P9-AQK-502

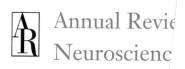

Annual Review
Neuroscience

Annual Review of
Neuroscience

Volume 34, 2011

Steven E. Hyman, *Editor*
Harvard University

Thomas M. Jessell, *Associate Editor*
Columbia University

Carla J. Shatz, *Associate Editor*
Stanford University

Charles F. Stevens, *Associate Editor*
Salk Institute for Biological Studies

Huda Y. Zoghbi, *Associate Editor*
Baylor College of Medicine

www.annualreviews.org • science@annualreviews.org • 650-493-4400

Annual Reviews
4139 El Camino Way • P.O. Box 10139 • Palo Alto, California 94303-0139

Annual Reviews
Palo Alto, California, USA

International Standard Serial Number: 0147-006X
International Standard Book Number: 978-0-8243-2434-6

TYPESET BY APTARA
PRINTED AND BOUND BY SHERIDAN BOOKS, INC., CHELSEA, MICHIGAN

Annual Review of
Neuroscience

Volume 34, 2011

Contents

Indexes

Errata

An online log of corrections to *Annual Review of Neuroscience* articles may be found at
http://neuro.annualreviews.org/

Related Articles

From the ***Annual Review of Biochemistry***, Volume 80 (2011)

Amyloid Structure: Conformational Diversity and Consequences
Brandon H. Toyama and Jonathan S. Weissman

Development of Probes for Cellular Functions Using Fluorescent Proteins
and Fluorescence Resonance Energy Transfer
Atsushi Miyawaki

Emerging In Vivo Analyses of Cell Function Using Fluorescene Imaging
Jennifer Lippincott-Schwartz

From the ***Annual Review of Biomedical Engineering***, Volume 12 (2010)

Image-Guided Interventions: Technology Review and Clinical Applications
Kevin Cleary and Terry M. Peters

Topography, Cell Response, and Nerve Regeneration
Diane Hoffman-Kim, Jennifer A. Mitchel, and Ravi V. Bellamkonda

From the ***Annual Review of Cell and Developmental Biology***, Volume 26 (2010)

The Diverse Functions of Oxysterol-Binding Proteins
Sumana Raychaudhuri and William A. Prinz

Ubiquitination in Postsynaptic Function and Plasticity
Angela M. Mabb and Michael D. Ehlers

α-Synuclein: Membrane Interactions and Toxicity in Parkinson's Disease
Pavan K. Auluck, Gabriela Caraveo, and Susan Lindquist

Novel Research Horizons for Presenilins and γ-Secretases in Cell Biology
and Disease
Bart De Strooper and Wim Annaert

Assembly of Fibronectin Extracellular Matrix
Purva Singh, Cara Carraher, and Jean E. Schwarzbauer

Common Factors Regulating Patterning of the Nervous and Vascular Systems
Mariana Melani and Brant M. Weinstein

From the ***Annual Review of Clinical Psychology***, Volume 6 (2010)

The Genetics of Mood Disorders
Jennifer Y.F. Lau and Thalia C. Eley

Collision Detection as a Model for Sensory-Motor Integration

Haleh Fotowat[1] and Fabrizio Gabbiani[2,3]

[1] Department of Biology, McGill University, Montreal, Quebec, H3A-1B1, Canada; email: haleh.fotowat@mcgill.ca

[2] Department of Neuroscience, Baylor College of Medicine, Houston, Texas 77030; email: gabbiani@bcm.edu

[3] Department of Computational and Applied Mathematics, Rice University, Houston, Texas 77005

Annu. Rev. Neurosci. 2011. 34:1–19

First published online as a Review in Advance on March 10, 2011

The *Annual Review of Neuroscience* is online at neuro.annualreviews.org

This article's doi: 10.1146/annurev-neuro-061010-113632

Keywords

single neuron computation, looming, neuroethology, escape behavior, LGMD, DCMD

Abstract

Visually guided collision avoidance is critical for the survival of many animals. The execution of successful collision-avoidance behaviors requires accurate processing of approaching threats by the visual system and signaling of threat characteristics to motor circuits to execute appropriate motor programs in a timely manner. Consequently, visually guided collision avoidance offers an excellent model with which to study the neural mechanisms of sensory-motor integration in the context of a natural behavior. Neurons that selectively respond to approaching threats and brain areas processing them have been characterized across many species. In locusts in particular, the underlying sensory and motor processes have been analyzed in great detail: These animals possess an identified neuron, called the LGMD, that responds selectively to approaching threats and conveys that information through a second identified neuron, the DCMD, to motor centers, generating escape jumps. A combination of behavioral and in vivo electrophysiological experiments has unraveled many of the cellular and network mechanisms underlying this behavior.

Contents

INTRODUCTION

How does an animal brain use the sensory information it acquires from the outside world to generate appropriate motor actions? When multiple behavioral choices are available, which neural mechanisms underlie making a particular decision? These questions have been the subject of intense research for decades. For the most part, sensory and motor physiology have been studied separately to characterize the input and output stages of the nervous system. However, owing to the inherent variability in both sensory and motor responses, their relation and the neural transformations that occur between them cannot be fully appreciated unless they are studied simultaneously.

Behaviors that are critical for survival provide a favorable model for sensory-motor integration because many animals have evolved specialized neural circuitry to execute them. Studying their underlying neural mechanisms provides an opportunity to understand basic principles of sensory-motor integration, which may generalize to other behaviors as well. In this review, we focus on visually evoked collision-avoidance and escape behaviors. Their investigation is particularly feasible in the laboratory because approaching objects can be effectively simulated using two-dimensional

projections on a computer screen, called looming stimuli.

In the following sections, we first discuss what is known about the psychophysics of collision-avoidance behaviors across different animal species, from humans to insects. Next, we present a comparative overview of the structure and function of specialized looming-sensitive neurons and their hypothesized role in generating escape responses. Finally, we focus on the locust, the model system in which the neural mechanisms underlying visually evoked escape behaviors are best understood. Despite the relatively small size of their brains, these animals are champions in executing such behaviors. At the sensory-motor interface, they possess identified neurons that detect looming stimuli and are part of a compact neural network that transforms sensory signals to the motor commands required for collision avoidance. Thus, in the locust, investigators can undertake detailed studies of the neural processing that occurs between the sight of a threat and the execution of an escape behavior.

VISUALLY EVOKED COLLISION-AVOIDANCE BEHAVIORS

Virtually all animals endowed with spatial vision exhibit an avoidance response to objects approaching on a collision course. To be effective, such responses should be executed in a timely manner and only if the approaching object might be a real threat. Thus the properties of the approaching object must be monitored in real time for the animal to decide whether, at which moment, and in which direction, to generate an escape response.

The psychophysics of visually evoked collision-avoidance behaviors has been studied for many decades, across different animal species (e.g., humans: Ball & Tronick 1971, King et al. 1992; monkeys: Schiff et al. 1962; pigeons: Wang & Frost 1992; turtles: Hayes & Saiff 1967; frogs: Ingle & Hoff 1990, King & Comer 1996, King et al. 1999, Yamamoto et al. 2003; goldfish: Preuss et al. 2006; crabs:

Looming stimulus: the two-dimensional expanding shadow associated with an approaching object by central projection on a screen

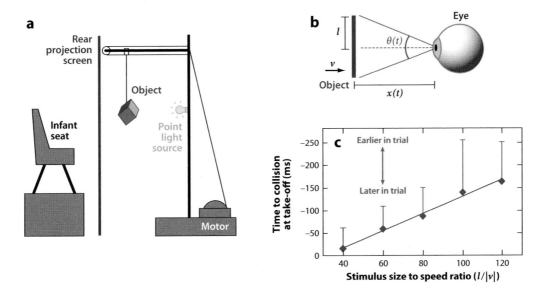

Figure 1

Looming stimuli and evoked behavior. (*a*) Schematics of the experiments of Ball & Tronick (1971). The subject is seated in a chair and views either a real approaching object (without projection screen) or the object's expanding shadow projected on the screen (looming stimulus). Reactions to the two types of stimuli are indistinguishable. (*b*) Kinematic variables characterizing a looming stimulus viewed monocularly. A solid disc is characterized by its half-size, l, and its constant approach speed, v. The distance to the eye is $x(t)$ and the subtended angle, $\theta(t)$. The time course of θ is fully determined by the size-to-speed ratio, $l/|v|$ (see sidebar, Approach Kinematics and Angular Threshold). (*c*) In locusts, the time of jump relative to collision as a function of the looming stimulus size-to-speed ratio is well fitted by a line [$\rho = 0.6$, data points are mean and standard deviation pooled over 13 animals; adapted from Fotowat & Gabbiani (2007)].

Hemmi 2005, Oliva et al. 2007, Sztarker & Tomsic 2008; crayfish: Wine & Krasne 1972, Glantz 1974; flies: Holmqvist & Srinivasan 1991, Tammero & Dickinson 2002, Hammond & O'Shea 2007, Card & Dickinson 2008, Fotowat et al. 2009; locusts: Robertson & Johnson 1993, Santer et al. 2005a,b, Fotowat & Gabbiani 2007). In response to impending collision, most animals generate a motor response that moves them away from the threat, covers the most vulnerable parts of their body, or generates a fast, unpredictable movement trajectory. For example, Ball & Tronick (1971) reported that human infants, 2–11 weeks old, presented with approaching objects, moved their head back and away from the object and brought their arms to their face. They found that such collision-avoidance behaviors were extremely robust and invariant with age, suggesting that they either are innate or require minimal learning. Additionally, the authors found that the two-dimensional shadow of an

approaching object was equally effective in generating avoidance responses (**Figure 1a**), and they concluded that either human infants are not capable of extracting depth information or such information is not necessary for generating the avoidance response. Similarly, both infant and adult rhesus monkeys show a "persistent fear response" to the symmetrically expanding shadows of physically approaching objects (Schiff et al. 1962). Thus, the symmetrical two-dimensional expansion of a silhouette is sufficient to simulate the approach of an object on a collision course and to generate avoidance responses. For such two-dimensional stimuli, referred to as looming stimuli, the "time of collision" is defined as the moment when the angular size subtended by the silhouette on the retina reaches 180° (Gibson 1958). In the particular case of monocular stimulation, the temporal dynamics of the retinal image of an object with half size l, approaching on a collision course with a constant approach speed, v,

APPROACH KINEMATICS AND ANGULAR THRESHOLD

In reference to **Figure 1b**, let $x > 0$ be the object's position with respect to the subject's eye; i.e., $x = 0$ at collision. Define $t = 0$ as the time of expected collision and $t < 0$ before collision. Consequently, the object's speed, v, is < 0 when the object is approaching and $-v = |v|$, the absolute value of v. The object's position is given by $x(t) = vt$ and by trigonometry,

$$\tan\theta(t)/2 = l/(vt) \quad \text{or} \quad \theta(t) = 2\tan^{-1}[l/(vt)]. \qquad 1.$$

Assume that the time of an event relative to collision depends linearly on $l/|v|$, as is the case for jump time in **Figure 1c**. Then

$$-t_{event} = \alpha l/|v| - \delta \quad (> 0 \text{ before collision}), \qquad 2.$$

where α is the slope of this linear relation (dimensionless) and δ the intercept with the ordinate axis (in units of ms). Equivalently, the angle, θ_{thres}, subtended by the looming stimulus δ ms before the event (i.e., at time $t_{event} - \delta$) is independent of the stimulus size-to-speed ratio, $l/|v|$. To see this, use successively Equations 1 and 2,

$$\tan\theta_{thres}/2 = l/[v(t_{event} - \delta)] = l/[-v\alpha l/|v|] = \alpha^{-1}$$

which is indeed independent of $l/|v|$. Furthermore, the threshold half-angle's tangent, $\tan\theta_{thres}/2$, is the inverse of the slope, α.

can be fully characterized by its size-to-speed ratio, $l/|v|$ (**Figure 1b**) (Gabbiani et al. 1999). Therefore, for a given object size, a faster approach speed implies a faster expansion rate and a smaller $l/|v|$ ratio. Varying the $l/|v|$ ratio thus allows investigators to manipulate the temporal dynamics of the looming stimulus to study which of its various aspects are used by the animal to guide its behavior.

For example, in response to looming stimuli, locusts jump and fly away, and the time of take-off relative to collision varies linearly with $l/|v|$ (**Figure 1c**). This linear relationship shows that the jump does not occur at a fixed time before collision, but implies rather that it occurs at a fixed delay after the stimulus reaches a threshold angular size on the retina (see sidebar, Approach Kinematics and Angular Threshold). Retinal image size has also been reported to be a critical stimulus parameter for

triggering visually evoked escape responses in several other species (frogs: Ingle & Hoff 1990, Yamamoto et al. 2003; goldfish: Preuss et al. 2006; fruit fly: Fotowat et al. 2009). An estimate of time to collision, on the other hand, may be relatively easily extracted from looming stimuli (Lee 1976) and, a priori, seems better suited than retinal image size for triggering collision-avoidance behaviors because it is not confounded by object size. There is, however, little evidence implicating time to collision in responses to approaching objects. Time to collision, however, has been linked to other types of avoidance behaviors, e.g., the triggering of wing folding in diving gannets (Lee & Reddish 1981), landing in the house fly (Wagner 1982), and various interception tasks in humans (Tresilian 1999). In the case of landing, other visual variables may also be used (Borst & Bahde 1988, Srinivasan et al. 2000), and it is likely that separate neural pathways drive collision-avoidance and landing behaviors (Tammero & Dickinson 2002).

Another interesting aspect of visually guided escape responses is their preparatory phase. For example, locusts (Santer et al. 2005b) and fruit flies (Card & Dickinson 2008) use their middle legs to displace their center of mass laterally and thus tilt the direction of their jump away from the approaching object. Additionally, fruit flies raise their wings prior to take-off (Hammond & O'Shea 2007, Fotowat et al. 2009), and locusts prepare for take-off by storing the required energy in the elastic elements of their hind legs (Heitler 1974). Whether the same neural pathways control collision-avoidance preparation and its final execution remains unknown.

In many animals, looming-evoked escape responses are probabilistic and habituate over repeated stimulus presentations (Hayes & Saiff 1967, Holmqvist & Srinivasan 1991, Yamamoto et al. 2003, Fotowat & Gabbiani 2007, Hammond & O'Shea 2007, Oliva et al. 2007, Fotowat et al. 2009). Interestingly, in locusts, the dynamics of habituation have been linked to the social lifestyle of the animals (solitary or swarming; Matheson et al. 2004).

Because visually evoked escape responses are natural, robust, probabilistic, and often multi-staged, planned behaviors, they provide an excellent context for studying the neural mechanisms of sensory-motor integration. Efficient detection of approaching threats is the critical first step for generating an escape response; hence many animal species possess neurons specialized in this task. In the next section, we discuss the properties of such looming-sensitive neurons and the different ways by which they could convey information about approaching objects to downstream motor centers.

LOOMING-SENSITIVE NEURONS

Electrophysiological recordings, electrical and pharmacological stimulation, and lesion studies have been carried out in many different animal species, revealing classes of neurons and neuronal pathways preferentially activated by objects approaching on a collision course in the context of visually guided collision-avoidance or escape behaviors. These neurons and pathways are commonly located in regions of the brain thought to be involved in sensory-motor integration, such as the superior colliculus and some of its target nuclei (monkeys: King & Cowey 1992; rats: Sahibzada et al. 1986, Mitchell et al. 1988a,b, Redgrave et al. 1988, Dean et al. 1989, Westby et al. 1990; hamsters: Northmore et al. 1988). In nonmammalian vertebrates, the optic tectum is homologous to the superior colliculus, and looming-sensitive neurons are found in midbrain nuclei along the tectofugal pathway in pigeons (Wang & Frost 1992, Wu et al. 2005), the optic tectum of frogs (King et al. 1999, Nakagawa & Hongjian 2010), and the nucleus isthmi in fish (Gallagher & Northmore 2006). The nucleus isthmi in birds may also be implicated in detecting looming stimuli in the context of spatial visual attention (Asadollahi et al. 2010). In mice, looming-sensitive neurons have been reported as early as in the retina (Münch et al. 2009). In the frog retina, synchronized activity of dimming-detectors plays an important role in generating looming-evoked escape behaviors (Lettvin

et al. 1968, Ishikane et al. 2005). Investigators have also identified looming-sensitive neurons in the brains of crustaceans, teleost fish, and insects [lobula giant neurons of crabs: Medan et al. 2007, Oliva et al. 2007, Sztarker & Tomsic 2008; medial giant neurons of crayfish: Wine & Krasne 1972; Mauthner cell of goldfish: Preuss et al. 2006; lobula giant movement detector (LGMD) of locusts: Schlotterer 1977, Rind & Simmons 1992, Hatsopoulos et al. 1995; neurons in the moth optic lobe: Wicklein & Strausfeld 2000).

Despite many differences in brain structure and behavioral repertoire, these neurons and pathways seem to share some common features. For example, many looming-sensitive neurons possess large dendritic fields, consistent with wide-field integration of sensory inputs (O'Shea & Williams 1974, Wicklein & Strausfeld 2000, Medan et al. 2007). They are sensitive to collision-bound trajectories across a substantial fraction of the visual field (Wang & Frost 1992, Gabbiani et al. 2001, Medan et al. 2007, Rogers et al. 2010) and respond very weakly, if at all, to optic flow stimuli, such as those generated by self-motion (O'Shea & Rowell 1975a, Wang & Frost 1992, Gabbiani et al. 2002, Medan et al. 2007). Additionally, these neurons are often multimodal. For example, looming-sensitive neurons in goldfish and locusts also respond to auditory stimuli (Zottoli 1977, O'Shea 1975, Zottoli & Faber 2000), and three of the four identified classes of looming-sensitive neurons in crabs respond to mechanical stimuli (Medan et al. 2007). In rats, the neural projection from the superior colliculus to the cuneiform area, which is implicated in generating collision avoidance to looming stimuli, is multimodal as well (Westby et al. 1990). Electrical and pharmacological stimulation of those neuronal populations leads to changes in respiration and heart rate, as well as blood pressure (Keay et al. 1988, 1990). Similarly, Wang & Frost (1992) reported that increases in heart rate correlate with increases in the activity of looming-sensitive neurons in pigeons. Thus, looming-sensitive neurons are likely to provide input to the autonomic nervous system in

LGMD: lobula giant movement detector

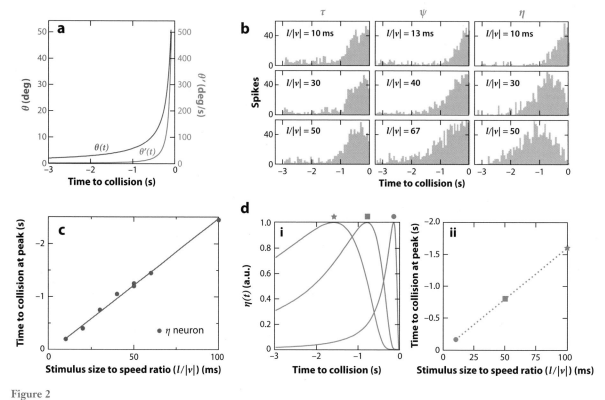

Figure 2

Kinematics of looming stimuli and associated neural responses. (*a*) During looming, both stimulus angular size, $\theta(t)$ (*blue curve*), and speed, $\theta'(t)$ (*red curve*), grow nonlinearly with time. (*b*) Responses of neurons in the nucleus rotundus of pigeons to looming stimuli (sum of 5 trials; 50 ms bins). A representative of the first class (τ, *left*) starts firing at a fixed time to collision, independent of $l/|v|$. Neurons of the ψ class (*middle*) start to fire when the stimulus angular speed exceeds a threshold and neurons of the η class (*right*) exhibit a peak that shifts relative to collision time with $l/|v|$. (*c*) The time of peak firing rate of η neurons is linearly related to $l/|v|$. (*d*) Their firing rate time course is well described by the η function described in the text, (*i*), whose peak time varies linearly with $l/|v|$ as well, (*ii*). Symbols mark the correspondence to different $l/|v|$ values. Panels *a* and *d* adapted from Laurent and Gabbiani (1998); panels *b* and *c* adapted from Sun & Frost (1998).

vertebrates, in addition to their involvement in motor aspects of collision-avoidance behaviors.

Which kinematic variables of an approaching object are extracted by looming-sensitive neurons in different animal species, and what are the similarities and differences in the computations they perform? Using the time course of retinal image expansion, these neurons could, in theory, compute several time-varying quantities associated with an approaching object, e.g., angular size and speed, time remaining to collision, or various combinations thereof. A defining characteristic of looming stimuli is that, as collision becomes imminent, both the angular size, $\theta(t)$, and speed, $\theta'(t)$, grow nonlinearly

(**Figure 2*a***), whereas time remaining to collision, τ, decreases linearly.

Among all vertebrate species, looming-sensitive neurons have been best characterized in the pigeon, where three distinct classes of neurons sensitive to distinct kinematic variables have been reported in the midbrain nucleus rotundus (Sun & Frost 1998). Neurons of the first class always initiate their response at the same time remaining to collision, i.e., after τ reaches a threshold. Thus, their response onset time relative to collision does not vary with the stimulus size or speed (**Figure 2*b***, τ). The time course of the firing rate of neurons in the second class differs because it starts earlier relative to collision for larger or slower objects (**Figure 2b**, ψ), and its onset is well described by a threshold in the angular speed of the stimulus. After response onset, the temporal slope of the firing rate is not different for stimuli with different size-to-speed ratios, suggesting that these neurons do not track the stimulus angular speed per se, but rather begin responding after it reaches a threshold. The response profile of the third class of neurons is similar to the second one in that it initiates earlier for larger or slower objects, but it is different in that it decreases when the stimulus reaches large angular sizes showing a distinct peak, which also occurs earlier for larger or slower objects (**Figure 2b**, η). Furthermore, the timing of the peak varies linearly with the stimulus size-to-speed ratio (**Figure 2c**). Just as in the case of locust jump escape behaviors, this linear relationship indicates that the peak occurs a fixed delay after the stimulus reaches a threshold angular size on the retina (Gabbiani et al. 1999). Such a pattern of neural activity could be achieved if the neurons in this class were following a nonlinear function of the stimulus angular size, $\theta(t)$, and speed, $\theta'(t)$, which first increases and then decreases toward the end of the approach. A simple multiplication of angular speed with a negative exponential of angular size fulfills this requirement: $\eta(t) = C\theta'(t - \delta)exp(-\alpha\theta(t - \delta))$ (**Figure 2d**). Here, the constant, C (converting angular velocity to firing rate), as well as α (related to the threshold angular size), and δ (implementing neural delays), can be fitted to experimental data. This function initially increases in parallel with angular speed, but eventually decreases as the negative exponential of size overwhelms it. Thus, $\eta(t)$ exhibits a peak that occurs earlier relative to collision for larger or slower objects and varies linearly with the object's size-to-speed ratio, $l/|v|$. In fact, this function had first been proposed to fit the time course of activity of an identified, looming-sensitive neuron in the locust, the LGMD (Hatsopoulos et al. 1995). This multiplicative combination of angular speed and a negative exponential of size is essentially unique in being able to reproduce the characteristic linear relation between peak firing rate and $l/|v|$ (appendix 3 of Gabbiani et al. 1999) and was later found to fit the time course of looming-evoked excitatory postsynaptic potentials in the Mauthner cell of goldfish, as well (Preuss et al. 2006).

The locust LGMD is perhaps the most extensively studied looming-sensitive neuron (O'Shea & Rowell 1976, Rowell et al. 1977, Schlotterer 1977, Rind & Simmons 1992, Hatsopoulos et al. 1995, Judge & Rind 1997, Gabbiani et al. 1999, Gabbiani et al. 2002, Matheson et al. 2004, Krapp & Gabbiani 2005, Guest & Gray 2006, Peron et al. 2009, Peron & Gabbiani 2009). It makes a strong synapse with the descending contralateral movement detector (DCMD) (O'Shea & Williams 1974, Rind 1984, Killmann & Schürmann 1985), whose large axon travels down the animal's contralateral nerve cord and contacts motor and interneurons involved in generating jumps and flight steering (Burrows & Rowell 1973, O'Shea et al. 1974, Simmons 1980). Therefore, the LGMD-DCMD system provides an excellent framework for studying the sensory-motor transformations that occur during looming-evoked collision-avoidance behaviors. In the following sections, we first review the known anatomical and physiological properties of the LGMD neuron and next describe the role of the LGMD-DCMD system in the sensory-motor integration mechanisms that underlie visually evoked collision-avoidance behaviors.

DCMD: descending contralateral movement detector

THE LOBULA GIANT MOVEMENT DETECTOR NEURON OF LOCUSTS

The LGMD of locusts is an identified neuron located in the lobula neuropil of each of their optic lobes. As in other insects, the two bilaterally symmetric optic lobes consist of a photoreceptor layer, the retina, and three hierarchically organized neuropils, the lamina, medulla, and lobula, connected by two optic chiasms. Thus, the lobula is three synapses away from the photoreceptors. Each LGMD neuron receives visual input through three distinct dendritic fields (**Figure 3a**). The largest one, field A, is ellipsoidal in shape, just like the eye, and receives ~15,000 excitatory retinotopic inputs from the entire visual hemifield (O'Shea & Williams 1974, Strausfeld & Nässel 1981, Krapp & Gabbiani 2005, Peron et al. 2007). The retinotopic organization of this projection is remarkably precise and preserved down to the level of single ommatidia or facets (Peron et al. 2009). In contrast, the LGMD's other two smaller dendritic fields, B and C, each receive ~500 nonretinotopic, feedforward inhibitory inputs, which are best activated by large, transient changes in luminance (Palka 1967,

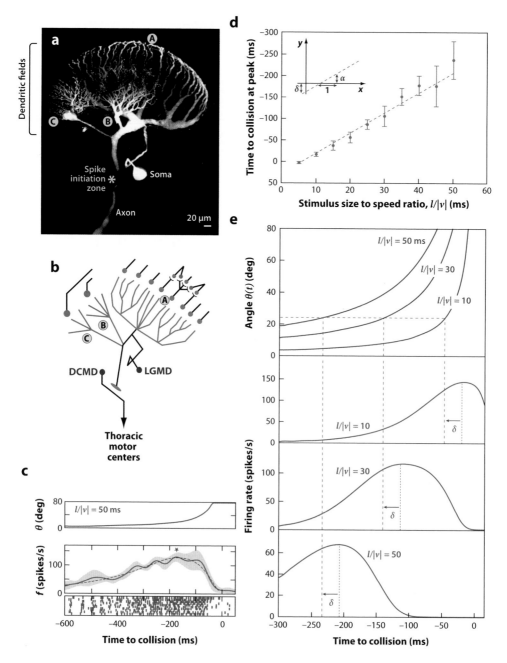

Rowell et al. 1977, Hatsopoulos et al. 1995, Gabbiani et al. 2005). In addition, the excitatory pathway is endowed with a lateral inhibitory network activated by wide-field motion that protects the LGMD's responses from habituation (**Figure 3***b*) (O'Shea & Rowell 1975a).

The LGMD neuron responds to small stimuli translating in its visual receptive field but is maximally activated by objects approaching on a collision course with the animal or by looming stimuli (Rowell et al. 1977, Schlotterer 1977, Rind & Simmons 1992, Hatsopoulos et al. 1995). The selectivity of the LGMD for looming versus translating stimuli is due in part to spike frequency adaptation mediated by SK-like (small K$^+$ conductance) potassium channels located immediately adjacent to the LGMD's spike initiation zone (Gabbiani & Krapp 2006, Peron & Gabbiani 2009). This conductance acts as a veto mechanism, effectively shutting off synaptic excitation when it is not maximally activated. In response to looming stimuli, the firing rate of the neuron increases as the stimulus angular size increases but then declines as it exceeds a threshold (**Figure 3***c*). This decline is caused by the feed-forward inhibitory inputs onto dendritic fields B and C. Feedforward excitation and inhibition are activated concurrently during a looming stimulus, with excitation slightly leading inhibition. As a result, the firing rate peak occurs well before excitation or inhibition has ceased (Gabbiani et al. 2002, 2005). The peak time of LGMD activity is linearly related to $l/|v|$, occurring earlier relative to collision for larger size-to-speed ratios (**Figure 3***d*) (Gabbiani et al. 1999). Again, this finding indicates that the peak occurs a fixed delay after the stimulus reaches a threshold angular size on the retina (**Figure 3***e*), where the delay and threshold angular size can be simply calculated from the slope and intercept of the linear fit (see sidebar, Approach Kinematics and Angular Threshold). This angular size threshold is invariant to changes in the shape, texture, and direction of the approaching object (Gabbiani et al. 2001, Rogers et al. 2010), presumably enabling the locust to respond similarly to the wide range of predators it experiences in the wild (Kuitert & Connin 1952, Preston-Mafham 1990).

In response to looming stimuli, the time course of the LGMD firing rate can be described as a function of the angular speed of stimulus expansion multiplied by a negative exponential of its angular size (Hatsopoulos et al. 1995, Gabbiani et al. 2002), similar to the η function used to fit the response of looming-sensitive neurons in pigeons. Experimental evidence suggests that this multiplicative operation is carried out within the LGMD itself rather than presynaptically (Gabbiani et al. 2002). How could the LGMD biophysically compute the η function using the visual inputs it receives? Information about the angular speed of the stimulus is conveyed by the excitatory input to the LGMD (from

---

Figure 3

Anatomy, circuit properties, and responses of the LGMD to looming stimuli. (*a*) Stain of the LGMD neuron illustrating its dendritic fields A, B, and C, and the location of the spike initiation zone (*). (*b*) Schematics of the input-output circuitry of the LGMD. Field A receives retinotopic, feedforward excitatory synaptic input from the entire visual hemifield (*light blue*). A lateral inhibitory network (*red dots*) between adjacent excitatory inputs to field A lies presynaptic to the LGMD (*purple dot*). Fields B and C receive nonretinotopic feedforward inhibition (*red*). The LGMD makes a synaptic contact with the DCMD neuron (*purple dot*) in the protocerebrum, which in turn sends its axon toward thoracic motor centers. (*c*) Response to a looming stimulus ($l/|v| = 30$ ms). The angular size subtended by the stimulus at the retina is illustrated on top (*dark blue curve*). The firing rate is illustrated in the middle panel (*dark yellow line* and *gray fill*, indicating standard deviation), as well as the multiplicative model discussed in the text (*dashed gray line*). Ten spike rasters representing ten different trials are illustrated at the bottom. Time of peak is indicated by a star. (*d*) Plot of peak time as a function of the size-to-speed ratio of the looming stimulus. The relation is close to linear and has an intercept δ with the y-axis and a slope α (*top left inset*). Data in panels *c* and *d* are from different animals. (*e*) Diagram illustrating the significance of the parameters α and δ derived from the linear fit in panel *d*. The time course of the angle $\theta(t)$ is depicted on top for three $l/|v|$ values. The bottom three panels illustrate the mean firing rate of the LGMD in response to these three looming stimuli. Note that in each case, the angular size of the object is equal to $24°$ at a delay $\delta = 27$ ms before the peak DCMD activity (*horizontal and vertical dashed lines*). Adapted from Gabbiani et al. (1999, 2002).

dendritic subfield A; Krapp & Gabbiani 2005), whereas feedforward inhibition conveys information about stimulus size (from subfields B or C; Gabbiani et al. 2005). Thus, if the excitatory input were proportional to the logarithm of angular speed, $\log \theta'$, and the inhibitory input to angular size, $-\alpha\theta$, these two signals could be added and exponentiated within the LGMD to yield the desired η function. A complete verification of this hypothesis requires a quantitative description of the time course of excitation and inhibition along the feedforward pathways leading to the LGMD and their interaction within its dendritic tree. In support of this hypothesis, the transformation between membrane potential and firing rate at the spike initiation zone of the LGMD is well approximated by a third-order power law, close to the exponentiation postulated above (Gabbiani et al. 2002). Thus, the locust LGMD is an example of a single neuron within a compact nervous system able to carry out a complex nonlinear computation within its dendritic tree. Mammalian neurons are endowed with many types of active dendritic conductances (Johnston & Narayanan 2008) potentially allowing them to carry out complex, nonlinear computations (Polsky et al. 2004), although their relation to sensory and motor processing remains less well understood.

In the brain, each LGMD neuron makes a mixed chemical and electrical synapse (Killmann & Schürmann 1985) with the DCMD neuron (O'Shea et al. 1974). The LGMD-DCMD synapse is so strong that every LGMD spike generates a spike in the DCMD (O'Shea & Rowell 1975b, Rind 1984). The DCMD axon crosses the midline in the brain, enters the contralateral nerve cord, and eventually reaches downstream motor ganglia in the thorax. Because the LGMD-DCMD spikes are 1:1, the DCMD acts as a faithful relay of the LGMD activity to motor centers. The DCMD axon is located peripherally in the dorso-medial segment of the nerve cord and has one of the largest diameters (15–17 µm; O'Shea et al. 1974), making it amenable to high signal-to-noise ratio extracellular recordings.

In the thorax, the DCMD axon forms a single branch within the first or prothoracic ganglion and three branches in the second (meso-) and third (meta-) thoracic ganglia, where it makes synapses with identified motor and interneurons implicated in the generation of jump and flight (Burrows & Rowell 1973, O'Shea et al. 1974, O'Shea & Williams 1974, Pearson et al. 1980, Simmons 1980, Pearson & Robertson 1981). Because of its selectivity to approaching objects and its postsynaptic targets, the DCMD neuron has long been thought to play an important role in triggering visually evoked escape behaviors. In the next section, we describe recent findings on the role played by the DCMD in triggering escape jumping.

ROLE OF THE DCMD IN LOOMING-EVOKED ESCAPE BEHAVIORS

The DCMD neuron is thought to be involved in generating escape jumps in response to looming stimuli (Fotowat & Gabbiani 2007, Santer et al. 2008), as well as in-flight collision-avoidance behaviors (Santer et al. 2006). Here, we focus on looming-evoked jump escape behaviors for which electrophysiological data are available in freely behaving animals (Fotowat et al. 2011).

The locust jump is a ballistic movement, requiring about ten times more power than what can be provided by the extensor muscles of the hind legs (Bennet-Clark 1975). Locusts, therefore, prepare to jump by storing mechanical energy in the elastic elements of their hind legs. The jump motor program consists of three phases (**Figure 4a**) (Godden 1975, Heitler & Burrows 1977, Burrows 1995, Burrows & Morris 2001). First, the flexor tibiae motor neurons become active. After its full flexion, the tibia becomes locked into the femur through an engaging lump near the joint. Next, the extensor motor neurons become active as well, resulting in the isometric cocontraction of flexors and extensors and the distortion of spring-like elastic elements of the joints, which act as energy storage devices (Burrows & Morris 2001).

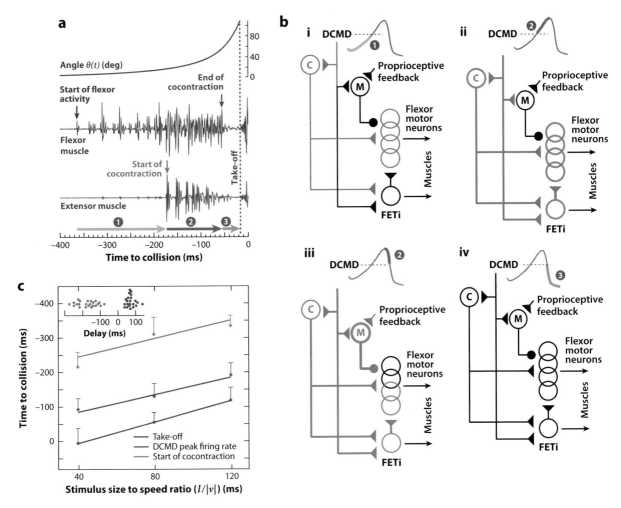

Figure 4

DCMD's involvement in the motor sequence leading to a jump. (*a*) In response to a looming stimulus (top, $l/|v| = 40$ ms), flexor motor neurons start to fire (*green arrow*), causing the tibia to flex and align itself with the femur. Next, the fast extensor motor neuron starts to fire, initiating the cocontraction phase (*red arrow*). Finally, the flexors become inhibited, signaling the end of the cocontraction phase (*blue arrow*). This leads shortly thereafter to take-off (*dashed blue vertical line*). The corresponding three jump phases are indicated by three colored arrows at bottom and three dark yellow circled numerals. (*b*) (*i–iv*) Simplified diagram of the circuitry involved in jump escape behaviors in response to looming stimuli and the postulated relation between DCMD firing phases, the activation of the circuit generating the jump, and motor phases of the behavior. Triangles and the circle indicate excitatory and inhibitory synapses, respectively. See text for details. Thick and thin red lines indicate strong and weakly activated pathways, respectively, at a time point within the given time interval (*dark yellow circled numeral*). Colors on schematized DCMD firing rate profile correspond with arrows indicating motor phases in panel *a*. (*i*) The rising phase of the DCMD activity coincides with the initial flexor activity until the DCMD firing rate crosses a threshold (*dashed horizontal line*), after which the cocontraction starts. (*ii*) Following this threshold, the DCMD maximally excites the fast extensor tibiae (FETi) motor neuron during the cocontraction phase. (*iii*) Around the DCMD peak activity, contingent upon the presence of sufficient proprioceptive feedback input, the M interneuron becomes activated, contributing to the inhibition of flexor motor neurons. (*iv*) The decline in the DCMD firing rate contributes further to the silencing of flexors as well as the FETi motor neuron, resulting in the end of cocontraction and take-off. (*c*) Time of start of cocontraction (*red*), DCMD peak firing rate (*gray*) and take-off (*blue*) as a function of stimulus size to speed ratio (mean and standard deviation across multiple trials and animals). The inset illustrates representative delays between DCMD peak and cocontraction onset (*red*) and between peak and take-off (*blue*) in single trials. Positive delays correspond to events after the peak. Thus the time of peak firing always follows cocontraction onset and precedes take-off. Modified and adapted from Fotowat & Gabbiani (2007) and Fotowat et al. (2011).

FETi: fast extensor tibiae (muscle and corresponding motor neuron)

Finally, once the required level of energy storage is achieved, the flexor tibiae motor neurons become inhibited, the lock gets released, and the tibia extends, leading to take-off.

Many of the identified motor neurons and interneurons that contribute to the generation of this motor pattern receive excitatory input from the DCMD (**Figure 4b**) (Burrows 1996). Each DCMD makes an excitatory connection with the C (Cocking) interneurons, a bilaterally symmetric pair located in the second (meso-) thoracic ganglion (Pearson & Robertson 1981). The C interneurons in turn make excitatory connections with both the fast extensor of the tibia (FETi) and flexor motor neurons. They can evoke spikes in both, provided that the tibia is in a fully flexed position, and therefore contribute to the cocontraction phase of the jump. Each DCMD also directly excites the FETi motor neuron of both hind legs, which in turn makes excitatory connections with flexor motor neurons. This highly unusual connection between antagonistic motor neurons, found only at the level of the hind legs, is most likely active during cocontraction as well (Hoyle & Burrows 1973, Burrows et al. 1989). In the third thoracic ganglion, each DCMD innervates a pair of interneurons called M (Multimodal). These interneurons, which also receive proprioceptive feedback from the hind legs and make inhibitory connections with the flexor motor neurons, are thought to contribute to the release of cocontraction, allowing take-off (Pearson et al. 1980; Steeves & Pearson 1982; Gynther & Pearson 1986, 1989). Therefore, on the basis of these anatomical connections, the DCMD neuron could contribute to both triggering and release of cocontraction.

Tethered locusts respond to looming stimuli by flexing their hind legs, and this behavior was shown to coincide with the time when the DCMD firing activity is largest [>150 spikes (spk)/s, Santer et al. 2008]. However, simultaneous recordings obtained from the DCMD and the FETi motor neuron in fixed locusts have shown that spiking in the DCMD is typically insufficient to activate the FETi (Burrows & Rowell 1973, Rogers et al. 2007), leaving unsolved the role of the DCMD in triggering the cocontraction and take-off (Burrows 1996).

In freely behaving locusts, however, the firing rate of the DCMD neuron can reach much higher values (>250 spk/s), and the cocontraction is triggered after the DCMD firing rate exceeds a threshold [**Figure 4b(ii)**] (Fotowat et al. 2011). The timing of such a threshold also varies linearly with $l/|v|$ (Gabbiani et al. 2002), thus the cocontraction phase is triggered after the approaching object reaches a threshold angular size. It is important to note, however, that although a DCMD threshold firing rate appears necessary for triggering cocontraction, it is not sufficient (Fotowat et al. 2011). For example, cocontraction cannot start before the hind leg is fully flexed (Burrows & Pflüger 1988). After the onset of cocontraction, the FETi motor neuron spikes closely follow those of the DCMD (Fotowat et al. 2011) and thus the onset of cocontraction appears to act as a switch that promotes the DCMD input to a major source of excitation for this neuron.

The DCMD's characteristic peak in firing rate occurs after the onset of cocontraction and well before take-off (**Figure 4c**), rendering the DCMD maximally active during cocontraction. Furthermore, the timing of take-off can be predicted with high accuracy from the timing of the DCMD peak firing rate, suggesting that the postsynaptic targets of the DCMD control the time of take-off on the basis of the time of DCMD peak activity, most likely through at least two parallel pathways. First, the postsynaptic M interneuron, which has a high firing threshold (Pearson et al. 1980), is likely to be maximally excitable around the DCMD peak firing time. Thus, given concurrent proprioceptive feedback excitation that signals the level of stored energy, the M interneuron could eventually shut off the flexor motor neurons [**Figure 4b(iii)**] (Pearson et al. 1980, Gynther & Pearson 1989). Second, the end time of cocontraction will be influenced by the DCMD peak time through the subsequent reduction in DCMD activity, resulting in decreased excitation to the flexors and extensors [**Figure 4b(iv)**]. This complex sequence of events illustrates how the elements of a local neuronal network may use different aspects of a sensory neuron's activity to generate various components of a motor behavior.

Because looming-evoked escapes in freely behaving locusts are probabilistic, they provide the opportunity to study the contribution of sensory and motor neuron activity to the final decision to escape. Recordings in freely behaving locusts have revealed that in trials where locusts do not jump, the cocontraction is still triggered, though significantly later. Additionally, in those trials, the number of DCMD spikes fired after cocontraction onset is significantly lower, whereas the timing of the DCMD peak firing rate remains unchanged. A corollary of these observations is that the occurrence of a jump can be predicted on a trial-by-trial basis from both sensory and motor aspects of the discharge patterns evoked by looming stimuli (Fotowat et al. 2011). It would be useful to identify the sources of sensory response variability that ultimately influence the behavioral outcome. One possible source of such variability is the modulation of the DCMD activity by identified octopaminergic neurons in the medulla, whose axons arborize in the optic lobe and respond to stimuli that increase locust arousal (Rowell 1971b, Bacon et al. 1995, Rind et al. 2008). Therefore, in an aroused locust, increased levels of octopamine could increase the excitability of the DCMD and, consequently, the excitatory drive to the downstream motor circuits.

Is the DCMD the sole source of looming information to the jump motor circuitry, or are there other parallel visual pathways that could generate escapes, as reported in other species (zebrafish: Liu & Fetcho 1999; fruit fly: Fotowat et al. 2009)? In fact, each locust nerve cord contains the axon of a neuron that carries to motor centers a nearly identical copy of the information provided by the DCMD. This neuron is called the descending ipsilateral movement detector (DIMD), and its axon runs in the nerve cord ipsilateral to the eye from which it receives input, i.e., opposite to that of the DCMD (Rowell 1971a, Burrows &

Rowell 1973). This redundancy allows a locust with one ablated nerve cord to still react and jump in response to stimuli presented to either eye (Santer et al. 2008, Fotowat et al. 2011). The DIMD has yet to be identified anatomically but is thought to have a large axon like the DCMD. Furthermore, its spikes are thought to summate with those of the DCMD at the level of the FETi neuron (Rowell 1971a, Burrows & Rowell 1973). The looming-evoked responses in the DCMD and DIMD are very similar, with indistinguishable peak time and amplitude (Fotowat et al. 2011). In addition to the DIMD, laser ablation of the DCMD has revealed other contralateral descending neurons that respond to looming; however, their peak activity occurs much later than that of the DCMD or DIMD. In the absence of these two neurons, locusts still flex their hind legs in preparation for cocontraction; however, they rarely take off, and when they do, it is after projected collision (Fotowat et al. 2011). These results suggest that the activity of the DCMD or DIMD is likely not necessary for the initial flexion stage [**Figure 4***b*(*i*)] but plays a critical role in generating correctly timed jump take-offs.

In summary, studies in freely escaping locusts illustrate how different aspects of the activity of single sensory neurons, e.g., firing rate threshold, spike count, peak firing rate, or decreases in firing rate following a peak, could be used in parallel by downstream neural networks to trigger distinct stages of complex behaviors. Moreover, sensory response variability itself can contain ample information about the variability in behavior, as well as its final outcome, i.e., to jump or not.

DISCUSSION AND FUTURE DIRECTIONS

Recent evidence increasingly indicates that sensory neurons tuned to looming stimuli share many physiological features across widely different animal species, including fruit flies, fish, frogs, locusts, and pigeons. These features include (*a*) specific tuning for approaching, as opposed to, e.g., translating, object motion;

DIMD: descending ipsilateral movement detector

(*b*) encoding of an angular threshold size in their peak firing rate; (*c*) the ability to segment effectively an approaching object from whole field optic flow, caused, e.g., by ego-motion; and (*d*) invariant responses to many aspects of looming stimuli, such as texture, shape, contrast, or approach angle. Elucidation of the underlying algorithms and of the biophysical implementation that gives rise to these features will bring us closer to understanding how collision-avoidance behaviors are generated and, more generally, will shed light on how the brain processes sensory information in a natural context, at the levels of both single neurons and neural circuits. This program is being pursued in the locust visual system by taking advantage of its identified neurons, the relatively compact neural circuitry involved, and the ability to carry out many types of electrophysiological experiments in vivo. Other species will undoubtedly offer complementary advantages to progress toward the same goal, for example, the use of genetic techniques in fruit flies or zebrafish (Pfeiffer et al. 2008, Scott 2009).

In locusts, the emerging picture suggests that distinct aspects of a single neuron's time-varying activity could be used by different elements of downstream networks during motor planning and execution. Moreover, proprioceptive feedback plays a role in shaping sensory-motor integration and the final behavioral outcome. Simultaneous monitoring of sensory input and proprioceptive feedback will thus be important in unraveling details about the process of sensory-motor integration.

On the basis of studies in fixed animals, the DCMD had long been thought incapable of driving visually evoked escape behaviors. Yet, in freely behaving animals, the DCMD firing rate reaches significantly higher levels and can drive spiking in motor neurons such as the FETi. These results call for the study of sensory-motor integration in freely behaving animals (Fotowat et al. 2011, Maimon et al. 2010).

Finally, the role of neuromodulators in shaping the outcome of collision-avoidance behaviors remains largely open to investigation. Looming-sensitive neurons often show pronounced habituation, or fluctuations in overall responsiveness, usually attributed to neuromodulation. The fact that trial-by-trial variations in LGMD-DCMD activity are related to the final outcome of collision-avoidance behaviors indicates that neuromodulators are an integral part of the sensory-motor circuitry that mediates them. Mapping out the associated neural components and characterizing their impact would thus constitute an important step in understanding how sensory-motor integration contributes to decision making in the context of collision-avoidance behaviors.

Although there are striking similarities regarding the sensory processing of looming stimuli between animals such as locusts, fish, and pigeons, they possess quite distinct motor networks and muscular machinery for generating escape behaviors. Relatively little is known about how different motor networks use sensory activity for generating complex behaviors and whether common sensory-motor transformation rules are exploited in different species that possess similar sensory processing stages. Comparative studies are thus necessary to draw general conclusions about the biophysical implementations of these neural computations.

DISCLOSURE STATEMENT

The authors are not aware of any affiliations, memberships, funding, or financial holdings that might be perceived as affecting the objectivity of this review.

ACKNOWLEDGMENTS

We gratefully acknowledge support for our work over the years from the AFRL, HSFP, NIMH, NSF, as well as the Sloan and Gillson Longenbaugh Foundations. We thank Mr. P.W. Jones, as well as Drs. N. Chen, N.J. Cowan, R.B. Dewell and R. Krahe for comments.

LITERATURE CITED

Asadollahi A, Mysore SP, Knudsen EI. 2010. Stimulus-driven competition in a cholinergic midbrain nucleus. *Nat. Neurosci.* 13:889–95

Bacon JP, Thompson KS, Stern M. 1995. Identified octopaminergic neurons provide an arousal mechanism in the locust brain. *J. Neurophysiol.* 74:2739–43

Ball W, Tronick E. 1971. Infant responses to impending collision: optical and real. *Science* 171:818–20

Bennet-Clark HC. 1975. The energetics of the jump of the locust *Schistocerca gregaria*. *J. Exp. Biol.* 63:53–83

Borst A, Bahde S. 1988. Spatio-temporal integration of motion: a simple strategy for safe landing in flies. *Naturwissenschaften* 75:265–67

Burrows M. 1995. Motor patterns during kicking movements in the locust. *J. Comp. Physiol. A* 176:289–305

Burrows M. 1996. *The Neurobiology of an Insect Brain*. Oxford: Oxford Univ. Press

Burrows M, Morris G. 2001. The kinematics and neural control of high-speed kicking movements in the locust. *J. Exp. Biol.* 204:3471–81

Burrows M, Pflüger H. 1988. Positive feedback loops from proprioceptors involved in leg movements of the locust. *J. Comp. Physiol. A* 163:425–40

Burrows M, Rowell CHF. 1973. Connections between descending visual interneurons and metathoracic motoneurons in the locust. *J. Comp. Physiol.* 85:221–34

Burrows M, Watson AHD, Brunn DE. 1989. Physiological and ultrastructural characterization of a central synaptic connection between identified motor neurons in the locust. *Eur. J. Neurosci.* 1:111–26

Card G, Dickinson MH. 2008. Visually mediated motor planning in the escape response of *Drosophila*. *Curr. Biol.* 18:1300–7

Dean P, Redgrave P, Westby GWM. 1989. Event or emergency? Two response systems in the mammalian superior colliculus. *Trends Neurosci.* 12:137–47

Fotowat H, Fayyazuddin A, Bellen HJ, Gabbiani F. 2009. A novel neuronal pathway for visually guided escape in *Drosophila melanogaster*. *J. Neurophysiol.* 102:875–85

Fotowat H, Gabbiani F. 2007. Relationship between the phases of sensory and motor activity during a looming-evoked multistage escape behavior. *J. Neurosci.* 27:10047–59

Fotowat H, Harrison RR, Gabbiani F. 2011. Multiplexing of motor information in the discharge of a collision detecting neuron during escape behaviors. *Neuron* 69:147–58

Gabbiani F, Cohen I, Laurent G. 2005. Time-dependent activation of feed-forward inhibition in a looming-sensitive neuron. *J. Neurophysiol.* 94:2150–61

Gabbiani F, Krapp HG. 2006. Spike-frequency adaptation and intrinsic properties of an identified, looming-sensitive neuron. *J. Neurophysiol.* 96:2951–62

Gabbiani F, Krapp HG, Koch C, Laurent G. 2002. Multiplicative computation in a visual neuron sensitive to looming. *Nature* 21:320–24

Gabbiani F, Krapp HG, Laurent G. 1999. Computation of object approach by a wide-field, motion-sensitive neuron. *J. Neurosci.* 19:1122–41

Gabbiani F, Mo C, Laurent G. 2001. Invariance of angular threshold computation in a wide-field looming-sensitive neuron. *J. Neurosci.* 21:314–29

Gallagher SP, Northmore DPM. 2006. Responses of teleostean nucleus isthmi to looming objects and other moving stimuli. *Vis. Neurosci.* 23:209–19

Gibson JJ. 1958. Visually controlled locomotion and visual orientation in animals. *Br. J. Psychol.* 49:182–94

Glantz RM. 1974. Defense reflex and motion detector responsiveness to approaching targets: the motion detector trigger to the defense reflex pathway. *J. Comp. Physiol.* 95:297–314

Godden D. 1975. The neural basis for locust jumping. *Comp. Biochem. Physiol. A* 51:351–60

Guest B, Gray J. 2006. Responses of a looming-sensitive neuron to compound and paired object approaches. *J. Neurophysiol.* 95:1428–41

Gynther IC, Pearson KG. 1986. Intracellular recordings from interneurones and motoneurones during bilateral kicks in the locust: implications for mechanisms controlling the jump. *J. Exp. Biol.* 122:323–43

Gynther IC, Pearson KG. 1989. An evaluation of the role of identified interneurons in triggering kicks and jumps in the locust. *J. Neurophysiol.* 61:45–57

Hammond S, O'Shea M. 2007. Escape flight initiation in the fly. *J. Comp. Physiol. A* 193:471–76

Hatsopoulos N, Gabbiani F, Laurent G. 1995. Elementary computation of object approach by wide-field visual neuron. *Science* 270:1000–3

Hayes WN, Saiff EI. 1967. Visual alarm reactions in turtles. *Anim. Behav.* 15:102–6

Heitler WJ. 1974. The locust jump. *J. Comp. Physiol. A* 89:93–104

Heitler W, Burrows M. 1977. The locust jump. I. The motor programme. *J. Exp. Biol.* 66:203–19

Hemmi J. 2005. Predator avoidance in fiddler crabs: 1. Escape decisions in relation to the risk of predation. *Anim. Behav.* 69:603–14

Holmqvist M, Srinivasan M. 1991. A visually evoked escape response of the housefly. *J. Comp. Physiol. A* 169:451–59

Hoyle G, Burrows M. 1973. Neural mechanisms underlying behavior in locust *Schistocerca gregaria*. 1. Physiology of identified motorneurons in methathoracic ganglion. *J. Neurobiol.* 4:3–41

Ingle DJ, Hoff KV. 1990. Visually elicited evasive behavior in frogs. *BioScience* 40:284–91

Ishikane H, Gangi M, Honda S, Tachibana M. 2005. Synchronized retinal oscillations encode essential information for escape behavior in frogs. *Nat. Neurosci.* 8:1087–95

Johnston D, Narayanan R. 2008. Active dendrites: colorful wings of the mysterious butterflies. *Trends Neurosci.* 31:309–16

Judge S, Rind F. 1997. The locust DCMD, a movement-detecting neurone tightly tuned to collision trajectories. *J. Exp. Biol.* 200:2209–16

Keay KA, Dean P, Redgrave P. 1990. N-methyl D-aspartate (NMDA) evoked changes in blood pressure and heart rate from the rat superior colliculus. *Exp. Brain Res.* 80:148–56

Keay KA, Redgrave P, Dean P. 1988. Cardiovascular and respiratory changes elicited by stimulation of rat superior colliculus. *Brain Res. Bull.* 20:13–26

Killmann F, Schürmann F. 1985. Both electrical and chemical transmission between the 'lobula giant movement detector' and the 'descending contralateral movement detector' neurons of locusts are supported by electron microscopy. *J. Neurocytol.* 14:637–52

King JG, Lettvin JY, Gruberg ER. 1999. Selective, unilateral, reversible loss of behavioral responses to looming stimuli after injection of tetrodotoxin or cadmium chloride into the frog optic nerve. *Brain. Res.* 841:20–26

King JR, Comer CM. 1996. Visually elicited turning behavior in *Rana pipiens*: comparative organization and neural control of escape and prey capture. *J. Comp Physiol. A* 178:293–305

King SM, Cowey A. 1992. Defensive responses to looming visual stimuli in monkeys with unilateral striate cortex ablation. *Neuropsychologia* 30:1017–24

King SM, Dykeman C, Redgrave P, Dean P. 1992. Use of a distracting task to obtain defensive head movements to looming visual stimuli by human adults in a laboratory setting. *Perception* 21:245–59

Krapp HG, Gabbiani F. 2005. Spatial distribution of inputs and local receptive field properties of a wide-field, looming sensitive neuron. *J. Neurophysiol.* 93:2240–53

Kuitert LC, Connin RV. 1952. Biology of the American grasshopper in the southeastern United States. *Florida Entomol.* 35:22–33

Laurent G, Gabbiani F. 1998. Collision-avoidance: nature's many solutions. *Nat. Neurosci.* 1:261–63

Lee DN. 1976. A theory of visual control of braking based on information about time-to-collision. *Perception* 5:437–59

Lee DN, Reddish PE. 1981. Plummeting gannets: a pradigm of ecological optics. *Nature* 293:293–94

Lettvin JY, Maturana H, McCulloch W, Pitts W. 1968. What the frog's eye tells the frog's brain. In *The Mind: Biological Approaches to its Function*, ed. WC Corning, M Balaban, 7:233–58. New York: Interscience. 321 pp.

Liu KS, Fetcho JR. 1999. Laser ablations reveal functional relationships of segmental hindbrain neurons in zebrafish. *Neuron* 23:325–35

Maimon G, Straw AD, Dickinson MH. 2010. Active flight increases the gain of visual motion processing in *Drosophila*. *Nat. Neurosci.* 13:393–99

Matheson T, Rogers SM, Krapp HG. 2004. Plasticity in the visual system is correlated with a change in lifestyle of solitarious and gregarious locusts. *J. Neurophysiol.* 91:1–12

Medan V, Oliva D, Tomsic D. 2007. Characterization of lobula giant neurons responsive to visual stimuli that elicit escape behaviors in the crab Chasmagnathus. *J. Neurophysiol.* 98:2414–28

Mitchell IJ, Redgrave P, Dean P. 1988a. Plasticity of behavioural response to repeated injection of glutamate in cuneiform area of rat. *Brain Res.* 460:394–97

Mitchell IJ, Dean P, Redgrave P. 1988b. The projection from superior colliculus to cuneiform area in the rat. II. Defence-like responses to stimulation with glutamate in cuneiform nucleus and surrounding structures. *Exp. Brain Res.* 72:626–39

Münch TA, da Silveira RA, Siegert S, Viney TJ, Awatramani GB, Roska B. 2009. Approach sensitivity in the retina processed by a multifunctional neural circuit. *Nat. Neurosci.* 12:1308–16

Nakagawa H, Hongjian K. 2010. Collision-sensitive neurons in the optic tectum of the bullfrog, *Rana catesbeiana.* *J. Neurophysiol.* 104:2487–99

Northmore DP, Levine ES, Schneider GE. 1988. Behavior evoked by electrical stimulation of the hamster superior colliculus. *Exp. Brain Res.* 73:595–605

O'Shea M. 1975. Two sites of axonal spike initiation in a bimodal interneuron. *Brain Res.* 96:93–98

O'Shea M, Rowell CHF. 1975a. Protection from habituation by lateral inhibition. *Nature* 254:53–55

O'Shea M, Rowell CHF. 1975b. A spike-transmitting electrical synapse between visual interneurones in the locust movement detector system. *J. Comp. Physiol.* 97:143–58

O'Shea M, Rowell CHF. 1976. The neuronal basis of a sensory analyser, the acridid movement detector system. II. Response decrement, convergence, and the nature of the excitatory afferents to the fan-like dendrites of the LGMD. *J. Exp. Biol.* 65:289–308

O'Shea M, Rowell CHF, Williams JLD. 1974. The anatomy of a locust visual interneurone; the descending contralateral movement detector. *J. Exp. Biol.* 60:1–12

O'Shea M, Williams JLD. 1974. The anatomy and output connection of a locust visual interneurone; the lobula giant movement detector (LGMD) neurone. *J. Comp. Physiol.* 91:257–66

Oliva D, Medan V, Tomsic D. 2007. Escape behavior and neuronal responses to looming stimuli in the crab *Chasmagnathus granulatus* (Decapoda: Grapsidae). *J. Exp. Biol.* 210:865–80

Palka J. 1967. An inhibitory process influencing visual responses in a fibre of the ventral nerve cord of locusts. *J. Insect Physiol.* 13:235–48

Pearson KG, Robertson RM. 1981. Interneurons coactivating hindleg flexor and extensor motoneurons in the locust. *J. Comp. Physiol. A* 144:391–400

Pearson KG, Heitler WJ, Steeves JD. 1980. Triggering of locust jump by multimodal inhibitory interneurons. *J. Neurophysiol.* 43:257–78

Peron SP, Gabbiani F. 2009. Spike frequency adaptation mediates looming stimulus selectivity in a collision-detecting neuron. *Nat. Neurosci.* 12:318–26

Peron SP, Jones PW, Gabbiani F. 2009. Precise subcellular input retinotopy and its computational consequences in an identified visual interneuron. *Neuron* 63:830–42

Peron SP, Krapp HG, Gabbiani F. 2007. Influence of electrotonic structure and synaptic mapping on the receptive field properties of a collision-detecting neuron. *J. Neurophysiol.* 97:159–77

Pfeiffer BD, Jenett A, Hammonds AS, Ngo TT, Misra S, et al. 2008. Tools for neuroanatomy and neurogenetics in *Drosophila*. *Proc. Natl. Acad. Sci. USA* 105:9715–20

Polsky A, Mel BW, Schiller J. 2004. Computational subunits in thin dendrites of pyramidal cells. *Nat. Neurosci.* 7:621–27

Preston-Mafham K. 1990. *Grasshoppers and Mantids of the World*. London: Blandford

Preuss T, Osei-Bonsu PE, Weiss SA, Wang C, Faber DS. 2006. Neural representation of object approach in a decision-making motor circuit. *J. Neurosci.* 26:3454–64

Redgrave P, Dean P, Mitchell IJ, Odekunle A, Clark A. 1988. The projection from superior colliculus to cuneiform area in the rat. I. Anatomical studies. *Exp. Brain Res.* 72:611–25

Rind FC, Simmons PJ. 1992. Orthopteran DCMD neuron: a reevaluation of responses to moving objects. I. Selective responses to approaching objects. *J. Neurophysiol.* 68:1654–66

Rind FC. 1984. A chemical synapse between two motion detecting neurones in the locust brain. *J. Exp. Biol.* 110:143–67

Rind FC, Santer RD, Wright GA. 2008. Arousal facilitates collision avoidance mediated by a looming sensitive visual neuron in a flying locust. *J. Neurophysiol.* 100:670–80

Robertson RM, Johnson AG. 1993. Retinal image size triggers obstacle avoidance in flying locusts. *Naturwissenschaften* 80:176–78

Rogers SM, Harston GW, Kilburn-Toppin F, Matheson T, Burrows M, et al. 2010. Spatiotemporal receptive field properties of a looming-sensitive neuron in solitarious and gregarious phases of the desert locust. *J. Neurophysiol.* 103:779–92

Rogers SM, Krapp HG, Burrows M, Matheson T. 2007. Compensatory plasticity at an identified synapse tunes a visuomotor pathway. *J. Neurosci.* 27:4621–33

Rowell CHF. 1971a. The orthopteran descending movement detector (DMD) neurones: a characterisation and review. *J. Comp. Physiol. A* 73:167–94

Rowell CHF. 1971b. Variable responsiveness of a visual interneurone in the free-moving locust, and its relation to behaviour and arousal. *J. Exp. Biol.* 55:727–47

Rowell CHF, O'Shea M, Williams JL. 1977. The neuronal basis of a sensory analyser, the acridid movement detector system. IV. The preference for small field stimuli. *J. Exp. Biol.* 68:157–85

Sahibzada N, Dean P, Redgrave P. 1986. Movements resembling orientation or avoidance elicited by electrical stimulation of the superior colliculus in rats. *J. Neurosci.* 6:723–33

Santer RD, Rind FC, Stafford R, Simmons PJ. 2006. Role of an identified looming-sensitive neuron in triggering a flying locust's escape. *J. Neurophysiol.* 95:3391–400

Santer RD, Simmons PJ, Rind FC. 2005a. Gliding behaviour elicited by lateral looming stimuli in flying locusts. *J. Comp. Physiol. A* 191:61–73

Santer RD, Yamawaki Y, Rind FC, Simmons PJ. 2005b. Motor activity and trajectory control during escape jumping in the locust *Locusta migratoria*. *J. Comp. Physiol. A* 191:965–75

Santer RD, Yamawaki Y, Rind FC, Simmons PJ. 2008. Preparing for escape: an examination of the role of the DCMD neuron in locust escape jumps. *J. Comp. Physiol. A* 194:69–77

Schiff W, Caviness JA, Gibson JJ. 1962. Persistent fear responses in rhesus monkeys to optical stimulus of "looming". *Science* 136:982–83

Schlotterer GR. 1977. Response of the locust descending movement detector neuron to rapidly approaching and withdrawing visual stimuli. *Can. J. Zool.* 55:1372–76

Scott EK. 2009. The Gal4/UAS toolbox in zebrafish: new approaches for defining behavioral circuits. *J. Neurochem.* 110:441–56

Simmons PJ. 1980. Connexions between a movement-detecting visual interneurone and flight motoneurones of a locust. *J. Exp. Biol.* 86:87–97

Srinivasan MV, Zhang SW, Chahl JS, Barth E, Venkatesh S. 2000. How honeybees make grazing landings on flat surfaces. *Biol. Cybern.* 83:171–83

Steeves JD, Pearson KG. 1982. Proprioceptive gating of inhibitory pathways to hindleg flexor motoneurons in the locust. *J. Comp. Physiol. A* 146:507–15

Strausfeld NJ, Nässel DR. 1981. Neural architecture serving compound eyes of crustacea and insects. In *Handbook of Sensory Physiology, Comparative Physiology and Evolution of Vision of Invertebrates, B: Invertebrate Visual Centers and Behavior*, ed. H Autrum, 7:1–132. Berlin: Springer Verlag

Sun H, Frost BJ. 1998. Computation of different optical variables of looming objects in pigeon nucleus rotundus neurons. *Nat. Neurosci.* 1:296–303

Sztarker J, Tomsic D. 2008. Neuronal correlates of the visually elicited escape response of the crab Chasmagnathus upon seasonal variations, stimuli changes and perceptual alterations. *J. Comp. Physiol. A* 194:587–96

Tammero LF, Dickinson MH. 2002. Collision-avoidance and landing responses are mediated by separate pathways in the fruit fly *Drosophila melanogaster*. *J. Exp. Biol.* 205:2785–98

Tresilian JR. 1999. Visually timed action: time-out for 'tau'? *Trends Cog. Sci.* 3:301–10

Wagner H. 1982. Flow-field variables trigger landing in flies. *Nature* 297:147–48

Wang Y, Frost BJ. 1992. Time to collision is signalled by neurons in the nucleus rotundus of pigeons. *Nature* 356:236–38

Westby GW, Keay KA, Redgrave P, Dean P, Bannister M. 1990. Output pathways from the rat superior colliculus mediating approach and avoidance have different sensory properties. *Exp. Brain Res.* 81:626–38

Wicklein M, Strausfeld NJ. 2000. Organization and significance of neurons that detect change of visual depth in the hawk moth *Manduca sexta*. *J. Comp. Neurol.* 424:356–76

Wine JJ, Krasne FB. 1972. The organization of escape behaviour in the crayfish. *J. Exp. Biol.* 56:1–18

Wu LQ, Niu YQ, Yang J, Wang SR. 2005. Tectal neurons signal impending collision of looming objects in the pigeon. *Eur. J. Neurosci.* 22:2325–31

Yamamoto K, Nakata M, Nakagawa H. 2003. Input and output characteristics of collision avoidance behavior in the frog *Rana catesbeiana*. *Brain Behav. Evol.* 62:201–11

Zottoli SJ. 1977. Correlation of the startle reflex and Mauthner cell auditory responses in unrestrained goldfish. *J. Exp. Biol.* 66:243–54

Zottoli SJ, Faber DS. 2000. The Mauthner cell: What has it taught us? *Neuroscientist* 6:26–38

Myelin Regeneration: A Recapitulation of Development?

Stephen P.J. Fancy,[1] Jonah R. Chan,[2]
Sergio E. Baranzini,[2] Robin J.M. Franklin,[4]
and David H. Rowitch[1,3]

[1]Departments of Pediatrics and Neurosurgery, Eli and Edyth Broad Institute for Stem Cell Research and Regeneration Medicine and Howard Hughes Medical Institute, [2]Department of Neurology, and [3]Division of Neonatology, University of California, San Francisco, California 94143; email: rowitchd@peds.ucsf.edu

[4]MRC Center for Stem Cell Biology and Regenerative Medicine and Department of Veterinary Medicine, University of Cambridge, Cambridge CB3 0ES, United Kingdom

Annu. Rev. Neurosci. 2011. 34:21–43

The *Annual Review of Neuroscience* is online at neuro.annualreviews.org

This article's doi:
10.1146/annurev-neuro-061010-113629

Keywords

myelination, remyelination, multiple sclerosis, oligodendrocyte, CNS development

Abstract

The developmental process of myelination and the adult regenerative process of remyelination share the common objective of investing nerve axons with myelin sheaths. A central question in myelin biology is the extent to which the mechanisms of these two processes are conserved, a concept encapsulated in the recapitulation hypothesis of remyelination. This question also has relevance for translating myelin biology into a better understanding of and eventual treatments for human myelin disorders. Here we review the current evidence for the recapitulation hypothesis and discuss recent findings in the development and regeneration of myelin in the context of human neurological disease.

Contents

OPCs:
oligodendrocyte
precursor cells

OLs:
oligodendrocytes

MS: multiple sclerosis

INTRODUCTION

The recapitulation hypothesis of myelin regeneration holds that mechanisms that underlie remyelination after injury are essentially a rerunning of a developmental myelination program (Franklin & Hinks 1999). Rather than providing a comprehensive review of the literature of developmental myelination and remyelination, which have been reviewed extensively elsewhere (Chong & Chan 2010, Dubois-Dalcq et al. 2008, Franklin & Ffrench-Constant 2008, Rowitch 2004), our purpose here is to focus on current evidence that supports or refutes the recapitulation hypothesis. An ever-increasing appreciation of the relationship between these two processes will help investigators translate glial developmental studies to understanding human myelin diseases.

Why is understanding processes of developmental myelination and its possible conservation in myelin regeneration important? One reason is that aberrations in the functions of oligodendrocyte precursor cells (OPCs) and oligodendrocytes (OLs) are a significant cause of human neurologic disease.

Demyelinating disease appears in many forms, of which multiple sclerosis (MS) is the most prevalent in young adults (Compston & Coles 2002). MS typically presents in a relapsing/remitting form, with remyelination and functional recovery following on from acute demyelinating episodes (**Figure 1**). However, a significant contributor to the deterioration seen in MS is the axonal loss that characterizes the progressive phase of the disease (Bjartmar et al. 2003). Several lines of evidence link axonal loss to the failure of remyelination, which occurs as the disease progresses. For example, axonal loss is less in areas of remyelination than in areas of chronic demyelination (**Figure 1c,d**) (Kornek et al. 2000), and also secondary disruption of axonal integrity is associated with primary defects in myelinating OLs (Edgar et al. 2004, Garbern et al. 2002, Griffiths et al. 1998, Lappe-Siefke et al. 2003). Therefore, understanding the mechanisms of remyelination is an important objective; it allows investigators to determine why remyelination fails and, hence, to develop potential therapeutic targets (Franklin & Ffrench-Constant 2008).

Periventricular leukomalacia (PVL), which is increasingly referred to by the term white matter injury, is a significant acquired cause

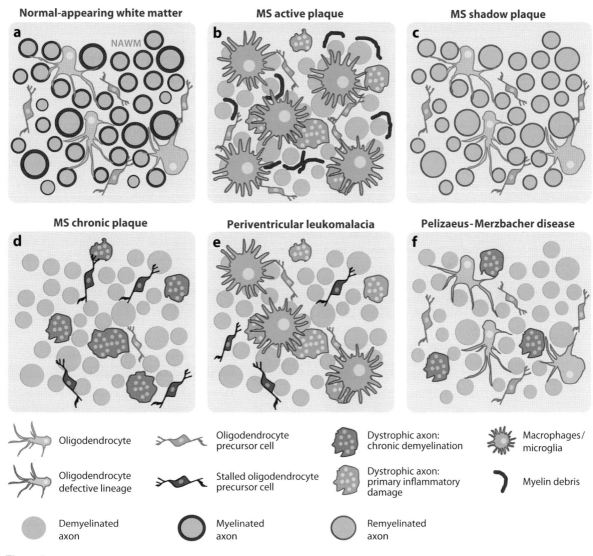

Figure 1

Cartoon illustrating significant differences observed among myelin sheaths, axons, and inflammatory cells in various pathological conditions. Transverse sections through white matter axons (*blue circles*) of (*a*) normal-appearing white matter (NAWM), (*b*) acutely injured white matter (active plaque), (*c*) myelin regeneration in a shadow plaque, or (*d*) chronic demyelination (chronic plaque) in an inflammatory disease such as multiple sclerosis (MS). Note that in MS, myelin sheaths surrounding axons in NAWM are thicker than in remyelinated axons (*compare black outer rings, a, c*) and (*b*) acute inflammation (macrophages/microglia, *green*) is thought to cause axonal damage and clear inhibitors (e.g., myelin debris). In chronically demyelinated lesions (*d*), stalled OPCs (*dark blue*) can be demonstrated, which suggests that these cells are blocked from effective repair. (*e*) Similar cells are observed in periventricular leukomalacia (PVL). (*f*) Contrasting these demyelinating conditions is the congenital leukodystrophy Pelizaeus-Merzbacher disease (PMD), in which oligodendrocytes typically are defective in myelin production, causing a failure of developmental myelination (primary phase of disease), which ultimately leads to massive axonal damage (secondary phase of disease).

PVL: periventricular
leukomalacia

of cerebral palsy in developed countries that provide advanced neonatal care to extremely low birth weight (ELBW) infants born at <28 weeks gestation (normal is 40 weeks) (Volpe 2001, Yoon et al. 2003). This disease is characterized by cystic/focal as well as diffuse gliotic damage to developing white matter around the lateral ventricles of the brain, eventually leading to fixed, demyelinated lesions. Kinney & Back (1998) hypothesize that the white matter damage results from intrauterine or perinatal hypoxic-ischemic events during peak periods of oligodendroglial development, leading to excitotoxic glutamate death of oligodendroglia (**Figure 1e**). However, more recent studies have identified OPCs within lesions of PVL, suggesting that failure of OPC differentiation accounts in part for fixed demyelination (Billiards et al. 2008). The increasing impact of PVL in the United States (due to rising rates of ELBW survival) is a compelling reason to redouble efforts to understand this complex disorder (Back & Rivkees 2004).

A large class of congenital myelin diseases results from abnormal production/composition of myelin, owing to mutations in at least 20 genes with roles in development of oligodendroglia (Inoue et al. 1999, Touraine et al. 2000) or myelination (Inoue et al. 2004). The archetype for this class is the leukodystrophy Pelizaeus-Merzbacher disease (**Figure 1f**), which can result mostly from point mutations or duplications of the *Phospholipid Protein 1* (*PLP1*) gene, encoding a protein component of myelin (Mimault et al. 1999). Mouse models of the disease (e.g., *jimpy*) exhibit a phenotype more severe than can simply be explained by structural abnormalities of myelin proteins alone, and these models suggest that *PLP* function may be required for OLs to develop and mature in postnatal stages (Campagnoni & Skoff 2001, Schneider et al. 1992).

In the following sections, we briefly review (*a*) the process of developmental myelination, (*b*) instances in which observations from development may or may not show conservation in remyelination, and (*c*) future directions and priorities for myelin regeneration research, discussing translational insights and therapy for human neurological diseases.

OVERVIEW OF OLIGODENDROCYTE DEVELOPMENT

The innovation of myelinating OLs in the CNS and Schwann cells in the peripheral nervous system (PNS) comprised a major evolutionary advance, permitting more complex nervous system structure and function. Although invertebrates have ensheathing glia, and some even produce myelin (Roots 2008, Zalc & Colman 2000), the myelinating OL is a cell type uniformly found in vertebrates above the jawless fishes. In addition to ensheathment, which facilitates electrical conduction, OLs fulfill diverse functional roles such as maintenance of axonal integrity and participation in signaling networks with neurons (Bergles et al. 2000, Karadottir et al. 2008, Lin & Bergles 2004).

Pattern Formation and Embryonic Oligodendrocyte Specification

Embryonic oligodendroglial specification shares mechanistic features with motor neurons of the ventral neural tube (Kessaris et al. 2001, Rowitch 2004). A critical step is the establishment of distinct neuronal progenitor cell domains during the pattern formation process (Jessell 2000, Marquardt & Pfaff 2001). Important patterning molecules include the ventralizing signal, Sonic hedgehog (Shh), and dorsally secreted bone morphogenetic proteins (BMPs). Such positional cues, secreted from organizing centers (e.g., floor plate and roof plate), instruct nearby progenitor cells to adopt identities unique to their anterior/posterior and dorsal/ventral position in the embryonic axes. This positional information is reflected in the expression of specific sets of transcription factors (TFs) that regulate subsequent cell fate decisions and progeny subtype specialization (**Figure 2a**).

In the embryonic spinal cord, OPCs are specified in a ventral progenitor domain, which

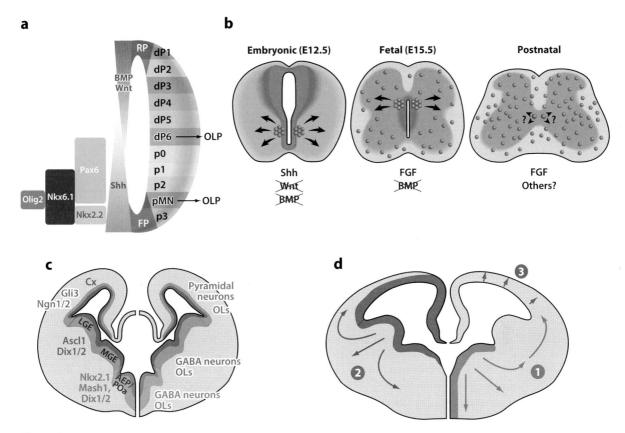

Figure 2

Key events in oligodendrocyte precursor cell (OPC) development. (*a*) Patterning of the embryonic neural tube by organizing signals (Shh, Wnt, BMP) is reflected in the expression of bHLH (Olig2) and homeodomain (Nkx2.2, Nkx6.1, Pax6) transcription factors in restricted progenitor domains (pMN, p0-p3, dP1-dP6) that give rise to specific neuron and glial subtypes. (*b*) Embryonic OPCs arise from the ventral pMN domain under control of Shh, whereas at fetal stages, a dorsal OPC-generating domain is active that is Shh-independent. OPCs from ventral and dorsal regions are presumably intermixed in the spinal cord at birth, although the later OPCs appear to occupy a more dorsal distribution. (*c*) In the forebrain, distinct domains demarcated by expression of specific transcription factors, such as Dlx1/2, Ascl1, and Olig2, give rise to neurons and glia. (*d*) Multiple waves of OPC production occur from embryonic to postnatal stages, emerging in a ventral-to-dorsal progression. The later dorsally derived cortical OPCs distribute to all parts of the brain and entirely compensate on a functional level for early ventral OPCs.

also produces motor neurons, called the pMN. Shh signaling, acting through activation of Olig2, is essential for motor neuron and OL development (Lu et al. 2002). Conversely, dorsally derived BMPs and Wnts antagonize OL development, while supporting astrocyte development (Agius et al. 2004, Shimizu et al. 2005). Therefore, the relative intensity of Shh, BMP, and Wnt signals is a critical determinant of early glial specification. Together, these findings indicate that embryonic OL and astrocyte precursors undergo initial specification in mutually

exclusive and restricted progenitor domains of the ventral neural tube (Rowitch 2004, Zhou & Anderson 2002) as opposed to arising from a common glial-restricted precursor cell present at all dorsoventral levels of the spinal cord (Noble et al. 2004, Rao & Mayer-Proschel 1997).

Temporal Waves of Oligodendrocyte Production

pMN progenitors develop in stages, with complete motor neuron formation by E10.5 in mice, followed by a phase of OL production (blue cells, **Figure 2b**) (commencing at E12.5) and exhaustion of cycling pMN progenitors during embryonic stages at E14.5. Several lines of evidence indicate that the pMN contains independently segregating populations of neuroblasts and glioblasts, rather than a bi-potent neuronglial precursor (Wu et al. 2006).

OPCs derived from the pMN continue to proliferate after specification, and they migrate laterally and dorsally to occupy all areas of the spinal cord. Additionally, at later (fetal, commencing ~E15.5) stages of spinal cord development, dorsal sources of OPC production emerge (red cells, **Figure 2b**) (Cai et al. 2005, Fogarty et al. 2005, Vallstedt et al. 2005). The overall contribution of dorsal- versus ventral-derived OPC is estimated to be <20%–30% of the total OPCs seen in wild-type animals (Vallstedt et al. 2005). Chandran et al. (2003) initially showed that, unlike their ventral counterparts, dorsally derived OPCs do not require Shh for development in vitro and arise in a fibroblast growth factor (FGF)-dependent manner (**Figure 2b**). Cai et al. (2005) subsequently confirmed and extended this finding by demonstrating in vivo that, although ventral OPCs do not develop in $Shh^{-/-}$ mice, OPCs begin appearing in more dorsal regions of the neural tube at 14.5 dpc, indicating that dorsal OPC development does not require Shh.

An analogous, if not more complex, situation exists in forebrain where cre recombinase fate-mapping experiments have demonstrated multiple waves of OPC production from embryonic to postnatal stages that emerge in a ventral-to-dorsal progression (Kessaris et al. 2006) (**Figure 2d**). The later dorsally derived cortical OPCs distribute to all parts of the brain and entirely compensate on a functional level for early ventral OPCs. The multiple sites of OPC development have caused investigators to speculate that OPC progeny are diversified in molecular or functional terms, but such differences have not been confirmed.

Once specified, OPCs exhibit multidirectional migration from their origins in the ventricular zone (VZ) to distant sites under control of both repulsive and attractive cues (Simpson & Armstrong 1999; Tsai et al. 2002, 2003). OPCs are marked by expression of platelet-derived growth factor receptor alpha (PDGFRα), Nkx2.2, and NG2, whereas mature OLs express *adenomatous polyposis coli* (APC), proteolipid protein (PLP), myelin basic protein (MBP), and myelin oligodendrocyte glycoprotein (MOG) (**Figure 3**), among others.

Regulation of Oligodendrocyte Maturation

Whereas some OPCs find targets and differentiate into myelinating cells early on, others are reserved as immature cells throughout the parenchyma, presumably for purposes of replacing OLs at a constant rate of turnover or at an accelerated level for myelin regeneration (Kang et al. 2010, Tripathi et al. 2010). However, emerging information on the electrical properties of these cells, such as their ability to engage in synaptic interactions (De Biase et al. 2010), suggests that they also perform physiological roles in the intact adult CNS.

Developmental myelination in the CNS depends on the coordination of two processes that are intimately associated with one another: terminal differentiation of OPCs into OLs and the wrapping of the relevant axons. Complicated by the heterogeneous spatiotemporal origin of OPCs and the lack of inductive cues that promote differentiation and myelination, uncoupling these two cellular processes has been difficult. In recent years, much effort has been devoted to understanding the transcriptional program necessary for differentiation of OLs. This process has led to the identification of the *Olig* genes (Lu et al. 2000, Zhou et al. 2000), basic helix-loop-helix (bHLH) TFs that play multiple roles in determining the oligodendroglial lineage, and in the differentiation of OLs (Li et al. 2007, Lu et al. 2002, Qi et al. 2002, Takebayashi et al. 2002, Tekki-Kessaris et al. 2001, Xin et al. 2005, Zhou & Anderson 2002). Olig2 knockout mice fail to develop cells of the oligodendroglial lineage (Lu et al. 2002, Zhou & Anderson 2002), whereas Olig1

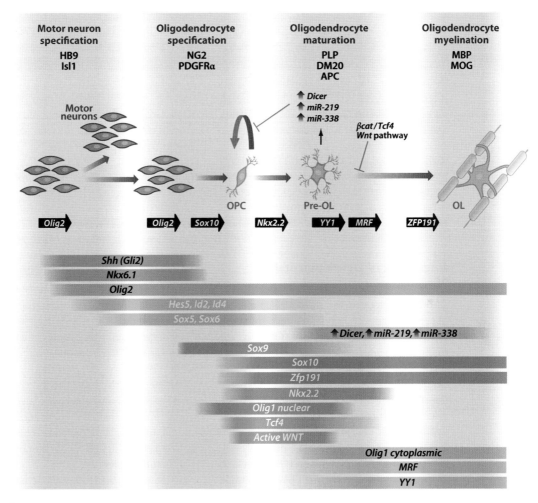

Figure 3

Schematic illustrating multiple mechanisms influencing oligodendrocyte development and differentiation at various developmental stages. The diagram summarizes the interactions that are implicated in the neuron-glial switch and oligodendrocyte precursor cell (OPC) maturation modeled on progeny produced from the ventral pMN domain of the neural tube. OPCs are marked by PDGFRα and NG2 expression, whereas oligodendrocytes express *adenomatous polyposis coli* (APC) and proteolipid protein (PLP). Myelinating oligodendrocytes express myelin basic protein (MBP) and myelin oligodendrocyte glycoprotein (MOG) and make compact myelin (demonstrated, for example, by ultra structure). During embryogenesis, ongoing exposure to Shh is required until the time of oligodendrocyte precursor (OPC) migration from the ventricular zone, reflecting, in part, essential roles for *Gli2*, Nkx6.1, and Nkx6.2, as well as temporal-dependent cell fate specification mechanisms (Briscoe & Novitch 2008). These activities are required so that the pMN is competent to produce motor neurons and oligodendrocytes. The switch of progeny cell production seems to involve downregulation of neurogenic factors (e.g., Ngn2) at the end of motor neuron production and maintenance of glial precursors until the second wave by Delta–Notch signaling (and downstream Hes factors), Id2/4 and proglial activity of Sox9. Later phases of OPC maturation are Shh-independent and require Sox10, Nkx2.2, and Olig1 function. Several key regulators of developmental myelination have recently been reported, including transcription factors YY1, MRF, ZFP191, and Tcf4, intracellular signaling pathways such as Wnt (mediated by β-catenin/Tcf4), and also posttranscriptional control via miRNAs. *Solid black arrows* represent OL-lineage transitions that are dependent on specific transcription factors (based on mouse knockout experiments), and *colored gradient bars* represent predicted temporal expression patterns. Our understanding of how these multifaceted transcription factors, intracellular signaling pathways, and posttranscriptional mechanisms interact in an integrated fashion to coordinate myelin generation and regeneration represents an important future area of integrative research.

is thought to function later in development (Lu et al. 2002, Xin et al. 2005) and myelin regeneration (Arnett et al. 2004). In forebrain, *Olig2* function is similarly required for OPC development, and *Dlx1/2* function regulates interneuron versus OL cell fate by repressing *Olig2* (Petryniak et al. 2007). Loss-of-function analysis shows general and essential roles for *Nkx2.2* in differentiation of OLs, but not in their specification (Qi et al. 2001). Stolt et al. (2002) made similar observations in the CNS of *Sox10*$^{-/-}$ null mutants, in which a global delay of mature marker acquisition is observed but initial specification of OPCs is unaffected.

Regulation of Developmental Myelination

Although investigators have described numerous transcriptional regulators that influence differentiation of OPCs into OLs, recent findings provide evidence for an additional maturation step in the transition of OLs into myelin-generating cells. Myelin gene regulatory factor (MRF) (Emery et al. 2009), the zinc finger protein 191 (Zfp191) (Howng et al. 2010), and the TF Ying Yang 1 (YY1) (He et al. 2007) are all thought to function after cell cycle exit and terminal differentiation. Mice lacking MRF in the oligodendroglial lineage continue to generate OLs; however, these cells do not fully mature and display defects in myelin gene expression and myelin internode formation (Emery et al. 2009). Similarly, mice expressing a mutation in *Zfp191* or a loss-of-function of *YY1* generate OLs that do not fully mature to form proper myelin segments (He et al. 2007, Howng et al. 2010).

YY1 is thought to repress transcriptional inhibitors by recruiting histone deacetylases (HDACs) (He et al. 2007). HDACs normally remove acetyl groups from histones to allow for chromatin compaction, which subsequently silences transcription. Recent evidence suggests that this mechanism may be responsible for the repression of pathways that normally prevent OL differentiation (Li et al. 2009, Shen & Casaccia-Bonnefil 2008).

We and others recently demonstrated that canonical Wnt signaling is a potent regulator of OL differentiation (Fancy et al. 2009, Feigenson et al. 2009, Ye et al. 2009). Classically, the Wnt pathway acts to stabilize β-catenin protein levels, which then can complex with T-cell factor/lymphoid enhancer factor (Tcf/LEF) family TFs to activate or repress gene targets (Behrens et al. 1996, Molenaar et al. 1996). In OLs lacking *HDAC1/2* function, there is stabilization of β-catenin. Moreover, HDAC1/2 directly associates with Tcf4 to inhibit β-catenin-Tcf4 interactions. Thus, *HDAC1/2* null OLs have increased Wnt tone (Ye et al. 2009, Zhao et al. 2010). Constitutive expression of stabilized ß-catenin impairs OPC differentiation and re-myelination (Fancy et al. 2009, Ye et al. 2009). Heterozygous loss of *adenomatous polyposis coli* (*APC*), encoding a β-catenin antagonist, results in a similar phenotype (Fancy et al. 2009). Taken together, these findings suggest a role for the Wnt pathway, representing an environmental cue that may influence the complexity of transcriptional and epigenetic mechanisms for the precise spatiotemporal coordination required for OL development. The cellular source of Wnt signals in development and repair is still unclear.

OL development is dependent on precise regulation of transcriptional programs, although miRNAs also play essential roles at multiple stages. Conditional ablation of *Dicer*, the processing enzyme for miRNAs, results in defects in OL cell cycle arrest, differentiation, and myelination. These findings are dependent on the temporal expression of the promoter driving the expression of the cre recombinase (Dugas et al. 2010, Shin et al. 2009, Zhao et al. 2010). By identifying targets for a subset of the miRNAs, investigators uncovered a number of genes that are important for OPC maintenance. These target genes include the TFs Hes5, Sox6, Zfp238, and FoxJ3, as well as PDGFRα, essential for promoting OPC proliferation (Dugas et al. 2010, Zhao et al. 2010). Additionally, the finding that very long chain fatty acids protein 7 (Elolv7) is a target of miRNA in postmitotic OLs suggests roles in lipid metabolism regulation and myelin membrane maintenance (Shin et al. 2009).

Because some axons are myelinated whereas others are not, it seems likely that axon-derived signals regulate myelination. In fact, axon diameter may represent a crucial regulator of myelination, as evidenced by experimentally increasing the size of an axonal target, which is sufficient to promote a corresponding increase in axon diameter and subsequent myelination of previously unmyelinated axons (Voyvodic 1989). Although this experiment was performed on peripheral axons, changes in axon diameter also likely regulate CNS myelination.

However, more recent studies suggest that an increase in axon diameter may not represent the only axonal factor responsible for myelination by OLs. Identifying extrinsic inhibitors of myelination, including the Notch pathway (Genoud et al. 2002, Zhang et al. 2009), PSA-NCAM (Charles et al. 2002), hyaluronan (Back et al. 2005), myelin (Miller 1999), and LINGO-1 (Li et al. 2007, Mi et al. 2005), suggests that endogenous inhibitors of developmental myelination may ensure myelination of the proper axons. With the exception of hyaluronan, these inhibitory cues represent axon-derived membrane-bound factors, implying that cell–cell contact is required to prevent differentiation and/or myelination. These findings indicate that even though relevant axons require myelination, establishing a permissive axonal environment does not occur by default.

Investigators have identifed various intracellular signaling pathways and mechanisms that transduce signals to promote and/or inhibit OL differentiation and myelination. Several lines of evidence indicate that the PI-3 kinase/Akt/mTOR signaling pathway is required to promote myelination (Flores et al. 2008, Narayanan et al. 2009). Forced expression of constitutively active Akt or inactivation of *phosphatase and tensin homolog (PTEN)* results in hypermyelination of axons, which is dependent on activating the Akt substrate mTOR (Flores et al. 2008, Harrington et al. 2010, Narayanan et al. 2009). Further studies

demonstrate that the membrane proteases γ- and β-secretase function during OL development. Myelination is delayed and myelin extent is reduced in β-secretase null mice (Hu et al. 2006), whereas γ-secretase inhibition enhances OL differentiation and myelination in vitro (Watkins et al. 2008). These opposing effects illustrate the presence of endogenous inductive and inhibitory cues because β-secretase is downstream of Akt signaling and γ-secretase is responsible for inhibitory Notch signaling. Integrating various exogenous cues and balanced inductive and inhibitory signaling may lead to the proper timing of differentiation and myelination (discussed below).

OVERVIEW OF REMYELINATION

Remyelination is the regenerative process during which new myelin sheaths are restored to axons following pathological demyelination. This process is sometimes called myelin repair, although this term carries the inaccurate connotation that damage to myelin is patched up rather than that lost myelin is being replaced. The remaining sections review processes associated with effective myelin regeneration and provide a critical assessment of the recapitulation hypothesis as a means to generate perspective on its underlying molecular mechanisms.

Remyelination and Maintenance of Axon Health

Until relatively recently, the sole benefit of remyelination was thought to be the restoration of saltatory conduction and hence restoration of lost function. Although this remains an important role of remyelination (Duncan et al. 2009, Jeffery & Blakemore 1997, Smith et al. 1979), remyelination may also play an important role in maintaining axonal integrity. This view is based on several lines of evidence suggesting that the metabolic stress on neurons maintaining long axons is offset, in part, by local trophic support from myelinating cells (Nave 2010, Nave & Trapp 2008). For example, OL-specific deletions in myelin-associated genes *PLP*, *myelin associated glycoprotein* (*MAG*), and *2', 3'-cyclic nucleotide 3'-phosphodiesterase* (*CNPase*) in mice do not cause any obvious defect in myelination but eventually lead to axonal pathology (Griffiths et al. 1998, Lappe-Siefke et al. 2003, Li et al. 1994). Similarly, substituting PLP with the peripheral myelin protein P_0 in mice leads to axonal dystrophy, implying an as yet uncharacterized trophic role for PLP in axonal maintenance (Yin et al. 2006). This newly appreciated role of myelin (*a*) provides a compelling explanation for the progressive atrophy of chronically demyelinated axons that underpins the irreversible and progressive decline that characterizes the later stages of MS and (*b*) implies that therapeutic promotion of remyelination could effectively prevent the neurodegenerative component of demyelinating disease. Interestingly, patients with severe PMD lack myelin (**Figure 1f**) yet may show a prolonged period (e.g., 2 years) before there is grossly evident loss of axon tracts by cranial scans [e.g., magnetic resonance imaging (MRI)].

Figure 1 illustrates some of the differences between normal-appearing white matter (NAWM), acutely injured white matter, and the ultimate outcomes of remyelination or chronic demyelination in an inflammatory disease such as MS. Note that in NAWM, the myelin sheath thickens with increasing axon diameter, whereas in remyelination the myelin sheaths are uniformly thin regardless of axon diameter (compare **Figures 1a,c**). The innate immune component of acute inflammation (**Figure 1b**) is necessary for remyelination because inflammatory mediators likely activate OPCs (Glezer et al. 2006), whereas macrophages play an important role in removing differentiation inhibitors within myelin debris (Kotter et al. 2006). In chronically demyelinated lesions (**Figure 1d**), OPCs can be present in high numbers but fail to differentiate into remyelinating OLs (Chang et al. 2002, Kuhlmann et al. 2008, Wolswijk 1998). A similar situation seems to pertain to PVL (**Figure 1e**), where OPCs, but not myelinating OLs (Billiards et al. 2008), occur. Axon loss can be extensive in PVL

(Kinney & Back 1998). In contrast to these acquired demyelinating conditions, in the inherited myelin disorders (leukodystrophies) such as Pelizaeus-Merzbacher disease (**Figure 1f**), normal myelination fails because OLs die or are dysfunctional as a result of an inherent genetic defect. This may be associated with extensive secondary axonal loss.

Conserved Mechanisms in Myelination and Remyelination: The Recapitulation Hypothesis

Myelination and remyelination share a common objective of investing axons with myelin sheaths and establishing the correct molecular architecture at nodes and paranodes that allow for saltatory conduction. They may therefore be expected, as a consequence of evolutionary economy, to use the same mechanisms. This concept is encapsulated in the "recapitulation hypothesis" of remyelination (Franklin & Hinks 1999). Both developmental myelination and remyelination involve the same key stages of OPC expansion within and toward regions containing naked axons by migration and proliferation and the subsequent differentiation and maturation into myelin-forming OLs. Although the literature is unbalanced, such that more is known about regulation of developmental myelination than remyelination, we go on to discuss several examples that either support or refute the recapitulation hypothesis.

Myelin-axon relationship in myelination and remyelination. The most obvious difference between myelination and remyelination is in the relationship between axon diameter and the length and thickness of the myelin sheath (or internode). In developmental myelination, a close association exists, whereby larger diameter axons are invested with proportionally thicker myelin and longer internodes. In contrast, following remyelination, the myelin sheath length and thickness remain roughly constant regardless of the axon diameter (**Figure 1**).

With small-diameter axons close to the threshold of myelination, this difference is less apparent, making it difficult to identify unequivocally remyelination in tracts of small-diameter axons such as the corpus callosum. However, as the axon size increases, the relative thinness of the remyelinated internode, which is still the defining morphological feature of remyelination, becomes more obvious. This finding raises two questions: First, which factors regulate myelin sheath parameters during myelination, and second, why does the close relationship between axon and myelin size break down in remyelination?

A straightforward mechanism accounting for the relationship between axon size and myelin size in the PNS has emerged on the basis of the levels of axonal expression of neuregulin (Nrg) 1-type III (Michailov et al. 2004). The situation in the CNS appears more complex: Whereas $Nrg1^{+/-}$ heterozygote mice show reduced myelin sheath thickness in the corpus callosum (Taveggia et al. 2008), mice lacking the *Nrg1* gene or the Nrg receptors Erb3/4 have normal myelin (Brinkmann et al. 2008). However, transgenic overexpression of *Nrg1* types I and III results in increased myelin sheath thickness (hypermyelination). This Nrg1-induced stimulation in myelin sheath thickness does not occur following demyelination where the myelin sheath parameters of OL remyelination are unchanged compared with wildtype controls (although the CNS remyelination by Schwann cells shows the same hypermyelination seen in the PNS) (Brinkmann et al. 2008).

Together, these data suggest that other mechanisms help regulate the precise relationship between axon and OL in CNS myelination. One such signal is the Akt/mTOR pathway, which, when overactivated, dramatically increases relative myelin sheath thickness (Flores et al. 2008, Narayanan et al. 2009, Tyler et al. 2009). A similar effect is seen in conditional knockout of the PI-3K pathway inhibitor *PTEN* in OLs (Harrington et al. 2010). However, as with the Nrg overexpression experiments, this ability to enhance myelin sheath thickness is limited to developmental myelination because the remyelinated myelin sheaths

in the conditional *PTEN* knockout remain as thin as controls (Harrington et al. 2010). Thus, the powerful mechanisms described thus far, which control myelin parameters in developmental myelination, appear dispensable during remyelination, at least in the particular demyelinating models studied.

Adult OPCs: role in remyelination. Remyelination involves the generation of new mature OLs because the number of OLs within an area of remyelination is higher than in equivalent areas of myelination, and remyelination occurs within areas depleted of OLs (Sim et al. 2002). Recent evidence using a cre-lox fate-mapping strategy in transgenic mice to express the marker protein YFP in OPCs has confirmed that "local" OPCs are the source of most remyelinating OLs (Zawadzka et al. 2010). The stem (type B cells) and NG2 progenitor cells of the adult subventricular zone (SVZ) can also contribute to remyelination of demyelinated axons in its near vicinity, such as within the corpus callosum (Menn et al. 2006, Nait-Oumesmar et al. 1999).

In striking contrast to developmental myelination, CNS remyelination can also be mediated by Schwann cells, the myelin-forming cells of the PNS. Schwann cell involvement in this process occurs in several demyelination experimental animal models, as well as in human demyelinating disease, and is especially associated with areas of demyelination in which gliosis (astrocyte hypertrophy/proliferation) is absent. These remyelinating Schwann cells within the CNS were generally thought to migrate from PNS sources such as spinal and cranial roots, meningeal fibers, or autonomic nerves following a breach in the *glia limitans* [the converse of this has also been reported (Coulpier et al. 2010)]. However, transgenic fate-mapping studies have revealed that very few of these cells originate in the PNS: Instead the overwhelming majority are derived from OPCs, revealing a surprising capacity of CNS precursors to generate a cell type that normally develops from the embryonic neural crest and is restricted to the PNS (Zawadzka et al. 2010). What causes

CNS-resident precursors to become Schwann cells is not fully understood. One possibility is that OPCs are exposed to extrinsic cues within demyelinated lesions that induce Schwann cell differentiation. Alternatively, OPCs may be a heterogeneous population of cells, some of which are intrinsically more disposed to acquire a Schwann cell phenotype.

The topic of OPC heterogeneity, however, is not well established. Early studies sought to distinguish OPCs on the basis of molecular expression characteristics and responses to growth factors in culture (Wolswijk & Noble 1989). More recently, two distinct types of OPCs have been distinguished on the basis of sodium channel expression, synaptic input, and ability to fire action potentials (Karadottir et al. 2008), although whether this could relate to the developmental potential of OPCs is unclear. As described above, OPCs are generated in several temporal waves, reflecting different spatial origins in the forebrain (Kessaris et al. 2006). Adult OPCs differ from their perinatal forebears in many ways, including antigenic markers, growth factor responsiveness, and basal motility rates and cell cycle. Thus, an interesting question for the future is whether tempo-spatially distinct classes of OPCs could have different capabilities in remyelination and Schwann cell production. Similarly, whether the functional subtypes of OPC serve other roles in remyelination requires further analysis, given the finding that OPCs responding to demyelination receive synaptic inputs (Etxeberria et al. 2010).

The stages of remyelination. Following white matter injury, OPCs (recognized by expression of NG2, PDGFRα, and Olig2; see **Figures 3** and **4**) in and around the damaged area rapidly undergo change from an essentially quiescent state to one of proliferative expansion. This activation involves hypertrophy and, of more functional relevance, an upregulation of several genes, many associated with the generation of OLs during development, such as those encoding the TFs Olig2, Nkx2.2, MyT1, and Sox2 (Fancy et al. 2004, Shen et al. 2008, Sim et al. 2002). OPC activation is likely to be in response to acute injury and associated changes in microglia and astrocytes (both of which detect disturbances in tissue homeostasis and undergo activation). Activated microglia and astrocytes are major sources of OPC mitogens and chemotaxic factors, enabling these cells to rapidly colonize areas of damage in a stage of remyelination called the recruitment phase (Carroll & Jennings 1994, Glezer et al. 2006, Rhodes et al. 2006). In general, the recruitment phase of remyelination involves mechanisms that support the recapitulation hypothesis. For example, platlet-derived growth factor (PDGF) and FGF are both major mitogens in the developmental and regenerative processes.

The final phase of remyelination is where recruited OPCs exit the cell cycle and differentiate into remyelinating OLs. As in myelination, this phase encompasses several distinct steps: (*a*) establishment of contact with the axon to be remyelinated, (*b*) expression of myelin genes, (*c*) generation of myelin membrane, and finally (*d*) wrapping and compaction to form the sheath. These points are illustrated in **Figure 4a**. Although some evidence indicates that remyelination failure can occur because of OPC recruitment failure, in which expression of semaphorin 3A and 3F is lacking (Williams et al. 2007), several lines of evidence, both experimental and clinical, suggest that differentiation failure represents a major cause of remyelination failure (Chang et al. 2002, Kuhlmann et al. 2008, Wolswijk 1998).

The efficiency of remyelination decreases with age. Indeed, for a chronic illness such as MS, age may be a primary contributor to remyelination failure owing to a decrease in the ability of recruited cells to differentiate (Shen et al. 2008, Sim et al. 2002, Woodruff et al. 2004). If so, the age-related factors may account for the finding of OPCs that fail to differentiate within chronic MS plaques (Chang et al. 2002, Kuhlmann et al. 2008, Wolswijk 1998). Differentiation failure is also associated with myelin regeneration failure in neonatal ischemic white matter injury such as PVL (Billiards et al. 2008), which clearly does not arise for age-related reasons. Nevertheless,

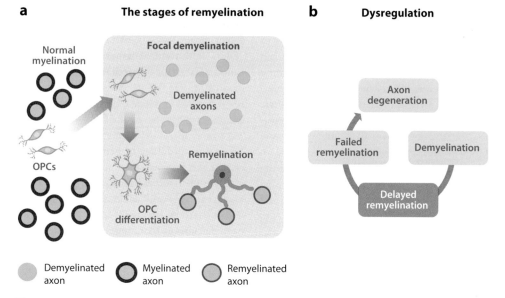

a The stages of remyelination   **b** Dysregulation

Normal myelination

Focal demyelination

Demyelinated axons

Remyelination

OPCs

OPC differentiation

Axon degeneration

Failed remyelination

Demyelination

Delayed remyelination

Demyelinated axon   Myelinated axon   Remyelinated axon

Figure 4

The stages of remyelination and dysregulation in disease. (*a*) In an initial stage of remyelination, resting oligodendrocyte precursor cells (OPCs) from surrounding normal myelinated tissue are activated from a quiescent state to proliferate and migrate. They then enter the area of focal demyelination in a recruitment stage (*gray gradient arrow*). OPCs subsequently exit the cell cycle and undergo terminal differentiation (*blue arrows*) into mature remyelinating oligodendrocytes (OLs), firstly contacting the demyelinated axon, then expressing myelin genes and generating the myelin membrane, and subsequently wrapping and compacting the myelin sheath. In contrast with developmental myelination (in which a close association exists whereby the larger the axon diameter the longer and thicker the myelin internode), the myelin sheath thickness and internode length during remyelination remain roughly constant regardless of the axon diameter. (*b*) If any of the stages of remyelination are dysregulated or delayed, OL lineage cells may miss a critical time window when the signaling environment is conducive to successful remyelination. Remyelination is thereby stalled and axons are left chronically demyelinated, where they are more prone to degeneration because of the lack of trophic support from the myelin sheath. In cases of demyelinating injury (e.g., PVL, MS), where there is also primary axon damage, failure of remyelination of demyelinated axons can exacerbate the extent of axon degeneration.

identifying pathways involved in the regulation of OPC differentiation in remyelination might translate into new therapeutic strategies for enhancing remyelination and axonal protection in MS, PVL, and other myelin disorders.

Extrinsic regulators of remyelination. The processes of OPC differentiation into myelinating OLs in development and during regeneration share many similarities but also differ in several important ways. Here we focus exclusively on factors that have been tested for roles in developmental myelination and remyelination. For example, both PDGF and insulin-like

growth factor 1 (IGF-1) are OPC mitogens in development and remyelination (Fruttiger et al. 1999, Woodruff et al. 2004, Carson et al. 1993, Mason et al. 2003), whereas FGF plays a key role in inhibiting differentiation as well as promoting recruitment and thereby regulates the correct transition from the recruitment to the differentiation phases (Armstrong et al. 2002). The ability of anti-Lingo1 antibodies to enhance remyelination suggests that Lingo-1 plays a negative regulatory role in remyelination similar to that in myelination, although whether the exact same mechanisms pertain to both is not yet known (Mi et al. 2005, 2009).

A critical difference that exists between myelination and remyelination is that the latter is associated with inflammation generated (*a*) in response to demyelination (myelin debris) and (*b*) in the case of autoimmune-demyelinating disease, such as MS, as the cause of demyelination itself. A critical role played by phagocytic macrophages is the removal of myelin debris generated during demyelination because CNS myelin contains proteins inhibitory to OPC differentiation both in vitro and during remyelination. Demyelination generates vast amounts of myelin debris as the compacted myelin unravels. Injection of myelin debris into experimentally induced areas of demyelination in young animals causes remyelination failure because precursor differentiation does not occur (Kotter et al. 2006). Although the concept of uncleared myelin debris as a cause of remyelination failure derives exclusively from experimental studies and has yet to be clearly demonstrated in MS tissue, these observations nevertheless raise important questions about the existence of pathways in precursor cells that inhibit differentiation that may offer new therapeutic targets. Both the fyn-rho-ROCK and protein kinase C (PKC) signaling pathways have been identified as being critical mediators of myelin-mediated inhibition (Baer et al. 2009). Although microglia cells may contribute to developmental myelination in a manner that has yet to be fully explored, the inflammatory context in which remyelination occurs is undoubtedly greater and makes a significant contribution to the environmental signals by which the behavior of OPCs during remyelination is governed. Several studies have highlighted the role of the innate immune response to demyelination in creating an environment conducive to remyelination (Arnett et al. 2001, Kotter et al. 2001, Mason et al. 2001). We have already discussed the key role that microglia play in responding to damage and activating OPCs to promote remyelination. Indeed, disrupting the innate immune response can impair remyelination, as shown by experiments involving depletion or pharmacological inhibition of macrophages and deletion of proinflammatory

cytokines IL-1β (Mason et al. 2001), TNF-α (Arnett et al. 2001), lymphotoxin-β receptor (Plant et al. 2007), or MHCII (Arnett et al. 2003). The adaptive immune system, which is not a feature of developmental myelination, may also contribute beneficially because remyelination is impaired in the absence of T lymphocytes (Bieber et al. 2003).

Intrinsic regulators of remyelination. Several of the key pathways and TFs known to regulate OPC differentiation in development have also been studied in the context of remyelination. In general, these studies confirm the notion that similar mechanisms are employed in developmental myelination and remyelination.

Notch pathway. During development, activation of Notch receptors on OPCs by the Notch ligand expressed from axons has an inhibitory regulatory effect on differentiation, leading Blaschuk & Ffrench-Constant (1998) to speculation that persistent signaling via the Notch pathway following demyelination would lead to chronic demyelination. Notch receptor is expressed by adult OPCs following demyelination in animals and within MS lesions, whereas the ligand Jagged and TF Hairy/Enhancer of Split (HES) are expressed predominantly in inflammatory cells and astrocytes (John et al. 2002). However, both Notch and Jagged are expressed in lesions that undergo complete remyelination, suggesting that their presence does not block, but rather may regulate the kinetics of repair (Seifert et al. 2007, Stidworthy et al. 2004). When Notch1 is specifically deleted following demyelination using an inducible cre-lox approach in which cre activity was driven by either *PLP* (Stidworthy et al. 2004) or Olig1 (Zhang et al. 2009) regulatory sequences, the predicted premature differentiation is slight and does not accelerate remyelination. Thus, the canonical Notch pathway seems to play a relatively minor role in the negative regulation of OPC differentiation in development and repair. This may be due, in part, to competitive activation of noncanonical Notch signaling in OPCs that is involved in the induction of OPC differentiation (Hu et al. 2003, Nakahara et al. 2009).

The Wnt Pathway. We recently carried out a genome-wide screen of TF-encoding genes expressed during remyelination (Fancy et al. 2009), which identified ∼50 such genes with dynamic regulation during myelin regeneration. Among these, the HMGbox containing TF called T cell transcription factor-4 (Tcf4, also called TCF7L2) is a critical intranuclear component of canonical Wnt signaling, forming a nucleoprotein complex with dephosphorylated β-catenin at the final gene-regulatory stage of the pathway. Tcf4 is expressed specifically within OPCs during postnatal developmental myelination, and in the adult, it was expressed by OPCs within remyelinating lesions—but not normal—adult white matter. A gain-of-function transgenic strategy showed that activation of the canonical Wnt pathway in OL lineage cells results in delayed kinetics of OPC differentiation in both myelination and remyelination.

How might a negative regulator of OPC differentiation, such as Wnt signaling in remyelination, determine the success or failure of remyelination? We have already highlighted several stages needed for effective myelin regeneration to occur (**Figure 4a**). We have also argued, based on findings from human chronic demyelinated lesions of MS and PVL that contain OPCs, that differentiation block is a likely stage that remyelination fails. In this respect, dysregulation (hyperactivation) of an inhibitory pathway such as Wnt could account for a (potentially catastrophic) delay in differentiation. Equally, OPCs must not exit the cell cycle too early and differentiate before an adequate number of cells has been generated (Casaccia-Bonnefil & Liu 2003). Thus, differentiation and expansion of OPC pools must be carefully balanced.

Transcription factors. Above, we have described the role of certain TFs as key regulators of oligodendrogliogenesis and developmental myelination (Emery et al. 2009, He et al. 2007,

Rowitch 2004). However, roles of such TFs in remyelination have not generally been assessed. One exception is the bHLH TF Olig1, which has key roles in early OPC development (Xin et al. 2005) as well as in OPC differentiation in remyelination (Arnett et al. 2004). *Olig1* null animals showed a near total failure of myelin regeneration despite adequate numbers of recruited OPCs into lesions caused by lysolethicin or cuprizone toxic injury (Arnett et al. 2004). Studying the role of many other developmentally significant transcriptional regulators such as YY1, Ascl1, and MRF in remyelination might provide important new insights.

Remyelination failure. The recapitulation hypothesis generally provides a useful framework for understanding the relationship between myelination and remyelination. Although notable differences between the two processes exist (e.g., myelin thickness–axon diameter relationship, roles for inflammation), certain key regulatory pathways appear to be well conserved.

Franklin (2002) proposed a hypothesis for remyelination failure in which the signaling environment becomes inappropriately regulated or "dysregulated." According to this hypothesis, remyelination fails because the complex and finely tuned mechanism of myelin regeneration loses precise coordination. For example, one could conceive of situations in which prolonged inhibition of OPC differentiation would result in missing a critical window of opportunity for remyelination. In this scenario, chronic demyelination results from failure of remyelination during an acute phase rather than because inhibitory signals were sustained in the chronic phase (**Figure 4b**).

In addition to the signaling factors discussed above such as activated Wnt, dysregulation of remyelination could also involve epigenetic mechanisms. For example, genes encoding the TFs HES1, HES5, Id2, Id4, and Sox2 must be downregulated for timely OPC differentiation. Such gene regulation involves acetylation of histones in the promoter regions that determine chromatin conformation. In CNS precursors, this process is mediated by members of the class of HDAC enzymes (Shen et al. 2005). As animals age, the efficiency of HDAC recruitment to suppress expression of inhibitory TF genes declines, resulting in cycling precursor cells that remain refractory to differentiation-inducing factors longer than those in young animals (Shen et al. 2008). Thus, OPCs within lesions in older animals could miss a critical window for differentiation because of an epigenetic mechanism and interactions with HDACs and signaling pathways such as Wnt (Ye et al. 2009). Further work is needed to understand how to regulate the precise timing of precursor recruitment and differentiation necessary for efficient remyelination and how these processes could become pathologically dysregulated, causing myelin regeneration to fail.

PRIORITIES FOR FUTURE RESEARCH FOR TRANSLATIONAL INSIGHTS INTO HUMAN CNS MYELIN DISORDERS

In this section, we highlight several areas for future research that may provide further insights into fundamental mechanisms of myelin development and regeneration, as well as clues to better understand human white matter disorders that target OLs.

Continuing Validation of Developmental Studies in the Setting of Regeneration

We have reviewed some of the recent and remarkable advances in our understanding of OL lineage development and function. To gain further translational insights into remyelination, such developmental insights must be directly tested for conserved roles in repair. Investigators have recently described several key players that regulate developmental myelination, including TFs YY1 (He et al. 2007), MRF (Emery et al. 2009) and ZFP191 (Howng et al. 2010), as well as new receptors (Chen et al. 2009) that influence OL differentiation

and myelination. miRNA posttranscriptional control during development is also emerging as an important mechanism for controlling OL differentiation. Although researchers often assume that mechanisms critical to development will also be involved in injury repair in the adult, a crucial future goal is the specific validation of these developmental paradigms within myelin regeneration postinjury.

Toward Understanding an Integrated Program Controlling Remyelination

Although significant steps forward have been made toward identifying individual factors critical to biology, we lack clear understanding of how extracellular factors, intracellular signaling pathways, and transcriptional, posttranscriptional, and epigenetic mechanisms work together to coordinate myelin regeneration. For example, TFs are known to operate together in CNS precursors, yet our understanding of the dynamic changes in TF complexes in OPCs during remyelination remains rudimentary. The integration of extrinsic signals and levels of cross-talk between intracellular signaling pathways is also poorly understood. Further research should integrate current knowledge to develop a unified conceptual framework for the complex factors at play during remyelination.

Improved Access to Human Tissue

To establish translational relevance of animal model findings to human disease, investigators must have access to high-quality human tissues to analyze at histological, protein, mRNA, miRNA, and epigenetic levels. Although tissue banks for MS have been set up, access to these resources may still be challenging. To evaluate human brain development, and consequently how white matter is affected in newborn brain injuries that lead to cerebral palsy, human neonatal tissue banks are similarly critical. With improved access to human tissues and marker development for use in the human brain, establishing conservation of mechanisms across species should be achievable.

Integration of Glial Biology and Genetic/Epigenetic Studies for MS and Other Human White Matter Disorders

MS is characterized by moderate, but complex, risk heritability. Integrating our knowledge of biology with that of MS genome-wide association studies (GWAS) is necessary to improve our understanding of disease pathogenesis. The first genomic scans in MS patients identified the *human leukocyte antigen* (HLA) as the primary susceptibility region (Haines et al. 1996). Recent large-scale genomic studies have revealed new heritable high-risk alleles for MS development (Gregory et al. 2007, Hafler et al. 2007) and identified several immune-related genes as genetic determinants of disease susceptibility (De Jager et al. 2009). Might as-yet-unidentified genetic determinants relate to regulation of the disease repair phase? Scrutiny of appropriately powered whole-genome data sets for candidate factors involved in myelin regeneration could reveal genetic or epigenetic markers associated with disease course. Indeed, given that MS in some cases has a nongenetic basis (Baranzini et al. 2010), understanding the role of epigenetic regulation of remyelination is a high priority.

CONCLUSION

The focus of this review has been to evaluate the recapitulation hypothesis as a relevant means to improve understanding of myelin regeneration. Several studies now support the recapitulation hypothesis, but others highlight differences between development and repair; these differences may reflect the significant variation in the inflammatory environment and/or signaling from axons during remyelination. The study of developmental myelination continues to reveal key mechanisms that may translate to our understanding of remyelination and human myelin regeneration. However, much more work must be done to establish to what extent development is mimicked in myelin regeneration. Such investigation will yield, as desirable by-products, information on which to base rational design of cell-based and small-molecule therapies to promote repair in human myelin disorders.

DISCLOSURE STATEMENT

The authors are not aware of any affiliations, memberships, funding, or financial holdings that might be perceived as affecting the objectivity of this review.

ACKNOWLEDGMENTS

The authors apologize to investigators whose work we could not cite owing to space limitations. This work was supported by grants from the National Multiple Sclerosis Society (to R.J.M.F. and D.H.R.), the United Kingdom Multiple Sclerosis Society (to R.J.M.F.) and the National Institute of Health (to D.H.R. and J.R.C.). J.R.C. and S.E.B. are Harry Weaver Neuroscience Scholars of the NMSS. D.H.R. is a Howard Hughes Medical Institute Investigator.

LITERATURE CITED

Agius E, Soukkarieh C, Danesin C, Kan P, Takebayashi H, et al. 2004. Converse control of oligodendrocyte and astrocyte lineage development by Sonic hedgehog in the chick spinal cord. *Dev. Biol.* 270:308–21

Armstrong RC, Le TQ, Frost EE, Borke RC, Vana AC. 2002. Absence of fibroblast growth factor 2 promotes oligodendroglial repopulation of demyelinated white matter. *J. Neurosci.* 22:8574–85

Arnett HA, Fancy SPJ, Alberta JA, Zhao C, Plant SR, et al. 2004. bHLH transcription factor Olig1 is required to repair demyelinated lesions in the CNS. *Science* 306:2111–15

Arnett HA, Mason J, Marino M, Suzuki K, Matsushima GK, Ting JP. 2001. TNF alpha promotes proliferation of oligodendrocyte progenitors and remyelination. *Nat. Neurosci.* 4:1116–22

Arnett HA, Wang Y, Matsushima GK, Suzuki K, Ting JP. 2003. Functional genomic analysis of remyelination reveals importance of inflammation in oligodendrocyte regeneration. *J. Neurosci.* 23:9824–32

Back SA, Rivkees SA. 2004. Emerging concepts in periventricular white matter injury. *Semin. Perinatol.* 28:405–14

Back SA, Tuohy TM, Chen H, Wallingford N, Craig A, et al. 2005. Hyaluronan accumulates in demyelinated lesions and inhibits oligodendrocyte progenitor maturation. *Nat. Med.* 11:966–72

Baer AS, Syed YA, Kang SU, Mitteregger D, Vig R, et al. 2009. Myelin-mediated inhibition of oligodendrocyte precursor differentiation can be overcome by pharmacological modulation of Fyn-RhoA and protein kinase C signalling. *Brain* 132:465–81

Baranzini SE, Mudge J, van Velkinburgh JC, Khankhanian P, Khrebtukova I, et al. 2010. Genome, epigenome and RNA sequences of monozygotic twins discordant for multiple sclerosis. *Nature* 464:1351–56

Behrens J, von Kries JP, Kuhl M, Bruhn L, Wedlich D, et al. 1996. Functional interaction of beta-catenin with the transcription factor LEF-1. *Nature* 382:638–42

Bergles DE, Roberts JD, Somogyi P, Jahr CE. 2000. Glutamatergic synapses on oligodendrocyte precursor cells in the hippocampus. *Nature* 405:187–91

Bieber AJ, Kerr S, Rodriguez M. 2003. Efficient central nervous system remyelination requires T cells. *Ann. Neurol.* 53:680–84

Billiards SS, Haynes RL, Folkerth RD, Borenstein NS, Trachtenberg FL, et al. 2008. Myelin abnormalities without oligodendrocyte loss in periventricular leukomalacia. *Brain Pathol.* 18:153–63

Bjartmar C, Wujek JR, Trapp BD. 2003. Axonal loss in the pathology of MS: consequences for understanding the progressive phase of the disease. *J. Neurol. Sci.* 206:165–71

Blaschuk KL, Ffrench-Constant C. 1998. Developmental neurobiology: notch is tops in the developing brain. *Curr. Biol.* 8:R334–37

Brinkmann BG, Agarwal A, Sereda MW, Garratt AN, Muller T, et al. 2008. Neuregulin-1/ErbB signaling serves distinct functions in myelination of the peripheral and central nervous system. *Neuron* 59:581–95

Briscoe J, Novitch BG. 2008. Regulatory pathways linking progenitor patterning, cell fates and neurogenesis in the ventral neural tube. *Philos. Trans. R Soc. Lond. B Biol. Sci.* 363:57–70

Cai J, Qi Y, Hu X, Tan M, Liu Z, et al. 2005. Generation of oligodendrocyte precursor cells from mouse dorsal spinal cord independent of Nkx6 regulation and Shh signaling. *Neuron* 45:41–53

Campagnoni AT, Skoff RP. 2001. The pathobiology of myelin mutants reveal novel biological functions of the MBP and PLP genes. *Brain Pathol.* 11:74–91

Carroll WM, Jennings AR. 1994. Early recruitment of oligodendrocyte precursors in CNS demyelination. *Brain* 117(Pt. 3):563–78

Carson MJ, Behringer RR, Brinster RL, McMorris FA. 1993. Insulin-like growth factor I increases brain growth and central nervous system myelination in transgenic mice. *Neuron* 10:729–40

Casaccia-Bonnefil P, Liu A. 2003. Relationship between cell cycle molecules and onset of oligodendrocyte differentiation. *J. Neurosci. Res.* 72:1–11

Chandran S, Kato H, Gerreli D, Compston A, Svendsen CN, Allen ND. 2003. FGF-dependent generation of oligodendrocytes by a hedgehog-independent pathway. *Development* 130:6599–609

Chang A, Tourtellotte WW, Rudick R, Trapp BD. 2002. Premyelinating oligodendrocytes in chronic lesions of multiple sclerosis. *N. Engl. J. Med.* 346:165–73

Charles P, Reynolds R, Seilhean D, Rougon G, Aigrot MS, et al. 2002. Re-expression of PSA-NCAM by demyelinated axons: an inhibitor of remyelination in multiple sclerosis? *Brain* 125:1972–79

Chen Y, Wu H, Wang S, Koito H, Li J, et al. 2009. The oligodendrocyte-specific G protein-coupled receptor GPR17 is a cell-intrinsic timer of myelination. *Nat. Neurosci.* 12:1398–406

Chong SY, Chan JR. 2010. Tapping into the glial reservoir: cells committed to remaining uncommitted. *J. Cell Biol.* 188:305–12

Compston A, Coles A. 2002. Multiple sclerosis. *Lancet* 359:1221–31

Coulpier F, Decker L, Funalot B, Vallat JM, Garcia-Bragado F, et al. 2010. CNS/PNS boundary transgression by central glia in the absence of Schwann cells or Krox20/Egr2 function. *J. Neurosci.* 30:5958–67

De Biase LM, Nishiyama A, Bergles DE. 2010. Excitability and synaptic communication within the oligodendrocyte lineage. *J. Neurosci.* 30(10):3600–11

De Jager PL, Jia X, Wang J, de Bakker PI, Ottoboni L, et al. 2009. Meta-analysis of genome scans and replication identify CD6, IRF8 and TNFRSF1A as new multiple sclerosis susceptibility loci. *Nat. Genet.* 41:776–82

Dubois-Dalcq M, Williams A, Stadelmann C, Stankoff B, Zalc B, Lubetzki C. 2008. From fish to man: understanding endogenous remyelination in central nervous system demyelinating diseases. *Brain* 131:1686–700

Dugas JC, Cuellar TL, Scholze A, Ason B, Ibrahim A, et al. 2010. Dicer1 and miR-219 are required for normal oligodendrocyte differentiation and myelination. *Neuron* 65:597–611

Duncan ID, Brower A, Kondo Y, Curlee JF Jr, Schultz RD. 2009. Extensive remyelination of the CNS leads to functional recovery. *Proc. Natl. Acad. Sci. USA* 106:6832–36

Edgar JM, McLaughlin M, Yool D, Zhang SC, Fowler JH, et al. 2004. Oligodendroglial modulation of fast axonal transport in a mouse model of hereditary spastic paraplegia. *J. Cell Biol.* 166:121–31

Emery B. 2010. Transcriptional and post-transcriptional control of CNS myelination. *Curr. Opin. Neurobiol.* 20:601–7

Emery B, Agalliu D, Cahoy JD, Watkins TA, Dugas JC, et al. 2009. Myelin gene regulatory factor is a critical transcriptional regulator required for CNS myelination. *Cell* 138:172–85

Etxeberria A, Mangin JM, Aguirre A, Gallo V. 2010. Adult-born SVZ progenitors receive transient synapses during remyelination in corpus callosum. *Nat. Neurosci.* 13:287–89

Fancy SPJ, Baranzini SE, Zhao C, Yuk DI, Irvine KA, et al. 2009. Dysregulation of the Wnt pathway inhibits timely myelination and remyelination in the mammalian CNS. *Genes Dev.* 23:1571–85

Fancy SPJ, Zhao C, Franklin RJM. 2004. Increased expression of Nkx2.2 and Olig2 identifies reactive oligodendrocyte progenitor cells responding to demyelination in the adult CNS. *Mol. Cell Neurosci.* 27:247–54

Feigenson K, Reid M, See J, Crenshaw EB 3rd, Grinspan JB. 2009. Wnt signaling is sufficient to perturb oligodendrocyte maturation. *Mol. Cell Neurosci.* 42:255–65

Flores AI, Narayanan SP, Morse EN, Shick HE, Yin X, et al. 2008. Constitutively active Akt induces enhanced myelination in the CNS. *J. Neurosci.* 28:7174–83

Fogarty M, Richardson WD, Kessaris N. 2005. A subset of oligodendrocytes generated from radial glia in the dorsal spinal cord. *Development* 132:1951–59

Franklin RJM. 2002. Why does remyelination fail in multiple sclerosis? *Nat. Rev. Neurosci.* 3:705–14

Franklin RJM, Ffrench-Constant C. 2008. Remyelination in the CNS: from biology to therapy. *Nat. Rev. Neurosci.* 9:839–55

Franklin RJM, Hinks GL. 1999. Understanding CNS remyelination: clues from developmental and regeneration biology. *J. Neurosci. Res.* 58:207–13

Fruttiger M, Karlsson L, Hall AC, Abramsson A, Calver AR, et al. 1999. Defective oligodendrocyte development and severe hypomyelination in PDGF-A knockout mice. *Development* 126:457–67

Garbern JY, Yool DA, Moore GJ, Wilds IB, Faulk MW, et al. 2002. Patients lacking the major CNS myelin protein, proteolipid protein 1, develop length-dependent axonal degeneration in the absence of demyelination and inflammation. *Brain* 125:551–61

Genoud S, Lappe-Siefke C, Goebbels S, Radtke F, Aguet M, et al. 2002. Notch1 control of oligodendrocyte differentiation in the spinal cord. *J. Cell Biol.* 158:709–18

Glezer I, Lapointe A, Rivest S. 2006. Innate immunity triggers oligodendrocyte progenitor reactivity and confines damages to brain injuries. *FASEB J.* 20:750–52

Gregory SG, Schmidt S, Seth P, Oksenberg JR, Hart J, et al. 2007. Interleukin 7 receptor alpha chain (IL7R) shows allelic and functional association with multiple sclerosis. *Nat. Genet.* 39:1083–91

Griffiths I, Klugmann M, Anderson T, Yool D, Thomson C, et al. 1998. Axonal swellings and degeneration in mice lacking the major proteolipid of myelin. *Science* 280:1610–13

Hafler DA, Compston A, Sawcer S, Lander ES, Daly MJ, et al. 2007. Risk alleles for multiple sclerosis identified by a genomewide study. *N. Engl. J. Med.* 357:851–62

Haines JL, Ter-Minassian M, Bazyk A, Gusella JF, Kim DJ, et al. 1996. A complete genomic screen for multiple sclerosis underscores a role for the major histocompatability complex. The Multiple Sclerosis Genetics Group. *Nat. Genet.* 13:469–71

Harrington EP, Zhao C, Fancy SPJ, Kaing S, Franklin RJM, Rowitch DHR. 2010. Oligodendrocyte PTEN required for myelin and axonal integrity not remyelination. *Ann. Neurol.* 68:703–16

He Y, Dupree J, Wang J, Sandoval J, Li J, et al. 2007. The transcription factor Yin Yang 1 is essential for oligodendrocyte progenitor differentiation. *Neuron* 55:217–30

Howng SY, Avila RL, Emery B, Traka M, Lin W, et al. 2010. ZFP191 is required by oligodendrocytes for CNS myelination. *Genes Dev.* 24:301–11

Hu QD, Ang BT, Karsak M, Hu WP, Cui XY, et al. 2003. F3/contactin acts as a functional ligand for Notch during oligodendrocyte maturation. *Cell* 115:163–75

Hu X, Hicks CW, He W, Wong P, Macklin WB, et al. 2006. Bace1 modulates myelination in the central and peripheral nervous system. *Nat. Neurosci.* 9:1520–25

Inoue K, Khajavi M, Ohyama T, Hirabayashi S, Wilson J, et al. 2004. Molecular mechanism for distinct neurological phenotypes conveyed by allelic truncating mutations. *Nat. Genet.* 36:361–69

Inoue K, Tanabe Y, Lupski JR. 1999. Myelin deficiencies in both the central and the peripheral nervous systems associated with a SOX10 mutation. *Ann. Neurol.* 46:313–18

Jeffery ND, Blakemore WF. 1997. Locomotor deficits induced by experimental spinal cord demyelination are abolished by spontaneous remyelination. *Brain* 120(Pt. 1):27–37

Jessell TM. 2000. Neuronal specification in the spinal cord: inductive signals and transcriptional codes. *Nat. Rev. Genet.* 1:20–29

John GR, Shankar SL, Shafit-Zagardo B, Massimi A, Lee SC, et al. 2002. Multiple sclerosis: re-expression of a developmental pathway that restricts oligodendrocyte maturation. *Nat. Med.* 8:1115–21

Karadottir R, Hamilton NB, Bakiri Y, Attwell D. 2008. Spiking and nonspiking classes of oligodendrocyte precursor glia in CNS white matter. *Nat. Neurosci.* 11:450–56

Kessaris N, Fogarty M, Iannarelli P, Grist M, Wegner M, Richardson WD. 2006. Competing waves of oligodendrocytes in the forebrain and postnatal elimination of an embryonic lineage. *Nat. Neurosci.* 9:173–79

Kessaris N, Pringle N, Richardson WD. 2001. Ventral neurogenesis and the neuron-glial switch. *Neuron* 31:677–80

Kinney HC, Back SA. 1998. Human oligodendroglial development: relationship to periventricular leukomalacia. *Semin. Pediatr. Neurol.* 5:180–89

Kornek B, Storch MK, Weissert R, Wallstroem E, Stefferl A, et al. 2000. Multiple sclerosis and chronic autoimmune encephalomyelitis: a comparative quantitative study of axonal injury in active, inactive, and remyelinated lesions. *Am. J. Pathol.* 157:267–76

Kotter MR, Li WW, Zhao C, Franklin RJM. 2006. Myelin impairs CNS remyelination by inhibiting oligodendrocyte precursor cell differentiation. *J. Neurosci.* 26:328–32

Kotter MR, Setzu A, Sim FJ, Van Rooijen N, Franklin RJM. 2001. Macrophage depletion impairs oligodendrocyte remyelination following lysolecithin-induced demyelination. *Glia* 35:204–12

Kuhlmann T, Miron V, Cui Q, Wegner C, Antel J, Bruck W. 2008. Differentiation block of oligodendroglial progenitor cells as a cause for remyelination failure in chronic multiple sclerosis. *Brain* 131:1749–58

Lappe-Siefke C, Goebbels S, Gravel M, Nicksch E, Lee J, et al. 2003. Disruption of Cnp1 uncouples oligodendroglial functions in axonal support and myelination. *Nat. Genet.* 33:366–74

Li C, Tropak MB, Gerlai R, Clapoff S, Abramow-Newerly W, et al. 1994. Myelination in the absence of myelin-associated glycoprotein. *Nature* 369:747–50

Li H, He Y, Richardson WD, Casaccia P. 2009. Two-tier transcriptional control of oligodendrocyte differentiation. *Curr. Opin. Neurobiol.* 19:479–85

Li QM, Tep C, Yune TY, Zhou XZ, Uchida T, et al. 2007. Opposite regulation of oligodendrocyte apoptosis by JNK3 and Pin1 after spinal cord injury. *J. Neurosci.* 27:8395–404

Lin SC, Bergles DE. 2004. Synaptic signaling between GABAergic interneurons and oligodendrocyte precursor cells in the hippocampus. *Nat. Neurosci.* 7:24–32

Lu QR, Sun T, Zhu Z, Ma N, Garcia M, et al. 2002. Common developmental requirement for Olig function indicates a motor neuron/oligodendrocyte connection. *Cell* 109:75–86

Lu QR, Yuk D, Alberta JA, Zhu Z, Pawlitzky I, et al. 2000. Sonic hedgehog–regulated oligodendrocyte lineage genes encoding bHLH proteins in the mammalian central nervous system. *Neuron* 25:317–29

Marquardt T, Pfaff SL. 2001. Cracking the transcriptional code for cell specification in the neural tube. *Cell* 106:651–54

Mason JL, Suzuki K, Chaplin DD, Matsushima GK. 2001. Interleukin-1beta promotes repair of the CNS. *J. Neurosci.* 21:7046–52

Mason JL, Xuan S, Dragatsis I, Efstratiadis A, Goldman JE. 2003. Insulin-like growth factor (IGF) signaling through type 1 IGF receptor plays an important role in remyelination. *J. Neurosci.* 23:7710–18

Menn B, Garcia-Verdugo JM, Yaschine C, Gonzalez-Perez O, Rowitch D, Alvarez-Buylla A. 2006. Origin of oligodendrocytes in the subventricular zone of the adult brain. *J. Neurosci.* 26:7907–18

Mi S, Miller RH, Lee X, Scott ML, Shulag-Morskaya S, et al. 2005. LINGO-1 negatively regulates myelination by oligodendrocytes. *Nat. Neurosci.* 8:745–51

Mi S, Miller RH, Tang W, Lee X, Hu B, et al. 2009. Promotion of central nervous system remyelination by induced differentiation of oligodendrocyte precursor cells. *Ann. Neurol.* 65:304–15

Michailov GV, Sereda MW, Brinkmann BG, Fischer TM, Haug B, et al. 2010. Axonal neuregulin-1 regulates myelin sheath thickness. *Neuron* 68(4):668–81

Miller RH. 1999. Contact with central nervous system myelin inhibits oligodendrocyte progenitor maturation. *Dev. Biol.* 216:359–68

Mimault C, Giraud G, Courtois V, Cailloux F, Boire JY, et al. 1999. Proteolipoprotein gene analysis in 82 patients with sporadic Pelizaeus-Merzbacher Disease: duplications, the major cause of the disease, originate more frequently in male germ cells, but point mutations do not. The Clinical European Network on Brain Dysmyelinating Disease. *Am. J. Hum. Genet.* 65:360–69

Molenaar M, van de Wetering M, Oosterwegel M, Peterson-Maduro J, Godsave S, et al. 1996. XTcf-3 transcription factor mediates beta-catenin-induced axis formation in Xenopus embryos. *Cell* 86:391–99

Nait-Oumesmar B, Decker L, Lachapelle F, Avellana-Adalid V, Bachelin C, Van Evercooren AB. 1999. Progenitor cells of the adult mouse subventricular zone proliferate, migrate and differentiate into oligo-dendrocytes after demyelination. *Eur. J. Neurosci.* 11:4357–66

Nakahara J, Kanekura K, Nawa M, Aiso S, Suzuki N. 2009. Abnormal expression of TIP30 and arrested nucleocytoplasmic transport within oligodendrocyte precursor cells in multiple sclerosis. *J. Clin. Invest.* 119:169–81

Narayanan SP, Flores AI, Wang F, Macklin WB. 2009. Akt signals through the mammalian target of rapamycin pathway to regulate CNS myelination. *J. Neurosci.* 29:6860–70

Nave KA. 2010. Myelination and the trophic support of long axons. *Nat. Rev. Neurosci.* 11:275–83

Nave KA, Trapp BD. 2008. Axon-glial signaling and the glial support of axon function. *Annu. Rev. Neurosci.* 31:535–61

Noble M, Proschel C, Mayer-Proschel M. 2004. Getting a GR(i)P on oligodendrocyte development. *Dev. Biol.* 265:33–52

Petryniak MA, Potter GB, Rowitch DH, Rubenstein JL. 2007. Dlx1 and Dlx2 control neuronal versus oligo-dendroglial cell fate acquisition in the developing forebrain. *Neuron* 55:417–33

Plant SR, Iocca HA, Wang Y, Thrash JC, O'Connor BP, et al. 2007. Lymphotoxin beta receptor (Lt betaR): dual roles in demyelination and remyelination and successful therapeutic intervention using Lt betaR-Ig protein. *J. Neurosci.* 27:7429–37

Qi Y, Cai J, Wu Y, Wu R, Lee J, et al. 2001. Control of oligodendrocyte differentiation by the Nkx2.2 homeodomain transcription factor. *Development* 128:2723–33

Qi Y, Stapp D, Qiu M. 2002. Origin and molecular specification of oligodendrocytes in the telencephalon. *Trends Neurosci.* 25:223–25

Rao MS, Mayer-Proschel M. 1997. Glial-restricted precursors are derived from multipotent neuroepithelial stem cells. *Dev. Biol.* 188:48–63

Rhodes KE, Raivich G, Fawcett JW. 2006. The injury response of oligodendrocyte precursor cells is induced by platelets, macrophages and inflammation-associated cytokines. *Neuroscience* 140:87–100

Roots BI. 2008. The phylogeny of invertebrates and the evolution of myelin. *Neuron. Glia Biol.* 4:101–9

Rowitch DH. 2004. Glial specification in the vertebrate neural tube. *Nat. Rev. Neurosci.* 5:409–19

Schneider A, Montague P, Griffiths I, Fanarraga M, Kennedy P, et al. 1992. Uncoupling of hypomyelination and glial cell death by a mutation in the proteolipid protein gene. *Nature* 358:758–61

Seifert T, Bauer J, Weissert R, Fazekas F, Storch MK. 2007. Notch1 and its ligand Jagged1 are present in remyelination in a T-cell- and antibody-mediated model of inflammatory demyelination. *Acta Neuropathol.* 113:195–203

Shen S, Casaccia-Bonnefil P. 2008. Post-translational modifications of nucleosomal histones in oligodendrocyte lineage cells in development and disease. *J. Mol. Neurosci.* 35:13–22

Shen S, Li J, Casaccia-Bonnefil P. 2005. Histone modifications affect timing of oligodendrocyte progenitor differentiation in the developing rat brain. *J. Cell Biol.* 169:577–89

Shen S, Sandoval J, Swiss VA, Li J, Dupree J, et al. 2008. Age-dependent epigenetic control of differentiation inhibitors is critical for remyelination efficiency. *Nat. Neurosci.* 11:1024–34

Shimizu T, Kagawa T, Wada T, Muroyama Y, Takada S, Ikenaka K. 2005. Wnt signaling controls the timing of oligodendrocyte development in the spinal cord. *Dev. Biol.* 282:397–410

Shin D, Shin JY, McManus MT, Ptacek LJ, Fu YH. 2009. Dicer ablation in oligodendrocytes provokes neuronal impairment in mice. *Ann. Neurol.* 66:843–57

Sim FJ, Zhao C, Penderis J, Franklin RJM. 2002. The age-related decrease in CNS remyelination efficiency is attributable to an impairment of both oligodendrocyte progenitor recruitment and differentiation. *J. Neurosci.* 22:2451–59

Simpson PB, Armstrong RC. 1999. Intracellular signals and cytoskeletal elements involved in oligodendrocyte progenitor migration. *Glia* 26:22–35

Smith EJ, Blakemore WF, McDonald WI. 1979. Central remyelination restores secure conduction. *Nature* 280:395–96

Stidworthy MF, Genoud S, Li WW, Leone DP, Mantei N, et al. 2004. Notch1 and Jagged1 are expressed after CNS demyelination, but are not a major rate-determining factor during remyelination. *Brain* 127:1928–41

Stolt CC, Rehberg S, Ader M, Lommes P, Riethmacher D, et al. 2002. Terminal differentiation of myelin-forming oligodendrocytes depends on the transcription factor Sox10. *Genes. Dev.* 16:165–70

Takebayashi H, Nabeshima Y, Yoshida S, Chisaka O, Ikenaka K. 2002. The basic helix-loop-helix factor olig2 is essential for the development of motoneuron and oligodendrocyte lineages. *Curr. Biol.* 12:1157–63

Taveggia C, Thaker P, Petrylak A, Caporaso GL, Toews A, et al. 2008. Type III neuregulin-1 promotes oligodendrocyte myelination. *Glia* 56:284–93

Tekki-Kessaris N, Woodruff R, Hall AC, Gaffield W, Kimura S, et al. 2001. Hedgehog-dependent oligodendrocyte lineage specification in the telencephalon. *Development* 128:2545–54

Touraine RL, Attie-Bitach T, Manceau E, Korsch E, Sarda P, et al. 2000. Neurological phenotype in Waardenburg syndrome type 4 correlates with novel SOX10 truncating mutations and expression in developing brain. *Am. J. Hum. Genet.* 66:1496–503

Tripathi RB, Rivers LE, Young KM, Jamen F, Richardson WD. 2010. NG2 glia generate new oligodendrocytes but few astrocytes in a murine experimental autoimmune encephalomyelitis model of demyelinating disease. *J. Neurosci.* 30:16383–90

Tsai HH, Frost E, To V, Robinson S, Ffrench-Constant C, et al. 2002. The chemokine receptor CXCR2 controls positioning of oligodendrocyte precursors in developing spinal cord by arresting their migration. *Cell* 110:373–83

Tsai HH, Tessier-Lavigne M, Miller RH. 2003. Netrin 1 mediates spinal cord oligodendrocyte precursor dispersal. *Development* 130:2095–105

Tyler WA, Gangoli N, Gokina P, Kim HA, Covey M, et al. 2009. Activation of the mammalian target of rapamycin (mTOR) is essential for oligodendrocyte differentiation. *J. Neurosci.* 29:6367–78

Vallstedt A, Klos JM, Ericson J. 2005. Multiple dorsoventral origins of oligodendrocyte generation in the spinal cord and hindbrain. *Neuron* 45:55–67

Volpe JJ. 2001. Neurobiology of periventricular leukomalacia in the premature infant. *Pediatr. Res.* 50:553–62

Voyvodic JT. 1989. Target size regulates calibre and myelination of sympathetic axons. *Nature* 342:430–33

Watkins TA, Emery B, Mulinyawe S, Barres BA. 2008. Distinct stages of myelination regulated by gamma-secretase and astrocytes in a rapidly myelinating CNS coculture system. *Neuron* 60:555–69

Williams A, Piaton G, Aigrot MS, Belhadi A, Theaudin M, et al. 2007. Semaphorin 3A and 3F: key players in myelin repair in multiple sclerosis? *Brain* 130:2554–65

Wolswijk G. 1998. Chronic stage multiple sclerosis lesions contain a relatively quiescent population of oligo-dendrocyte precursor cells. *J. Neurosci.* 18:601–9

Wolswijk G, Noble M. 1989. Identification of an adult-specific glial progenitor cell. *Development* 105:387–400

Woodruff RH, Fruttiger M, Richardson WD, Franklin RJM. 2004. Platelet-derived growth factor regulates oligodendrocyte progenitor numbers in adult CNS and their response following CNS demyelination. *Mol. Cell Neurosci.* 25:252–62

Wu S, Wu Y, Capecchi MR. 2006. Motoneurons and oligodendrocytes are sequentially generated from neural stem cells but do not appear to share common lineage-restricted progenitors in vivo. *Development* 133:581–90

Xin M, Yue T, Ma Z, Wu FF, Gow A, Lu QR. 2005. Myelinogenesis and axonal recognition by oligodendro-cytes in brain are uncoupled in Olig1-null mice. *J. Neurosci.* 25:1354–65

Ye F, Chen Y, Hoang T, Montgomery RL, Zhao XH, et al. 2009. HDAC1 and HDAC2 regulate oligoden-drocyte differentiation by disrupting the beta-catenin-TCF interaction. *Nat. Neurosci.* 12:815–27

Yin X, Baek RC, Kirschner DA, Peterson A, Fujii Y, et al. 2006. Evolution of a neuroprotective function of central nervous system myelin. *J. Cell Biol.* 172:469–78

Yoon BH, Park CW, Chaiworapongsa T. 2003. Intrauterine infection and the development of cerebral palsy. *BJOG* 110(Suppl. 20):124–27

Zalc B, Colman DR. 2000. Origins of vertebrate success. *Science* 288:271–72

Zawadzka M, Rivers LE, Fancy SPJ, Zhao C, Tripathi R, et al. 2010. CNS-resident glial progenitor/stem cells produce Schwann cells as well as oligodendrocytes during repair of CNS demyelination. *Cell Stem Cell* 6:578–90

Zhang Y, Argaw AT, Gurfein BT, Zameer A, Snyder BJ, et al. 2009. Notch1 signaling plays a role in regulating precursor differentiation during CNS remyelination. *Proc. Natl. Acad. Sci. USA* 106:19162–67

Zhao X, He X, Han X, Yu Y, Ye F, et al. 2010. MicroRNA-mediated control of oligodendrocyte differentiation. *Neuron* 65:612–26

Zhou Q, Anderson DJ. 2002. The bHLH transcription factors OLIG2 and OLIG1 couple neuronal and glial subtype specification. *Cell* 109:61–73

Zhou Q, Wang S, Anderson DJ. 2000. Identification of a novel family of oligodendrocyte lineage-specific basic helix-loop-helix transcription factors. *Neuron* 25:331–43

Neural Representations for Object Perception: Structure, Category, and Adaptive Coding

Zoe Kourtzi[1] and Charles E. Connor[2]

[1] School of Psychology, University of Birmingham, Edgbaston, Birmingham, B15 2TT, United Kingdom; email: z.kourtzi@bham.ac.uk

[2] Krieger Mind/Brain Institute and Department of Neuroscience, Johns Hopkins University, Baltimore, Maryland 21218, USA; email: connor@jhu.edu

Annu. Rev. Neurosci. 2011. 34:45–67

First published online as a Review in Advance on March 24, 2011

The *Annual Review of Neuroscience* is online at neuro.annualreviews.org

This article's doi: 10.1146/annurev-neuro-060909-153218

Keywords

shape, ventral pathway, recognition, learning

Abstract

Object perception is one of the most remarkable capacities of the primate brain. Owing to the large and indeterminate dimensionality of object space, the neural basis of object perception has been difficult to study and remains controversial. Recent work has provided a more precise picture of how 2D and 3D object structure is encoded in intermediate and higher-level visual cortices. Yet, other studies suggest that higher-level visual cortex represents categorical identity rather than structure. Furthermore, object responses are surprisingly adaptive to changes in environmental statistics, implying that learning through evolution, development, and also shorter-term experience during adulthood may optimize the object code. Future progress in reconciling these findings will depend on more effective sampling of the object domain and direct comparison of these competing hypotheses.

Contents

INTRODUCTION

Object perception is critical for understanding and interacting with the world. Our ability to perceive objects is amazingly rapid, robust, and accurate, given the extreme computational difficulty of extracting object information from natural images (Dickinson 2009). The neural coding mechanisms underlying this remarkable ability have been a subject of intense study for half a century. Yet the fundamental principles of object processing in the brain remain uncertain and controversial. In contrast, other aspects of visual perception have been satisfyingly explained at a mechanistic level. For example, scholars widely accept that visual motion is represented by populations of neurons tuned for direction and speed in areas MT (middle temporal) and MST (middle superior temporal) and in other parts of the dorsal visual pathway (McCool & Britten 2007). But, in the ventral visual pathway (Ungerleider & Mishkin 1982, Felleman & Van Essen 1991), we have no comparable consensus on the coding dimensionality for objects. In fact, studies of ventral pathway function often avoid the question of what specific information is encoded by neural responses, somewhat comparable to studying MT neurons without knowing about direction tuning.

The reason for this gap in understanding is the difficulty of adequately sampling the enormous input domain for ventral pathway neurons. Object space is simply too high dimensional to study in the same way as other visual subdomains. Motion coding can be studied by sampling neural responses to stimuli along a few obvious dimensions such as direction and speed. These responses can be fit with mathematical tuning functions that capture the motion information conveyed by the neural responses. This basic ability to characterize the information encoded by neurons has been the foundation for spectacular work on perceptual causality and decision-making in the dorsal motion pathway (McCool & Britten 2007).

This basic approach cannot be applied in the same way to the ventral object pathway. The dimensionality of the object domain is too high to sample comprehensively, and it is unknown: There is no single, obvious way to represent a complex object with neural responses. As a result, the standard approach has been to sample object space randomly, with arbitrary sets of real or photographic objects. Such experiments have provided seminal insights into ventral pathway function, including the discovery of face-processing neurons (Desimone et al. 1984) and the description of columnar organization in inferotemporal cortex (Fujita et al. 1992). But because sampling is sparse and incomplete, these experiments cannot elucidate the specific information conveyed by neural responses; they cannot determine the coding dimensions of object-selective neurons, and they cannot constrain mathematical models of neural tuning in those dimensions.

This review describes three recent trends in the ongoing effort to grapple with the high dimensionality of object space. First, investigators have recently attempted to parameterize object structure and quantify neural tuning in structural dimensions. Second, others have attempted to quantify the relationship of neural responses to object categories. Third, recent studies have addressed the dynamic nature of ventral pathway coding, in the hope that object representation can be understood in terms of

Ventral pathway:
one of the two main pathways in the primate visual cortical hierarchy; the ventral pathway processes object-related information, including shape, color, and texture

the learning mechanisms that generate neural codes during development and that recalibrate coding on shorter timescales.

STRUCTURAL CODING

The classic approach to neural coding is to parameterize stimuli along one or more dimensions, to sample neural responses comprehensively along those dimensions, and to fit those responses with mathematical functions to describe how neurons encode information along those dimensions. In the object domain, this approach is problematic because the dimensionality of objects is (*a*) vast, necessitating the use of very large stimulus sets to cover the domain with some level of completeness, and (*b*) indeterminate, requiring novel experimental and analytical designs to test hypotheses about neural coding dimensions for objects. Nevertheless, progress has been made in quantifying neural tuning in structural dimensions across large sets of parametrically varying object stimuli.

Boundary Fragment Coding in Intermediate Cortex

Problems of sampling and stimulus parameterization are more tractable at intermediate processing stages such as area V4 because receptive fields are smaller and thus the complexity of object information encoded by neurons is correspondingly lower. Attempts to understand object coding in area V4 have partially extrapolated from what is known about structural representation in early visual cortex. Thus, V4 has been studied with grating stimuli (Gallant et al. 1993), contour stimuli (Pasupathy & Connor 1999, 2001), and natural object photographs (David et al. 2006). In all three cases, the scale and complexity of stimuli have been increased commensurate with V4 receptive field sizes, which are on the same order as retinal eccentricity (i.e., at 3° eccentricity, receptive field diameter is roughly 3°).

Responses in early visual cortex can be well characterized with tuning models in the orientation/spatial frequency domain that account for phase invariance (David & Gallant 2005). Such models capture less variance at the V4 level (David et al. 2006), which suggests that additional dimensions are represented in V4. A number of studies have shown that V4 neurons are sensitive not only to orientation but also to curvature (Gallant et al. 1993; Pasupathy & Connor 1999, 2001), which is the derivative or rate of change of orientation with respect to contour length. This finding makes sense because contrast edges in natural scenes (typically produced by object boundaries) are more likely to change orientation within the larger image windows encompassed by V4 receptive fields. Curvature is also a salient quality in human perception (Andrews et al. 1973, Treisman & Gormican 1988, Wilson et al. 1997, Wolfe et al. 1992, Ben-Shahar 2006). Thus, explicit coding of curvature in V4 is an effective way to represent important boundary elements of natural objects (Connor et al. 2007).

Another tuning dimension that appears in intermediate ventral pathway cortex is relative position. Absolute, retinotopic position coding deteriorates as receptive fields grow larger through progressively higher processing stages in the ventral pathway (Felleman & Van Essen 1991). Yet information about the positional arrangement of structural elements is critical for recognizing objects and perceiving their physical structure. Hence, it is not surprising that neurons at the V4 level and higher are acutely sensitive to the position of structural elements relative to each other and to the object as a whole (Connor et al. 2007).

Figure 1*a* exemplifies V4 tuning for curvature and relative position of object boundary fragments. This particular neuron responded to objects with acute convex curvature near the top. This response pattern can be captured with a two-dimensional (2D) Gaussian function on the curvature/angular position domain (**Figure 1***b*). The response pattern remained consistent across changes in absolute, retinotopic position (Pasupathy & Connor, 2001). Also, tuning for convexity near the top remained consistent across wide variations in

Area V4: a major intermediate stage in the primate ventral visual pathway

2D: two dimensional

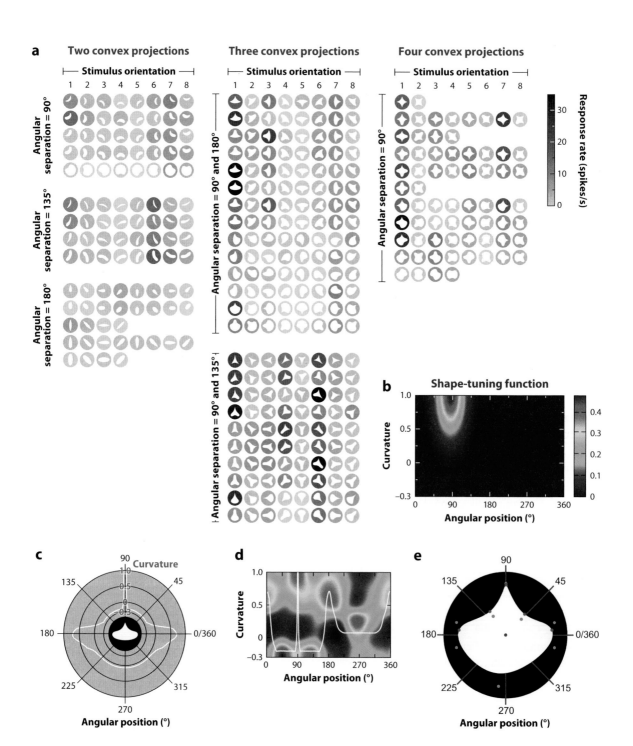

a

Two convex projections

├── **Stimulus orientation** ──┤

1 2 3 4 5 6 7 8

Angular separation = 90°

Angular separation = 135°

Angular separation = 180°

Three convex projections

├── **Stimulus orientation** ──┤

1 2 3 4 5 6 7 8

Angular separation = 90° and 180°

Angular separation = 90° and 135°

Four convex projections

├── **Stimulus orientation** ──┤

1 2 3 4 5 6 7 8

Angular separation = 90°

Response rate (spikes/s)

30

20

10

0

b **Shape-tuning function**

Curvature

1.0

0.5

0

−0.3

0 90 180 270 360

Angular position (°)

0.4

0.3

0.2

0.1

0

c

90 Curvature

1.0

135 0.5 45

0

−0.3

180 0/360

225 315

270

Angular position (°)

d

Curvature

1.0

0.5

0

−0.3

0 90 180 270 360

Angular position (°)

e

90

135 45

180 0/360

225 315

270

Angular position (°)

global object shape (**Figure 1a**). This is a critical prediction of structural coding theories that depend on representation by components (Hubel & Wiesel 1959, Selfridge 1959, Sutherland 1968, Barlow 1972, Milner 1974, Marr & Nishihara 1978, Hoffman & Richards 1984, Biederman 1987, Dickinson et al. 1992): Component signals from a given neuron must have the same information value regardless of shape variations elsewhere in the object. In agreement with this prediction, neurons in V4 (Pasupathy & Connor 2001) and higher-level processing stages in inferotemporal (IT) cortex (Brincat & Connor 2004, Yamane et al. 2008) respond at maximal levels to a wide variety of global shapes sharing some spatially localized structural element(s).

The range of different V4 tuning functions is broad and comprehensive enough to serve as a basis set for representing global shape at the population level. This is demonstrated in **Figure 1c–e**, where a single shape from the stimulus set (**Figure 1c**) is reconstructed from the neural population response to that shape. Each neuron's tuning function (e.g., **Figure 1b**) was weighted by its response to the shape and summed into the overall pattern in **Figure 1d**. The local maxima in this pattern correspond to the curvatures and positions of the boundary fragments that make up the shape. These local maxima can be used to reconstruct the approximate shape of the original stimulus (**Figure 1e**). All stimuli were approximately recoverable in this fashion, showing that V4 neurons carry relatively complete information

about the structure of 2D object boundaries at the population level (Pasupathy & Connor 2002). These analyses provide a neural confirmation of the theory of representation by components.

Configural Coding in Higher-Level Cortex

Beyond V4, neurons with larger receptive fields integrate information across entire objects, and as a result the dimensionality of object space becomes much less tractable. Two-dimensional object structure can be parameterized and tested comprehensively at a level of moderate complexity with the use of very large stimulus sets, on the order of 10^3, which is near the practical limits of neural recording experiments (Brincat & Connor 2004). But this approach becomes unworkable for three-dimensional (3D) object structure, which would require stimulus sets on the order of 10^4 or 10^5 to address object representation at a comparable level of complexity.

Although random and systematic sampling are inadequate at this level of structural complexity, a promising alternative is adaptive sampling, i.e., search through object space guided by neural responses. One version of this idea was pioneered by Tanaka and colleagues (1991). Beginning with a test of IT neural responses to randomly selected objects, the object evoking the strongest response was deconstructed into simpler components. The end point for each neuron was the simplest pattern that still evoked

Representation by components: a phrase coined by Irving Biederman (1987) to describe representation of objects in terms of their component parts

Inferotemporal (IT) cortex: a general anatomical label for the more anterior, higher-level stages in the primate ventral visual pathway

3D: three dimensional

Figure 1

Boundary fragment coding in intermediate ventral pathway cortex. (*a*) Responses of an individual V4 neuron to two-dimensional (2D) silhouette stimuli, recorded from a macaque monkey performing a fixation task. Stimuli were flashed at the cell's receptive field center. Average responses across 5 presentations are represented by gray levels surrounding each stimulus icon (see scale bar). (*b*) Gaussian function describing the response pattern in part *a*. The vertical axis represents boundary curvature (squashed to a scale from −1 to 1), and the horizontal axis represents angular position of boundary fragments with respect to the shape's center of mass. The color scale on the right indicates normalized predicted response. The tuning peak corresponds to sharp convex curvature (1.0) near the top of the shape (84.6°). (*c*) Curvature/angular position function for a single stimulus, plotted in polar coordinates to illustrate correspondence with the stimulus outline. (*d*) Estimated V4 population response across the curvature/angular position domain (colored surface, plotted in Cartesian coordinates) with the veridical curvature function (*white line*) superimposed. A Cartesian plot is used here because a polar plot would distort peak width in the population response. (*e*) Reconstruction of the stimulus shape based on the population response surface in part *d*. Modified from Pasupathy & Connor 2002.

near-maximal responses. This approach was a critical tool for demonstrating the columnar organization of IT (Fujita et al. 1992). However, because the method is strictly convergent, the final, simplified structure is limited to whatever existed in the original set of random objects, and the single end point cannot constrain a quantitative model of neural tuning.

A divergent, evolutionary method for adaptively sampling object space was recently tested by Yamane and colleagues (2008). The example experiment on an IT neuron presented in **Figure 2** began with two sets of 50 random 3D shapes (**Figure 2a**, Run 1 and Run 2). Three-dimensionality was conveyed by shading cues (visible in **Figure 2a**) and binocular disparity cues. The average responses of the neuron to each stimulus (indicated by background color, according to the scale bar in **Figure 2a**) were used as feedback to a probabilistic algorithm for defining subsequent generations of stimuli. Subsequent generations emphasized partially morphed versions of high-response stimuli from previous generations, which ensured that structural components eliciting neural responses propagated, evolved, and recombined, producing dense sampling in the most relevant region of object space. For this example neuron, both runs evolved high-response stimuli characterized by a specific configuration of sharp convex projections and concave indentations in the upper right quadrant of the objects (**Figure 2b,c**).

This configuration of surface fragments was well described with models based on structural tuning for surface curvature, surface orientation, and 3D relative position. The model shown here is based on two Gaussian tuning functions in the curvature/orientation/position domain (**Figure 2d**). Projection of these tuning functions onto the surface of example high-response stimuli (**Figure 2e**) shows that the cyan function captured the sharp convexities and the magenta function captured the interleaved concavities. Successful cross-prediction of responses between runs (**Figure 2e**) demonstrates that the adaptive search algorithm converged on the same result from different starting points.

Across the IT population, neurons exhibit a wide range of tuning for surface fragment configurations (**Figure 3a**). Tuning for configurations, as opposed to individual structural elements, has been a consistent finding in IT cortex in previous 2D shape experiments as well (Brincat & Connor 2004). Tuning for configurations develops gradually over the course of ~60 ms following initial responses to individual components (Brincat & Connor 2006). Configural tuning may represent a coding optimum between the extremes of component-level representation (as in V4, see **Figure 1**)

Figure 2

Adaptive sampling of object structure space. Neural responses were recorded from a single cell in IT of a macaque monkey performing a fixation task. Stimuli were flashed at the center of gaze for 750 ms each. Two independent stimulus lineages (Run 1 and Run 2) are shown in the left and right columns, respectively. Background color (see scale bar) indicates the average response to each stimulus across five presentations. (*a*) Initial generations of 50 randomly constructed 3D shape stimuli. Stimuli are ordered from top left to bottom right according to average response strength. (*b*) Partial family trees showing how stimulus shape and response strength evolved across successive generations. (*c*) Highest-response stimuli across 10 generations (500 stimuli) in each lineage. (*d*) Response models based on two Gaussian tuning functions. The Gaussian functions describe tuning for surface fragment geometry, defined in terms of curvature (principal, i.e., maximum and minimum, cross-sectional curvatures), orientation (of a surface normal vector, projected onto the *x/y* and *y/z* planes), and position (relative to object center of mass in x/y/z coordinates). The curvature scale is squashed to a range between –1 (concave) and 1 (convex). The 1.0 standard deviation boundaries of the two Gaussians (*magenta and cyan*) are shown projected onto different combinations of these dimensions. The equations show the overall response models, with fitted weights for the two Gaussians, the product or interaction term, and the baseline response. (*e*) The two Gaussian functions are shown projected onto the surface of a high-response stimulus from each run. The stimulus surface is tinted according to the tuning amplitude in the corresponding region of the model domain. The scatterplots show the relationship between observed responses and responses predicted by the model. In each case, self-prediction by the model is illustrated by the stimulus/scatterplot pair on the left and cross-prediction by the pair on the right. Reproduced from Yamane et al. (2008).

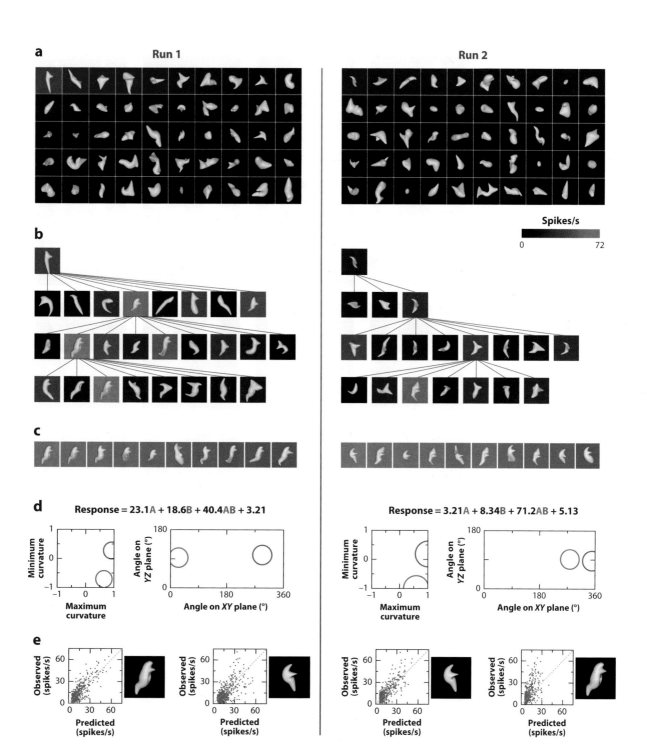

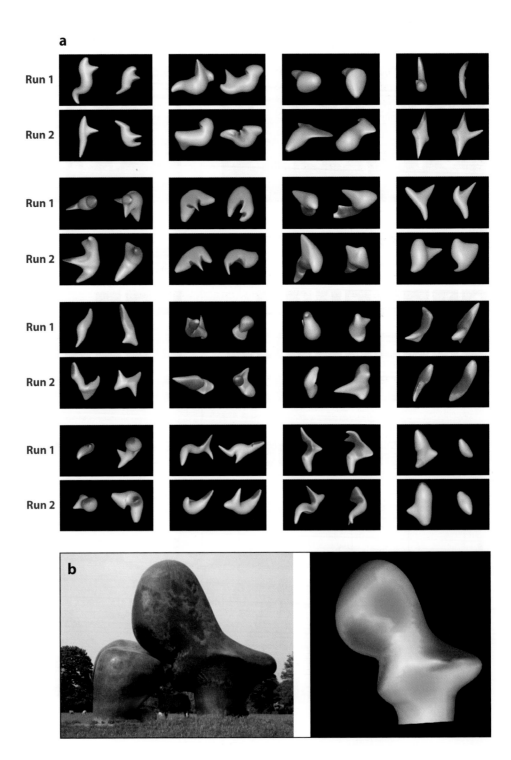

and holistic representation. Component-level representation is combinatorial and therefore highly productive, suitable for representing the virtual infinity of potential object shapes. Holistic representation schemes, in which individual neurons signal information about global shape, have more potential for sparse, efficient representation. Configural coding may represent a compromise between productivity and sparseness.

Conceivably, IT neurons tuned for surface fragment configurations serve as basis functions for representing complete 3D object structure. This idea is represented diagrammatically in **Figure 3b**, where a detail from a Henry Moore sculpture is approximated with a computer rendering. Tuning functions from Yamane et al. (2008) are projected onto the surface to suggest how the complete shape could be represented by a neural ensemble signaling its constituent surface fragment configurations. This coding scheme would provide a compact, explicit representation of the kind of 3D object structure we experience perceptually.

Representation of Face Structure

Another way to tackle the enormous dimensionality of object space is to restrict investigation to a well-defined subspace of objects. This approach makes sense for neurons that operate primarily within such a subspace, as with neurons in face-processing regions of the ventral pathway, which show remarkable selectivity for face stimuli over other natural categories of objects (Tsao et al. 2006). Given this restricted coding context, investigators can explore the relevant input space densely

and comprehensively. Freiwald and colleagues (2009) did this by parameterizing cartoon faces in terms of size, shape, and relative positions of eyes, brows, nose, and mouth. Neurons in the middle face-processing region of monkey IT exhibited tuning for configurations of parts defined according to these dimensions. The range of tuning patterns suggested that this face patch contains a complete basis function representation of a facial structure space.

Other groups have taken a different theoretical approach inspired by psychophysical results, suggesting that faces are represented in terms of holistic structural similarity and organized with respect to a grand geometric average over all faces encountered through time (Rhodes et al. 1987, Mauro & Kubovy 1992, Leopold et al. 2001, Webster et al. 2004). Loffler and colleagues (2005) provided evidence in favor of this average face principle by showing strong fMRI cross-adaptation in human fusiform face area to stimuli lying along the same morph direction from the average face. Leopold and colleagues (2006) provided parallel evidence for tuning along such morph lines at the level of individual neurons in macaque monkey IT. It would be interesting to see the holistic similarity hypothesis tested directly against the component structure hypothesis with a suitable stimulus set parameterized in both domains simultaneously.

CATEGORICAL CODING

The main alternative to structural object representation is categorical representation. In both words and actions, we group objects into categories on the basis of characteristics

fMRI: functional magnetic resonance imaging

Figure 3

Configural coding in higher-level ventral pathway cortex. (*a*) Surface configuration tuning for 16 example IT neurons. In each case, two high-response stimuli are shown from the first run (*top row*) and the second run (*bottom row*). Models were fit as described in **Figure 2** and the two Gaussian tuning functions were projected onto the surface of the stimuli. (*b*) Hypothetical example of configural coding of 3D object structure. Five 2-Gaussian tuning models (*red, green, blue, cyan, magenta*) from Yamane et al. (2008) are projected onto a 3D rendering (*right*) of the larger figure in Henry Moore's "Sheep Piece" (1971–1972, *left*; reproduced by permission of the Henry Moore Foundation, **http://www.henry-moore-fdn.co.uk**). Reproduced from Yamane et al. (2008).

that are often partially or wholly nonstructural: animacy, behavior, utility, and especially association, either episodic or conceptual. It seems certain that both structure and category must be represented somewhere in the brain and that those representations must interact in some way. But there is potential controversy over which domain provides the most fundamental explanation of object coding in the ventral pathway and, by extension, underlies our perceptual experience of objects.

Categorical representation of objects has long been studied at the qualitative level (Desimone et al. 1984, Vogels 1999). Recently, researchers have begun to use quantitative analyses to study categorical representation in functionally homologous regions of the ventral pathway cortex (Denys et al, 2004). Kiani and colleagues (2007) analyzed categorical representation in a massive data set of 674 neurons recorded from monkey anterior IT, each studied with more than 1000 natural object photographs. Multidimensional scaling (MDS), applied to the distances between objects in neural response space, revealed an overarching division between animate and inanimate objects, with further subdivision of animate objects into subcategories that included human faces, monkey faces, nonprimate faces, hands, human bodies, and quadrupeds. The higher-level divisions, between animate and inanimate and between faces and bodies, have been replicated for human inferior temporal visual cortex by analyzing fMRI voxel response patterns (Kriegeskorte et al. 2008) (**Figure 4**). Analyses for reconstruction of natural images from

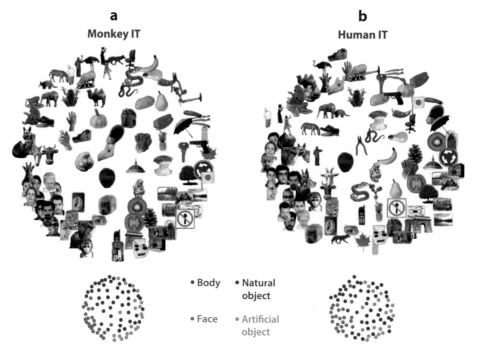

Figure 4

Categorical coding in higher-level ventral pathway cortex. Ninety-two object photographs were presented to monkeys and humans performing a fixation task. Responses were recorded from 674 IT neurons in two monkeys. Responses of IT cortex in four humans were measured with high-resolution fMRI. For both data sets, multidimensional scaling techniques were used to produce the stimulus arrangements shown here, in which distance between stimuli corresponds approximately to distance in neural (monkey) or voxel (human) response space (i.e., dissimilarity of response patterns). Both stimulus arrangements show that faces, bodies, and other objects fall into separate response clusters. Reproduced from Kriegeskorte et al. (2008).

fMRI voxel response patterns also indicate the existence of category information in anterior human visual cortex (Naselaris et al. 2009).

These findings pose an interesting challenge to structural coding hypotheses. Apparent structural tuning may only be a reflection of selectivity for object categories, which are definable to some extent by their structural characteristics. Conversely, apparent selectivity for an object category could reflect more fundamental tuning for structural characteristics of that category. These alternatives could be differentiated by studies that simultaneously analyze categorical and structural tuning and contrast their explanatory power. Freedman and colleagues (2003) did this for a single, learned categorical distinction (between cat-like and dog-like stimuli) and found that the amount of category information in monkey IT was no greater than that expected on the basis of structural tuning. Similar analyses remain to be done for the naturalistic categories identified in the studies cited above.

ADAPTIVE CODING

In the search for neural codes, we typically measure responses to input alone (e.g., objects, faces) without accounting for context in space (i.e., scene configuration) or time (i.e., previous experiences with a given object). However, accumulating evidence suggests an adaptive neural code that is dynamically shaped by experience. Here, we summarize work showing that experience plays a critical role in shaping structural and categorical coding for object perception. That is, learning optimizes the neural processes that mediate binding of local elements and parts into objects, recognition of objects across image changes that preserve identity (e.g., position, orientation, clutter), and selection of behaviorally relevant features for object categorization. We propose that similar learning mechanisms may mediate long-term optimization through evolution and development, tune the visual system to fundamental principles of feature binding, and shape structure and category representations.

Learning to See Objects

Evolution and development shape the organization of the visual system and facilitate visual recognition in cluttered scenes (Gilbert et al. 2001, Simoncelli & Olshausen 2001). Recent studies suggest that the primate brain is sensitive to regularities that occur frequently in natural scenes (e.g., orientation similarity in neighboring elements) and has developed a network of connections that mediate integration of object features based on these correlations (Sigman et al. 2001, Geisler 2008). However, long-term experience is not the only means by which visual processes become optimized. Learning through everyday experiences in adulthood plays a key role in facilitating the detection and recognition of targets in cluttered scenes (Dosher & Lu 1998, Goldstone 1998, Schyns et al. 1998, Gold et al. 1999, Sigman & Gilbert 2000, Gilbert et al. 2001, Brady & Kersten 2003). Observers are shown to learn distinctive target features by using image regularities to integrate relevant object features and by suppressing background noise (Dosher & Lu 1998, Gold et al. 1999, Brady & Kersten 2003, Li et al. 2004).

Here, we propose that long-term experience and short-term training interact to shape the optimization of visual recognition processes. Whereas long-term experience through evolution and development hones the principles of organization that mediate feature grouping for object recognition, short-term training in adulthood may establish new principles for interpreting natural scenes. For example, long-term experience with the high prevalence of collinear edges in natural environments (Sigman et al. 2001, Geisler 2008) has resulted in enhanced sensitivity for detecting collinear contours in clutter. However, short-term training alters the behavioral relevance of image regularities that violate the typical principles of contour linking (Sigman et al. 2001, Simoncelli & Olshausen 2001, Geisler 2008). Although collinearity is a prevalent principle for perceptual integration in natural scenes, recent evidence (Schwarzkopf & Kourtzi 2008)

Statistical learning:
learning of regularities
by mere exposure

Recurrent
processing:
processing based on
horizontal and
feedback connections

suggests that the brain can learn to exploit other image regularities (i.e., orthogonal alignments) that typically signify discontinuities for contour linking. Furthermore, both infants and adults learn fast and without explicit feedback to extract and exploit novel spatial and temporal regularities that appear frequently in visual scenes. Examples of this type of statistical learning comprise parsing speech into meaningful language streams (Saffran et al. 1996, Pena et al. 2002), integrating shapes across space (Fiser & Aslin 2001, Baker et al. 2004, Turk-Browne et al. 2009), combining object views across time (Kourtzi & Shiffrar 1997, Wallis & Bulthoff 2001), grouping objects into spatial configurations and visual scenes (Fiser & Aslin 2005, Orban et al. 2008), and abstracting visual categories (Brady & Oliva 2008).

Which are the neural mechanisms that mediate our ability to extract statistical regularities and learn novel principles of perceptual organization for object detection and recognition? Recent neurophysiology and imaging studies implicate recurrent processing between local integration mechanisms that tune image statistics in visual cortex and top-down fronto-parietal mechanisms that mediate the formation and flexible selection of behaviorally relevant rules and features. Consistent with a theoretical model of attention-gated reinforcement learning (Roelfsema & van Ooyen 2005), learning enhances responses in fronto-parietal circuits (Schwarzkopf & Kourtzi 2008). These gain effects may relate to a global reinforcement mechanism that is important for identifying salient image regions and detecting objects in clutter. Goal-directed attentional mechanisms may then optimize visual processing within these salient regions and change the neural sensitivity to the relevant object features rather than spurious image correlations. Thus, learning may support efficient target detection by enhancing the salience of targets through increased correlation of neuronal signals related to the target features and decorrelation of signals related to target and background features (Jagadeesh et al. 2001, Li et al. 2008). That is, feedback from higher fronto-parietal regions may change neural processing (i.e., neural selectivity or local correlations) in higher occipitotemporal circuits that support shape integration and recognition (Kourtzi et al. 2005, Sigman et al. 2005, Schwarzkopf et al. 2009).

Recent studies combining behavioral and brain-imaging measurements (Zhang & Kourtzi 2010) propose two routes to visual learning in clutter (**Figure 5**). These studies show that long-term experience with statistical regularities (i.e., collinearity)

Figure 5

Learning statistical regularities. (*a*) Examples of stimuli: Collinear contours in which elements are aligned along the contour path and orthogonal contours in which elements are oriented at 90° to the contour path. For demonstration purposes only, two rectangles illustrate the position of the two contour paths in each stimulus. (*b*) Average behavioral performance across subjects (percent correct) before and after supervised training (i.e., observers received feedback on a contour detection task) or exposure (i.e., observers performed an irrelevant contrast discrimination task) to collinear or orthogonal contours. Before training, detection was difficult for both collinear and orthogonal contours. After training, the observers' performance in detecting orthogonal contours improved significantly following supervised training but not following mere exposure. In contrast, for collinear contours, observers showed similar improvement in detection performance following supervised training or exposure. These learning effects were specific to the trained contour orientation for orthogonal contours, whereas they generalized to untrained orientations for collinear contours. (*c*) fMRI responses for observers trained with orthogonal versus collinear contours. fMRI data (percent signal change for contour minus random stimuli) are shown for trained contour orientations before and after supervised training on orthogonal (*upper panel*) versus collinear (*lower panel*) contours. Training enhanced responses in intraparietal regions for orthogonal contours while in higher occipitotemporal regions for collinear contours. Taken together, the behavioral and fMRI findings demonstrate that opportunistic learning of statistical regularities (i.e., collinear contours) may occur by frequent exposure and is mediated by occipitotemporal areas, whereas bootstrap-based learning of discontinuities (i.e., orthogonal contours) requires extensive training and is mediated by intraparietal regions. Adapted from Zhang & Kourtzi (2010).

may facilitate opportunistic learning (i.e., learning to exploit image cues), whereas learning to integrate discontinuities (i.e., elements orthogonal to contour paths) entails bootstrap-based training (i.e., learning new features) for detecting contours in clutter. Learning to integrate collinear contours occurs simply through frequent exposure, generalizes across untrained stimulus features, and shapes processing in higher occipitotemporal regions implicated in the representation of global forms. In contrast, learning to integrate discontinuities (i.e., elements orthogonal to contour paths) required task-specific training

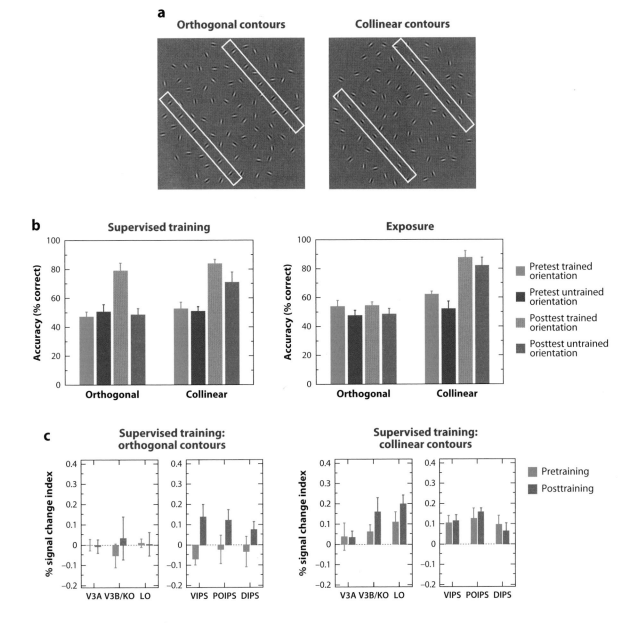

(bootstrap-based learning), was stimulus dependent, and enhanced processing in intraparietal regions implicated in attention-gated learning. Similarly, recent neuroimaging studies suggest that a ventral cortex region becomes specialized through experience and development for letter integration and word recognition (Dehaene et al. 2005), whereas parietal regions are recruited for recognizing words presented in unfamiliar formats (Cohen et al. 2008). Taken together, these findings propose that opportunistic learning of statistical regularities shapes bottom-up object processing in occipitotemporal areas, whereas learning new features and rules for perceptual integration recruits parietal regions involved in the attentional gating of recognition processes.

Learning Object Structure

How does the brain construct structural object representations that are sensitive to subtle differences in object identity so we can discriminate between similar objects while being tolerant of image changes that preserve object identity, enabling us to recognize different presentations of the same object? Recent neurophysiological studies propose that although individual neurons contain highly selective information for image features, connections across neural populations may support object recognition across image changes. In particular, neural populations in higher temporal areas may contain information about object identity that may generalize across image changes (e.g., Rolls 2000, Grill-Spector & Malach 2004, Hung et al. 2005, Quiroga et al. 2005). Computational models (Fukushima 1980, Riesenhuber & Poggio 1999, Ullman & Soloviev 1999) propose that the brain builds these robust object representations using neuronal connections that group together similar image features across image transformations. Furthermore, recent neurophysiological studies (Zoccolan et al. 2007) show that temporal cortex neurons with high object selectivity have low invariance. These studies suggest that connections between neurons selective for similar features are critical for the binding of feature configurations and the robust representation of object identity.

But how does the brain know which neurons to connect or which connections across neural populations to strengthen to build robust object representations? Experience and training may be a solution to this problem (Foldiak 1991, Wallis & Rolls 1997, Ullman & Soloviev 1999, Wallis & Bulthoff 2001) by enhancing the sparseness and clustering of the neural code. fMRI studies show that at the level of large neural populations training results in differential responses to trained compared with untrained object categories (see sidebar, Adaptive Coding Across Temporal Scales). In particular, learning changes the distribution of voxel preferences for the trained stimuli, suggesting altered sensitivity to stimulus features rather than simply gain modulations that would preserve the spatial distribution of activity (Op de Beeck et al. 2006, Schwarzkopf et al. 2009).

At the single-neuron level, training with novel object configurations and combinations

ADAPTIVE CODING ACROSS TEMPORAL SCALES

A range of fMRI studies using learning or repetition suppression paradigms (i.e., when a stimulus is presented repeatedly) show similar effects for long-term training, rapid learning, and priming, which depend on the nature of the stimulus representation. In particular, enhanced responses have been observed when learning engages processes necessary for new representations to form, as in the case of unfamiliar (Schacter et al. 1995, Gauthier et al. 1999, Henson et al. 2000), degraded (Tovee et al. 1996, Dolan et al. 1997, George et al. 1999), masked unrecognizable (Grill-Spector et al. 2000, James et al. 2000), or noise-embedded (Kourtzi et al. 2005) targets. In contrast, when the stimulus perception is unambiguous (e.g., familiar, undegraded, recognizable targets presented in isolation), training results in more efficient processing of the stimulus features indicated by attenuated neural responses (Henson et al. 2000, James et al. 2000, Jiang et al. 2000, van Turennout et al. 2000, Koutstaal et al. 2001, Chao et al. 2002, Kourtzi et al. 2005).

of object parts or mere experience with novel objects in the animals' living environment tunes temporal cortex neurons to novel objects and supports some generalization to neighboring object views (Miyashita & Chang 1988, Logothetis et al. 1995, Rolls 1995, Kobatake et al. 1998, Baker et al. 2002). Furthermore, training enhances not only the selectivity but also the clustering of IT neurons, with similar object selectivity enabling stronger local interactions (Erickson et al. 2000). Temporal continuity enhances the binding of disparate images into the same object representation (Kourtzi & Shiffrar 1997, Wallis & Bulthoff 2001, Cox et al. 2005). For example, recent work (Li & DiCarlo 2008) has shown that IT neurons learn to bind into the same object features that are presented at different retinal locations but in temporal correlation, supporting position-invariant object representations.

Taken together, neurophysiology and imaging studies provide evidence for learning-dependent plasticity mechanisms in the temporal cortex that mediate robust representations of object structure. However, whether learning results in long-term changes in neural properties or optimizes the readout signals in IT remains an open question. Recent neurophysiology studies showing that learning enhances the selectivity of the most informative neurons for a feature discrimination task (Raiguel et al. 2006) suggest that learning optimizes the readout of IT neurons. In particular, learning is thought to operate via top-down mechanisms that originate at decision stages, determine the relevance of object features, and reweight neural selectivity in sensory areas in a task-dependent manner (Dosher & Lu 1998, Ahissar & Hochstein 2004, Roelfsema & van Ooyen 2005, Law & Gold 2008). Accumulating evidence for such mechanisms comes from studies showing task-dependent learning effects in visual cortex (Gilbert et al. 2001, Kourtzi et al. 2005, Sigman et al. 2005). Thus, learning shapes robust object representations by enhancing the processing of feature detectors in local circuits using top-down knowledge about the relevant task dimensions and demands.

Learning Object Category

Extensive behavioral work on visual categorization (e.g., Goldstone et al. 2001) suggests that the brain learns the relevance of visual features for categorical decisions rather than simply representing physical similarity. That is, learning may reduce object space dimensionality by reweighting feature representations on the basis of their behavioral relevance in the context of a task.

Although a large network of brain areas has been implicated in visual category learning (see sidebar, Brain Networks for Category Learning), the role of temporal cortex in the learning and representation of visual categories remains controversial. Recent imaging studies have revealed a distributed pattern of activations for object categories in the temporal cortex (Haxby et al. 2001), including regions specialized for categories of biological importance (e.g., faces, bodies, places) (Reddy & Kanwisher 2006). However, some neurophysiological studies propose that the temporal cortex represents primarily the visual similarity between stimuli (Op de Beeck et al. 2001,

BRAIN NETWORKS FOR CATEGORY LEARNING

A large network of cortical and subcortical areas has been implicated in visual category learning (e.g. Vogels et al 2002; for reviews, see Keri 2003, Ashby & Maddox 2005). In particular, areas in the prefrontal cortex have been implicated in rule-based tasks in which the category structure is determined by a single stimulus dimension. This is consistent with the role of the prefrontal cortex in guiding visual attention to select behaviorally relevant information (for reviews, see Miller 2000, Duncan 2001). In contrast, the basal ganglia have been implicated primarily in information-integration tasks that require combining information from different stimulus dimensions for making categorical decisions. Furthermore, the medial temporal cortex has been implicated in category-learning tasks that rely on memorization. Finally, prototype-distortion tasks during which participants compare category exemplars to prototypical visual stimuli engage occipitotemporal regions.

Thomas et al. 2001, Freedman et al. 2003, Jiang et al. 2007, Op de Beeck et al. 2008), whereas others suggest that it represents learned stimulus categories (Meyers et al. 2008) and diagnostic stimulus dimensions for categorization (Sigala & Logothetis 2002, Mirabella et al. 2007). Furthermore, recent work suggests that the representations of object categories in the temporal cortex are modulated by task demands (Koida & Komatsu 2007) and experience (e.g., Op de Beeck et al. 2006, Gillebert et al. 2009).

Understanding the mechanisms that mediate adaptive coding of object categories is critical to understanding our ability to make flexible perceptual decisions. Here, we propose that adaptive categorical coding is implemented by interactions between top-down mechanisms related to the formation of rules and local processing of task-relevant object features. For example, recent neuroimaging studies (Li et al. 2007) using multivariate analysis methods provide evidence that learning shapes feature and object representations in a network of areas with dissociable roles in visual categorization (**Figure 6**). In particular, observers were trained to categorize dynamic shape configurations on the basis of single stimulus dimension

(form versus motion) or feature conjunctions. Temporal and parietal areas encode the perceived similarity in form and motion features, respectively. In contrast, frontal areas and the striatum represent task-relevant conjunctions of spatio-temporal features critical for forming more complex categorization rules. These findings suggest that neural representations in these areas are shaped by the behavioral relevance of sensory features and by previous experience to reflect the perceptual (categorical) rather than the physical similarities between stimuli. This notion is consistent with neurophysiological evidence for recurrent processes that modulate selectivity for perceptual categories along the behaviorally relevant stimulus dimensions in a top-down manner (Freedman et al. 2003, Smith et al. 2004, Mirabella et al. 2007) resulting in enhanced selectivity for the relevant stimulus features in visual areas.

Further evidence for recurrent processing for flexible categorical representations comes from recent work (Li et al. 2009) showing that category learning shapes decision-related processes in frontal and higher occipitotemporal regions rather than signal detection or response execution in primary visual or motor areas.

Figure 6

Learning rules for categorical decisions. (*a*) Five sample frames of a prototypical stimulus depicting a dynamic figure. Each stimulus comprised ten dots that were configured in a skeleton arrangement and moved in a biologically plausible manner (i.e., sinusoidal motion trajectories). (*b*) Stimuli were generated by applying spatial morphing (steps of percent stimulus B) between prototypical trajectories (e.g., A–B) and temporal warping (steps of time warping constant). Stimuli were assigned to one of four groups: A fast-slow (AFS), A slow-fast (ASF), B fast-slow (BFS), and B slow-fast (BSF). For the simple categorization task (*left panel*), the stimuli were categorized according to their spatial similarity: Category 1 (*red dots*) consisted of AFS, ASF, and Category 2 (*blue dots*) of BFS, BSF. For the complex task (*right panel*), the stimuli were categorized on the basis of their spatial and temporal similarity: Category 1 (*red dots*) consisted of ASF, BFS, and Category 2 (*blue dots*) of AFS, BSF. (*c*) Multivariate pattern analysis (MVPA) of fMRI data: Prediction accuracy (i.e., probability with which the presented and perceived stimuli are correctly predicted from brain activation patterns using a linear support vector machine classifier (SVM) for the spatial similarity (*blue line*) and complex (*green line*) classification schemes across categorization tasks (simple, complex task). Prediction accuracies for these MVPA rules are compared with accuracy for the shuffling rule (baseline prediction accuracy, *dotted line*). Interactions of prediction accuracy across tasks in dorsolateral prefrontal cortex (DLPFC) and lateral occipital (LO) regions indicate that the categories perceived by the observers are reliably decoded from fMRI responses in these areas. In contrast, the lack of a significant interaction in V1 shows that the stimuli are represented on the basis of their physical similarity rather than on the rule used by the observers for categorization. Adapted from Li et al. (2007).

In particular, in prefrontal circuits, learning shapes the estimation of the decision criterion only in the context of the categorization task. In contrast, in higher occipitotemporal regions, the representations of perceived categories are sustained after training independent of the task and may serve as selective readout signals for optimal decisions (**Figure 7**).

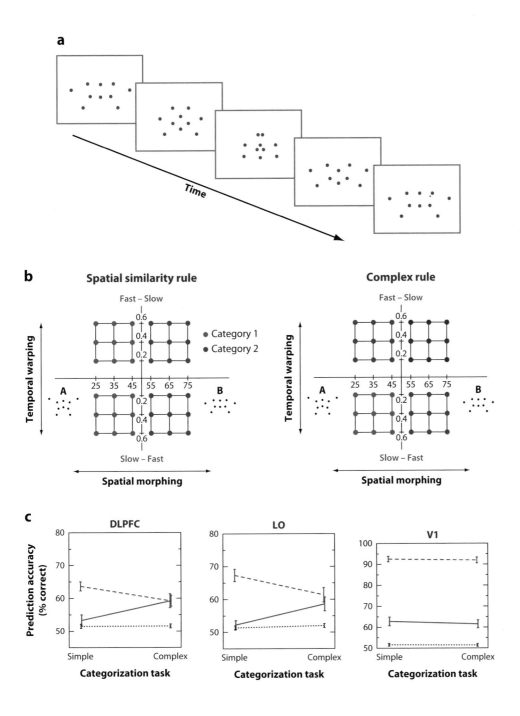

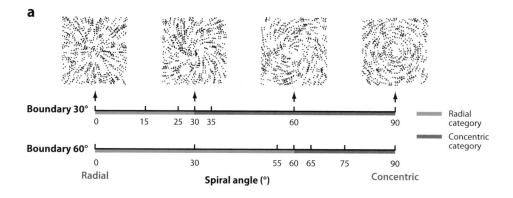

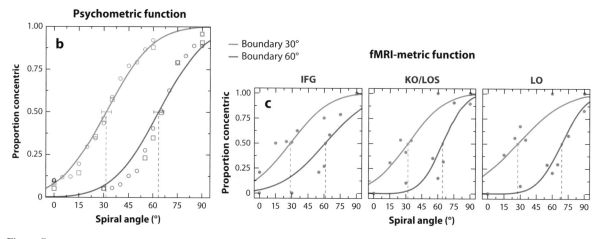

Figure 7

Learning shapes behavioral choice. (*a*) Observers were trained to categorize global form patterns as radial or concentric. Four example Glass pattern stimuli (100% signal) are shown at spiral angles of 0°, 30°, 60°, and 90°. Before training (pretraining test), the mean categorization boundary (50% point on the psychometric function) was close to the mean of the physical stimulus space (45° spiral angle). Observers were then trained with feedback to assign stimuli into categories on the basis of two different category boundaries: 30°, 60° spiral angle. The two tested boundaries and spiral angles that indicate the categorical membership of the stimuli for each boundary are shown (*blue bar: stimuli that resemble radial; red bar: stimuli that resemble concentric*). Observers were first trained on one of the two boundaries and then retrained on the other. (*b*) Testing the observers without feedback after training demonstrated that training had shifted the observers' criteria for categorization to the trained boundary (i.e., criterion of psychometric functions). (*c*) A linear support vector machine classifier (SVM) was trained to classify fMRI signals on the basis of the observer's behavioral choice (radial versus concentric) on each trial and tested for accuracy in predicting the observers' choice on an independent data set. For each observer, the mean performance of the classifier (proportion of trials classified as concentric for each stimulus condition) was calculated across cross-validations (fMR-metric functions). Comparing the classifier's choices with the observer's choices showed that fMR-metric functions in frontal and higher occipitotemporal areas resemble psychometric functions, suggesting a link between behavioral and neural responses. Adapted from Li et al. (2009).

CONCLUSION

Object vision is a remarkable perceptual capacity that has remained largely unexplained at the level of neural coding mechanisms. A primary obstacle has been the high, unknown dimensionality of objects, which precludes comprehensive sampling of the relevant input space. We have reviewed recent approaches to this problem: quantitative modeling of structural coding, adaptive sampling of object space, quantitative evaluation of categorical representation, and measurement of adaptive changes in object coding. Results from these different approaches are compelling, but they do not obviously cohere within a single framework. Structure is a conceptually different domain from category, and it is not clear which domain provides more fundamental explanations or how the two might interrelate. Both structural and categorical coding require some level of stability, a principle that is challenged by the strong adaptability of object responses. Future progress will depend in part on addressing these different themes within the same experimental contexts. The high dimensionality of object space will remain an enormous challenge, demanding further innovation in experimental and analytical design.

DISCLOSURE STATEMENT

The authors are not aware of any affiliations, memberships, funding, or financial holdings that might be perceived as affecting the objectivity of this review.

LITERATURE CITED

Ahissar M, Hochstein S. 2004. The reverse hierarchy theory of visual perceptual learning. *Trends Cogn. Sci.* 8:457–64

Andrews DP, Butcher AK, Buckley BR. 1973. Acuities for spatial arrangement in line figures: human and ideal observers compared. *Vision Res.* 13:599–620

Ashby FG, Maddox WT. 2005. Human category learning. *Annu. Rev. Psychol.* 56:149–178

Baker CI, Behrmann M, Olson CR. 2002. Impact of learning on representation of parts and wholes in monkey inferotemporal cortex. *Nat. Neurosci.* 5:1210–16

Baker CI, Olson CR, Behrmann M. 2004. Role of attention and perceptual grouping in visual statistical learning. *Psychol. Sci.* 15:460–66

Barlow HB. 1972. Single units and sensation: a neuron doctrine for perceptual psychology? *Perception* 1:371–94

Ben-Shahar O. 2006. Visual saliency and texture segregation without feature gradient. *Proc. Natl. Acad. Sci. USA* 103:15704–9

Biederman I. 1987. Recognition-by-components: a theory of human image understanding. *Psychol. Rev.* 94:115–47

Brady MJ, Kersten D. 2003. Bootstrapped learning of novel objects. *J. Vis.* 3:413–22

Brady TF, Oliva A. 2008. Statistical learning using real-world scenes: extracting categorical regularities without conscious intent. *Psychol. Sci.* 19:678–85

Brincat SL, Connor CE. 2004. Underlying principles of visual shape selectivity in posterior inferotemporal cortex. *Nat. Neurosci.* 7:880–86

Brincat SL, Connor CE. 2006. Dynamic shape synthesis in posterior inferotemporal cortex. *Neuron* 49:17–24

Chao LL, Weisberg J, Martin A. 2002. Experience-dependent modulation of category-related cortical activity. *Cereb. Cortex* 12:545–51

Cohen L, Dehaene S, Vinckier F, Jobert A, Montavont A. 2008. Reading normal and degraded words: contribution of the dorsal and ventral visual pathways. *Neuroimage* 40:353–66

Connor CE, Brincat SL, Pasupathy A. 2007. Transformation of shape information in the ventral pathway. *Curr. Opin. Neurobiol.* 17:140–47

Cox DD, Meier P, Oertelt N, DiCarlo JJ. 2005. 'Breaking' position-invariant object recognition. *Nat. Neurosci.* 8:1145–47

David SV, Gallant JL. 2005. Predicting neuronal responses during natural vision. *Network* 16:239–60

David SV, Hayden BY, Gallant JL. 2006. Spectral receptive field properties explain shape selectivity in area V4. *J. Neurophysiol.* 96:3492–505

Dehaene S, Cohen L, Sigman M, Vinckier F. 2005. The neural code for written words: a proposal. *Trends Cogn. Sci.* 9:335–41

Denys K, Vanduffel W, Fize D, Nelissen K, Peuskens H, et al. 2004. The processing of visual shape in the cerebral cortex of human and nonhuman primates: a functional magnetic resonance imaging study. *J Neurosci.* 24:2551–65

Desimone R, Albright TA, Gross CG, Bruce C. 1984. Stimulus-selective properties of inferior temporal neurons in the macaque. *J. Neurosci.* 4:2051–62

Dickinson SJ. 2009. The evolution of object categorization and the challenge of image abstraction. In *Object Categorization: Computer and Human Vision Perspectives*, ed. SJ Dickinson, A Leonardis, B Schiele, MJ Tarr, pp. 1–32. Cambridge, UK: Cambridge Univ. Press

Dickinson SJ, Pentland AP, Rosenfeld A. 1992. From volumes to views: an approach to 3-D object recognition. *CVGIP: Image Underst.* 55:130–54

Dolan RJ, Fink GR, Rolls E, Booth M, Holmes A, et al. 1997. How the brain learns to see objects and faces in an impoverished context. *Nature* 389:596–99

Dosher BA, Lu ZL. 1998. Perceptual learning reflects external noise filtering and internal noise reduction through channel reweighting. *Proc. Natl. Acad. Sci. USA* 95:13988–93

Duncan J. 2001. An adaptive coding model of neural function in prefrontal cortex. *Nat. Rev. Neurosci.* 2:820–29

Erickson CA, Jagadeesh B, Desimone R. 2000. Clustering of perirhinal neurons with similar properties following visual experience in adult monkeys. *Nat. Neurosci.* 3:1143–48

Felleman DJ, Van Essen DC. 1991. Distributed hierarchical processing in the primate cerebral cortex. *Cereb. Cortex* 1:1–47

Fiser J, Aslin RN. 2001. Unsupervised statistical learning of higher-order spatial structures from visual scenes. *Psychol. Sci.* 12:499–504

Fiser J, Aslin RN. 2005. Encoding multielement scenes: statistical learning of visual feature hierarchies. *J. Exp. Psychol. Gen.* 134:521–37

Foldiak P. 1991. Learning invariance from transformation sequences. *Neural Comput.* 3:194–200

Freedman DJ, Riesenhuber M, Poggio T, Miller EK. 2003. A comparison of primate prefrontal and inferior temporal cortices during visual categorization. *J. Neurosci.* 23:5235–46

Freiwald WA, Tsao DY, Livingstone MS. 2009. A face feature space in the macaque temporal lobe. *Nat. Neurosci.* 12:1187–96

Fujita I, Tanaka K, Ito M, Cheng K. 1992. Columns for visual features of objects in monkey inferotemporal cortex. *Nature* 360:343–46

Fukushima K. 1980. Neocognitron: a self organizing neural network model for a mechanism of pattern recognition unaffected by shift in position. *Biol. Cybern.* 36:193–202

Gallant JL, Braun J, Van Essen DC. 1993. Selectivity for polar, hyperbolic and Cartesian gratings in macaque visual cortex. *Science* 259:100–3

Gauthier I, Tarr MJ, Anderson AW, Skudlarski P, Gore JC. 1999. Activation of the middle fusiform 'face area' increases with expertise in recognizing novel objects. *Nat. Neurosci.* 2:568–73

Geisler WS. 2008. Visual perception and the statistical properties of natural scenes. *Annu. Rev. Psychol.* 59:167–92

George N, Dolan RJ, Fink GR, Baylis GC, Russell C, Driver J. 1999. Contrast polarity and face recognition in the human fusiform gyrus. *Nat. Neurosci.* 2:574–80

Gilbert CD, Sigman M, Crist RE. 2001. The neural basis of perceptual learning. *Neuron* 31:681–97

Gillebert CR, Op de Beeck HP, Panis S, Wagemans J. 2009. Subordinate categorization enhances the neural selectivity in human object-selective cortex for fine shape differences. *J. Cogn. Neurosci.* 21:1054–64

Gold J, Bennett PJ, Sekuler AB. 1999. Signal but not noise changes with perceptual learning. *Nature* 402:176–78

Goldstone RL. 1998. Perceptual learning. *Annu. Rev. Psychol.* 49:585–612

Goldstone RL, Lippa Y, Shiffrin RM. 2001. Altering object representations through category learning. *Cognition* 78:27–43

Grill-Spector K, Kushnir T, Hendler T, Malach R. 2000. The dynamics of object-selective activation correlate with recognition performance in humans. *Nat. Neurosci.* 3:837–43

Grill-Spector K, Malach R. 2004. The human visual cortex. *Annu. Rev. Neurosci.* 27:649–77

Haxby JV, Gobbini MI, Furey ML, Ishai A, Schouten JL, Pietrini P. 2001. Distributed and overlapping representations of faces and objects in ventral temporal cortex. *Science* 293:2425–30

Henson R, Shallice T, Dolan R. 2000. Neuroimaging evidence for dissociable forms of repetition priming. *Science* 287:1269–72

Hoffman DD, Richards WA. 1984. Parts of recognition. *Cognition* 18:65–96

Hubel DH, Wiesel TN. 1959. Receptive fields of single neurons in the cat's striate cortex. *J. Physiol. (Lond.)* 148:574–91

Hung CP, Kreiman G, Poggio T, DiCarlo JJ. 2005. Fast readout of object identity from macaque inferior temporal cortex. *Science* 310:863–66

Jagadeesh B, Chelazzi L, Mishkin M, Desimone R. 2001. Learning increases stimulus salience in anterior inferior temporal cortex of the macaque. *J. Neurophysiol.* 86:290–303

James TW, Humphrey GK, Gati JS, Menon RS, Goodale MA. 2000. The effects of visual object priming on brain activation before and after recognition. *Curr. Biol.* 10:1017–24

Jiang X, Bradley E, Rini RA, Zeffiro T, Vanmeter J, Riesenhuber M. 2007. Categorization training results in shape- and category-selective human neural plasticity. *Neuron* 53:891–903

Jiang Y, Haxby JV, Martin A, Ungerleider LG, Parasuraman R. 2000. Complementary neural mechanisms for tracking items in human working memory. *Science* 287:643–46

Keri S. 2003. The cognitive neuroscience of category learning. *Brain Res. Brain Res. Rev.* 43:85–109

Kiani R, Esteky H, Mirpour K, Tanaka K. 2007. Object category structure in response patterns of neuronal population in monkey inferior temporal cortex. *J. Neurophysiol.* 97:4296–309

Kobatake E, Wang G, Tanaka K. 1998. Effects of shape-discrimination training on the selectivity of inferotemporal cells in adult monkeys. *J. Neurophysiol.* 80:324–30

Koida K, Komatsu H. 2007. Effects of task demands on the responses of color-selective neurons in the inferior temporal cortex. *Nat. Neurosci.* 10:108–16

Kourtzi Z, Betts LR, Sarkheil P, Welchman AE. 2005. Distributed neural plasticity for shape learning in the human visual cortex. *PLoS Biol.* 3:e204

Kourtzi Z, Shiffrar M. 1997. One-shot view invariance in a moving world. *Psychol. Sci.* 8:461–66

Koutstaal W, Wagner AD, Rotte M, Maril A, Buckner RL, Schacter DL. 2001. Perceptual specificity in visual object priming: functional magnetic resonance imaging evidence for a laterality difference in fusiform cortex. *Neuropsychologia* 39:184–99

Kriegeskorte N, Marieke M, Ruff DA, Kiani R, Bodurka J, et al. 2008. Matching categorical object representations in inferior temporal cortex of man and monkey. *Neuron* 60:1126–41

Law CT, Gold JI. 2008. Neural correlates of perceptual learning in a sensory-motor, but not a sensory, cortical area. *Nat. Neurosci.* 11:505–13

Leopold DA, Bondar IV, Giese MA. 2006. Norm-based face encoding by single neurons in the monkey inferotemporal cortex. *Nature* 442:572–75

Leopold DA, O'Toole AJ, Vetter T, Blanz V. 2001. Prototype-referenced shape encoding revealed by high-level aftereffects. *Nat. Neurosci.* 4:89–94

Li N, DiCarlo JJ. 2008. Unsupervised natural experience rapidly alters invariant object representation in visual cortex. *Science* 321:1502–7

Li RW, Levi DM, Klein SA. 2004. Perceptual learning improves efficiency by re-tuning the decision 'template' for position discrimination. *Nat. Neurosci.* 7:178–83

Li S, Mayhew SD, Kourtzi Z. 2009. Learning shapes the representation of behavioral choice in the human brain. *Neuron* 62:441–52

Li S, Ostwald D, Giese M, Kourtzi Z. 2007. Flexible coding for categorical decisions in the human brain. *J. Neurosci.* 27:12321–30

Li W, Piech V, Gilbert CD. 2008. Learning to link visual contours. *Neuron* 57:442–51

Loffler G, Yourganov G, Wilkinson F, Wilson HR. 2005. fMRI evidence for the neural representation of faces. *Nat. Neurosci.* 8:1386–90

Logothetis NK, Pauls J, Poggio T. 1995. Shape representation in the inferior temporal cortex of monkeys. *Curr. Biol.* 5:552–63

Marr D, Nishihara HK. 1978. Representation and recognition of the spatial organization of three-dimensional shapes. *Proc. R. Soc. Lond. B Biol. Sci.* 200:269–94

Mauro R, Kubovy M. 1992. Caricature and face recognition. *Mem. Cognit.* 20:433–40

McCool CH, Britten KH. 2007. Cortical processing of visual motion. In *The Senses: A Comprehensive Reference*, ed. MC Bushnell, DV Smith, GK Beauchamp, SJ Firestein, P Dallos, et al., 2:157–87. New York: Academic

Meyers EM, Freedman DJ, Kreiman G, Miller EK, Poggio T. 2008. Dynamic population coding of category information in inferior temporal and prefrontal cortex. *J. Neurophysiol.* 100:1407–19

Miller EK. 2000. The prefrontal cortex and cognitive control. *Nat. Rev. Neurosci.* 1:59–65

Milner PM. 1974. A model for visual shape recognition. *Psychol. Rev.* 81:521–35

Mirabella G, Bertini G, Samengo I, Kilavik BE, Frilli D, et al. 2007. Neurons in area V4 of the macaque translate attended visual features into behaviorally relevant categories. *Neuron* 54:303–18

Miyashita Y, Chang HS. 1988. Neuronal correlate of pictorial short-term memory in the primate temporal cortex. *Nature* 331:68–70

Naselaris T, Prenger RJ, Kay KN, Oliver M, Gallant JL. 2009. Bayesian reconstruction of natural images from human brain activity. *Neuron* 63:902–15

Op de Beeck H, Wagemans J, Vogels R. 2001. Inferotemporal neurons represent low-dimensional configurations of parameterized shapes. *Nat. Neurosci.* 4:1244–52

Op de Beeck HP, Baker CI, DiCarlo JJ, Kanwisher NG. 2006. Discrimination training alters object representations in human extrastriate cortex. *J. Neurosci.* 26:13025–36

Op de Beeck HP, Torfs K, Wagemans J. 2008. Perceived shape similarity among unfamiliar objects and the organization of the human object vision pathway. *J. Neurosci.* 28:10111–23

Orban G, Fiser J, Aslin RN, Lengyel M. 2008. Bayesian learning of visual chunks by human observers. *Proc. Natl. Acad. Sci. USA* 105:2745–50

Pasupathy A, Connor CE. 1999. Responses to contour features in macaque area V4. *J. Neurophysiol.* 82:2490–502

Pasupathy A, Connor CE. 2001. Shape representation in area V4: position-specific tuning for boundary conformation. *J. Neurophysiol.* 86:2505–19

Pasupathy A, Connor CE. 2002. Population coding of shape in area V4. *Nat. Neurosci.* 5:1332–38

Pena M, Bonatti LL, Nespor M, Mehler J. 2002. Signal-driven computations in speech processing. *Science* 298:604–7

Quiroga RQ, Reddy L, Kreiman G, Koch C, Fried I. 2005. Invariant visual representation by single neurons in the human brain. *Nature* 435:1102–7

Raiguel S, Vogels R, Mysore SG, Orban GA. 2006. Learning to see the difference specifically alters the most informative V4 neurons. *J. Neurosci.* 26:6589–602

Reddy L, Kanwisher N. 2006. Coding of visual objects in the ventral stream. *Curr. Opin. Neurobiol.* 16:408–14

Rhodes G, Brennan S, Carey S. 1987. Identification and ratings of caricatures: implications for mental representations of faces. *Cogn. Psychol.* 19:473–97

Riesenhuber M, Poggio T. 1999. Hierarchical models of object recognition in cortex. *Nat. Neurosci.* 2:1019–25

Roelfsema PR, van Ooyen A. 2005. Attention-gated reinforcement learning of internal representations for classification. *Neural. Comput.* 17:2176–214

Rolls ET. 1995. Learning mechanisms in the temporal lobe visual cortex. *Behav. Brain Res.* 66:177–85

Rolls ET. 2000. Functions of the primate temporal lobe cortical visual areas in invariant visual object and face recognition. *Neuron* 27:205–18

Saffran JR, Aslin RN, Newport EL. 1996. Statistical learning by 8-month-old infants. *Science* 274:1926–28

Schacter DL, Reiman E, Uecker A, Polster MR, Yun LS, Cooper LA. 1995. Brain regions associated with retrieval of structurally coherent visual information. *Nature* 376:587–90

Schwarzkopf DS, Kourtzi Z. 2008. Experience shapes the utility of natural statistics for perceptual contour integration. *Curr. Biol.* 18:1162–67

Schwarzkopf DS, Zhang J, Kourtzi Z. 2009. Flexible learning of natural statistics in the human brain. *J. Neurophysiol.* 102:1854–67

Schyns PG, Goldstone RL, Thibaut JP. 1998. The development of features in object concepts. *Behav. Brain Sci.* 21:1–17

Selfridge OG. 1959. Pandemonium: a paradigm for learning. In *Mechanization of Thought Processes: Proceedings of a Symposium Held at the National Physical Laboratory*, pp. 513–26. London: HMSO

Sigala N, Logothetis NK. 2002. Visual categorization shapes feature selectivity in the primate temporal cortex. *Nature* 415:318–20

Sigman M, Cecchi G, Gilbert C, Magnasco M. 2001. On a common circle: natural scenes and Gestalt rules. *Proc. Natl. Acad. Sci. USA* 98:1935–40

Sigman M, Gilbert CD. 2000. Learning to find a shape. *Nat. Neurosci.* 3:264–69

Sigman M, Pan H, Yang Y, Stern E, Silbersweig D, Gilbert CD. 2005. Top-down reorganization of activity in the visual pathway after learning a shape identification task. *Neuron* 46:823–35

Simoncelli EP, Olshausen BA. 2001. Natural image statistics and neural representation. *Annu. Rev. Neurosci.* 24:1193–216

Smith ML, Gosselin F, Schyns PG. 2004. Receptive fields for flexible face categorizations. *Psychol. Sci.* 15:753–61

Sutherland NS. 1968. Outlines of a theory of visual pattern recognition in animals and man. *Proc. R. Soc. Lond. B Biol. Sci.* 171:297–317

Tanaka K, Saito H, Fukada Y, Moriya M. 1991. Coding visual images of objects in the inferotemporal cortex of the macaque monkey. *J. Neurophysiol.* 66:170–89

Thomas E, Van Hulle MM, Vogels R. 2001. Encoding of categories by noncategory-specific neurons in the inferior temporal cortex. *J. Cogn. Neurosci.* 13:190–200

Tovee MJ, Rolls ET, Ramachandran VS. 1996. Rapid visual learning in neurones of the primate temporal visual cortex. *Neuroreport* 7:2757–60

Treisman A, Gormican S. 1988. Feature analysis in early vision: evidence from search asymmetries. *Psychol. Rev.* 95:15–48

Tsao DY, Freiwald WA, Tootell RB, Livingstone MS. 2006. A cortical region consisting entirely of face-selective cells. *Science* 311:670–74

Turk-Browne NB, Scholl BJ, Chun MM, Johnson MK. 2009. Neural evidence of statistical learning: efficient detection of visual regularities without awareness. *J. Cogn. Neurosci.* 21:1934–45

Ullman S, Soloviev S. 1999. Computation of pattern invariance in brain-like structures. *Neural. Netw.* 12:1021–36

Ungerleider L, Mishkin M. 1982. Two cortical visual systems. In *Analysis of Visual Behavior*, ed. DJ Ingle, MA Goodale, RJW Mansfield, pp. 549–86. Cambridge, MA: MIT Press

van Turennout M, Ellmore T, Martin A. 2000. Long-lasting cortical plasticity in the object naming system. *Nat. Neurosci.* 3:1329–34

Vogels R. 1999. Categorization of complex visual images by rhesus monkeys. Part 2: single-cell study. *Eur J Neurosci.* 11:1239–55

Vogels R, Sary G, Dupont P, Orban GA. 2002. Human brain regions involved in visual categorization. *Neuroimage* 16:401–14

Wallis G, Bulthoff HH. 2001. Effects of temporal association on recognition memory. *Proc. Natl. Acad. Sci. USA* 98:4800–4

Wallis G, Rolls ET. 1997. Invariant face and object recognition in the visual system. *Prog. Neurobiol.* 51:167–94

Webster MA, Kaping D, Mizokami Y, Duhamel P. 2004. Adaptation to natural facial categories. *Nature* 428:557–61

Wilson HR, Wilkinson F, Asaad W. 1997. Concentric orientation summation in human form vision. *Vision Res.* 37:2325–30

Wolfe JM, Yee A, Friedman-Hill SR. 1992. Curvature is a basic feature for visual search tasks. *Perception* 21:465–80

Yamane Y, Carlson ET, Bowman KC, Wang Z, Connor CE. 2008. A neural code for three-dimensional object shape in macaque inferotemporal cortex. *Nat. Neurosci.* 11:1352–60

Zhang J, Kourtzi Z. 2010. Learning-dependent plasticity with and without training in the human brain. *Proc. Natl. Acad. Sci. USA* 107:13503–8

Zoccolan D, Kouh M, Poggio T, DiCarlo JJ. 2007. Trade-off between object selectivity and tolerance in monkey inferotemporal cortex. *J. Neurosci.* 27:12292–307

Gender Development and the Human Brain

Melissa Hines

Department of Social and Developmental Psychology, University of Cambridge, Cambridge, CB2 3RQ, United Kingdom; email: mh504@cam.ac.uk

Annu. Rev. Neurosci. 2011. 34:69–88

First published online as a Review in Advance on March 25, 2011

The *Annual Review of Neuroscience* is online at neuro.annualreviews.org

This article's doi:
10.1146/annurev-neuro-061010-113654

Keywords

sex difference, testosterone, behavior, sex differentiation, androgen, play

Abstract

Convincing evidence indicates that prenatal exposure to the gonadal hormone, testosterone, influences the development of children's sex-typical toy and activity interests. In addition, growing evidence shows that testosterone exposure contributes similarly to the development of other human behaviors that show sex differences, including sexual orientation, core gender identity, and some, though not all, sex-related cognitive and personality characteristics. In addition to these prenatal hormonal influences, early infancy and puberty may provide additional critical periods when hormones influence human neurobehavioral organization. Sex-linked genes could also contribute to human gender development, and most sex-related characteristics are influenced by socialization and other aspects of postnatal experience, as well. Neural mechanisms underlying the influences of gonadal hormones on human behavior are beginning to be identified. Although the neural mechanisms underlying experiential influences remain largely uninvestigated, they could involve the same neural circuitry as that affected by hormones.

Contents

INTRODUCTION

Males and females differ both behaviorally and neurally. Indeed, the existence of behavioral sex differences implies the existence of neural sex differences, given that behavior depends on the nervous system. Contemporary research shows that gendered behavior results from a complex interplay of genes, gonadal hormones, socialization, and cognitive development related to gender identification. This article focuses on the role of gonadal hormones, particularly testosterone, during early development. This focus has been chosen because extensive experimental research in nonhuman mammals shows that testosterone exerts powerful

Gonadal hormones: gonads' products, including androgens, produced mainly by the testes, and estrogens and progesterone, produced mainly by the ovaries

Testosterone: the major androgenic hormone produced by the testes

influences on both gender-related behavior and the developing brain and because recent research provides convincing evidence that testosterone exerts similar influences on human development. The article critically reviews the evidence of prenatal hormonal influences on human neurobehavioral sexual differentiation and contextualizes these hormonal effects with genetic, social, and cognitive influences on gender development. It also critically reviews the evidence regarding possible neural changes underlying hormonal influences on human behavior and suggests that neural systems similar to those influenced by hormones underlie other types of influences on gendered behavior.

GENERAL PRINCIPLES OF SEXUAL DIFFERENTIATION

Gender development begins at conception with the union of two X chromosomes (genetic female) or an X and a Y chromosome (genetic male). The main role of these sex chromosomes in human sexual differentiation is to determine whether the gonads become testes or ovaries (Arnold 2009). Genetic information on the Y chromosome leads to testicular differentiation (Wilson et al. 1981), whereas without the Y chromosome, ovaries develop instead of testes. The testes begin to produce testosterone prenatally, and the ovaries do not (Wilson et al. 1981). Consequently, male and female fetuses differ in the amount of testosterone to which they are exposed. This sex difference appears to be maximal between about weeks 8 and 24 of human gestation, with testosterone in males tapering off before birth (Carson et al. 1982, Reyes et al. 1973). In nonhuman mammals at comparable stages of early development, testosterone and hormones produced from testosterone influence neural survival, neuroanatomical connectivity, and neurochemical specification, producing sex differences in brain structure and function (McCarthy et al. 2009). These effects of testosterone on neural development provide powerful mechanisms for influencing behavior across the life span.

ANIMAL MODELS OF HORMONE EFFECTS

The influences of early testosterone exposure on neurobehavioral development were first documented in a landmark study by Phoenix et al. (1959). They showed that administering testosterone to pregnant guinea pigs produced female offspring who showed increased capacity for male-typical sexual behavior and decreased capacity for female-typical sexual behavior in adulthood (Phoenix et al. 1959). Phoenix et al. contrasted these early, and permanent, effects of hormones, which they called organizational because they were thought to reflect changes in the organization of neural systems, with the later, and transient, effects of hormones after puberty, which they called activational because they were thought to reflect transient activation of the previously organized systems. This organizational/activational distinction has stood up well in the subsequent 50 some years (Arnold 2009), and thousands of studies on numerous species, including not only guinea pigs, but also rats, mice, hamsters, gerbils, ferrets, dogs, sheep, and marmoset and rhesus monkeys, have documented the early organizing effects of testosterone on a wide variety of behaviors that show sex differences (Hines 2004, 2009; McCarthy et al. 2009). For instance, the female offspring of rhesus macaques treated with testosterone during pregnancy show increased male-typical, and reduced female-typical, sexual behavior in adulthood, and increased male-typical, rough-and-tumble play as juvenile animals.

The organizing influences of testosterone on behavioral development were originally thought to reflect subtle neural changes (Phoenix et al. 1959). Subsequent research, however, has shown that early hormone manipulations produce dramatic changes in the structure of neural regions with the relevant hormone receptors. The first dramatic neural sex difference described in the rodent brain was the sexually dimorphic nucleus of the preoptic area (SDN-POA). This region of the anterior hypothalamic/preoptic area (AH/POA) is several fold larger in the adult male rat than in the adult female rat, and its volume can be altered by manipulating testosterone during early development (Gorski et al. 1978, 1980). Administering testosterone to developing female animals increases the volume of the SDN-POA, and removing testosterone from developing males reduces its volume (Dohler et al. 1984, Jacobson et al. 1981). Other neural regions in addition to the SDN-POA show sex differences, and in these regions too, the size of the sex difference is influenced by the early hormone environment. For instance, a second region of the preoptic area, the anteroventral paraventricular nucleus, is larger and contains more neurons in female rats than in males, and these characteristics are reduced by early testosterone treatment (Ito et al. 1986, Sumida et al. 1993). Similar neural sex differences have been reported in other rodent species, including gerbils, hamsters, mice, and guinea pigs, as well as in ferrets, sheep, and rhesus monkeys, and studies investigating early hormone influences have found similar results to those seen in rats, in other rodent species, and in ferrets (Bleier et al. 1982, Byne 1998, Hines et al. 1987, Roselli et al. 2004, Simerly et al. 1997, Tobet et al. 1986, Ulibarri & Yahr 1988).

Some general principles can be derived from the extensive experimental work in nonhuman mammals, and these principles have informed hypotheses regarding possible hormonal influences on human brain and behavior (see Hines 2009, for a review). First, during early development, estrogens generally do not promote female-typical development. Instead, female-typical development occurs in the absence of testicular hormones. Thus, exposure to high levels of estrogen is not expected to femininize neurobehavioral development. Second, the effects of testosterone on development are graded and linear; the more hormone the animal is exposed to, the more male-typical its behavior and brain structure become. An implication of this principle is that gonadal hormones can contribute to individual differences within each sex, as well as to differences between the sexes. Third, neurobehavioral sexual differentiation is a multidimensional process. The many

Rough-and-tumble play: juvenile behavior characterized by overall body contact or playful aggression; more common in males than in females

SDN-POA: sexually dimorphic nucleus of the preoptic area

behaviors and neural systems that differ for males and females can be influenced by hormones during slightly different critical periods, or can be sensitive to different doses of hormone, or to different metabolites of testosterone, or can involve different downstream mechanisms such as cofactors. Implications of this principle include an expectation that the many human behaviors and brain structures that differ by sex may not relate in a uniform way to one another and that individuals can develop complicated patterns of sex-typed behavior, being masculine in some respects and feminine in others. Fourth, the effects of hormones can differ somewhat from one species to another. For instance, behaviors that are influenced in one species may not be influenced in all others. Similarly, brain regions that differ for males and females in one species may not show a sex difference in another. Thus the specific effects of gonadal steroids seen on the brain and behavior of nonhuman mammals cannot be automatically generalized to humans, as well. Instead, hypothesized neural and behavioral influences of testosterone during early development must be evaluated directly in humans. Fifth, the behaviors and neural features that are influenced by gonadal hormones are those that show sex differences, meaning that they differ on average for males and females. Therefore, the characteristics that are likely to be influenced in humans are also those that show sex differences.

HUMAN RESEARCH

Ethical considerations generally preclude experimental manipulations of gonadal hormones in humans during early development. However, information from genetic syndromes that produce fetal hormone abnormality, as well as from situations in which pregnant women have been prescribed hormones, and studies relating normal variability in hormones early in life to normal variability in subsequent behavior all suggest that hormones contribute to human gender development. The most convincing evidence of these influences has come from studies of childhood play.

Why Study Children's Play?

Girls and boys differ in their toy, playmate, and activity preferences (Hines, 2010a). For example, boys tend to prefer toy vehicles, whereas girls tend to prefer dolls. Girls and boys also generally prefer playmates of their own sex, and boys spend more time in rough-and-tumble play than girls do. Children's sex-typed play behavior is the aspect of human gender development that has been studied most extensively in relation to the early hormone environment. This focus on childhood play reflects several considerations. First, children spend most of their time playing, and play is thought to be essential for healthy cognitive and emotional development (Ginsburg et al. 2007, Piaget 1970, Vygotsky 1976). Second, children's sex-typed play behavior can be assessed readily and reliably. Third, large sex differences exist in children's play, larger than those in cognitive abilities or personality characteristics (Hines 2010b), providing scope for detecting hormonal influences. Fourth, sex differences in children's play are evident early in life and relate to other behaviors that show sex differences, including sexual orientation and gender identification (Bailey & Zucker 1995, Green 1985, Hines et al. 2004). Fifth, play can be assessed during a period of hormonal quiescence, allowing examination of the early and permanent organizational influences of hormones on brain development, prior to the addition of the transient, activational influences of hormones that occur after puberty.

Several types of studies provide convergent evidence that testosterone concentrations prenatally influence children's subsequent sex-typed toy, playmate, and activity preferences. Studies of girls exposed to unusually high levels of testosterone and other androgens before birth, because they have the genetic disorder known as classic congenital adrenal hyperplasia (CAH), consistently find that these girls show increased male-typical play and reduced female-typical play (Berenbaum & Hines 1992, Dittmann et al. 1990, Ehrhardt et al. 1968, Ehrhardt & Baker 1974, Hall et al.

2004, Nordenstrom et al. 2002, Pasterski et al. 2005). Similarly, children whose mothers took androgenic progestins during pregnancy have shown increased male-typical toy and activity preferences, whereas the opposite occurs in children whose mothers took antiandrogenic progestins (Ehrhardt et al. 1977, Ehrhardt & Money 1967).

Is the Behavioral Alteration Caused by Hormones Acting on the Developing Brain?

The external genitalia, as well as the brain, contain androgen receptors, and girls with CAH, as well as those whose mothers took androgenic progestins, are typically born with varying degrees of genital virilization (enlarged clitoris, fused labia). Those skeptical of gonadal hormone influences on human neurobehavioral development suggest that the abnormal genital appearance, rather than the neural influences of androgens, could cause behavioral masculinization (Fausto-Sterling 1992, Jordan-Young 2010). Specifically, they suggest that parents may treat their daughters differently because of the girls' external virilization at birth and that this difference in parental treatment could alter sex-typed behavior. In addition, they suggest that virilized genitalia could reduce self-identification as female, which could in turn cause increased male-typical behavior.

Some evidence suggests that parents can influence the development of children's gender-typical behavior. For instance, parents generally encourage sex-typical play (Fagot 1978, Langlois & Downs 1980, Pasterski et al. 2005), and the amount of such encouragement has been found to correlate with the amount of sex-typed toy play, at least among typically developing children (Pasterski et al. 2005). However, parents have been found to offer more, rather than less, encouragement of sex-typical play to their daughters with CAH than to their daughters who do not have the disorder (Pasterski et al. 2005), suggesting that parental encouragement is not responsible for cross-gendered toy choices in girls with CAH.

Similarly, although gender identification plays a role in children's acquisition of gender-related behavior, at least in typically developing children (Hines 2010a, Ruble et al. 2006), it is unlikely that the male-typical behavior in girls with CAH results solely from altered gender identity based on genital virilization at birth. Evidence arguing against this explanation comes from studies relating normal variability in prenatal testosterone exposure to normal variability in subsequent behavior. Testosterone concentrations in maternal blood samples taken during pregnancy or in amniotic fluid from normally developing fetuses relate positively to male-typical childhood behavior (Auyeung et al. 2009b, Hines et al. 2002). Because the children in these studies have normal appearing genitalia, it is unlikely that differential parental socialization or changes in gender identification based on genital appearance account for the observed relationships between prenatal testosterone and postnatal behavior.

Researchers have also looked at species in which children's toys are novel objects and therefore not subject to the socialization histories or processes of gender identification thought to explain sex-typed toy preferences in children. Two studies of nonhuman primates have reported sex-typed toy preferences similar to those seen in children. Male vervet monkeys (Alexander & Hines 2002) have been found to spend more time than females contacting toys that are typically preferred by boys (e.g., a car) and less time contacting toys that are typically preferred by girls (e.g., a doll) (**Figure 1**). Similarly, male rhesus monkeys have been found to prefer toys normally preferred by boys (wheeled toys) to plush toys (Hassett et al. 2008). These findings show that sex-typed toy preferences can arise independent of the social and cognitive processes involved in gender development.

Rethinking Children's Preferences for Sex-Typed Toys

Children's sex-typical toy preferences have been widely assumed to result from socialization and other postnatal factors and to

Androgens: substances, including testosterone, that promote masculinization. Produced by the testes, adrenal glands, and ovaries, with the testes the largest source

Virilized genitalia: masculinized genitalia, typically involving an enlarged clitoris and partially fused labia

Figure 1

Examples of a male and a female vervet monkey contacting human children's sex-typed toys. The female animal (*left*) appears to be inspecting the doll in a manner similar to that in which vervet monkeys inspect infant vervets. The male animal (*right*) appears to be moving the car along the ground as a child might do. Reproduced by permission from Alexander & Hines (2002).

provide rehearsals for adult sex-typed social roles. Evidence of inborn influences has led researchers to reevaluate this perspective and to investigate the object features that make certain toys more or less interesting to brains exposed prenatally to different amounts of testosterone. Although boys' toys and girls' toys differ in shape and color (boys' toys tend to be angular and blue, whereas girls' toys tend to be rounded and pink), sex differences in toy preferences are present in very young infants (Alexander et al. 2009b, Campbell et al. 2000, Jadva et al. 2010, Serbin et al. 2001), before sex differences in color or shape preferences are seen (Jadva et al. 2010), suggesting that the object preferences do not result from the color or shape preferences. Another possibility is that boys like toys that can be moved in space,

and prenatal androgen exposure may increase interest in watching things move in space (Alexander 2003, Alexander & Hines 2002, Hines 2004), perhaps by altering development of the visual system (Alexander 2003).

Early Hormone Influences on Sexual Orientation and Core Gender Identity

Adult behaviors that show sex differences, including sexual orientation and core gender identity, also appear to be influenced by prenatal testosterone exposure. Women with CAH not only recall more male-typical childhood behavior, but also show reduced heterosexual orientation as adults, and these two outcomes correlate (Hines et al. 2004; see Meyer-Bahlburg et al. 2008 for a review of additional studies of CAH and sexual orientation). Normal variability in testosterone prenatally, e.g., from maternal blood or amniotic fluid, has not yet been related to sexual orientation, but a characteristic that is thought to provide an indirect measure of prenatal testosterone exposure, the ratio of the second to the fourth digit of the hand (2D:4D), which is greater in females than in males, has been linked. A study of more than 200,000 individuals, who measured their own 2D:4D and reported their sexual orientation online, found that 2D:4D related as predicted to sexual orientation in males, but not in females (Collaer et al. 2007). A meta-analysis that did not include this large study reached a somewhat different conclusion, however, finding that 2D:4D related as predicted to sexual orientation in females, but not in males (Grimbos et al. 2010). Finger ratios are probably a weak correlate of prenatal testosterone exposure, perhaps explaining the somewhat inconsistent results.

Women with CAH not only show reduced heterosexual interest, but also show diminished identification with the female gender, and this too correlates with their recalled childhood sex-typical behavior (Hines et al. 2004; see Hines 2010a for a review of studies of gender identity in females with CAH). About 3% of women with CAH express a desire to live as men in adulthood, despite having been reared as girls, in contrast with ~0.005% of all women (Dessens et al. 2005). Although 3% may seem small, it indicates that women with CAH are ~600 times more likely than women in general to experience severe gender dysphoria. Girls with other disorders involving exposure to unusually high levels of androgens prenatally also show increased gender dysphoria (Slijper et al. 1998). Additionally, even when not gender dysphoric, or wishing to change sex, girls and women with CAH show somewhat reduced satisfaction with the female sex assignment (Ehrhardt et al. 1968, Ehrhardt & Baker 1974, Hines et al. 2004). No evidence thus far has linked normal variability in the early hormone environment to gender dysphoria. In addition, research attempting to link 2D:4D to gender identification has produced inconsistent findings (Kraemer et al. 2009, Schneider et al. 2006, Wallien et al. 2008), again perhaps because of the weak relationship between 2D:4D and prenatal androgen exposure.

Early Hormone Influences on Personality and Cognition

Sex differences in personality characteristics and in cognitive ability are smaller than are sex differences in children's sex-typed activities, sexual orientation, or gender identity (Hines 2010b). Nevertheless, they also have been examined for evidence of early hormonal influence.

Some personality characteristics that show sex differences relate to prenatal testosterone exposure. For instance, empathy, which is higher on average in females than in males, appears to be reduced by testosterone exposure before birth. Females with CAH show reduced empathy (Mathews et al. 2009), and testosterone measured in amniotic fluid relates negatively to empathy in both boys and girls (Chapman et al. 2006). Tendencies toward physical aggression, which are higher on average in males than in females, also relate to prenatal testosterone exposure, with prenatal testosterone exposure increasing aggression. Girls and women with CAH show increased physical aggression (Mathews et al. 2009, Pasterski et al. 2007), as do children exposed prenatally to androgenic progestins (Reinisch 1981). Not all personality dimensions that show average sex differences relate to prenatal testosterone exposure, however. For instance, the study that reported increased aggression and reduced empathy in females with CAH also considered the personality dimension of dominance/assertiveness, which is higher on average in males than in females. Despite seeing the expected sex difference in healthy controls, no difference in dominance/assertiveness was seen between females with and without CAH.

Cognitive and motor abilities that show sex differences also have been examined for influences of prenatal testosterone exposure (reviewed in Hines 2010a). One study found that females with CAH showed more male-typical behavior in the form of increased accuracy in throwing balls and darts at targets (Hines et al. 2003), a result that was not accounted for by increased muscle strength (Collaer et al. 2009). Some studies have also found that females with CAH resemble males in showing enhanced mental rotations performance, but other studies have not corroborated these results (Hines et al. 2003). Two studies found that males with CAH show reduced performance on mental rotations or other visuo-spatial tasks (Hampson et al. 1998, Hines et al. 2003), results that had not been predicted. Several studies have also found that both males and females with CAH show impaired performance on arithmetic and mathematical tests (Baker & Ehrhardt 1974, Perlman 1973, Sinforiani et al. 1994), despite males generally being viewed as better than females at mathematics. Studies relating amniotic fluid testosterone to spatial and mathematical performance have also produced inconsistent and largely negative results (Finegan et al. 1992, Grimshaw et al. 1995). Most studies have found no alterations in individuals with CAH on tasks at which females excel, such as verbal fluency or perceptual speed, although one study suggests reduced female-typical behavior in females with CAH in the form of impaired fine motor performance (Collaer et al. 2009). Perhaps prenatal testosterone exposure has a clearer impact on motor abilities that show sex differences (e.g., targeting and fine motor performance) than on cognitive abilities assessed with paper-and-pencil tests.

Socialization and Sex Differences in Cognitive Performance

Substantial evidence supports social and cultural influences on some cognitive sex differences (see Hines 2010a for a review). For instance, sex differences on certain measures of cognitive abilities appear to have declined over time (Feingold 1988). For the SAT Mathematics, in particular, the sex ratio among those scoring at the upper extreme has declined from

13 boys to one girl in 1982 to 2.8 boys to 1 girl more recently (Halpern et al. 2007). There are also large national differences in mathematical and science performance, differences that are many fold larger than the sex difference within any nation (Mullis et al. 2008). Additionally, the magnitude of the sex difference in mathematics performance within a nation relates to the role of women. Nations where women and men are similar in regard to variables such as representation in the legislature show more equal mathematics performance (Guiso et al. 2008).

Sex-Related Psychiatric Disorders

Some psychiatric disorders are more common in one sex or the other, and testosterone could contribute here, as well. For example, prenatal testosterone exposure has been suggested to contribute to autistic spectrum conditions (ASC) (Baron-Cohen 2002) and to obsessive compulsive disorder (OCD) and Tourette Syndrome (Alexander & Peterson 2004), and to be protective against eating disorders (Culbert et al. 2008, Klump et al. 2006). For OCD and Tourette syndrome, evidence that individuals with these disorders are more male-typical in other respects, such as childhood play behavior, has been interpreted to support a link to testosterone (Alexander & Peterson 2004). For ASC (Auyeung et al. 2009a, Chapman et al. 2006, Knickmeyer et al. 2006) and for eating disorders (Culbert et al. 2008, Klump et al. 2006), behaviors in the normal range that are similar to those seen in the disorders (e.g., empathy for ASC, disordered eating for eating disorders) have been linked to prenatal androgens, although for disordered eating, some studies have failed to replicate these results (Raevuori et al. 2008). In addition, for ASC and for eating disorders, studies have not shown that variability in the early hormone environment leads to the disorder itself, as opposed to behaviors in the normal range that resemble those that characterize the disorder. For instance, although a study of females exposed to high levels of androgens prenatally, because of CAH, found increased scores on an inventory of traits related to ASC, none of the women with CAH scored high enough to suggest a clinical diagnosis (Knickmeyer et al. 2006). The proposed link between prenatal testosterone and ASC has also been questioned by evidence indicating that both males and females with gender identity disorder, rather than females only, are at increased risk of ASC (de Vries et al. 2010) and by the larger male predominance for the less severe ASC, Asperger syndrome, than for the more severe ASC, classical autism. One possibility is that prenatal androgen exposure contributes to individual differences within the normal range in behaviors that show sex differences and that some of these resemble behaviors associated with developmental disorders, such as ASC. As a consequence, exposure to testosterone before birth, when added to other risk factors, could contribute to some individuals crossing a threshold for diagnosis. However, developmental disorders are one area in which direct genetic effects (Skuse 2006), particularly those of genes encoded in the X and Y chromosomes (Reinius et al. 2008, Reinius & Jazin 2009, Skuse 2006), may play an important role.

Sex Differences in Brain Structure and Function

There are numerous reports of sex differences in human brain structure or function (reviewed by Cahill 2009, Hines 2009). For instance, total brain volume, like body size, is larger in males than females. In addition, the amygdala is larger in males, whereas the hippocampus is larger in females (Goldstein et al. 2001). Women also show greater cortical thickness than men do in many regions (Luders et al. 2006). Perhaps in compensation for the smaller brain, women also show greater gyrification in parts of frontal and parietal cortex and perhaps more efficient use of white matter (Gur et al. 1999). There are many reports of sex differences in the human brain, particularly in its function, and many of these are as yet unreplicated. Because males and females are routinely compared in studies, and positive results are more readily published than

negative results, some findings of neural sex differences may prove to be spurious.

In addition, although many neural sex differences have been described, few have been linked to behavioral sex differences. In fact, many differences in brain function have been noted during equivalent performance by the sexes. For instance, men and women show different patterns of asymmetry of function when performing certain phonological tasks, despite showing no sex difference in task performance (Shaywitz et al. 1995). Similarly, for men and women matched for mathematical ability, mathematical performance correlates with temporal lobe activation in men but not in women (Haier & Benbow 1995), and for women, performance on intelligence tests that do not differ by sex correlates with gray and white matter in frontal regions, whereas for men the correlation is with parietal regions (Haier et al. 2005). Indeed, neural sex differences may sometimes, or even commonly, exist to produce similar behavior in males and females, rather than to produce differences (De Vries & Sodersten 2009, McCarthy et al. 2009). Additionally, it appears that during performance of many tasks, male and female brains function similarly (Frost et al. 1999, Halari et al. 2005, Mansour et al. 1996).

Despite the many neural sex similarities and the many neural sex differences that do not relate to any behavioral sex difference, neural and behavioral sex differences have been linked in some instances. Much research in this area has focused on neural differences related to sexual orientation, particularly in men. The only finding in this area that has been independently replicated, at least as of yet, involves the third interstitial nucleus of the anterior hypothalamus (INAH-3). INAH-3 is thought to be the human homolog of the rodent SDN-POA, and four different research groups have reported that INAH-3, like the SDN-POA, is larger in males than in females (Allen et al. 1989, Byne et al. 2001, Garcia-Falgueras & Swaab, 2008, LeVay 1991). INAH-3 is also smaller (i.e., more female-typical) in homosexual than heterosexual men (Byne et al. 2001, LeVay 1991), although the number of neurons in the nucleus

INAH 1 to 4: interstitial nuclei of the anterior hypothalamus, numbers 1 to 4

appears similar for these two groups (Byne et al. 2001). The volumetric sex difference does not appear to relate to disease processes (e.g., HIV status) or to hormone use in adulthood (see Hines 2009 for discussion). Because the sex difference in SDN-POA volume in other mammals results from early testosterone exposure, differences in INAH-3 volume in humans may relate to the early hormone environment, as well. This possibility has not yet been directly investigated, however.

Heterosexual and homosexual men also differ in corpus callosum anatomy; the isthmus, in particular, is significantly larger in right-handed homosexual compared with right-handed heterosexual men (Witelson et al. 2008). Patterns of cerebral asymmetry and functional cortical connectivity have also been linked to sexual orientation in both men and women (Savic & Lindstrom 2008).

Researchers have also searched for neural correlates of gender identity disorder. One group has reported that the central subregion of the BNST (BNSTc) is smaller in women and in male-to-female transsexuals than in nontranssexual men (Zhou et al. 1995). Interpretation of this finding is complicated, however, because the sex difference in BNSTc does not appear until after puberty (Chung et al. 2002), whereas most transsexual individuals recall feeling strongly cross-gendered from early childhood. Thus, the difference in BNSTc may be the result of experience (Hines, 2009) or of the adult hormone treatment associated with changing sex (Lawrence 2009). This same research group also reported that INAH-3 is smaller and contains fewer neurons in male-to-female transsexuals than in control males (Garcia-Falgueras & Swaab 2008).

In the realm of cognitive and motor sex differences, the midsagittal area of posterior callosal regions, particularly the splenium, relates negatively to language lateralization and positively to verbal fluency in women (Hines et al. 1992b). These findings suggest a correspondence between female-typical brain structure and female-typical cognitive function, given that language lateralization is reduced in women compared to men (McGlone, 1980, Voyer, 1996), whereas verbal fluency is greater in women than in men (Hyde & Linn 1988, Kolb & Wishaw 1985, Spreen & Strauss 1991) and posterior callosal regions tend to be larger in women than in men as well (de LaCoste-Utamsing & Holloway 1982, Witelson 1985).

Gron et al. (2000) have also described links between sex differences in brain function and navigational performance (Gron et al. 2000). In both men and women, navigating through a virtual maze, on which males perform better on average than do females, is accompanied by neural activity in the medial occipital gyri, medial and lateral superior parietal lobules, posterior cingulate and parahippocampal gyri, and the right hippocampus proper. However, women show more activity than men do in the right prefrontal cortex at Brodmann's areas 46/9, the right inferior parietal lobule, and the right superior parietal lobule, whereas men show significantly more activity than women do in the left hippocampus proper, the right parahippocampal gyrus, and the left posterior cingulate. Women with CAH have been found to perform better than healthy women on a different virtual maze task (Mueller et al. 2008), but there is, as yet, no evidence regarding neural activation in women with and without CAH while performing navigational tasks.

Effects of Experience on the Brain

Sex differences in brain structure or even function are often interpreted to imply inborn differences between males and females. This leap is inappropriate, however. Behavioral differences must be accompanied by neural differences, so the observation of a neural sex difference on its own tells us little to nothing about how the difference developed. This is true not only for differences in brain function, but also, at least in some cases, for differences in brain structure. For instance, experience can change the mammalian brain throughout the life span, and even neurogenesis in some brain regions can continue in adulthood (Juraska 1998, Maguire et al. 2006, Ming & Song 2005).

Hence, the existence of a neural sex difference, even one that relates to a behavior known to be influenced by early androgen exposure, does not prove that the hormone exposure caused the neural difference. A more direct strategy for identifying links between early hormones and the brain could be to look at neural structure or function in individuals with early hormone abnormality or in individuals for whom the early hormone environment has been measured. Although very little information of this type is available, some neural differences have been described in individuals with CAH (Hines 2009). Most notably, both males and females with CAH show decreased amygdala volume (Merke et al. 2003), and females with CAH show increased amygdala activation to negative facial emotions and, in this respect, resemble healthy males (Ernst et al. 2007). These findings fit well with expectations based on experimental work in other species because the amygdala, or some of its subregions, is larger in males than in females, contains receptors for androgen, is influenced by early manipulations of testosterone, and is involved in behaviors that show sex differences, including rough-and-tumble play and aggression (Cooke et al. 2007, Hines et al. 1992a).

Other Potential Critical Periods for Hormone Influences on Gender Development

In addition to the difference in testosterone in male and female fetuses, males and females differ in gonadal hormone levels neonatally. Shortly after birth, testosterone surges in boys (Forest et al. 1974), and estrogen surges in girls (Bidlingmaier et al. 1974, 1987). The testosterone surge has been called "mini-puberty" and may play a role in development of the gonads and external genitalia in infant boys (Quigley 2002). Human brain development, particularly cortical development, continues rapidly for the first two years after birth and reacts to experience (de Graaf-Peters & Hadders-Algra 2005, Huttenlocher 2002). Thus, this early postnatal period could provide

a time when gonadal hormones and experience interact to shape the brain and behavioral propensities.

The early postnatal hormone surges, particularly the testosterone surge in boys, are a focus of current research activity. Men with anorchia (missing testes) but with normal penile development, who apparently experience normal testosterone levels prenatally, but who lack testosterone after birth, resemble controls in terms of sexual orientation, core gender identity, and questionnaire measures of personality characteristics viewed as masculine or feminine (Poomthavorn et al. 2009). Other characteristics may be influenced by the postnatal hormone surge, however. For example, men who lack the postnatal testosterone surge because they have idiopathic hypogonadotropic hypogonadism show reduced spatial abilities, and this condition is not reversed by subsequent testosterone replacement (Hier & Crowley 1982). In addition, females who do not experience the early postnatal surge of gonadal steroids because they have Turner syndrome show evidence of reduced performance on tasks at which males excel, as well as on tasks at which females excel, but not on sex-neutral tasks (Collaer et al. 2002). Evidence from healthy infants also suggests that the postnatal testosterone surge may play a role in gender development. Initial evidence suggests that testosterone during early infancy relates to infants' visual preferences for social stimuli (Alexander et al. 2009a), to neural organization for language processing (Friederici et al. 2008), and to sex-related development of the visual system (Held et al. 1996). Although these initial reports are somewhat inconsistent and require replication, they provide intriguing glimpses through a potential new window on early gonadal hormone contributions to human gender development.

Contemporary research is also focusing on possible hormonal influences on neurobehavioral sexual differentiation at puberty. The hormonal changes of puberty produce dramatic changes in the human body, and experimental research in rodents suggests that they produce

Idiopathic hypogonadotropic hypogonadism: involves gonadotropin deficiency and impaired gonadal steroid production after birth in affected males

Turner syndrome: absent or imperfect second X chromosome causes ovarian regression, typically before birth, impairing or eliminating ovarian hormone production

an additional wave of neural and behavioral organization as well (Schulz et al. 2009).

Puberty is also a time of great change in human behavior, characterized by increased sexual interest and activity and the emergence of some types of behavioral problems, including higher rates of depression in females than in males (Halpern et al. 1993, Hyde et al. 2008). Evidence supports the existence of sex differences in the timing of some neural changes that accompany puberty, and these seem to parallel the earlier puberty experienced by girls compared with boys. Total cerebral volume peaks earlier in girls than in boys (at about age 10.5 years versus 14.5 years), and although both adolescent girls and adolescent boys show an inverted-U-shaped pattern of change in cortical and subcortical gray matter, the peak occurs one to two years earlier in girls than in boys (Lenroot et al. 2007). Studies have also proposed links to hormones at this time. One study found that among girls global gray matter volume related negatively to estradiol levels, but among boys the same variable related positively to testosterone levels (Peper et al. 2009). Similarly, neural sex differences in adolescent girls and boys have been found to relate to circulating testosterone levels (Neufang et al. 2009). These data are correlational, so investigators do not know if hormones, or other associated developmental processes, are the causal factors. Another study, however, suggests that testosterone may play a role in the growth of white matter in the adolescent brain. In this study, white matter increased at different rates in girls and boys, and testosterone levels and androgen receptor type interacted in relation to this sex difference. The association between male-typical brain development and testosterone levels was significantly stronger in boys with the more efficient type of androgen receptor than in boys with the less efficient type (Perrin et al. 2008). Like the investigations of possible organizational influences of hormones during neonatal development, research on puberty as an additional time of brain organization in relation to gender-linked behavior is in its early

stages, but it offers promise for understanding the dramatic behavioral changes that occur at this time of adolescent development.

CONCLUDING REMARKS

The prenatal hormone environment clearly contributes to the development of sex-related variation in human behavior and plays a role in the development of individual differences in behavior within each sex, as well as differences between the sexes. Thus, early hormone differences appear to be part of the answer to questions such as why males and females differ behaviorally and neurally, as well as why some of us are more sex-typical than others. In other species, the early hormone environment exerts its enduring effects on behavior by altering neural development. Similar neural changes are thought to underlie associations between the early hormone environment and human behavior, but the specific neural changes involved are just beginning to be identified. Many sex differences have been described in the human brain, but only a subset of these has been related to behavioral sex differences and still fewer have been linked to the early hormone environment. Steroid-sensitive regions, including regions of the hypothalamus and the amygdala, are implicated, as are interhemispheric connections, but establishing firm links between early hormones, brain development, and behavior is a primary area for future research.

Although this review has focused on hormonal influences on gender-related brain development and behavior, it has also discussed direct genetic influences that may contribute, in particular, to developmental disabilities, such as autistic spectrum conditions. In addition, the role of socialization, culture, and cognitive developmental processes in the development of behavioral differences between males and females has been noted. Although hormones contribute to behavioral sex differences, other factors contribute, as well. In addition, gender development is multidimensional, and developmental processes involved in each dimension are likely to differ somewhat. A good example of the numerous types of factors that can influence human gender development comes from research on children's play. Here, evidence clearly shows that prenatal testosterone exposure plays a role in sex differences and individual differences, promoting male-typical toy, playmate, and activity interests. After birth, the early surges of testosterone or estrogen may be important, too, but socialization factors also gain in importance, as parents and then peers and eventually teachers encourage children to engage in gender-typed play (Fagot 1978, Langlois & Downs 1980, Pasterski et al. 2005). The child also begins to develop the understanding that he or she is male or female, and this knowledge produces motivation to imitate the behavior of others of the same sex and to respond to information that things, such as toys or activities, are for girls or for boys by choosing the things that they have been told are for their own sex (Bussey & Bandura 1999, Martin et al. 2002, Masters et al. 1979, Perry & Bussey 1979). These social and cognitive developmental influences on children's activities could engage the same neural circuitry as underlies the effects of the early hormone environment. Thus, identifying the brain systems influenced by early androgen exposure could help elucidate systems involved in other types of influences on the same behavioral outcomes, as well.

SUMMARY POINTS

1. Human gender development begins before birth and is influenced by levels of testosterone prenatally, and perhaps neonatally. Sex-typed play in childhood relates to levels of testosterone before birth, and evidence indicates that the prenatal hormone environment also contributes to variability in sexual orientation, gender identity, and some, but not all, personality traits that differ on average for males and females.

2. Other types of influences on neurobehavioral gender development include direct genetic effects of the sex chromosomes and postnatal socialization and cognitive understanding of gender.

3. Gender development is multidimensional, and the combinations of factors that influence the many different dimensions of gender appear to differ. Early hormonal influences appear to play a larger role, for example, in children's toy preferences than they do in cognitive abilities that show sex differences, where social and cultural influences appear to be more important.

FUTURE ISSUES

1. Which neural changes can be associated with the early hormone environment, either in individuals with disorders that cause hormone abnormality or in healthy individuals for whom measures of the early hormone environment are available?

2. Does the early hormone environment contribute to the development of psychological disorders that are more common in one sex or the other?

3. Will early infancy and puberty prove to be critical periods when hormones exert permanent influences on human gendered behavior, as has been shown for prenatal development?

4. Are the neural systems associated with hormone-induced changes in behaviors that show sex differences also the systems that respond to experiential effects on the same behavioral outcomes?

DISCLOSURE STATEMENT

The author is not aware of any affiliations, memberships, funding, or financial holdings that might be perceived as affecting the objectivity of this review.

ACKNOWLEDGMENTS

The author's research described in this article was supported in part by the United States Public Health Service (HD24542) and the Wellcome Trust (069606).

LITERATURE CITED

Alexander GM. 2003. An evolutionary perspective of sex-typed toy preferences: pink, blue, and the brain. *Arch. Sex. Behav.* 32:7–14

Alexander GM, Hines M. 2002. Sex differences in response to children's toys in nonhuman primates (cercopithecus aethiops sabaeus). *Evol. Hum. Behav.* 23:467–79

Alexander GM, Peterson BS. 2004. Testing the prenatal hormone hypothesis of tic-related disorders: gender identity and gender role behavior. *Dev. Psychopathol.* 16:407–20

Alexander GM, Wilcox T, Farmer ME. 2009a. Hormone-behavior associations in early infancy. *Horm. Behav.* 56:498–502

Alexander GM, Wilcox T, Woods R. 2009b. Sex differences in infants' visual interest in toys. *Arch. Sex. Behav.* 38:427–33

Allen LS, Hines M, Shryne JE, Gorski RA. 1989. Two sexually dimorphic cell groups in the human brain. *J. Neurosci.* 9:497–506

Arnold AP. 2009. The organizational-activational hypothesis as the foundation for a unified theory of sexual differentiation of all mammalian tissues. *Horm. Behav.* 55:570–78

Auyeung B, Baron-Cohen S, Ashwin E, Knickmeyer R, Taylor K, Hackett G. 2009a. Fetal testosterone and autistic traits. *Br. J. Psychol.* 100:1–22

Auyeung B, Baron-Cohen S, Chapman E, Knickmeyer R, Taylor K, et al. 2009b. Fetal testosterone predicts sexually differentiated childhood behavior in girls and in boys. *Psychol. Sci.* 20:144–48

Bailey JM, Zucker KJ. 1995. Childhood sex-typed behavior and sexual orientation: a conceptual analysis and quantitative review. *Dev. Psychol.* 31:43–55

Baker SW, Ehrhardt AA. 1974. Prenatal androgen, intelligence and cognitive sex differences. See Friedman et al. 1974, pp. 53–76

Baron-Cohen S. 2002. The extreme male brain theory of autism. *Trends Cogn. Sci.* 6:248–54

Berenbaum SA, Hines M. 1992. Early androgens are related to childhood sex-typed toy preferences. *Psychol. Sci.* 3:203–6

Bidlingmaier F, Strom TM, Dörr G, Eisenmenger W, Knorr D. 1987. Estrone and estradiol concentrations in human ovaries, testes, and adrenals during the first two years of life. *J. Clin. Endocrinol. Metab.* 65:862–67

Bidlingmaier F, Versmold H, Knorr D. 1974. Plasma estrogens in newborns and infants. In *Sexual Endocrinology of the Perinatal Period*, ed. M Forest, J Bertrand, pp. 299–314. Paris: Inserm

Bleier R, Byne W, Siggelkow I. 1982. Cytoarchitectonic sexual dimorphisms of the medial preoptic and anterior hypothalamic areas in guinea pig, rat, hamster, and mouse. *J. Comp. Neurol.* 212:118–30

Bussey K, Bandura A. 1999. Social cognitive theory of gender development and differentiation. *Psychol. Rev.* 106:676–713

Byne W. 1998. The medial preoptic and anterior hypothalamic regions of the rhesus monkey: cytoarchitectonic comparison with the human and evidence for sexual dimorphism. *Brain Res.* 793:346–50

Byne W, Tobet SA, Mattiace LA, Lasco MS, Kemether E, et al. 2001. The interstitial nuclei of the human anterior hypothalamus: an investigation of variation with sex, sexual orientation, and HIV status. *Horm. Behav.* 40:86–92

Cahill L. 2009. Sex differences in human brain structure and function: relevance to learning and memory. See Pfaff et al. 2009, pp. 2307–15

Campbell A, Shirley L, Heywood C. 2000. Infants' visual preference for sex-congruent babies, children, toys, and activities: a longitudinal study. *Br. J. Dev. Psychol.* 18:479–98

Carson DJ, Okuno A, Lee PA, Stetten G, Didolkar SM, Migeon CJ. 1982. Amniotic fluid steroid levels: fetuses with adrenal hyperplasia, 46,XXY fetuses, and normal fetuses. *Am. J. Dis. Child.* 136:218–22

Chapman E, Baron-Cohen S, Auyeung B, Knickmeyer R, Taylor K, Hackett G. 2006. Fetal testosterone and empathy: evidence from the Empathy Quotient (EQ) and the "reading the mind in the eyes" test. *Soc. Neurosci.* 1:135–48

Chung WCJ, De Vries GJ, Swaab D. 2002. Sexual differentation of the bed nucleus of the stria terminalis in humans may extend into adulhood. *J. Neurosci.* 22(3):1027–33

Collaer ML, Brook CDG, Conway GS, Hindmarsh PC, Hines M. 2009. Motor development in individuals with congenital adrenal hyperplasia: strength, targeting, and fine motor skill. *Psychoneuroendocrinology* 34:249–58

Collaer ML, Geffner M, Kaufman FR, Buckingham B, Hines M. 2002. Cognitive and behavioral characteristics of Turner syndrome: exploring a role for ovarian hormones in female sexual differentiation. *Horm. Behav.* 41:139–55

Collaer ML, Hines M. 1995. Human behavioral sex differences: a role for gonadal hormones during early development? *Psychol. Bull.* 118:55–107

Collaer ML, Reimers S, Manning J. 2007. Visuospatial performance on an Internet line judgment task and potential hormonal markers: sex, sexual orientation, and 2D:4D. *Arch. Sex. Behav.* 36:177–92

Cooke BM, Stokas MR, Woolley CS. 2007. Morphological sex differences and laterality in the prepubertal medial amygdala. *J. Comp. Neurol.* 501:904–15

Culbert KM, Breedlove SM, Burt SA, Klump KL. 2008. Prenatal hormone exposure and risk for eating disorders. *Arch. Gen. Psychiatry* 65:329–36

de Graaf-Peters VP, Hadders-Algra M. 2005. Ontogeny of the human central nervous system: What is happening when? *Early Hum. Dev.* 82:257–66

de Lacoste-Utamsing C, Holloway RL. 1982. Sexual dimorphism in the human corpus callosum. *Science* 216:1431–32

Dessens AB, Slijper FME, Drop SLS. 2005. Gender dysphoria and gender change in chromosomal females with congenital adrenal hyperplasia. *Arch. Sex. Behav.* 34:389–97

de Vries ALC, Noens ILJ, Cohen-Kettenis PT, Berckelaer-Onnes IA, Doreleijers TA. 2010. Autism spectrum disorders in gender dysphoric children and adolescents. *J. Autism Dev. Disord.* 40:930–36

De Vries GJ, Sodersten P. 2009. Sex differences in the brain: the relation between structure and function. *Horm. Behav.* 55:589–96

Dittmann RW, Kappes MH, Kappes ME, Börger D, Stegner H, et al. 1990. Congenital adrenal hyperplasia I: gender-related behavior and attitudes in female patients and sisters. *Psychoneuroendocrinology* 15:401–20

Dohler KD, Coquelin A, Davis F, Hines M, Shryne JE, Gorski RA. 1984. Pre- and postnatal influence of testosterone propionate and diethylstilbestrol on differentiation of the sexually dimorphic nucleus of the preoptic area in male and female rats. *Brain Res.* 302:291–95

Ehrhardt AA, Baker SW. 1974. Fetal androgens, human central nervous system differentiation, and behavior sex differences. See Friedman et al. 1974, pp. 33–52

Ehrhardt AA, Epstein R, Money J. 1968. Fetal androgens and female gender identity in the early-treated adrenogenital syndrome. *Johns Hopkins Med. J.* 122:160–67

Ehrhardt AA, Grisanti GC, Meyer-Bahlburg HFL. 1977. Prenatal exposure to medroxyprogesterone acetate (MPA) in girls. *Psychoneuroendocrinology* 2:391–98

Ehrhardt AA, Money J. 1967. Progestin-induced hermaphroditism: IQ and psychosexual identity in a study of ten girls. *J. Sex Res.* 3:83–100

Ernst M, Maheu FS, Schroth E, Hardin J, Golan LS, et al. 2007. Amygdala function in adolescents with congenital adrenal hyperplasia: a model for the study of early steroid abnormalities. *Neuropsychologia* 45:2104–13

Fagot BI. 1978. The influence of sex of child on parental reactions to toddler children. *Child Dev.* 49:459–65

Fausto-Sterling A. 1992. *Myths of Gender*. New York: Basic Books

Feingold A. 1988. Cognitive gender differences are disappearing. *Am. Psychol.* 43:95–103

Finegan JK, Niccols GA, Sitarenios G. 1992. Relations between prenatal testosterone levels and cognitive abilities at 4 years. *Dev. Psychol.* 28:1075–89

Forest MG, Sizonenko PC, Cathiard AM, Bertrand J. 1974. Hypophyso-gonadal function in humans during the first year of life. I. Evidence for testicular activity in early infancy. *J. Clin. Invest.* 53:819–28

Friederici AD, Pannekamp A, Partsch C-J, Ulmen U, Oehler K et al. 2008. Sex hormone testosterone affects language organization in the infant brain. *Neuroreport* 19:283–86

Friedman RC, Richart RN, Vande Wiele RL, eds. 1974. *Sex Differences in Behavior*. New York: Wiley

Frost JA, Binder JR, Springer JA, Hammeke TA, Bellgowan SF, et al. 1999. Language processing is strongly left lateralized in both sexes—evidence from functional MRI. *Brain* 122:199–208

Garcia-Falgueras A, Swaab DF. 2008. A sex difference in the hypothalamic uncinate nucleus: relationship to gender identity. *Brain* 131:3132–46

Ginsburg KR, Committee on Communications, Committee on Psychosocial Aspects of Child and Family Health. 2007. The importance of play in promoting healthy child development and maintaining strong parent-child bonds. *Pediatrics* 119:182–91

Goldstein JM, Seidman LJ, Horton NJ, Makris N, Kennedy DN, et al. 2001. Normal sexual dimorphism of the adult human brain assessed by in vivo magnetic resonance imaging. *Cereb. Cortex* 11:490–97

Gorski RA, Gordon JH, Shryne JE, Southam AM. 1978. Evidence for a morphological sex difference within the medial preoptic area of the rat brain. *Brain Res.* 148:333–46

Gorski RA, Harlan RE, Jacobson CD, Shryne JE, Southam AM. 1980. Evidence for the existence of a sexually dimorphic nucleus in the preoptic area of the rat. *J. Comp. Neurol.* 193:529–39

Green R. 1985. Gender identity in childhood and later sexual orientation. *Am. J. Psychiatry* 142:339–41

Grimbos T, Dawood K, Burriss RP, Zucker KJ, Puts DA. 2010. Sexual orientation and the second to fourth finger length ratio: a meta-analysis in men and women. *Behav. Neurosci.* 124:278–87

Grimshaw GM, Sitarenios G, Finegan JK. 1995. Mental rotation at 7 years: relations with prenatal testosterone levels and spatial play experiences. *Brain Cogn.* 29:85–100

Gron G, Wunderlich AP, Spitzer M, Tomczak R, Riepe MW. 2000. Brain activation during human navigation: gender-different neural networks as substrate of performance. *Nat. Neurosci.* 3:404–8

Guiso L, Monte F, Sapienza P, Zingales L. 2008. Culture, gender and math. *Science* 320:1164–65

Gur RC, Turetsky BI, Matsui M, Yan M, Bilker W, et al. 1999. Sex differences in gray and white matter in healthy young adults: correlations with cognitive performance. *J. Neurosci.* 19:4065–72

Haier RJ, Benbow CP. 1995. Sex differences and lateralization in temporal lobe glucose metabolism during mathematical reasoning. *Dev. Neuropsychol.* 11:405–14

Haier RJ, Jung RE, Yeo RA, Head K, Alkire MT. 2005. The neuroanatomy of general intelligence: Sex matters. *Neuroimage* 25:320–27

Halari R, Sharma T, Hines M, Andrew C, Simmons A, Kumari V. 2005. Comparable fMRI activity with differential behavioural performance on mental rotation and overt verbal fluency tasks in healthy men and women. *Exp. Brain Res.* 169:1–14

Hall CM, Jones JA, Meyer-Bahlburg HFL, Dolezal C, Coleman M, et al. 2004. Behavioral and physical masculinization are related to genotype in girls with congenital adrenal hyperplasia. *J. Clin. Endocrinol. Metab.* 89:419–24

Halpern CT, Udry JR, Campbell B, Suchindran C. 1993. Testosterone and pubertal development as predictors of sexual activity: a panel analysis of adolescent males. *Psychosom. Med.* 55:436–47

Halpern DF, Benbow CP, Geary DC, Gur RC, Hyde JS, Gernsbacher MA. 2007. The science of sex differences in science and mathematics. *Psychol. Sci. Public Interest* 8:1–51

Hampson E, Rovet J, Altmann D. 1998. Spatial reasoning in children with congenital adrenal hyperplasia due to 21-hydroxylase deficiency. *Dev. Neuropsychol.* 14(2):299–320

Hassett JM, Siebert ER, Wallen K. 2008. Sex differences in rhesus monkey toy preferences parallel those of children. *Horm. Behav.* 54:359–64

Held R, Thorn F, Gwiazda J, Bauer J. 1996. Development of binocularity and its sexual differentiation. In *Infant Vision*, ed. F Vital-Durant, J Atkinson, OJ Braddick, pp. 265–74. New York: Oxford Univ. Press

Hier DB, Crowley WF. 1982. Spatial ability in androgen-deficient men. *N. Engl. J. Med.* 306:1202–5

Hines M. 2004. *Brain Gender*. New York: Oxford Univ. Press

Hines M. 2009. Gonadal hormones and sexual differentiation of human brain and behavior. See Pfaff et al. 2009, pp. 1869–909

Hines M. 2010a. Gendered behavior across the lifespan. In *Life-Span Development*, Vol. 2, ed. RM Lerner, ME Lamb, A Freund, pp. 341–78. New York: Wiley

Hines M. 2010b. Sex-related variation in human behavior and the brain. *Trends Cogn. Sci.* 14:448–56

Hines M, Allen LS, Gorski RA. 1992a. Sex differences in subregions of the medial nucleus of the amygdala and the bed nucleus of the stria terminalis of the rat. *Brain Res.* 579:321–26

Hines M, Alsum P, Roy M, Gorski RA, Goy RW. 1987. Estrogenic contributions to sexual differentiation in the female guinea pig: influences of diethylstilbestrol and tamoxifen on neural, behavioral and ovarian development. *Horm. Behav.* 21:402–17

Hines M, Brook C, Conway GS. 2004. Androgen and psychosexual development: core gender identity, sexual orientation and recalled childhood gender role behavior in women and men with congenital adrenal hyperplasia (CAH). *J. Sex Res.* 41:75–81

Hines M, Chiu L, McAdams LA, Bentler PM, Lipcamon J. 1992b. Cognition and the corpus callosum: verbal fluency, visuospatial ability and language lateralization related to midsagittal surface areas of callosal subregions. *Behav. Neurosci.* 106:3–14

Hines M, Fane BA, Pasterski VL, Mathews GA, Conway GS, Brook C. 2003. Spatial abilities following prenatal androgen abnormality: targeting and mental rotations performance in individuals with congenital adrenal hyperplasia (CAH). *Psychoneuroendocrinology* 28:1010–26

Hines M, Golombok S, Rust J, Johnston K, Golding J, ALSPAC Study Team. 2002. Testosterone during pregnancy and childhood gender role behavior: a longitudinal population study. *Child Dev.* 73:1678–87

Huttenlocher PR. 2002. *Neural Plasticity*. Cambridge, MA: Harvard Univ. Press

Hyde JS, Mezulis AH, Abramson LY. 2008. The ABCs of depression: integrating affective, biological, and cognitive models to explain the emergence of the gender difference in depression. *Psychol. Rev.* 115:291–313

Hyde JS, Linn MC. 1988. Gender differences in verbal ability: a meta-analysis. *Psychol. Bull.* 104:53–69

Ito S, Murakami S, Yamanouchi K, Arai Y. 1986. Perinatal androgen exposure decreases the size of the sexually dimorphic medial preoptic nucleus in the rat. *Proc. Japan Acad.* 62(B):408–11

Jacobson CD, Csernus VJ, Shryne JE, Gorski RA. 1981. The influence of gonadectomy, androgen exposure, or a gonadal graft in the neonatal rat on the volume of the sexually dimorphic nucleus of the preoptic area. *J. Neurosci.* 1:1142–47

Jadva V, Golombok S, Hines M. 2010. Infants' preferences for toys, colors and shapes. *Arch. Sex. Behav.* 39:1261–73

Jordan-Young RM. 2010. *Brainstorm: The Flaws in the Science of Sex Differences*. Cambridge, MA: Harvard Univ. Press

Juraska JM. 1998. Neural plasticity and the development of sex differences. *Annu. Rev. Sex Res.* 9:20–38

Klump KL, Gobrogge KL, Perkins PS, Thorne D, Sisk CL, Breedlove SM. 2006. Preliminary evidence that gonadal hormones organize and activate disordered eating. *Psychol. Med.* 36:539–46

Knickmeyer R, Baron-Cohen S, Fane B, Wheelwright S, Mathews GA, et al. 2006. Androgen and autistic traits: a study of individuals with congenital adrenal hyperplasia. *Horm. Behav.* 50:148–53

Kolb B, Whishaw IQ. 1985. *Fundamentals of Human Neuropsychology*. New York: W.H. Freeman and Co.

Kraemer B, Noll T, Delsignore A, Milos G, Schnyder U, Hepp U. 2009. Finger length ratio (2D:4D) in adults with gender identity disorder. *Arch. Sex. Behav.* 38:359–63

Langlois JH, Downs AC. 1980. Mothers, fathers and peers as socialization agents of sex-typed play behaviors in young children. *Child Dev.* 51:1217–47

Lawrence AA. 2009. Parallels between gender identity disorder and body integrity identity disorder: a review and update. In *Body Integrity Identity Disorder: Psychological, Neurobiological, Ethical, and Legal Aspects*, ed. A Stirn, A Thiel, S Oddo, pp. 154–72. Lengerich, Germ.: Pabst

Lenroot RK, Gogtay N, Greenstein DK, Wells EM, Wallace GL, et al. 2007. Sexual dimorphism of brain developmental trajectories during childhood and adolescence. *Neuroimage* 36:1065–73

LeVay S. 1991. A difference in hypothalamic structure between heterosexual and homosexual men. *Science* 253:1034–37

Luders E, Narr Kl, Thompson PM, Rex DE, Woods RP, et al. 2006. Gender effects on cortical thickness and the influence of scaling. *Hum. Brain Mapp.* 27:314–24

Maguire EA, Woollett K, Spiers HJ. 2006. London taxi drivers and bus drivers: a structural MRI and neuropsychological analysis. *Hippocampus* 16:1091–101

Mansour CS, Haier RJ, Buchsbaum MS. 1996. Gender comparisons of cerebral glucose metabolic rate in healthy adults during a cognitive task. *Pers. Individ. Differ.* 20:183–91

Martin CL, Ruble DN, Szkrybalo J. 2002. Cognitive theories of early gender development. *Psychol. Bull.* 128:903–33

Masters JC, Ford ME, Arend R, Grotevant HD, Clark LV. 1979. Modeling and labelling as integrated determinants of children's sex-typed imitative behavior. *Child Dev.* 50:364–71

Mathews GA, Fane BA, Conway GS, Brook C, Hines M. 2009. Personality and congenital adrenal hyperplasia: possible effects of prenatal androgen exposure. *Horm. Behav.* 55:285–91

McCarthy MM, De Vries GJ, Forger NG. 2009. Sexual differentiation of the brain: mode, mechanisms, and meaning. See Pfaff et al. 2009, pp. 1707–44

McGlone J. 1980. Sex differences in human brain asymmetry: a critical survey. *Behav. Brain Sci.* 3:215–63

Merke DP, Fields JD, Keil MF, Vaituzis AC, Chroussos GP, Giedd JN. 2003. Children with classic congenital adrenal hyperplasia have decreased amygdala volume: potential prenatal and postnatal hormonal effects. *J. Clin. Endocrinol. Metab.* 88(4):1760–65

Meyer-Bahlburg HFL, Dolezal C, Baker SW, New MI. 2008. Sexual orientation in women with classical or non-classical congenital adrenal hyperplasia as a function of degree of prenatal androgen excess. *Arch. Sex. Behav.* 37:85–99

Ming G, Song H. 2005. Adult neurogenesis in the mammalian central nervous system. *Annu. Rev. Neurosci.* 28:223–50

Mueller SC, Temple V, Oh E, VanRyzin C, Williams A, et al. 2008. Early androgen exposure modulates spatial cognition in congenital adrenal hyperplasia (CAH). *Psychoneuroendocrinology* 33:973–80

Mullis IVS, Martin MO, Foy P. 2008. *TIMSS 2007 International Maths Report.* Boston: Int. Assoc. Eval. Educ. Achiev.

Neufang S, Specht K, Hausmann M, Gunturkun O, Herpertz-Dahlmann B, et al. 2009. Sex differences and the impact of steroid hormones on the developing human brain. *Cereb. Cortex* 19:464–73

Nordenstrom A, Servin A, Bohlin G, Larsson A, Wedell A. 2002. Sex-typed toy play behavior correlates with the degree of prenatal androgen exposure assessed by CYP21 genotype in girls with congenital adrenal hyperplasia. *J. Clin. Endocrinol. Metab.* 87:5119–24

Pasterski VL, Geffner ME, Brain C, Hindmarsh P, Brook C, Hines M. 2005. Prenatal hormones and postnatal socialization by parents as determinants of male-typical toy play in girls with congenital adrenal hyperplasia. *Child Dev.* 76:264–78

Pasterski VL, Hindmarsh P, Geffner M, Brook C, Brain C, Hines M. 2007. Increased aggression and activity level in 3- to 11-year-old girls with congenital adrenal hyperplasia (CAH). *Horm. Behav.* 52:368–74

Peper JS, Brouwer RM, Schnack HG, van Baal GC, van Leeuwen M, Van Den Berg SM, et al. 2009. Sex steroids and brain structure in pubertal boys and girls. *Psychoneuroendocrinology* 34:332–42

Perlman SM. 1973. Cognitive abilities of children with hormone abnormalities: screening by psychoeducational tests. *J. Learn. Disabil.* 6:21–29

Perrin JS, Herve P-Y, Leonard G, Perron M, Pike GB, et al. 2008. Growth of white matter in the adolescent brain: role of testosterone and androgen receptor. *J. Neurosci.* 28:9519–24

Perry DG, Bussey K. 1979. The social learning theory of sex difference: Imitation is alive and well. *J. Pers. Soc. Psychol.* 37:1699–712

Pfaff DW, Arnold AP, Etgen AM, Fahrbach SE, Rubin RT, eds. 2009. *Hormones, Brain and Behavior.* San Diego: Academic

Phoenix CH, Goy RW, Gerall AA, Young WC. 1959. Organizing action of prenatally administered testosterone propionate on the tissues mediating mating behavior in the female guinea pig. *Endocrinology* 65:163–96

Piaget J. 1970. *Science of Education and the Psychology of the Child*. New York: Orion

Poomthavorn P, Stargatt R, Zacharin M. 2009. Psychosexual and psychosocial functions of anorchid young adults. *J. Clin. Endocrinol. Metab.* 94:2502–5

Quigley CA. 2002. The postnatal gonadotrophin and sex steroid surge-insights from the androgen insensitivity syndrome. *Clin. Endocrinol. Metab.* 87(1):24–28

Raevuori A, Kaprio J, Hoek HW, Sihvola E, Rissanen A, Keski-Rahkonen A. 2008. Anorexia and bulimia nervosa in same-sex and opposite-sex twins: lack of an association with twin type in a nationwide study of Finnish twins. *Am. J. Psychiatry* 165:1604–10

Reinisch JM. 1981. Prenatal exposure to synthetic progestins increases potential for aggression in humans. *Science* 211:1171–73

Reinius B, Jazin E. 2009. Prenatal sex differences in the human brain. *Mol. Psychiatry* 14:988–89

Reinius B, Saetre P, Leonard JA, Blekhman R, Merino-Martinez R, et al. 2008. An evolutionarily conserved sexual signature in the primate brain. *PLoS Genet.* 4:e1000100

Reyes FI, Winter JSD, Faiman C. 1973. Studies on human sexual development. I. Fetal gonadal and adrenal sex steroids. *J. Clin. Endocrinol. Metab.* 37:74–78

Roselli CE, Larkin K, Resko JA, Stellflug JN, Stromshak F. 2004. The volume of a sexually dimorphic nucleus in the ovine medial preoptic area/anterior hypothalamus varies with sexual partner preference. *Endocrinology* 145(2):478–83

Ruble DN, Martin CL, Berenbaum SA. 2006. Gender development. In *Handbook of Child Psychology*, Vol. 3: *Social, Emotional and Personality Development*, ed. W Damon, RM Lerner, N Eisenberg, pp. 858–932. New York: Wiley

Savic I, Lindstrom P. 2008. PET and MRI show differences in cerebral asymmetry and functional connectivity between homo- and heterosexual subjects. *Proc. Natl. Acad. Sci. USA* 105:9403–8

Schneider HJ, Pickel J, Stalla GK. 2006. Typical female 2nd-4th finger length (2D:4D) ratios in male-to-female transsexuals—possible implications for prenatal androgen exposure. *Psychoneuroendocrinology* 31:265–69

Schulz KM, Molenda-Figueira HA, Sisk CL. 2009. Back to the future: the organizational-activational hypothesis adapted to puberty and adolescence. *Horm. Behav.* 55:597–604

Serbin LA, Poulin-Dubois D, Colbourne KA, Sen MG, Eichstedt JA. 2001. Gender stereotyping in infancy: visual preferences for and knowledge of gender-stereotyped toys in the second year. *Int. J. Behav. Dev.* 25:7–15

Shaywitz BA, Shaywitz SE, Pugh KR, Constable RT, Skudlarski P, et al. 1995. Sex differences in the functional organization of the brain for language. *Nature* 373:607–9

Simerly RB, Zee MC, Pendleton JW, Lubahn DB, Korach KS. 1997. Estrogen receptor-dependent sexual differentiation of dopaminergic neurons in the preoptic region of the mouse. *Proc. Natl. Acad. Sci. USA* 94:14077–82

Sinforiani E, Livieri C, Mauri M, Bisio P, Sibilla L, et al. 1994. Cognitive and neuroradiological findings in congenital adrenal hyperplasia. *Psychoneuroendocrinology* 19:55–64

Skuse DH. 2006. Sexual dimorphism in cognition and behaviour: the role of X-linked genes. *Eur. J. Endocrinol.* 155:S99–106

Slijper FME, Drop SLS, Molenaar JC, de Muinck Keizer-Schrama SMPF. 1998. Long-term psychological evaluation of intersex children. *Arch. Sex. Behav.* 27:125–44

Spreen O, Strauss E. 1991. *A Compendium of Neuropsychological Tests*. New York: Oxford Univ. Press

Sumida H, Nishizuka M, Kano Y, Arai Y. 1993. Sex differences in the anteroventral periventricular nucleus of the preoptic area and in the related effects of androgen in prenatal rats. *Neurosci. Lett.* 151:41–44

Tobet SA, Zahniser DJ, Baum MJ. 1986. Differentiation in male ferrets of a sexually dimorphic nucleus of the preoptic/anterior hypothalamic area requires prenatal estrogen. *Neuroendocrinology* 44:299–308

Ulibarri C, Yahr P. 1988. Role of neonatal androgens in sexual differentiation of brain structure, scent marking and gonadotropin secretion in gerbils. *Behav. Neural Biol.* 49:27–44

Voyer D. 1996. On the magnitude of laterality effects and sex differences in functional lateralities. *Laterality* 1:51–83

Vygotsky LS. 1976. Play and its role in the mental development of the child. *Sov. Psychol.* 5:6–18

Wallien MSC, Zucker KJ, Steensma TD, Cohen-Kettenis PT. 2008. 2D:4D finger-length ratios in children and adults with gender identity disorder. *Horm. Behav.* 54:450–54

Wilson JD, George FW, Griffin JE. 1981. The hormonal control of sexual development. *Science* 211:1278–84

Witelson SF. 1989. Hand and sex differences in the isthmus and genu of the human corpus callosum: a postmortem morphological study. *Brain* 112:799–835

Witelson SF, Kigar DL, Scamvougeras A, Kideckel DM, Buck B, et al. 2008. Corpus callosum anatomy in right-handed homosexual and heterosexual men. *Arch. Sex. Behav.* 37:857–63

Zhou J, Hofman MA, Gooren LJG, Swaab DF. 1995. A sex difference in the human brain and its relation to transsexuality. *Nature* 378:68–70

Too Many Cooks? Intrinsic and Synaptic Homeostatic Mechanisms in Cortical Circuit Refinement

Gina Turrigiano

Department of Biology, Center for Complex Systems, and Center for Behavioral Genomics, Brandeis University, Waltham, Massachusetts 02454; email: Turrigiano@brandeis.edu

Annu. Rev. Neurosci. 2011. 34:89–103

First published online as a Review in Advance on March 25, 2011

The *Annual Review of Neuroscience* is online at neuro.annualreviews.org

This article's doi:
10.1146/annurev-neuro-060909-153238

0147-006X/11/0721-0089$20.00

Keywords

synaptic scaling, intrinsic plasticity, homeostatic plasticity, critical period

Abstract

Maintaining the proper balance between excitation and inhibition is critical for the normal function of cortical circuits. This balance is thought to be maintained by an array of homeostatic mechanisms that regulate neuronal and circuit excitability, including mechanisms that target excitatory and inhibitory synapses, and mechanisms that target intrinsic neuronal excitability. In this review, I discuss where and when these mechanisms are used in complex microcircuits, what is currently known about the signaling pathways that underlie them, and how these different ways of achieving network stability cooperate and/or compete. An important challenge for the field of homeostatic plasticity is to assemble our understanding of these individual mechanisms into a coherent view of how microcircuit stability is maintained during experience-dependent circuit refinement.

Contents

INTRODUCTION

Epilepsy is a disorder of circuit excitability that affects 1%–2% of the population, often to devastating effect. What is extraordinary about this incidence to a neuroscientist who studies cortical microcircuits is not that it is so high, but rather that most people, most of the time, are not epileptic. This observation is surprising because many of the circuits within our cerebral cortex are composed of highly unstable networks with extensive positive feedback (Burkhalter 2008), where even small changes in the balance between excitation and inhibition can set off uncontrolled seizure-like activity. Yet despite the existence of many forces that continuously perturb the balance between excitation and inhibition, such as learning-related or developmental changes in synapse number and strength, somehow, most of the time, our brains manage to compensate for these changes and maintain stable function. Our brains appear to be constructed so that the flexibility that enables us to adapt and learn is balanced by stabilizing mechanisms that preserve overall network function, and these forces

of plasticity and stability are able to coexist and cooperate without interfering with each other.

How do our brains achieve this remarkable feat? In the past two decades, major inroads have been made into elucidating the mechanisms that allow neurons and circuits to maintain stable function in the face of these ongoing perturbations. Our brains employ an array of classic homeostatic negative feedback mechanisms that allow neurons and/or circuits to sense how active they are and to adjust their excitability to keep this activity within some target range (Davis 2006, Marder & Goaillard 2006, Turrigiano & Nelson 2004), and collectively these stabilizing mechanisms have been termed homeostatic plasticity. To implement homeostatic plasticity neurons need to sense some aspect of "activity," and when this measure deviates from a target value, a force must be generated that adjusts excitability to move neuronal activity back toward this target. In principle, if individual neurons can stabilize their own firing, then overall network activity will also be stabilized; however, depending on network architecture, the rules for homeostatic regulation are likely to be tuned for the function of particular neurons within the circuit. For example, one might predict that excitatory and inhibitory neurons would use distinct homeostatic rules. Although much has been learned about the cellular and synaptic mechanisms of homeostatic plasticity in reduced preparations such as neuronal cultures, little is currently known about how homeostatic plasticity is implemented in complex, highly recurrent microcircuits such as those of the neocortex, where many different cell types subserve distinct functions and likely express unique forms of plasticity.

Neuronal firing arises from the interplay between synaptic currents and the intrinsic firing properties of a neuron. Thus one can imagine two fundamentally different ways that neurons could homeostatically regulate excitability (**Figure 1**). First, they could slowly adjust synaptic strengths up or down in the right direction to stabilize average neuronal firing rates (Turrigiano et al. 1998). Conversely, instead of regulating synaptic strengths, they

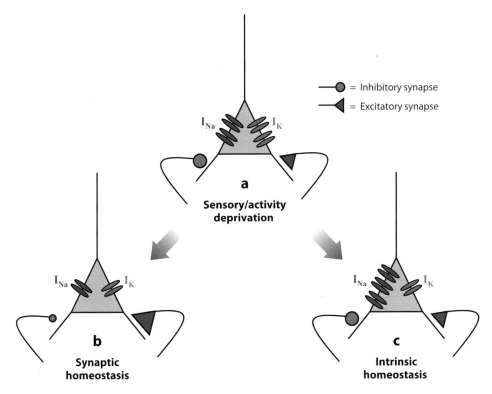

Figure 1

Two fundamentally different mechanisms for the homeostatic regulation of neuronal firing. (*a*) Neuronal activity is determined both by the strength of excitatory and inhibitory synaptic inputs and by the balance of inward and outward voltage-dependent conductances that regulate intrinsic excitability, here illustrated as the relative number of Na (*blue*) and K (*red*) channels. Neurons can compensate for reduced sensory drive either by using synaptic mechanisms to modify the balance between excitatory and inhibitory inputs (*b*) or by using intrinsic mechanisms to modify the balance of inward and outward voltage-dependent currents (*c*).

could modulate intrinsic excitability to shift the relationship between synaptic input and firing rate (their "input-output curve") (Desai et al. 1999, Turrigiano et al. 1994). In principle, both of these processes could work, and many neurons appear able to undergo homeostatic regulation of firing via either mechanism (Desai et al. 1999, Maffei & Turrigiano 2008b, Turrigiano et al. 1998). This raises the questions of why neurons sometimes use one method and sometimes another, whether important functional differences exist between these two forms of homeostatic plasticity, and whether there are hierarchical rules for their engagement. In this review, I focus on recent work examining the interactions between intrinsic and synaptic homeostasis,

using cortical networks as a major example. I begin by reviewing the evidence for these two forms of homeostatic regulation, explore what is currently known about their interaction in both reduced preparations and in vivo, and speculate about their function. Understanding the rules that underlie network homeostasis is likely to shed important light onto disease processes to which imbalances in excitation and inhibition contribute, such as epilepsy, schizophrenia, and autism.

HOMEOSTATIC REGULATION OF NEURONAL FIRING

Compelling evidence from a variety of systems, both in vivo and in vitro, indicates that circuit activity is homeostatically regulated to maintain firing rates and/or firing patterns within certain functional boundaries. An early example came from studies of invertebrate central pattern generators (CPGs), where investigators observed that perturbations that made these networks arrhythmic resulted in compensatory changes in intrinsic neuronal properties that, over time, restored rhythmicity (Haedo & Golowasch 2006; Thoby-Brisson & Simmers 1998; Turrigiano et al. 1994, 1995); Gonzalez-Islas & Wenner (2006) found a similar phenomenon in developing vertebrate spinal cord central pattern generators. Similarly, central neurons in dissociated cultures are able to maintain average firing rates around a homeostatic set point. When cortical or hippocampal neurons are induced to fire more than normal, over many hours, firing returns to baseline levels, and if neuronal firing is reduced over time neurons also compensate and again firing is restored to baseline (Burrone et al. 2002, Turrigiano et al. 1998). These studies lend strong support to the idea that neuronal circuits possess mechanisms that maintain firing around a homeostatic stable point. Because fluctuations in firing are the currency of information transfer in the brain, it may seem at first glance to be highly problematic for neurons to maintain stable average firing rates without impairing information flow. Most forms of homeostatic compensation in central neurons are slow and operate over hours to days, many orders of magnitude slower than the moment-to-moment fluctuations in firing that transmit information (Turrigiano & Nelson 2004). Thus the temporal characteristics of firing rate homeostasis appear designed to prevent interference with the business of information transfer.

In contrast with the strong evidence for firing rate (or firing pattern) homeostasis cited above, for many in vivo vertebrate circuits the evidence is less direct. In both visual tectum and visual cortex, studies have shown that neuronal response amplitudes remain roughly constant following visual deprivation, suggesting that homeostatic compensation has occurred (Chandrasekaran et al. 2007, Mrsic-Flogel et al.

2007). Furthermore, lowering sensory drive in both visual and auditory cortex can generate compensatory changes in synaptic and/or intrinsic network properties that enhance circuit excitability when measured ex vivo (Maffei et al. 2004, 2006; Maffei & Turrigiano 2008b; Vale & Sanes 2002), but it is not yet clear whether these changes result in true homeostasis, that is, act to maintain an activity set point.

Further complicating matters, in both sensory cortex and hippocampus the mechanisms and sites of homeostatic compensation are strongly developmentally regulated (Desai et al. 2002, Echegoyen et al. 2007). Research has demonstrated this point most clearly in the primary visual cortex, where studies in rodent have shown that compensation for lowered visual drive is implemented in a layer- and cell-type-specific manner. In layer 4 (the primary cortical input layer), homeostatic compensation is present early in development but turns off at the opening of the classical visual system critical period (CP) (Desai et al. 2002; Maffei et al. 2004, 2006). In contrast, in the upper cortical layers (which mediate extensive lateral interactions between functionally related cortical regions), compensation is absent early, turns on at the onset of the CP, and remains active into adulthood (Goel & Lee 2007, Maffei & Turrigiano 2008b). Thus not all neurons, nor indeed all local microcircuits, are subject to homeostatic regulation at all times in an animal's life. A major challenge for the field is to identify the rules that guide the placement and timing of homeostatic mechanisms within complex neural circuits.

SYNAPTIC HOMEOSTASIS

Central neurons are embedded in complex networks composed of many distinct cell types, including both excitatory neurons and a rich variety of inhibitory neurons with distinct morphologies and functions. In most networks, small changes in the balance between excitation and inhibition (the E/I balance) can have a major impact on ongoing activity, and compelling evidence indicates that the E/I balance is tightly regulated (Atallah & Scanziani 2009, Pouille et al. 2009, Shu et al. 2003). Given this complexity, the ability of networks to compensate for external or internal perturbations and to maintain stable firing is not trivial and likely requires mechanisms that can adjust both excitatory and inhibitory synaptic strengths in a cell-type-specific manner. Indeed, experimenters have now uncovered a rich variety of homeostatic synaptic mechanisms that operate on both excitatory and inhibitory synapses, which I describe below in turn.

Synaptic Scaling of Excitatory Synapses

Investigators have identified several forms of homeostatic plasticity of excitatory synapses. For example, there is evidence for both "global" mechanisms such as synaptic scaling that operate on all a neuron's synapses and "local" mechanisms that act on individual or small groups of synapses (Turrigiano 2008, Yu & Goda 2009). Similarly, some forms of synaptic homeostasis occur through presynaptic and others through postsynaptic changes in function (Davis & Bezprozvanny 2001). Currently, the best studied form of homeostatic plasticity at central excitatory synapses is global synaptic scaling, which I will focus on here both because of space limitations and because of strong evidence that this form of homeostatic plasticity is important for the in vivo function of cortical networks; several recent reviews provide an excellent discussion of presynaptic and/or local forms of compensatory synaptic plasticity (Thiagarajan et al. 2007, Turrigiano 2008, Yu & Goda 2009).

Synaptic scaling was first identified in cultured neocortical neurons, where investigators observed that pharmacological manipulations of activity induced compensatory and bidirectional changes in the unit strength of glutamatergic synapses (Turrigiano et al. 1998). By measuring miniature excitatory postsynaptic currents (minis) mediated by 2-amino-3-(5-methyl-3-oxo-1,2- oxazol-4-yl) propanoic acid (AMPA) and N-methyl D-

aspartate-type glutamate receptors, researchers found that modulating network activity induced uniform increases or decreases in the entire mini amplitude distribution, in effect scaling postsynaptic strength up or down (Desai et al. 2002, Gainey et al. 2009, Turrigiano et al. 1998). These changes in mini amplitude translate into changes in the amplitude of evoked transmission, with little or no change in short-term synaptic dynamics (Maffei et al. 2004, Watt et al. 2000, Wierenga et al. 2005). Such a postsynaptic scaling process is predicted to stabilize activity without changing the relative strength of synaptic inputs, thus avoiding disrupting information-storage mechanisms that rely on differences in synaptic weights. Synaptic scaling has now been demonstrated in a variety of central neurons both in vitro and in vivo, including neocortical and hippocampal pyramidal neurons and spinal neurons (Desai et al. 2002, Goel & Lee 2007, Kim & Tsien 2008, Knogler et al. 2010, O'Brien et al. 1998, Stellwagen & Malenka 2006, Turrigiano et al. 1998). A fascinating and still unanswered question is the nature of the biophysical mechanism that allows neurons to scale synaptic strengths up and down proportionally.

How do neurons sense perturbations in activity during synaptic scaling? Two recent studies have provided strong evidence that synaptic scaling is a cell-autonomous process in which neurons sense changes in their own activity through changes in firing/depolarization and calcium influx. For example, selectively blocking firing in an individual cortical pyramidal neuron scales up that neuron's synaptic strengths to the same degree as does blockade of network activity, through a process that requires a drop in somatic calcium influx, reduced activation of calcium/calmodulin dependent (CaM) Kinase Kinase (CaMKK) and CaM Kinase IV (CaMKIV), and transcription (Ibata et al. 2008). This signaling pathway then leads to enhanced accumulation of AMPA-type glutamate receptors (AMPAR) in the postsynaptic membrane at all excitatory synapses, thus scaling up mini amplitude and enhancing evoked transmission. This global enhancement of AMPAR abundance in response to activity blockade requires sequences on the C-terminus of the GluR2 subunit of the AMPAR (Gainey et al. 2009), which distinguishes synaptic scaling from other forms of synaptic enhancement such as long-term potentiation (LTP) that require sequences on the GluR1 subunit (Malenka & Bear 2004). Thus synaptic scaling up is fundamentally different from LTP: It operates over a longer temporal scale (hours) and a wider spatial scale (global) and utilizes trafficking steps that target the GluR2 subunit to enhance AMPAR abundance at synapses.

Like scaling up in neocortical neurons, scaling down in hippocampal slice cultures in response to enhanced activity (using channel-rhodopsin and optical stimulation) can also be induced by cell-autonomous changes in calcium influx, and this process also involves CaMKK/CaMKIV signaling and transcription and requires the GluR2 subunit for its expression (Goold & Nicoll 2010). Unlike CaMKK, CaMKIV was found to be necessary but not sufficient to trigger a reduction in synaptic strength, suggesting that CaMKK activates several parallel signaling pathways that cooperate to reduce synaptic strength. In hippocampal neurons, driving individual neurons to fire induces synapse loss as well as reduced quantal amplitude (Goold & Nicoll 2010), something not seen in young neocortical or spinal neurons in response to elevated network activity (O'Brien et al. 1998, Turrigiano et al. 1998); it is not clear whether this discrepancy is due to differences in activation method, neuron type, or neuron age between studies.

We do not currently understand the entire sequence of events that lead from cell-autonomous changes in calcium influx to bidirectional changes in AMPAR abundance, and a number of parallel signaling pathways, and dozens of molecules, likely contribute to synaptic scaling. For example, there is evidence that the neurotrophin brain-derived neurotrophic factor (BDNF) (Rutherford et al. 1998), the immediate early gene *Arc* (Shepherd et al. 2006), the cytokine TNFα (Steinmetz & Turrigiano 2010, Stellwagen & Malenka

2006), the immune molecule MHC1 (Goddard et al. 2007), Beta3 integrins (Cingolani et al. 2008), and the polo-like kinase 2 (Plk2)-CDK5 signaling pathway (Seeburg et al. 2008), among others, are all involved in or essential for synaptic scaling. Several of these molecules are known to regulate AMPA receptor trafficking; for example, *Arc* interacts with the endocytic machinery that removes AMPAR from the membrane (Chowdhury et al. 2006), TNFα directly increases synaptic AMPAR accumulation (Beattie et al. 2002; Stellwagen et al. 2005), and Beta3 integrins regulate AMPAR surface expression (Cingolani et al. 2008). Some of these molecules are involved in only one branch of synaptic scaling (either scaling up or scaling down), indicating that although some signaling elements (such as CaMKK and CaMKIV) are shared during bidirectional scaling (Goold & Nicoll 2010, Ibata et al. 2008), others are not (Rutherford et al. 1998, Shepherd et al. 2006, Stellwagen et al. 2005). Many of these signaling molecules are likely to play permissive rather than instructive roles in synaptic scaling, as has recently been shown for TNFα (Steinmetz & Turrigiano 2010).

Synapse-Type Specificity of Excitatory Synaptic Scaling

Neural circuits are composed of many excitatory and inhibitory cell types interconnected in highly specific ways, and it would clearly be counterproductive from a homeostatic point of view to scale all synapses up or down together without regard for the function of the postsynaptic neuron, and indeed there is evidence that the rules for scaling excitatory synapses are cell-type specific. In cultured cortical and hippocampal neurons, excitatory synapses onto pyramidal neurons are scaled up by activity blockade, whereas excitatory synapses onto γ-Aminobutyric acid (GABA)-ergic interneurons are either unaffected (Rutherford et al. 1998) or reduced (Chang et al., 2010), possibly depending on GABAergic cell type. Conversely, enhancing network activity increases excitatory transmission onto GABAergic interneurons

(Chang et al. 2010, Rutherford et al. 1998) through a process that involves the activity-dependent regulation of the immediate early gene *Narp*. *Narp* appears to be secreted by presynaptic pyramidal neurons and accumulates preferentially at excitatory synapses into parvalbumin-positive interneurons (Chang et al. 2010), suggesting that homeostatic regulation of excitatory synapses onto these neurons is a noncell-autonomous process that depends on pyramidal neuron activity, a theme I revisit below in the discussion of inhibitory synapse scaling.

Not all excitatory neurons express synaptic scaling at all times during development. As discussed above, in visual cortex the expression of synaptic scaling is strongly developmentally regulated and is expressed by layer 4 pyramidal neurons early in postnatal development, but then turns off in layer 4 and turns on in layer 2/3 pyramidal neurons around the opening of the classical visual system CP (Desai et al. 2002; Goel & Lee 2007; Maffei et al. 2004, 2006; Maffei & Turrigiano 2008b). Similarly, activity blockade in hippocampal networks scales up CA1 but not CA3 excitatory synapses, suggesting that the rules for expression of scaling in hippocampus are cell-type specific (Kim & Tsien 2008). These studies underscore the point that not all cell types or networks are designed to maintain homeostasis of firing at all periods of development. Rather, they suggest that synaptic scaling is specifically expressed when and where it is needed.

An interesting and unanswered question is whether a given postsynaptic neuron can preferentially scale one subtype of excitatory synapse while leaving others unaffected. In cortical neurons in dissociated culture and in vivo, it is thought that all excitatory synapses are affected equally during synaptic scaling in response to a drop in activity, based on the observation that the entire distribution of mini amplitudes is scaled up or down proportionally. However, if a synapse type that represented only a small fraction of a neuron's synapses was not affected, this analysis is unlikely to be sensitive enough to detect the resulting deviation from pure scaling. Conversely, if synapse-specific and global synaptic plasticity mechanisms (that affect quantal amplitude) are activated simultaneously by a given activity manipulation, then the net change in mini distribution may not follow a simple scaling rule even though synaptic scaling of all excitatory synapses has occurred. Thus the presence of proportional scaling of the quantal amplitude distribution does not rule out some synapse specificity, nor does its absence necessarily rule out that synaptic scaling has occurred. Changes in the mini amplitude distribution induced by manipulations of network activity should thus be interpreted with due caution.

Homeostatic Regulation of Inhibitory Synapses

A powerful way to stabilize network activity is to reciprocally regulate the relative strengths of excitatory and inhibitory synapses, and a long literature shows that inhibition is regulated by long-lasting changes in activity and/or sensory drive. Early work in primate and rodent visual cortex demonstrated that visual deprivation or inhibition of retinal activity with tetrodotoxin (TTX) decreased immunoreactivity for GABA (Benevento et al. 1995; Hendry et al. 1994; Hendry & Jones 1986, 1988) and reduced inhibition and inhibitory synapse number in cortical and hippocampal cultures (Marty et al. 1997, Rutherford et al. 1997), leading to a reduction in the amount of functional inhibition (Rutherford et al. 1997). These studies raised the possibility that inhibitory synaptic strength is regulated homeostatically in the opposite direction from excitatory synapses.

Indeed, the same paradigm that scales up miniature excitatory postsynaptic currents onto pyramidal neurons in culture scales down the amplitude of miniature inhibitory postsynaptic currents through a mechanism that can involve both changes in accumulation of postsynaptic GABAA receptors and a reduction in presynaptic GABAergic markers, such as GAD65 (Hartman et al. 2006, Kilman et al. 2002). Both in vitro and in vivo studies have suggested that homeostatic regulation of inhibition can occur via a constellation of changes in postsynaptic strength, synapse number, and GABA packaging and release in various combinations (Hartman et al. 2006, Kilman et al. 2002, Maffei et al. 2004); this variability in expression mechanism could reflect several distinct inhibitory plasticity mechanisms (as is the case for homeostasis at excitatory synapses) or perhaps the great diversity of inhibitory synapse types in hippocampus and cortex.

The distinct and opposing plasticity rules at excitatory and inhibitory synapses appear to be designed to stabilize the firing of principal (in cortex, and hippocampus, pyramidal) neurons, suggesting that, from a network point of view, it is the activity of the principal neurons that is homeostatically constrained. This raises the question of whose activity matters in the regulation of inhibition: the presynaptic inhibitory neuron or the postsynaptic pyramidal neuron. This question was recently addressed in hippocampal cultures by preventing firing in either the postsynaptic pyramidal neuron or the presynaptic inhibitory neuron, while measuring inhibitory synapses onto the pyramidal neuron; the answer was that neither manipulation was sufficient to mimic the effects of blocking network firing. This result argues that, in contrast to synaptic scaling of excitatory synapses, homeostatic regulation of inhibition is a noncell-autonomous process that either requires changes in both pre- and postsynaptic activity simultaneously or is triggered by global changes in network activity.

Interneurons come in a wide range of functional varieties, and although in neocortical networks the net effects of changes in excitation and inhibition appear to be homeostatic (Maffei et al. 2004, Rutherford et al. 1998, Turrigiano et al. 1998), different classes of inhibitory synapse are regulated differently by lowered activity. When sensory drive to primary visual cortex is lowered in vivo, connections from fast-spiking basket cells onto layer 4 pyramidal neurons are reduced in amplitude, whereas connections from another class of interneuron become sparser but stronger (Maffei et al. 2004). Similarly, activity

blockade with TTX in neocortical slice cultures differentially regulates different classes of inhibitory synapses (Bartley et al. 2008), whereas in hippocampal networks, activity blockade reduces net inhibition but increases the strength of a specific, endocannabinoid-sensitive class of inhibitory input (Kim & Alger 2010). Under some conditions, net inhibition in hippocampal networks can change in the same direction as excitation (Echegoyen et al. 2007), but whether this acts to enhance or oppose stability is not entirely clear. All these studies strongly support the idea that inhibitory synapses are regulated in a subtype-specific manner, presumably because the molecular machinery that subserves plasticity at inhibitory synapses is different at different synapse types.

Several released factors, including BDNF and endocannabinoids, have been implicated in the homeostatic regulation of inhibition (Kim & Alger 2010, Rutherford et al. 1997, Swanwick et al. 2006). Different subclasses of interneuron have receptors for endocannabinoids and BDNF, suggesting a mechanistic basis for cell-type specificity in the homeostatic regulation of inhibition.

HOMEOSTASIS OF INTRINSIC EXCITABILITY

Changes in intrinsic excitability that alter a neuron's input-output function can strongly affect network behavior, and there is mounting evidence for activity-dependent plasticity of intrinsic excitability in a variety of neurons (Marder & Goaillard 2006, Zhang & Linden 2003). Just as synaptic plasticity comes in several flavors and can be induced through a variety of signaling cascades, intrinsic plasticity also exhibits great diversity and can be either destabilizing or homeostatic. For example, the classic stimuli used to induce hippocampal synaptic plasticity also induce intrinsic plasticity, and these changes can either boost the effects of synaptic plasticity through local changes in dendritic excitability or serve a homeostatic function by regulating somatic spike generation (Fan et al. 2005, Frick et al. 2004, Narayanan et al. 2010). Similarly, both destabilizing and homeostatic forms of intrinsic plasticity have now been well-documented in neocortical neurons (Breton & Stuart 2009, Cudmore et al. 2010, Cudmore & Turrigiano 2004, Desai et al. 1999, Nataraj et al. 2010); furthermore, as for synaptic homeostatic mechanisms, there appear to be a variety of intrinsic plasticity processes that operate over distinct spatial and temporal scales to modulate neuronal activity (Daoudal & Debanne 2003, Zhang & Linden 2003). A fascinating and largely unanswered question is whether homeostatic forms of intrinsic and synaptic plasticity serve redundant or distinct functions within neuronal networks.

Many organisms live for decades, whereas the ion channels that subserve neuronal excitability turn over on a timescale of days to weeks. How, then, do neurons maintain stability in their intrinsic firing characteristics? An idea that has emerged over the past two decades is that neurons regulate their intrinsic excitability in a homeostatic manner by using some signal (such as intracellular calcium) to trigger changes in the balance of inward and outward currents (Marder & Goaillard 2006). This process has been beautifully documented in invertebrate central pattern generators, and both theoretical and experimental work has lent considerable support to the idea that these neurons can compensate intrinsically for changes in modulatory drive and so maintain their ability to fire in bursts (Golowasch et al. 1999, LeMasson et al. 1993, Marder & Prinz 2002, Turrigiano et al. 1994). In vertebrate neurons, the regulation of neuronal input/output curves serves as a gain-control mechanism underlying adaptive plasticity of the vestibulo-ocular reflex (Gittis & du Lac 2006, Nelson et al. 2005) and contributes to the activity-dependent development of the *Xenopus* retinotectal system (Aizenman et al. 2003, Pratt & Aizenman 2007). In cultured neocortical pyramidal neurons, the same activity-deprivation paradigm that leads to scaling up of synaptic strengths also enhances intrinsic excitability so that neurons fire more to the same synaptic input (Desai et al. 1999). As for invertebrate neurons, this process occurs

through the reciprocal regulation of inward and outward voltage-dependent currents (generally with no change in passive electrical properties), although the exact currents targeted by intrinsic plasticity depend on the identity and function of the targeted neuron (Breton & Stuart 2009, Desai et al. 1999, Nelson et al. 2005). It was recently shown that in addition to changes in inward and outward current densities, enhanced firing (by changes to calcium influx) can regulate the location of the axon initial segment (AIS) so that it moves further from the soma (Grubb & Burrone 2010). Conversely, auditory deprivation can increase the length of the AIS in auditory brain stem neurons (Kuba et al. 2010), suggesting that modifications in the location and size of the AIS may be a bidirectional and fairly general neuronal response to changes in activity. The exact contribution of these changes in AIS to neuronal excitability has not been determined, but they are predicted to alter firing threshold and so could play an important role in the homeostatic regulation of neuronal excitability.

Although the phenomenon of homeostatic intrinsic plasticity has now been widely documented, very little is known about the underlying induction and expression mechanisms. The cell biological processes that regulate the abundance and trafficking of glutamate and other neurotransmitter receptors are likely to apply to voltage-gated ion channels as well, but it remains unclear whether the same signaling pathways target voltage- and ligand-gated channels in parallel during homeostatic plasticity or whether synaptic and intrinsic plasticity are triggered by distinct signaling pathways. Understanding the mechanisms of intrinsic plasticity will be key for illuminating whether synaptic and intrinsic plasticity cooperate or compete during experience-dependent plasticity.

INTERPLAY BETWEEN SYNAPTIC AND INTRINSIC PLASTICITY: WHAT TO USE WHEN?

The discussion above highlights the intriguing point that neural circuits have a variety of homeostatic mechanisms to chose from: When faced with a destabilizing perturbation, they could respond by regulating inhibition, excitation, intrinsic excitability, or all the above. This raises the perplexing questions of why neurons should possess both synaptic and intrinsic homeostatic mechanisms, whether these two forms of homeostasis are simply redundant or subserve distinct functions, and whether they are generally induced in parallel or whether there is some kind of hierarchy that regulates when and where they are brought into play. Some insights into these questions are being generated by recent studies into the cellular mechanisms of experience-dependent plasticity within visual cortex.

Visual cortex has been used extensively as a model system for exploring the role of experience in refining cortical function, and much work has gone into determining the cellular plasticity mechanisms that underlie various phases of cortical circuit refinement, using a venerable sensory deprivation paradigm pioneered by Hubel and Wiesel decades ago (Hubel & Wiesel 1970, Wiesel & Hubel 1963). Although visual cortical plasticity was initially largely ascribed to Hebbian forms of synaptic plasticity, it has recently become clear that sensory deprivation (like activity deprivation in vitro) engages a rich array of plasticity mechanisms, including the entire cast of homeostatic characters identified above (Feldman 2009, Nelson & Turrigiano 2008). Moreover, these mechanisms are employed in a cell-type-specific, layer-specific, and developmentally regulated manner, making for a degree of complexity that can seem bewildering. For example, in rodents, the functional effects of visual deprivation and the underlying plasticity mechanisms are different during the pre-CP just after eye opening and during the classical visual system CP that begins a week later (Maffei & Turrigiano 2008a); I discuss the role of synaptic and intrinsic homeostatic plasticity during both developmental periods in turn below.

During the pre-CP, visual deprivation induces compensatory changes in local circuit excitability in layer 4 by scaling up excitatory synapses onto pyramidal neurons and reducing inhibition (Desai et al. 2002, Maffei et al. 2004). However, these synaptic changes are not accompanied by changes in intrinsic excitability, indicating that in layer 4 during the pre-CP synaptic changes are used preferentially to compensate for reduced sensory drive. This finding is in marked contrast with experiments on young visual cortical neurons in culture, where activity deprivation (using TTX) induces compensatory changes in synaptic strengths and intrinsic excitability in parallel (Desai 2003). This raises the possibility that in neocortical circuits synaptic homeostatic mechanisms might be used first, and intrinsic mechanisms might be engaged only when synaptic mechanisms do not suffice—for example, when firing is blocked in culture using TTX so that neurons cannot restore firing no matter how strongly they regulate synaptic strengths (Desai et al. 1999).

The functional effects of visual deprivation change abruptly at the beginning of the classical visual system CP and so do the expression patterns of various forms of plasticity within the microcircuits of V1. In rodents, the response to visual deprivation during the CP follows a biphasic time course: There is an initial loss of responsiveness to the deprived eye, followed more slowly (over several days) by a gain of responsiveness to both eyes. The net effect of these two processes is to shift the relative drive to the two eyes to favor the open eye [and so to shift ocular dominance (OD)] (Frenkel & Bear 2004, Kaneko et al. 2008, Mrsic-Flogel et al. 2007). The potentiation phase also occurs in monocular cortex or in response to binocular deprivation in the binocular cortex, indicating that it does not require competition between the two eyes and raising the possibility that it is induced by a homeostatic rather than a Hebbian mechanism (Kaneko et al. 2008, Mrsic-Flogel et al. 2007).

Manipulations that block synaptic scaling in vitro (namely, blocking TNFα or Arc signaling) also block the delayed potentiation during visual deprivation, raising the possibility that synaptic scaling underlies response potentiation during the CP (Kaneko et al.

2008, McCurry et al. 2010). However, there is currently little direct evidence for this hypothesis, and these manipulations likely interfere with other plasticity mechanisms as well as synaptic scaling. Furthermore, although synaptic scaling has not been examined in binocular cortex, in monocular cortex, lid suture (the method of deprivation generally used to induce OD plasticity) does not scale up mEPSC amplitude in layer 2/3 pyramidal neurons, although retinal blockade and dark-rearing do (Desai et al. 2002, Goel & Lee 2007, Maffei & Turrigiano 2008b). Instead, lid suture triggers an enhancement of intrinsic excitability (Maffei & Turrigiano 2008b), raising the possibility that intrinsic homeostatic plasticity rather than synaptic scaling might underlie delayed response potentiation. Resolving this issue will require further experiments in binocular cortex to examine the time course and induction requirements of these two forms of homeostatic plasticity. Why homeostatic compensation in layer 2/3 is delayed in response to lid suture is unclear. Also unanswered is why retinal activity block and lid suture induce different forms of homeostatic plasticity within layer 2/3. One possibility is that lid suture decorrelates sensory drive to cortex more than does blocking retinal activity (Linden et al. 2009), and this decorrelation drives very strong synaptic depression in layer 2/3 (Maffei & Turrigiano 2008b, Rittenhouse et al. 1999). It may be that this depression exceeds the ability of synaptic scaling to compensate, which then triggers intrinsic homeostatic plasticity. In this view, there is a hierarchy between synaptic scaling and intrinsic plasticity, with intrinsic plasticity coming into play only when synaptic mechanisms cannot fully compensate for perturbations in drive.

In hippocampal networks, activity deprivation can also activate both intrinsic and synaptic homeostasis, but the hierarchical relationship between them appears to be different from that suggested above by work in neocortex. In hippocampal slice cultures, for example, synaptic and intrinsic mechanisms are also dissociable, but intrinsic mechanisms seem to come into play first before synaptic

mechanisms kick in (Karmarkar & Buonomano 2006). Furthermore, activity blockade in vivo was observed to induce hyperexcitability in the CA1 region, which was accompanied by both synaptic and intrinsic changes in young, but only intrinsic changes in old, CA1 pyramidal neurons (Echegoyen et al. 2007). These studies demonstrate that in hippocampus, as in neocortex, a complex and developmentally regulated interplay occurs between these two forms of homeostatic plasticity. In contrast to visual cortex and hippocampus, the retinotectal system in *Xenopus* seems to use intrinsic mechanisms preferentially for homeostatic compensation. Reducing synaptic drive leads to enhanced intrinsic excitability, but not vice versa: When neuronal excitability is suppressed by overexpressing a K^+ channel, neurons respond by upregulating Na^+ currents to bring intrinsic excitability back up (Pratt & Aizenman 2007). Thus the rules by which intrinsic and synaptic mechanisms are applied to achieve homeostatic compensation seem to vary across systems according to rules we have not yet fathomed.

CONCLUDING REMARKS

Neural circuits utilize a complex mix of synaptic and intrinsic mechanisms to optimize their function and/or adapt to a changing environment (Feldman 2009, Nelson & Turrigiano 2008). These include an array of homeostatic mechanisms that allow circuits to regulate differentially the strength of excitation, inhibition, and intrinsic excitability. Functionally, intrinsic and synaptic plasticity well may not have identical effects on information transfer. Synaptic scaling adjusts the gain of the input, rather than the gain of the output, and does so in a manner that can reciprocally change excitation and inhibition, thus modifying the E/I balance. In contrast, intrinsic excitability modifies the contribution of a neuron to circuit function without changing synaptic currents, but the functional effects of this modulation will depend on how excitability is modified. For example, if intrinsic plasticity affects the slope of the neuronal input-output function without affecting threshold, then the neuron becomes more or less sensitive to both excitation and inhibition, and this form of gain control will not involve changes in the effective E/I ratio. In contrast, if excitability is regulated by sliding the input-output function left or right to modify the firing threshold, then the relative effectiveness of excitation and inhibition will be modified by intrinsic plasticity; moving the threshold left makes excitation more effective at firing the neuron and inhibition less able to prevent firing, and vice versa. One great challenge currently facing the field of cortical plasticity is to elucidate how the appropriate homeostatic mechanism is selected by particular firing patterns or activity levels to achieve the appropriate outcome for the circuit.

DISCLOSURE STATEMENT

The author is not aware of any affiliations, memberships, funding, or financial holdings that might be perceived as affecting the objectivity of this review.

LITERATURE CITED

Aizenman CD, Akerman CJ, Jensen KR, Cline HT. 2003. Visually driven regulation of intrinsic neuronal excitability improves stimulus detection in vivo. *Neuron* 39:831–42

Atallah BV, Scanziani M. 2009. Instantaneous modulation of gamma oscillation frequency by balancing excitation with inhibition. *Neuron* 62:566–77

Bartley AF, Huang ZJ, Huber KM, Gibson JR. 2008. Differential activity-dependent, homeostatic plasticity of two neocortical inhibitory circuits. *J. Neurophysiol.* 100:1983–94

Beattie EC, Stellwagen D, Morishita W, Bresnahan JC, Ha BK, et al. 2002. Control of synaptic strength by glial TNFalpha. *Science* 295:2282–85

Benevento LA, Bakkum BW, Cohen RS. 1995. gamma-aminobutyric acid and somatostatin immunoreactivity in the visual cortex of normal and dark-reared rats. *Brain Res.* 689:172–82

Breton JD, Stuart GJ. 2009. Loss of sensory input increases the intrinsic excitability of layer 5 pyramidal neurons in rat barrel cortex. *J. Physiol.* 587:5107–19

Burkhalter A. 2008. Many specialists for suppressing cortical excitation. *Front. Neurosci.* 2:155–67

Burrone J, O'Byrne M, Murthy VN. 2002. Multiple forms of synaptic plasticity triggered by selective suppression of activity in individual neurons. *Nature* 420:414–18

Chandrasekaran AR, Shah RD, Crair MC. 2007. Developmental homeostasis of mouse retinocollicular synapses. *J. Neurosci.* 27:1746–55

Chang MC, Park JM, Pelkey KA, Grabenstatter HL, Xu D, et al. 2010. Narp regulates homeostatic scaling of excitatory synapses on parvalbumin-expressing interneurons. *Nat. Neurosci.* 13:1090–97

Chowdhury S, Shepherd JD, Okuno H, Lyford G, Petralia RS, et al. 2006. Arc/Arg3.1 interacts with the endocytic machinery to regulate AMPA receptor trafficking. *Neuron* 52:445–59

Cingolani LA, Thalhammer A, Yu LM, Catalano M, Ramos T, et al. 2008. Activity-dependent regulation of synaptic AMPA receptor composition and abundance by beta3 integrins. *Neuron* 58:749–62

Cudmore RH, Fronzaroli-Molinieres L, Giraud P, Debanne D. 2010. Spike-time precision and network synchrony are controlled by the homeostatic regulation of the D-type potassium current. *J. Neurosci.* 30:12885–95

Cudmore RH, Turrigiano GG. 2004. Long-term potentiation of intrinsic excitability in LV visual cortical neurons. *J. Neurophysiol.* 92:341–48

Daoudal G, Debanne D. 2003. Long-term plasticity of intrinsic excitability: learning rules and mechanisms. *Learn. Mem.* 10:456–65

Davis GW. 2006. Homeostatic control of neural activity: from phenomenology to molecular design. *Annu. Rev. Neurosci.* 29:307–23

Davis GW, Bezprozvanny I. 2001. Maintaining the stability of neural function: a homeostatic hypothesis. *Annu. Rev. Physiol.* 63:847–69

Desai NS. 2003. Homeostatic plasticity in the CNS: synaptic and intrinsic forms. *J. Physiol. Paris* 97:391–402

Desai NS, Cudmore RH, Nelson SB, Turrigiano GG. 2002. Critical periods for experience-dependent synaptic scaling in visual cortex. *Nat. Neurosci.* 5:783–89

Desai NS, Rutherford LC, Turrigiano GG. 1999. Plasticity in the intrinsic excitability of cortical pyramidal neurons. *Nat. Neurosci.* 2:515–20

Echegoyen J, Neu A, Graber KD, Soltesz I. 2007. Homeostatic plasticity studied using in vivo hippocampal activity-blockade: synaptic scaling, intrinsic plasticity and age-dependence. *PLoS One* 2:e700

Fan Y, Fricker D, Brager DH, Chen X, Lu HC, et al. 2005. Activity-dependent decrease of excitability in rat hippocampal neurons through increases in I(h). *Nat. Neurosci.* 8:1542–51

Feldman DE. 2009. Synaptic mechanisms for plasticity in neocortex. *Annu. Rev. Neurosci.* 32:33–55

Frenkel MY, Bear MF. 2004. How monocular deprivation shifts ocular dominance in visual cortex of young mice. *Neuron* 44:917–23

Frick A, Magee J, Johnston D. 2004. LTP is accompanied by an enhanced local excitability of pyramidal neuron dendrites. *Nat. Neurosci.* 7:126–35

Gainey MA, Hurvitz-Wolff JR, Lambo ME, Turrigiano GG. 2009. Synaptic scaling requires the GluR2 subunit of the AMPA receptor. *J. Neurosci.* 29:6479–89

Gittis AH, du Lac S. 2006. Intrinsic and synaptic plasticity in the vestibular system. *Curr. Opin. Neurobiol.* 16:385–90

Goddard CA, Butts DA, Shatz CJ. 2007. Regulation of CNS synapses by neuronal MHC class I. *Proc. Natl. Acad. Sci. USA* 104:6828–33

Goel A, Lee HK. 2007. Persistence of experience-induced homeostatic synaptic plasticity through adulthood in superficial layers of mouse visual cortex. *J. Neurosci.* 27:6692–700

Golowasch J, Casey M, Abbott LF, Marder E. 1999. Network stability from activity-dependent regulation of neuronal conductances. *Neural. Comput.* 11:1079–96

Gonzalez-Islas C, Wenner P. 2006. Spontaneous network activity in the embryonic spinal cord regulates AMPAergic and GABAergic synaptic strength. *Neuron* 49:563–75

Goold CP, Nicoll RA. 2010. Single-cell optogenetic excitation drives homeostatic synaptic depression. *Neuron* 68:512–28

Grubb MS, Burrone J. 2010. Activity-dependent relocation of the axon initial segment fine-tunes neuronal excitability. *Nature* 465:1070–4

Haedo RJ, Golowasch J. 2006. Ionic mechanism underlying recovery of rhythmic activity in adult isolated neurons. *J. Neurophysiol.* 96:1860–76

Hartman KN, Pal SK, Burrone J, Murthy VN. 2006. Activity-dependent regulation of inhibitory synaptic transmission in hippocampal neurons. *Nat. Neurosci.* 9:642–49

Hendry SH, Huntsman MM, Vinuela A, Mohler H, de Blas AL, Jones EG. 1994. GABAA receptor subunit immunoreactivity in primate visual cortex: distribution in macaques and humans and regulation by visual input in adulthood. *J. Neurosci.* 14:2383–401

Hendry SH, Jones EG. 1986. Reduction in number of immunostained GABAergic neurones in deprived-eye dominance columns of monkey area 17. *Nature* 320:750–53

Hendry SH, Jones EG. 1988. Activity-dependent regulation of GABA expression in the visual cortex of adult monkeys. *Neuron* 1:701–12

Hubel DH, Wiesel TN. 1970. The period of susceptibility to the physiological effects of unilateral eye closure in kittens. *J. Physiol.* 206:419–36

Ibata K, Sun Q, Turrigiano GG. 2008. Rapid synaptic scaling induced by changes in postsynaptic firing. *Neuron* 57:819–26

Kaneko M, Stellwagen D, Malenka RC, Stryker MP. 2008. Tumor necrosis factor-alpha mediates one component of competitive, experience-dependent plasticity in developing visual cortex. *Neuron* 58:673–80

Karmarkar UR, Buonomano DV. 2006. Different forms of homeostatic plasticity are engaged with distinct temporal profiles. *Eur. J. Neurosci.* 23:1575–84

Kilman V, van Rossum MC, Turrigiano GG. 2002. Activity deprivation reduces miniature IPSC amplitude by decreasing the number of postsynaptic GABA(A) receptors clustered at neocortical synapses. *J. Neurosci.* 22:1328–37

Kim J, Alger BE. 2010. Reduction in endocannabinoid tone is a homeostatic mechanism for specific inhibitory synapses. *Nat. Neurosci.* 13:592–600

Kim J, Tsien RW. 2008. Synapse-specific adaptations to inactivity in hippocampal circuits achieve homeostatic gain control while dampening network reverberation. *Neuron* 58:925–37

Knogler LD, Liao M, Drapeau P. 2010. Synaptic scaling and the development of a motor network. *J. Neurosci.* 30:8871–81

Kuba H, Oichi Y, Ohmori H. 2010. Presynaptic activity regulates Na^+ channel distribution at the axon initial segment. *Nature* 465:1075–78

LeMasson G, Marder E, Abbott LF. 1993. Activity-dependent regulation of conductances in model neurons. *Science* 259:1915–17

Linden ML, Heynen AJ, Haslinger RH, Bear MF. 2009. Thalamic activity that drives visual cortical plasticity. *Nat. Neurosci.* 12:390–92

Maffei A, Nataraj K, Nelson SB, Turrigiano GG. 2006. Potentiation of cortical inhibition by visual deprivation. *Nature* 443:81–84

Maffei A, Nelson SB, Turrigiano GG. 2004. Selective reconfiguration of layer 4 visual cortical circuitry by visual deprivation. *Nat. Neurosci.* 7:1353–59

Maffei A, Turrigiano G. 2008a. The age of plasticity: developmental regulation of synaptic plasticity in neo-cortical microcircuits. *Prog. Brain Res.* 169:211–23

Maffei A, Turrigiano GG. 2008b. Multiple modes of network homeostasis in visual cortical layer 2/3. *J. Neurosci.* 28:4377–84

Malenka RC, Bear MF. 2004. LTP and LTD: an embarrassment of riches. *Neuron* 44:5–21

Marder E, Goaillard JM. 2006. Variability, compensation and homeostasis in neuron and network function. *Nat. Rev. Neurosci.* 7:563–74

Marder E, Prinz AA. 2002. Modeling stability in neuron and network function: the role of activity in homeostasis. *Bioessays* 24:1145–54

Marty S, Berzaghi Mda P, Berninger B. 1997. Neurotrophins and activity-dependent plasticity of cortical interneurons. *Trends Neurosci.* 20:198–202

McCurry CL, Shepherd JD, Tropea D, Wang KH, Bear MF, Sur M. 2010. Loss of Arc renders the visual cortex impervious to the effects of sensory experience or deprivation. *Nat. Neurosci.* 13:450–57

Mrsic-Flogel TD, Hofer SB, Ohki K, Reid RC, Bonhoeffer T, Hubener M. 2007. Homeostatic regulation of eye-specific responses in visual cortex during ocular dominance plasticity. *Neuron* 54:961–72

Narayanan R, Dougherty KJ, Johnston D. 2010. Calcium store depletion induces persistent perisomatic increases in the functional density of h channels in hippocampal pyramidal neurons. *Neuron* 68:921–35

Nataraj K, Le Roux N, Nahmani M, Lefort S, Turrigiano G. 2010. Visual deprivation suppresses L5 pyramidal neuron excitability by preventing the induction of intrinsic plasticity. *Neuron* 68:750–62

Nelson AB, Gittis AH, du Lac S. 2005. Decreases in CaMKII activity trigger persistent potentiation of intrinsic excitability in spontaneously firing vestibular nucleus neurons. *Neuron* 46:623–31

Nelson SB, Turrigiano GG. 2008. Strength through diversity. *Neuron* 60:477–82

O'Brien RJ, Kamboj S, Ehlers MD, Rosen KR, Fischbach GD, Huganir RL. 1998. Activity-dependent modulation of synaptic AMPA receptor accumulation. *Neuron* 21:1067–78

Pouille F, Marin-Burgin A, Adesnik H, Atallah BV, Scanziani M. 2009. Input normalization by global feedforward inhibition expands cortical dynamic range. *Nat. Neurosci.* 12:1577–85

Pratt KG, Aizenman CD. 2007. Homeostatic regulation of intrinsic excitability and synaptic transmission in a developing visual circuit. *J. Neurosci.* 27:8268–77

Rittenhouse CD, Shouval HZ, Paradiso MA, Bear MF. 1999. Monocular deprivation induces homosynaptic long-term depression in visual cortex. *Nature* 397:347–50

Rutherford LC, DeWan A, Lauer HM, Turrigiano GG. 1997. Brain-derived neurotrophic factor mediates the activity-dependent regulation of inhibition in neocortical cultures. *J. Neurosci.* 17:4527–35

Rutherford LC, Nelson SB, Turrigiano GG. 1998. BDNF has opposite effects on the quantal amplitude of pyramidal neuron and interneuron excitatory synapses. *Neuron* 21:521–30

Seeburg DP, Feliu-Mojer M, Gaiottino J, Pak DT, Sheng M. 2008. Critical role of CDK5 and Polo-like kinase 2 in homeostatic synaptic plasticity during elevated activity. *Neuron* 58:571–83

Shepherd JD, Rumbaugh G, Wu J, Chowdhury S, Plath N, et al. 2006. Arc/Arg3.1 mediates homeostatic synaptic scaling of AMPA receptors. *Neuron* 52:475–84

Shu Y, Hasenstaub A, McCormick DA. 2003. Turning on and off recurrent balanced cortical activity. *Nature* 423:288–93

Steinmetz CC, Turrigiano GG. 2010. Tumor necrosis factor-alpha signaling maintains the ability of cortical synapses to express synaptic scaling. *J. Neurosci.* 30:14685–90

Stellwagen D, Beattie EC, Seo JY, Malenka RC. 2005. Differential regulation of AMPA receptor and GABA receptor trafficking by tumor necrosis factor-alpha. *J. Neurosci.* 25:3219–28

Stellwagen D, Malenka RC. 2006. Synaptic scaling mediated by glial TNF-alpha. *Nature* 440:1054–59

Swanwick CC, Murthy NR, Kapur J. 2006. Activity-dependent scaling of GABAergic synapse strength is regulated by brain-derived neurotrophic factor. *Mol. Cell. Neurosci.* 31:481–92

Thiagarajan TC, Lindskog M, Malgaroli A, Tsien RW. 2007. LTP and adaptation to inactivity: overlapping mechanisms and implications for metaplasticity. *Neuropharmacology* 52:156–75

Thoby-Brisson M, Simmers J. 1998. Neuromodulatory inputs maintain expression of a lobster motor pattern-generating network in a modulation-dependent state: evidence from long-term decentralization in vitro. *J. Neurosci.* 18:2212–25

Turrigiano G, Abbott LF, Marder E. 1994. Activity-dependent changes in the intrinsic properties of cultured neurons. *Science* 264:974–77

Turrigiano G, LeMasson G, Marder E. 1995. Selective regulation of current densities underlies spontaneous changes in the activity of cultured neurons. *J. Neurosci.* 15:3640–52

Turrigiano GG. 2008. The self-tuning neuron: synaptic scaling of excitatory synapses. *Cell* 135:422–35

Turrigiano GG, Leslie KR, Desai NS, Rutherford LC, Nelson SB. 1998. Activity-dependent scaling of quantal amplitude in neocortical neurons. *Nature* 391:892–96

Turrigiano GG, Nelson SB. 2004. Homeostatic plasticity in the developing nervous system. *Nat. Rev. Neurosci.* 5:97–107

Vale C, Sanes DH. 2002. The effect of bilateral deafness on excitatory and inhibitory synaptic strength in the inferior colliculus. *Eur. J. Neurosci.* 16:2394–404

Watt AJ, van Rossum MC, MacLeod KM, Nelson SB, Turrigiano GG. 2000. Activity coregulates quantal AMPA and NMDA currents at neocortical synapses. *Neuron* 26:659–70

Wierenga CJ, Ibata K, Turrigiano GG. 2005. Postsynaptic expression of homeostatic plasticity at neocortical synapses. *J. Neurosci.* 25:2895–905

Wiesel TN, Hubel DH. 1963. Effects of visual deprivation on morphology and physiology of cells in the cats lateral geniculate body. *J. Neurophysiol.* 26:978–93

Yu LM, Goda Y. 2009. Dendritic signalling and homeostatic adaptation. *Curr. Opin. Neurobiol.* 19:327–35

Zhang W, Linden DJ. 2003. The other side of the engram: experience-driven changes in neuronal intrinsic excitability. *Nat. Rev. Neurosci.* 4:885–900

Reward, Addiction, Withdrawal to Nicotine

Mariella De Biasi and John A. Dani

Department of Neuroscience, Center on Addiction, Learning, Memory, Baylor College of Medicine, Houston, Texas 77030; email: jdani@bcm.edu; debiasi@bcm.edu

Annu. Rev. Neurosci. 2011. 34:105–30

First published online as a Review in Advance on March 25, 2011

The *Annual Review of Neuroscience* is online at neuro.annualreviews.org

This article's doi: 10.1146/annurev-neuro-061010-113734

Keywords

synaptic plasticity, dopamine, ventral tegmental area, nucleus accumbens, habenula, interpeduncular nucleus

Abstract

Nicotine is the principal addictive component that drives continued tobacco use despite users' knowledge of the harmful consequences. The initiation of addiction involves the mesocorticolimbic dopamine system, which contributes to the processing of rewarding sensory stimuli during the overall shaping of successful behaviors. Acting mainly through nicotinic receptors containing the $\alpha4$ and $\beta2$ subunits, often in combination with the $\alpha6$ subunit, nicotine increases the firing rate and the phasic bursts by midbrain dopamine neurons. Neuroadaptations arise during chronic exposure to nicotine, producing an altered brain condition that requires the continued presence of nicotine to be maintained. When nicotine is removed, a withdrawal syndrome develops. The expression of somatic withdrawal symptoms depends mainly on the $\alpha5$, $\alpha2$, and $\beta4$ (and likely $\alpha3$) nicotinic subunits involving the epithalamic habenular complex and its targets. Thus, nicotine taps into diverse neural systems and an array of nicotinic acetylcholine receptor (nAChR) subtypes to influence reward, addiction, and withdrawal.

Contents

INTRODUCTION AND SUMMARY

Tobacco use is the leading cause of preventable death in developed countries (Benowitz 2008, Mathers & Loncar 2006, Peto et al. 1996), and nicotine is the main addictive component (Benowitz 2009, Dani et al. 2009, Dani & Heinemann 1996, Mansvelder & McGehee 2002). Nicotine is a tertiary amine alkaloid. In its charged form, nicotine binds to diverse nicotinic acetylcholine receptor (nAChR) subtypes that have unique expression patterns in the central nervous system. Like the cholinergic neurotransmitter systems, nAChRs are widely distributed, and they participate in cholinergic signaling in nearly every neural area (Changeux & Edelstein 2005, Dani & Bertrand 2007, Woolf 1991). In its uncharged form, nicotine is membrane permeable. Therefore, nicotine can influence intracellular processes indirectly via nAChRs and directly by entering the cytoplasm (Lester et al. 2009, Rezvani et al. 2007).

nAChR: nicotinic acetylcholine receptor

DA: dopamine

VTA: ventral tegmental area

PFC: prefrontal cortex

NAc: nucleus accumbens

Nicotine initiates addiction by impinging directly on neural circuitry that normally reinforces behaviors that lead to rewarding goals. Since the earliest studies of intracranial electrical self-stimulation, investigators have identified cortical and limbic brain structures as mediating reward (Olds 1958, Olds & Milner 1954). In particular, the mesocorticolimbic dopamine (DA) system plays an important role in self-stimulation and in processing environmental reward. Likewise, this DA system plays a critical role in the acquisition of behaviors that are inappropriately reinforced by psychostimulant drugs, including nicotine (Balfour 2004, Corrigall 1999, Dani et al. 2009, Dani & Heinemann 1996, Di Chiara 2000, Mansvelder & McGehee 2002). One important dopaminergic pathway originates in the ventral tegmental area (VTA) of the midbrain and projects to the prefrontal cortex (PFC) as well as to limbic and striatal structures, including the nucleus accumbens (NAc). Acting mainly via midbrain nAChRs composed of the $\beta 2$ subunit in combination with the $\alpha 4$ and/or $\alpha 6$ subunits, nicotine increases the firing rate and phasic burst activity of midbrain DA neurons and elevates DA in the PFC, NAc, and other targets (Imperato et al. 1986, Grenhoff et al. 1986, Mameli-Engvall et al. 2006, Pons et al. 2008, Zhang et al. 2009b). Those decisive mechanistic actions underlie nicotine's ability to enhance reward from brain stimulation, reinforce conditioned place preference, and support self-administration (Corrigall 1999, Dani et al. 2009, Dani & Harris 2005, Mameli-Engvall et al. 2006, Picciotto et al. 1998, Stolerman & Shoaib 1991).

After chronic use of nicotine, removal of nicotine produces a withdrawal syndrome that can be relieved by nicotine replacement therapy (Dani et al. 2009, Dani & Heinemann 1996, De Biasi & Salas 2008, Di Chiara 2000, Mansvelder & McGehee 2002, Stolerman & Shoaib 1991). The withdrawal syndrome is not mediated exactly by the same mechanisms or by the same neural circuits that initiate addiction or dependency. The epithalamic habenular complex and its targets appear to be critical for the withdrawal syndrome. The medial habenula

(MHb) and one of its primary targets, the interpeduncular nucleus (IPN), richly express β4 and α5 nAChR subunits (and α2 only in the rodent IPN), which are necessary for the neuroadaptations that lead to somatic withdrawal symptoms during nicotine abstinence (Salas et al. 2004b, 2009). Recent research shows that nicotine-induced reward, addiction, and withdrawal involve a wide range of nAChR subtypes expressed in diverse neural systems.

REWARD AND INITIATION OF ADDICTION

Rewarding stimuli promote learning of goal-directed behaviors, generate positive emotions, and subsequently stimulate repetition of those learned behaviors (Schultz 2010, Thorndike 1911). Intrinsic neural responses to reward evolved to ensure successful behaviors that perpetuate the genetic material of the individual and its species. Therefore, reward mechanisms are observed across species, from insects to primates (Hodos 1961, Lau et al. 2006, McClure et al. 2007, Qin & Wheeler 2007, Wise 2006). For humans, rewards can be more complex and abstract than food and sex and can take the form of monetary, aesthetic, cognitive, or social stimuli (Montague et al. 2006, O'Doherty et al. 2001). The effectiveness or value of rewards can be estimated by measuring the effort a subject expends on behaviors that achieve the rewarded goal (Hodos 1961, Schultz 2006).

The overall circuitry that mediates reward is broad and diverse, but particular fiber tracts and nuclei are now recognized as being especially important. The study of the brain reward circuitry was spearheaded in the 1950s by Olds and Milner (Olds 1958, Olds & Milner 1954), who showed that rats work incessantly to self-administer intracranial electrical stimulation to certain regions of the brain (Conover et al. 1994, Routtenberg & Lindy 1965, Spies 1965). Early investigations showed that the midbrain dopamine (DA) systems have an important role in the response to salient stimuli and to brain self-stimulation (Phillips & Olds 1969; Fibiger 1978, Wise 1978). Self-stimulation of the medial forebrain bundle facilitates DA release (Garris et al. 1999, Gratton et al. 1988, Phillips et al. 1989), and DA receptor antagonists or DA neuron lesions inhibit brain self-stimulation (Fouriezos & Wise 1976, Lippa et al. 1973, Stellar & Corbett 1989). Those early studies emphasized the importance of the mesocorticolimbic DA pathways during reward-motivated behaviors. The DA efferents originating from the midbrain VTA and particularly targeting the PFC and NAc became recognized as paramount neural structures shaping reward-related behavior (Schultz et al. 1997; Wise 2004, 2009; Wise & Rompre 1989). However, the role of the DA projections from the substantia nigra compacta (SNc) and wide-ranging DA targets have important, but less well-defined, roles (e.g., see Packard & Knowlton 2002).

The midbrain DA area expresses diverse nAChR subtypes (Grady et al. 2010, Klink et al. 2001, Pidoplichko et al. 1997, Woolltorton et al. 2003) and receives afferent cholinergic innervation from the nearby pedunculopontine tegmentum (PPT) and the laterodorsal tegmentum (LDT), which are a loose collection of cholinergic neurons interspersed with γ-aminobutyric acid (GABA)ergic and glutamatergic neurons (Omelchenko & Sesack 2005). Although the VTA receives a strong excitatory glutamate input from the PFC, that excitation is mainly onto DA neurons that project back to the cortex, not to the NAc (Carr & Sesack 2000). Rather, the PPT/LDT neurons provide potent glutamatergic excitation to the DA neurons projecting to the NAc (Omelchenko & Sesack 2005). In addition, the PPT/LDT cholinergic afferents have important endogenous influences over DA neurons by acting via nAChRs (Maskos 2008, 2010). The PPT/LDT cholinergic afferents boost glutamate afferent transmission via presynaptic nAChRs (Pidoplichko et al. 2004, Schilstrom et al. 2000) and provide some excitatory drive to GABA projection neurons and interneurons, mainly via β2-containing (β2*) nAChRs (Mansvelder et al. 2002, Pidoplichko et al. 2004, Pons et al. 2008). The PPT/LDT contribute to events associated

IPN: interpeduncular nucleus

SNc: substantia nigra compacta

PPT: pedunculopontine tegmentum

LDT: laterodorsal tegmentum

GABA: γ-aminobutyric acid

with drug taking (Picciotto & Corrigall 2002), as exemplified by lesions in the PPT reducing nicotine self-administration (Lanca et al. 2000). This influence of the PPT/LDT arises, at least in part, because these areas contribute to the phasic burst firing of DA neurons (Floresco et al. 2003, Lodge & Grace 2006).

In the midbrain, nAChRs comprising the β2 nAChR subunit in combination with α4 and/or α6 subunits are essential for nicotine-induced reinforcement and the DA signal (Drenan et al. 2010, Mameli-Engvall et al. 2006, Picciotto et al. 1998, Pons et al. 2008, Tapper et al. 2004). Nicotine directly activates DA neurons via those and other nAChR subtypes (Pidoplichko et al. 1997), and nicotine modifies the firing modes and firing frequency of DA neurons through excitatory and inhibitory inputs and synaptic plasticity (Mameli-Engvall et al. 2006; Schilstrom et al. 2000, 2003; Zhang et al. 2009b).

DA neurons display a number of different firing modes that for convenience will be grouped into two different contributions: tonic and phasic. Tonic firing is at a relatively low frequency (roughly 4 Hz) that generally consists of individual action potentials. Phasic firing is usually at a higher average frequency, and it includes high-frequency burst discharges (roughly 20-80 Hz) (Grace et al. 2007; Grace & Bunney 1983, 1984; Hyland et al. 2002; Robinson et al. 2004). Although phasic bursts can be rather short, sometimes containing only one or two extra spikes, they are thought to convey or differentiate behaviorally relevant information. For example, phasic burst firing is induced by unpredicted reward or unanticipated cues that have been conditioned to a known reward (Schultz 2007b, Schultz et al. 1997). Disrupting those bursts impairs conditioned behavioral responses, diminishes the ability to learn cues about reward and salience, and disrupts the overall processing of reward (Schultz 2007b). Phasic burst firing by DA neurons and the consequent downstream phasic DA signals are vital for processing reward-related information and, thus, are important for the initial phase of addiction.

While increasing the firing of DA neurons, nicotine potently increases the number and length of phasic bursts that mediate the processing of reward (**Figure 1**) (Grenhoff et al. 1986, Mameli-Engvall et al. 2006, Zhang et al. 2009b). The phasic burst firing induced by nicotine causes greater extracellular DA release compared with tonic, single-spike firing (Floresco et al. 2003, Gonon 1988, Grace 1991, Zhang et al. 2009b). This burst firing mode normally mediates the phasic behavioral response to salient environmental stimuli (Maskos et al. 2005, Schultz et al. 1997). Nicotine, however, acts directly on this circuitry, as if a reward-related sensory input has been received. In mice lacking the β2 nAChR subunit (β2 -/-), nicotine does not induce phasic bursts by DA neurons (Mameli-Engvall et al. 2006), nicotine does not induce DA release, and nicotine does not support self-administration (Picciotto et al. 1998).

The irrefutable role of β2* nAChRs in the reinforcing properties of nicotine is reflected in the analysis of nAChR null mice in combination with nAChR subunit re-expression with lentiviral techniques (Mameli-Engvall et al. 2006, Pons et al. 2008). Nicotine self-administration is reinstated by re-expression of the β2 subunit in the VTA but not in the neighboring SNc (Avale et al. 2008), confirming the specific role of the VTA in reward. Nicotine self-administration is also reinstated by the re-expression of either the α4 or the α6 nAChR subunit, consistent with the expression of α4β2 and α4α6β2 nAChRs in the VTA (Pons et al. 2008). Thus, β2* nAChRs in combinations with both α4 and α6 subunits are important for the reinforcing properties of nicotine, and the α4 and α6 subunits cannot completely substitute for each other even though they are abundantly expressed in VTA neurons (Champtiaux et al. 2003, Salminen et al. 2007).

Interestingly, local re-expression of α7 in the VTA did not reinstate self-administration in α7 -/- mice (Pons et al. 2008), consistent with the pharmacological evidence indicating that the α7 subunit does not influence chronic nicotine self-administration in

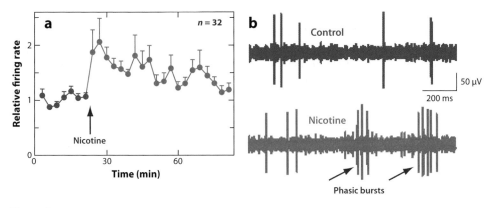

Figure 1

Nicotine increases the action potential firing rate and phasic burst firing of putative midbrain dopamine (DA) neurons in freely moving rats. (*a*) The normalized average firing rate of DA neurons in response to administration of nicotine (0.4–0.5 mg/kg, i.p., free-base equivalent). (*b*) In vivo, chronic tetrode recording from a putative DA neuron, indicating that the DA neuron fires more phasic bursts (*gray arrows*) after nicotine administration (*red trace*) than before (*blue control trace*). Adapted from Zhang et al. (2009b).

trained rats (Grottick & Higgins 2000) nor nicotine-induced conditioned place preference (Stolerman et al. 2004, Walters et al. 2006). The roles of $\alpha 7^*$ nAChRs in nicotine addiction may be more subtle and widely distributed in the CNS, and these receptors may work appropriately only in combination with other nAChR subtypes. For instance, the $\alpha 7$ nAChRs seem to fine-tune nicotine-induced DA neuron firing only after $\beta 2^*$ nAChRs have been activated (Mameli-Engvall et al. 2006).

DOWNSTREAM FROM DOPAMINE NEURON PHASIC BURSTS

If phasic burst firing by DA neurons is critical for nicotine self-administration, what are the downstream consequences of those bursts? The targets innervated by DA projections are broad and diverse, and their influences on reward-related behaviors have focused attention onto limbic and cortical sites, including the ventral and dorsal striatum, the amygdala, the hippocampus, the PFC, and the orbitofrontal cortex. Those DA projection target areas are involved in reinforcement (ventral striatum), learning and declarative memory (hippocampus), emotional memory (amygdala), and habit

formation (dorsal striatum), as well as executive functions and working memory (PFC and orbitofrontal cortex) (Schultz 2007b, Laviolette 2007, Seamans & Yang 2004). Therefore, the dopaminergic system broadly influences neural processing that underlies reward-based, as well as other, mechanisms for memory and behavior. The NAc of the ventral striatum has garnered special interest in processing reward, beginning with the early intracranial self-stimulation experiments. Electrodes placed along fibers leading to the NAc potently mediate electrical self-stimulation (Corbett & Wise 1979, Olds & Olds 1969).

The NAc and portions of the olfactory tubercle comprise the ventral striatum, which is mainly limbic related. It receives extensive excitatory innervation from the PFC, the hippocampus, and the amygdala (Haber et al. 2000, Heimer 2000, Pennartz et al. 1994). The primary DA innervation arises from the VTA and, to a much lesser degree, from the SNc (van Domburg & ten Donkelaar 1991). Among its functions, the NAc serves as the limbic-motor or motivation-action interface, and those functions are part of the NAc participation in reward-based learning and addiction (Haber et al. 2000, Mogenson et al. 1980). Addictive drugs (including nicotine) increase the basal DA concentration in the NAc shell as measured by microdialysis, but that background increase is smaller or not detected in the dorsolateral striatum (Di Chiara 1999, Pidoplichko et al. 2004, Pontieri et al. 1996, Zhang et al. 2009b). This increase in basal or background DA measured by microdialysis in the NAc shell is considered indicative of a drug's addictive influence.

Nicotine, at the concentration obtained from tobacco, induces phasic bursts from VTA/SNc DA neurons (**Figure 1b**) that innervate much of the dorsal to ventral extent of the striatum (Grenhoff et al. 1986, Pidoplichko et al. 1997, Schilstrom et al. 2003, Zhang et al. 2009b), suggesting that the DA signal should be comparable throughout the striatum. However, nicotinic and other local striatal mechanisms regulate the frequency dependence of DA release (Cragg 2003, Rice & Cragg 2004, Zhang et al. 2009a, Zhang & Sulzer 2004, Zhou et al. 2001). Nicotinic receptors are expressed on DA neuron fibers and terminals, and the activity of those nAChRs, particularly containing the β2, α4, and α6 subunits (Pons et al. 2008), is highly regulated by ongoing striatal cholinergic

interneuron discharges and dense acetylcholine esterase activity (Zhou et al. 2001, 2003).

Although only ~2% of the striatal neurons are cholinergic, they have extremely dense axonal arbors that provide denser cholinergic innervation than is seen anywhere else in the brain (Butcher & Woolf 1984, Zhou et al. 2001, 2002). The background firing rate of striatal cholinergic neurons is ~6 Hz, providing ongoing acetylcholine (ACh) release via direct synaptic transmission and volume transmission (Dani & Bertrand 2007, de Rover et al. 2002, Koos & Tepper 2002, Wilson 2004). The dense striatal acetylcholine esterase rapidly hydrolyzes the released ACh, preventing the appearance of long-lasting ACh signals that would strongly desensitize the high-affinity subtypes of nAChRs (Dani et al. 2000, Grady et al. 1994, Wooltorton et al. 2003). This ongoing ACh release and breakdown produce a background of cholinergic activity in the striatum that has a number of functions, including regulation of DA release (Exley & Cragg 2008, Zhang et al. 2009a, Zhang & Sulzer 2004).

The frequency dependence and nicotinic modulation of DA release differ in the dorsolateral striatum and NAc shell (Chergui et al. 1994; Cragg 2003; Cragg et al. 2002; Zhang et al. 2009a,b). Direct measures of DA release using fast-scan cyclic voltammetry indicate that the transfer function relating DA neuron activity and DA release differs in those (and likely other) locations (**Figure 2**) (Cragg 2003, Rice & Cragg 2004, Zhang et al. 2009a, Zhang & Sulzer 2004, Zhou et al. 2001). Presynaptic or axonal nAChRs enhance DA release evoked at low firing frequencies (Zhou et al. 2001). The ongoing activity of striatal cholinergic interneurons (Bennett & Wilson 1999) excites presynaptic nAChRs on the DA terminals (Jones et al. 2001), producing an increase of intraterminal calcium that enhances DA release (Grady et al. 1997, Wonnacott et al. 2000). However, when nicotine is added to striatal brain slices, a significant proportion of presynaptic and axonal nAChRs is desensitized. As a consequence, low-frequency, tonic DA release is inhibited (Rice & Cragg 2004, Zhang &

ACh: acetylcholine

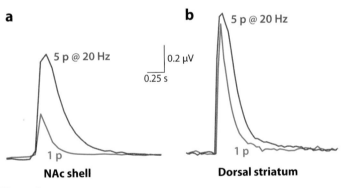

Figure 2

Different dopamine (DA) release properties in the dorsal striatum and NAc shell. DA release was evoked by a single electrical stimulus pulse (1p) or by a stimulus train of 5 pulses delivered at 20 Hz (5p @ 20Hz) to brain slices. The traces represent DA concentration measured in brain slices by carbon-fiber voltammetry. Example measurements are shown of the DA signal evoked by 1p (*green traces*) or by 5p at 20 Hz (*gray traces*). In the NAc shell, the DA signal evoked by 1p is much smaller than that evoked by 5p (as measured by the area under the curve). In the dorsal striatum, the DA signal evoked by 1p is only slightly smaller than that evoked by a 5p train. Adapted from data within Zhang et al. (2009b).

Sulzer 2004, Zhou et al. 2001), but phasic DA release is regulated differently, such that phasic bursts induce large DA signals (Cragg 2003; Rice & Cragg 2004; Zhang et al. 2009a,b).

The difference in DA release between the NAc shell and the dorsal striatum is exemplified in **Figure 2**, which shows DA-concentration traces arising from electrical stimulation applied to DA afferents (Zhang et al. 2009a,b). In the NAc shell, phasic burst discharges produce a greater increase in DA release than in the dorsal striatum, where the DA release from a single-pulse stimulus is comparable to the release from a phasic-burst stimulus. These results arise, in part, because the probability of DA release to a single-action potential is greater in the dorsal striatum than the NAc shell (Zhang et al. 2009a,b; Zhang & Sulzer 2004). In the dorsal striatum, DA release probability is already high in response to a single-tonic stimulus. Therefore, a phasic burst is not able to increase DA release much further. On the other hand, the NAc shell has a low probability of release in response to a single-tonic stimulus and, thus, has further to increase DA release in response to a phasic burst.

More telling was the experiment examining the effect of nicotine over DA release (**Figure 3**). A stimulus train representing in vivo DA neuron firing in the absence or presence of nicotine evokes DA release differently from a dorsal striatal slice (**Figure 3a**) and from a NAc shell slice (**Figure 3b**). The stimulus trains are shown below the DA-concentration traces. With identical DA neuron bursting, nicotine increases the average DA release more in the NAc shell than in the dorsal striatum (**Figure 3c**). The details of the DA signaling change in both anatomical areas, but as judged by the area under the curves (**Figure 3a,b**), the data suggest that the background DA concentration would increase in the NAc shell and not the dorsal striatum (Zhang et al. 2009b). These results are consistent with the direct measurements of background DA concentrations made by microdialysis (Zhang et al. 2009b).

Measured in vivo from the VTA, nicotine administration to a rat causes (some) DA neurons to fire at a higher rate and to fire more and longer bursts (**Figure 1**). The downstream DA signaling is altered, but the transfer function from DA neuron spikes to DA release is dependent on the target and time because regulatory mechanisms change with local conditions (Zhang et al. 2009a). DA axon terminals in the dorsolateral striatum and NAc shell respond differently to incoming action potentials,

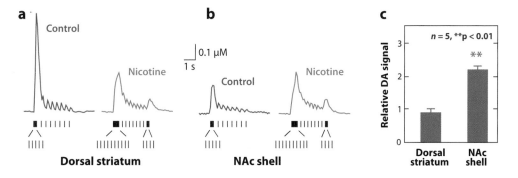

Figure 3

The burst firing by dopamine (DA) neurons measured in vivo that is induced by nicotine causes a greater increase of DA release in the NAc shell than in the dorsal striatum. The traces represent DA concentration measured in brain slices by carbon-fiber voltammetry. Electrical stimulus patterns used to evoke DA release were designed to mimic the in vivo firing patterns (*vertical tick marks below traces*) of DA neurons measured from freely moving rats before (control) and after nicotine injections. Patterned stimulus trains based on the in vivo DA-unit recordings are shown below the evoked DA release in the absence (control) or presence of nicotine in the dorsal striatum (*a*) or the NAc shell (*b*). (*c*) The relative DA signal (calculated as the area under the curve) was unchanged by nicotine in the dorsal striatum but was increased in the NAc shell. Adapted from Zhang et al. (2009b).

consistent with phasic bursts inducing a greater increase of DA release in the NAc shell (Cragg 2003; Rice & Cragg 2004; Zhang et al. 2009a,b; Zhang & Sulzer 2004), which has a more important role in mediating the rewarding properties of nicotine that drive the early phases of addiction. Given the properties and topology of DA fiber tracts, it is likely that DA release in cortical areas commonly responds to DA firing frequency more as is seen in the NAc shell rather than as is seen in the dorsal striatum.

Despite these potent effects of nicotine in the striatum, it is the nAChRs and nicotinic mechanisms within and impinging on the midbrain DA centers that are required for the addictive drive of nicotine (Mameli-Engvall et al. 2006, Maskos et al. 2005, Pons et al. 2008). Synaptic plasticity within the midbrain and onto DA neurons is an important early step in the nicotine addiction process (Dani et al. 2001, Mansvelder et al. 2002, Mansvelder & McGehee 2000, Pidoplichko et al. 2004). The extensive cholinergic innervation and nicotinic effects throughout the striatum suggest, however, that the direct and indirect actions of nicotine on the striatal (and other) targets also serve underappreciated roles in the overall nicotine addiction process.

AVERSIVE SENSORY SIGNALING IN THE MESOCORTICOLIMBIC SYSTEM

Nicotine in both humans and animals follows an inverted U-shaped dose-response curve, with aversive effects coming into play at higher nicotine concentrations (Ashton et al. 1980, Fattore et al. 2002, Herskovic et al. 1986, Paterson et al. 2003). Aside from effects on the peripheral nervous system, this dose-response relationship arises from nicotine acting on a heterogeneous population of nAChR subtypes in different neuronal circuits that mediate rewarding and aversive effects. Smokers titrate their nicotine intake to experience the rewarding while avoiding the aversive actions (Benowitz 2001, Benowitz & Jacob 1985, Dani et al. 2009,

Hutchison & Riley 2008, Kassel et al. 2007, Simpson & Riley 2005, Wise et al. 1976).

Although investigators originally postulated that the DA released from the VTA mediates only the hedonic or pleasurable effects of natural reinforcers and drugs in the NAc, evidence indicates that DA can signal events of opposite hedonic valence based on novelty or salience. Reward omission induces phasic inhibition of VTA neurons (Schultz 2007a,b), and aversive stimuli excite ventral (Brischoux et al. 2009) and inhibit dorsal (Ungless et al. 2004) VTA neurons. VTA inhibition has been proposed to signal the motivational value of unexpected, aversive events whereas VTA excitation is thought to signal salience to help respond to environmental changes (Bromberg-Martin et al. 2010). In the NAc of rats, intraoral sucrose (i.e., rewarding) infusion increased DA levels, whereas quinine (i.e., aversive) infusion decreased DA levels (Roitman et al. 2008). Owing to afferents mediating sensory input, NAc neurons decrease their firing to appetitive stimuli and increase their firing to aversive stimuli (Roitman et al. 2005). Because the activity of DA neurons and the resulting DA release on its targets depend on the balance between excitatory and inhibitory inputs (Marinelli et al. 2006), understanding how DA neurons are inhibited provides insight into behavior and the complex effects of nicotine.

Aversive, negative sensory input is processed by the habenular complex, an epithalamic structure involved in fear, anxiety, depression, and stress (Geisler et al. 2008, Ikemoto 2010, Winter et al. 2010). The habenula is divided into the MHb and the two divisions of the lateral nucleus (LHb). It receives massive afferents from the medial frontal cortex, the NAc, the olfactory bulb, the septum, and the striatum via the stria medullaris thalami and sends projections to the IPN, the VTA, the SNc, the medial raphe complex, the locus coeruleus, and the periaqueductal gray (Geisler et al. 2008; Greatrex & Phillipson 1982; Herkenham & Nauta 1977, 1979; Kim & Chang 2005; Scheibel 1997; Sutherland 1982). Whereas the MHb receives inputs primarily from the limbic system and

LHb: lateral habenula

sends outputs mainly to the IPN, the LHb receives inputs primarily from the basal ganglia and sends outputs mainly to dopaminergic and serotonergic neurons (Hikosaka 2010).

An aversive sensory input administered to monkeys elicits LHb activity that indirectly inhibits DA neurons (Hikosaka et al. 2008; Ikemoto 2010; Matsumoto & Hikosaka 2007, 2009). Functional MRI studies in humans are consistent with those results. A human decision-making error that causes a negative sensory feedback or the absence of an expected sensory reward produces strong activity within the habenular complex (Salas et al. 2010, Shepard et al. 2006, Ullsperger & von Cramon 2003). The firing patterns of glutamatergic LHb neurons negatively correlate with the firing patterns of DA neurons. For example, LHb neurons increase their firing in the absence of predicted reward and decrease their firing upon delivery of reward, which is opposite from the usual firing of DA neurons (Hikosaka et al. 2008, Matsumoto & Hikosaka 2007). Furthermore, electrical stimulation of the LHb inhibits most DA neurons (Christoph et al. 1986, Ji & Shepard 2007, Matsumoto & Hikosaka 2007). In contrast, habenular lesions increase DA turnover in the NAc and PFC, reflecting an activation of the dopaminergic system (Lisoprawski et al. 1980, Nishikawa et al. 1986).

Recent studies suggest that the LHb, rather than projecting directly to the VTA, exerts its effects through the stimulation of neurons in the newly characterized rostromedial tegmental nucleus (RMTg), a region at the tail of the VTA (Perrotti et al. 2005). The RMTg sends GABAergic projections to the VTA and SNc and inhibits DA cells (Ikemoto 2010, Jhou et al. 2009a). As in the LHb, RMTg neurons are phasically activated by aversive stimuli and inhibited by natural rewards such as food or reward-predictive stimuli, and conversely, RMTg neurons are stimulated by aversive stimuli (Jhou et al. 2009b). RMTg neurons are also modulated by drugs of abuse, such as cocaine and methamphetamine. Nicotine increases the RMTg neuron activity (Lecca et al. 2010). Because physiological doses of nicotine commonly

stimulate DA neurons (Grenhoff et al. 1986, Mameli-Engvall et al. 2006, Pidoplichko et al. 1997, Zhang et al. 2009b), nicotine-induced RMTg activity may contribute to the inverted U-shaped dose-response curve. Furthermore, a transient inhibition of some DA neurons can be detected before the longer-lasting, nicotine-induced excitation (Erhardt et al. 2002).

The MHb densely expresses nAChRs and receives rich cholinergic innervation, but its role in regulating catecholamine transmission is not known. Although projections have been identified from the MHb to the LHb (Kim & Chang 2005), the MHb sends most of its projections to the IPN (Klemm 2004). The IPN, in turn, sends projections to the serotonergic raphé nuclei and to the dopaminergic VTA (Groenewegen et al. 1986, Klemm 2004, Montone et al. 1988). Therefore, the MHb, at least, influences monoaminergic transmission via its connections to the IPN.

NICOTINE-INDUCED NEUROADAPTATIONS

The addiction process produces cellular adaptations that influence widely distributed neural circuits, particularly including the mesocorticolimbic system (Koob & Le Moal 2001). Acting via distinct anatomical distributions and cellular locations of nAChR subtypes, nicotine produces neuroadaptations through its influence over intracellular pathways (Wonnacott et al. 2005). Diverse nAChR subtypes are expressed on DA, GABA, and Glu neurons of the midbrain, and nAChRs are expressed on afferent projections arriving from diverse areas, including the cortex, the PPT/LDT, and the NAc (Kalivas 1993, Steffensen et al. 1998, Walaas & Fonnum 1980). Each of these neuronal subpopulations expresses nAChRs of differing subunit compositions and in differing cellular localizations (Calabresi et al. 1989, Klink et al. 2001, Wooltorton et al. 2003).

One of the most direct and important neuroadaptations induced by chronic nicotine use is the subtype-specific upregulation of nAChRs (Buisson & Bertrand 2001; Changeux et al.

1984; Dani & Heinemann 1996; Flores et al. 1997; Mansvelder et al. 2002; Marks et al. 1983, 1992; Perry et al. 1999; Rezvani et al. 2007, 2009; Rowell & Wonnacott 1990; Schwartz & Kellar 1983). The mechanisms invoked for nAChR upregulation are multiple and likely act in parallel. Because nicotine is not hydrolyzed by acetylcholine esterase, as ACh is, nicotine's long-lasting presence favors excessive nAChR desensitization. The homeostatic response to desensitized (turned-off) receptors is upregulation (Dani & Heinemann 1996, Fenster et al. 1999, Picciotto et al. 2008). Other mechanisms that likely contribute to upregulation are decreased surface receptor turnover (Peng et al. 1997) and increased receptor assembly at the endoplasmic reticulum (Corringer et al. 2006; Darsow et al. 2005; Rezvani et al. 2007, 2010; Sallette et al. 2005). Isomerization of surface nAChRs to high-affinity nicotinic sites has also been proposed (Buisson & Bertrand 2002, Vallejo et al. 2005).

Nicotinic receptor upregulation differs among the diverse subtypes, varies among brain regions for the same nAChR subtype, and depends on the contingency of nicotine administration (Gaimarri et al. 2007, Marks et al. 1983, McCallum et al. 2006, Metaxas et al. 2010, Nguyen et al. 2004, Pauly et al. 1996, Schwartz & Kellar 1983). Because nicotinic mechanisms are distributed throughout the brain, changes in nAChR subtype expression densities could have influences over diverse neuronal circuitry.

In parallel with and, in part, arising from nAChR upregulation, nicotine also produces heterologous neuroadaptations. Nicotine-induced synaptic plasticity increases the α-amino-3-hydroxy-5-methyl-4-isoxazole propionic acid/N-methyl-D-aspartate current ratio of glutamate receptors (GluRs) at locations mediating drug-associated memory (Dani et al. 2001, Kauer & Malenka 2007, Tang & Dani 2009), including at DA neurons (Gao et al. 2010, Placzek et al. 2009, Saal et al. 2003). Nicotine exposure also increases high-affinity DA D2 receptors in the NAc (Novak et al. 2010), an effect, also produced by cocaine, that causes DA hypersensitivity (Briand et al.

2008, Grigoriadis & Seeman 1986). Beyond the well-studied events of synaptic plasticity, chronic nicotine exposure leads to further remodeling of synapses, producing changes in scaffolding proteins, such as PSD95 and Shank (Hwang & Li 2006, Rezvani et al. 2007). These changes derive, at least in part, from nicotine-induced changes in protein turnover and from partial inhibition of proteasomal function (Rezvani et al. 2007). Because proteasomes are a fundamental component of the protein degradation machinery, nicotine entrance into the cytoplasm that influences proteasomes has far-reaching effects even on cells that do not express nAChRs (Rezvani et al. 2007).

Chronic nicotine exposure also has indirect effects on motivational systems by altering synthesis and release of opioid peptides in a time-dependent and peptide-specific fashion (Berrendero et al. 2010). The endogenous opioid system influences positive and negative motivational and affective states (Steiner & Gerfen 1998) and participates in the behavioral effects of several addictive drugs, including nicotine (Berrendero et al. 2005, Hadjiconstantinou & Neff 2010, Trigo et al. 2009). Opioid peptides affect DA function in the VTA and the striatum. In particular, dynorphins decrease and enkephalins increase the release of DA (Devine et al. 1993; Di Chiara & Imperato 1988; Longoni et al. 1991; Pentney & Gratton 1991; Spanagel et al. 1990a,b, 1992). In addition, DA controls the synthesis of striatal dynorphin and enkephalin by affecting mRNA levels (Angulo & McEwen 1994, Hadjiconstantinou & Neff 2010). These changes could have broad influence over circuits influencing reward-related processes and, thus, drug use and withdrawal.

More generally, nAChRs affect the release of virtually every major neurotransmitter (Alkondon et al. 1997, Dani & Bertrand 2007, Hadjiconstantinou & Neff 2010, Kenny 2010, McGehee et al. 1995, Gray et al. 1996, Role & Berg 1996). Therefore, neuroadaptations arising from chronic nicotine exposure may cause widespread alterations in brain neurotransmission. Consequently, the concept that only certain brain areas narrowly and rigidly mediate reward, aversion, and addiction to nicotine is becoming obsolete. The overall neuroadaptations contribute to the mechanisms that maintain nicotine consumption, including mechanisms underlying associative learning (Tang & Dani, 2009), and they also participate in the nicotine-withdrawal syndrome (De Biasi & Salas 2008, Hadjiconstantinou & Neff 2010).

BEHAVIORAL MANIFESTATIONS OF NICOTINE WITHDRAWAL

As the brain adapts to chronic nicotine exposure, a new homeostatic condition is achieved by the brain's circuitry that requires the presence of nicotine to be maintained. When access to nicotine is not available, these homeostatic neuroadaptations are no longer appropriate, which contributes to the withdrawal syndrome (Koob & Volkow 2009).

Nicotine withdrawal is a collection of somatic and affective symptoms that is observed within a few hours after discontinuation of nicotine intake (De Biasi & Salas 2008). The symptoms reflect the imbalance in brain neurochemistry created by the absence of nicotine in the addicted brain. The most prominent withdrawal symptoms are irritability, anxiety, anger, difficulty concentrating, sleep disturbance, increased appetite, and weight gain (APA 2000, Hogle et al. 2010, Hughes 2007). Other quantifiable manifestations include reduced heart rates, altered neurohormonal profiles, disrupted electroencephalographic theta power, and perturbations of learned behaviors (Buchhalter et al. 2008, Koob & Le Moal 2005). Powerful cravings also accompany withdrawal, and they may be precipitated by drug-reinforced memories, such as the sight of a cigarette and the place associated with smoking (Dalley & Everitt 2009, Dani & Montague 2007, Shiffman et al. 2002, Smolka et al. 2005, Volkow et al. 2002).

Withdrawal is also associated with dysregulation of the hypothalamic-pituitary-adrenal (HPA) axis and the stress-response system (Koob 2008). Acute withdrawal from many drugs elevates corticosterone and corticotropin-releasing factor (CRF) (Koob 2008, Koob & Kreek 2007). During nicotine withdrawal, corticotropin-releasing factor receptor CRF1 influences anxiety and brain reward functions (Bruijnzeel et al. 2009, George et al. 2007). The endogenous opioid system, dynorphin in particular, engages (Hadjiconstantinou & Neff 2010). Nicotine cessation increases the levels of prodynorphin mRNA in NAc dynorphinergic neurons, possibly as a compensatory mechanism to increased dynorphinergic tone (Isola et al. 2008).

The emergence of negative affective symptoms such as dysphoria, anxiety, and irritability (Koob & Le Moal 2001) and, to a lesser extent, the somatic manifestations of withdrawal, signal dependency and underlie drug-seeking behavior (Koob & Le Moal 2005, Koob & Volkow 2009). The negative emotional state produced by the combination of withdrawal symptoms causes enough distress to become a deterrent to abstinence and a drive to relapse (Bruijnzeel & Gold 2005, Koob & Le Moal 2005, West et al. 1989). Drug taking is then maintained through negative reinforcement mechanisms.

Withdrawal symptoms can be assessed in animals chronically exposed to nicotine by the sudden discontinuation of nicotine treatment or administration of nAChR antagonists (De Biasi & Salas 2008). Mecamylamine, which has a slightly higher affinity for $\alpha 3\beta 4^*$ nAChRs than for $\beta 2^*$ receptors, is the best antagonist at precipitating nicotine withdrawal (Damaj et al. 2003). Like humans, rodents manifest both somatic and nonsomatic signs of withdrawal (Malin & Goyarzu 2009). In rats, somatic signs of physical dependency include teeth-chattering, chewing, gasps, palpebral ptosis, tremors, shakes, and yawns (Malin et al. 1992). Similar symptoms are observed in mice, including shaking, grooming, and scratching, which become repetitive and long lived. Jumping, which is not normally observed, also appears in mice undergoing nicotine withdrawal (Damaj et al. 2003, De Biasi & Salas 2008, Isola et al. 1999).

Anhedonia, conditioned place aversion, anxiety-related behavior, and conditioned fear are four major affective manifestations of nicotine withdrawal. Anhedonia is quantifiable in animals as an increase in stimulus threshold needed to induce intracranial electrical self-stimulation to reward pathways (Paterson & Markou 2007). Increased brain-stimulation reward thresholds reflect a decreased sensitivity to the rewarding effect of the electrical stimulation. Both spontaneous (Epping-Jordan et al. 1998, Harrison et al. 2001) and mecamylamine-precipitated (Watkins et al. 2000) nicotine withdrawal induce increases in self-stimulation thresholds. In the conditioned place-aversion paradigm, the animal learns to avoid the compartment in which nicotine withdrawal is experienced (Jackson et al. 2009, Kenny & Markou 2001). A previously neutral environment becomes aversive because of the association of the place with the negative affective state of withdrawal (Suzuki et al. 1996).

Another nonsomatic manifestation of nicotine withdrawal is increased anxiety-like behavior in the elevated plus maze (Pellow et al. 1985). Mice and rats undergoing nicotine withdrawal show increased anxiety-like behavior in the elevated plus maze (Damaj et al. 2003, Irvine et al. 2001). This phenomenon is similar to the anxiety reported by humans experiencing nicotine abstinence and suggests that the sensitivity to anxiety influences the degree of withdrawal signs. These properties may contribute to the more severe withdrawal symptoms commonly experienced by those suffering from depression or anxiety disorders (Dani & Harris 2005, Pomerleau et al. 2005). The conditioned fear paradigm can be used to measure the cognitive symptoms of withdrawal (Davis et al. 2005) as nicotine withdrawal produces deficits in contextual fear conditioning and affects the acquisition of learned responses (Portugal et al. 2008).

MOLECULAR MECHANISMS OF WITHDRAWAL

Withdrawal reflects the abrupt disruption of the homeostasis maintained in the presence of nicotine, and withdrawal triggers new neuroadaptations that counteract the negative state. Spontaneous or mecamylamine-precipitated nicotine withdrawal is associated with a decrease in extracellular DA levels in the NAc (Carboni et al. 2000, Gaddnas et al. 2002, Hildebrand et al. 1998, Rada et al. 2001, Rahman et al. 2004). This phenomenon is also observed during ethanol, morphine, cocaine, and amphetamine withdrawal (Rossetti et al. 1992, Weiss et al. 1992) and is a trigger of the withdrawal syndrome. The decrease in extracellular DA concentrations reflects changes in both DA release and reuptake (Duchemin et al. 2009). Nicotine abstinence is accompanied by increased DA uptake into synaptosomes and elevated DA transporters (DATs) in the SNc/VTA (Hadjiconstantinou & Neff 2010, Rahman et al. 2004). Changes in DA release and uptake are short lived and normalize within 48 h after nicotine cessation (Hadjiconstantinou & Neff 2010).

DAT upregulation and decreased DA overflow may contribute to withdrawal symptoms because those DA signaling changes temporally coincide with the emergence of somatic and affective signs of withdrawal (Hadjiconstantinou & Neff 2010). Although buproprion inhibits nAChRs, it is mainly a norepinephrine/DA reuptake blocker. Buproprion may aid smoking cessation, in part, by normalizing NAc DA levels, and those DA levels may in turn attenuate anhedonia and somatic withdrawal signs (Cryan et al. 2003, Paterson & Markou 2007).

During withdrawal, the NAc displays a DA deficit, whereas the PFC has an increase in DA output (Carboni et al. 2000). Because enhanced DA transmission in the PFC occurs during stressful and aversive stimuli and is implicated in anxiety, the elevated PFC DA during nicotine withdrawal may contribute to the aversive manifestations (Bradberry et al. 1991, Broersen et al. 2000, Inglis & Moghaddam 1999, Kawasaki et al. 2001, Thierry et al. 1976). Besides DA, other neurotransmitters, such as serotonin and norepinephrine, known to mediate the manifestations of withdrawal from other drugs may play a role (Bruijnzeel et al. 2010, Fletcher et al. 2008, Semenova & Markou 2010, Slotkin & Seidler 2007).

DISTINCT NICOTINIC RECEPTOR SUBTYPES MEDIATE WITHDRAWAL

The use of mouse models has clarified the specific brain area that mediates physical dependency. Salas et al. (2003a, 2004a) observed that mice null for the α5 and β4 nAChR subunits are less sensitive to the acute effects of nicotine measured as hypolocomotion in the open field arena and nicotine-induced seizure activity. Attenuation of nicotine-induced seizure is also present in mice heterozygous for the α3 subunit, which is found in the same gene cluster that contains α5 and β4 (Boulter et al. 1990, Salas et al. 2004a). Absence of β4 abolishes the somatic signs of withdrawal precipitated by systemic injection of mecamylamine (Salas et al. 2004b) and prevents withdrawal-induced hyperalgesia (Salas et al. 2004b). This phenotype is in sharp contrast with that of β2 -/- mice, which display normal somatic signs of withdrawal (Besson et al. 2006, Salas et al. 2004b). Physical signs of dependency are also absent in α5 -/- mice in which nicotine withdrawal either occurs spontaneously or upon nicotine cessation or is precipitated by mecamylamine injection (Salas et al. 2009). Both α5 -/- and β4 -/- strains also display reduced anxiety-related behaviors (Gangitano et al. 2009, Salas et al. 2003b), in contrast with β2 -/- mice, which show normal anxiety-like responses (Maskos et al. 2005). The latter is an important result, considering the role of anxiety and stress in nicotine withdrawal and relapse (De Biasi & Salas 2008). Another nicotinic subunit that contributes to the somatic signs of withdrawal is the α2 subunit (Salas et al. 2009), which is selectively expressed in rodent IPN and olfactory bulb (Ishii et al. 2005, Whiteaker et al. 1999). The α7 subunit also plays some role in mecamylamine-precipitated somatic signs of withdrawal as α7 -/- mice showed an intermediate withdrawal phenotype (Salas et al. 2007). The same mice manifested decreased hyperalgesia upon spontaneous nicotine withdrawal (Grabus et al. 2005).

In the mouse, the α5, α2, and β4 nAChR subunits are expressed at high levels in either MHb and/or IPN (De Biasi & Salas 2008, Grady et al. 2009), and no mRNA encoding for nAChR subunits can be detected in the LHb by in situ hybridization techniques. Therefore, investigators postulated that the MHb/IPN axis may be involved in the somatic signs of withdrawal. Indeed, microinjection of the nAChR antagonist mecamylamine into the Hb and IPN was sufficient to precipitate nicotine withdrawal symptoms in mice chronically treated with nicotine. Conversely, mecamylamine injection into the VTA, the hippocampus, or the cortex did not trigger somatic signs of withdrawal (Salas et al. 2009), establishing the MHb and IPN as mediators of somatic withdrawal from nicotine.

Other data in the literature suggest additional involvements of the MHb/IPN axis with brain reward areas. First, electrical stimulation of the MHb and the fasciculus retroflexus produces rewarding effects (Sutherland & Nakajima 1981). Second, most stimulant drugs of abuse cause axonal degeneration in the LHb and the fasciculus retroflexus, and nicotine causes degeneration of neurons in the portion of the fasciculus retroflexus that connects the MHb to the IPN (Ellison 2002, Ellison et al. 1996). Third, a derivative of the alkaloid ibogaine, 18-methoxycoronaridine (18-MC) (Vastag 2005), is a potent antagonist of β4* nAChRs (Glick et al. 2002) with significant antiaddictive properties (Maisonneuve & Glick 2003). In animal models, 18-MC reduced self-administration of nicotine, morphine, cocaine, and methamphetamine; reduced oral intake of alcohol and nicotine; and decreased signs of opioid withdrawal (Maisonneuve & Glick 2003). Fourth, a recent report shows that α5 -/- mice self-administer nicotine at doses that elicit strong aversion in wild-type mice. This result suggests that α5-containing nAChRs in the MHb are key to controlling the amounts of nicotine self-administered (Fowler et al. 2011).

The MHb/IPN axis has a prominent role mediating the aversive effects of nicotine and the somatic symptoms of withdrawal. The data suggest the existence of reward and withdrawal circuits with partially overlapping functions. Smokers finely control nicotine intake by regulating how they puff on a cigarette and how deeply they inhale. In that way, they achieve the most desirable nicotine dose, at the top of the inverted U dose-response curve. The primary circuits underlying this exquisitely regulated dosing are mainly the mesocorticolimbic DA system and the epithalamic habenular circuits: One supports the rising arm, the other the falling arm of the inverted U dose-response curve.

Recent findings on the influence of gene variants in the *CHRNA5-CHRNA3-CHRNB4* gene cluster on nicotine addiction (Bierut et al. 2008, Thorgeirsson et al. 2008) show that a single nucleotide polymorphism (SNP) within CHRNA5, rs16969968, seems to correlate with nicotine dependency risk, heavy smoking, and the pleasurable sensation produced by a cigarette (Berrettini et al. 2008, Bierut et al. 2008, Saccone et al. 2007, Thorgeirsson et al. 2008). This nonsynonymous SNP, located in the second intracellular loop of the α5 subunit, leads to a reduction in receptor function (Bierut et al. 2008). Because the nAChR subunits comprised in the *CHRNA5-CHRNA3-CHRNB4* reduce the aversive effects of nicotine and drive its consumption, it is tempting to speculate that people with SNP rs16969968 smoke more and become addicted at a younger age because they lack sufficient functional α5* nAChRs. The presence of fewer aversive effects (even at higher nicotine doses) during the initial contact with the drug would promote the hedonic drive, thereby promoting the transition from use to abuse and dependency. New mouse models examining the role of nAChR gene variants will help investigators explore the role of gene polymorphisms in nicotine addiction and will provide insight into cessation therapies tailored for specific subpopulations of smokers.

DISCLOSURE STATEMENT

The authors are not aware of any affiliations, memberships, funding, or financial holdings that might be perceived as affecting the objectivity of this review.

ACKNOWLEDGMENTS

We thank William Doyon and Michael Paolini for helpful comments and Dang Q. Dao and Erika E. Perez for assistance with the figures. We are supported by grants from the Cancer Prevention and Research Institute of Texas and the National Institutes of Health (NINDS NS21229 and NIDA DA09411, DA017173, DA029157).

LITERATURE CITED

Alkondon M, Pereira EF, Barbosa CT, Albuquerque EX. 1997. Neuronal nicotinic acetylcholine receptor activation modulates gamma-aminobutyric acid release from CA1 neurons of rat hippocampal slices. *J. Pharmacol. Exp. Ther.* 283:1396–411

Am. Psychiatr. Assoc. (APA). 2000. *American Psychiatric Association, Diagnostic and Statistical Manual—IVTR.* Washington, DC: APA

Angulo JA, McEwen BS. 1994. Molecular aspects of neuropeptide regulation and function in the corpus striatum and nucleus accumbens. *Brain Res. Brain Res. Rev.* 19:1–28

Ashton H, Marsh VR, Millman JE, Rawlins MD, Telford R, Thompson JW. 1980. Biphasic dose-related responses of the CNV (contingent negative variation) to I.V. nicotine in man. *Br. J. Clin. Pharmacol.* 10:579–89

Avale ME, Faure P, Pons S, Robledo P, Deltheil T, et al. 2008. Interplay of beta2* nicotinic receptors and dopamine pathways in the control of spontaneous locomotion. *Proc. Natl. Acad. Sci. USA* 105:15991–96

Balfour DJ. 2004. The neurobiology of tobacco dependence: a preclinical perspective on the role of the dopamine projections to the nucleus. *Nicotine Tob. Res.* 6:899–912

Bennett BD, Wilson CJ. 1999. Spontaneous activity of neostriatal cholinergic interneurons in vitro. *J. Neurosci.* 19:5586–96

Benowitz NL. 2001. Compensatory smoking of low yield cigarettes. In *Risks Associated with Smoking Cigarettes with Low Machine-Measured Yields of Tar and Nicotine*, ed. DR Shopland, pp. 39–64. NIH Publ. No. 02–5074. Bethesda, MD: NIH

Benowitz NL. 2008. Clinical pharmacology of nicotine: implications for understanding, preventing, and treating tobacco addiction. *Clin. Pharmacol. Ther.* 83:531–41

Benowitz NL. 2009. Pharmacology of nicotine: addiction, smoking-induced disease, and therapeutics. *Annu. Rev. Pharmacol. Toxicol.* 49:57–71

Benowitz NL, Jacob P 3rd. 1985. Nicotine renal excretion rate influences nicotine intake during cigarette smoking. *J. Pharmacol. Exp. Ther.* 234:153–55

Berrendero F, Mendizabal V, Robledo P, Galeote L, Bilkei-Gorzo A, et al. 2005. Nicotine-induced antinociception, rewarding effects, and physical dependence are decreased in mice lacking the preproenkephalin gene. *J. Neurosci.* 25:1103–12

Berrendero F, Robledo P, Trigo JM, Martin-Garcia E, Maldonado R. 2010. Neurobiological mechanisms involved in nicotine dependence and reward: participation of the endogenous opioid system. *Neurosci. Biobehav. Rev.* 35:220–31

Berrettini W, Yuan X, Tozzi F, Song K, Francks C, et al. 2008. Alpha-5/alpha-3 nicotinic receptor subunit alleles increase risk for heavy smoking. *Mol. Psychiatry* 13:368–73

Besson M, David V, Suarez S, Cormier A, Cazala P, et al. 2006. Genetic dissociation of two behaviors associated with nicotine addiction: beta-2 containing nicotinic receptors are involved in nicotine reinforcement but not in withdrawal syndrome. *Psychopharmacology* 187:189–99

Bierut LJ, Stitzel JA, Wang JC, Hinrichs AL, Grucza RA, et al. 2008. Variants in nicotinic receptors and risk for nicotine dependence. *Am. J. Psychiatry* 165:1163–71

Boulter J, O'Shea-Greenfield A, Duvoisin RM, Connolly JG, Wada E, et al. 1990. Alpha 3, alpha 5, and beta 4: three members of the rat neuronal nicotinic acetylcholine receptor-related gene family form a gene cluster. *J. Biol. Chem.* 265:4472–82

Bradberry CW, Lory JD, Roth RH. 1991. The anxiogenic beta-carboline FG 7142 selectively increases dopamine release in rat prefrontal cortex as measured by microdialysis. *J. Neurochem.* 56:748–52

Briand LA, Flagel SB, Seeman P, Robinson TE. 2008. Cocaine self-administration produces a persistent increase in dopamine D2 High receptors. *Eur. Neuropsychopharmacol.* 18:551–56

Brischoux F, Chakraborty S, Brierley DI, Ungless MA. 2009. Phasic excitation of dopamine neurons in ventral VTA by noxious stimuli. *Proc. Natl. Acad. Sci. USA* 106:4894–99

Broersen LM, Abbate F, Feenstra MG, de Bruin JP, Heinsbroek RP, Olivier B. 2000. Prefrontal dopamine is directly involved in the anxiogenic interoceptive cue of pentylenetetrazol but not in the interoceptive cue of chlordiazepoxide in the rat. *Psychopharmacology* 149:366–76

Bromberg-Martin ES, Matsumoto M, Hikosaka O. 2010. Dopamine in motivational control: rewarding, aversive, and alerting. *Neuron* 68:815–34

Bruijnzeel AW, Bishnoi M, van Tuijl IA, Keijzers KF, Yavarovich KR, et al. 2010. Effects of prazosin, clonidine, and propranolol on the elevations in brain reward thresholds and somatic signs associated with nicotine withdrawal in rats. *Psychopharmacology* 212:485–99

Bruijnzeel AW, Gold MS. 2005. The role of corticotropin-releasing factor-like peptides in cannabis, nicotine, and alcohol dependence. *Brain Res. Brain Res. Rev.* 49:505–28

Bruijnzeel AW, Prado M, Isaac S. 2009. Corticotropin-releasing factor-1 receptor activation mediates nicotine withdrawal-induced deficit in brain reward function and stress-induced relapse. *Biol. Psychiatry* 66:110–17

Buchhalter AR, Fant RV, Henningfield JE. 2008. Novel pharmacological approaches for treating tobacco dependence and withdrawal: current status. *Drugs* 68:1067–88

Buisson B, Bertrand D. 2001. Chronic exposure to nicotine upregulates the human (alpha)4((beta)2 nicotinic acetylcholine receptor function. *J. Neurosci.* 21:1819–29

Buisson B, Bertrand D. 2002. Nicotine addiction: the possible role of functional upregulation. *Trends Pharmacol. Sci.* 23:130–36

Butcher LL, Woolf NJ. 1984. Histochemical distribution of acetylcholinesterase in the central nervous system: clues to the localization of cholinergic neurons. In *Handbook of Chemical Neuroanatomy*, Vol. 3: *Classical Transmitters and Transmitter Receptors in the CNS, Part II*, ed. A Björklund, T Hökfelt, MJ Kuhar, pp. 1–50. Amsterdam: Elsevier Biomed.

Calabresi P, Lacey MG, North RA. 1989. Nicotinic excitation of rat ventral tegmental neurones in vitro studied by intracellular recording. *Br. J. Pharmacol.* 98:135–40

Carboni E, Bortone L, Giua C, Di Chiara G. 2000. Dissociation of physical abstinence signs from changes in extracellular dopamine in the nucleus accumbens and in the prefrontal cortex of nicotine dependent rats. *Drug Alcohol Depend.* 58:93–102

Carr DB, Sesack SR. 2000. Projections from the rat prefrontal cortex to the ventral tegmental area: target specificity in the synaptic associations with mesoaccumbens and mesocortical neurons. *J. Neurosci.* 20:3864–73

Champtiaux N, Gotti C, Cordero-Erausquin M, David DJ, Przybylski C, et al. 2003. Subunit composition of functional nicotinic receptors in dopaminergic neurons investigated with knock-out mice. *J. Neurosci.* 23:7820–29

Changeux JP, Devillers-Thiery A, Chemouilli P. 1984. Acetylcholine receptor: an allosteric protein. *Science* 225:1335–45

Changeux JP, Edelstein SJ. 2005. *Nicotinic Acetylcholine Receptors: From Molecular Biology to Cognition*. New York: Odile Jacob

Chergui K, Suaud-Chagny MF, Gonon F. 1994. Nonlinear relationship between impulse flow, dopamine release and dopamine elimination in the rat brain in vivo. *Neuroscience* 62:641–45

Christoph GR, Leonzio RJ, Wilcox KS. 1986. Stimulation of the lateral habenula inhibits dopamine-containing neurons in the substantia nigra and ventral tegmental area of the rat. *J. Neurosci.* 6:613–19

Conover KL, Woodside B, Shizgal P. 1994. Effects of sodium depletion on competition and summation between rewarding effects of salt and lateral hypothalamic stimulation in the rat. *Behav. Neurosci.* 108:549–58

Corbett D, Wise RA. 1979. Intracranial self-stimulation in relation to the ascending noradrenergic fiber systems of the pontine tegmentum and caudal midbrain: a moveable electrode mapping study. *Brain Res.* 177:423–36

Corrigall WA. 1999. Nicotine self-administration in animals as a dependence model. *Nicotine Tob. Res.* 1:11–20

Corringer PJ, Sallette J, Changeux JP. 2006. Nicotine enhances intracellular nicotinic receptor maturation: a novel mechanism of neural plasticity? *J. Physiol.* 99:162–71

Cragg SJ. 2003. Variable dopamine release probability and short-term plasticity between functional domains of the primate striatum. *J. Neurosci.* 23:4378–85

Cragg SJ, Hille CJ, Greenfield SA. 2002. Functional domains in dorsal striatum of the nonhuman primate are defined by the dynamic behavior of dopamine. *J. Neurosci.* 22:5705–12

Cryan JF, Bruijnzeel AW, Skjei KL, Markou A. 2003. Bupropion enhances brain reward function and reverses the affective and somatic aspects of nicotine withdrawal in the rat. *Psychopharmacology* 168:347–58

Dalley JW, Everitt BJ. 2009. Dopamine receptors in the learning, memory and drug reward circuitry. *Semin. Cell Dev. Biol.* 20:403–10

Damaj MI, Kao W, Martin BR. 2003. Characterization of spontaneous and precipitated nicotine withdrawal in the mouse. *J. Pharmacol. Exp. Ther.* 307:526–34

Dani JA, Bertrand D. 2007. Nicotinic acetylcholine receptors and nicotinic cholinergic mechanisms of the central nervous system. *Annu. Rev. Pharmacol. Toxicol.* 47:699–729

Dani JA, Harris RA. 2005. Nicotine addiction and comorbidity with alcohol abuse and mental illness. *Nat. Neurosci.* 8:1465–70

Dani JA, Heinemann S. 1996. Molecular and cellular aspects of nicotine abuse. *Neuron* 16:905–8

Dani JA, Ji D, Zhou FM. 2001. Synaptic plasticity and nicotine addiction. *Neuron* 31:349–52

Dani JA, Kosten TR, Benowitz NL. 2009. The pharmacology of nicotine and tobacco. In *Principles of Addiction Medicine*, ed. RK Ries, DA Fiellin, SC Miller, R Saitz, pp. 179–91. Philadelphia, PA: Wolters Kluwer/Lippincott, Williams & Wilkins

Dani JA, Montague PR. 2007. Disrupting addiction through the loss of drug-associated internal states. *Nat. Neurosci.* 10:403–4

Dani JA, Radcliffe KA, Pidoplichko VI. 2000. Variations in desensitization of nicotinic acetylcholine receptors from hippocampus and midbrain dopamine areas. *Eur. J. Pharmacol.* 393:31–38

Darsow T, Booker TK, Pina-Crespo JC, Heinemann SF. 2005. Exocytic trafficking is required for nicotine-induced up-regulation of alpha 4 beta 2 nicotinic acetylcholine receptors. *J. Biol. Chem.* 280:18311–20

Davis JA, James JR, Siegel SJ, Gould TJ. 2005. Withdrawal from chronic nicotine administration impairs contextual fear conditioning in C57BL/6 mice. *J. Neurosci.* 25:8708–13

De Biasi M, Salas R. 2008. Influence of neuronal nicotinic receptors over nicotine addiction and withdrawal. *Exp. Biol. Med.* 233:917–29

de Rover M, Lodder JC, Kits KS, Schoffelmeer AN, Brussaard AB. 2002. Cholinergic modulation of nucleus accumbens medium spiny neurons. *Eur. J. Neurosci.* 16:2279–90

Devine DP, Leone P, Pocock D, Wise RA. 1993. Differential involvement of ventral tegmental mu, delta and kappa opioid receptors in modulation of basal mesolimbic dopamine release: in vivo microdialysis studies. *J. Pharmacol. Exp. Ther.* 266:1236–46

Di Chiara G. 1999. Drug addiction as dopamine-dependent associative learning disorder. *Eur. J. Pharmacol.* 375:13–30

Di Chiara G. 2000. Role of dopamine in the behavioural actions of nicotine related to addiction. *Eur. J. Pharmacol.* 393:295–314

Di Chiara G, Imperato A. 1988. Opposite effects of mu and kappa opiate agonists on dopamine release in the nucleus accumbens and in the dorsal caudate of freely moving rats. *J. Pharmacol. Exp. Ther.* 244:1067–80

Drenan RM, Grady SR, Steele AD, McKinney S, Patzlaff NE, et al. 2010. Cholinergic modulation of locomotion and striatal dopamine release is mediated by alpha6alpha4* nicotinic acetylcholine receptors. *J. Neurosci.* 30:9877–89

Duchemin AM, Zhang H, Neff NH, Hadjiconstantinou M. 2009. Increased expression of VMAT2 in dopaminergic neurons during nicotine withdrawal. *Neurosci. Lett.* 467:182–86

Ellison G. 2002. Neural degeneration following chronic stimulant abuse reveals a weak link in brain, fasciculus retroflexus, implying the loss of forebrain control circuitry. *Eur. Neuropsychopharmacol.* 12:287–97

Ellison G, Irwin S, Keys A, Noguchi K, Sulur G. 1996. The neurotoxic effects of continuous cocaine and amphetamine in Habenula: implications for the substrates of psychosis. *NIDA Res. Monogr.* 163:117–45

Epping-Jordan MP, Watkins SS, Koob GF, Markou A. 1998. Dramatic decreases in brain reward function during nicotine withdrawal. *Nature* 393:76–79

Erhardt S, Schwieler L, Engberg G. 2002. Excitatory and inhibitory responses of dopamine neurons in the ventral tegmental area to nicotine. *Synapse* 43:227–37

Exley R, Cragg SJ. 2008. Presynaptic nicotinic receptors: a dynamic and diverse cholinergic filter of striatal dopamine neurotransmission. *Br. J. Pharmacol.* 153(Suppl. 1):S283–97

Fattore L, Cossu G, Martellotta MC, Fratta W. 2002. Baclofen antagonizes intravenous self-administration of nicotine in mice and rats. *Alcohol Alcohol.* 37:495–98

Fenster CP, Whitworth TL, Sheffield EB, Quick MW, Lester RA. 1999. Upregulation of surface alpha4beta2 nicotinic receptors is initiated by receptor desensitization after chronic exposure to nicotine. *J. Neurosci.* 19:4804–14

Fibiger HC. 1978. Drugs and reinforcement mechanisms: a critical review of the catecholamine theory. *Annu. Rev. Pharmacol. Toxicol.* 18:37–56

Fletcher PJ, Le AD, Higgins GA. 2008. Serotonin receptors as potential targets for modulation of nicotine use and dependence. *Prog. Brain Res.* 172:361–83

Flores CM, Davila-Garcia MI, Ulrich YM, Kellar KJ. 1997. Differential regulation of neuronal nicotinic receptor binding sites following chronic nicotine administration. *J. Neurochem.* 69:2216–19

Floresco SB, West AR, Ash B, Moore H, Grace AA. 2003. Afferent modulation of dopamine neuron firing differentially regulates tonic and phasic dopamine transmission. *Nat. Neurosci.* 6:968–73

Fouriezos G, Wise RA. 1976. Pimozide-induced extinction of intracranial self-stimulation: response patterns rule out motor or performance deficits. *Brain Res.* 103:377–80

Fowler CD, Lu Q, Johnson PM, Marks MJ, Kenny PJ. 2011. Habenular alpha5 nicotinic receptor subunit signalling controls nicotine intake. *Nature.* In Press

Gaddnas H, Piepponen TP, Ahtee L. 2002. Mecamylamine decreases accumbal dopamine output in mice treated chronically with nicotine. *Neurosci. Lett.* 330:219–22

Gaimarri A, Moretti M, Riganti L, Zanardi A, Clementi F, Gotti C. 2007. Regulation of neuronal nicotinic receptor traffic and expression. *Brain Res. Rev.* 55:134–43

Gangitano D, Salas R, Teng Y, Perez E, De Biasi M. 2009. Progesterone modulation of alpha5 nAChR subunits influences anxiety-related behavior during estrus cycle. *Genes Brain Behav.* 8:398–406

Gao M, Jin Y, Yang K, Zhang D, Lukas RJ, Wu J. 2010. Mechanisms involved in systemic nicotine-induced glutamatergic synaptic plasticity on dopamine neurons in the ventral tegmental area. *J. Neurosci.* 30:13814–25

Garris PA, Kilpatrick M, Bunin MA, Michael D, Walker QD, Wightman RM. 1999. Dissociation of dopamine release in the nucleus accumbens from intracranial self-stimulation. *Nature* 398:67–69

Geisler S, Marinelli M, Degarmo B, Becker ML, Freiman AJ, et al. 2008. Prominent activation of brainstem and pallidal afferents of the ventral tegmental area by cocaine. *Neuropsychopharmacology* 33:2688–700

George O, Ghozland S, Azar MR, Cottone P, Zorrilla EP, et al. 2007. CRF-CRF1 system activation mediates withdrawal-induced increases in nicotine self-administration in nicotine-dependent rats. *Proc. Natl. Acad. Sci. USA* 104:17198–203

Glick SD, Maisonneuve IM, Kitchen BA, Fleck MW. 2002. Antagonism of alpha 3 beta 4 nicotinic receptors as a strategy to reduce opioid and stimulant self-administration. *Eur. J. Pharmacol.* 438:99–105

Gonon FG. 1988. Nonlinear relationship between impulse flow and dopamine released by rat midbrain dopaminergic neurons as studied by in vivo electrochemistry. *Neuroscience* 24:19–28

Grabus SD, Martin BR, Imad Damaj M. 2005. Nicotine physical dependence in the mouse: involvement of the alpha7 nicotinic receptor subtype. *Eur. J. Pharmacol.* 515:90–93

Grace AA. 1991. Phasic versus tonic dopamine release and the modulation of dopamine system responsivity: a hypothesis for the etiology of schizophrenia. *Neuroscience* 41:1–24

Grace AA, Bunney BS. 1983. Intracellular and extracellular electrophysiology of nigral dopaminergic neurons–1. Identification and characterization. *Neuroscience* 10:301–15

Grace AA, Bunney BS. 1984. The control of firing pattern in nigral dopamine neurons: single spike firing. *J. Neurosci.* 4:2866–76

Grace AA, Floresco SB, Goto Y, Lodge DJ. 2007. Regulation of firing of dopaminergic neurons and control of goal-directed behaviors. *Trends Neurosci.* 30:220–27

Grady SR, Grun EU, Marks MJ, Collins AC. 1997. Pharmacological comparison of transient and persistent [3H]dopamine release from mouse striatal synaptosomes and response to chronic L-nicotine treatment. *J. Pharmacol. Exp. Ther.* 282:32–43

Grady SR, Marks MJ, Collins AC. 1994. Desensitization of nicotine-stimulated [3H]dopamine release from mouse striatal synaptosomes. *J. Neurochem.* 62:1390–98

Grady SR, Moretti M, Zoli M, Marks MJ, Zanardi A, et al. 2009. Rodent habenulo-interpeduncular pathway expresses a large variety of uncommon nAChR subtypes, but only the alpha3beta4* and alpha3beta3beta4* subtypes mediate acetylcholine release. *J. Neurosci.* 29:2272–82

Grady SR, Salminen O, McIntosh JM, Marks MJ, Collins AC. 2010. Mouse striatal dopamine nerve terminals express alpha4alpha5beta2 and two stoichiometric forms of alpha4beta2*-nicotinic acetylcholine receptors. *J. Mol. Neurosci.* 40:91–95

Gratton A, Hoffer BJ, Gerhardt GA. 1988. Effects of electrical stimulation of brain reward sites on release of dopamine in rat: an in vivo electrochemical study. *Brain Res. Bull.* 21:319–24

Gray R, Rajan AS, Radcliffe KA, Yakehiro M, Dani JA. 1996. Hippocampal synaptic transmission enhanced by low concentrations of nicotine. *Nature* 83:713–16

Greatrex RM, Phillipson OT. 1982. Demonstration of synaptic input from prefrontal cortex to the habenula in the rat. *Brain Res.* 238:192–97

Grenhoff J, Aston-Jones G, Svensson TH. 1986. Nicotinic effects on the firing pattern of midbrain dopamine neurons. *Acta Physiol. Scand.* 128:351–58

Grigoriadis D, Seeman P. 1986. [3H]-domperidone labels only a single population of receptors which convert from high to low affinity for dopamine in rat brain. *Naunyn-Schmiedebergs Arch. Pharmakol.* 332:21–25

Groenewegen HJ, Ahlenius S, Haber SN, Kowall NW, Nauta WJ. 1986. Cytoarchitecture, fiber connections, and some histochemical aspects of the interpeduncular nucleus in the rat. *J. Comp. Neurol.* 249:65–102

Grottick AJ, Higgins GA. 2000. Effect of subtype selective nicotinic compounds on attention as assessed by the five-choice serial reaction time task. *Behav. Brain Res.* 117:197–208

Haber SN, Fudge JL, McFarland NR. 2000. Striatonigrostriatal pathways in primates form an ascending spiral from the shell to the dorsolateral striatum. *J. Neurosci.* 20:2369–82

Hadjiconstantinou M, Neff NH. 2011. Nicotine and endogenous opioids: neurochemical and pharmacological evidence. *Neuropharmacology.* In Press

Harrison AA, Liem YT, Markou A. 2001. Fluoxetine combined with a serotonin-1A receptor antagonist reversed reward deficits observed during nicotine and amphetamine withdrawal in rats. *Neuropsychopharmacology* 25:55–71

Heimer L. 2000. Basal forebrain in the context of schizophrenia. *Brain Res. Brain Res. Rev.* 31:205–35

Herkenham M, Nauta WJ. 1977. Afferent connections of the habenular nuclei in the rat. A horseradish peroxidase study, with a note on the fiber-of-passage problem. *J. Comp. Neurol.* 173:123–46

Herkenham M, Nauta WJ. 1979. Efferent connections of the habenular nuclei in the rat. *J. Comp. Neurol.* 187:19–47

Herskovic JE, Rose JE, Jarvik ME. 1986. Cigarette desirability and nicotine preference in smokers. *Pharmacol. Biochem. Behav.* 24:171–75

Hikosaka O. 2010. The habenula: from stress evasion to value-based decision-making. *Nat. Rev. Neurosci.* 11:503–13

Hikosaka O, Bromberg-Martin E, Hong S, Matsumoto M. 2008. New insights on the subcortical representation of reward. *Curr. Opin. Neurobiol.* 18:203–8

Hildebrand BE, Nomikos GG, Hertel P, Schilstrom B, Svensson TH. 1998. Reduced dopamine output in the nucleus accumbens but not in the medial prefrontal cortex in rats displaying a mecamylamine-precipitated nicotine withdrawal syndrome. *Brain Res.* 779:214–25

Hodos W. 1961. Progressive ratio as a measure of reward strength. *Science* 134:943–44

Hogle JM, Kaye JT, Curtin JJ. 2010. Nicotine withdrawal increases threat-induced anxiety but not fear: neuroadaptation in human addiction. *Biol. Psychiatry* 68:719–25

Hughes JR. 2007. Effects of abstinence from tobacco: valid symptoms and time course. *Nicotine Tob. Res.* 9:315–27

Hutchison MA, Riley AL. 2008. Adolescent exposure to nicotine alters the aversive effects of cocaine in adult rats. *Neurotoxicol. Teratol.* 30:404–11

Hwang YY, Li MD. 2006. Proteins differentially expressed in response to nicotine in five rat brain regions: identification using a 2-DE/MS-based proteomics approach. *Proteomics* 6:3138–53

Hyland BI, Reynolds JN, Hay J, Perk CG, Miller R. 2002. Firing modes of midbrain dopamine cells in the freely moving rat. *Neuroscience* 114:475–92

Ikemoto S. 2010. Brain reward circuitry beyond the mesolimbic dopamine system: a neurobiological theory. *Neurosci. Biobehav. Rev.* 35:129–50

Imperato A, Mulas A, Di Chiara G. 1986. Nicotine preferentially stimulates dopamine release in the limbic system of freely moving rats. *Eur. J. Pharmacol.* 132:337–38

Inglis FM, Moghaddam B. 1999. Dopaminergic innervation of the amygdala is highly responsive to stress. *J. Neurochem.* 72:1088–94

Irvine EE, Cheeta S, File SE. 2001. Tolerance to nicotine's effects in the elevated plus-maze and increased anxiety during withdrawal. *Pharmacol. Biochem. Behav.* 68:319–25

Ishii K, Wong JK, Sumikawa K. 2005. Comparison of alpha2 nicotinic acetylcholine receptor subunit mRNA expression in the central nervous system of rats and mice. *J. Comp. Neurol.* 493:241–60

Isola R, Vogelsberg V, Wemlinger TA, Neff NH, Hadjiconstantinou M. 1999. Nicotine abstinence in the mouse. *Brain Res.* 850:189–96

Isola R, Zhang H, Tejwani GA, Neff NH, Hadjiconstantinou M. 2008. Dynorphin and prodynorphin mRNA changes in the striatum during nicotine withdrawal. *Synapse* 62:448–55

Jackson KJ, Kota DH, Martin BR, Damaj MI. 2009. The role of various nicotinic receptor subunits and factors influencing nicotine conditioned place aversion. *Neuropharmacology* 56:970–74

Jhou TC, Fields HL, Baxter MG, Saper CB, Holland PC. 2009a. The rostromedial tegmental nucleus (RMTg), a GABAergic afferent to midbrain dopamine neurons, encodes aversive stimuli and inhibits motor responses. *Neuron* 61:786–800

Jhou TC, Geisler S, Marinelli M, Degarmo BA, Zahm DS. 2009b. The mesopontine rostromedial tegmental nucleus: a structure targeted by the lateral habenula that projects to the ventral tegmental area of Tsai and substantia nigra compacta. *J. Comp. Neurol.* 513:566–96

Ji H, Shepard PD. 2007. Lateral habenula stimulation inhibits rat midbrain dopamine neurons through a GABA(A) receptor-mediated mechanism. *J. Neurosci.* 27:6923–30

Jones IW, Bolam JP, Wonnacott S. 2001. Presynaptic localisation of the nicotinic acetylcholine receptor beta2 subunit immunoreactivity in rat nigrostriatal dopaminergic neurones. *J. Comp. Neurol.* 439:235–47

Kalivas PW. 1993. Neurotransmitter regulation of dopamine neurons in the ventral tegmental area. *Brain Res. Brain Res. Rev.* 18:75–113

Kassel JD, Greenstein JE, Evatt DP, Wardle MC, Yates MC, et al. 2007. Smoking topography in response to denicotinized and high-yield nicotine cigarettes in adolescent smokers. *J. Adolesc. Health* 40:54–60

Kauer JA, Malenka RC. 2007. Synaptic plasticity and addiction. *Nat. Rev. Neurosci.* 8:844–58

Kawasaki H, Kaufman O, Damasio H, Damasio AR, Granner M, et al. 2001. Single-neuron responses to emotional visual stimuli recorded in human ventral prefrontal cortex. *Nat. Neurosci.* 4:15–16

Kenny PJ. 2010. Tobacco dependence, the insular cortex and the hypocretin connection. *Pharmacol. Biochem. Behav.* 97:700–7

Kenny PJ, Markou A. 2001. Neurobiology of the nicotine withdrawal syndrome. *Pharmacol. Biochem. Behav.* 70:531–49

Kim U, Chang SY. 2005. Dendritic morphology, local circuitry, and intrinsic electrophysiology of neurons in the rat medial and lateral habenular nuclei of the epithalamus. *J. Comp. Neurol.* 483:236–50

Klemm WR. 2004. Habenular and interpeduncularis nuclei: shared components in multiple-function networks. *Med. Sci. Monit.* 10:RA261–73

Klink R, de Kerchove d'Exaerde A, Zoli M, Changeux JP. 2001. Molecular and physiological diversity of nicotinic acetylcholine receptors in the midbrain dopaminergic nuclei. *J. Neurosci.* 21:1452–63

Koob G, Kreek MJ. 2007. Stress, dysregulation of drug reward pathways, and the transition to drug dependence. *Am. J. Psychiatry* 164:1149–59

Koob GF. 2008. A role for brain stress systems in addiction. *Neuron* 59:11–34

Koob GF, Le Moal M. 2001. Drug addiction, dysregulation of reward, and allostasis. *Neuropsychopharmacology* 24:97–129

Koob GF, Le Moal M. 2005. Plasticity of reward neurocircuitry and the 'dark side' of drug addiction. *Nat. Neurosci.* 8:1442–44

Koob GF, Volkow ND. 2009. Neurocircuitry of addiction. *Neuropsychopharmacology* 35:217–38

Koos T, Tepper JM. 2002. Dual cholinergic control of fast-spiking interneurons in the neostriatum. *J. Neurosci.* 22:529–35

Lanca AJ, Adamson KL, Coen KM, Chow BL, Corrigall WA. 2000. The pedunculopontine tegmental nucleus and the role of cholinergic neurons in nicotine self-administration in the rat: a correlative neuroanatomical and behavioral study. *Neuroscience* 96:735–42

Lau B, Bretaud S, Huang Y, Lin E, Guo S. 2006. Dissociation of food and opiate preference by a genetic mutation in zebrafish. *Genes Brain Behav.* 5:497–505

Laviolette SR. 2007. Dopamine modulation of emotional processing in cortical and subcortical neural circuits: evidence for a final common pathway in schizophrenia? *Schizophr. Bull.* 33:971–81

Lecca S, Melis M, Luchicchi A, Ennas MG, Castelli MP, et al. 2010. Effects of drugs of abuse on putative rostromedial tegmental neurons, inhibitory afferents to midbrain dopamine cells. *Neuropsychopharmacology* 36:589–602

Lester HA, Xiao C, Srinivasan R, Son CD, Miwa J, et al. 2009. Nicotine is a selective pharmacological chaperone of acetylcholine receptor number and stoichiometry. Implications for drug discovery. *AAPS J.* 11:167–77

Lippa AS, Antelman SM, Fisher AE, Canfield DR. 1973. Neurochemical mediation of reward: a significant role for dopamine? *Pharmacol. Biochem. Behav.* 1:23–28

Lisoprawski A, Herve D, Blanc G, Glowinski J, Tassin JP. 1980. Selective activation of the mesocortico-frontal dopaminergic neurons induced by lesion of the habenula in the rat. *Brain Res.* 183:229–34

Lodge DJ, Grace AA. 2006. The laterodorsal tegmentum is essential for burst firing of ventral tegmental area dopamine neurons. *Proc. Natl. Acad. Sci. USA* 103:5167–72

Longoni R, Spina L, Mulas A, Carboni E, Garau L, et al. 1991. (D-Ala2)deltorphin II: D1-dependent stereo-typies and stimulation of dopamine release in the nucleus accumbens. *J. Neurosci.* 11:1565–76

Maisonneuve IM, Glick SD. 2003. Anti-addictive actions of an iboga alkaloid congener: a novel mechanism for a novel treatment. *Pharmacol. Biochem. Behav.* 75:607–18

Malin DH, Goyarzu P. 2009. Rodent models of nicotine withdrawal syndrome. *Handb. Exp. Pharmacol.* 2009:401–34

Malin DH, Lake JR, Newlin-Maultsby P, Roberts LK, Lanier JG, et al. 1992. Rodent model of nicotine abstinence syndrome. *Pharmacol. Biochem. Behav.* 43:779–84

Mameli-Engvall M, Evrard A, Pons S, Maskos U, Svensson TH, et al. 2006. Hierarchical control of dopamine neuron-firing patterns by nicotinic receptors. *Neuron* 50:911–21

Mansvelder HD, Keath JR, McGehee DS. 2002. Synaptic mechanisms underlie nicotine-induced excitability of brain reward areas. *Neuron* 33:905–19

Mansvelder HD, McGehee DS. 2000. Long-term potentiation of excitatory inputs to brain reward areas by nicotine. *Neuron* 27:349–57

Mansvelder HD, McGehee DS. 2002. Cellular and synaptic mechanisms of nicotine addiction. *J. Neurobiol.* 53:606–17

Marinelli M, Rudick CN, Hu XT, White FJ. 2006. Excitability of dopamine neurons: modulation and physiological consequences. *CNS Neurol. Disord. Drug Targets* 5:79–97

Marks MJ, Burch JB, Collins AC. 1983. Effects of chronic nicotine infusion on tolerance development and nicotinic receptors. *J. Pharmacol. Exp. Ther.* 226:817–25

Marks MJ, Pauly JR, Gross SD, Deneris ES, Hermans-Borgmeyer I, et al. 1992. Nicotine binding and nicotinic receptor subunit RNA after chronic nicotine treatment. *J. Neurosci.* 12:2765–84

Maskos U. 2008. The cholinergic mesopontine tegmentum is a relatively neglected nicotinic master modulator of the dopaminergic system: relevance to drugs of abuse and pathology. *Br. J. Pharmacol.* 153(Suppl. 1):S438–45

Maskos U. 2010. Role of endogenous acetylcholine in the control of the dopaminergic system via nicotinic receptors. *J. Neurochem.* 114:641–6

Maskos U, Molles BE, Pons S, Besson M, Guiard BP, et al. 2005. Nicotine reinforcement and cognition restored by targeted expression of nicotinic receptors. *Nature* 436:103–7

Mathers CD, Loncar D. 2006. Projections of global mortality and burden of disease from 2002 to 2030. *PLoS Med.* 3:e442

Matsumoto M, Hikosaka O. 2007. Lateral habenula as a source of negative reward signals in dopamine neurons. *Nature* 447:1111–15

Matsumoto M, Hikosaka O. 2009. Two types of dopamine neuron distinctly convey positive and negative motivational signals. *Nature* 459:837–41

McCallum SE, Parameswaran N, Bordia T, Fan H, McIntosh JM, Quik M. 2006. Differential regulation of mesolimbic alpha 3/alpha 6 beta 2 and alpha 4 beta 2 nicotinic acetylcholine receptor sites and function after long-term oral nicotine to monkeys. *J. Pharmacol. Exp. Ther.* 318:381–88

McClure SM, Ericson KM, Laibson DI, Loewenstein G, Cohen JD. 2007. Time discounting for primary rewards. *J. Neurosci.* 27:5796–804

McGehee DS, Heath MJ, Gelber S, Devay P, Role LW. 1995. Nicotine enhancement of fast excitatory synaptic transmission in CNS by presynaptic receptors. *Science* 269:1692–96

Metaxas A, Bailey A, Barbano MF, Galeote L, Maldonado R, Kitchen I. 2010. Differential region-specific regulation of alpha4beta2* nAChRs by self-administered and non-contingent nicotine in C57BL/6J mice. *Addict. Biol.* 15:464–79

Mogenson GJ, Jones DL, Yim CY. 1980. From motivation to action: functional interface between the limbic system and the motor system. *Prog. Neurobiol.* 14:69–97

Montague PR, King-Casas B, Cohen JD. 2006. Imaging valuation models in human choice. *Annu. Rev. Neurosci.* 29:417–48

Montone KT, Fass B, Hamill GS. 1988. Serotonergic and nonserotonergic projections from the rat interpeduncular nucleus to the septum, hippocampal formation and raphe: a combined immunocytochemical and fluorescent retrograde labelling study of neurons in the apical subnucleus. *Brain Res. Bull.* 20:233–40

Nguyen HN, Rasmussen BA, Perry DC. 2004. Binding and functional activity of nicotinic cholinergic receptors in selected rat brain regions are increased following long-term but not short-term nicotine treatment. *J. Neurochem.* 90:40–49

Nishikawa T, Fage D, Scatton B. 1986. Evidence for, and nature of, the tonic inhibitory influence of habenu-lointerpeduncular pathways upon cerebral dopaminergic transmission in the rat. *Brain Res.* 373:324–36

Novak G, Seeman P, Le Foll B. 2010. Exposure to nicotine produces an increase in dopamine D2(High) receptors: a possible mechanism for dopamine hypersensitivity. *Int. J. Neurosci.* 120:691–97

O'Doherty J, Kringelbach ML, Rolls ET, Hornak J, Andrews C. 2001. Abstract reward and punishment representations in the human orbitofrontal cortex. *Nat. Neurosci.* 4:95–102

Olds J. 1958. Self-stimulation of the brain: its use to study local effects of hunger, sex, and drugs. *Science* 127:315–24

Olds J, Milner P. 1954. Positive reinforcement produced by electrical stimulation of septal area and other regions of rat brain. *J. Comp. Physiol. Psychol.* 47:419–27

Olds ME, Olds J. 1969. Effects of lesions in medial forebrain bundle on self-stimulation behavior. *Am. J. Physiol.* 217:1253–64

Omelchenko N, Sesack SR. 2005. Laterodorsal tegmental projections to identified cell populations in the rat ventral tegmental area. *J. Comp. Neurol.* 483:217–35

Packard MG, Knowlton BJ. 2002. Learning and memory functions of the basal ganglia. *Annu. Rev. Neurosci.* 25:563–93

Paterson NE, Markou A. 2007. Animal models and treatments for addiction and depression co-morbidity. *Neurotox. Res.* 11:1–32

Paterson NE, Semenova S, Gasparini F, Markou A. 2003. The mGluR5 antagonist MPEP decreased nicotine self-administration in rats and mice. *Psychopharmacology* 167:257–64

Pauly JR, Marks MJ, Robinson SF, van de Kamp JL, Collins AC. 1996. Chronic nicotine and mecamylamine treatment increase brain nicotinic receptor binding without changing alpha 4 or beta 2 mRNA levels. *J. Pharmacol. Exp. Ther.* 278:361–69

Pellow S, Chopin P, File SE, Briley M. 1985. Validation of open:closed arm entries in an elevated plus-maze as a measure of anxiety in the rat. *J. Neurosci. Methods* 14:149–67

Peng X, Gerzanich V, Anand R, Wang F, Lindstrom J. 1997. Chronic nicotine treatment up-regulates alpha3 and alpha7 acetylcholine receptor subtypes expressed by the human neuroblastoma cell line SH-SY5Y. *Mol. Pharmacol.* 51:776–84

Pennartz CM, Groenewegen HJ, Lopes da Silva FH. 1994. The nucleus accumbens as a complex of functionally distinct neuronal ensembles: an integration of behavioural, electrophysiological and anatomical data. *Prog. Neurobiol.* 42:719–61

Pentney RJ, Gratton A. 1991. Effects of local delta and mu opioid receptor activation on basal and stimulated dopamine release in striatum and nucleus accumbens of rat: an in vivo electrochemical study. *Neuroscience* 45:95–102

Perrotti LI, Bolanos CA, Choi KH, Russo SJ, Edwards S, et al. 2005. DeltaFosB accumulates in a GABAergic cell population in the posterior tail of the ventral tegmental area after psychostimulant treatment. *Eur. J. Neurosci.* 21:2817–24

Perry E, Walker M, Grace J, Perry R. 1999. Acetylcholine in mind: a neurotransmitter correlate of consciousness? *Trends Neurosci.* 22:273–80

Peto R, Lopez AD, Boreham J, Thun M, Heath C Jr, Doll R. 1996. Mortality from smoking worldwide. *Br. Med. Bull.* 52:12–21

Phillips AG, Blaha CD, Fibiger HC. 1989. Neurochemical correlates of brain-stimulation reward measured by ex vivo and in vivo analyses. *Neurosci. Biobehav. Rev.* 13:99–104

Phillips MI, Olds J. 1969. Unit activity: motiviation-dependent responses from midbrain neurons. *Science* 165:1269–71

Picciotto MR, Addy NA, Mineur YS, Brunzell DH. 2008. It is not "either/or": activation and desensitization of nicotinic acetylcholine receptors both contribute to behaviors related to nicotine addiction and mood. *Prog. Neurobiol.* 84:329–42

Picciotto MR, Corrigall WA. 2002. Neuronal systems underlying behaviors related to nicotine addiction: neural circuits and molecular genetics. *J. Neurosci.* 22:3338–41

Picciotto MR, Zoli M, Rimondini R, Lena C, Marubio LM, et al. 1998. Acetylcholine receptors containing the beta2 subunit are involved in the reinforcing properties of nicotine. *Nature* 391:173–77

Pidoplichko VI, DeBiasi M, Williams JT, Dani JA. 1997. Nicotine activates and desensitizes midbrain dopamine neurons. *Nature* 390:401–4

Pidoplichko VI, Noguchi J, Areola OO, Liang Y, Peterson J, et al. 2004. Nicotinic cholinergic synaptic mechanisms in the ventral tegmental area contribute to nicotine addiction. *Learn. Mem.* 11:60–69

Placzek AN, Zhang TA, Dani JA. 2009. Age dependent nicotinic influences over dopamine neuron synaptic plasticity. *Biochem. Pharmacol.* 78:686–92

Pomerleau OF, Pomerleau CS, Mehringer AM, Snedecor SM, Ninowski R, Sen A. 2005. Nicotine dependence, depression, and gender: characterizing phenotypes based on withdrawal discomfort, response to smoking, and ability to abstain. *Nicotine Tob. Res.* 7:91–102

Pons S, Fattore L, Cossu G, Tolu S, Porcu E, et al. 2008. Crucial role of alpha4 and alpha6 nicotinic acetylcholine receptor subunits from ventral tegmental area in systemic nicotine self-administration. *J. Neurosci.* 28:12318–27

Pontieri FE, Tanda G, Orzi F, Di Chiara G. 1996. Effects of nicotine on the nucleus accumbens and similarity to those of addictive drugs. *Nature* 382:255–57

Portugal GS, Kenney JW, Gould TJ. 2008. Beta2 subunit containing acetylcholine receptors mediate nicotine withdrawal deficits in the acquisition of contextual fear conditioning. *Neurobiol. Learn. Mem.* 89:106–13

Qin J, Wheeler AR. 2007. Maze exploration and learning in *C. elegans. Lab Chip* 7:186–92

Rada P, Jensen K, Hoebel BG. 2001. Effects of nicotine and mecamylamine-induced withdrawal on extracellular dopamine and acetylcholine in the rat nucleus accumbens. *Psychopharmacology* 157:105–10

Rahman S, Zhang J, Engleman EA, Corrigall WA. 2004. Neuroadaptive changes in the mesoaccumbens dopamine system after chronic nicotine self-administration: a microdialysis study. *Neuroscience* 129:415–24

Rezvani AH, Slade S, Wells C, Petro A, Lumeng L, et al. 2010. Effects of sazetidine-A, a selective alpha4beta2 nicotinic acetylcholine receptor desensitizing agent on alcohol and nicotine self-administration in selectively bred alcohol-preferring (P) rats. *Psychopharmacology* 211:161–74

Rezvani K, Teng Y, De Biasi M. 2009. The ubiquitin-proteasome system regulates the stability of neuronal nicotinic acetylcholine receptors. *J. Mol. Neurosci.* 40:177–84

Rezvani K, Teng Y, Shim D, De Biasi M. 2007. Nicotine regulates multiple synaptic proteins by inhibiting proteasomal activity. *J. Neurosci.* 27:10508–19

Rice ME, Cragg SJ. 2004. Nicotine amplifies reward-related dopamine signals in striatum. *Nat. Neurosci.* 7:583–84

Robinson S, Smith DM, Mizumori SJ, Palmiter RD. 2004. Firing properties of dopamine neurons in freely moving dopamine-deficient mice: effects of dopamine receptor activation and anesthesia. *Proc. Natl. Acad. Sci. USA* 101:13329–34

Roitman MF, Wheeler RA, Carelli RM. 2005. Nucleus accumbens neurons are innately tuned for rewarding and aversive taste stimuli, encode their predictors, and are linked to motor output. *Neuron* 45:587–97

Roitman MF, Wheeler RA, Wightman RM, Carelli RM. 2008. Real-time chemical responses in the nucleus accumbens differentiate rewarding and aversive stimuli. *Nat. Neurosci.* 11:1376–77

Role LW, Berg DK. 1996. Nicotinic receptors in the development and modulation of CNS synapses. *Neuron* 16:1077–85

Rossetti ZL, Hmaidan Y, Gessa GL. 1992. Marked inhibition of mesolimbic dopamine release: a common feature of ethanol, morphine, cocaine and amphetamine abstinence in rats. *Eur. J. Pharmacol.* 221:227–34

Routtenberg A, Lindy J. 1965. Effects of the availability of rewarding septal and hypothalamic stimulation on bar pressing for food under conditions of deprivation. *J. Comp. Physiol. Psychol.* 60:158–61

Rowell PP, Wonnacott S. 1990. Evidence for functional activity of up-regulated nicotine binding sites in rat striatal synaptosomes. *J. Neurochem.* 55:2105–10

Saal D, Dong Y, Bonci A, Malenka RC. 2003. Drugs of abuse and stress trigger a common synaptic adaptation in dopamine neurons. *Neuron* 37:577–82

Saccone SF, Hinrichs AL, Saccone NL, Chase GA, Konvicka K, et al. 2007. Cholinergic nicotinic receptor genes implicated in a nicotine dependence association study targeting 348 candidate genes with 3713 SNPs. *Hum. Mol. Genet.* 16:36–49

Salas R, Baldwin P, De Biasi M, Montague PR. 2010. BOLD responses to negative reward prediction errors in human habenula. *Front. Hum. Neurosci.* 4:36

Salas R, Cook KD, Bassetto L, De Biasi M. 2004a. The alpha3 and beta4 nicotinic acetylcholine receptor subunits are necessary for nicotine-induced seizures and hypolocomotion in mice. *Neuropharmacology* 47:401–7

Salas R, Main A, Gangitano D, De Biasi M. 2007. Decreased withdrawal symptoms but normal tolerance to nicotine in mice null for the alpha7 nicotinic acetylcholine receptor subunit. *Neuropharmacology* 53:863–69

Salas R, Orr-Urtreger A, Broide RS, Beaudet A, Paylor R, De Biasi M. 2003a. The nicotinic acetylcholine receptor subunit alpha 5 mediates short-term effects of nicotine in vivo. *Mol. Pharmacol.* 63:1059–66

Salas R, Pieri F, De Biasi M. 2004b. Decreased signs of nicotine withdrawal in mice null for the beta4 nicotinic acetylcholine receptor subunit. *J. Neurosci.* 24:10035–39

Salas R, Pieri F, Fung B, Dani JA, De Biasi M. 2003b. Altered anxiety-related responses in mutant mice lacking the beta4 subunit of the nicotinic receptor. *J. Neurosci.* 23:6255–63

Salas R, Sturm R, Boulter J, De Biasi M. 2009. Nicotinic receptors in the habenulo-interpeduncular system are necessary for nicotine withdrawal in mice. *J. Neurosci.* 29:3014–18

Sallette J, Pons S, Devillers-Thiery A, Soudant M, Prado de Carvalho L, et al. 2005. Nicotine upregulates its own receptors through enhanced intracellular maturation. *Neuron* 46:595–607

Salminen O, Drapeau JA, McIntosh JM, Collins AC, Marks MJ, Grady SR. 2007. Pharmacology of alpha-conotoxin MII-sensitive subtypes of nicotinic acetylcholine receptors isolated by breeding of null mutant mice. *Mol. Pharmacol.* 71:1563–71

Scheibel AB. 1997. The thalamus and neuropsychiatric illness. *J. Neuropsychiatry Clin. Neurosci.* 9:342–53

Schilstrom B, Fagerquist MV, Zhang X, Hertel P, Panagis G, et al. 2000. Putative role of presynaptic alpha7* nicotinic receptors in nicotine stimulated increases of extracellular levels of glutamate and aspartate in the ventral tegmental area. *Synapse* 38:375–83

Schilstrom B, Rawal N, Mameli-Engvall M, Nomikos GG, Svensson TH. 2003. Dual effects of nicotine on dopamine neurons mediated by different nicotinic receptor subtypes. *Int. J. Neuropsychopharmacol.* 6:1–11

Schultz W. 2006. Behavioral theories and the neurophysiology of reward. *Annu. Rev. Psychol.* 57:87–115

Schultz W. 2007a. Behavioral dopamine signals. *Trends Neurosci.* 30:203–10

Schultz W. 2007b. Multiple dopamine functions at different time courses. *Annu. Rev. Neurosci.* 30:259–88

Schultz W. 2010. Dopamine signals for reward value and risk: basic and recent data. *Behav. Brain Funct.* 6:24

Schultz W, Dayan P, Montague PR. 1997. A neural substrate of prediction and reward. *Science* 275:1593–99

Schwartz RD, Kellar KJ. 1983. Nicotinic cholinergic receptor binding sites in the brain: regulation in vivo. *Science* 220:214–16

Seamans JK, Yang CR. 2004. The principal features and mechanisms of dopamine modulation in the prefrontal cortex. *Prog. Neurobiol.* 74:1–58

Semenova S, Markou A. 2010. The alpha2 adrenergic receptor antagonist idazoxan, but not the serotonin-2A receptor antagonist M100907, partially attenuated reward deficits associated with nicotine, but not amphetamine, withdrawal in rats. *Eur. Neuropsychopharmacol.* 20:731–46

Shepard PD, Holcomb HH, Gold JM. 2006. Schizophrenia in translation: the presence of absence: habenular regulation of dopamine neurons and the encoding of negative outcomes. *Schizophr. Bull.* 32:417–21

Shiffman S, Gwaltney CJ, Balabanis MH, Liu KS, Paty JA, et al. 2002. Immediate antecedents of cigarette smoking: an analysis from ecological momentary assessment. *J. Abnorm. Psychol.* 111:531–45

Simpson GR, Riley AL. 2005. Morphine preexposure facilitates morphine place preference and attenuates morphine taste aversion. *Pharmacol. Biochem. Behav.* 80:471–79

Slotkin TA, Seidler FJ. 2007. A unique role for striatal serotonergic systems in the withdrawal from adolescent nicotine administration. *Neurotoxicol. Teratol.* 29:10–16

Smolka MN, Bühler M, Klein S, Zimmermann U, Mann K, et al. 2005. Severity of nicotine dependence modulates cue-induced brain activity in regions involved in motor preparation and imagery. *Psychopharmacology* 184:1–12

Spanagel R, Herz A, Shippenberg TS. 1990a. The effects of opioid peptides on dopamine release in the nucleus accumbens: an in vivo microdialysis study. *J. Neurochem.* 55:1734–40

Spanagel R, Herz A, Shippenberg TS. 1990b. Identification of the opioid receptor types mediating beta-endorphin-induced alterations in dopamine release in the nucleus accumbens. *Eur. J. Pharmacol.* 190:177–84

Spanagel R, Herz A, Shippenberg TS. 1992. Opposing tonically active endogenous opioid systems modulate the mesolimbic dopaminergic pathway. *Proc. Natl. Acad. Sci. USA* 89:2046–50

Spies G. 1965. Food versus intracranial self-stimulation reinforcement in food-deprived rats. *J. Comp. Physiol. Psychol.* 60:153–57

Steffensen SC, Svingos AL, Pickel VM, Henriksen SJ. 1998. Electrophysiological characterization of GABAergic neurons in the ventral tegmental area. *J. Neurosci.* 18:8003–15

Steiner H, Gerfen CR. 1998. Role of dynorphin and enkephalin in the regulation of striatal output pathways and behavior. *Exp. Brain Res.* 123:60–76

Stellar JR, Corbett D. 1989. Regional neuroleptic microinjections indicate a role for nucleus accumbens in lateral hypothalamic self-stimulation reward. *Brain Res.* 477:126–43

Stolerman IP, Chamberlain S, Bizarro L, Fernandes C, Schalkwyk L. 2004. The role of nicotinic receptor alpha 7 subunits in nicotine discrimination. *Neuropharmacology* 46:363–71

Stolerman IP, Shoaib M. 1991. The neurobiology of tobacco addiction. *Trends Pharmacol. Sci.* 12:467–73

Sutherland RJ. 1982. The dorsal diencephalic conduction system: a review of the anatomy and functions of the habenular complex. *Neurosci. Biobehav. Rev.* 6:1–13

Sutherland RJ, Nakajima S. 1981. Self-stimulation of the habenular complex in the rat. *J. Comp. Physiol. Psychol.* 95:781–91

Suzuki T, Ise Y, Tsuda M, Maeda J, Misawa M. 1996. Mecamylamine-precipitated nicotine-withdrawal aversion in rats. *Eur. J. Pharmacol.* 314:281–84

Tang J, Dani JA. 2009. Dopamine enables in vivo synaptic plasticity associated with the addictive drug nicotine. *Neuron* 63:673–82

Tapper AR, McKinney SL, Nashmi R, Schwarz J, Deshpande P, et al. 2004. Nicotine activation of alpha4* receptors: sufficient for reward, tolerance, and sensitization. *Science* 306:1029–32

Thierry AM, Tassin JP, Blanc G, Glowinski J. 1976. Selective activation of mesocortical DA system by stress. *Nature* 263:242–44

Thorgeirsson TE, Geller F, Sulem P, Rafnar T, Wiste A, et al. 2008. A variant associated with nicotine dependence, lung cancer and peripheral arterial disease. *Nature* 452:638–42

Thorndike EL, ed. 1911. *Animal Intelligence*. New York: Macmillan

Trigo JM, Martin-Garcia E, Berrendero F, Robledo P, Maldonado R. 2009. The endogenous opioid system: a common substrate in drug addiction. *Drug Alcohol Depend.* 108:183–94

Ullsperger M, von Cramon DY. 2003. Error monitoring using external feedback: specific roles of the habenular complex, the reward system, and the cingulate motor area revealed by functional magnetic resonance imaging. *J. Neurosci.* 23:4308–14

Ungless MA, Magill PJ, Bolam JP. 2004. Uniform inhibition of dopamine neurons in the ventral tegmental area by aversive stimuli. *Science* 303:2040–42

Vallejo YF, Buisson B, Bertrand D, Green WN. 2005. Chronic nicotine exposure upregulates nicotinic receptors by a novel mechanism. *J. Neurosci.* 25:5563–72

van Domburg PHMF, ten Donkelaar HJ. 1991. The human substantia nigra and ventral tegmental area. *Adv. Anat. Embryol. Cell Biol.* 121:1–132

Vastag B. 2005. Addiction research. Ibogaine therapy: a 'vast, uncontrolled experiment'. *Science* 308:345–46

Volkow ND, Fowler JS, Wang GJ, Goldstein RZ. 2002. Role of dopamine, the frontal cortex and memory circuits in drug addiction: insight from imaging studies. *Neurobiol. Learn. Mem.* 78:610–24

Walaas I, Fonnum F. 1980. Biochemical evidence for gamma-aminobutyrate containing fibres from the nucleus accumbens to the substantia nigra and ventral tegmental area in the rat. *Neuroscience* 5:63–72

Walters CL, Brown S, Changeux JP, Martin B, Damaj MI. 2006. The beta2 but not alpha7 subunit of the nicotinic acetylcholine receptor is required for nicotine-conditioned place preference in mice. *Psychopharmacology* 184:339–44

Watkins SS, Stinus L, Koob GF, Markou A. 2000. Reward and somatic changes during precipitated nicotine withdrawal in rats: centrally and peripherally mediated effects. *J. Pharmacol. Exp. Ther.* 292:1053–64

Weiss F, Markou A, Lorang MT, Koob GF. 1992. Basal extracellular dopamine levels in the nucleus accumbens are decreased during cocaine withdrawal after unlimited-access self-administration. *Brain Res.* 593:314–18

West RJ, Hajek P, Belcher M. 1989. Severity of withdrawal symptoms as a predictor of outcome of an attempt to quit smoking. *Psychol. Med.* 19:981–85

Whiteaker P, Davies AR, Marks MJ, Blagbrough IS, Potter BV, et al. 1999. An autoradiographic study of the distribution of binding sites for the novel alpha7-selective nicotinic radioligand [3H]-methyllycaconitine in the mouse brain. *Eur. J. Neurosci.* 11:2689–96

Wilson CJ. 2004. Basal ganglia. In *The Synaptic Organization of the Brain*, ed. GM Shepherd, pp. 361–413. Oxford, UK: Oxford Univ. Press

Winter C, Vollmayr B, Djodari-Irani A, Klein J, Sartorius A. 2010. Pharmacological inhibition of the lateral habenula improves depressive-like behavior in an animal model of treatment resistant depression. *Behav. Brain Res.* 216:463–65

Wise RA. 1978. Catecholamine theories of reward: a critical review. *Brain Res.* 152:215–47

Wise RA. 2004. Dopamine, learning and motivation. *Nat. Rev. Neurosci.* 5:483–94

Wise RA. 2006. Role of brain dopamine in food reward and reinforcement. *Philos. Trans. R. Soc. Lond. B* 361:1149–58

Wise RA. 2009. Roles for nigrostriatal—not just mesocorticolimbic—dopamine in reward and addiction. *Trends Neurosci.* 32:517–24

Wise RA, Rompre PP. 1989. Brain dopamine and reward. *Annu. Rev. Psychol.* 40:191–225

Wise RA, Yokel RA, DeWit H. 1976. Both positive reinforcement and conditioned aversion from amphetamine and from apomorphine in rats. *Science* 191:1273–75

Wonnacott S, Kaiser S, Mogg A, Soliakov L, Jones IW. 2000. Presynaptic nicotinic receptors modulating dopamine release in the rat striatum. *Eur. J. Pharmacol.* 393:51–58

Wonnacott S, Sidhpura N, Balfour DJ. 2005. Nicotine: from molecular mechanisms to behaviour. *Curr. Opin. Pharmacol.* 5:53–59

Woolf NJ. 1991. Cholinergic systems in mammalian brain and spinal cord. *Prog. Neurobiol.* 37:475–524

Wooltorton JR, Pidoplichko VI, Broide RS, Dani JA. 2003. Differential desensitization and distribution of nicotinic acetylcholine receptor subtypes in midbrain dopamine areas. *J. Neurosci.* 23:3176–85

Zhang H, Sulzer D. 2004. Frequency-dependent modulation of dopamine release by nicotine. *Nat. Neurosci.* 7:581–82

Zhang L, Doyon WM, Clark JJ, Phillips PE, Dani JA. 2009a. Controls of tonic and phasic dopamine transmission in the dorsal and ventral striatum. *Mol. Pharmacol.* 76:396–404

Zhang T, Zhang L, Liang Y, Siapas AG, Zhou FM, Dani JA. 2009b. Dopamine signaling differences in the nucleus accumbens and dorsal striatum exploited by nicotine. *J. Neurosci.* 29:4035–43

Zhou FM, Liang Y, Dani JA. 2001. Endogenous nicotinic cholinergic activity regulates dopamine release in the striatum. *Nat. Neurosci.* 4:1224–29

Zhou FM, Wilson C, Dani JA. 2003. Muscarinic and nicotinic cholinergic mechanisms in the mesostriatal dopamine systems. *Neuroscientist* 9:23–36

Zhou FM, Wilson CJ, Dani JA. 2002. Cholinergic interneuron characteristics and nicotinic properties in the striatum. *J. Neurobiol.* 53:590–605

Neuronal Intrinsic Mechanisms of Axon Regeneration

Kai Liu, Andrea Tedeschi, Kevin Kyungsuk Park, and Zhigang He

F.M. Kirby Neurobiology Center, Children's Hospital, and Department of Neurology, Harvard Medical School, Boston, Massachusetts 02115; email: zhigang.he@childrens.harvard.edu

Annu. Rev. Neurosci. 2011. 34:131–52

First published online as a Review in Advance on March 25, 2011

The *Annual Review of Neuroscience* is online at neuro.annualreviews.org

This article's doi: 10.1146/annurev-neuro-061010-113723

0147-006X/11/0721-0131$20.00

Keywords

axon regeneration, spinal cord injury, optic nerve injury, growth cone, axon transport

Abstract

Failure of axon regeneration after central nervous system (CNS) injuries results in permanent functional deficits. Numerous studies in the past suggested that blocking extracellular inhibitory influences alone is insufficient to allow the majority of injured axons to regenerate, pointing to the importance of revisiting the hypothesis that diminished intrinsic regenerative ability critically underlies regeneration failure. Recent studies in different species and using different injury models have started to reveal important cellular and molecular mechanisms within neurons that govern axon regeneration. This review summarizes these observations and discusses possible strategies for stimulating axon regeneration and perhaps functional recovery after CNS injury.

Contents

INTRODUCTION

About 30 years ago, Aguayo and his colleagues demonstrated that some injured axons can regrow into the sciatic nerve grafts transplanted to the lesion sites in the adult central nervous system (CNS) (Aguayo et al. 1981, David & Aguayo 1981, Richardson et al. 1984). Sciatic nerve, a part of the peripheral nervous system (PNS), is considered a permissive substrate that supports axon regeneration of peripheral neurons. Thus, different from Ramon y Cajal's (1928) statement that "in adult centers the nerve paths are something fixed, ended, im-

mutable. Everything may die, nothing may be regenerated" (p. 750), these observations suggested that mature neurons in the adult CNS retain the intrinsic ability to grow axons and that the major reason for regeneration failure must be the nonpermissive local environment in the CNS. Based on this hypothesis, an optimistic view has proposed that the removal or blockade of environmental inhibitory activities should be sufficient to promote regeneration of injured axons and perhaps even the recovery of functions after injury. Thus numerous studies have focused on characterizing the molecular identities and signaling mechanisms of extracellular inhibitors for axon outgrowth (Case & Tessier-Lavigne 2005, Filbin 2006, Fitch & Silver 2008, Harel & Strittmatter 2006, Schwab & Bartholdi 1996, Yiu & He 2006). As the result, a number of molecules associated with the CNS myelin, the glial scar formed by reactive astrocytes, or CNS neurons themselves have been shown to possess potent inhibitory activities for neurite growth in cultured neurons. In addition, the signaling mechanisms for their inhibitory influences have also been extensively investigated (Filbin 2006, Fitch & Silver 2008, Harel & Strittmatter 2006, Yiu & He 2006).

However, removal or blockade of various extracellular inhibitory activities has thus far failed to achieve expected axon regeneration in experimental CNS injury models. Although it is still possible that certain key inhibitors remain to be identified, a more likely possibility is that most adult CNS neurons have lost their intrinsic growth ability. In fact, in previous transplantation studies, only limited subsets of CNS axons were able to grow into such permissive grafts, whereas others, such as axons of the corticospinal tract (CST), or most axons in the optic nerve, failed to do so (David & Aguayo 1981, Richardson et al. 1984). Among these, CST axons are highly refractory to regeneration. These axons form the longest descending projections in the CNS and originate in the corticospinal neurons in layer V of the cerebral cortex. In rodents, CST fibers enter the spinal cord white matter around the neonatal stage (Bareyre et al. 2005, Gribnau et al. 1986, Schreyer & Jones

PNS: peripheral nervous system

CST: corticospinal tract

1982). When the neonatal rat spinal cord is injured, at which point the CST has not fully developed, those CST axons that have not yet reached the level of the injury can grow past the lesion site and innervate the appropriate sublesional areas (Bregman et al. 1989). However, after the CST is fully formed, no CST axons can regrow past the lesion after a spinal cord injury (Bregman et al. 1989). Thus, mature axons in the adult CNS have likely reduced or lost their regenerative ability after injury.

In principle, intrinsic regenerative ability determines how neurons respond to an axotomy insult. Young neurons in the early developmental stage possess robust axon growth and regenerative ability. Furthermore, PNS axons in adult mammals and even CNS axons in lower organisms such as amphibians can still regenerate. In this review, we first describe different aspects of neuronal responses to axotomy, and then we discuss cellular and molecular insights obtained from different experimental paradigms.

INJURY RESPONSES

Cell Body Responses

In neuronal cell bodies, axotomy-triggered morphological changes have been referred to as chromatolysis (Brodal 1981, Kreutzberg 1982, Lieberman 1971) (**Figure 1**). Acute reactions are quite similar in both regeneration-competent neurons (such as PNS neurons) and regeneration-incompetent neurons (such as CNS neurons), and these include dispersal of the Nissl substance, displacement of the nucleus to the cell's periphery, swelling of the

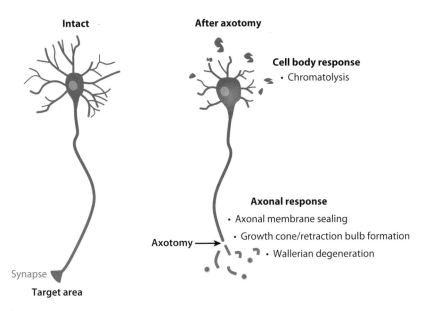

Figure 1

Neuronal responses to axotomy. After an axotomy occurs in the adult, both the injured axon and its cell body show distinct responses. The injured axon terminal reseals. The distal axonal segment undergoes self-destructive Wallerian degeneration. Depending on the neuronal types, the proximal end forms a growth cone–like structure or a retraction bulb and initiates regenerative growth (PNS neurons) or abortive sprouts (CNS neurons). Distant from the axotomy site, the cell body often shows chromatolysis to different extents, depending on the neuronal types, the distance between the injury site and the cell body, and the numbers of remaining collaterals. Overall, regeneration-competent neuronal cell bodies appear hypertrophic, whereas regeneration-incompetent neurons become atrophic after injury.

cell body, and loss or retraction of synaptic terminals. The extent of these changes depends on the distance of the injury site to the cell body and the number of remaining axonal collaterals. However, the long-term cell body responses induced by axotomy vary considerably. In the regeneration-competent neurons, such as spinal motor neurons, cell bodies remain hypertrophic and show signs of increased metabolism and protein synthesis, such as increased free ribosomes and other intracellular organelles (Kreutzberg 1982, Lieberman 1971). Whereas in regeneration-incompetent CNS neurons, these hypertrophic changes do not occur. Instead, many of these injured neurons appear atrophic, displaying reduced cell volume and dendritic arborization. In neurons such as Purkinje cells, cell bodies show little change after axotomy, and their axons do not regenerate (Rossi et al. 1995).

Axonal Responses

Axotomy also triggers drastic responses in axonal compartments (**Figure 1**). While the distal part undergoes Wallerian degeneration, the proximal lesion site reseals the damaged axonal membrane (Fishman & Bittner 2003, Schlaepfer & Bunge 1973). A critical step for axon regeneration is the transformation of the cut axonal end into a growth cone–like structure that can integrate extracellular signals and orchestrate the use of cellular materials for axon regrowth (Tessier-Lavigne & Goodman 1996, Yu & Bargmann 2001). In the CNS of most adults, a cut axonal end often fails to transform into a growth cone apparatus, but instead forms an end bulb or retraction bulb (Hill et al. 2001, Li & Raisman 1995). For example, after spinal cord injury in adult rodents, damaged CST axons rostral to the injury form retraction bulbs and undergo considerable dying-back away from the lesion site (Bernstein & Stelzner 1983, Bregman et al. 1989, Hill et al. 2001). Often these injured CST axons also make initial attempts to sprout and regrow. However, such sprouting responses are always transient and abortive

(Bregman et al. 1989; Joosten et al. 1995; Li et al. 1997, 1998; Ramon-Cueto et al. 2000).

Nonneuronal Responses

As do neuronal cell bodies and axons, nonneuronal components show varied responses after axotomy in PNS and CNS. In the PNS, axon and myelin debris could be efficiently cleared (George & Griffin 1994), whereas such processes are significantly slower in the CNS (Vargas & Barres 2007). Endogenous antibodies are likely required for robust PNS myelin clearance and axon regeneration (Vargas et al. 2010). B-cell knockout the targeted deletion of J_H locus (JHD) mice display a significant delay in macrophage influx, myelin clearance, and axon regeneration. Rapid clearance of myelin debris could be restored in mutant JHD mice by passively transferring antibodies from naïve wild-type mice or by using an anti-PNS myelin antibody, which suggests that the immune-privileged status of the CNS may contribute to CNS myelin clearance failure and axon regeneration failure after injury (Vargas et al. 2010). In contrast, an injury in the adult CNS initiates specific glial responses, such as the upregulation of chondroitin sulfate proteoglycans expression and the formation of the glial scar, which restrict axon regeneration both physically and chemically (Fitch & Silver 2008, Properzi et al. 2003).

Thus, to survive an axotomy, an injured axon needs to reseal the damaged membrane quickly. Local cytoskeletal rearrangement often occurs in the axonal terminal, causing the growth cone to reform and perhaps sprout. However, to sustain lengthy axon regeneration, a cell body response appears to be required, such as activating an axon regrowth program to ensure the synthesis of raw materials for axon growth and transport and assembly of axonal components along the axonal shaft and at the terminal. All these steps must be coordinated to ensure successful axon regrowth. Depending on the types of neurons and injury, the hurdles for successful axon regeneration could vary. Recent studies in different models have begun to reveal possible rate-limiting steps in axon regeneration, which we discuss below.

DEVELOPMENT-DEPENDENT DECLINE OF AXON GROWTH ABILITY

During early development, axon growth is likely to be supported by a set of trophic factors and extracellular matrix molecules. When an axon reaches its targets, it must stop elongating and transform its growth cone into a presynaptic terminal. Correspondingly, neuronal function will also switch axon growth to synaptic growth and maturation. Further synaptic activity could further consolidate this terminal differentiated state. Thus, one hypothesis proposes that the loss of regenerative ability in mature neurons could be developmentally programmed and associated with synapse formation and function.

Which mechanisms account for axon growth during development? In PNS neurons, axon growth requires a group of polypeptide growth factors such as nerve growth factor (NGF) and other neurotrophins (NT) (Glebova & Ginty 2005, Goldberg & Barres 2000, Reichardt 2006, Segal & Greenberg 1996). The neurotrophin family members NGF, brain-derived neurotrophic factor (BDNF), NT3, and NT4 act via their corresponding receptor tyrosine kinases (TrkA, TrkB, and TrkC). Upon activation by neurotrophins, these receptor tyrosine kinases trigger a number of intracellular signaling pathways (Reichardt 2006, Segal & Greenberg 1996, Zhou & Snider 2006). Among different signaling pathways engaged by neurotrophins, two of them play important roles in stimulating axon growth: the classical phosphoinositide 3-kinases (PI3K)/serine/threonine protein kinase B (AKT) pathway and the Ras-activated rapidly accelerated fibrosarcoma (RAF)/extracellular signal-regulated kinases (ERK) kinase cascade (Liu & Snider 2001, Markus et al. 2002).

In CNS neurons, neuronal activity, in addition to trophic factors, is also important for neuronal survival and axon growth. Cultured retinal ganglion cells (RGCs) do not extend axons by default. Forced expression of Bcl-2 is sufficient to keep purified RGCs alive, but these neurons extend axons only in the presence of trophic growth factors. Furthermore, physiological levels of electric activity could significantly enhance axonal growth stimulated by trophic factors (Goldberg et al. 2002a). One mechanism could be related to the observation that depolarization and cyclic adenosine monophosphate (cAMP) elevation could rapidly recruit TrkB to the plasma membrane by stimulating the exocytosis of TrkB-carrying vesicles (Meyer-Franke et al. 1998).

Evidence suggests that mature neurons, in contrast with immature neurons, in the CNS possess reduced or lost capacity for axon growth. For RGCs, a dramatic reduction in axon elongation ability occurs around birth (Chen et al. 1995, Goldberg et al. 2002a). Embryonic RGCs extend axons ~10 times faster than do the mature ones (Goldberg et al. 2002a, Jhaveri et al. 1991). Signals from other cell types such as amacrine cells are thought to trigger the switch in RGC growth capacity (Goldberg et al. 2002b). However, another hypothesis proposes that the loss of axon growth ability could be a by-product of neurons exiting the cell cycle. The anaphase-promoting complex (APC), an E3 ubiquitin ligase complex, is highly expressed in postmitotic neurons. Konishi et al. (2004) showed that inhibition of APC and its activator Cdh1 could promote axon growth in cerebellar granule cells, possibly by affecting several substrates of APC, such as transcription factor Id2 (Lasorella et al. 2006). However, the contribution of this pathway to axon regeneration in vivo remains unknown. In sum, loss of axon growth ability over the course of development is likely the result of interactions between extracellular signals and neuronal intrinsic programs.

What could be the molecular basis for this development-dependent reduction of axon growth ability? Bcl-2 was one of the first molecules implicated in this process (Chen et al. 1997). In lower vertebrates, such as fish and frog, CNS neurons upregulate Bcl-2 after

RGCs: retinal ganglion cells

cAMP: cyclic adenosine monophosphate

injury, which correlates with axon regeneration (Cristino et al. 2000). In rodents, Bcl-2 expression also parallels axon extension during development, and overexpression of Bcl-2 appears to increase axonal growth ability of RGCs (Cho et al. 2005).

Recent studies suggest that Kruppel-like factors (KLFs), a set of zinc-finger transcription factors, are another potential type of development-dependent molecular regulators of axon growth ability. These molecules were initially implicated in regulating cell-cycle exit and terminal differentiation of nonneuronal cells. Among these different members, KLF4 is one of the four transcription factors that are sufficient to transform fibroblasts into pluripotent stem cells (Takahashi et al. 2007). The first line of evidence indicating KLF involvement in axon growth and regeneration came from studies in zebrafish (Veldman et al. 2007). Unlike mammals, teleost fish can mount an efficient and robust regenerative response following optic nerve injury. To identify genes involved in successful nerve regeneration, Veldman et al. (2007) analyzed gene expression in zebrafish RGCs following optic nerve injury and found that KLF6a and KLF7a are upregulated after injury, and their expression is also necessary for robust RGC axon regrowth. Further gene ontological analysis of the data set suggests that regenerating neurons upregulate genes associated with RGC development. However, not all regeneration-associated genes are expressed in developing RGCs, indicating that regeneration is not simply a recapitulation of development.

In an independent study, Moore et al. (2009) identified another KLF, KLF4, as a potent inhibitory gene for axon growth of embryonic hippocampal neurons and RGCs. In a survey of more than 100 neuronal genes whose expression in rat RGCs changes around birth, KLF4 was the most potent inhibitor of axonal growth when tested in rat hippocampal neurons (Moore et al. 2009). Conversely, targeted deletion of the KLF4 gene increases the number and length of regenerating RGC axons after optic nerve injury. Notably, different KLFs possess temporally distinct expression patterns during development. In comparison with embryonic RGCs, KLF6/7 are downregulated and KLF4/9 are upregulated in adult. Overexpression of these different KLFs leads to opposite effects on neurite growth in cortical neurons. However, we still do not know how different KLFs and other transcription factors control the expression of extrinsic and/or neuronal intrinsic regulators of growth ability.

In addition to Bcl-2 and KLFs, the mammalian target of rapamycin (mTOR) levels also show a development-dependent decline in both RGCs (Park et al. 2008) and cortical neurons (Liu et al. 2010). Thus, it is likely that multiple molecules and pathways account for development-dependent decline of axon growth ability in CNS neurons.

CONDITIONING EFFECTS OF A PERIPHERAL LESION IN SENSORY NEURONS

The preconditioning lesion of primary sensory neurons from dorsal root ganglia (DRGs) has been an extensively explored model to study mechanisms regulating axon regeneration. The primary sensory neurons are unique because they have axons projecting in both the PNS and the CNS. Each DRG neuron has two major branches stemming from a unipolar axon: a peripheral axon that innervates peripheral targets and a central axon that relays the information to the spinal cord. However, these two branches of sensory axons differ with regard to their ability to regenerate after axotomy. Whereas the peripheral axon can regenerate back to its original target after lesion, the central branch from the same DRG neuron engages in only abortive sprouting after axotomy within the CNS. A first injury at the peripheral branch, called a conditioning lesion, can dramatically increase the regenerative responses to a second lesion occurring at either the peripheral or the central branches (Neumann & Woolf 1999, Richardson & Verge 1986). Richardson & Issa (1984) initially discovered that DRG neurons could regenerate their central axons into peripheral nerve grafts if the peripheral axons had been previously cut. Woolf and colleagues (Neumann & Woolf 1999) subsequently showed that injured central axons from DRG neurons with a conditioning lesion could regenerate across the hostile spinal cord injury sites. Because adult DRG neurons possess the competence for axon regeneration, the conditioning effects provide a unique model to investigate how the regenerative program is turned on or off.

The effects of the conditioning lesion require gene transcription alteration. In this regard, a peripheral injury could trigger a recapitulation of the developmental patterns of expression for growth-associated proteins (GAP-43, CAP23, SPRR1A, and cytoskeleton components) and a decrease in the neurofilament proteins. Other upregulated genes include transcription factors (such as ATF-3, c-Jun, Sox11, and Smad1), translation regulators, and arginase 1 (a rate-limiting enzyme in polyamine biosynthesis). The downregulated genes include ion channels and protein involved in neurotransmitter synthesis (reviewed by Hoffman 2010 and Richardson et al. 2009).

To transform a peripheral lesion into axonal regeneration program activation, retrograde signals must be involved (Cragg 1970, Rishal & Fainzilber 2010) (**Figure 2**). Such signals could be negative (e.g., interruption of the

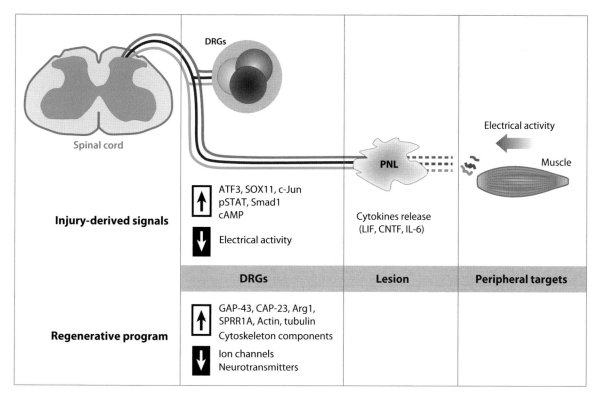

Figure 2

Both positive and negative signals contribute to the conditioning effects of a peripheral nerve lesion (PNL). In the adult dorsal root ganglia (DRG) neurons, a PNL activates the axon regenerative program, indicated by the upregulation of regeneration-associated genes such as GAP-43, Smad1, and cytoskeleton components, as well as by the down-regulation of other genes such as ion channels and neurotransmitters. This activation will allow a robust axon regeneration to occur if the same prelesioned neurons receive a second axotomy even in the central branch. Many factors likely contribute to this conditioning effect, including injury-induced positive regulators such as activation of the Janus kinase/signal transducers and activators of transcription (JAK/STAT) pathway as the result of locally released cytokines, and the removal of negative influence from peripheral targets, such as electrical activity.

transport of peripheral target-derived signals) and/or positive (e.g., new signals generated from the stump of the injury). The concept of injury-elicited positive retrograde signals was elegantly demonstrated in *Aplysia*, whereby injection of axoplasm from injured sensory neurons could induce regenerative responses in uninjured neurons (Ambron et al. 1996, Zhang & Ambron 2000). Recent studies point to several potentially important mediators in the effects of a conditioning lesion on axon regeneration.

Neuropoietic Cytokines

Several cytokines, such as leukemia inhibitory factor (LIF), interleukin-6 (IL-6), and ciliary neurotrophic factor (CNTF), released locally at the peripheral lesion sites (Cao et al. 2006, Sendtner et al. 1992, Subang & Richardson 2001) (**Figure 2**) could play important roles in activating the regenerative program. Through gp130-containing receptor complexes, these cytokines could activate intracellular signaling pathways such as the JAK-STAT pathway. Consistent with an involvement of this pathway in the conditioning effect, the phospho-Stat3 signal is accumulated in the nucleus of DRG neurons after a peripheral but not a central injury (Qiu et al. 2005, Schwaiger et al. 2000). Functionally, IL-6 or overexpression of active Stat3 could promote the neurite growth of cultured DRG neurons (Cao et al. 2006, Miao et al. 2006). In culture, while inhibitors of ERK or PI3K could block neurite growth in embryonic DRG neurons, a specific Jak2 inhibitor could efficiently abolish the neurite growth of DRG neurons with a conditioning lesion (Liu & Snider 2001, Miao et al. 2006). However, IL-6 knockout mice still show normal axon regeneration in the PNS (Cao et al. 2006). Thus, this pathway's contribution to the conditioning effect still remains unknown.

Cyclic AMP (cAMP)

Because multiple inhibitors are associated with CNS myelin and the glial scar, injured ascending dorsal column axons from DRG neurons with a conditioning lesion must be able to overcome these local inhibitory activities. An important underlying mechanism could be mediated by the elevation of cAMP levels in these neurons with a conditioning lesion. Early studies in cultured *Xenopus* spinal neurons suggested that elevating cAMP levels could allow these neurons to convert from repulsion to attraction in response to inhibitory molecules such as myelin-associated glycoprotein (MAG) (Song et al. 1997, 1998). In mammalian neurons, the ability of embryonic, but not mature, axons to elongate in the presence of MAG correlates with their intracellular cAMP levels; cAMP concentration is much higher in embryonic neurons (Cai et al. 2001). A conditioning lesion could transiently increase cAMP concentration in adult DRG neurons, which correlates with their increased axon growth ability (Qiu et al. 2002). How cAMP affects axonal growth is not entirely clear yet, but both CREB [cAMP response element–binding protein (Cai et al. 2001, Cao et al. 2006)] and arginase 1 (Cai et al. 2002, Deng et al. 2009) have been implicated as important mediators.

Axonal Transport

Although intraganglionic injection of a cAMP analog induces some growth-promoting changes, it does not fully mimic the effects of the conditioning lesion on long-distance regeneration of dorsal column axons in a peripheral nerve graft (Cai et al. 2002, Han et al. 2004, Lu et al. 2004). A possible explanation could be that a conditioning lesion, but not a cAMP analog, could promote the enhancement of anterograde axonal transport (Han et al. 2004). Considering the much longer axonal lengths in the adult compared with that during development, delivery of molecules synthesized in the cell body to regenerating axonal growth cones could be a formidable challenge for axon regeneration in the adult.

In general, axonal transport involves the ATP-dependent interactions of motor proteins and axonal microtubules, which are oriented with their plus ends toward the axonal terminals. Pulse-labeling studies showed that cytoskeletal polymers and soluble proteins are transported much slower than are membrane-associated proteins (Lasek 1968). The rate of regeneration is comparable to the velocity of such slow axonal transports (Lasek & Hoffman 1976, Wujek & Lasek 1983). A conditioning lesion could increase this slow axon transport (McQuarrie et al. 1977); however, the underlying mechanisms are still unknown. The correlation between axon transport and regeneration is not limited to sensory neurons. McKerracher and colleagues have shown that after optic nerve injury, axonal transport is dramatically decreased (McKerracher et al. 1990, McKerracher & Hirscheimer 1992). Application of a Rho antagonist could promote local sprouting but not axonal transport and lengthy regeneration (Bertrand et al. 2005), consistent with the importance of axonal transport to long-distance axon regeneration.

In addition to axonal transport, a possible strategy to bypass the long distance between the soma and the axonal terminal is to employ local protein synthesis (reviewed by Gumy et al. 2010). Dissociated sensory neurons from DRGs with a conditioning lesion have increased mRNAs in the axonal compartments (Zheng et al. 2001), suggesting a potentially important source for synthesizing axonal components such as β-actin and growth-associated tubulin isotypes (Yoo et al. 2010). In addition, locally synthesized proteins could also facilitate the retrograde transport of injury signals in the activating regenerative program in the neuronal cell bodies (Hanz et al. 2003, Perlson et al. 2005). For example, local translated vimentin and importin-β could allow the transport of activated mitogen-associated protein (MAP) kinase (ERK1/2) from the injury site to the cell bodies (Hanz et al. 2003, Perlson et al. 2005).

Silencing of Electrical Activity

In addition to the positive factors discussed above, evidence also suggests that the peripheral targets of sensory neurons convey a suppressive signal to inhibit axon growth.

For example, the interruption of retrograde transport in the peripheral axons by transection or ligation can enhance axonal growth ability of affected sensory neurons (Lankford et al. 1998). Thus, a conditioning lesion could result at least partially from relieving sensory neurons from such inhibitory influence. Recent studies suggest that a possible contributor is the silencing of the electrical activity by the peripheral lesion (Enes et al. 2010). Assayed by in vivo electrophysiological recording of sensory neurons, Enes et al. (2010) showed that a lesion at the peripheral, but not a central, branch dramatically decreases the sensory-evoked firing rate of DRG neurons, consistent with the fact that the peripheral axon transmits sensory inputs to DRG neurons. In cultured adult DRG neurons, Enes et al. (2010) found that treatments of increasing neuronal electrical activity reduce neurite initiation and elongation, reminiscent of the observations that electrical stimulation of cultured DRG neurons leads to rapid growth cone collapse (Fields et al. 1990). Further analysis suggests that such effects are mediated by the peripheral lesion-triggered downregulation of Cav1.2, a component of L-type voltage-gated calcium channel (VGCC) (Enes et al. 2010). However, its different patterns of electrical activity could have much more complicated roles in modulating neuronal growth and differentiation. In fact, Udina et al. (2008) demonstrated that electrical stimulation of the sciatic nerve could actually promote regeneration of the injured central axons of the stimulated sensory neurons. Thus, it is clear that multiple mechanisms could contribute to conditioning effects, and it is a major challenge to dissect out their relative contributions.

AXON REGENERATION REGULATORS IN INVERTEBRATES

Robust axon regeneration after injury and powerful forward genetics make invertebrates excellent models for analyzing cellular and molecular pathways involved in axon regeneration. This analysis has been greatly facilitated

DLK-1: dual leucine
zipper kinase 1

by recent developments of a precise laser axotomy method, which allows investigators to lesion single axons and to monitor their regeneration in live *C. elegans* (Yanik et al. 2004). By this method, Yan et al. (2009) identified the DLK (dual leucine zipper kinase 1) MAP kinase pathway as an important regulator of axon regeneration. Independently, Hammarlund et al. (2007) took advantage of an available Unc-70 mutant in which axons break as animals move with continuous regeneration. A large-scale screen for Unc-70 mutant animals with diminished regeneration identified a requirement for DLK-1 in axon regeneration. Time-lapse analysis revealed that, although filopodia extend from severed axons, in DLK-1 mutants the motile growth cone fails to form (Hammarlund et al. 2009). Overexpression of DLK-1 in a wild-type background increases the success of growth cone formation and the extent of axon regeneration. Initial axon growth appears normal in DLK-1 mutants, suggesting that DLK-1 is required for axon regeneration but not for axonal development (Hammarlund et al. 2009), consistent with the results by an independent study showing that distinct pathways mediate initial axon development and adult-stage axon regeneration (Gabel et al. 2008).

DLK-1 is a component of a conserved mitogen-activated protein kinase (MAPK) cascade that includes the MAP kinase kinase MKK-4 and the p38 MAP kinase PMK-3. Loss-of-function mutations of the *dlk-1*, *mkk-4*, or *pmk-3* genes all impair axon regeneration in *C. elegans*, suggesting that this entire pathway is required for axon regeneration (Hammarlund et al. 2009). A possible mechanism is that DLK-1 functions by signaling via a MAPKAP kinase, MAK-2, to stabilize the mRNA encoding CEBP-1, a bZip protein related to CCAAT/enhancer-binding protein (C/EBP), and induces translation of CEBP-1 in axons (Yan et al. 2009). In the mammalian PNS, C/EBP is also induced by injury, and murine C/EBPβ can activate the transcription of the α-tubulin gene associated with injury responses (Korneev et al. 1997, Nadeau et al. 2005).

However, because the DLK-1 pathway is required for growth cone formation ∼7 h after axotomy, Hammarlund et al. (2009) proposed that this pathway affects the process of microtubule dynamics during growth cone formation (**Figure 3**). Previous studies have implicated a critical role of microtubule reorganization in growth cone formation after injury. Imaging studies in *Aplysia* neurons revealed that axotomy leads to reorientation of the microtubule polarity (Erez et al. 2007). Different from a regeneration-competent growth cone, a prominent feature of a regeneration-incompetent retraction bulb is the disorganized microtubules in the terminal. In addition, in both cultured DRG neurons (Chierzi et al. 2005) and *Aplysia* neurons (Kamber et al. 2009), local calcium is critical for growth cone formation after transection. In *C. elegans*, both calcium and cAMP promote growth cone formation in a DLK-1 kinase-dependent manner (Ghosh-Roy et al. 2010). Together, these observations highlight growth cone formation as an important step in the initiation of axon regeneration. It would be interesting to examine how these pathways regulate microtubule dynamics during growth cone formation. Some components of this and related pathways have been implicated in injury response and axon regeneration in mammals. For example, we know that axotomy could activate Jun N-terminal kinase (JNK) (Cavalli et al. 2005, Lindwall et al. 2004) and c-Jun knockout mice show defects in PNS axon regeneration (Raivich et al. 2004, Zhong et al. 1999). Whether these pathways also regulate growth cone formation or other steps of axon regeneration remains unknown.

REGULATORS OF AXON REGENERATIVE ABILITY IN MAMMALIAN CNS

As discussed above, only some injured axons could regrow into the transplanted permissive grafts after a CNS lesion. For example, some propriospinal axons and brain stem–derived axons, but not CST, could grow into the grafts transplanted to the injured spinal cord (David

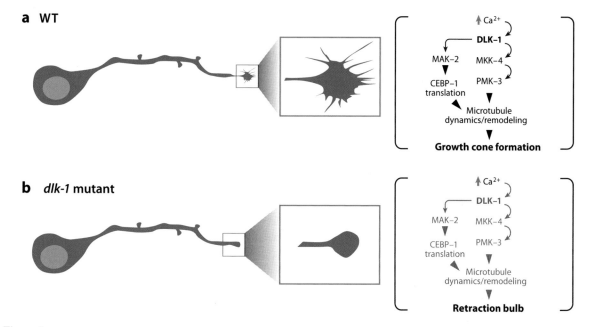

a WT

b *dlk-1* mutant

Figure 3

The DLK-1 pathway is required for growth cone formation and axon regeneration in *C. elegans*. After an axotomy in wild-type animals, the resealed terminals of injured axons elaborate filopodia and form growth cones for axon regeneration. In *dlk-1* mutants, injured axons still extend filopodia but fail to form growth cones and regenerate injured axon. Mutants of several DLK-1 downstream genes such as MKK-4 and PMK-3 have similar phenotypes, suggesting that this pathway is required for axon regeneration. In addition, evidence also suggests that MAK-2 and CEBP-1 local translation is involved.

& Aguayo 1981, Richardson et al. 1984), which suggests that different white matter tracts differ in their intrinsic regenerative ability. Even within the same types of axons, the distance between the neuronal soma and the injury site could influence the regenerative responses. In axotomized CNS neurons, the upregulation of regeneration-related genes fails to occur or happens transiently if the neurons are injured at greater distances from their cell bodies (Fernandes et al. 1999). For example, only after axotomy at the cervical but not thoracic spinal cord level, rubrospinal neurons increase growth-associated protein 43 (GAP-43) expression, and injured rubrospinal axons can regenerate into the transplanted permissive grafts (Fernandes et al. 1999). Thus, these observations suggest that, similar to PNS neurons, increasing neuronal responses to axotomy in the cell bodies are critical for axon regeneration. However, for sustained axon regeneration cell

bodies and also axonal terminals and shafts must be involved. Several approaches have been utilized to explore the underlying mechanisms.

Reactivating Trophic Responses

On the basis of extensive evidence establishing roles of neurotrophins in promoting axon growth during development (Reichardt 2006, Zhou & Snider 2006, Zweifel et al. 2005), efforts have been made to activate growth-promoting pathways in injured CNS neurons by exogenously providing trophic factors or overexpressing signaling molecules. These approaches have generated mixed results. In optic nerve injury models, intravitreal application of BDNF could promote survival of axotomized RGCs, whereas CNTF could increase the axon growth to some extent (Leaver et al. 2006, Nakazawa et al. 2002, Park et al. 2004, Pernet & Di Polo 2006, Smith et al. 2009). For the CST,

NT-3, but not NGF or BDNF, could increase the sprouting of injured CST axons rostral to a spinal cord lesion, but all fail to stimulate the re-growth of severed CST axons across the lesion site (Schnell et al. 1994) or even into permissive grafts (Blits et al. 2000, Grill et al. 1997). More recently, in purified corticospinal neurons from early postnatal stages, Ozdinler & Macklis (2006) found that insulin-like growth factor (IGF) and BDNF could promote axon elongation or branching, respectively. However, IGF delivery in the adult could promote corticospinal neuronal survival but not axon regeneration after CNS injury (Hollis et al. 2009b). Thus, it appears that neuronal responses to trophic factors may vary even at different developmental stages or conditions. A potential explanation for limited success of growth factors in promoting axon regeneration is the reduced or altered responsiveness of mature neurons to these trophic factors. In fact, Shen et al. (1999) showed that isolated (injured) RGCs do not respond to growth factors unless they are depolarized or their intracellular cAMP levels are elevated.

Additional efforts have been made to activate specific signaling pathways directly. For example, lentivirus-mediated TrkB expression in cortical neurons could increase the regrowth of corticospinal axons after a subcortical injury (Hollis et al. 2009a). In this study, evidence also suggested an important role for ERK activation because a mutant Trk with a selective mutation of the Shc/FRS-2 activation domain fails to elicit axon regrowth. In this regard, mammalian sterile 20-like kinase-3b (Mst3b, encoded by Stk24), an ERK downstream-signaling molecule, could promote axon growth (Lorber et al. 2009). However, over-expression of active ERK1/2 could promote neuronal survival but not axon regeneration after optic nerve injury (Pernet et al. 2005). As we describe below, the mTOR activity, which is regulated by different trophic factors, appears to be an important limiting factor for neuronal regenerative responses in the adult CNS.

Activation of Growth Ability by Inflammation

Although application of trophic factors was based on their effects on immature neurons during development, certain stimuli may exist specifically for promoting neuronal responses to injury in the adult CNS. By transplanting a fragment of a peripheral nerve into the vitreous humor, Berry et al. (1996) found that this procedure can promote axon regeneration after optic nerve crush. Although this mechanism was initially interpreted as the result of trophic factors from Schwann cells, later studies suggested that primarily the infiltrated inflammatory cells such as macrophages are responsible for activating axon regeneration. Other procedures of activating macrophages, such as lens injury and intravitreous injection of zymozan, could also promote optic nerve regeneration (Fischer et al. 2001, Leon et al. 2000). Isolated sensory neurons from the DRG with zymosan injection also show increased neurite growth (Steinmetz et al. 2005), suggesting that inflammation in the cell bodies could activate the intrinsic regenerative program of both CNS and PNS neurons.

What could be the underlying mechanisms? A possible contributor is oncomodulin, a macrophage-derived growth factor for RGCs (Yin et al. 2006). However, evidence also supports the involvement of injury-induced cytokine induction (Hauk et al. 2008, Muller et al. 2007). In particular, Leibinger et al. (2009) showed that the effect of lens injury is reduced in CNTF-/- mice and completely blocked in CNTF-/-LIF-/- double mutant mice, suggesting a critical role of injury-induced cytokines in inflammation-induced activation of axon regeneration.

Increasing Regenerative Response by Modulating PTEN/mTOR

To explain the difference between embryonic-neuron and adult-neuron axon growth ability, a simple hypothesis holds that axon growth ability is regulated by the same mechanisms that control cellular growth and size in other cell types. Several molecular pathways have been implicated in preventing cell overgrowth (Weinberg 2007). Park et al. (2008) tested involvement of these pathways in axon regeneration responses of RGCs after optic nerve injury models. In so doing, investigators used a conditional gene knockout approach, in which a particular tumor suppressor gene was targeted for deletion by injecting adeno-associated virus (AAV)-Cre into the retina of floxed mice. Eliminating the gene encoding the tumor suppressor p53 enhanced RGC survival, but not axon regeneration, which is consistent with the notion that axon regeneration could be operated by survival-independent mechanisms. However, eliminating the gene encoding the tumor suppressor phosphatase and tensin homolog (PTEN) not only prevented apoptosis of induced RGCs, but also promoted robust axon extension (Park et al. 2008).

In antagonizing PI3K activity, PTEN is a phosphatase that could convert phosphatidylinositol (3,4,5) trisphosphate (PIP_3) to phosphatidylinositol (4,5) bisphosphate (PIP_2). PTEN deletion leads to PIP_3 accumulation, which in turn recruits and activates phosphatidylinositol-dependent kinase 1/2 (PDK1/2), resulting in AKT activation (Carnero 2010, Guertin & Sabatini 2007, Luo et al. 2003, Park et al. 2010). Among the multiple downstream targets of this pathway, mTOR controls cap-dependent protein translation by regulating the activity of ribosomal S6 kinase and eukaryotic initiation factor 4E binding protein (Guertin & Sabatini 2007). Applying rapamycin, an inhibitor of mTOR, abolished the regenerative effect of PTEN deficiency (Park et al. 2008), suggesting that axon regeneration induced by PTEN deletion is dependent on the mTOR pathway.

Further analysis suggests that mTOR activity is an important indicator of axon regenerative ability in RGCs. During development, the mTOR level decreases in most RGCs, which may provide a plausible explanation for development-dependent decline of axon growth ability. In addition, axotomy triggers a stress response, resulting in further reduction of the mTOR levels in axotomized RGCs (Park et al. 2008). Because of the central role of mTOR in regulating protein translation, diminished mTOR activity in axotomized RGCs could be an important mechanism of regeneration failure.

However, mTOR is unlikely to be the only downstream effector of PTEN deletion-triggered axon regeneration. Genetic ablation of a negative mTOR regulator, tuberous sclerosis complex 1 (TSC1), partially but not completely mimicked the effects of PTEN deficiency on axon regeneration (Park et al. 2008), indicating that other PTEN-regulated pathways could be involved in controlling axon growth. In this respect, $3'$ phosphorylated phosphoinositides ($3'$ PIs), the direct products of PTEN, regulate asymmetric signaling and orientating polarized outgrowth during axon formation (Adler et al. 2006). In addition, glycogen synthesis kinases (GSK-3) may also be involved. Zhou & Snider (2005) suggest that GSK-3 is a master kinase that relays convergent signals from extracellular factors in controlling different aspects of microtubule assembly in the initiation and extension of a growth cone. By regulating collapsin response mediator protein 2 (CRMP2) and perhaps other α/β tubulin dimer-binding proteins, GSK-3β may regulate microtubule polymerization (Fukata et al. 2002). On the other hand, by phosphorylating APC, a microtubule plus end binding protein, GSK-3 reduces the stability of microtubule in the axon terminals (Zhou et al. 2004). In addition to these primed substrates of GSK-3s for which a prior phosphorylation event in the vicinity of GSK-3 phosphorylation sites is required (Woodgett 2001), GSK-3s have nonprimed substrates such as a major microtubule binding protein MAP1b, which can also affect microtubule dynamics (Trivedi et al. 2005). However, the participation of these mTOR-independent pathways in axon regeneration remains to be determined.

Increased axon regeneration after PTEN deletion is not limited to RGCs. Liu et al. (2010) showed that mTOR activity is also an important regulator of axon regrowth (both compensatory

PTEN: phosphatase and tensin homolog

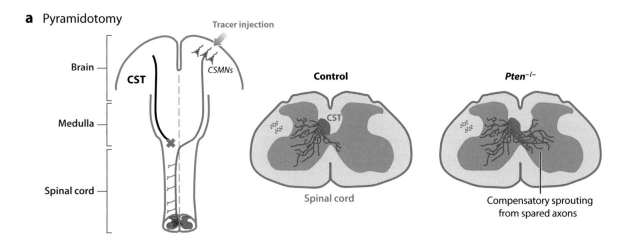

a Pyramidotomy

b Spinal cord injury

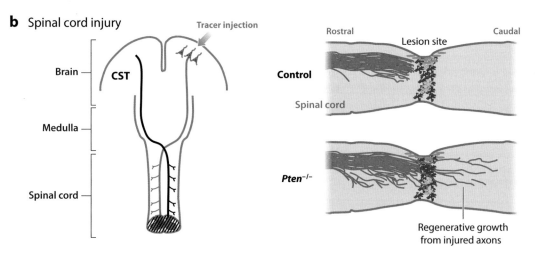

Figure 4

Two types of CST axon regrowth. In assessing the intrinsic growth ability of corticospinal motoneurons (CSMNs), one can examine both regenerative growth of injured axons and compensatory sprouting of intact axons. The scheme illustrated in panel *a* will assess the compensatory sprouting of intact CST axons after one side of CST axons is transected prior to decussation at the pyramid by a unilateral pyramidotomy. An anterograde tracer will be injected into the contralateral side of motor cortex to label intact CST axons (*red*). Transverse sections of the spinal cord could be used to determine the extent of compensatory sprouting. In control mice, most CST axons innervate the contralateral spinal cord, thus few CST axons could cross the midline and project to the contralateral side of the spinal cord. However, PTEN deletion in the CSMNs increases their intrinsic growth ability, and more CST axons could cross the midline after unilateral pyramidotomy. However, a spinal cord injury that transects CST axons, illustrated in panel *b*, could be used to examine the regenerative growth of injured CST axons. Whereas in control mice, injured CST axons often retract from the lesion site, PTEN deletion in CSMNs could promote robust regenerative growth in the spinal cord both proximal and caudal to the lesion site.

sprouting from spared axons and regenerative growth from injured axons) (**Figure 4**) of CST axons. As occurs with RGCs, a development-dependent reduction of mTOR activity occurs in cortical neurons, and axotomy could further reduce this activity in injured corticospinal neurons. While maintaining mTOR activity after injury, PTEN deletion in these cortical neurons promotes robust sprouting of CST axons rostral to the lesion sites. In addition, significant numbers of regrowing CST axons could pass the lesion site in different spinal cord injury models (Liu et al. 2010). These regenerating CST axons can form synapses in the spinal cord caudal to the lesion sites (Liu et al. 2010); however, whether these regenerating axons could mediate functional recovery is still unknown. Not only in cortical neurons and RGCs, deletion of PTEN or TSC2 in DRG neurons also increases axon growth and regeneration (Abe et al. 2010, Christie et al. 2010), suggesting that PTEN/mTOR is a common regulator of axon regeneration.

SOCS3-Dependent Regulation of Axon Regeneration

Using the in vivo AAV-mediated conditional knockout approach discussed above, Smith et al. (2009) found significant optic nerve regeneration after deletion of suppressor of cytokine signaling 3 (SOCS3) in adult RGCs. SOCS3 is a known negative regulator of the gp130-dependent JAK/STAT pathway (Baker et al. 2009). No axon regeneration was observed when both SOCS3 and gp130 were deleted, indicating that the regeneration of SOCS3 mutant axons is dependent on cytokine ligands of gp130 (Smith et al. 2009). Exogenous application of CNTF to SOCS3 deleted mice could dramatically increase the extent of axon regeneration (Smith et al. 2009). An interesting observation was that downregulation of SOCS3 might contribute to cAMP-induced effects on axon growth in RGCs (Park et al. 2009). Together with the data of PTEN deletion (Christie et al. 2010, Liu et al. 2010, Park et al. 2008), these new observations highlight the dominant role of negative regulators of growth signaling pathways in restricting axon regeneration in adult neurons. A challenge for future studies is to understand how mature neurons acquire these intrinsic barriers of axon regeneration over the course of development.

SUMMARY AND PERSPECTIVES

As discussed above, successful regeneration requires the coordinated action of the axonal terminal, the shaft, and the cell bodies. An injured axonal terminal needs to reform a growth cone after injury and assemble appropriate cytoskeleton structures for axon extension. The cell body is responsible for most of the synthesis of axonal building blocks, which need to be transported along the axonal shaft for axonal regrowth. From these studies in different models, it appears that obvious rate-limiting steps in achieving axon regeneration are associated with the formation of the growth cone, with retrograde transport of the injury signal to the cell bodies, and with responsiveness of the cell body to axonal injury.

The formation of regeneration-competent growth cone is likely the result of the interaction between the local cytoskeleton and the surrounding environment. Several molecular players such as DLK-1 have been identified, but how they interact with other pathways to reorient the microtubules and permit the extension of regrowing axons needs to be further explored.

Several models such as conditioning lesion in sensory neurons and distance-dependent activation of regeneration-related genes in CNS neurons suggest that relaying the injury signal to the cell body of axotomized neurons is an important step in initiating a regenerative response. Injury signals could be generated in the lesion sites, such as cytokines in Schwann cells after sciatic nerve injury (Cao et al. 2006, Sendtner et al. 1992, Subang & Richardson 2001) or around the cell bodies, such as CNTF upregulation in retinal astrocytes after optic nerve injury (Muller et al. 2007). However, the links between axotomy and induction of these signals still remain a mystery. Observations of cytokines as regeneration-promoting factors in both PNS and CNS neurons and identification of SOCS3 as an intrinsic blocker for axon regeneration in adult RGCs suggested that even mature CNS neurons still possess a natural tendency to regenerate injured axons. A possible major difference between CNS and PNS neurons is their ability to translate (propagate or suppress) such signals. Understanding the underlying mechanisms and identifying

SOCS3: suppressor of cytokine signaling 3

additional intrinsic brakes should be another dimension for future studies.

Although local cytoskeleton rearrangements might be sufficient for a transient sprouting response, an obvious limiting factor for axon regeneration is the neuronal competence to sustain long-distance regrowth. As an important regulator of protein translation (Sarbassov et al. 2005), mTOR could be a potential determinant for axon regeneration. mTOR downregulation occurs during development and after injury (Liu et al. 2010, Park et al. 2008), suggesting the importance of understanding the regulatory mechanisms of neuronal mTOR activity, which may reveal important targets for potential strategies of promoting axon regeneration. In addition, reduced axon regeneration observed after tuberous sclerosis protein 1 (TSC1) deletion (in comparison to PTEN deletion) also highlighted the contribution of other mTOR-independent mechanisms such as axonal transport and cytoskeleton assembly in axon regeneration. Further studies on these aspects should enhance our understanding of axon regeneration.

Although neonatal organisms share some mechanisms with developmental axon growth, certain features of adult organisms create unique challenges for axon regeneration. For example, the nervous system is significantly larger in the adult than during development. Despite observed axon regeneration after different surgical or genetic manipulations as discussed above, the distance needed for regenerating axons to reach their functional targets in the adult is still a formidable challenge. Thus, in addition to promoting the initiation of axon regeneration, increasing the speed of axon regeneration should be important for functional recovery. Possible clues could be learned from the maturation process of the nervous system because most axons continue to grow even after establishing their synaptic connections during early development. This "tethered" growth of integrated axon tracts occurs in a growth cone–independent manner and could be recapitulated in vitro (Smith 2009). Understanding these mechanisms may allow us to design novel strategies to speed up axon regeneration.

We have focused on the intrinsic mechanisms regulating axon regeneration. However, a successful regenerative strategy is likely to require combinatorial modulations of both intrinsic and extrinsic mechanisms. Methods that could remove extracellular inhibitory influences, promote the remyelination of regenerating axons, and stimulate the synaptic integration of regenerating axons into the pre-existing circuits for neuronal functions must all be considered. Recent success in allowing regenerating axons to pass the lesion sites should set up an important basis for these future developments toward functional regeneration.

DISCLOSURE STATEMENT

Z. He is a cofounder of Axonis, a biotech company.

LITERATURE CITED

Abe N, Borson SH, Gambello MJ, Wang F, Cavalli V. 2010. Mammalian target of rapamycin (mTOR) activation increases axonal growth capacity of injured peripheral nerves. *J. Biol. Chem.* 285:28034–43

Adler CE, Fetter RD, Bargmann CI. 2006. UNC-6/Netrin induces neuronal asymmetry and defines the site of axon formation. *Nat. Neurosci.* 9:511–18

Aguayo AJ, David S, Bray GM. 1981. Influences of the glial environment on the elongation of axons after injury: transplantation studies in adult rodents. *J. Exp. Biol.* 95:231–40

Ambron RT, Zhang XP, Gunstream JD, Povelones M, Walters ET. 1996. Intrinsic injury signals enhance growth, survival, and excitability of Aplysia neurons. *J. Neurosci.* 16:7469–77

Baker BJ, Akhtar LN, Benveniste EN. 2009. SOCS1 and SOCS3 in the control of CNS immunity. *Trends Immunol.* 30:392–400

Bareyre FM, Kerschensteiner M, Misgeld T, Sanes JR. 2005. Transgenic labeling of the corticospinal tract for monitoring axonal responses to spinal cord injury. *Nat. Med.* 11:1355–60

Bernstein DR, Stelzner DJ. 1983. Plasticity of the corticospinal tract following midthoracic spinal injury in the postnatal rat. *J. Comp. Neurol.* 221:382–400

Berry M, Carlile J, Hunter A. 1996. Peripheral nerve explants grafted into the vitreous body of the eye promote the regeneration of retinal ganglion cell axons severed in the optic nerve. *J. Neurocytol.* 25:147–70

Bertrand J, Winton MJ, Rodriguez-Hernandez N, Campenot RB, McKerracher L. 2005. Application of Rho antagonist to neuronal cell bodies promotes neurite growth in compartmented cultures and regeneration of retinal ganglion cell axons in the optic nerve of adult rats. *J. Neurosci.* 25:1113–21

Blits B, Dijkhuizen PA, Boer GJ, Verhaagen J. 2000. Intercostal nerve implants transduced with an adenoviral vector encoding neurotrophin-3 promote regrowth of injured rat corticospinal tract fibers and improve hindlimb function. *Exp. Neurol.* 164:25–37

Bregman BS, Kunkel-Bagden E, McAtee M, O'Neill A. 1989. Extension of the critical period for developmental plasticity of the corticospinal pathway. *J. Comp. Neurol.* 282:355–70

Brodal A. 1981. *Neurological Anatomy in Relation to Clinical Medicine.* Oxford, UK: Oxford Univ. Press

Cai D, Deng K, Mellado W, Lee J, Ratan RR, Filbin MT. 2002. Arginase I and polyamines act downstream from cyclic AMP in overcoming inhibition of axonal growth MAG and myelin in vitro. *Neuron* 35:711–19

Cai D, Qiu J, Cao Z, McAtee M, Bregman BS, Filbin MT. 2001. Neuronal cyclic AMP controls the developmental loss in ability of axons to regenerate. *J. Neurosci.* 21:4731–39

Cao Z, Gao Y, Bryson JB, Hou J, Chaudhry N, et al. 2006. The cytokine interleukin-6 is sufficient but not necessary to mimic the peripheral conditioning lesion effect on axonal growth. *J. Neurosci.* 26:5565–73

Carnero A. 2010. The PKB/AKT pathway in cancer. *Curr. Pharm. Des.* 16:34–44

Case LC, Tessier-Lavigne M. 2005. Regeneration of the adult central nervous system. *Curr. Biol.* 15:R749–53

Cavalli V, Kujala P, Klumperman J, Goldstein LS. 2005. Sunday Driver links axonal transport to damage signaling. *J. Cell Biol.* 168:775–87

Chen DF, Jhaveri S, Schneider GE. 1995. Intrinsic changes in developing retinal neurons result in regenerative failure of their axons. *Proc. Natl. Acad. Sci. USA* 92:7287–91

Chen DF, Schneider GE, Martinou JC, Tonegawa S. 1997. Bcl-2 promotes regeneration of severed axons in mammalian CNS. *Nature* 385:434–39

Chierzi S, Ratto GM, Verma P, Fawcett JW. 2005. The ability of axons to regenerate their growth cones depends on axonal type and age, and is regulated by calcium, cAMP and ERK. *Eur. J. Neurosci.* 21:2051–62

Cho KS, Yang L, Lu B, Feng Ma H, Huang X, et al. 2005. Re-establishing the regenerative potential of central nervous system axons in postnatal mice. *J. Cell Sci.* 118:863–72

Christie KJ, Webber CA, Martinez JA, Singh B, Zochodne DW. 2010. PTEN inhibition to facilitate intrinsic regenerative outgrowth of adult peripheral axons. *J. Neurosci.* 30:9306–15

Cragg BG. 1970. What is the signal for chromatolysis? *Brain Res.* 23:1–21

Cristino L, Pica A, Della Corte F, Bentivoglio M. 2000. Co-induction of nitric oxide synthase, bcl-2 and growth-associated protein-43 in spinal motoneurons during axon regeneration in the lizard tail. *Neuroscience* 101:451–58

David S, Aguayo AJ. 1981. Axonal elongation into peripheral nervous system "bridges" after central nervous system injury in adult rats. *Science* 214:931–33

Deng K, He H, Qiu J, Lorber B, Bryson JB, Filbin MT. 2009. Increased synthesis of spermidine as a result of upregulation of arginase I promotes axonal regeneration in culture and in vivo. *J. Neurosci.* 29:9545–52

Enes J, Langwieser N, Ruschel J, Carballosa-Gonzalez MM, Klug A, et al. 2010. Electrical activity suppresses axon growth through Ca(v)1.2 channels in adult primary sensory neurons. *Curr. Biol.* 20:1154–64

Erez H, Malkinson G, Prager-Khoutorsky M, De Zeeuw CI, Hoogenraad CC, Spira ME. 2007. Formation of microtubule-based traps controls the sorting and concentration of vesicles to restricted sites of regenerating neurons after axotomy. *J. Cell Biol.* 176:497–507

Fernandes KJ, Fan DP, Tsui BJ, Cassar SL, Tetzlaff W. 1999. Influence of the axotomy to cell body distance in rat rubrospinal and spinal motoneurons: differential regulation of GAP-43, tubulins, and neurofilament-M. *J. Comp. Neurol.* 414:495–510

Fields RD, Neale EA, Nelson PG. 1990. Effects of patterned electrical activity on neurite outgrowth from mouse sensory neurons. *J. Neurosci.* 10:2950–64

Fischer D, Heiduschka P, Thanos S. 2001. Lens-injury-stimulated axonal regeneration throughout the optic pathway of adult rats. *Exp. Neurol.* 172:257–72

Filbin MT. 2006. Recapitulate development to promote axonal regeneration: good or bad approach? *Philos. Trans. R. Soc. Lond. B Biol. Sci.* 361:1565–74

Fishman HM, Bittner GD. 2003. Vesicle-mediated restoration of a plasmalemmal barrier in severed axons. *News Physiol. Sci.* 18:115–8

Fitch MT, Silver J. 2008. CNS injury, glial scars, and inflammation: Inhibitory extracellular matrices and regeneration failure. *Exp. Neurol.* 209:294–301

Fukata Y, Itoh TJ, Kimura T, Menager C, Nishimura T, et al. 2002. CRMP-2 binds to tubulin heterodimers to promote microtubule assembly. *Nat. Cell Biol.* 4:583–91

Gabel CV, Antoine F, Chuang CF, Samuel AD, Chang C. 2008. Distinct cellular and molecular mechanisms mediate initial axon development and adult-stage axon regeneration in *C. elegans. Development* 135:1129–36

George R, Griffin JW. 1994. Delayed macrophage responses and myelin clearance during Wallerian degeneration in the central nervous system: the dorsal radiculotomy model. *Exp. Neurol.* 129:225–36

Ghosh-Roy A, Wu Z, Goncharov A, Jin Y, Chisholm AD. 2010. Calcium and cyclic AMP promote axonal regeneration in *Caenorhabditis elegans* and require DLK-1 kinase. *J. Neurosci.* 30:3175–83

Glebova NO, Ginty DD. 2005. Growth and survival signals controlling sympathetic nervous system development. *Annu. Rev. Neurosci.* 28:191–222

Goldberg JL, Barres BA. 2000. The relationship between neuronal survival and regeneration. *Annu. Rev. Neurosci.* 23:579–612

Goldberg JL, Espinosa JS, Xu Y, Davidson N, Kovacs GT, Barres BA. 2002a. Retinal ganglion cells do not extend axons by default: promotion by neurotrophic signaling and electrical activity. *Neuron* 33:689–702

Goldberg JL, Klassen MP, Hua Y, Barres BA. 2002b. Amacrine-signaled loss of intrinsic axon growth ability by retinal ganglion cells. *Science* 296:1860–4

Gribnau AA, de Kort EJ, Dederen PJ, Nieuwenhuys R. 1986. On the development of the pyramidal tract in the rat. II. An anterograde tracer study of the outgrowth of the corticospinal fibers. *Anat. Embryol. (Berl.)* 175:101–10

Grill RJ, Blesch A, Tuszynski MH. 1997. Robust growth of chronically injured spinal cord axons induced by grafts of genetically modified NGF-secreting cells. *Exp. Neurol.* 148:444–52

Guertin DA, Sabatini DM. 2007. Defining the role of mTOR in cancer. *Cancer Cell* 12:9–22

Gumy LF, Tan CL, Fawcett JW. 2010. The role of local protein synthesis and degradation in axon regeneration. *Exp. Neurol.* 223:28–37

Hammarlund M, Jorgensen EM, Bastiani MJ. 2007. Axons break in animals lacking beta-spectrin. 176:269–75

Hammarlund M, Nix P, Hauth L, Jorgensen EM, Bastiani M. 2009. Axon regeneration requires a conserved MAP kinase pathway. *Science* 323:802–6

Han PJ, Shukla S, Subramanian PS, Hoffman PN. 2004. Cyclic AMP elevates tubulin expression without increasing intrinsic axon growth capacity. *Exp. Neurol.* 189:293–302

Hanz S, Perlson E, Willis D, Zheng JQ, Massarwa R, et al. 2003. Axoplasmic importins enable retrograde injury signaling in lesioned nerve. *Neuron* 40:1095–104

Harel NY, Strittmatter SM. 2006. Can regenerating axons recapitulate developmental guidance during recovery from spinal cord injury? *Nat. Rev. Neurosci.* 7:603–16

Hauk TG, Muller A, Lee J, Schwendener R, Fischer D. 2008. Neuroprotective and axon growth promoting effects of intraocular inflammation do not depend on oncomodulin or the presence of large numbers of activated macrophages. *Exp. Neurol.* 209:469–82

Hill CE, Beattie MS, Bresnahan JC. 2001. Degeneration and sprouting of identified descending supraspinal axons after contusive spinal cord injury in the rat. *Exp. Neurol.* 171:153–69

Hoffman PN. 2010. A conditioning lesion induces changes in gene expression and axonal transport that enhance regeneration by increasing the intrinsic growth state of axons. *Exp. Neurol.* 223:11–18

Hollis ER 2nd, Jamshidi P, Low K, Blesch A, Tuszynski MH. 2009a. Induction of corticospinal regeneration by lentiviral trkB-induced Erk activation. *Proc. Natl. Acad. Sci. USA* 106:7215–20

Hollis ER 2nd, Lu P, Blesch A, Tuszynski MH. 2009b. IGF-I gene delivery promotes corticospinal neuronal survival but not regeneration after adult CNS injury. *Exp. Neurol.* 215:53–59

Jhaveri S, Edwards MA, Schneider GE. 1991. Initial stages of retinofugal axon development in the hamster: evidence for two distinct modes of growth. *Exp. Brain Res.* 87:371–82

Joosten EA, Bar PR, Gispen WH. 1995. Directional regrowth of lesioned corticospinal tract axons in adult rat spinal cord. *Neuroscience* 69:619–26

Kamber D, Erez H, Spira ME. 2009. Local calcium-dependent mechanisms determine whether a cut axonal end assembles a retarded endbulb or competent growth cone. *Exp. Neurol.* 219:112–25

Konishi Y, Stegmuller J, Matsuda T, Bonni S, Bonni A. 2004. Cdh1-APC controls axonal growth and patterning in the mammalian brain. *Science* 303:1026–30

Korneev S, Fedorov A, Collins R, Blackshaw SE, Davies JA. 1997. A subtractive cDNA library from an identified regenerating neuron is enriched in sequences up-regulated during nerve regeneration. *Invert. Neurosci.* 3:185–92

Kreutzberg GW. 1982. Acute neural reaction to injury. In *Repair and Regeneration of the Nervous System, Life Science Research Report 24*, ed. JG Nicholls, pp. 57–69. Berlin: Springer

Lankford KL, Waxman SG, Kocsis JD. 1998. Mechanisms of enhancement of neurite regeneration in vitro following a conditioning sciatic nerve lesion. *J. Comp. Neurol.* 391:11–29

Lasek RJ. 1968. Axoplasmic transport of labeled proteins in rat ventral motoneurons. *Exp. Neurol.* 21:41–51

Lasek RJ, Hoffman PN. 1976. The neuronal cytoskeleton, axonal transport and axonal growth. In *Cold Spring Harbor Laboratory Conference on Cell Proliferation, Cell Motility*. Vol. 3, ed. R Goldman, T Pollard, J Rosenbaum, pp. 1021–49. Cold Spring Harbor, NY: Cold Spring Harbor Lab. Press

Lasorella A, Stegmüller J, Guardavaccaro D, Liu G, Carro MS, et al. 2006. Degradation of Id2 by the anaphase-promoting complex couples cell cycle exit and axonal growth. *Nature* 442:471–74

Leaver SG, Cui Q, Plant GW, Arulpragasam A, Hisheh S, et al. 2006. AAV-mediated expression of CNTF promotes long-term survival and regeneration of adult rat retinal ganglion cells. *Gene Ther.* 13:1328–41

Leibinger M, Muller A, Andreadaki A, Hauk TG, Kirsch M, Fischer D. 2009. Neuroprotective and axon growth-promoting effects following inflammatory stimulation on mature retinal ganglion cells in mice depend on ciliary neurotrophic factor and leukemia inhibitory factor. *J. Neurosci.* 29:14334–41

Leon S, Yin Y, Nguyen J, Irwin N, Benowitz LI. 2000. Lens injury stimulates axon regeneration in the mature rat optic nerve. *J. Neurosci.* 20:4615–26

Li Y, Field PM, Raisman G. 1997. Repair of adult rat corticospinal tract by transplants of olfactory ensheathing cells. *Science* 277:2000–2

Li Y, Field PM, Raisman G. 1998. Regeneration of adult rat corticospinal axons induced by transplanted olfactory ensheathing cells. *J. Neurosci.* 18:10514–24

Li Y, Raisman G. 1995. Sprouts from cut corticospinal axons persist in the presence of astrocytic scarring in long-term lesions of the adult rat spinal cord. *Exp. Neurol.* 134:102–11

Lieberman AR. 1971. The axon reaction: a review of the principal features of perikaryal responses to axon injury. *Int. Rev. Neurobiol.* 14:49–124

Lindwall C, Dahlin L, Lundborg G, Kanje M. 2004. Inhibition of c-Jun phosphorylation reduces axonal outgrowth of adult rat nodose ganglia and dorsal root ganglia sensory neurons. *Mol. Cell Neurosci.* 27:267–79

Liu K, Lu Y, Lee JK, Samara R, Willenberg R, et al. 2010. PTEN deletion enhances the regenerative ability of adult corticospinal neurons. *Nat. Neurosci.* 13:1075–81

Liu RY, Snider WD. 2001. Different signaling pathways mediate regenerative versus developmental sensory axon growth. *J. Neurosci.* 21:RC164

Lorber B, Howe ML, Benowitz LI, Irwin N. 2009. Mst3b, an Ste20-like kinase, regulates axon regeneration in mature CNS and PNS pathways. *Nat. Neurosci.* 12:1407–14

Lu P, Yang H, Jones LL, Filbin MT, Tuszynski MH. 2004. Combinatorial therapy with neurotrophins and cAMP promotes axonal regeneration beyond sites of spinal cord injury. *J. Neurosci.* 24:6402–9

Luo J, Manning BD, Cantley LC. 2003. Targeting the PI3K-Akt pathway in human cancer: rationale and promise. *Cancer Cell* 4:257–62

Markus A, Zhong J, Snider WD. 2002. Raf and akt mediate distinct aspects of sensory axon growth. *Neuron* 35:65–76

McKerracher L, Hirscheimer A. 1992. Slow transport of the cytoskeleton after axonal injury. *J. Neurobiol.* 23:568–78

McKerracher L, Vidal-Sanz M, Essagian C, Aguayo AJ. 1990. Selective impairment of slow axonal transport after optic nerve injury in adult rats. *J. Neurosci.* 10:2834–41

McQuarrie IG, Grafstein B, Gershon MD. 1977. Axonal regeneration in the rat sciatic nerve: effect of a conditioning lesion and of dbcAMP. *Brain Res.* 132:443–53

Meyer-Franke A, Wilkinson GA, Kruttgen A, Hu M, Munro E, et al. 1998. Depolarization and cAMP elevation rapidly recruit TrkB to the plasma membrane of CNS neurons. *Neuron* 21:681–93

Miao T, Wu D, Zhang Y, Bo X, Subang MC, et al. 2006. Suppressor of cytokine signaling-3 suppresses the ability of activated signal transducer and activator of transcription-3 to stimulate neurite growth in rat primary sensory neurons. *J. Neurosci.* 26:9512–19

Moore DL, Blackmore MG, Hu Y, Kaestner KH, Bixby JL, et al. 2009. KLF family members regulate intrinsic axon regeneration ability. *Science* 326:298–301

Muller A, Hauk TG, Fischer D. 2007. Astrocyte-derived CNTF switches mature RGCs to a regenerative state following inflammatory stimulation. *Brain* 130:3308–20

Nadeau S, Hein P, Fernandes KJ, Peterson AC, Miller FD. 2005. A transcriptional role for C/EBP beta in the neuronal response to axonal injury. *Mol. Cell Neurosci.* 29:525–35

Nakazawa T, Tamai M, Mori N. 2002. Brain-derived neurotrophic factor prevents axotomized retinal ganglion cell death through MAPK and PI3K signaling pathways. *Invest. Ophthalmol. Vis. Sci.* 43:3319–26

Neumann S, Woolf CJ. 1999. Regeneration of dorsal column fibers into and beyond the lesion site following adult spinal cord injury. *Neuron* 23:83–91

Ozdinler PH, Macklis JD. 2006. IGF-I specifically enhances axon outgrowth of corticospinal motor neurons. *Nat. Neurosci.* 9:1371–81

Park K, Luo JM, Hisheh S, Harvey AR, Cui Q. 2004. Cellular mechanisms associated with spontaneous and ciliary neurotrophic factor-cAMP-induced survival and axonal regeneration of adult retinal ganglion cells. *J. Neurosci.* 24:10806–15

Park KK, Hu Y, Muhling J, Pollett MA, Dallimore EJ, et al. 2009. Cytokine-induced SOCS expression is inhibited by cAMP analogue: impact on regeneration in injured retina. *Mol. Cell Neurosci.* 41:313–24

Park KK, Liu K, Hu Y, Kanter JL, He Z. 2010. PTEN/mTOR and axon regeneration. *Exp. Neurol.* 223:45–50

Park KK, Liu K, Hu Y, Smith PD, Wang C, et al. 2008. Promoting axon regeneration in the adult CNS by modulation of the PTEN/mTOR pathway. *Science* 322:963–66

Perlson E, Hanz S, Ben-Yaakov K, Segal-Ruder Y, Seger R, Fainzilber M. 2005. Vimentin-dependent spatial translocation of an activated MAP kinase in injured nerve. *Neuron* 45:715–26

Pernet V, Di Polo A. 2006. Synergistic action of brain-derived neurotrophic factor and lens injury promotes retinal ganglion cell survival, but leads to optic nerve dystrophy in vivo. *Brain* 129:1014–26

Pernet V, Hauswirth WW, Di Polo A. 2005. Extracellular signal-regulated kinase 1/2 mediates survival, but not axon regeneration, of adult injured central nervous system neurons in vivo. *J. Neurochem.* 93:72–83

Properzi F, Asher RA, Fawcett JW. 2003. Chondroitin sulfate proteoglycans in the central nervous system: changes and synthesis after injury. *Biochem. Soc. Trans.* 31:335–36

Qiu J, Cafferty WB, McMahon SB, Thompson SW. 2005. Conditioning injury-induced spinal axon regeneration requires signal transducer and activator of transcription 3 activation. *J. Neurosci.* 25:1645–53

Qiu J, Cai D, Dai H, McAtee M, Hoffman PN, et al. 2002. Spinal axon regeneration induced by elevation of cyclic AMP. *Neuron* 34:895–903

Raivich G, Bohatschek M, Da Costa C, Iwata O, Galiano M, et al. 2004. The AP-1 transcription factor c-Jun is required for efficient axonal regeneration. *Neuron* 43:57–67

Ramon-Cueto A, Cordero MI, Santos-Benito FF, Avila J. 2000. Functional recovery of paraplegic rats and motor axon regeneration in their spinal cords by olfactory ensheathing glia. *Neuron* 25:425–35

Ramon y Cajal S, ed. 1928. *Degeneration and Regeneration of the Nervous System*. London: Oxford Univ. Press

Reichardt LF. 2006. Neurotrophin-regulated signaling pathways. *Philos. Trans. R. Soc. Lond. B Biol. Sci.* 361:1545–64

Richardson PM, Issa VM. 1984. Peripheral injury enhances central regeneration of primary sensory neurones. *Nature* 309:791–93

Richardson PM, Issa VM, Aguayo AJ. 1984. Regeneration of long spinal axons in the rat. *J. Neurocytol.* 13:165–82

Richardson PM, Miao T, Wu D, Zhang Y, Yeh J, Bo X. 2009. Responses of the nerve cell body to axotomy. *Neurosurgery* 65:A74–79

Richardson PM, Verge VM. 1986. The induction of a regenerative propensity in sensory neurons following peripheral axonal injury. *J. Neurocytol.* 15:585–94

Rishal I, Fainzilber M. 2010. Retrograde signaling in axonal regeneration. *Exp. Neurol.* 223:5–10

Rossi F, Jankovski A, Sotelo C. 1995. Differential regenerative response of Purkinje cell and inferior olivary axons confronted with embryonic grafts: environmental cues versus intrinsic neuronal determinants. *J. Comp. Neurol.* 359:663–77

Sarbassov DD, Ali SM, Sabatini DM. 2005. Growing roles for the mTOR pathway. *Curr. Opin. Cell Biol.* 17:596–603

Schlaepfer WW, Bunge RP. 1973. Effects of calcium ion concentration on the degeneration of amputated axons in tissue culture. *J. Cell Biol.* 59:456–70

Schnell L, Schneider R, Kolbeck R, Barde YA, Schwab ME. 1994. Neurotrophin-3 enhances sprouting of corticospinal tract during development and after adult spinal cord lesion. *Nature* 367:170–73

Schreyer DJ, Jones EG. 1982. Growth and target finding by axons of the corticospinal tract in prenatal and postnatal rats. *Neuroscience* 7:1837–53

Schwab ME, Bartholdi D. 1996. Degeneration and regeneration of axons in the lesioned spinal cord. *Physiol. Rev.* 76:319–70

Schwaiger FW, Hager G, Schmitt AB, Horvat A, Streif R, et al. 2000. Peripheral but not central axotomy induces changes in Janus kinases (JAK) and signal transducers and activators of transcription (STAT). *Eur. J. Neurosci.* 12:1165–76

Segal RA, Greenberg ME. 1996. Intracellular signaling pathways activated by neurotrophic factors. *Annu. Rev. Neurosci.* 19:463–89

Sendtner M, Stockli KA, Thoenen H. 1992. Synthesis and localization of ciliary neurotrophic factor in the sciatic nerve of the adult rat after lesion and during regeneration. *J. Cell Biol.* 118:139–48

Shen S, Wiemelt AP, McMorris FA, Barres BA. 1999. Retinal ganglion cells lose trophic responsiveness after axotomy. *Neuron* 23:285–95

Smith DH. 2009. Stretch growth of integrated axon tracts: extremes and exploitations. *Prog. Neurobiol.* 89:231–39

Smith PD, Sun F, Park KK, Cai B, Wang C, et al. 2009. SOCS3 deletion promotes optic nerve regeneration in vivo. *Neuron* 64:617–23

Song H, Ming G, He Z, Lehmann M, McKerracher L, et al. 1998. Conversion of neuronal growth cone responses from repulsion to attraction by cyclic nucleotides. *Science* 281:1515–18

Song HJ, Ming GL, Poo MM. 1997. cAMP-induced switching in turning direction of nerve growth cones. *Nature* 388:275–79

Steinmetz MP, Horn KP, Tom VJ, Miller JH, Busch SA, et al. 2005. Chronic enhancement of the intrinsic growth capacity of sensory neurons combined with the degradation of inhibitory proteoglycans allows functional regeneration of sensory axons through the dorsal root entry zone in the mammalian spinal cord. *J. Neurosci.* 25:8066–76

Subang MC, Richardson PM. 2001. Synthesis of leukemia inhibitory factor in injured peripheral nerves and their cells. *Brain Res.* 900:329–31

Takahashi K, Tanabe K, Ohnuki M, Narita M, Ichisaka T, et al. 2007. Induction of pluripotent stem cells from adult human fibroblasts by defined factors. *Cell* 131:861–72

Tessier-Lavigne M, Goodman CS. 1996. The molecular biology of axon guidance. *Science* 274:1123–33

Trivedi N, Marsh P, Goold RG, Wood-Kaczmar A, Gordon-Weeks PR. 2005. Glycogen synthase kinase-3beta phosphorylation of MAP1B at Ser1260 and Thr1265 is spatially restricted to growing axons. *J. Cell Sci.* 118:993–1005

Udina E, Furey M, Busch S, Silver J, Gordon T, Fouad K. 2008. Electrical stimulation of intact peripheral sensory axons in rats promotes outgrowth of their central projections. *Exp. Neurol.* 210:238–47

Vargas ME, Barres BA. 2007. Why is Wallerian degeneration in the CNS so slow? *Annu. Rev. Neurosci.* 30:153–79

Vargas ME, Watanabe J, Singh SJ, Robinson WH, Barres BA. 2010. Endogenous antibodies promote rapid myelin clearance and effective axon regeneration after nerve injury. *Proc. Natl. Acad. Sci. USA* 107:11993–98

Veldman MB, Bemben MA, Thompson RC, Goldman D. 2007. Gene expression analysis of zebrafish retinal ganglion cells during optic nerve regeneration identifies KLF6a and KLF7a as important regulators of axon regeneration. *Dev. Biol.* 312:596–612

Weinberg RA. 2007. *The Biology of Cancer*. New York: Garland Sci.

Woodgett JR. 2001. Judging a protein by more than its name: GSK-3. *Sci. STKE* 2001:re12

Wujek JR, Lasek RJ. 1983. Correlation of axonal regeneration and slow component B in two branches of a single axon. *J. Neurosci.* 3:243–51

Yan D, Wu Z, Chisholm AD, Jin Y. 2009. The DLK-1 kinase promotes mRNA stability and local translation in *C. elegans* synapses and axon regeneration. *Cell* 138:1005–18

Yanik MF, Cinar H, Cinar HN, Chisholm AD, Jin Y, Ben-Yakar A. 2004. Neurosurgery: functional regeneration after laser axotomy. *Nature* 432:822

Yin Y, Henzl MT, Lorber B, Nakazawa T, Thomas TT, et al. 2006. Oncomodulin is a macrophage-derived signal for axon regeneration in retinal ganglion cells. *Nat. Neurosci.* 9:843–52

Yiu G, He Z. 2006. Glial inhibition of CNS axon regeneration. *Nat. Rev. Neurosci.* 7:617–27

Yoo SK, Deng Q, Cavnar PJ, Wu YI, Hahn KM, Huttenlocher A. 2010. Differential regulation of protrusion and polarity by PI3K during neutrophil motility in live zebrafish. *Dev. Cell* 18:226–36

Yu TW, Bargmann CI. 2001. Dynamic regulation of axon guidance. *Nat. Neurosci.* 4(Suppl.):1169–76

Zhang XP, Ambron RT. 2000. Positive injury signals induce growth and prolong survival in Aplysia neurons. *J. Neurobiol.* 45:84–94

Zheng JQ, Kelly TK, Chang B, Ryazantsev S, Rajasekaran AK, et al. 2001. A functional role for intra-axonal protein synthesis during axonal regeneration from adult sensory neurons. *J. Neurosci.* 21:9291–303

Zhong J, Dietzel ID, Wahle P, Kopf M, Heumann R. 1999. Sensory impairments and delayed regeneration of sensory axons in interleukin-6-deficient mice. *J. Neurosci.* 19:4305–13

Zhou FQ, Snider WD. 2005. Cell biology: GSK-3beta and microtubule assembly in axons. *Science* 308:211–14

Zhou FQ, Snider WD. 2006. Intracellular control of developmental and regenerative axon growth. *Philos. Trans. R. Soc. Lond. B Biol. Sci.* 361:1575–92

Zhou FQ, Zhou J, Dedhar S, Wu YH, Snider WD. 2004. NGF-induced axon growth is mediated by localized inactivation of GSK-3beta and functions of the microtubule plus end binding protein APC. *Neuron* 42:897–912

Zweifel LS, Kuruvilla R, Ginty DD. 2005. Functions and mechanisms of retrograde neurotrophin signalling. *Nat. Rev. Neurosci.* 6:615–25

Transcriptional Control of the Terminal Fate of Monoaminergic Neurons

Nuria Flames[1,2,3] and Oliver Hobert[1]

[1] Department of Biochemistry and Molecular Biophysics, Howard Hughes Medical Institute, Columbia University Medical Center, New York, New York 10032; email: or38@columbia.edu

[2] Genes & Disease Program, Center for Genomic Regulation (CRG), Barcelona, Spain E-08003; email: nflamesb@gmail.com

[3] Present address: Instituto de Biomedicina de Valencia IBV-CSIC, E-46010 Valencia, Spain

Annu. Rev. Neurosci. 2011. 34:153–84

The *Annual Review of Neuroscience* is online at neuro.annualreviews.org

This article's doi:
10.1146/annurev-neuro-061010-113824

0147-006X/11/0721-0153$20.00

Keywords

transcription factors, terminal differentiation, serotonin, dopamine, noradrenaline, gene regulation

Abstract

Monoaminergic neurons are critical functional components of all nervous systems across phylogeny. The terminally differentiated state of individual types of monoaminergic neurons is defined by the coordinated expression of a battery of genes that instructs the synthesis and transport of specific monoamines, such as serotonin or dopamine. Dysfunction or deregulation of several of these enzymes and transporter system has been proposed to be the underlying basis of several pathological conditions. We review here the state of knowledge of the nature of the transcriptional regulatory programs that control the expression of what we term monoaminergic gene batteries (enzymes and transporters for specific monoamines) and thereby define the terminally differentiated state of monoaminergic neurons. We review several case studies in vertebrate and invertebrate model systems and propose that the coordinated expression of the genes that define individual monoaminergic cell types may be brought about by transcriptional coregulatory strategies.

Contents

INTRODUCTION

Monoamines, also called biogenic amines, are neurotransmitters synthesized from aromatic amino acids such as phenylalanine, tyrosine, and tryptophan and include the catecholamines (dopamine, noradrenaline, and adrenaline), the tryptamines (serotonin and

melatonin), tyramine, octopamine, and histamine (**Figure 1**). Monoamines act through ionotropic and metabotropic receptors and are involved in a vast array of different brain functions in which they often serve critical modulatory roles. Defects in monoaminergic transmission—conferred by malfunction or deregulation of genes involved in synthesizing and transporting monoamines—result in a number of pathological conditions, including schizophrenia, depression, addictive behaviors, and many more (Webster 2001).

In their terminally differentiated state, monoaminergic neuron types are defined by the coordinated expression of batteries of specific enzymes and transporters that synthesize a specific monoamine, package the monoamine into synaptic vesicles, and reuptake the monoamine into the neuron after release (**Figure 2**). To understand how different types of monoaminergic neurons develop, one needs to understand how the coordinated coexpression of the genes that encode these enzymes and transporters is brought about. In this review, we discuss the present state of knowledge regarding the regulation of the expression of the defining features of monoaminergic neuron types. We first give an overview of the battery of proteins that define individual monoaminergic neuron types; we then provide an overview of the features of monoaminergic systems in distinct animal phyla, focusing entirely on neurons and leaving aside nonneuronal sources of monoamine, and then provide a summary of our current knowledge of gene regulatory strategies and, specifically, transcriptional control mechanisms that define these neuron types across phylogeny.

MONOAMINERGIC CELL TYPES ARE DEFINED BY COEXPRESSION OF SPECIFIC ENZYMES AND TRANSPORTERS

Catecholaminergic Synthesis: Dopamine, Noradrenaline, and Adrenaline

All catecholamines are synthesized from the amino acid tyrosine in a common biosynthetic

---

Monoamines or biogenic amines: neurotransmitters synthesized from aromatic amino acids

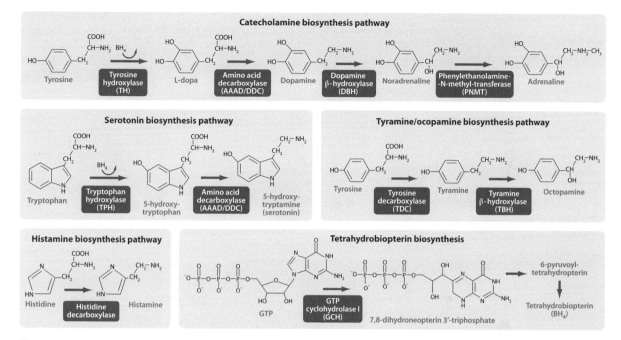

Figure 1

Biosynthesis of monoamines. Besides the core monoaminergic synthesis pathways, the tetrahydrobiopterin (BH$_4$) synthesis pathway is also shown because BH$_4$ is a cofactor present in both catecholamine and serotonin pathways. Two of the enzymes involved in the BH$_4$ pathway are broadly expressed, whereas GTP-cyclohydrolase expression is confined to monoaminergic neurons.

pathway that first generates dopamine (**Figure 1**). Norepinephrinergic neurons continue to process dopamine via dopamine β-hydroxylase (DBH) to norepinephrine (also known as noradrenaline). In adrenergic cells, phenyl-ethanolamine-*N*-methyl transferase (PNMT) methylates norepinephrine to form epinephrine (also known as adrenaline) (**Figure 1**). A vesicular monoamine transporter concentrates the neurotransmitter into vesicles, and upon release, a specific reuptake transporter allows the neurotransmitter to clear from the synaptic cleft (**Figure 2**). Reuptake leads to the recycling of the neurotransmitter, or finally, catecholamines are degraded by monoamine oxidases (MAO) and catechol-*O*-methyltransferase (COMT) (Mannisto & Kaakkola 1999, Youdim et al. 2006).

Serotonin Synthesis Pathway

Tryptophan hydroxylation is the first step of the serotonin biosynthesis pathway (**Figure 1**). Catecholaminergic and serotonergic pathways share some enzymes, and the same vesicular monoamine transporter also packages serotonin. A specific serotonin transporter (SERT) reuptakes the neurotransmitter from the synaptic cleft (**Figure 2**). Finally, serotonin is degraded by MAO (Youdim et al. 2006).

Tyramine and Octopamine Synthesis Pathway

Invertebrates lack both DBH and PNMT; consequently, they cannot convert tyrosine (via dopamine) into adrenaline or noradrenaline. Instead, invertebrate-specific enzymes convert tyrosine into the neurotransmitters octopamine and tyramine (**Figure 1**). Tyramine functions both as an octopamine precursor and as a neurotransmitter in its own right (**Figure 2**). The vesicular monoamine transporter packages tyramine and octopamine into vesicles for its release (**Figure 2**). The octopamine transporter

Terminally differentiated state: the hard-wired expression of a unique combination of genes expressed throughout the life of a mature, nondividing neuron

MAO: monoamine oxidase

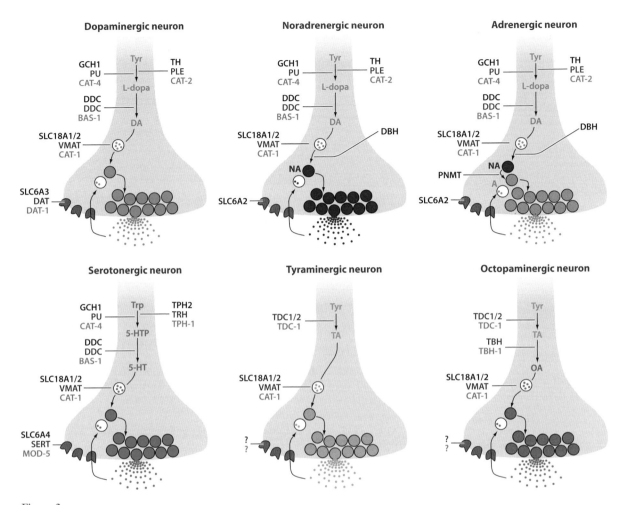

Figure 2

Monoaminergic gene batteries that define monoaminergic identity. For each neuron subtype, the biosynthetic intermediaries appear in the color of the corresponding neurotransmitter (*blue* for dopamine, *purple* for noradrenaline, *pink* for adrenaline, *green* for serotonin, *orange* for tyramine, and *red* for octopamine). For each neuron subtype, the mouse (*black*), *D. melanogaster* (*purple*) and *C. elegans* (*red*) protein names involved in every step (**Figure 1**) are shown. Gene batteries that define individual monoaminergic phenotypes have also been called "monoaminergic pathway genes"; e.g., "dopamine pathway genes." Abbreviations: ?, unidentified gene; 5HT, serotonin; 5HTP, 5 hydroxytryptophan; A, adrenaline; BAS-1, biogenic amine synthesis related 1; CAT, abnormal catecholamine distribution gene; DA, dopamine; DAT, dopamine transporter; DBH, dopamine-β-hydroxylase; DDC, dopa decarboxylase [also called amino acid decarboxylase (AAAD)]; GCH1, GTP cyclohydrolase 1; MOD-5, modulation of locomotion defective; NA, noradrenaline; OA, octopamine; PLE, pale; PNMT, phenylethanolamine-N-methyl-transferase; PU, punch; SERT, serotonin transporter; SLC, solute carrier; TA, tyramine; TBH, tyramine-β-hydroxylase; TDC, tyrosine decarboxylase; TH, tyrosine hydroxylase; TPH, tryptophan hydroxylase; TRH, tryptophan hydroxylase; Trp, tryptophan; Tyr, tyrosine; VMAT, vesicular monoamine transporter.

(OAT) responsible for the selective reuptake of octopamine and tyramine has been identified in some insects but is absent from *D. melanogaster* or *C. elegans* (Caveney et al. 2006, Roeder 2005). Octopamine is inactivated by methylation and acetylation (Roeder et al. 2003).

Other Biosynthetic Monoamines

Other monoamines function as neurotransmitters (Hilakivi 1987, Weiner & Ganong 1978), the most prominent being histamine, which is a neurotransmitter in both invertebrates and vertebrates. It is synthesized from histidine by decarboxylation through a histidine decarboxylase (**Figure 1**).

THE STRUCTURAL AND FUNCTIONAL ANATOMY OF MONOAMINERGIC SYSTEMS ACROSS PHYLOGENY

Monoaminergic neurons are broadly distributed in all nervous systems across phylogeny. The different mammalian monoaminergic cell groups were initially identified by using formaldehyde-induced-fluorescence (FIF) staining, which converts catecholamines and serotonin into intensely fluorescent products (Carlsson et al. 1961). Groups of stained cells were named with letters, "A" for catecholaminergic and "B" for serotonergic and numbers corresponding to the anterior-posterior location (Carlsson et al. 1962, Dahlstrom & Fuxe 1964). Later, antibody staining, directed against the neurotransmitter or toward a specific enzyme of the biosynthetic pathway (**Figure 2**), corroborated the existence of the different monoaminergic nuclei and their neurotransmitter identity. Below, we first briefly describe monoaminergic systems in the most commonly used model species.

Mouse Monoaminergic Systems

In mammals, such as the mouse, dopaminergic neurons are present in telencephalic, diencephalic, and mesencephalic nuclei. Dopaminergic neuron groups have been classically named A8 to A17 (**Figure 3**). In the mesencephalon, the dopaminergic neurons in the substantia nigra (SN) are involved in motor control and are lost in Parkinson's disease, whereas dopaminergic signaling from the ventral tegmental area (VTA) regulates behaviors such as reward, motivation, cognition, and drug addiction (Kandel et al. 1991).

Following the same nomenclature, serotonergic neurons are designated as B1 to B9 groups (Kandel et al. 1991). All mammalian serotonergic neurons arise from the hindbrain (**Figure 3**). Serotonin function regulates sleep, mood, and feeding behaviors. In the peripheral nervous system (PNS), some enteric neurons release serotonin (Blaugrund et al. 1996).

Mouse brain adrenergic nuclei are named A1 to A7. The medulla noradrenergic neurons (A1 and A2) project to the hypothalamus and control cardiovascular and endocrine functions. In the pons, A5 and A7 groups modulate autonomic reflexes and pain sensation, and A6 is part of the locus coeruleus and maintains vigilance and responsiveness to unexpected stimuli. The A3 and A4 regions, initially identified by FIF staining, could not be confirmed to be monoaminergic by immunostaining (**Figure 3**) (Kandel et al. 1991). In the PNS, there is an important noradrenergic component: the sympathetic neurons. Noradrenergic sympathetic neurons derive from neural crest cells, which also give rise to parasympathetic, enteric, and sensory neurons (Howard 2005). There is also a small adrenergic component in the brain, C1 to C3 groups, which reside in the medulla (**Figure 3**) (Kandel et al. 1991).

Zebrafish Monoaminergic Systems

The distribution of zebrafish dopaminergic neurons slightly differs from mammals because zebrafish have some extra telencephalic and diencephalic dopaminergic nuclei, whereas mesencephalic dopaminergic neurons are absent (Rink & Wullimann 2002) (**Figure 4**). Serotonergic neurons are also more diverse in zebrafish than in mammals. Apart from the raphe nuclei, serotonergic cells are also found in the diencephalon (Kaslin & Panula 2001) (**Figure 4**). Noradrenergic neurons are organized in a similar way as in mammals and can be found in the locus coeruleus, in the medulla (**Figure 4**), and in the sympathetic ganglia of the PNS (Rink & Wullimann 2002). Zebrafish is also unusual in its complement of enzymes that define monoaminergic neurons. Genome duplication at the base of teleost evolution resulted in *tyrosine hydroxylase* (*Th*), *GTP cyclohydrolase* (*Gch*), and *serotonin transporter* (*Sert*) genes being duplicated; in addition, three *Tph* genes show nonoverlapping expression in the brain. The expression of some of these genes has been analyzed (Bellipanni et al. 2002; Chen et al. 2009; Filippi et al. 2010; Holzschuh et al. 2001; Teraoka et al. 2004; Yamamoto et al. 2010, 2011).

Drosophila Melanogaster Monoaminergic Systems

Drosophila has orthologs for all mouse dopamine and serotonin pathway genes. TH, AAAD, tyrosine decarboxylase (TDC), and TBH antibody staining and Gal4 lines, as well as serotonin immunostaining, have been used to identify monoaminergic groups in the fly (Vomel & Wegener 2008). There are ~80 dopaminergic neurons in the *Drosophila* third larval instar located in both the brain and the ventral ganglion (**Figure 5**) (Lundell & Hirsh 1994, Monastirioti 1999, Vomel & Wegener 2008). *Drosophila* dopaminergic

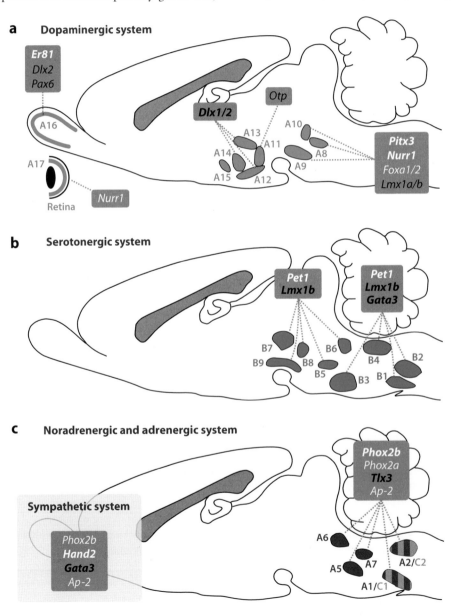

neurons arise from different progenitors in development. Dopaminergic signaling in flies controls arousal, sexual behavior, learning, and memory (Birman 2005, Keene & Waddell 2005, Rothenfluh & Heberlein 2002).

Similarly to the dopaminergic neurons, serotonergic neurons are located in the brain and ventral ganglion (Monastirioti 1999) (**Figure 5**). Serotonin modulates locomotion, learning and memory, circadian rhythms, reproduction, feeding, heart rate, and locomotion (Monastirioti 1999, Neckameyer et al. 2007). Octopaminergic neurons in the larvae are located in the ventral ganglion and absent from the brain (Monastirioti 1999) (**Figure 5**), whereas in the adult they are present in both structures (Monastirioti 1999, Roeder 2005). Tyraminergic neurons have been described in both the larvae ventral ganglia and the brain (Nagaya et al. 2002). Functionally, the octopamine and tyramine system in *Drosophila* has been compared with the vertebrate adrenergic system because both are involved in the fight or flight response. Furthermore, octopamine signaling has been described in learning and memory (Roeder 2005).

Caenorhabditis Elegans Monoaminergic Systems

The 302 neuron-containing nervous system of *C. elegans* employs dopamine, serotonin, tyramine, and octopamine as neurotransmitters (Alkema et al. 2005, Horvitz et al. 1982, Sulston et al. 1975). The *C. elegans* hermaphrodite contains four pairs of dopaminergic neurons, which fall into three morphologically distinct classes (**Figure 6**). Unlike in any other system, the lineage history of every *C. elegans* cell is known, revealing that each individual dopaminergic neuron pair arises from different progenitors in development (**Figure 6**) but nevertheless shares the expression of all the biosynthesis enzymes and transporters that allow the use of dopamine as a neurotransmitter (**Figure 2**). This diversity in origin is similar to the diverse developmental origins of fly and vertebrate monoaminergic neurons (**Figures 3–5**). The *C. elegans* dopamine biosynthetic pathway is virtually identical to the one in vertebrates, and nematode mutants can be rescued by introducing the corresponding human ortholog (Duerr et al. 1999). All *C. elegans* dopaminergic neurons are mechanosensory ciliated neurons and regulate different behaviors, including locomotion, learning, habituation, and male mating behavior (Chase & Koelle 2007, Liu & Sternberg 1995, Rose & Rankin 2001, Sawin et al. 2000).

Anti-serotonin staining labels six different neuronal classes in the *C. elegans* hermaphrodite (**Figure 6**) (Horvitz et al. 1982). Serotonergic neurons, like dopaminergic neurons, arise from very different progenitors in development. However, in contrast with

---

Figure 3

CNS monoaminergic systems and their regulation in mouse. (*a*) Dopamine system: In the midbrain, A8–A10 groups correspond to the retrorubral field, the substantia nigra (SN), and ventral tegmental area (VTA), respectively. Diencephalic neurons constitute A11–A15 groups. The hypothalamic dopaminergic system is distributed through A11–A14. Groups A15 (olfactory tubercle) and A16 (olfactory bulb) are part of the olfactory system, and finally A17 is the amacrine dopaminergic neurons of the retina. (*b*) Serotonin system: B1–B4 are located in the raphe nucleus and constitute the caudal division. B5–B9 constitute the rostral division and are located in the pons and midbrain. (*c*) Noradrenergic (A) and adrenergic (C) system: In the medulla, A1/C1 corresponds to the nucleus ambiguous and A2/C2 to the nucleus of the solitary tract and dorsal motor vagal nucleus. A5 and A7 are located in the pons, and A6 is part of the locus coeruleus. Transcription factors in the yellow boxes are candidate terminal selectors of the corresponding monoaminergic group, based on the following evidence: The transcription factor is expressed postmitotically and/or maintained in the adult, and the mutant phenotype shows loss of monoaminergic markers. Bold means it has been determined that, in the mutant, cells are generated and express neuronal or other features but not monoamine pathway genes. White text means there are data for direct binding and/or activation of monoamine pathway genes.

Adult lateral view **6-day-old dorsal view**

a Dopaminergic system

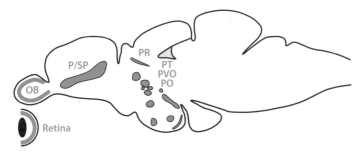
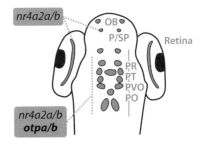

b Serotonergic system

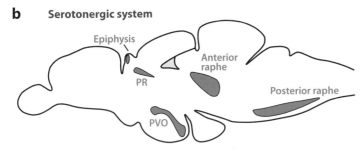
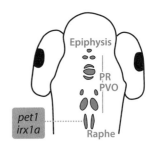

c Noradrenergic system

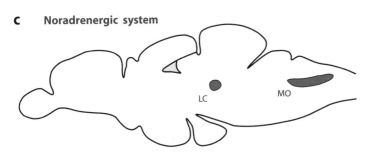
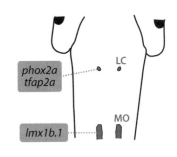

Figure 4

CNS monoaminergic systems and their regulation in zebrafish. Adult and larval anatomy of the main monoaminergic groups in zebrafish. At six days old, all the components of the monoaminergic systems are already present. (*a*) Dopaminergic system. (*b*) Serotonergic system. (*c*) Noradrenergic system. See **Figure 3** legend for explanation of how gene regulatory factors are labeled. LC, locus coeruleus; MO, medulla oblongata; OB, olfactory bulb; P/SP, pallial/subpallial system; PO, preoptic area; PR, pretectum; PT, posterior tuberculum; PVO, paraventricular area.

dopaminergic neurons, serotonergic neurons belong to different functional classes, including chemosensory and neurosecretory neurons, interneurons, and motorneurons. Serotonin signaling has been linked to a plethora of behaviors, including chemosensation, feeding, egg laying, and male mating behavior (Chase & Koelle 2007, Horvitz et al. 1982, Sawin et al. 2000, Waggoner et al. 1998, Weinshenker et al. 1995).

Although tyramine was initially thought to be an intermediate for octopamine synthesis, it is now clear that both function as neurotransmitters in invertebrates. In *C. elegans*, tyramine is synthesized in only two types of neurons, RIC (ring interneuron C) and RIM (ring

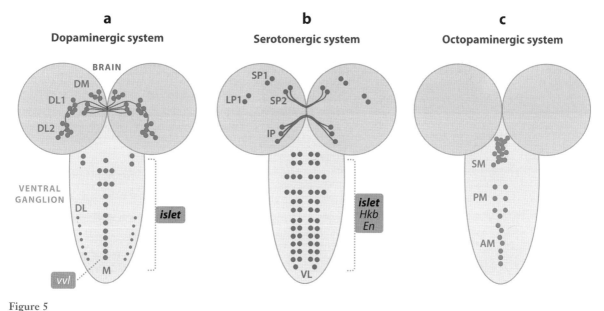

a
Dopaminergic system

BRAIN
DM
DL1
DL2
VENTRAL
GANGLION
DL
islet
M
vvl

b
Serotonergic system

SP1
LP1
SP2
IP
islet
Hkb
En
VL

c
Octopaminergic system

SM
PM
AM

Figure 5

Monoaminergic systems and their regulation in *Drosophila*. Representation of the dopaminergic (*a*), serotonergic (*b*), and octopaminergic (*c*) systems in the CNS of a third larval instar. See legend to **Figure 3** for explanation of how gene regulatory factors are labeled.

interneuron M) (Alkema et al. 2005). In the RIC neuron, TBH regulates the conversion of tyramine to octopamine; thus, RIM is considered a tyraminergic neuron, whereas RIC is an octopaminergic neuron (**Figure 2**). Octopamine function regulates behaviors such as movement, egg laying, and pharyngeal pumping (Horvitz et al. 1982), whereas tyramine has been implicated in egg laying and escape response (Alkema et al. 2005, Pirri et al. 2009). Both neurotransmitters also modulate dopamine- and serotonin-regulated behaviors (Horvitz et al. 1982, Suo et al. 2006, Wragg et al. 2007).

REGULATORY PROGRAMS THAT CONTROL TERMINAL DIFFERENTIATION OF MONOAMINERGIC NEURONS

The molecular composition of different types of monoaminergic neurons, i.e., the specific battery of enzymes and transporters that each monoaminergic neuron expresses (**Figure 2**), poses the question of how the expression of this gene battery is regulated. Answering this question holds the key to understanding terminal monoaminergic neuron differentiation. Although monoaminergic gene batteries are clearly the key identifying features of individual monoaminergic neurons, specific monoaminergic neuron types naturally also display other defining features, such as their defined position, projection patterns, and connectivity features. However, there are far fewer molecular descriptors of these additional features, thereby lending themselves less readily to investigate molecular and gene regulatory mechanisms.

The most commonly used traditional strategy to decipher regulatory programs that generate monoaminergic neurons is to ask which genes impinge on expression of defined fate markers (such as TH for dopamine neurons) or to examine the presence of the amine itself (serotonin staining, for example, which is a reflection of the presence of the appropriate biosynthetic enzymes). These types of approaches have yielded, as expected, two classes of regulatory factors: Early-acting regulatory factors, which act during a restricted time window in progenitor or immature cells

Gene battery: set of coexpressed genes that together define a specific property, such as the terminal identity of a neuron

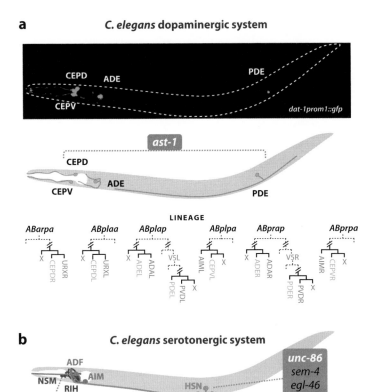

a   ***C. elegans* dopaminergic system**

CEPD   ADE   PDE

CEPV   *dat-1prom1::gfp*

ast-1

CEPD

ADE

CEPV   PDE

LINEAGE

| ABarpa | ABplaa | ABplap | ABplpa | ABprap | ABprpa |

X CEPDR / URXR   X CEPDL / URXL   X ADEL / ADAL   V5L PDEL / X PVDL   X CEPVL / AIML   X ADER / ADAR   V5R PDER / X PVDR   X AIMR / CEPVR

b   ***C. elegans* serotonergic system**

ADF

NSM   AIM

RIH

RIM

HSN

VC4 VC5

unc-86
sem-4
egl-46
hlh-3

unc-86

c   ***C. elegans* tyraminergic/octopaminergic system**

RIM   RIC

Figure 6

Monoaminergic systems and their regulation in *C. elegans* (*a*) Picture of a transgenic worm expressing GFP (green fluorescent protein) under the control of the 5′ upstream region of the dopamine transporter shows the 4 bilateral pairs of dopamine neurons. Schematic representation of the dopaminergic system and representation of the worm lineage to show the widespread origin of dopamine neurons. (*b*) Serotonergic system schematic representation; different colors represent different neuron subtypes, based on morphology and function. (*c*) Tyraminergic (RIM)/octopaminergic (RIC) system. See legend to **Figure 3** for explanation of how gene regulatory factors are labeled.

controllers of neuronal identity, and we refer to them as "terminal selectors," a term we define in the sidebar, Terminal Selectors.

We first review which gene-regulatory factors have been implicated in the terminal differentiation of monoaminergic neurons in distinct species; then in the final section, we attempt to carve out some concepts and common principles.

Specification of Monoaminergic Systems in the Mouse

Dopaminergic neuron specification in the mouse

Mesencephalic dopaminergic neuron specification. Midbrain dopaminergic neurons arise from the most ventral part of the neural tube, with the floor plate being the major source of dopaminergic neurons (Ono et al. 2007). Because midbrain dopaminergic neurons are clinically relevant, the factors involved in the patterning and specification of the midbrain have been extensively studied. A combination of SHH, FGFs (fibroblast growth factors) and WNT morphogens specifies dopaminergic progenitors of the midbrain, which are characterized by the expression of different transcription factors, including OTX2, EN1/2, NGN2, LMX1a, MSX1, and FOXA2 (reviewed in Abeliovich & Hammond 2007, Smidt & Burbach 2007). These progenitors generate postmitotic cells that will mature into dopaminergic neurons. Several factors have been described to be specifically involved in the terminal differentiation of the mesencephalic dopaminergic neurons; among them, NURR1 and PITX3 stand out as the best candidates for being involved in directly regulating the expression of the dopamine gene battery (**Figure 3**).

NURR1 is an orphan nuclear hormone receptor transcription factor expressed during development in different brain regions, including the midbrain, where it is expressed in postmitotic cells. Its expression is maintained in adult stages, thereby perfectly matching with TH-expressing cells (Backman et al. 1999, Zetterstrom et al. 1996).

(including signaling molecules), and late-acting regulatory factors, which directly control the activation (and often the sustained presence) of the monoaminergic pathway genes. These late-acting regulatory factors, therefore, are direct

Nurr1 mutants die at birth and completely lack TH-expressing neurons in the midbrain (Zetterstrom et al. 1997). Several observations indicate that *Nurr1* has a specific role in terminal differentiation of the dopaminergic phenotype. (*a*) In accordance with the postmitotic expression of *Nurr1*, *Nurr1* mutants show normal patterning of the floor plate (Wallen et al. 1999). (*b*) At embryonic day e11.5, when the first TH-expressing cells appear in the wild type, postmitotic cells are generated in *Nurr1* mutants. These cells partially differentiate, they express general neuronal markers such as TUJ1, but they show reduced or lack of expression of TH, DAT (dopamine transporter), VMAT, and AAAD/DDC. Later in development, these undifferentiated cells die and are completely absent by birth (Saucedo-Cardenas et al. 1998, Smits et al. 2003, Wallen et al. 1999). This finding seems to indicate that NURR1 is necessary for the correct acquisition of neurotransmitter identity in the midbrain dopaminergic neurons and that it may also have a direct role in cell survival. Alternatively, cell loss could be a secondary consequence of lack of proper differentiation. (*c*) NURR1 binds the upstream regions of the dopamine pathway genes in vivo (Jacobs et al. 2009b), and in some contexts and cell lines, NURR1 is sufficient to activate the dopamine pathway genes, although with low efficiency (Kim et al. 2003, Sakurada et al. 1999). (*d*) Finally, NURR1 activity is continuously required to maintain dopaminergic fate (Kadkhodaei et al. 2009). However, some of these points were questioned in an earlier report (Witta et al. 2000).

The limited effect of NURR1 ectopic expression together with its normal expression in nondopaminergic neurons suggests that other transcription factors must act together with NURR1 to confer specificity. The best candidate for such a cofactor is the PITX3 homeodomain transcription factor, which within the brain is exclusively expressed in the midbrain dopaminergic neurons and whose expression is maintained in the adult (Smidt et al. 1997). In *Pitx3* mutant mice, at early stages of

TERMINAL SELECTORS

A few decades ago, Garcia-Bellido coined the term selector genes for genes that affect the identity of organs and developing fields in an organism (Garcia-Bellido 1975). Building on this concept, the term terminal selector gene was recently proposed for transcription factors that directly control the terminal identity of individual cell types in the nervous system (and beyond) (Hobert 2008). Terminal identity is defined by the expression of terminal differentiation genes, that is, genes that code for nuts and bolts identity features of a neuron, such as ion channels and neurotransmitter-synthesizing enzymes. Terminal differentiation genes are continuously expressed throughout the life of a neuron and their regulatory sequences contain binding sites for terminal selector transcription factors. An individual neuron type may not contain a large number of parallel acting regulatory routines, meaning, a large number of different terminal selectors for different terminal features. Rather, many terminal differentiation genes appear to be coregulated through one common terminal selector (or a common combination of several transcription factors). Transcription factors that directly control the expression of the monoaminergic identity determinants discussed in this review classify as terminal selectors. These terminal selectors regulate not just the expression of monoaminergic identity, but also many other identity features. For a more detailed discussion of terminal selectors, see Hobert (2008) and Hobert et al. (2010).

development, dopaminergic cells are generated and correctly specified, but later on, cells that are generated fail to express TH; these undifferentiated cells eventually die (Hwang et al. 2003, Jacobs et al. 2009b, Maxwell et al. 2005, Nunes et al. 2003, Smidt et al. 2004, van den Munckhof et al. 2003). Biochemical and functional analyses have demonstrated that PITX3 and NURR1 cooperate to bind and activate *Th* and *Dat* expression directly (Cazorla et al. 2000, Jacobs et al. 2009b, Martinat et al. 2006). Via the regulation of a miRNA that negatively controls its own expression, PITX3 is also involved in a negative feedback loop that presumably gauges protein levels to the exact required levels (Kim et al. 2007). PITX3 signaling has also been linked to the retinoic acid (RA) pathway. In addition to the early role of RA in the patterning of the midbrain, RA synthesis is maintained in adult dopaminergic neurons. *Pitx3* mutants show defects in expression of the enzyme responsible for the synthesis of RA: the aldehyde dehydrogenase RALDH1. Maternal supplementation of RA partially rescues the loss of TH-expressing cells in *Pitx3* mutants, suggesting a role of RA signaling in dopamine differentiation (Jacobs et al. 2007).

Four other transcription factors, FOXA1/FOXA2 and LMX1a/LMX1b, are also expressed in adult dopaminergic neurons (Kittappa et al. 2007, Zou et al. 2009). During development, these factors are expressed in dopaminergic progenitors, they are required for midbrain dopaminergic specification, and mutant analysis has placed them upstream of *Pitx3* and *Nurr1* (Andersson et al. 2006, Chung et al. 2009, Ferri et al. 2007, Kittappa et al. 2007, Roybon et al. 2008, Smidt et al. 2000). But since FOXA2 and FOXA1 directly bind a *cis*-regulatory motif in the *Th* gene (Lee et al. 2010, Lin et al. 2009) and since their expression is maintained in dopaminergic neurons, they could also be part of the NURR1/PITX3 combinatorial code of terminal selectors of midbrain dopaminergic fate (**Figure 3**).

Olfactory bulb dopaminergic neuron specification. Dopaminergic neurons of the olfactory bulb (A16) are interneurons, which also produce GABA (gamma-amino butyric acid) and are located in the periglomerular layer where they modulate the activity of the output neurons, the mitral cells (Cave & Baker 2009). Three different classes of GABAergic periglomerular interneurons have been identified to date: the dopaminergic, the calbindin (CB)-expressing, and the calretinin (CR)-expressing interneurons. All periglomerular interneurons, plus the granular GABAergic neurons, are continuously generated throughout the animal's life (Luskin 1993). Different morphogens and transcription factors are required for the specification of the olfactory bulb progenitors both in development and in the adult (reviewed in Cave & Baker 2009, Lledo et al. 2008). A number of factors may

have a direct role in terminal differentiation of these DA neurons, among them ER81, an ETS transcription factor expressed in all TH- and some CB-positive cells of the periglomerular layer (Allen et al. 2007, Flames & Hobert 2009, Saino-Saito et al. 2007). *Er81* mutants show decreased numbers of TH-expressing cells in the olfactory bulb and ER81 can bind and activate *Th* expression (Cave et al. 2010, Flames & Hobert 2009), suggesting it has a direct role in terminal differentiation. Other dopamine battery genes seem unaffected in *Er81* mutants (Cave et al. 2010); however, as in the case of NURR1, ER81 is not exclusively expressed in the dopaminergic neurons (Stenman et al. 2003). Thus, a number of additional transcription factors may act together with ER81 to regulate terminal dopamine neuron differentiation, and these factors may compensate for the lack of ER81. Among these factors may be members of the Distalless (*Dlx*) homeobox gene family (**Figure 3**). During development, DLX1, -2, -5, and -6 are expressed in the progenitors, in the migrating cells, and in the mature olfactory bulb neurons (Eisenstat et al. 1999), and the same expression pattern is maintained in the adult (Allen et al. 2007, Brill et al. 2008, Saino-Saito et al. 2003). In *Dlx2* mutant mice, the periglomerular layer appears organized normally, but there is a dramatic loss in the number of TH-expressing cells (Qiu et al. 1995). Retroviral injections of DLX2 in adult mice induce neurogenesis and differentiation toward TH-expressing cells, whereas a dominant-negative form of DLX2 has opposite effects (Brill et al. 2008). DLX2 needs the presence of the paired-type homeobox gene PAX6 to induce TH cells, and the two transcription factors molecularly interact (Brill et al. 2008), (**Figure 3**). PAX6 is expressed in the olfactory bulb interneuron progenitors, in a subpopulation of the migrating neuroblasts, and in the glomerular layer, where it is preferentially expressed in dopamine neurons (Haba et al. 2009, Stoykova & Gruss 1994). Partial or complete loss of PAX6 results in a loss of dopaminergic olfactory neurons (Dellovade et al. 1998, Hack et al. 2005, Hill et al. 1991, Kohwi et al. 2005). PAX6 also has a direct role in controlling the survival of the differentiated dopaminergic neurons by activating crystallin α (Ninkovic et al. 2010). As in the case of *Dlx2*, it remains to be determined if *Pax6* has a direct role in postmitotic cells to instruct dopaminergic differentiation, if it directly binds to the regulatory control region of dopamine pathway genes, and if it is required to maintain the dopaminergic phenotype.

Mice lacking *Sal3* (zinc finger transcription factor), *Id2* (HLH transcription factor), or *Arx* (homeobox gene) also show defects in olfactory bulb dopaminergic neuron specification; however, because they also affect other neuronal populations in the olfactory bulb, it appears most likely that these genes act at earlier progenitor stages (Harrison et al. 2008, Havrda et al. 2008, Yoshihara et al. 2005). Last, *Nurr1* is also expressed in the dopaminergic olfactory bulb neurons, where its expression is maintained throughout adulthood (Backman et al. 1999, Zetterstrom et al. 1996); although *Nurr1* mutants show normal TH staining, dopamine levels are decreased by 60% in the olfactory bulb (Le et al. 1999).

Specification of other dopaminergic neurons. The factors described above may have roles in other dopaminergic neurons, as well. Amacrine dopaminergic neurons are probably the least characterized dopaminergic neurons of the brain. All adult amacrine dopaminergic neurons express DLX2 (de Melo et al. 2003); however, a role for DLX2 in retinal dopaminergic specification has not been tested. Besides its role in midbrain dopamine neuron specification, NURR1 is also necessary for amacrine dopaminergic specification (**Figure 3**) (Jiang & Xiang 2009). NURR1 is postmitotically expressed in a subgroup of GABAergic amacrine cells, including all the dopaminergic neurons and its expression is maintained in the adult. Because NURR1 is also expressed in other amacrine subpopulations that are also affected in *Nurr1* conditional mutants (Jiang & Xiang 2009), NURR1 may work with other factors to specify amacrine dopaminergic neurons. Unlike in midbrain dopamine neurons, this cofactor cannot be PITX3, which is not expressed in the retina (Smidt et al. 1997). Taken together, NURR1 may have a very broad role in dopamine differentiation.

As part of the forebrain dopaminergic neurons, diencephalic dopaminergic neurons (A11–A15) share some characteristics with the olfactory bulb dopaminergic neurons. Adult *Dlx1* mutants show decreased numbers in A12 and A14 TH-expressing cells, and in *Dlx1/2* mutant, embryos A13 postmitotic neurons are generated (assessed by the expression of TUJ1) but fail to express TH (Andrews et al. 2003); these findings indicate that Dlx factors may act as terminal selectors for both olfactory bulb and diencephalic dopaminergic neurons (**Figure 3**). Finally, another homeodomain transcription factor, *orthopedia* (*Otp*), is necessary for dopaminergic A11 nucleus differentiation (**Figure 3**). OTP is expressed in A11 dopaminergic neurons, and *Otp* mutants show no TH or AAAD/DDC staining in the A11 region (Ryu et al. 2007). The role of OTP in diencephalic dopaminergic differentiation seems to be evolutionary conserved (Ryu et al. 2007) (for more detail, see the zebrafish section below). However, it remains to be determined if *Otp* acts at the terminal level or upstream of other regulatory factors.

Serotonergic neuron specification in the mouse. Brain serotonergic progenitors are located in the ventral hindbrain. Visceral motor neurons (VMNs) and serotonergic neurons arise sequentially from the same progenitors. Two rhombomeres are the exception to this rule: r1, which produces only serotonergic neurons and will constitute more than half of the total serotonergic population, and r4, which never produces serotonergic neurons. At embryonic stage E10.5, rhombomeres r2, r3, and r5–7 switch from VMN generation to serotonergic neuron production by inhibiting the expression of the homeodomain transcription factor *Phox2b* (paired-like home-odomain protein 2b), which at this level specifies VMNs (Pattyn et al. 2003). In accordance with its role in inhibiting serotonergic fate, *Phox2b* is never expressed in r1, and it is not turned off in r4. In *Phox2b* mutants, VMNs are lost and serotonergic neurons are precociously generated in rhombomeres 2–7, including r4 (Pattyn et al. 2003). As we discuss below, at other locations of the hindbrain, as well as in the PNS, PHOX2b induces noradrenergic fate.

Several transcription factors are known to be expressed in the progenitors of the serotonergic neurons and instruct serotonergic fate. Early steps of monoaminergic differentiation are not the scope of this review, and readers are referred to previous reviews (Alenina et al. 2006, Goridis & Rohrer 2002, Scott & Deneris 2005). Briefly, FGF8, FGF4, and SHH are early inducers of the hindbrain pattern (Ye et al. 1998), and progenitor expression of *Mash1*, *Foxa2*, and *Nkx2.2* is necessary to generate serotonergic neurons (Briscoe et al. 1999, Cheng et al. 2003, Jacob et al. 2007, Pattyn et al. 2004, Simon et al. 2005, Wassarman et al. 1997). These genes act upstream and regulate the expression of the *Gata3*, *Lmx1b*, and *Pet1* transcription factors (**Figure 3**). *Lmx1b* and *Pet1* are expressed in postmitotic serotonergic neurons before the onset of *Tph2* expression, and their expression is maintained throughout the animal's life (Asbreuk et al. 2002, Ding et al. 2003, Hendricks et al. 1999). *Pet1* and *Lmx1b* mutants show defects in expression of the serotonin pathway genes (*Tph2*, *Sert*, *Vmat2*, *Aaad/Ddc*) but show no defects in the expression of earlier markers. In both mutants, postmitotic cells are still generated, and at least in the case of *Pet1* mutants, they still express panneuronal features and do not switch their fate to that of another neuron type (Cheng et al. 2003, Ding et al. 2003, Hendricks et al. 2003). This phenotype is very similar to the ones described for *Nurr1* and *Pitx3* for the dopaminergic specification. PET1 and LMX1b have other characteristics typical of terminal selectors. PET1 directly binds to regulatory elements in all the serotonin pathway genes and also autoregulates

its own maintenance of expression, and both PET1 and LMX1b are required to maintain the serotonergic phenotype (Hendricks et al. 1999, Liu et al. 2010, Scott et al. 2005, Song et al. 2011, Zhao et al. 2006).

Ectopic coexpression of both LMX1b and PET1 is not sufficient to turn on expression of the serotonin pathway genes (Cheng et al. 2003), suggesting that an unidentified factor is part of the combinatorial code of terminal selectors for serotonergic fate. The GATA3 transcription factor is a candidate for being part of the code because it is expressed in all the caudal serotonergic neurons and in a subset of the anterior serotonergic cells (**Figure 3**). In *Gata3* mutants, postmitotic cells are normally generated at caudal levels but fail to synthesize serotonin. Rostral serotonergic neurons are not affected in *Gata3* mutants (van Doorninck et al. 1999). This observation indicates the presence of slightly different regulatory mechanisms between anterior and posterior serotonergic populations, a notion also made by transcriptional profiling (Wylie et al. 2010). It also indicates that other terminal selectors remain to be identified for serotonergic specification. *Gata3* expression is maintained in adult serotonergic neurons (Zhao et al. 2008); however, its role in the adult remains to be studied. *Gata3* is also expressed in the noradrenergic sympathetic neurons, where it acts as a survival factor (Tsarovina et al. 2010).

Noradrenergic neuron specification in the mouse. Noradrenergic and adrenergic neurons in the brain represent a minor population of cells located mainly in the locus coeruleus (A6) and in some other groups of cells (A1, A2, A5, A7, C1, and C2). *Phox2a* and *Phox2b*, two homeodomain transcription factors, are expressed in all noradrenergic and adrenergic nuclei in the brain, and *Phox2a* expression is maintained postnatally (Pattyn et al. 1997, 2000; Tiveron et al. 1996; Valarché et al. 1993). In the locus coeruleus, *Phox2a* starts its expression in postmitotic cells, followed by *Phox2b* expression (Pattyn et al. 1997). *Phox2a* and *Phox2b* mutant mice lack noradrenergic neurons in the locus coeruleus and other brain nor/adrenergic groups (Morin et al. 1997; Pattyn et al. 1997, 2000). In *Phox2a* mutants, *Phox2b*, *Th*, and *Dbh* expression is not detected (Morin et al. 1997, Pattyn et al. 1997); thus *Phox2a* could have a role in generation, survival, or specification of central noradrenergic cells. *Phox2b* mutants initially show expression of *Phox2a* but lack *Th* or *Dbh* expression. These cells are later undetectable, suggesting that they died (Pattyn et al. 2000). Thus, PHOX2a is not sufficient to drive noradrenergic differentiation in the absence of PHOX2b: Swap experiments between the *Phox2a* and *Phox2b* loci show that *Phox2b* expressed under the *Phox2a* locus can substitute for *Phox2a* in the generation of noradrenergic cells in the locus coeruleus but not vice versa, which supports a major role of *Phox2b* in CNS noradrenergic specification (Coppola et al. 2005). However, both PHOX2a and PHOX2b bind and activate *Th* and *Dbh* upstream regions in different cell lines (Kim et al. 1998, Yang et al. 1998, Zellmer et al. 1995), and only *Phox2a* expression is maintained in the adult. Therefore, *Phox2b* may establish noradrenergic terminal differentiation, whereas *Phox2a* could have a role in noradrenergic maintenance (**Figure 3**). Postmitotic removal of *Phox2a* function has no effect on noradrenergic fate maintenance because there is a compensatory upregulation of *Phox2b*, whereas postmitotic depletion of both factors leads to loss of noradrenergic markers; it remains unclear whether the loss in noradrenergic markers corresponds to a loss of the expression of differentiated properties or to cell death (Coppola et al. 2010).

Another homeodomain transcription factor, *Tlx3* (also known as *Rnx*), is transiently expressed in the noradrenergic neurons in the brain. *Tlx3* mutants show defects in *Dbh* and *Th* expression, but *Phox2a* and *Phox2b* expression is not affected, which suggests that both factors act in parallel pathways (**Figure 3**) (Qian et al. 2001). *Phox2b* and *Tlx3* are coexpressed in other non-noradrenergic neurons in the brain, which suggests that other factors are required to specify noradrenergic fate.

The main noradrenergic population is not in the brain but in the autonomic nervous system. The sympathetic neurons derive from the neural crest cells that migrate to the dorsal aorta, where they are specified by bone morphogenetic proteins (BMPs). Neural crest cells start expressing different transcription factors such as *Mash1*, *Phox2b*, *Phox2a*, *Hand2*, and *Gata2/3*, which play important roles in sympathetic neuron proliferation, differentiation, and survival (reviewed in Goridis & Rohrer 2002, Howard 2005). Similar to their role in CNS noradrenergic differentiation, *Phox2a* and *Phox2b* are also involved in sympathetic neuron specification. Contrary to CNS noradrenergic neurons, PNS noradrenergic neurons express *Phox2b* earlier than *Phox2a* (Pattyn et al. 1997, Tiveron et al. 1996, Valarché et al. 1993), and *Phox2b* expression is not lost in *Phox2a* mutants (Morin et al. 1997, Pattyn et al. 1997). Consequently, *Phox2a* mutants show normal *Dbh* and *Th* expression, whereas *Phox2b* mutants lack sympathetic noradrenergic neurons (Morin et al. 1997; Pattyn et al. 1997, 1999). In sympathoblasts, *Phox2b* functions both in specification and in survival because the early ganglionic cells never switch on *Th* or *Dbh* and die soon after their aggregation (Pattyn et al. 1999). Conversely, ectopic PHOX2a/b can induce *Dbh* and *Th* expression in chick (Stanke et al. 1999). As already mentioned, both transcription factors directly bind *cis*-regulatory elements in the *Dbh* and *Th* genes and activate their expression, arguing for a role in terminal differentiation. PHOX2a/b function is also required to maintain the sympathetic noradrenergic differentiated state (Coppola et al. 2010).

Phox2a and *Phox2b* are not exclusively expressed in the sympathetic neurons, which means other factors must provide specificity. Two other transcription factors could participate in such specificity: the bHLH transcription factor HAND2 and the GATA3 transcription factor (**Figure 3**). *Hand2* is absent from the CNS and in the PNS is expressed in the sympathoadrenal and enteric systems (Cserjesi et al. 1995). *Hand2* mutants have reduced *Th* and *Dbh* expression (Hendershot et al. 2008,

Morikawa et al. 2007). Although analysis of these mutant mice shows contrasting results, it seems that panneuronal features are less affected than specific noradrenergic features (Hendershot et al. 2008, Morikawa et al. 2007) and that HAND2 also affects proliferation of a subpopulation of sympathetic neurons (Hendershot et al. 2008, Schmidt et al. 2009). HAND2 interacts with PHOX2b and enhances its ability to activate *Dbh* expression in vitro; this enhancement does not require DNA binding of HAND2 (Rychlik et al. 2003, Xu et al. 2003). *Hand2* expression is maintained in the adult sympathetic neurons, and its activity is continuously required to maintain *Dbh* and *Th* expression, but not their panneuronal features (Doxakis et al. 2008, Morikawa et al. 2005, Schmidt et al. 2009).

Gata3 is expressed in the sympathetic neurons but is absent from the CNS noradrenergic cells (**Figure 3**) (Lim et al. 2000). In *Gata3* mutants, sympathetic neurons are initially generated; they express *Phox2a*, *Phox2b*, and neuronal features but fail to express noradrenergic markers (*Dbh* and *Th*) (Lim et al. 2000), and later in development, these undifferentiated cells degenerate (Tsarovina et al. 2004). Ectopic expression of *Gata3* results in an increased number of TH-expressing neurons in primary neural crest stem cell culture (Hong et al. 2006). Removing *Gata3* function from differentiated noradrenergic neurons leads to cell death but does not seem to affect expression of the noradrenergic gene battery (Tsarovina et al. 2010). Taken together, *Gata3* may be part of the combinatorial code that drives sympathetic noradrenergic terminal differentiation.

AP-2β, a member of the AP-2 family of basic helix-span-helix transcription factors, is expressed in the sympathetic ganglion, and *Ap-2β* mutants show a decreased number of noradrenergic cells both in PNS and in CNS (**Figure 3**) (Hong et al. 2008). AP-2 binds to the 5′ upstream region of *Dbh* (Kim et al. 1998), suggesting it plays a role in terminal differentiation. Finally, INSM1, a zinc finger transcription factor, is involved in the generation of serotonergic neurons and CNS and PNS noradrenergic neurons, although its direct role in terminal differentiation is unclear (Jacob et al. 2009, Wildner et al. 2008).

Specification of Monoaminergic Systems in Zebrafish

Dopaminergic neuron specification in zebrafish. *Nurr1* and *Otp* seem to have evolutionarily conserved roles in dopaminergic differentiation. As mentioned above, zebrafish lack mesencephalic dopaminergic neurons. Zebrafish have two *Nurr1* genes—*Nr4a2a/b*, which are expressed in the retina, telencephalic, and diencephalic dopaminergic neurons—and morpholinos against these two genes decrease the number of TH-expressing cells (**Figure 4**) (Blin et al. 2008, Filippi et al. 2007, Luo et al. 2008). However, most of the ventral diencephalic neurons do not express *Nurr1*. The two zebrafish genes *Otpa* and *Otpb* are expressed in postmitotic and mature dopaminergic diencephalic cells, and disruption of these two genes decreases the number of TH- and DAT-expressing cells without affecting neurogenesis or apoptosis (**Figure 4**) (Del Giacco et al. 2006, Ryu et al. 2007). Other factors such as transcription elongation factor *Spt5*, the bHLH factor *Artn2*, and components of the transcriptional mediator complex play a role in dopaminergic neuron generation (Durr et al. 2006, Guo et al. 2000, Lohr et al. 2009). Regulatory elements for expression of zebrafish dopamine genes appear to be distributed over relatively large genomic intervals as it has so far not been possible to isolate small, discrete *cis*-regulatory elements required for expression in dopaminergic neurons (Bai & Burton 2009, Fujimoto et al. 2011).

Serotonergic neuron specification in zebrafish. Zebrafish *Pet1* is expressed in the raphe serotonergic neurons but not in the diencephalic serotonergic neurons (**Figure 4**) (Lillesaar et al. 2007). As in mammals, it regulates *tph-2* expression (Lillesaar et al. 2007). *Irx1a*, a TALE homeodomain transcription factor, is also expressed and required for serotonergic neurons to specify in the raphe (**Figure 4**) (Cheng et al. 2007). It is currently unknown if mammalian *Irx1* or other TALE homeobox genes could have similar roles. Finally, the zinc finger transcription factor *Fez* shows decreased numbers of both serotonergic and dopaminergic diencephalic neurons; however, owing to the cell nonautonomy of the phenotype, this gene's effect must be indirect (Levkowitz et al. 2003, Rink & Guo 2004).

Noradrenergic neuron specification in zebrafish. Several of the factors involved in mouse noradrenergic differentiation are also necessary in zebrafish. *Soulless*, a mutant that lacks noradrenergic neurons in the locus coeruleus, corresponds to a mutation in *Phox2a* (Guo et al. 1999a,b), and in the *Hand2* mutant *hands off*, sympathetic noradrenergic neurons fail to differentiate (Lucas et al. 2006). Finally, mutations in the zebrafish *Ap-2* gene (*tfap2a*) cause both CNS and PNS noradrenergic neurons not to differentiate, in the same way that *Ap-2β* and *Lmx1b.1* knockdown decreases the number of noradrenergic cells in the medulla (**Figure 4**) (Filippi et al. 2007, Holzschuh et al. 2003).

Specification of Monoaminergic Systems in *D. Melanogaster*

Dopaminergic neuron specification in *Drosophila*. Analysis of the regulatory region of the *Drosophila* AAAD enzyme, expressed by both dopamine and serotonin neurons (**Figure 2**), led to the identification of different *cis*-regulatory motifs necessary for *Aaad* expression in different tissues and neuronal subpopulations (Bray et al. 1988, Johnson et al. 1989, Johnson & Hirsh 1990, Lundell & Hirsh 1992). The POU homeodomain transcription factor *Ventral veins lacking* (*vvl*, a.k.a *Cf1a*) was reported to bind to the *Aaad cis*-regulatory motif, necessary for its expression in dopaminergic medial neurons of the ventral ganglion (**Figure 5**). However, this is not the only element required for dopaminergic *Aaad* expression because mutations of other regions not bound by *vvl* also abolish dopaminergic *Aaad* expression (Johnson & Hirsh 1990).

Islet, a LIM/homeodomain transcription factor (also called *Tail up* or *Tup*), may induce the terminal fate of a subpopulation of both dopamine and serotonin neurons in *Drosophila* (Thor & Thomas 1997). *Islet* is expressed in several postmitotic neurons, including all dopamine and serotonin neurons of the ventral ganglion, but it is absent from the brain monoaminergic populations (**Figure 5**). *Islet* mutants lose expression of *Th*, *Aaad*, and serotonin, and these defects can be rescued by postmitotic expression of *Islet* (Thor & Thomas 1997). Ectopic *Islet* expression induces *Th* and *Aaad* in other neurons of the ventral ganglion; however, no induction of the serotonin phenotype was observed (Thor & Thomas 1997). All the above observations suggest that *Islet* acts with other factors to induce both dopamine and serotonin fates and that the combination of factors is different for each monoaminergic subtype. The lack of expression of *Islet* in brain monoaminergic neurons indicates that, in *Drosophila*, different neuronal subtypes are regulated independently. *Islet* activity is not restricted to monoaminergic differentiation because *Islet* mutants also show defects in motorneuron specification and axon pathfinding (Thor & Thomas 1997).

Serotonergic neuron specification in *Drosophila*. The serotonergic interneurons of the larval ventral ganglion arise from the neuroblast NB 7–3, which divides 3 times during development and gives rise to three ganglion mother cells (GMCs). The first GMC divides to generate a serotonergic interneuron (EW1) and a motorneuron (GW1). Division of the second GMC generates a second serotonergic interneuron (EW2) and an apoptotic cell, and the third GMC generates a nonmonoaminergic interneuron (EW3) and an apoptotic cell (Dittrich et al. 1997, Isshiki et al. 2001, Novotny et al. 2002). In the most posterior segment A8, serotonergic neurons are a single cell instead of a pair (Lundell & Hirsh 1994, Monastirioti 1999, Vomel & Wegener 2008) (**Figure 5**). In addition to the already mentioned role of *Islet* in serotonergic specification,

Cis-regulatory motif: DNA sequence in a locus, of variable size, to which a regulatory factor binds to affect expression of the locus

three different transcription factors are known to specify the serotonergic lineage, *Eagle* (*Eg*), *Huckebein* (*Hkb*), and *Engrailed* (*En*) (**Figure 5**). *Eagle* is a zinc finger transcription factor that is expressed in the NB 7–3 and in three other neuroblasts; its expression is initially maintained in the progeny, including the serotonergic neurons, but it then fades from the entire CNS after stage 17 (Dittrich et al. 1997). In *Eg* mutants, the NB 7–3 divides to generate postmitotic cells, but the number of AAAD- and serotonin-expressing cells is dramatically reduced (Dittrich et al. 1997, Lundell & Hirsh 1998). However, a continuous role of *Eg* in serotonergic terminal differentiation seems unlikely because *Eg* expression is not maintained in the EW1 and EW2 serotonergic neurons (Dittrich et al. 1997). Presumably, *Eg* acts upstream of another, yet unidentified factor to activate the serotonergic gene battery.

Hkb is another zinc finger transcription factor that has been implicated in EW1 and EW2 specification. In the CNS, *Hkb* is expressed in eight neuroblasts, including NB 7–3 (Chu-LaGraff et al. 1995). *Hkb* expression is maintained in the postmitotic EW1 and EW2, at least until stage 18, and *Hkb* mutants show a reduction in the number of serotonin- and AAAD-expressing cells (Lundell et al. 1996). Although *Hkb* expression is maintained in the serotonergic neurons, there are no data about *Hkb* directly activating any of the serotonin pathway genes. Another factor involved in serotonergic differentiation is the homeodomain transcription factor *En*, which is coexpressed with *Hkb* specifically in the serotonergic neurons EW1 and EW2 (Lundell et al. 1996). *En* mutants, like *Hkb* and *Eg* mutants, show decreased numbers of AAAD- and serotonin-expressing cells, but it is not clear at which step *En* affects aminergic neuron differentiation.

None of the transcription factors described so far are expressed in brain serotonergic neurons, which indicates that, as in the case of the dopaminergic neurons, different serotonergic populations are regulated by independent mechanisms.

Tyraminergic and octopaminergic neuron specification in *Drosophila*. Despite the importance of octopamine and tyramine in controlling different behaviors (reviewed in Roeder 2005), no data about the molecular mechanisms that specify these neurons in *Drosophila* has been published.

Specification of Monoaminergic Systems in *C. elegans*

Dopaminergic neuron specification in *C. elegans*. The amenability of *C. elegans* to transgenic reporter gene studies allows investigators to approach the question of dopaminergic neuron specification from a bottom-up angle. Through the generation and subsequent mutagenesis of reporter gene constructs, we found that the five genes that code for the dopamine biosynthetic and transport machinery (**Figure 2**) are all coregulated through a common *cis*-regulatory logic (Flames & Hobert 2009). That is, all genes contain phylogenetically conserved ETS-domain binding sites that are required for the coordinated expression of all five genes in not only one but all eight dopaminergic neurons, despite their distinct lineage history (**Figure 6**). The *Pea3/Er81*-like ETS factor AST-1 is the *trans*-acting factor for this shared *cis*-regulatory motif. AST-1 is expressed in all dopaminergic neurons in the worm, and its expression is maintained throughout the animal's life. In *ast-1* mutants, the dopaminergic neurons are generated but fail to express all the dopamine pathway genes. In addition, ectopic expression of *ast-1* induces ectopic dopaminergic fate in some contexts (Flames & Hobert 2009). The mouse homolog of *ast-1*, *Er81/Etv1*, also appears to control at least some aspects of the dopaminergic fate in olfactory bulb neurons, as described above, suggesting a evolutionary conservation of the dopaminergic regulatory logic (Flames & Hobert 2009).

However, *ast-1* is expressed in neurons other than the dopaminergic neurons, which indicates that *ast-1* must act with other factors to specify the dopaminergic fate. Preliminary evidence suggests that AST-1 interacts with distinct homeodomain proteins (N. Flames and O. Hobert, unpublished data).

Serotonergic neuron specification in *C. elegans*. As in the case of dopaminergic neurons, *C. elegans* serotonergic neurons arise from very different lineages during development. However, contrary to the dopaminergic system in which *ast-1* coregulates all neuron subtypes, there is no transcription factor known to be involved in the differentiation of all *C. elegans* serotonergic neurons. Among the *C. elegans* serotonergic cells, the HSN (hermphroditic specific motor neuron), involved in egg laying, is the best characterized. HSN is born in the posterior part of the embryo and migrates during development to the middle of the body. At later larval stages, it matures and turns on the expression of all serotonin pathway genes. Using an extensive forward genetic screen, 38 distinct genetic loci affecting different aspects of HSN development and function were isolated (Desai et al. 1988, Desai & Horvitz 1989). *ham-2*, a zinc finger transcription factor, controls HSN migration (Baum et al. 1999), whereas *sem-4*, also a zinc finger transcription factor, controls HSN cellular morphology, axon pathfinding, and serotonin expression (**Figure 6**) (Basson & Horvitz 1996, Desai et al. 1988). However, it remains unclear whether *sem-4* directly activates serotonin pathway genes or if it acts upstream of other unknown factors. Worms lacking *egl-46*, another HSN-expressed zinc finger transcription factor, show multiple defects in HSN differentiation, including migration, axon projections, and serotonin production (**Figure 6**). As in the case of *sem-4*, it is unknown whether *egl-46* works at the terminal differentiation stage or if it is an upstream regulator of other factors (Wu et al. 2001). The best candidate for a terminal selector for HSN serotonergic differentiation, i.e., a transcription factor that directly controls differentiated features of HSN, is the POU transcription factor *unc-86*. HSN shows normal early differentiation in *unc-86* mutants but shows defects in terminal differentiation,

including lack of expression of the serotonin pathway genes and consequently serotonin synthesis (Sze et al. 2002). UNC-86 protein binds to the *tph-1* upstream region, arguing that regulation is direct (Sze et al. 2002). *unc-86* also regulates terminal differentiation of the serotonergic NSM (neurosecretory motorneuron); however, the amphid serotonergic neuron ADF does not express *unc-86* and, as expected, is not affected in *unc-86* mutants (**Figure 6**) (Sze et al. 2002). In the ADF neuron, the LIM homeobox gene *lim-4* appears to control the serotonergic phenotype (Zheng et al. 2005), but because *lim-4* is expressed only in the progenitor cell of ADF, it can only be upstream of a yet-to-be-identified terminal selector.

Another factor that could regulate HSN terminal differentiation is *hlh-3*. HLH-3 is a basic helix-loop-helix (bHLH) protein of the Achaete/Scute family; *hlh-3* is widely expressed during embryogenesis and its expression is maintained in HSN at least until the third larval stage. *hlh-3* mutants show normal generation and migration of HSN but fail to express *tph-1* and serotonin. Other serotonergic neurons are not affected in *hlh-3* mutants. Finally, *unc-86* expression is not affected in *hlh-3* mutants, which suggests that both factors act in parallel to specify HSN serotonergic fate (Doonan et al. 2008).

In summary, no single factor seems to be sufficient to specify serotonergic fate throughout the animal. Moreover, as in the case of *ast-1*, *unc-86* is expressed in many neurons apart from serotonergic neurons, indicating that *unc-86* must be one component of a combinatorial code of transcription factors that provides serotonergic specificity.

Tyraminergic and Octopaminergic specification in *C. elegans*. As in *Drosophila*, the factors that specify tyraminergic (RIM) and octopaminergic (RIC) neurons are unknown.

CONCLUSIONS

Despite the central role of monoaminergic signaling in the nervous system of all metazoans, the molecular mechanisms that directly activate the expression of the gene batteries necessary for the synthesis and release of these neurotransmitters remain incompletely understood at best. Even though a good number of transcription factors have been reported to affect the expression of specific enzymes in specific cell types, in most cases it is not known whether the regulation is direct or reflects on earlier activities of the transcription factor. In most cases, it is also not clear how extensively the fate of a monoaminergic class is affected.

There are a few well-characterized exceptions, and these exceptions may point to a common theme and principle in monoaminergic specification that will require further validation (**Table 1**). In *C. elegans ast-1* and *unc-86* mutants, in mouse Nurr1/Pitx3 mutants, and in mouse *Pet1* mutants, most if not all features of the respective monoaminergic gene batteries (as shown in **Figure 2**) are lost. Even though factors that act early in lineage specification would be expected to result in such a broad phenotype, in these specific cases, the loss of all markers appears rather to be the reflection of direct coregulation of monoaminergic gene batteries. In the example of dopaminergic neurons in worms, all dopaminergic identity genes are coregulated by an AST-1 binding site present in all five genes. Serotonergic neuron specification appears to occur by the same principle; all biosynthetic enzymes and transporters appear to be coregulated by the mouse ETS factor PET1 and *C. elegans* UNC-86. The NURR1/PITX3 transcription factor combination may act in the same way to coregulate all dopaminergic identity genes in the mouse midbrain (Jacobs et al. 2009b). These transcription factors, therefore, all classify as terminal selectors of the respective monoaminergic neuron identities.

The emerging theme of coregulation, schematically depicted in **Figure 7**, makes sense if one considers that the coordination of expression of these genes is critical. As in an assembly line, all the enzymes and transporters depend on each other to generate the final end product (a released monoamine). Coregulation of proteins acting in biosynthetic processes is a pervasive theme across phylogeny and can be observed down to the level of bacteria, which organize functionally related genes involved in specific biosynthetic processes into coregulated operons and regulons (Epstein & Beckwith 1968, Hobert et al. 2010). Coregulation of enzymes and transporters for classic neurotransmitters, e.g., GABA, has also been reported (Eastman et al. 1999). In the most extreme version of coregulation of neurotransmitter-related genes, the two genes required for synthesis and vesicular transport for acetylcholine are organized into an almost operon-like structure across the animal kingdom (from worms to vertebrates) (Erickson et al. 1994, Kitamoto et al. 1998, Rand 1989).

Coregulation of monoaminergic gene batteries may go beyond just those genes and include other identity determinants of the neurons as well (**Figure 7**), as indicated by the many additional targets of, e.g., NURR1/PITX3 (Jacobs et al. 2009a,b) or AST-1 (Flames & Hobert 2009). These ideas are important to keep in mind, particularly when one considers that, even though the synthesis and transport machinery for a specific monoamine are core identity determinants, other features define a specific monoaminergic neuron type—features such as, for example, the position and morphological properties of a specific monoaminergic neuron or its specific patterns of synaptic connectivity. Cases such as the terminal selectors *ast-1* or *Nurr1/Pitx3*, which control a plethora of distinct types of genes in their respective monoaminergic neuron types, may point to many identity determinants all being under coregulatory control (see sidebar, Terminal Selectors). Taken together, it is important to probe the hypothesis of coregulation more extensively, ideally through the dissection of the *cis*-regulatory landscape of monoaminergic gene battery members, as exemplified in *C. elegans* (Flames & Hobert 2009).

In some cases and in some organisms, all neurons that produce one specific monoamine may be regulated by a similar regulatory logic. For example, *ast-1* affects all *C. elegans*

Table 1 Summary of transcription factors involved in monoaminergic terminal differentiation[a]

| Gene name | M. musculus | | | | | | | | | | D. melanogaster | | | | | C. elegans | |
	Pitx3	Dlx1/2	Pax6	Phox2a/b	Lmx1b	Er81	Pet1	Nurr1	Hand2	Gata3	En	vvl	islet	Eg	bkb	unc-86	ast-1
TF family	HD	HD	Pax/HD	Prd-type HD	LIM/HD	ETS	ETS	NHR	bHLH	GATA	HD	POU/HD	LIM/HD	Zn finger	Zn finger	POU/HD	ETS
MA population	DA	DA	DA	NA	5HT[b]	DA	5HT	DA	NA	NA[c]	5HT	DA	DA and 5HT	5HT	5HT	5HT	DA
Exclusive expression	Yes	No	No	No	No	No	Yes	No	No	No	No	No	No	No	No	No	No
Maintained expression	Yes	Yes	Yes	Yes	Yes	Yes	Yes	Yes	Yes	Yes	?	?	?	No	Yes	Yes	Yes
Loss of MA differentiation in the mutant	Yes	Yes	Yes	Yes	Yes	Yes	Yes	Yes	Yes	Yes	Yes	Yes	Yes	Yes	Yes	Yes	Yes
Cell death in the mutant	Yes	?	Yes	Yes	No	No	No	Yes	No	Yes	?	?	No	?	?	No	No
Direct activation of MA pathway genes	Yes	?	?	Yes	?	Yes	Yes	Yes	Yes	?	?	Yes	?	?	?	Yes	Yes
Ectopic induction of MA pathway genes	?	Yes	Yes	Yes	Yes[d]	?	Yes[d]	?	Yes	Yes	?	?	Yes	Yes	?	?	Yes
Role in fate maintenance	?	?	No	Yes	Yes	?	Yes	Yes	Yes	No	?	?	?	No	?	?	Yes
Role in survival[e]	?	?	Yes	No[f]	Yes	?	No	Yes	No	Yes	?	?	?	No	?	?	?

[a] Abbreviations: 5HT, serotonin; ?, not reported; DA, dopamine; HD, homeodomain; MA, monoamine; NA, noradrenaline; NHR, nuclear hormone receptor; bHLH, basic helix loop helix; Prd-type, paired like; TF, transcription factor.

[b] Refers to its role in serotonergic cells. Lmx1b is also expressed in midbrain dopaminergic neurons.

[c] Refers to its role in sympathetic noradrenergic cell differentiation. Gata3 is also expressed in serotonergic cells.

[d] Induction of ectopic serotonergic fate only upon coexpression of Pet1, Lmx1b, and Nkx2.2. Nkx2.2 is expressed only in progenitor cells, thus it is not a factor directly involved in terminal differentiation and has not been included in the table.

[e] Refers specifically to survival of already differentiated cells.

[f] Refers to sympathetic neurons; in the brain, it remains to be determined if the role is in fate maintenance, survival, or both.

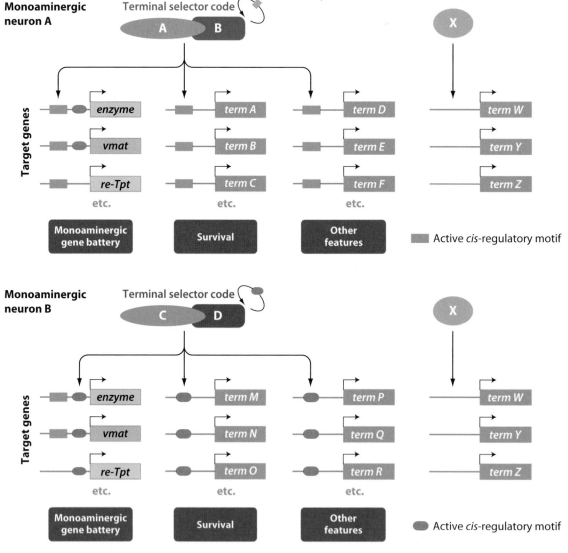

Figure 7

Concepts for monoaminergic specification. Monoaminergic gene batteries are genes encoding proteins shown in **Figure 2** (*vmat*, vesicular monoamine transporter; *re-Tpt*, reuptake transporter). In cases analyzed in sufficient detail, it appears that these genes are coregulated through a common regulatory logic. *Trans*-acting factors controlling the expression of the genes are called "terminal selectors" (see sidebar, Terminal Selectors). They often autoregulate their own expression to ensure the continuous presence of themselves and therefore also of their target genes. Examples for A and B or C and D are *Nurr1* and *Pitx3* for dopaminergic midbrain neurons or *ast-1* for worm DA neurons (plus as yet unknown factors) or *Pet1* for mouse serotonergic neurons. Terminal selectors may regulate survival of the cells and many other identity features of the respective neurons (indicated by "term"). There may also be parallel acting regulatory routines (X), such as those regulating pan-neuronal fate, which is often not affected by removal of terminal selectors. The blue square and red circle indicate that different *cis*-regulatory motifs (and different *trans*-acting factors) may be utilized for the same gene in distinct types of monoaminergic neurons or even in different populations of the same monoaminergic subtype.

dopaminergic neurons, and Pet1 affects the vast majority of mouse serotonergic neurons. But in most cases, neurons that produce one specific monoamine are too diverse to be all controlled by one regulatory logic. This is well exemplified in the mouse dopaminergic system. On the basis of the available knowledge of *trans*-acting factors involved in olfactory dopaminergic neurons versus midbrain dopaminergic neurons, it appears that the monoaminergic gene battery is controlled by distinct terminal selectors (**Figure 3**). It is possible that such independent regulation is reflected by members of the monoaminergic gene battery containing distinct *cis*-regulatory motifs that are activated in distinct monoaminergic neuron groups (**Figure 7**). The existence of parallel and independently acting *cis*-regulatory elements is a rather obvious hypothesis from the standpoint that some members of monoaminergic gene batteries are shared by different monoamines and yet are directly regulated by different transcription factors. For example, the *C. elegans* vesicular monoamine transporter *cat-1* is expressed in serotonin neurons and in dopaminergic neurons, and a *cis*-regulatory analysis of its gene-regulatory sequences has identified separable elements for driving expression in dopamine neurons (AST-1 binding site) and serotonin neurons (Flames & Hobert 2009). Even though distinct monoaminergic neuron groups may use distinct gene-regulatory factors, there may be some overlap. For example, *Nurr1*, but not *Pitx3*, may have a role in retinal dopaminergic neurons, similar to its role in midbrain dopamine neurons. In retinal neurons, NURR1 may pair up with a distinct *trans*-acting factor.

The common theme emerging from studying monoaminergic terminal differentiation is the use of combinatorial codes of regulatory genes (**Figure 7**). By acting in different combinations, regulatory genes can be employed in a distinct manner in distinct cell types, thereby allowing an organism to generate a plethora of different regulatory states with a limited number of regulatory genes. Even though some regulatory genes, such as the terminal selector *Pitx3*, may be expressed in only one neuron type, they nevertheless act with other transcription factors (*Nurr1* in the case of *Pitx3*), most likely because the binding specificity of one transcription factor alone may not suffice to select specific genes for regulation.

Several recent studies aimed to assess whether the mechanisms that induce the expression of terminal monoaminergic gene batteries are the same as those that maintain it throughout the life of the neuron. All presumptive terminal selectors discussed above are continuously expressed, and genetic evidence suggests that terminal selectors are indeed also continuously required to maintain such terminal batteries (**Table 1**) (Flames & Hobert 2009, Kadkhodaei et al. 2009, Liu et al. 2010, Schmidt et al. 2009, Song et al. 2011, Zhao et al. 2006). Another interesting observation is that most of the mouse terminal selectors are required to maintain the differentiated state and in some cases for the cells to survive, indicating that avoiding cell death may be a continuously active process that is part of the functional gene batteries controlled by the terminal selectors.

The next few years will surely see the identification of more regulatory factors involved in monoaminergic neuron specification. The use of relatively novel technology, such as chromatin-immunoprecipitation, will identify regulatory relationships in a more definitive manner, and through such accumulated knowledge, it may become possible to generate monoaminergic neurons in vitro.

DISCLOSURE STATEMENT

The authors are not aware of any affiliations, memberships, funding, or financial holdings that might be perceived as affecting the objectivity of this review.

ACKNOWLEDGMENTS

We apologize to those whose work we have not cited because of space constraints or our own ignorance. We are grateful to Jean-François Brunet and Marten P. Smidt for their valuable comments on this review. Work in the authors' lab is funded by the National Institutes of Health (R01NS039996-05; R01NS050266-03). N.F. has been funded by the NY Stem Cell Foundation, an EMBO fellowship, and a Marie Curie postdoctoral fellowship. O.H. is an Investigator of the Howard Hughes Medical Institute.

LITERATURE CITED

Abeliovich A, Hammond R. 2007. Midbrain dopamine neuron differentiation: factors and fates. *Dev. Biol.* 304:447–54

Alenina N, Bashammakh S, Bader M. 2006. Specification and differentiation of serotonergic neurons. *Stem Cell Rev.* 2:5–10

Alkema MJ, Hunter-Ensor M, Ringstad N, Horvitz HR. 2005. Tyramine functions independently of octopamine in the *Caenorhabditis elegans* nervous system. *Neuron* 46:247–60

Allen ZJ Jr, Waclaw RR, Colbert MC, Campbell K. 2007. Molecular identity of olfactory bulb interneurons: transcriptional codes of periglomerular neuron subtypes. *J. Mol. Histol.* 38:517–25

Andersson E, Tryggvason U, Deng Q, Friling S, Alekseenko Z, et al. 2006. Identification of intrinsic determinants of midbrain dopamine neurons. *Cell* 124:393–405

Andrews GL, Yun K, Rubenstein JL, Mastick GS. 2003. Dlx transcription factors regulate differentiation of dopaminergic neurons of the ventral thalamus. *Mol. Cell Neurosci.* 23:107–20

Asbreuk CH, Vogelaar CF, Hellemons A, Smidt MP, Burbach JP. 2002. CNS expression pattern of Lmx1b and coexpression with ptx genes suggest functional cooperativity in the development of forebrain motor control systems. *Mol. Cell Neurosci.* 21:410–20

Backman C, Perlmann T, Wallen A, Hoffer BJ, Morales M. 1999. A selective group of dopaminergic neurons express Nurr1 in the adult mouse brain. *Brain Res.* 851:125–32

Basson M, Horvitz HR. 1996. The *Caenorhabditis elegans* gene sem-4 controls neuronal and mesodermal cell development and encodes a zinc finger protein. *Genes Dev.* 10:1953–65

Baum PD, Guenther C, Frank CA, Pham BV, Garriga G. 1999. The *Caenorhabditis elegans* gene ham-2 links Hox patterning to migration of the HSN motor neuron. *Genes Dev.* 13:472–83

Bellipanni G, Rink E, Bally-Cuif L. 2002. Cloning of two tryptophan hydroxylase genes expressed in the diencephalon of the developing zebrafish brain. *Mech. Dev.* 119(Suppl. 1):S215–20

Birman S. 2005. Arousal mechanisms: Speedy flies don't sleep at night. *Curr. Biol.* 15:R511–13

Bai Q, Burton EA. 2009. Cis-acting elements responsible for dopaminergic neuron-specific expression of zebrafish slc6a3 (dopamine transporter) in vivo are located remote from the transcriptional start site. *Neuroscience* 164:1138–51

Blaugrund E, Pham TD, Tennyson VM, Lo L, Sommer L, et al. 1996. Distinct subpopulations of enteric neuronal progenitors defined by time of development, sympathoadrenal lineage markers and Mash-1-dependence. *Development* 122:309–20

Blin M, Norton W, Bally-Cuif L, Vernier P. 2008. NR4A2 controls the differentiation of selective dopaminergic nuclei in the zebrafish brain. *Mol. Cell Neurosci.* 39:592–604

Bray SJ, Johnson WA, Hirsh J, Heberlein U, Tjian R. 1988. A *cis*-acting element and associated binding factor required for CNS expression of the *Drosophila melanogaster* dopa decarboxylase gene. *EMBO J.* 7:177–88

Brill MS, Snapyan M, Wohlfrom H, Ninkovic J, Jawerka M, et al. 2008. A dlx2- and pax6-dependent transcriptional code for periglomerular neuron specification in the adult olfactory bulb. *J. Neurosci.* 28:6439–52

Briscoe J, Sussel L, Serup P, Hartigan-O'Connor D, Jessell TM, et al. 1999. Homeobox gene Nkx2.2 and specification of neuronal identity by graded Sonic hedgehog signalling. *Nature* 398:622–27

Carlsson A, Falck B, Hillarp NA. 1962. Cellular localization of brain monoamines. *Acta Physiol. Scand. Suppl.* 56:1–28

Carlsson A, Falck B, Hillarp NA, Thieme G, Torp A. 1961. A new histochemical method for visualization of tissue catechol amines. *Med. Exp. Int. J. Exp. Med.* 4:123–25

Cave JW, Akiba Y, Banerjee K, Bhosle S, Berlin R, Baker H. 2010. Differential regulation of dopaminergic gene expression by Er81. *J. Neurosci.* 30:4717–24

Cave JW, Baker H. 2009. Dopamine systems in the forebrain. *Adv. Exp. Med. Biol.* 651:15–35

Caveney S, Cladman W, Verellen L, Donly C. 2006. Ancestry of neuronal monoamine transporters in the metazoa. *J. Exp. Biol.* 209:4858–68

Cazorla P, Smidt MP, O'Malley KL, Burbach JP. 2000. A response element for the homeodomain transcription factor Ptx3 in the tyrosine hydroxylase gene promoter. *J. Neurochem.* 74:1829–37

Chase DL, Koelle MR. 2007. Biogenic amine neurotransmitters in *C. elegans*. In *WormBook*, ed. *C. elegans* Res. Community, WormBook, doi/10.1895/wormbook.1.132.1, **http://www.wormbook.org**

Chen YC, Priyadarshini M, Panula P. 2009. Complementary developmental expression of the two tyrosine hydroxylase transcripts in zebrafish. *Histochem. Cell Biol.* 132:375–81

Cheng CW, Yan CH, Choy SW, Hui MN, Hui CC, Cheng SH. 2007. Zebrafish homologue irx1a is required for the differentiation of serotonergic neurons. *Dev. Dyn.* 236:2661–67

Cheng L, Chen CL, Luo P, Tan M, Qiu M, et al. 2003. Lmx1b, Pet-1, and Nkx2.2 coordinately specify serotonergic neurotransmitter phenotype. *J. Neurosci.* 23:9961–67

Chu-LaGraff Q, Schmid A, Leidel J, Bronner G, Jackle H, Doe CQ. 1995. huckebein specifies aspects of CNS precursor identity required for motoneuron axon pathfinding. *Neuron* 15:1041–51

Chung S, Leung A, Han BS, Chang MY, Moon JI, et al. 2009. Wnt1-lmx1a forms a novel autoregulatory loop and controls midbrain dopaminergic differentiation synergistically with the SHH-FoxA2 pathway. *Cell Stem Cell* 5:646–58

Coppola E, d'Autreaux F, Rijli FM, Brunet JF. 2010. Ongoing roles of Phox2 homeodomain transcription factors during neuronal differentiation. *Development* 137:4211–20

Coppola E, Pattyn A, Guthrie SC, Goridis C, Studer M. 2005. Reciprocal gene replacements reveal unique functions for Phox2 genes during neural differentiation. *EMBO J.* 24:4392–403

Cserjesi P, Brown D, Lyons GE, Olson EN. 1995. Expression of the novel basic helix-loop-helix gene eHAND in neural crest derivatives and extraembryonic membranes during mouse development. *Dev. Biol.* 170:664–78

Dahlstrom A, Fuxe K. 1964. Localization of monoamines in the lower brain stem. *Experientia* 20:398–99

Del Giacco L, Sordino P, Pistocchi A, Andreakis N, Tarallo R, et al. 2006. Differential regulation of the zebrafish orthopedia 1 gene during fate determination of diencephalic neurons. *BMC Dev. Biol.* 6:50

Dellovade TL, Pfaff DW, Schwanzel-Fukuda M. 1998. Olfactory bulb development is altered in small-eye (Sey) mice. *J. Comp. Neurol.* 402:402–18

de Melo J, Qiu X, Du G, Cristante L, Eisenstat DD. 2003. Dlx1, Dlx2, Pax6, Brn3b, and Chx10 homeobox gene expression defines the retinal ganglion and inner nuclear layers of the developing and adult mouse retina. *J. Comp. Neurol.* 461:187–204

Desai C, Garriga G, McIntire SL, Horvitz HR. 1988. A genetic pathway for the development of the *Caenorhabditis elegans* HSN motor neurons. *Nature* 336:638–46

Desai C, Horvitz HR. 1989. *Caenorhabditis elegans* mutants defective in the functioning of the motor neurons responsible for egg laying. *Genetics* 121:703–21

Ding YQ, Marklund U, Yuan W, Yin J, Wegman L, et al. 2003. Lmx1b is essential for the development of serotonergic neurons. *Nat. Neurosci.* 6:933–38

Dittrich R, Bossing T, Gould AP, Technau GM, Urban J. 1997. The differentiation of the serotonergic neurons in the *Drosophila* ventral nerve cord depends on the combined function of the zinc finger proteins Eagle and Huckebein. *Development* 124:2515–25

Doonan R, Hatzold J, Raut S, Conradt B, Alfonso A. 2008. HLH-3 is a *C. elegans* Achaete/Scute protein required for differentiation of the hermaphrodite-specific motor neurons. *Mech. Dev.* 125:883–93

Doxakis E, Howard L, Rohrer H, Davies AM. 2008. HAND transcription factors are required for neonatal sympathetic neuron survival. *EMBO Rep.* 9:1041–47

Duerr JS, Frisby DL, Gaskin J, Duke A, Asermely K, et al. 1999. The cat-1 gene of *Caenorhabditis elegans* encodes a vesicular monoamine transporter required for specific monoamine-dependent behaviors. *J. Neurosci.* 19:72–84

Durr K, Holzschuh J, Filippi A, Ettl AK, Ryu S, et al. 2006. Differential roles of transcriptional mediator complex subunits Crsp34/Med27, Crsp150/Med14 and Trap100/Med24 during zebrafish retinal development. *Genetics* 174:693–705

Eastman C, Horvitz HR, Jin Y. 1999. Coordinated transcriptional regulation of the unc-25 glutamic acid decarboxylase and the unc-47 GABA vesicular transporter by the *Caenorhabditis elegans* UNC-30 homeodomain protein. *J. Neurosci.* 19:6225–34

Eisenstat DD, Liu JK, Mione M, Zhong W, Yu G, et al. 1999. DLX-1, DLX-2, and DLX-5 expression define distinct stages of basal forebrain differentiation. *J. Comp. Neurol.* 414:217–37

Epstein W, Beckwith J. 1968. Regulation of gene expression. *Annu. Rev. Biochem.* 37:411–36

Erickson JD, Varoqui H, Schafer MK, Modi W, Diebler MF, et al. 1994. Functional identification of a vesicular acetylcholine transporter and its expression from a "cholinergic" gene locus. *J. Biol. Chem.* 269:21929–32

Ferri AL, Lin W, Mavromatakis YE, Wang JC, Sasaki H, et al. 2007. Foxa1 and Foxa2 regulate multiple phases of midbrain dopaminergic neuron development in a dosage-dependent manner. *Development* 134:2761–69

Filippi A, Durr K, Ryu S, Willaredt M, Holzschuh J, Driever W. 2007. Expression and function of nr4a2, lmx1b, and pitx3 in zebrafish dopaminergic and noradrenergic neuronal development. *BMC Dev. Biol.* 7:135

Filippi A, Mahler J, Schweitzer J, Driever W. 2010. Expression of the paralogous tyrosine hydroxylase encoding genes th1 and th2 reveals the full complement of dopaminergic and noradrenergic neurons in zebrafish larval and juvenile brain. *J. Comp. Neurol.* 518:423–38

Flames N, Hobert O. 2009. Gene regulatory logic of dopamine neuron differentiation. *Nature* **458:885–89**

Fujimoto E, Stevenson TJ, Chien CB, Bonkowsky JL. 2011. Identification of a dopaminergic enhancer indicates complexity in vertebrate dopamine neuron phenotype specification. *Dev. Biol.* 352:393–404

Garcia-Bellido A. 1975. Genetic control of wing disc development in *Drosophila*. *Ciba Found. Symp.* 29:161–82

Goridis C, Rohrer H. 2002. Specification of catecholaminergic and serotonergic neurons. *Nat. Rev. Neurosci.* 3:531–41

Guo S, Brush J, Teraoka H, Goddard A, Wilson SW, et al. 1999a. Development of noradrenergic neurons in the zebrafish hindbrain requires BMP, FGF8, and the homeodomain protein soulless/Phox2a. *Neuron* 24:555–66

Guo S, Wilson SW, Cooke S, Chitnis AB, Driever W, Rosenthal A. 1999b. Mutations in the zebrafish unmask shared regulatory pathways controlling the development of catecholaminergic neurons. *Dev. Biol.* 208:473–87

Guo S, Yamaguchi Y, Schilbach S, Wada T, Lee J, et al. 2000. A regulator of transcriptional elongation controls vertebrate neuronal development. *Nature* 408:366–69

Haba H, Nomura T, Suto F, Osumi N. 2009. Subtype-specific reduction of olfactory bulb interneurons in Pax6 heterozygous mutant mice. *Neurosci. Res.* 65:116–21

Hack MA, Saghatelyan A, de Chevigny A, Pfeifer A, Ashery-Padan R, et al. 2005. Neuronal fate determinants of adult olfactory bulb neurogenesis. *Nat. Neurosci.* 8:865–72

Harrison SJ, Parrish M, Monaghan AP. 2008. Sall3 is required for the terminal maturation of olfactory glomerular interneurons. *J. Comp. Neurol.* 507:1780–94

Havrda MC, Harris BT, Mantani A, Ward NM, Paolella BR, et al. 2008. Id2 is required for specification of dopaminergic neurons during adult olfactory neurogenesis. *J. Neurosci.* 28:14074–86

Hendershot TJ, Liu H, Clouthier DE, Shepherd IT, Coppola E, et al. 2008. Conditional deletion of Hand2 reveals critical functions in neurogenesis and cell type-specific gene expression for development of neural crest-derived noradrenergic sympathetic ganglion neurons. *Dev. Biol.* 319:179–91

Hendricks T, Francis N, Fyodorov D, Deneris ES. 1999. The ETS domain factor Pet-1 is an early and precise marker of central serotonin neurons and interacts with a conserved element in serotonergic genes. *J. Neurosci.* **19:10348–56**

Hendricks TJ, Fyodorov DV, Wegman LJ, Lelutiu NB, Pehek EA, et al. 2003. Pet-1 ETS gene plays a critical role in 5-HT neuron development and is required for normal anxiety-like and aggressive behavior. *Neuron* 37:233–47

Extensive analysis of the *cis*-regulatory elements of all dopamine pathway genes in *C. elegans* that led to identification of the dopaminergic motif and its transacting factor *ast-1*.

In Hendricks et al. 1999 and 2003, *Pet1* is characterized as a terminal selector for serotonergic specification. PET1 directly binds to the regulatory regions of serotonin pathway genes and their expression is lost in *Pet1* mutants.

Hilakivi I. 1987. Biogenic amines in the regulation of wakefulness and sleep. *Med. Biol.* 65:97–104

Hill RE, Favor J, Hogan BL, Ton CC, Saunders GF, et al. 1991. Mouse small eye results from mutations in a paired-like homeobox-containing gene. *Nature* 354:522–25

Hobert O. 2008. Regulatory logic of neuronal diversity: terminal selector genes and selector motifs. *Proc. Natl. Acad. Sci. USA* 105:20067–71

Hobert O, Carrera I, Stefanakis N. 2010. The molecular and gene regulatory signature of a neuron. *Trends Neurosci.* 33:435–45

Holzschuh J, Barrallo-Gimeno A, Ettl AK, Durr K, Knapik EW, Driever W. 2003. Noradrenergic neurons in the zebrafish hindbrain are induced by retinoic acid and require tfap2a for expression of the neurotransmitter phenotype. *Development* 130:5741–54

Holzschuh J, Ryu S, Aberger F, Driever W. 2001. Dopamine transporter expression distinguishes dopaminergic neurons from other catecholaminergic neurons in the developing zebrafish embryo. *Mech. Dev.* 101:237–43

Hong SJ, Huh Y, Chae H, Hong S, Lardaro T, Kim KS. 2006. GATA-3 regulates the transcriptional activity of tyrosine hydroxylase by interacting with CREB. *J. Neurochem.* 98:773–81

Hong SJ, Lardaro T, Oh MS, Huh Y, Ding Y, et al. 2008. Regulation of the noradrenaline neurotransmitter phenotype by the transcription factor AP-2beta. *J. Biol. Chem.* 283:16860–67

Horvitz HR, Chalfie M, Trent C, Sulston JE, Evans PD. 1982. Serotonin and octopamine in the nematode *Caenorhabditis elegans*. *Science* 216:1012–14

Howard MJ. 2005. Mechanisms and perspectives on differentiation of autonomic neurons. *Dev. Biol.* 277:271–86

Hwang DY, Ardayfio P, Kang UJ, Semina EV, Kim KS. 2003. Selective loss of dopaminergic neurons in the substantia nigra of Pitx3-deficient aphakia mice. *Brain Res. Mol. Brain Res.* 114:123–31

Isshiki T, Pearson B, Holbrook S, Doe CQ. 2001. *Drosophila* neuroblasts sequentially express transcription factors which specify the temporal identity of their neuronal progeny. *Cell* 106:511–21

Jacob J, Ferri AL, Milton C, Prin F, Pla P, et al. 2007. Transcriptional repression coordinates the temporal switch from motor to serotonergic neurogenesis. *Nat. Neurosci.* 10:1433–39

Jacob J, Storm R, Castro DS, Milton C, Pla P, et al. 2009. Insm1 (IA-1) is an essential component of the regulatory network that specifies monoaminergic neuronal phenotypes in the vertebrate hindbrain. *Development* 136:2477–85

Jacobs FM, Smits SM, Noorlander CW, von Oerthel L, van der Linden AJ, et al. 2007. Retinoic acid counteracts developmental defects in the substantia nigra caused by Pitx3 deficiency. *Development* 134:2673–84

Jacobs FM, van der Linden AJ, Wang Y, von Oerthel L, Sul HS, et al. 2009a. Identification of Dlk1, Ptpru and Klhl1 as novel Nurr1 target genes in meso-diencephalic dopamine neurons. *Development* 136:2363–73

Jacobs FM, van Erp S, van der Linden AJ, von Oerthel L, Burbach JP, Smidt MP. 2009b. Pitx3 potentiates Nurr1 in dopamine neuron terminal differentiation through release of SMRT-mediated repression. *Development* 136:531–40

Jiang H, Xiang M. 2009. Subtype specification of GABAergic amacrine cells by the orphan nuclear receptor Nr4a2/Nurr1. *J. Neurosci.* 29:10449–59

Johnson WA, Hirsh J. 1990. Binding of a *Drosophila* POU-domain protein to a sequence element regulating gene expression in specific dopaminergic neurons. *Nature* 343:467–70

Johnson WA, McCormick CA, Bray SJ, Hirsh J. 1989. A neuron-specific enhancer of the *Drosophila* dopa decarboxylase gene. *Genes Dev.* 3:676–86

Kadkhodaei B, Ito T, Joodmardi E, Mattsson B, Rouillard C, et al. 2009. Nurr1 is required for maintenance of maturing and adult midbrain dopamine neurons. *J. Neurosci.* 29:15923–32

Kandel ER, Schwartz JM, Jessell TM, eds. 1991. *Principles of Neural Science*. Norwalk, CT: Appleton & Lange

Kaslin J, Panula P. 2001. Comparative anatomy of the histaminergic and other aminergic systems in zebrafish (*Danio rerio*). *J. Comp. Neurol.* 440:342–77

Keene AC, Waddell S. 2005. *Drosophila* memory: Dopamine signals punishment? *Curr. Biol.* 15:R932–34

Kim HS, Seo H, Yang C, Brunet JF, Kim KS. 1998. Noradrenergic-specific transcription of the dopamine beta-hydroxylase gene requires synergy of multiple *cis*-acting elements including at least two Phox2a-binding sites. *J. Neurosci.* 18:8247–60

Chip-on-chip analysis of NURR1 and PITX3 demonstrated their joint activity in controlling dopamine pathway genes as well as many other terminal identity features.

Kim J, Inoue K, Ishii J, Vanti WB, Voronov SV, et al. 2007. A microRNA feedback circuit in midbrain dopamine neurons. *Science* 317:1220–24

Kim KS, Kim CH, Hwang DY, Seo H, Chung S, et al. 2003. Orphan nuclear receptor Nurr1 directly transactivates the promoter activity of the tyrosine hydroxylase gene in a cell-specific manner. *J. Neurochem.* 85:622–34

Kitamoto T, Wang W, Salvaterra PM. 1998. Structure and organization of the *Drosophila* cholinergic locus. *J. Biol. Chem.* 273:2706–13

Kittappa R, Chang WW, Awatramani RB, McKay RD. 2007. The foxa2 gene controls the birth and spontaneous degeneration of dopamine neurons in old age. *PLoS Biol.* 5:e325

Kohwi M, Osumi N, Rubenstein JL, Alvarez-Buylla A. 2005. Pax6 is required for making specific subpopulations of granule and periglomerular neurons in the olfactory bulb. *J. Neurosci.* 25:6997–7003

Le W, Conneely OM, Zou L, He Y, Saucedo-Cardenas O, et al. 1999. Selective agenesis of mesencephalic dopaminergic neurons in Nurr1-deficient mice. *Exp. Neurol.* 159:451–58

Lee HS, Bae EJ, Yi SH, Shim JW, Jo AY, et al. 2010. Foxa2 and Nurr1 synergistically yield A9 nigral dopamine neurons exhibiting improved differentiation, function, and cell survival. *Stem Cells* 28:501–12

Levkowitz G, Zeller J, Sirotkin HI, French D, Schilbach S, et al. 2003. Zinc finger protein too few controls the development of monoaminergic neurons. *Nat. Neurosci.* 6:28–33

Lillesaar C, Tannhauser B, Stigloher C, Kremmer E, Bally-Cuif L. 2007. The serotonergic phenotype is acquired by converging genetic mechanisms within the zebrafish central nervous system. *Dev. Dyn.* 236:1072–84

Lim KC, Lakshmanan G, Crawford SE, Gu Y, Grosveld F, Engel JD. 2000. Gata3 loss leads to embryonic lethality due to noradrenaline deficiency of the sympathetic nervous system. *Nat. Genet.* 25:209–12

Lin W, Metzakopian E, Mavromatakis YE, Gao N, Balaskas N, et al. 2009. Foxa1 and Foxa2 function both upstream of and cooperatively with Lmx1a and Lmx1b in a feedforward loop promoting mesodiencephalic dopaminergic neuron development. *Dev. Biol.* 333:386–96

Liu C, Maejima T, Wyler SC, Casadesus G, Herlitze S, Deneris ES. 2010. Pet-1 is required across different stages of life to regulate serotonergic function. *Nat. Neurosci.* 13:1190–98

Liu KS, Sternberg PW. 1995. Sensory regulation of male mating behavior in *Caenorhabditis elegans*. *Neuron* 14:79–89

Lledo PM, Merkle FT, Alvarez-Buylla A. 2008. Origin and function of olfactory bulb interneuron diversity. *Trends Neurosci.* 31:392–400

Lohr H, Ryu S, Driever W. 2009. Zebrafish diencephalic A11-related dopaminergic neurons share a conserved transcriptional network with neuroendocrine cell lineages. *Development* 136:1007–17

Lucas ME, Müller F, Rüdiger R, Henion PD, Rohrer H. 2006. The bHLH transcription factor hand2 is essential for noradrenergic differentiation of sympathetic neurons. *Development* 133:4015–24

Lundell MJ, Chu-LaGraff Q, Doe CQ, Hirsh J. 1996. The engrailed and huckebein genes are essential for development of serotonin neurons in the *Drosophila* CNS. *Mol. Cell Neurosci.* 7:46–61

Lundell MJ, Hirsh J. 1992. The zfh-2 gene product is a potential regulator of neuron-specific dopa decarboxylase gene expression in *Drosophila*. *Dev. Biol.* 154:84–94

Lundell MJ, Hirsh J. 1994. Temporal and spatial development of serotonin and dopamine neurons in the *Drosophila* CNS. *Dev. Biol.* 165:385–96

Lundell MJ, Hirsh J. 1998. Eagle is required for the specification of serotonin neurons and other neuroblast 7–3 progeny in the *Drosophila* CNS. *Development* 125:463–72

Luo GR, Chen Y, Li XP, Liu TX, Le WD. 2008. Nr4a2 is essential for the differentiation of dopaminergic neurons during zebrafish embryogenesis. *Mol. Cell Neurosci.* 39:202–10

Luskin MB. 1993. Restricted proliferation and migration of postnatally generated neurons derived from the forebrain subventricular zone. *Neuron* 11:173–89

Mannisto PT, Kaakkola S. 1999. Catechol-O-methyltransferase (COMT): biochemistry, molecular biology, pharmacology, and clinical efficacy of the new selective COMT inhibitors. *Pharmacol. Rev.* 51:593–628

Martinat C, Bacci JJ, Leete T, Kim J, Vanti WB, et al. 2006. Cooperative transcription activation by Nurr1 and Pitx3 induces embryonic stem cell maturation to the midbrain dopamine neuron phenotype. *Proc. Natl. Acad. Sci. USA* 103:2874–79

Report of the direct binding to TH and DAT regulatory regions of NURR1 and PITX3 and their cooperative action in the generation of dopaminergic neurons from embryonic stem cells.

Maxwell SL, Ho HY, Kuehner E, Zhao S, Li M. 2005. Pitx3 regulates tyrosine hydroxylase expression in the substantia nigra and identifies a subgroup of mesencephalic dopaminergic progenitor neurons during mouse development. *Dev. Biol.* 282:467–79

Monastirioti M. 1999. Biogenic amine systems in the fruit fly *Drosophila melanogaster. Microsc. Res. Tech.* 45:106–21

Morikawa Y, Dai YS, Hao J, Bonin C, Hwang S, Cserjesi P. 2005. The basic helix-loop-helix factor Hand 2 regulates autonomic nervous system development. *Dev. Dyn.* 234:613–21

Morikawa Y, D'Autreaux F, Gershon MD, Cserjesi P. 2007. Hand2 determines the noradrenergic phenotype in the mouse sympathetic nervous system. *Dev. Biol.* 307:114–26

Morin X, Cremer H, Hirsch MR, Kapur RP, Goridis C, Brunet JF. 1997. Defects in sensory and autonomic ganglia and absence of locus coeruleus in mice deficient for the homeobox gene Phox2a. *Neuron* 18:411–23

Nagaya Y, Kutsukake M, Chigusa SI, Komatsu A. 2002. A trace amine, tyramine, functions as a neuromodulator in *Drosophila melanogaster. Neurosci. Lett.* 329:324–28

Neckameyer WS, Coleman CM, Eadie S, Goodwin SF. 2007. Compartmentalization of neuronal and peripheral serotonin synthesis in *Drosophila melanogaster. Genes Brain Behav.* 6:756–69

Ninkovic J, Pinto L, Petricca S, Lepier A, Sun J, et al. 2010. The transcription factor Pax6 regulates survival of dopaminergic olfactory bulb neurons via crystallin alphaA. *Neuron* 68:682–94

Novotny T, Eiselt R, Urban J. 2002. Hunchback is required for the specification of the early sublineage of neuroblast 7–3 in the *Drosophila* central nervous system. *Development* 129:1027–36

Nunes I, Tovmasian LT, Silva RM, Burke RE, Goff SP. 2003. Pitx3 is required for development of substantia nigra dopaminergic neurons. *Proc. Natl. Acad. Sci. USA* 100:4245–50

Ono Y, Nakatani T, Sakamoto Y, Mizuhara E, Minaki Y, et al. 2007. Differences in neurogenic potential in floor plate cells along an anteroposterior location: midbrain dopaminergic neurons originate from mesencephalic floor plate cells. *Development* 134:3213–25

Pattyn A, Goridis C, Brunet JF. 2000. Specification of the central noradrenergic phenotype by the homeobox gene Phox2b. *Mol. Cell Neurosci.* 15:235–43

Pattyn A, Morin X, Cremer H, Goridis C, Brunet JF. 1997. Expression and interactions of the two closely related homeobox genes Phox2a and Phox2b during neurogenesis. *Development* 124:4065–75

Pattyn A, Morin X, Cremer H, Goridis C, Brunet JF. 1999. The homeobox gene Phox2b is essential for the development of autonomic neural crest derivatives. *Nature* 399:366–70

Pattyn A, Simplicio N, van Doorninck JH, Goridis C, Guillemot F, Brunet JF. 2004. Ascl1/Mash1 is required for the development of central serotonergic neurons. *Nat. Neurosci.* 7:589–95

Pattyn A, Vallstedt A, Dias JM, Samad OA, Krumlauf R, et al. 2003. Coordinated temporal and spatial control of motor neuron and serotonergic neuron generation from a common pool of CNS progenitors. *Genes Dev.* 17:729–37

Pirri JK, McPherson AD, Donnelly JL, Francis MM, Alkema MJ. 2009. A tyramine-gated chloride channel coordinates distinct motor programs of a *Caenorhabditis elegans* escape response. *Neuron* 62:526–38

Qian Y, Fritzsch B, Shirasawa S, Chen CL, Choi Y, Ma Q. 2001. Formation of brainstem (nor)adrenergic centers and first-order relay visceral sensory neurons is dependent on homeodomain protein Rnx/Tlx3. *Genes Dev.* 15:2533–45

Qiu M, Bulfone A, Martinez S, Meneses JJ, Shimamura K, et al. 1995. Null mutation of Dlx-2 results in abnormal morphogenesis of proximal first and second branchial arch derivatives and abnormal differentiation in the forebrain. *Genes Dev.* 9:2523–38

Rand JB. 1989. Genetic analysis of the cha-1-unc-17 gene complex in *Caenorhabditis. Genetics* 122:73–80

Rink E, Guo S. 2004. The too few mutant selectively affects subgroups of monoaminergic neurons in the zebrafish forebrain. *Neuroscience* 127:147–54

Rink E, Wullimann MF. 2002. Development of the catecholaminergic system in the early zebrafish brain: an immunohistochemical study. *Brain Res. Dev. Brain Res.* 137:89–100

Roeder T. 2005. Tyramine and octopamine: ruling behavior and metabolism. *Annu. Rev. Entomol.* 50:447–77

Roeder T, Seifert M, Kahler C, Gewecke M. 2003. Tyramine and octopamine: antagonistic modulators of behavior and metabolism. *Arch. Insect Biochem. Physiol.* 54:1–13

Rose JK, Rankin CH. 2001. Analyses of habituation in *Caenorhabditis elegans. Learn Mem.* 8:63–69

Rothenfluh A, Heberlein U. 2002. Drugs, flies, and videotape: the effects of ethanol and cocaine on *Drosophila* locomotion. *Curr. Opin. Neurobiol.* 12:639–45

Roybon L, Hjalt T, Christophersen NS, Li JY, Brundin P. 2008. Effects on differentiation of embryonic ventral midbrain progenitors by Lmx1a, Msx1, Ngn2, and Pitx3. *J. Neurosci.* 28:3644–56

Rychlik JL, Gerbasi V, Lewis EJ. 2003. The interaction between dHAND and Arix at the dopamine beta-hydroxylase promoter region is independent of direct dHAND binding to DNA. *J. Biol. Chem.* 278:49652–60

Ryu S, Mahler J, Acampora D, Holzschuh J, Erhardt S, et al. 2007. Orthopedia homeodomain protein is essential for diencephalic dopaminergic neuron development. *Curr. Biol.* 17:873–80

Saino-Saito S, Berlin R, Baker H. 2003. Dlx-1 and Dlx-2 expression in the adult mouse brain: relationship to dopaminergic phenotypic regulation. *J. Comp. Neurol.* 461:18–30

Saino-Saito S, Cave JW, Akiba Y, Sasaki H, Goto K, et al. 2007. ER81 and CaMKIV identify anatomically and phenotypically defined subsets of mouse olfactory bulb interneurons. *J. Comp. Neurol.* 502:485–96

Sakurada K, Ohshima-Sakurada M, Palmer TD, Gage FH. 1999. Nurr1, an orphan nuclear receptor, is a transcriptional activator of endogenous tyrosine hydroxylase in neural progenitor cells derived from the adult brain. *Development* 126:4017–26

Analysis of the *Nurr1* mutant, which for the first time describes its role not in dopaminergic generation but in terminal differentiation and survival.

Saucedo-Cardenas O, Quintana-Hau JD, Le WD, Smidt MP, Cox JJ, et al. 1998. Nurr1 is essential for the induction of the dopaminergic phenotype and the survival of ventral mesencephalic late dopaminergic precursor neurons. *Proc. Natl. Acad. Sci. USA* 95:4013–18

Sawin ER, Ranganathan R, Horvitz HR. 2000. *C. elegans* locomotory rate is modulated by the environment through a dopaminergic pathway and by experience through a serotonergic pathway. *Neuron* 26:619–31

Schmidt M, Lin S, Pape M, Ernsberger U, Stanke M, et al. 2009. The bHLH transcription factor Hand2 is essential for the maintenance of noradrenergic properties in differentiated sympathetic neurons. *Dev. Biol.* 329:191–200

Scott MM, Deneris ES. 2005. Making and breaking serotonin neurons and autism. *Int. J. Dev. Neurosci.* 23:277–85

Scott MM, Krueger KC, Deneris ES. 2005. A differentially autoregulated Pet-1 enhancer region is a critical target of the transcriptional cascade that governs serotonin neuron development. *J. Neurosci.* 25:2628–36

Simon HH, Scholz C, O'Leary DD. 2005. Engrailed genes control developmental fate of serotonergic and noradrenergic neurons in mid- and hindbrain in a gene dose-dependent manner. *Mol. Cell Neurosci.* 28:96–105

Smidt MP, Asbreuk CH, Cox JJ, Chen H, Johnson RL, Burbach JP. 2000. A second independent pathway for development of mesencephalic dopaminergic neurons requires Lmx1b. *Nat. Neurosci.* 3:337–41

Smidt MP, Burbach JP. 2007. How to make a mesodiencephalic dopaminergic neuron. *Nat. Rev. Neurosci.* 8:21–32

Smidt MP, Smits SM, Bouwmeester H, Hamers FP, van der Linden AJ, et al. 2004. Early developmental failure of substantia nigra dopamine neurons in mice lacking the homeodomain gene Pitx3. *Development* 131:1145–55

Smidt MP, van Schaick HS, Lanctot C, Tremblay JJ, Cox JJ, et al. 1997. A homeodomain gene Ptx3 has highly restricted brain expression in mesencephalic dopaminergic neurons. *Proc. Natl. Acad. Sci. USA* 94:13305–10

Smits SM, Ponnio T, Conneely OM, Burbach JP, Smidt MP. 2003. Involvement of Nurr1 in specifying the neurotransmitter identity of ventral midbrain dopaminergic neurons. *Eur. J. Neurosci.* 18:1731–38

Song NN, Xiu JB, Huang Y, Chen JY, Zhang L, et al. 2011. Adult raphe-specific deletion of Lmx1b leads to central serotonin deficiency. *PLoS ONE* 6:e15998

Stanke M, Junghans D, Geissen M, Goridis C, Ernsberger U, Rohrer H. 1999. The Phox2 homeodomain proteins are sufficient to promote the development of sympathetic neurons. *Development* 126:4087–94

Stenman J, Toresson H, Campbell K. 2003. Identification of two distinct progenitor populations in the lateral ganglionic eminence: implications for striatal and olfactory bulb neurogenesis. *J. Neurosci.* 23:167–74

Stoykova A, Gruss P. 1994. Roles of Pax-genes in developing and adult brain as suggested by expression patterns. *J. Neurosci.* 14:1395–412

Sulston J, Dew M, Brenner S. 1975. Dopaminergic neurons in the nematode *Caenorhabditis elegans*. *J. Comp. Neurol.* 163:215–26

Suo S, Kimura Y, Van Tol HH. 2006. Starvation induces cAMP response element-binding protein-dependent gene expression through octopamine-Gq signaling in *Caenorhabditis elegans*. *J. Neurosci.* 26:10082–90

Sze JY, Zhang S, Li J, Ruvkun G. 2002. The *C. elegans* POU-domain transcription factor UNC-86 regulates the tph-1 tryptophan hydroxylase gene and neurite outgrowth in specific serotonergic neurons. *Development* 129:3901–11

Teraoka H, Russell C, Regan J, Chandrasekhar A, Concha ML, et al. 2004. Hedgehog and Fgf signaling pathways regulate the development of tphR-expressing serotonergic raphe neurons in zebrafish embryos. *J. Neurobiol.* 60:275–88

Thor S, Thomas JB. 1997. The *Drosophila* islet gene governs axon pathfinding and neurotransmitter identity. *Neuron* 18:397–409

Tiveron MC, Hirsch MR, Brunet JF. 1996. The expression pattern of the transcription factor Phox2 delineates synaptic pathways of the autonomic nervous system. *J. Neurosci.* 16:7649–60

Tsarovina K, Pattyn A, Stubbusch J, Müller F, van der Wees J, et al. 2004. Essential role of Gata transcription factors in sympathetic neuron development. *Development* 131:4775–86

Tsarovina K, Reiff T, Stubbusch J, Kurek D, Grosveld FG, et al. 2010. The Gata3 transcription factor is required for the survival of embryonic and adult sympathetic neurons. *J. Neurosci.* 30:10833–43

Valarché I, Tissier-Seta JP, Hirsch MR, Martinez S, Goridis C, Brunet JF. 1993. The mouse homeodomain protein Phox2 regulates Ncam promoter activity in concert with Cux/CDP and is a putative determinant of neurotransmitter phenotype. *Development* 119:881–96

van den Munckhof P, Luk KC, Ste-Marie L, Montgomery J, Blanchet PJ, et al. 2003. Pitx3 is required for motor activity and for survival of a subset of midbrain dopaminergic neurons. *Development* 130:2535–42

van Doorninck JH, van der Wees J, Karis A, Goedknegt E, Engel JD, et al. 1999. GATA-3 is involved in the development of serotonergic neurons in the caudal raphe nuclei. *J. Neurosci.* 19:RC12

Vomel M, Wegener C. 2008. Neuroarchitecture of aminergic systems in the larval ventral ganglion of *Drosophila melanogaster*. *PLoS One* 3:e1848

Waggoner LE, Zhou GT, Schafer RW, Schafer WR. 1998. Control of alternative behavioral states by serotonin in *Caenorhabditis elegans*. *Neuron* 21:203–14

Wallen A, Zetterstrom RH, Solomin L, Arvidsson M, Olson L, Perlmann T. 1999. Fate of mesencephalic AHD2-expressing dopamine progenitor cells in NURR1 mutant mice. *Exp. Cell Res.* 253:737–46

Wassarman KM, Lewandoski M, Campbell K, Joyner AL, Rubenstein JL, et al. 1997. Specification of the anterior hindbrain and establishment of a normal mid/hindbrain organizer is dependent on Gbx2 gene function. *Development* 124:2923–34

Webster R. 2001. *Neurotransmitters, Drugs and Brain Function*. Chichester, UK: Wiley

Weiner RI, Ganong WF. 1978. Role of brain monoamines and histamine in regulation of anterior pituitary secretion. *Physiol. Rev.* 58:905–76

Weinshenker D, Garriga G, Thomas JH. 1995. Genetic and pharmacological analysis of neurotransmitters controlling egg laying in *C. elegans*. *J. Neurosci.* 15:6975–85

Wildner H, Gierl MS, Strehle M, Pla P, Birchmeier C. 2008. Insm1 (IA-1) is a crucial component of the transcriptional network that controls differentiation of the sympatho-adrenal lineage. *Development* 135:473–81

Witta J, Baffi JS, Palkovits M, Mezey E, Castillo SO, Nikodem VM. 2000. Nigrostriatal innervation is preserved in Nurr1-null mice, although dopaminergic neuron precursors are arrested from terminal differentiation. *Brain Res. Mol. Brain Res.* 84:67–78

Wragg RT, Hapiak V, Miller SB, Harris GP, Gray J, et al. 2007. Tyramine and octopamine independently inhibit serotonin-stimulated aversive behaviors in *Caenorhabditis elegans* through two novel amine receptors. *J. Neurosci.* 27:13402–12

Wu J, Duggan A, Chalfie M. 2001. Inhibition of touch cell fate by egl-44 and egl-46 in *C. elegans*. *Genes Dev.* 15:789–802

Wylie CJ, Hendricks TJ, Zhang B, Wang L, Lu P, et al. 2010. Distinct transcriptomes define rostral and caudal serotonin neurons. *J. Neurosci.* 30:670–84

Xu H, Firulli AB, Zhang X, Howard MJ. 2003. HAND2 synergistically enhances transcription of dopamine-beta-hydroxylase in the presence of Phox2a. *Dev. Biol.* 262:183–93

Yamamoto K, Ruuskanen JO, Wullimann MF, Vernier P. 2010. Two tyrosine hydroxylase genes in vertebrates: new dopaminergic territories revealed in the zebrafish brain. *Mol. Cell Neurosci.* 43:394–402

Characterization of *unc-86* as a terminal selector for serotonergic fate in *C. elegans* based on the finding that *unc-86* mutants lose expression of different serotonin pathway genes and that *unc-86* directly interacts with the *tph* promoter.

Detailed description of the role of *tup* (*islet*) as a terminal differentiation factor for dopamine and serotonin neurons, as well as its role in controlling other features, such as axon pathfinding.

Yamamoto K, Ruuskanen JO, Wullimann MF, Vernier P. 2011. Differential expression of dopaminergic cell markers in the adult zebrafish forebrain. *J. Comp. Neurol.* 519:576–98

Yang C, Kim HS, Seo H, Kim CH, Brunet JF, Kim KS. 1998. Paired-like homeodomain proteins, Phox2a and Phox2b, are responsible for noradrenergic cell-specific transcription of the dopamine beta-hydroxylase gene. *J. Neurochem.* 71:1813–26

Ye W, Shimamura K, Rubenstein JL, Hynes MA, Rosenthal A. 1998. FGF and Shh signals control dopaminergic and serotonergic cell fate in the anterior neural plate. *Cell* 93:755–66

Yoshihara S, Omichi K, Yanazawa M, Kitamura K, Yoshihara Y. 2005. Arx homeobox gene is essential for development of mouse olfactory system. *Development* 132:751–62

Youdim MB, Edmondson D, Tipton KF. 2006. The therapeutic potential of monoamine oxidase inhibitors. *Nat. Rev. Neurosci.* 7:295–309

Zellmer E, Zhang Z, Greco D, Rhodes J, Cassel S, Lewis EJ. 1995. A homeodomain protein selectively expressed in noradrenergic tissue regulates transcription of neurotransmitter biosynthetic genes. *J. Neurosci.* 15:8109–20

Zetterstrom RH, Solomin L, Jansson L, Hoffer BJ, Olson L, Perlmann T. 1997. Dopamine neuron agenesis in Nurr1-deficient mice. *Science* 276:248–50

Zetterstrom RH, Williams R, Perlmann T, Olson L. 1996. Cellular expression of the immediate early transcription factors Nurr1 and NGFI-B suggests a gene regulatory role in several brain regions including the nigrostriatal dopamine system. *Brain Res. Mol. Brain Res.* 41:111–20

Zhao GY, Li ZY, Zou HL, Hu ZL, Song NN, et al. 2008. Expression of the transcription factor GATA3 in the postnatal mouse central nervous system. *Neurosci. Res.* 61:420–28

Zhao ZQ, Scott M, Chiechio S, Wang JS, Renner KJ, et al. 2006. Lmx1b is required for maintenance of central serotonergic neurons and mice lacking central serotonergic system exhibit normal locomotor activity. *J. Neurosci.* 26:12781–88

Zheng X, Chung S, Tanabe T, Sze JY. 2005. Cell-type specific regulation of serotonergic identity by the *C. elegans* LIM-homeodomain factor LIM-4. *Dev. Biol.* 286:618–28

Zou HL, Su CJ, Shi M, Zhao GY, Li ZY, et al. 2009. Expression of the LIM-homeodomain gene Lmx1a in the postnatal mouse central nervous system. *Brain Res. Bull.* 78:306–12

Amyloid Precursor Protein Processing and Alzheimer's Disease

Richard J. O'Brien[1] and Philip C. Wong[2]

[1]Department of Neurology, Johns Hopkins Bayview Medical Center

[2]Department of Pathology, Johns Hopkins University School of Medicine;
email: robrien@jhmi.edu

Annu. Rev. Neurosci. 2011. 34:185–204

First published online as a Review in Advance on
March 29, 2011

The *Annual Review of Neuroscience* is online at
neuro.annualreviews.org

This article's doi:
10.1146/annurev-neuro-061010-113613

Keywords

Neurodegeneration, dementia, BACE1, α-secretase, γ-secretase,
aging

Abstract

Alzheimer's disease (AD), the leading cause of dementia worldwide,
is characterized by the accumulation of the β-amyloid peptide (Aβ)
within the brain along with hyperphosphorylated and cleaved forms of
the microtubule-associated protein tau. Genetic, biochemical, and behavioral research suggest that physiologic generation of the neurotoxic
Aβ peptide from sequential amyloid precursor protein (APP) proteolysis is the crucial step in the development of AD. APP is a single-pass
transmembrane protein expressed at high levels in the brain and metabolized in a rapid and highly complex fashion by a series of sequential
proteases, including the intramembranous γ-secretase complex, which
also process other key regulatory molecules. Why Aβ accumulates in
the brains of elderly individuals is unclear but could relate to changes in
APP metabolism or Aβ elimination. Lessons learned from biochemical
and genetic studies of APP processing will be crucial to the development
of therapeutic targets to treat AD.

Contents

Neuritic plaques:
Large extracellular
aggregates of the
amyloid Aβ peptide
surrounded by
dystrophic neurites
(dendrites) containing
aggregated tau

APP: amyloid
precursor protein

Aβ: amyloid β
peptide

HISTORY OF ALZHEIMER'S DISEASE

In 1907, Alois Alzheimer reported the results of an autopsy on a 55-year-old woman named Auguste Deter, who had died from a progressive behavioral and cognitive disorder. Alzheimer noted the presence of two distinctive pathologies in Deter's brain: neurofibrillary tangles, which he correctly surmised were abnormal intracellular aggregates (and which were later shown to be composed of hyperphosphory-

lated and cleaved forms of the microtubule-associated protein tau), and neuritic plaques (which he called miliary foci), which were dystrophic neuronal processes surrounding a "special substance in the cortex" (Alzheimer et al. 1995). This "special substance" was isolated and purified in 1984 by Glenner & Wong (1984), who showed that it was a 4.2-kDa peptide, primarily 40 or 42 amino acids in length, which they speculated was cleaved from a larger precursor. Their prediction was verified in short order when the amyloid precursor protein (APP) was cloned in 1987 (Kang et al. 1987). The peptide isolated by Glenner & Wong has come to be known as the Aβ peptide, short for amyloid-β peptide.

Alzheimer was convinced that the case of Auguste Deter represented an unusual cause of dementia. It was not until the seminal work of Blessed, Tomlinson, and Roth (Blessed et al. 1968) that a relationship between the amount of neuritic Aβ plaques in the brains of elderly subjects and the risk of dementia was established. Alzheimer's disease (AD) is now recognized as a common dementing disorder of the elderly (see sidebar, Alzheimer's Disease: What's in a Name) with characteristic pathological findings (**Figure 1**). Young onset, frequently genetic, forms of the disease are a rare but important subset. AD currently afflicts 26 million people worldwide with projections of a fourfold increase in that number by 2050 (Brookmeyer et al. 2007).

RELATIONSHIP OF BRAIN Aβ ACCUMULATION TO DEMENTIA IN HUMAN PATHOLOGIC SPECIMENS

Cohorts of subjects who are followed with serial neuropsychological testing during life and then donate their brains to scientific research after death (Dolan et al. 2010a) have become a crucial tool for understanding the determinants of cognitive decline in older subjects. Most studies agree that the classical pathological criteria for AD, neuritic plaques and neurofibrillary tangles, can account for 40%–70% of the variance in cognition seen in elderly subjects, with additional pathologies such as cerebrovascular disease (Dolan et al. 2010b) and Lewy body pathology (Schneider et al. 2007) working together with AD pathology to account for an additional 20%–30% of dementia cases.

It was historically unclear whether the accumulating Aβ plaques in the brains of patients with dementia caused the dementia or simply indicated the presence of dying neurons. Indeed, studies of head trauma and brain ischemia in humans and animals demonstrate transient increases in brain Aβ deposition (Gentleman et al. 1997, Qi et al. 2007). However, studies of chronic brain injury and ischemia in humans do not suggest that Aβ is a nonspecific marker of neuronal injury (Dolan et al. 2010b, McKee et al. 2009), and there is no increase in the age-expected prevalence of Aβ pathology in patients with other neurodegenerative disorders such as Parkinson's disease and frontotemporal dementia.

STRUCTURE AND FUNCTION OF APP

The amyloid precursor protein (APP) is one member of a family of related proteins that includes the amyloid precursor-like proteins (APLP1 and APLP2) in mammals and the amyloid precursor protein-like (APPL) in Drosophila. All are single-pass transmembrane proteins with large extracellular domains (**Figure 2**), and all are processed in a manner similar to APP. Only APP generates an amyloidogenic fragment owing to sequence divergence at the internal Aβ site. Alternate splicing of the APP transcript generates 8 isoforms, of which 3 are most common: the 695 amino acid form, which is expressed predominantly in the CNS, and the 751 and 770 amino acid forms, which are more ubiquitously expressed (Bayer et al. 1999).

The precise physiological function of APP is not known and remains one of the vexing issues in the field. In most studies, APP overexpression shows a positive effect on cell health and growth. This effect is epitomized in transgenic

ALZHEIMER'S DISEASE: WHAT'S IN A NAME

Dementia is a clinical term that refers to the development of progressive cognitive deterioration associated with an inability to perform normal activities of daily living. When dementia afflicts the young, there is usually a single pathologic process present on autopsy such as the neuritic plaques and neurofibrillary tangles, which are indicative of Alzheimer's disease (AD). In the elderly, however, in whom dementia affects 1 out of 3 individuals, the brain pathology is usually mixed, with most demented individuals having a combination of AD pathology, atherosclerosis, and Lewy body pathology. Moreover, recent work with autopsy cohorts and neuroimaging studies has shown that many cognitively normal elderly subjects carry moderate amounts of AD or cerebrovascular pathology without symptoms as long as they do not have a second comorbid process. Thus, in the elderly, it seems more appropriate to use the term dementia to describe the clinical process and denote "Alzheimer-type pathology" as a risk factor for dementia. In the future, the relative amount of each of these pathologies in a symptomatic or asymptomatic individual will be quantifiable using neuroimaging and cerebrospinal fluid analysis, allowing the design of pathologically relevant clinical trials.

mice that overexpress wild-type APP and have enlarged neurons (Oh et al. 2009). In transiently transfected cell lines, APP modulates cell growth, motility, neurite outgrowth, and cell survival, functions that can be reproduced by the soluble ectodomain, which is released by cleavage of APP (**Figure 2**). These observations were extended in vivo by a recent study (Young-Pearse et al. 2007), which found neuronal migration abnormalities in embryonic rodents injected with APP RNAi. In adult animals, intracerebral injections of the APP ectodomain can improve cognitive function and synaptic density (Meziane et al. 1998, Roch et al. 1994). The sites most responsible for the bioactivity of the APP ectodomain appear to be its two heparin-binding domains (Mok et al. 1997). The second heparin-binding domain is also the site of binding F-spondin, the only potential ligand identified for APP (Ho & Sudhof 2004). Identifying a ligand-binding partner for APP is of some importance because APP has been compared with the developmental signaling

Dementia: a progressive decline in cognition associated with an inability to perform normal activities owing to the cognitive deterioration

Alzheimer's disease (AD): The most common underlying cause of dementia. Pathology shows frequent Aβ amyloid deposition and neurofibrillary tangles

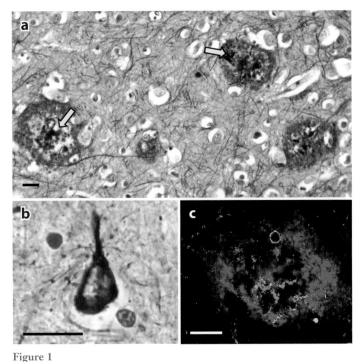

Figure 1

Pathology of Alzheimer's disease. (*a, b*) Brain sections from a patient with dementia are stained with silver, revealing neuritic plaques in panel *a* and a neurofibrillary tangle in panel *b*. The plaques in panel *a* consist of an amorphous reddish protein (Aβ) with dystrophic neurites (*yellow arrows, dark black material*). (*c*) An Aβ plaque stained with an anti-Aβ antibody (*red*) shows infiltrating microglia stained with an IBA1 antibody (*green*). Each line is 40 microns.

Death Receptor 6 (DR6). In this formulation, growth factor deprivation triggers cleavage of APP by the secretase BACE1, releasing the ectodomain, which then binds to DR6 and activates caspase 6 and caspase 3, causing axonal and cell body apoptotic degeneration, respectively. The growth factor deprivation, which triggers cleavage of APP, could be part of normal axonal pruning or could be a primary factor in neuronal degeneration. This work has raised the intriguing possibility that the APP ectodomain released by BACE1 cleavage (sAPPβ) has different properties than does the APP ectodomain released by α-cleavage (sAPPα), which has an extra 16 amino acids in its C-terminus.

In addition to a physiological role for APP, the Aβ peptide itself plays an important role in synaptic physiology, regulating synaptic scaling (Kamenetz et al. 2003) and synaptic vesicle release (Abramov et al. 2009).

It is disappointing, however, that deletion of APP in mice (and thus Aβ production) produces very little phenotype and does not suggest that a loss of APP or Aβ function is deleterious to the adult animal in any way. Triple knockouts involving APP, APLP1, and APLP2 (Herms et al. 2004) show scattered cortical migration abnormalities. Double knockout mice lacking APP and APLP2 exhibit a mismatch between presynaptic and postsynaptic markers at the neuromuscular junction along with excessive nerve terminal sprouting (Wang et al. 2005). This same phenotype is seen in DR6 knockout mice and in fly APPL deletions (Ashley et al. 2005), suggesting a role for APP and its family members in neuritic outgrowth and synaptic pruning. None of these, however, suggests a role for APP in the mature CNS, in which APP production is known to continue at a very high rate.

The intracellular C-terminus of APP is also important for its function and has been proposed to serve two roles, one as a transcriptional regulator and the other, related to its YENPTY amino acid domain (**Figure 2**), as a regulator of its own intracellular sorting. The YENPTY domain regulates clathrin-coated pit internalization (Chen et al. 1990) through a

Notch: Historically and developmentally important signaling protein with similar structure and processing as APP. The Notch intracellular domain regulates transcription

Secretase: protease designed to cleave transmembrane proteins to release bioactive forms or metabolize proteins prior to degradation

molecule Notch, which is structurally similar to APP. Notch proteolysis is triggered by binding to a member of the Delta, Serrate, Lag2 (DSL) family of ligands (Louvi & Artavanis-Tsakonas 2006), which causes Notch to be cleaved by many of the same secretases as APP, releasing a soluble ectodomain and an intracellular domain (NICD) that regulates nuclear target genes. Although F-spondin plays a role in neuronal development and repair, evidence demonstrating that APP is crucial to this process is lacking. Other potential N-terminal binding partners for APP include collagen, netrin-1, laminin, the Aβ peptide, and molecules that interact with APP when coexpressed in the same cell, including other members of the APP family and Notch itself (Chen et al. 2006).

Nikolaev et al. (2009) recently reported that the secreted APP ectodomain acts as a ligand for

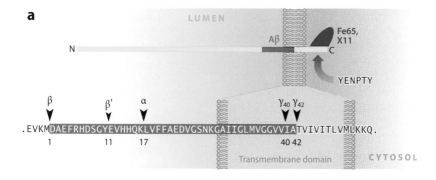

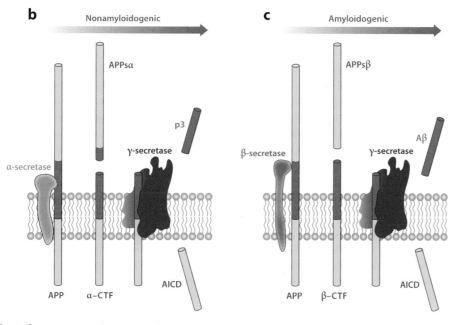

Figure 2

Sequential cleavage of the amyloid precursor protein (APP) occurs by two pathways. (*a*) The APP family of proteins has large, biologically active, N-terminal ectodomains as well as a shorter C-terminus that contains a crucial Tyrosine–Glutamic Acid-Asparagine-Proline-Threonine-Tyrosine (YENPTY) protein-sorting domain to which the adaptor proteins X11 and Fe65 bind. The Aβ peptide starts within the ectodomain and continues into the transmembrane region (*red*). (*b*) Nonamyloidogenic processing of APP involving α-secretase followed by γ-secretase is shown. (*c*) Amyloidogenic processing of APP involving BACE1 followed by γ-secretase is shown. Both processes generate soluble ectodomains (sAPPα and sAPPβ) and identical intracellular C-terminal fragments (AICD).

series of binding partners (see below). It is 100% conserved in all forms of APP from the fly to the human. Mutation at this site alters endocytosis of APP (Perez et al. 1999) and diminishes Aβ production (Ring et al. 2007). Phosphorylation of Thr 668, 14 amino acids away from the YENPTY domain, by cyclin-dependent kinase 5 interferes with at least some of the protein-protein interactions of the YENPTY domain (Ando et al. 2001), although mutation of this site did not alter brain Aβ accumulation in mice (Sano et al. 2006).

SorLA: sorting
protein–related
receptor, also called
LR11

Trans-Golgi
network (TGN):
final sorting stack of
the Golgi system from
which vesicles bud on
their way to the cell
surface, endosome,
and lysosome

Endosome: an
acidic transitional
compartment in
equilibrium with the
cell surface, TGN, and
lysosome, which
compartmentalizes
proteolytic function

Interaction with the YENPTY domain of APP requires the presence of a phosphotyrosine binding domain on the interacting protein. The two best characterized APP binding partners are X11 and Fe65, which were isolated using the yeast-two-hybrid screen. X11 contains one polypyrimidine-tract binding (PTB) domain as well as two postsynaptic density-DlgA-ZO1 (PDZ) domains (Feng & Zhang 2009), whereas Fe65 has two PTB domains, each with different binding specificities, and a tryptophan repeat domain that interacts with the actin cytoskeleton-associated proteins Mena and Evl (Borg et al. 1996, Lambrechts et al. 2000). Both X11 and Fe65 are highly expressed in brain and interact with all APLPs. Tissue culture studies have suggested that both binding partners couple APP to SorLA/LR11 in the trans-Golgi network (TGN), preventing APP from interacting with BACE1 (Pietrzik et al. 2004, Saito et al. 2008). When heterozygous X11 or Fe65 knockout mice are crossed with APP overexpressing mice, brain Aβ accumulation increases significantly (Saluja et al. 2009); moreover, mice that overexpress X11 or Fe65 have diminished brain Aβ accumulation (Lee et al. 2003, McLoughlin & Miller 2008).

Fe65 has also assumed a unique place in Alzheimer's research because it is known to bind the transcription factor complex CP2-LSF-LBP-1 and the histone deacetylase Tip60 via the non-APP-binding PTB domain (Cao & Sudhof 2001, Zambrano et al. 1998) and regulate transcription in cultured cell lines. Investigators initially felt that Fe65 acted in concert with the APP C-terminus to regulate transcription. More recent work has emphasized the independent role of Fe65 (Yang et al. 2006), suggesting that full-length APP may serve as a docking station to keep Fe65 out of the nucleus.

AMYLOID PRECURSOR PROTEIN PROCESSING

APP is produced in large quantities in neurons and is metabolized very rapidly (Lee et al. 2008). Multiple alternate pathways exist for APP proteolysis, some of which lead to generation of the Aβ peptide and some of which do not (**Figures 2** and **3**). After sorting in the endoplasmic reticulum and Golgi, APP is delivered to the axon, where it is transported by fast axonal transport to synaptic terminals (Koo et al. 1990). APP had been reported to function as a receptor for kinesin-1-mediated axonal transport (Kamal et al. 2000), but subsequent work has not confirmed the association of APP and kinesin-1 (Lazarov et al. 2005).

Crucial steps in APP processing occur at the cell surface and in the TGN (**Figure 3**). From the TGN, APP can be transported to the cell surface or directly to an endosomal compartment. Clathrin-associated vesicles mediate both these steps. On the cell surface, APP can be proteolyzed directly by α-secretase and then γ-secretase, a process that does not generate Aβ, or reinternalized in clathrin-coated pits into another endosomal compartment containing the proteases BACE1 and γ-secretase. The latter results in the production of Aβ, which is then dumped into the extracellular space following vesicle recycling or degraded in lysosomes. Although most APP must pass through the cell surface as part of its processing, this step is very rapid, as little APP is on the surface at any point in time. Why some surface APP is internalized into endosomes and some proteolyzed directly by α-secretase is unclear, although segregation of APP and BACE1 into lipid rafts may be a crucial element (Ehehalt et al. 2003). Finally, to complete the APP cycling loop, retrograde communication occurs between endosomal compartments and the TGN, mediated by a complex of molecules called retromers.

Experiments that have examined this model directly have found that 80% of Aβ release is blocked by preventing surface endocytosis (Koo & Squazzo 1994). Moreover, all the enzymes appear to be in the correct places. FRET analysis indicates that BACE1 interacts with APP predominantly in endosomes under native conditions (Kinoshita et al. 2003), whereas γ-secretase activity is present on the cell surface, where it complements α-secretase activity, and in endosomal compartments, where

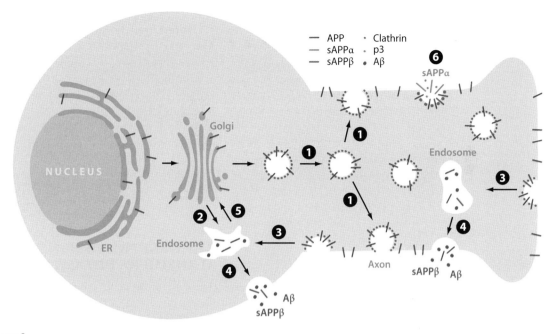

Figure 3

APP trafficking in neurons. Newly synthesized APP (*purple*) is transported from the Golgi down the axon (1) or into a cell body endosomal compartment (2). After insertion into the cell surface, some APP is cleaved by α-secretase (6) generating the sAPPα fragment, which diffuses away (*green*), and some is reinternalized into endosomes (3), where Aβ is generated (*blue*). Following proteolysis, the endosome recycles to the cell surface (4), releasing Aβ (*blue*) and sAPPβ. Transport from the endosomes to the Golgi prior to APP cleavage can also occur, mediated by retromers (5).

it complements BACE1 activity (Fukumori et al. 2006, Parvathy et al. 1999).

The enzymes that cleave APP have been extensively characterized. BACE1, a transmembrane aspartic protease, is directly involved in the cleavage of APP at the +1 (prior to amino acid 1) and +11 sites of Aβ. Neurons from BACE1-/- mice do not produce Aβ, confirming that BACE1 is the neuronal β-secretase (Cai et al. 2001). Following BACE1 cleavage and release of the sAPPβ ectodomain, the APP C-terminal fragment is cleaved by the γ-secretase complex at one of several sites varying from +40 to +44 to generate Aβ peptides (1–40 and 1–42 being most common) and the APP intracellular domain (**Figure 2**).

γ-secretase is a multiprotein complex composed of presenilin 1 (PS1) or presenilin 2 (PS2); nicastrin (Nct), a type I transmembrane glycoprotein; and Aph-1 and Pen-2, two multipass transmembrane proteins (Bergmans & De Strooper 2010). This complex is essential for the sequential intramembranous proteolysis of a variety of transmembrane proteins. PS1 and PS2 contain two aspartyl residues that play crucial roles in intramembranous cleavage; substitutions of these residues (D257 in TM 6 and at D385 in TM 7) reduces cleavage of APP and Notch1 (De Strooper et al. 1999, Wolfe et al. 1999). The functions of the various γ-secretase proteins and their interactions in the complex are not yet fully defined, but it has been suggested that the ectodomain of nicastrin recognizes and binds to the amino-terminal stubs of previously cleaved transmembrane proteins. Aph-1 aids the formation of a precomplex, which interacts with PS1 or PS2 while Pen-2 enters the complex to initiate the cleavage of PS1 or PS2 to form an N-terminal 28-kDa fragment and a C-terminal 18-kDa fragment, both of which are critical to the γ-secretase complex (Takasugi et al. 2003).

PS1: presenilin 1 protein

PS2: presenilin 2 protein

Several aspects of the standard model deserve comment. α-cleavage of APP (+17) is attributed to the ADAM (a disintegrin and metalloproteinase) family of proteases (Asai et al. 2003; Jorissen et al. 2010) and occurs, to a large extent, on the cell surface. However, there is some α-secretase activity in the trans-Golgi. This is of some significance because activation of protein kinase C (Mills & Reiner 1999) causes a significant increase in α-cleavage of APP by increasing transport of APP to the cell surface (Hung et al. 1993), by blocking access of cell surface APP to endosomes, and by stimulating α-cleavage in the TGN (Skovronsky et al. 2000). Because α-cleavage occurs within the Aβ sequence, it prevents Aβ generation. Indeed, increased expression of ADAM 10 or SIRT1, a regulator of ADAM 10 gene expression, in a mouse model of AD significantly attenuated Aβ deposition and cognitive deficits (Donmez et al. 2010; Postina et al. 2004).

Although the standard model suggests that little Aβ is generated outside of endosomal pathways, which (*a*) dump it outside the cell and (*b*) have a compulsory cell-surface transition prior to internalization, such is not necessarily the case. Shunting directly from the TGN to the endosome and back (**Figure 3**) is a potentially important pathway in APP processing. Indeed, early work in transfected cell lines had suggested that significant amounts of APP were processed to Aβ intracellularly (Greenfield et al. 1999), and evidence provides support for BACE1 in the TGN (Huse et al. 2002) and for intracellular Aβ accumulation in neurons from patients with early AD (Gouras et al 2000; LaFerla et al 2007). A key to intracellular generation of Aβ is the concept of retromer transport of APP and BACE1. Retromers are intracellular complexes that shuttle cargo predominantly but not exclusively from the endosome to the TGN. Adaptor proteins affix cargo to the retromer complex. SorLA, a member of the low density lipoprotein receptor superfamily, is one such adaptor protein and binds APP (Andersen et al. 2005) via its N-terminal complement-like domain and binds to the retromer complex via its vacuolar

protein-sorting domain (Jacobsen et al. 2001). C-terminal interactions also exist between APP, BACE1, and SorLA (Spoelgen et al. 2006) mediated through adaptor proteins such as Fe65 and X11 (Schmidt et al. 2007). Most current data suggest that SorLA keeps APP from interacting with BACE1 and promotes transport of APP to the Golgi and away from endosomes, reducing Aβ production. SorLA expression is diminished in neurons from patients with AD (Scherzer et al. 2004), whereas a reduction of SorLA in transgenic animals leads to Aβ accumulation (Andersen et al. 2005).

THE GENETICS OF ALZHEIMER'S DISEASE

The most important lessons to be learned from genetic cases of AD is that the pathology of autosomal dominant AD is very similar to sporadic AD (Shepherd et al. 2009), including the development of neurofibrillary tangles and microglial infiltration. This single observation is central to the concept that Aβ deposition is also the primary event in sporadic AD.

Autosomal Dominant Mutations Associated with AD

There are 32 *APP*, 179 *PSEN1* (presenilin 1 gene locus), and 14 *PSEN2* gene mutations that result in early-onset, autosomal dominant, fully penetrant AD. The mutations can be examined in detail at the Alzheimer Disease and Frontotemporal Dementia Mutation Database Web site online (**http://www.molgen.ua. ac.be/ADmutations/**). In APP, mutations cluster around the γ-secretase cleavage site, although the most famous APP mutation (APP-swe) causes a change in amino acids adjacent to the BACE1 cleavage site. *PSEN* gene mutations (which give rise to proteins called presenilins, PS1 and PS2) predominantly alter the amino acids in their nine transmembrane domains. The common thread to all these mutations is that they increase production of the less soluble and more toxic Aβ42 relative to Aβ40 (Shen & Kelleher 2007).

In Down's syndrome, overexpression of APP results in brain Aβ deposition when individuals are in their late 20s. Neurofibrillary tangles develop later and correlate with the onset of the mid-life cognitive decline that is common in these individuals (Hof et al. 1995).

APOE

Late-onset sporadic AD also has a significant genetic component, estimated at 50%–70%. Almost half that risk is conferred by the apolipoprotein E (*APOE*) allele (Avramopoulos 2009), although some recent data has challenged that association (Roses 2010). The *APOE* gene comes in three variants that encode proteins (designated ApoE2, ApoE3, and ApoE4) that differ at two amino acids. The *APOE E4* allele, which is present in 10%–20% of various populations (Singh et al. 2006), increases the risk for AD threefold in individuals carrying one copy and 15-fold for homozygous individuals. ApoE is the predominant apolipoprotein of the HDL complex in the brain. Although ApoE has many roles in brain physiology (Kim et al. 2009), the most compelling data regarding its role in the development of AD relates to its ability to bind Aβ.

Pathologically, the *APOE E4* allele is strongly associated with increased brain Aβ deposition (Tiraboschi et al. 2004), and ApoE2 and -E3 but not -E4 bind Aβ tightly (Tokuda et al. 2000). Human *APOE* genes expressed in mice diminish Aβ deposition except for *APOE E4*, which increases Aβ deposition (Holtzman et al. 1999). Thus, the effect of ApoE on AD risk may be entirely explained by its effect on Aβ deposition. It is appealing to think ApoE increases clearance or cellular uptake of Aβ via receptor-mediated binding, but the data are not yet clear on this issue (Kim et al. 2009).

MECHANISMS OF Aβ TOXICITY

The primacy of APP in the development of Alzheimer's disease depends on the toxicity of the Aβ peptide (**Figure 4**) because evidence does not show that loss of APP function is deleterious. In addition, Aβ toxicity must also explain other pathological aspects of AD including neurofibrillary tangles, inflammation, and oxidative damage.

Initial work in tissue culture showed that Aβ fibrils are acutely toxic to neurons (Yankner et al. 1989), resulting in complete death of all cells within 24 h of exposure. The mechanism of death in these cultured cells is likely to be apoptosis (Deshpande et al. 2006), triggered by oxidative effects of Aβ. Mutation of a single amino acid (methionine 35) of the Aβ peptide eliminates its ability to generate reactive oxidative species (Kanski et al. 2002). The form of the Aβ peptide that is toxic to neurons is controversial. Evidence exists for picomolar toxicity of an oligomeric assembly of Aβ, possibly a dimer (Shankar et al. 2008), as well as for a multimeric pore-like complex of Aβ monomers. However, there is still compelling evidence for toxicity caused by Aβ monomers, especially Aβ42 (Butterfield 2002), and by truncated, oxidized, and insoluble species of Aβ (Yankner & Lu 2009). There is also controversy about the relative toxicities of intracellular versus extracellular Aβ because intracellular injection of Aβ42 but not Aβ40 also kills neurons and intracellular Aβ is seen early in AD (LaFerla et al. 2007).

Work in vivo also supports the notion that Aβ is toxic to neurons. Mice that overexpress mutant human APPs develop Aβ deposition by 4 to 6 months and show evidence of subsequent neuronal injury. Studies show loss of synaptic terminals (Irizarry et al. 1997, Spires et al. 2005), synaptic dysfunction (Kamenetz et al. 2003, Shankar et al. 2008), abnormalities on spatial memory tests (Chen et al. 2000), and inflammation (El Khoury et al. 2007). These animals, along with human AD brains, show activation of multiple caspases, including caspases 3, 6, 7, 8, and 9, along with evidence of caspase cleavage of actin, fodrin, and the proteasome subunit p97 (Halawani et al. 2010, Rissman et al. 2004, Rohn & Head 2009). Moreover, caspase inhibitors (Rohn et al. 2009) and overexpression of the antiapoptotic protein Bcl-2 (Rohn et al. 2008) ameliorate Aβ toxicity in transgenic mice. One of the

ApoE4:
apolipoprotein E4 protein

APOE E4:
apolipoprotein E4 gene locus

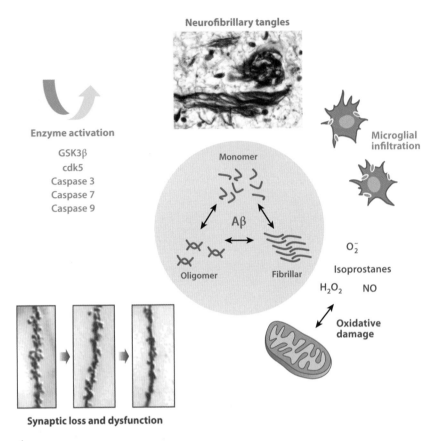

Figure 4

Aβ toxicity. An equilibrium between several species of extracellular and intracellular Aβ, including monomeric, oligomeric, and fibrillar forms, causes toxicity through several mechanisms including microglial infiltration, the generation of reactive oxygen species, and synaptic damage. Neurofibrillary tangles are generated by Aβ-induced tau phosphorylation and cleavage. Enzymes activated directly by extracellular Aβ include GSK3β, Cdk5, and multiple caspases, which activate tau cleavage and phosphorylation among their many deleterious effects.

Neurofibrillary tangles (NFT): Insoluble intracellular inclusions that stain darkly with silver and are composed primarily of hyperphosphorylated and cleaved forms of the microtubule-associated protein tau

more interesting aspects of caspase-induced cell damage is whether APP itself is a target for caspase cleavage, releasing a unique, potentially toxic, C-terminal intracellular fragment (Galvan et al. 2006, Gervais et al. 1999). Caspase-induced damage in AD may occur independent of apoptosis (Hyman 2011), to which mature neurons are resistant.

Soluble Aβ can also control cleavage and phosphorylation of tau, both of which are crucial for neurofibrillary tangle (NFT) generation. Phosphorylation of tau is regulated by several kinases, including GSK3β and cdk5, both of which are activated by extracellular Aβ

(Hernandez & Avila 2008, Lee et al. 2000). Pathways leading to tau cleavage including GSK3β, caspase 3, caspase 9, and calpain are also activated by soluble Aβ species (Cho & Johnson 2004, Chung et al. 2001). Moreover, it appears that tau is an important downstream mediator of Aβ toxicity. Triple transgenic mice with mutant APP, PS1, and tau proteins develop Aβ deposition prior to the appearance of NFT pathology (Oddo et al. 2003). Reducing levels of Aβ by immunotherapy prevents tau pathology from developing and abrogates spatial memory problems (Billings et al. 2005, Oddo et al. 2006). Moreover, when transgenic mice that overexpress APP are crossed with mice lacking tau, no detioration in spatial memory function is seen even though Aβ deposition is exuberant (Roberson et al. 2007).

WHY IS THERE AGE-RELATED BRAIN Aβ ACCUMULATION

Large amounts of APP are continuously metabolized to Aβ in the brain (Bateman et al. 2006). Early in the development of Alzheimer's disease, the concentration of Aβ42 in the cerebrospinal fluid (CSF) starts to fall (Shaw et al. 2009), while the concentration of Aβ42 in the brain is rising (Steinerman et al. 2008), suggesting a diminution in Aβ transport from the brain, a result strongly supported by a recent metabolic analysis of brain Aβ clearance in humans (Mawuenyega et al. 2010). Alternatively, a change in the ratio of Aβ42 to Aβ40 within the brain or any change in the production of CSF or the molecules that buffer Aβ in CSF, such as ApoE, may result in more Aβ aggregation and less CSF clearance.

One mechanism to eliminate extracellular Aβ in the brain includes the proteases neprilysin and insulin degrading enzyme (Selkoe 2001), the polyfunctional endothelial transport proteins P-glycoprotein, receptor for advanced glycation endproducts (RAGE), and low-density lipoprotein-like receptor (LRP1). Animal work has shown that each of these proteins can enhance either Aβ degradation (Leissring et al. 2003) or Aβ transport (Cirrito et al. 2005, Deane et al. 2004). However, the relevance of any one of these proteins to Aβ accumulation in the earliest stages of Alzheimer's disease is unclear.

One alternate explanation for Aβ accumulation in the elderly would be an alteration in the cleavage of APP. Excessive age-associated acetylation of the α-secretase gene may diminish nonamyloidogenic processing of APP (Donmez et al. 2010), whereas an increase in BACE1 activity, reported in early AD brain tissue, would increase amyloidogenic processing (Holsinger et al. 2002, Yang et al. 2003). Underlying the increase in brain BACE1

activity seen early in AD could be HIF1a, induced by oxidative stress (Guglielmotto et al. 2009). Several micro RNAs, altered early in AD, also regulate BACE1 RNA. Among these are miR-107, miR-298, miR-328, and the miR-29a/b-1 cluster (Hebert et al. 2008).

One very important emerging area of research is the role of environmental enrichment and exercise on brain APP metabolism and Aβ elimination. Enriched environments, including exercise, reduce brain Aβ in transgenic mice (Yuede et al. 2009) in part by increasing brain peptidases but also by stimulating synaptic processes and growth factors. This work is consistent with human research that shows intellectual stimulation and exercise are also protective against dementia (Rovio et al. 2005, Verghese et al. 2006). Combined with data showing that synaptic activity regulates Aβ production in neurons (Cirrito et al. 2008), there is powerful incentive for future work in this area.

POTENTIAL EVIDENCE AGAINST THE PRIMACY OF APP IN ALZHEIMER'S DISEASE

Aβ Immunization

Fortunately, we have now entered an era in which disease-modifying therapies for Alzheimer's disease are in human clinical trials (**http://www.alz.org/trialmatch**). These therapies target Aβ production, tau aggregation, oxidation, and inflammation. None has created the interest of Elan's initial AN1792 trial, which was an active immunization protocol against the full-length Aβ42 peptide (Schenk et al. 2005). Although the trial was stopped because encephalitis developed in some participants, eight subjects who received active immunization have undergone autopsies. Remarkably, Aβ pathology was virtually absent in three of the eight brains (Holmes et al. 2008). Unfortunately, all three of these subjects continued to have a relentlessly progressive disease course and neurofibrillary pathology was advanced at the time of death. Many explanations have been offered for these results, especially the failure to immunize patients early enough in the disease. However, other possible explanations for the results, such as the importance of intracellular Aβ or a bystander role for extracellular Aβ, need also be considered.

Abnormal Notch Processing as the Real Cause of AD

Besides cleaving APP, γ-secretase and BACE1 cleave other single-pass transmembrane proteins, including Notch (De Strooper et al. 1999), neuregulin, and E-cadherin (Parks & Curtis 2007). This raises the question of whether PS1 mutations or APP processing cause cognitive problems due to interference with the processing of other, more essential, substrates. Because almost every PS1 mutation that causes dementia is associated with significant Aβ deposition, this explanation seems unlikely. Moreover, although conditional PS1 deletions in mice result in impairments in memory and synaptic plasticity (Saura et al. 2004), more relevant reductions in PS1 activity have not had a similar effect.

Growth Factor and Hormone Deprivation as the Cause of Alzheimer's Disease

Several different growth factors and signaling molecules are linked to Alzheimer's disease. First, the levels of brain-derived neurotrophic factor (BDNF) in the brains of patients with AD decrease (Lee et al. 2005, Peng et al. 2005), and BDNF infusion can improve cognitive function in aged primates (Nagahara et al. 2009). However, BDNF polymorphisms, which have been associated with other cognitive syndromes, have not been consistently associated with AD (Zuccato & Cattaneo 2009).

Second, levels of the nerve growth factor (NGF) prohormone proNGF are significantly increased in early AD (Peng et al. 2004), potentially caused by a reduction in the expression of the NGF receptor TrkA (Counts et al. 2004). Cross talk occurs between APP processing and NGF signaling pathways such that Aβ generation may be altered by NGF signaling (Calissano et al. 2010). The most remarkable animal model of Alzheimer's disease is a mouse line engineered to express anti-NGF antibodies. At 15 months of age, these animals develop amyloid plaques and neurofibrillary tangles along with abnormalities in spatial learning (Capsoni et al. 2000). Combined with work suggesting growth factor deprivation can trigger APP cleavage by BACE1 and activate DR6 (Nikolaev et al. 2009), this line of investigation is a very interesting twist on the primacy of APP.

Third, a prospective study from the Framingham cohort (Lieb et al. 2009) showed that lower plasma levels of the hormone leptin were correlated with a significantly higher risk of developing Alzheimer's disease. Moreover, leptin supplementation in APP transgenic mice reduced Aβ accumulation and improved cognition (Greco et al. 2010). Pathways that may mediate this effect include leptin's ability to affect Aβ production and clearance (Fewlass et al. 2004, Greco et al. 2009b) and its ability to alter dendritic morphology and synaptic density (O'Malley et al. 2007). However it still remains to be shown that low leptin levels cause AD pathology and are not simply the result of those changes.

Systems Degeneration

Interesting recent research has suggested that very early AD may involve the degeneration of cortical areas coordinately active at rest called the "default mode network" (Seeley et al. 2009; Sorg et al. 2007). These are also the areas (posterior cingulate and parietal cortex) with the earliest Aβ accumulation (Sperling et al. 2009). This work suggests that there is a metabolic or synaptic component to the disease and that modulating neuronal activity as well as APP metabolism might be a useful approach. Alternative theories that AD pathology spreads as a synaptic contagion due to toxins or misfolded (prion-like) proteins have also been proposed.

CONCLUSIONS AND FUTURE DIRECTIONS

On the basis of the data presented in this review, there is much to suggest that abnormal processing of APP and the toxicity of the Aβ peptide are central to the development of dementia in the elderly. However, genetic data has suggested that targeting the components of APP processing as a pharmacologic strategy will not be without consequences. BACE1 null mice show altered performance on tests of cognition and emotion (Laird et al. 2005, Savonenko et al. 2008) and have abnormalities of myelination, reflecting alterations in the biology of neuregulin (Hu et al. 2006, Willem et al. 2006). Conditional PS1 deletions result in impairments in memory and in hippocampal synaptic plasticity (Saura et al. 2004), whereas nicastrin heterozygote knockout mice develop skin tumors (Li et al. 2007b). Other roadblocks also exist. The BACE1 catalytic site is quite large, and we do not know yet whether investigators can achieve adequate brain penetration of a compound of sufficient size to inhibit its activity.

γ-secretase activity is also an attractive target. Both genetic and pharmaceutical lowering of γ-secretase activity decrease production of Aβ (Li et al. 2007a,b). However, γ-secretase activity is also essential for processing Notch and a variety of other transmembrane proteins (Louvi & Artavanis-Tsakonas 2006). The γ-secretase inhibitor LY-411, for instance, reduces brain Aβ production but also has profound effects on T- and B- cell maturation (Barten et al. 2005).

Although challenges exist, significant progress has been made over the past 30 years. The future is likely to include multidrug regimens targeting several steps in Aβ production and clearance (Chow et al. 2010) given to individuals with asymptomatic Aβ accumulation detected by positron-emission topography (PET) scans or spinal fluid analysis. In this version of the future, our parents will be able to live out their lives with dignity and grace.

DISCLOSURE STATEMENT

The authors are not aware of any affiliations, memberships, funding, or financial holdings that might be perceived as affecting the objectivity of this review.

LITERATURE CITED

Abramov E, Dolev I, Fogel H, Ciccotosto GD, Ruff E, Slutsky I. 2009. Amyloid-beta as a positive endogenous regulator of release probability at hippocampal synapses. *Nat. Neurosci.* 12:1567–76

Alzheimer A, Stelzmann RA, Schnitzlein HN, Murtagh FR. 1995. An English translation of Alzheimer's 1907 paper, "Uber eine eigenartige Erkankung der Hirnrinde". *Clin Anat* 8:429–31

Andersen OM, Reiche J, Schmidt V, Gotthardt M, Spoelgen R, et al. 2005. Neuronal sorting protein-related receptor sorLA/LR11 regulates processing of the amyloid precursor protein. *Proc. Natl. Acad. Sci. USA* 102:13461–66

Ando K, Iijima KI, Elliott JI, Kirino Y, Suzuki T. 2001. Phosphorylation-dependent regulation of the interaction of amyloid precursor protein with Fe65 affects the production of beta-amyloid. *J. Biol. Chem.* 276:40353–61

Asai M, Hattori C, Szabo B, Sasagawa N, Maruyama K, et al. 2003. Putative function of ADAM9, ADAM10, and ADAM17 as APP alpha-secretase. *Biochem. Biophys. Res. Commun.* 301:231–35

Ashley J, Packard M, Ataman B, Budnik V. 2005. Fasciclin II signals new synapse formation through amyloid precursor protein and the scaffolding protein dX11/Mint. *J. Neurosci.* 25:5943–55

Avramopoulos D. 2009. Genetics of Alzheimer's disease: recent advances. *Genome Med.* 1:34.1–34.7

Barten DM, Guss VL, Corsa JA, Loo A, Hansel SB, et al. 2005. Dynamics of {beta}-amyloid reductions in brain, cerebrospinal fluid, and plasma of {beta}-amyloid precursor protein transgenic mice treated with a (Lilly)-secretase inhibitor. *J. Pharmacol. Exp. Ther.* 312:635–43

Bateman RJ, Munsell LY, Morris JC, Swarm R, Yarasheski KE, Holtzman DM. 2006. Human amyloid-beta synthesis and clearance rates as measured in cerebrospinal fluid in vivo. *Nat. Med.* 12:856–61

Bayer TA, Cappai R, Masters CL, Beyreuther K, Multhaup G. 1999. It all sticks together—the APP-related family of proteins and Alzheimer's disease. *Mol. Psychiatry* 4:524–28

Benn SC, Woolf CJ. 2004. Adult neuron survival strategies—slamming on the brakes. *Nat. Rev. Neurosci.* 5:686–700

Bergmans BA, De Strooper B. 2010. Gamma-secretases: from cell biology to therapeutic strategies. *Lancet Neurol.* 9:215–26

Billings LM, Oddo S, Green KN, McGaugh JL, LaFerla FM. 2005. Intraneuronal Abeta causes the onset of early Alzheimer's disease-related cognitive deficits in transgenic mice. *Neuron* 45:675–88

Blessed G, Tomlinson BE, Roth M. 1968. The association between quantitative measures of dementia and of senile change in the cerebral grey matter of elderly subjects. *Br. J. Psychiatry* 114:797–811

Borg JP, Ooi J, Levy E, Margolis B. 1996. The phosphotyrosine interaction domains of X11 and FE65 bind to distinct sites on the YENPTY motif of amyloid precursor protein. *Mol. Cell Biol.* 16:6229–41

Brookmeyer R, Johnson E, Ziegler-Graham K, Arrighi HM. 2007. Forecasting the global burden of Alzheimer's disease. *Alzheimers Dement.* 3:186–91

Butterfield DA. 2002. Amyloid beta-peptide (1–42)-induced oxidative stress and neurotoxicity: implications for neurodegeneration in Alzheimer's disease brain. A review. *Free Radic. Res.* 36:1307–13

Cai H, Wang Y, McCarthy D, Wen H, Borchelt DR, et al. 2001. BACE1 is the major beta-secretase for generation of Abeta peptides by neurons. *Nat. Neurosci.* 4:233–34

Calissano P, Matrone C, Amadoro G. 2010. Nerve growth factor as a paradigm of neurotrophins related to Alzheimer's disease. *Dev. Neurobiol.* 70:372–83

Cao X, Sudhof TC. 2001. A transcriptionally [correction of transcriptively] active complex of APP with Fe65 and histone acetyltransferase Tip60. *Science* 293:115–20

Capsoni S, Ugolini G, Comparini A, Ruberti F, Berardi N, Cattaneo A. 2000. Alzheimer-like neurodegeneration in aged antinerve growth factor transgenic mice. *Proc. Natl. Acad. Sci. USA* 97:6826–31

Chen CD, Oh SY, Hinman JD, Abraham CR. 2006. Visualization of APP dimerization and APP-Notch2 heterodimerization in living cells using bimolecular fluorescence complementation. *J. Neurochem.* 97:30–43

Chen G, Chen KS, Knox J, Inglis J, Bernard A, et al. 2000. A learning deficit related to age and beta-amyloid plaques in a mouse model of Alzheimer's disease. *Nature* 408:975–79

Chen WJ, Goldstein JL, Brown MS. 1990. NPXY, a sequence often found in cytoplasmic tails, is required for coated pit-mediated internalization of the low density lipoprotein receptor. *J. Biol. Chem.* 265:3116–23

Cho JH, Johnson GV. 2004. Glycogen synthase kinase 3 beta induces caspase-cleaved tau aggregation in situ. *J. Biol. Chem.* 279:54716–23

Chow VW, Savonenko AV, Melnikova T, Kim H, Price DL, et al. 2010. Modeling an anti-amyloid combination therapy for Alzheimer's disease. *Sci. Transl. Med.* 2:13ra1

Chung CW, Song YH, Kim IK, Yoon WJ, Ryu BR, et al. 2001. Proapoptotic effects of tau cleavage product generated by caspase-3. *Neurobiol. Dis.* 8:162–72

Cirrito JR, Deane R, Fagan AM, Spinner ML, Parsadanian M, et al. 2005. P-glycoprotein deficiency at the blood-brain barrier increases amyloid-beta deposition in an Alzheimer disease mouse model. *J. Clin. Invest.* 115:3285–90

Cirrito JR, Kang JE, Lee J, Stewart FR, Verges DK, et al. 2008. Endocytosis is required for synaptic activity-dependent release of amyloid-beta in vivo. *Neuron* 58:42–51

Counts SE, Nadeem M, Wuu J, Ginsberg SD, Saragovi HU, Mufson EJ. 2004. Reduction of cortical TrkA but not p75(NTR) protein in early-stage Alzheimer's disease. *Ann. Neurol.* 56:520–31

Deane R, Wu Z, Sagare A, Davis J, Du Yan S, et al. 2004. LRP/amyloid beta-peptide interaction mediates differential brain efflux of Abeta isoforms. *Neuron* 43:333–44

de Calignon A, Fox LM, Pitstick R, Carlson GA, Bacskai BJ, et al. 2010. Caspase activation precedes and leads to tangles. *Nature* 464:1201–4

Deshpande A, Mina E, Glabe C, Busciglio J. 2006. Different conformations of amyloid beta induce neurotoxicity by distinct mechanisms in human cortical neurons. *J. Neurosci.* 26:6011–18

De Strooper B, Annaert W, Cupers P, Saftig P, Craessaerts K, et al. 1999. A presenilin-1-dependent gamma-secretase-like protease mediates release of Notch intracellular domain. *Nature* 398:518–22

Dolan D, Troncoso J, Crain B, Resnick S, Zonderman A, O'Brien R. 2010a. Age, dementia and Alzheimer's disease in the BLSA. *Brain.* 133:2225–231

Dolan H, Troncoso J, Crain B, Resnick S, Zonderman A, O'Brien R. 2010b. Atherosclerosis and Alzheimer's disease in the Baltimore Longitudinal Study of Aging. *Ann. Neurol.* 68:231–40

Donmez G, Wang D, Cohen DE, Guarente L. 2010. SIRT1 suppresses beta-amyloid production by activating the alpha-secretase gene ADAM10. *Cell* 142:320–32

Ehehalt R, Keller P, Haass C, Thiele C, Simons K. 2003. Amyloidogenic processing of the Alzheimer beta-amyloid precursor protein depends on lipid rafts. *J. Cell Biol.* 160:113–23

El Khoury J, Toft M, Hickman SE, Means TK, Terada K, et al. 2007. Ccr2 deficiency impairs microglial accumulation and accelerates progression of Alzheimer-like disease. *Nat. Med.* 13:432–38

Feng W, Zhang M. 2009. Organization and dynamics of PDZ-domain-related supramodules in the postsynaptic density. *Nat. Rev. Neurosci.* 10:87–99

Fewlass DC, Noboa K, Pi-Sunyer FX, Johnston JM, Yan SD, Tezapsidis N. 2004. Obesity-related leptin regulates Alzheimer's Abeta. *FASEB J.* 18:1870–78

Fukumori A, Okochi M, Tagami S, Jiang J, Itoh N, et al. 2006. Presenilin-dependent gamma-secretase on plasma membrane and endosomes is functionally distinct. *Biochemistry* 45:4907–14

Galvan V, Gorostiza OF, Banwait S, Ataie M, Logvinova AV, et al. 2006. Reversal of Alzheimer's-like pathology and behavior in human APP transgenic mice by mutation of Asp664. *Proc. Natl. Acad. Sci. USA* 103:7130–35

Gentleman SM, Greenberg BD, Savage MJ, Noori M, Newman SJ, et al. 1997. A beta 42 is the predominant form of amyloid beta-protein in the brains of short-term survivors of head injury. *Neuroreport* 8:1519–22

Gervais FG, Xu D, Robertson GS, Vaillancourt JP, Zhu Y, et al. 1999. Involvement of caspases in proteolytic cleavage of Alzheimer's amyloid-beta precursor protein and amyloidogenic A beta peptide formation. *Cell* 97:395–406

Glenner GG, Wong CW. 1984. Alzheimer's disease: initial report of the purification and characterization of a novel cerebrovascular amyloid protein. *Biochem. Biophys. Res. Commun.* 120:885–90

Gouras GK, Tsai J, Naslund J, Vincent B, Edgar M, et al. 2000. Intraneuronal Abeta42 accumulation in human brain. *Am. J. Pathol.* 156:15–20

Greco SJ, Bryan KJ, Sarkar S, Zhu X, Smith MA, et al. 2009a. Chronic leptin supplementation ameliorates pathology and improves cognitive performance in a transgenic mouse model of Alzheimer's disease. *J. Alzheimers Dis.* Abstract

Greco SJ, Bryan KJ, Sarkar S, Zhu X, Smith MA, et al. 2010. Leptin reduces pathology and improves memory in a transgenic mouse model of Alzheimer's disease. *J Alzheimers Dis* 19:1155–67

Greco SJ, Sarkar S, Johnston JM, Tezapsidis N. 2009b. Leptin regulates tau phosphorylation and amyloid through AMPK in neuronal cells. *Biochem. Biophys. Res. Commun.* 380:98–104

Greenfield JP, Tsai J, Gouras GK, Hai B, Thinakaran G, et al. 1999. Endoplasmic reticulum and trans-Golgi network generate distinct populations of Alzheimer beta-amyloid peptides. *Proc. Natl. Acad. Sci. USA* 96:742–47

Guglielmotto M, Aragno M, Autelli R, Giliberto L, Novo E, et al. 2009. The up-regulation of BACE1 mediated by hypoxia and ischemic injury: role of oxidative stress and HIF1alpha. *J. Neurochem.* 108:1045–56

Halawani D, Tessier S, Anzellotti D, Bennett DA, Latterich M, LeBlanc AC. 2010. Identification of Caspase-6-mediated processing of the valosin containing protein (p97) in Alzheimer's disease: a novel link to dysfunction in ubiquitin proteasome system-mediated protein degradation. *J. Neurosci.* 30:6132–42

Hebert SS, Horre K, Nicolai L, Papadopoulou AS, Mandemakers W, et al. 2008. Loss of microRNA cluster miR-29a/b-1 in sporadic Alzheimer's disease correlates with increased BACE1/beta-secretase expression. *Proc. Natl. Acad. Sci. USA* 105:6415–20

Herms J, Anliker B, Heber S, Ring S, Fuhrmann M, et al. 2004. Cortical dysplasia resembling human type 2 lissencephaly in mice lacking all three APP family members. *EMBO J.* 23:4106–15

Hernandez F, Avila J. 2008. The role of glycogen synthase kinase 3 in the early stages of Alzheimers' disease. *FEBS Lett.* 582:3848–54

Ho A, Sudhof TC. 2004. Binding of F-spondin to amyloid-beta precursor protein: a candidate amyloid-beta precursor protein ligand that modulates amyloid-beta precursor protein cleavage. *Proc. Natl. Acad. Sci. USA* 101:2548–53

Hof PR, Bouras C, Perl DP, Sparks DL, Mehta N, Morrison JH. 1995. Age-related distribution of neuropathologic changes in the cerebral cortex of patients with Down's syndrome. Quantitative regional analysis and comparison with Alzheimer's disease. *Arch. Neurol.* 52:379–91

Holmes C, Boche D, Wilkinson D, Yadegarfar G, Hopkins V, et al. 2008. Long-term effects of Abeta42 immunisation in Alzheimer's disease: follow-up of a randomised, placebo-controlled phase I trial. *Lancet* 372:216–23

Holsinger RM, McLean CA, Beyreuther K, Masters CL, Evin G. 2002. Increased expression of the amyloid precursor beta-secretase in Alzheimer's disease. *Ann. Neurol.* 51:783–86

Holtzman DM, Bales KR, Wu S, Bhat P, Parsadanian M, et al. 1999. Expression of human apolipoprotein E reduces amyloid-beta deposition in a mouse model of Alzheimer's disease. *J. Clin. Invest.* 103:R15–21

Hu X, Hicks CW, He W, Wong P, Macklin WB, et al. 2006. Bace1 modulates myelination in the central and peripheral nervous system. *Nat. Neurosci.* 9:1520–25

Hung AY, Haass C, Nitsch RM, Qiu WQ, Citron M, et al. 1993. Activation of protein kinase C inhibits cellular production of the amyloid beta-protein. *J. Biol. Chem.* 268:22959–62

Huse JT, Liu K, Pijak DS, Carlin D, Lee VM, Doms RW. 2002. Beta-secretase processing in the trans-Golgi network preferentially generates truncated amyloid species that accumulate in Alzheimer's disease brain. *J. Biol. Chem.* 277:16278–84

Hyman BT. 2011. Caspase activation without apoptosis: insight into Abeta initiation of neurodegeneration. *Nat Neurosci* 14:5–6

Irizarry MC, Soriano F, McNamara M, Page KJ, Schenk D, et al. 1997. Abeta deposition is associated with neuropil changes, but not with overt neuronal loss in the human amyloid precursor protein V717F (PDAPP) transgenic mouse. *J. Neurosci.* 17:7053–59

Jacobsen L, Madsen P, Jacobsen C, Nielsen MS, Gliemann J, Petersen CM. 2001. Activation and functional characterization of the mosaic receptor SorLA/LR11. *J. Biol. Chem.* 276:22788–96

Jorissen E, Prox J, Bernreuther C, Weber S, Schwanbeck R, et al. 2010. The disintegrin/metalloproteinase ADAM10 is essential for the establishment of the brain cortex. *J. Neurosci.* 30:4833–44

Kamal A, Stokin GB, Yang Z, Xia CH, Goldstein LS. 2000. Axonal transport of amyloid precursor protein is mediated by direct binding to the kinesin light chain subunit of kinesin-I. *Neuron* 28:449–59

Kamenetz F, Tomita T, Hsieh H, Seabrook G, Borchelt D, et al. 2003. APP processing and synaptic function. *Neuron* 37:925–37

Kang J, Lemaire HG, Unterbeck A, Salbaum JM, Masters CL, et al. 1987. The precursor of Alzheimer's disease amyloid A4 protein resembles a cell-surface receptor. *Nature* 325:733–36

Kanski J, Aksenova M, Butterfield DA. 2002. The hydrophobic environment of Met35 of Alzheimer's Abeta(1–42) is important for the neurotoxic and oxidative properties of the peptide. *Neurotox. Res.* 4:219–23

Kim J, Basak JM, Holtzman DM. 2009. The role of apolipoprotein E in Alzheimer's disease. *Neuron* 63:287–303

Kinoshita A, Fukumoto H, Shah T, Whelan CM, Irizarry MC, Hyman BT. 2003. Demonstration by FRET of BACE interaction with the amyloid precursor protein at the cell surface and in early endosomes. *J. Cell Sci.* 116:3339–46

Koo EH, Sisodia SS, Archer DR, Martin LJ, Weidemann A, et al. 1990. Precursor of amyloid protein in Alzheimer disease undergoes fast anterograde axonal transport. *Proc. Natl. Acad. Sci. USA* 87:1561–65

Koo EH, Squazzo SL. 1994. Evidence that production and release of amyloid beta-protein involves the endocytic pathway. *J. Biol. Chem.* 269:17386–89

LaFerla FM, Green KN, Oddo S. 2007. Intracellular amyloid-beta in Alzheimer's disease. *Nat Rev Neurosci* 8:499–509

Laird FM, Cai H, Savonenko AV, Farah MH, He K, et al. 2005. BACE1, a major determinant of selective vulnerability of the brain to amyloid-beta amyloidogenesis, is essential for cognitive, emotional, and synaptic functions. *J. Neurosci.* 25:11693–709

Lambrechts A, Kwiatkowski AV, Lanier LM, Bear JE, Vandekerckhove J, et al. 2000. cAMP-dependent protein kinase phosphorylation of EVL, a Mena/VASP relative, regulates its interaction with actin and SH3 domains. *J. Biol. Chem.* 275:36143–51

Lazarov O, Morfini GA, Lee EB, Farah MH, Szodorai A, et al. 2005. Axonal transport, amyloid precursor protein, kinesin-1, and the processing apparatus: revisited. *J. Neurosci.* 25:2386–95

Lee J, Fukumoto H, Orne J, Klucken J, Raju S, et al. 2005. Decreased levels of BDNF protein in Alzheimer temporal cortex are independent of BDNF polymorphisms. *Exp. Neurol.* 194:91–96

Lee J, Retamal C, Cuitino L, Caruano-Yzermans A, Shin JE, et al. 2008. Adaptor protein sorting nexin 17 regulates amyloid precursor protein trafficking and processing in the early endosomes. *J. Biol. Chem.* 283:11501–8

Lee JH, Lau KF, Perkinton MS, Standen CL, Shemilt SJ, et al. 2003. The neuronal adaptor protein X11alpha reduces Abeta levels in the brains of Alzheimer's APPswe Tg2576 transgenic mice. *J. Biol. Chem.* 278:47025–29

Lee MS, Kwon YT, Li M, Peng J, Friedlander RM, Tsai LH. 2000. Neurotoxicity induces cleavage of p35 to p25 by calpain. *Nature* 405:360–64

Leissring MA, Farris W, Chang AY, Walsh DM, Wu X, et al. 2003. Enhanced proteolysis of beta-amyloid in APP transgenic mice prevents plaque formation, secondary pathology, and premature death. *Neuron* 40:1087–93

Li T, Wen H, Brayton C, Das P, Smithson LA, et al. 2007a. Epidermal growth factor receptor and notch pathways participate in the tumor suppressor function of gamma-secretase. *J. Biol. Chem.* 282:32264–73

Li T, Wen H, Brayton C, Laird FM, Ma G, et al. 2007b. Moderate reduction of gamma-secretase attenuates amyloid burden and limits mechanism-based liabilities. *J. Neurosci.* 27:10849–59

Lieb W, Beiser AS, Vasan RS, Tan ZS, Au R, et al. 2009. Association of plasma leptin levels with incident Alzheimer disease and MRI measures of brain aging. *JAMA* 302:2565–72

Louvi A, Artavanis-Tsakonas S. 2006. Notch signalling in vertebrate neural development. *Nat. Rev. Neurosci.* 7:93–102

Mawuenyega KG, Sigurdson W, Ovod V, Munsell L, Kasten T, et al. 2010. Decreased clearance of CNS beta-amyloid in Alzheimer's disease. *Science* 330:1774

McKee AC, Cantu RC, Nowinski CJ, Hedley-Whyte ET, Gavett BE, et al. 2009. Chronic traumatic encephalopathy in athletes: progressive tauopathy after repetitive head injury. *J. Neuropathol. Exp. Neurol.* 68:709–35

McLoughlin DM, Miller CC. 2008. The FE65 proteins and Alzheimer's disease. *J. Neurosci. Res.* 86:744–54

Meziane H, Dodart JC, Mathis C, Little S, Clemens J, et al. 1998. Memory-enhancing effects of secreted forms of the beta-amyloid precursor protein in normal and amnestic mice. *Proc. Natl. Acad. Sci. USA* 95:12683–88

Mills J, Reiner PB. 1999. Mitogen-activated protein kinase is involved in N-methyl-D-aspartate receptor regulation of amyloid precursor protein cleavage. *Neuroscience* 94:1333–38

Mok SS, Sberna G, Heffernan D, Cappai R, Galatis D, et al. 1997. Expression and analysis of heparin-binding regions of the amyloid precursor protein of Alzheimer's disease. *FEBS Lett.* 415:303–7

Nagahara AH, Merrill DA, Coppola G, Tsukada S, Schroeder BE, et al. 2009. Neuroprotective effects of brain-derived neurotrophic factor in rodent and primate models of Alzheimer's disease. *Nat. Med.* 15:331–37

Nikolaev A, McLaughlin T, O'Leary DD, Tessier-Lavigne M. 2009. APP binds DR6 to trigger axon pruning and neuron death via distinct caspases. *Nature* 457:981–89

Oddo S, Caccamo A, Kitazawa M, Tseng BP, LaFerla FM. 2003. Amyloid deposition precedes tangle formation in a triple transgenic model of Alzheimer's disease. *Neurobiol. Aging* 24:1063–70

Oddo S, Vasilevko V, Caccamo A, Kitazawa M, Cribbs DH, LaFerla FM. 2006. Reduction of soluble Abeta and tau, but not soluble Abeta alone, ameliorates cognitive decline in transgenic mice with plaques and tangles. *J. Biol. Chem.* 281:39413–23

Oh ES, Savonenko AV, King JF, Fangmark Tucker SM, Rudow GL, et al. 2009. Amyloid precursor protein increases cortical neuron size in transgenic mice. *Neurobiol. Aging* 30:1238–44

O'Malley D, MacDonald N, Mizielinska S, Connolly CN, Irving AJ, Harvey J. 2007. Leptin promotes rapid dynamic changes in hippocampal dendritic morphology. *Mol. Cell Neurosci.* 35:559–72

Parks AL, Curtis D. 2007. Presenilin diversifies its portfolio. *Trends Genet.* 23:140–50

Parvathy S, Hussain I, Karran EH, Turner AJ, Hooper NM. 1999. Cleavage of Alzheimer's amyloid precursor protein by alpha-secretase occurs at the surface of neuronal cells. *Biochemistry* 38:9728–34

Peng S, Wuu J, Mufson EJ, Fahnestock M. 2004. Increased proNGF levels in subjects with mild cognitive impairment and mild Alzheimer disease. *J. Neuropathol. Exp. Neurol.* 63:641–49

Peng S, Wuu J, Mufson EJ, Fahnestock M. 2005. Precursor form of brain-derived neurotrophic factor and mature brain-derived neurotrophic factor are decreased in the pre-clinical stages of Alzheimer's disease. *J. Neurochem.* 93:1412–21

Perez RG, Soriano S, Hayes JD, Ostaszewski B, Xia W, et al. 1999. Mutagenesis identifies new signals for beta-amyloid precursor protein endocytosis, turnover, and the generation of secreted fragments, including Abeta42. *J. Biol. Chem.* 274:18851–56

Pietrzik CU, Yoon IS, Jaeger S, Busse T, Weggen S, Koo EH. 2004. FE65 constitutes the functional link between the low-density lipoprotein receptor-related protein and the amyloid precursor protein. *J. Neurosci.* 24:4259–65

Postina R, Schroeder A, Dewachter I, Bohl J, Schmitt U, et al. 2004. A disintegrin-metalloproteinase prevents amyloid plaque formation and hippocampal defects in an Alzheimer disease mouse model. *J. Clin. Invest.* 113:1456–64

Qi JP, Wu H, Yang Y, Wang DD, Chen YX, et al. 2007. Cerebral ischemia and Alzheimer's disease: the expression of amyloid-beta and apolipoprotein E in human hippocampus. *J. Alzheimers Dis.* 12:335–41

Ring S, Weyer SW, Kilian SB, Waldron E, Pietrzik CU, et al. 2007. The secreted beta-amyloid precursor protein ectodomain APPs alpha is sufficient to rescue the anatomical, behavioral, and electrophysiological abnormalities of APP-deficient mice. *J. Neurosci.* 27:7817–26

Roses AD. 2010. An inherited variable poly-T repeat genotype in TOMM40 in Alzheimer disease. *Arch Neurol* 67:536–41

Rissman RA, Poon WW, Blurton-Jones M, Oddo S, Torp R, et al. 2004. Caspase-cleavage of tau is an early event in Alzheimer disease tangle pathology. *J. Clin. Invest.* 114:121–30

Roberson ED, Scearce-Levie K, Palop JJ, Yan F, Cheng IH, et al. 2007. Reducing endogenous tau ameliorates amyloid beta-induced deficits in an Alzheimer's disease mouse model. *Science* 316:750–54

Roch JM, Masliah E, Roch-Levecq AC, Sundsmo MP, Otero DA, et al. 1994. Increase of synaptic density and memory retention by a peptide representing the trophic domain of the amyloid beta/A4 protein precursor. *Proc. Natl. Acad. Sci. USA* 91:7450–54

Rohn TT, Head E. 2009. Caspases as therapeutic targets in Alzheimer's disease: Is it time to "cut" to the chase? *Int. J. Clin. Exp. Pathol.* 2:108–18

Rohn TT, Kokoulina P, Eaton CR, Poon WW. 2009. Caspase activation in transgenic mice with Alzheimer-like pathology: results from a pilot study utilizing the caspase inhibitor, Q-VD-OPh. *Int. J. Clin. Exp. Med.* 2:300–8

Rohn TT, Vyas V, Hernandez-Estrada T, Nichol KE, Christie LA, Head E. 2008. Lack of pathology in a triple transgenic mouse model of Alzheimer's disease after overexpression of the anti-apoptotic protein Bcl-2. *J. Neurosci.* 28:3051–59

Rovio S, Kareholt I, Helkala EL, Viitanen M, Winblad B, et al. 2005. Leisure-time physical activity at midlife and the risk of dementia and Alzheimer's disease. *Lancet Neurol.* 4:705–11

Saito Y, Sano Y, Vassar R, Gandy S, Nakaya T, et al. 2008. X11 proteins regulate the translocation of amyloid beta-protein precursor (APP) into detergent-resistant membrane and suppress the amyloidogenic cleavage of APP by beta-site-cleaving enzyme in brain. *J. Biol. Chem.* 283:35763–71

Saluja I, Paulson H, Gupta A, Turner RS. 2009. X11alpha haploinsufficiency enhances Abeta amyloid deposition in Alzheimer's disease transgenic mice. *Neurobiol. Dis.* 36:162–68

Sano Y, Nakaya T, Pedrini S, Takeda S, Iijima-Ando K, et al. 2006. Physiological mouse brain Abeta levels are not related to the phosphorylation state of threonine-668 of Alzheimer's APP. *PLoS One* 1:e51

Saura CA, Choi SY, Beglopoulos V, Malkani S, Zhang D, et al. 2004. Loss of presenilin function causes impairments of memory and synaptic plasticity followed by age-dependent neurodegeneration. *Neuron* 42:23–36

Savonenko AV, Melnikova T, Laird FM, Stewart KA, Price DL, Wong PC. 2008. Alteration of BACE1-dependent NRG1/ErbB4 signaling and schizophrenia-like phenotypes in BACE1-null mice. *Proc. Natl. Acad. Sci. USA* 105:5585–90

Schenk DB, Seubert P, Grundman M, Black R. 2005. A beta immunotherapy: lessons learned for potential treatment of Alzheimer's disease. *Neurodegener. Dis.* 2:255–60

Scherzer CR, Offe K, Gearing M, Rees HD, Fang G, et al. 2004. Loss of apolipoprotein E receptor LR11 in Alzheimer disease. *Arch. Neurol.* 61:1200–5

Schmidt V, Sporbert A, Rohe M, Reimer T, Rehm A, et al. 2007. SorLA/LR11 regulates processing of amyloid precursor protein via interaction with adaptors GGA and PACS-1. *J. Biol. Chem.* 282:32956–64

Schneider JA, Arvanitakis Z, Bang W, Bennett DA. 2007. Mixed brain pathologies account for most dementia cases in community-dwelling older persons. *Neurology* 69:2197–204

Seeley WW, Crawford RK, Zhou J, Miller BL, Greicius MD. 2009. Neurodegenerative diseases target large-scale human brain networks. *Neuron* 62:42–52

Selkoe DJ. 2001. Clearing the brain's amyloid cobwebs. *Neuron* 32:177–80

Shankar GM, Li S, Mehta TH, Garcia-Munoz A, Shepardson NE, et al. 2008. Amyloid-beta protein dimers isolated directly from Alzheimer's brains impair synaptic plasticity and memory. *Nat. Med.* 14:837–42

Shaw LM, Vanderstichele H, Knapik-Czajka M, Clark CM, Aisen PS, et al. 2009. Cerebrospinal fluid biomarker signature in Alzheimer's disease neuroimaging initiative subjects. *Ann. Neurol.* 65:403–13

Shen J, Kelleher RJ 3rd. 2007. The presenilin hypothesis of Alzheimer's disease: evidence for a loss-of-function pathogenic mechanism. *Proc. Natl. Acad. Sci. USA* 104:403–9

Shepherd C, McCann H, Halliday GM. 2009. Variations in the neuropathology of familial Alzheimer's disease. *Acta Neuropathol.* 118:37–52

Singh PP, Singh M, Mastana SS. 2006. APOE distribution in world populations with new data from India and the UK. *Ann. Hum. Biol.* 33:279–308

Skovronsky DM, Moore DB, Milla ME, Doms RW, Lee VM. 2000. Protein kinase C-dependent alpha-secretase competes with beta-secretase for cleavage of amyloid-beta precursor protein in the trans-Golgi network. *J. Biol. Chem.* 275:2568–75

Sorg C, Riedl V, Muhlau M, Calhoun VD, Eichele T, et al. 2007. Selective changes of resting-state networks in individuals at risk for Alzheimer's disease. *Proc Natl Acad Sci USA* 104:18760–65

Sperling RA, Laviolette PS, O'Keefe K, O'Brien J, Rentz DM, et al. 2009. Amyloid deposition is associated with impaired default network function in older persons without dementia. *Neuron* 63:178–88

Spires TL, Meyer-Luehmann M, Stern EA, McLean PJ, Skoch J, et al. 2005. Dendritic spine abnormalities in amyloid precursor protein transgenic mice demonstrated by gene transfer and intravital multiphoton microscopy. *J. Neurosci.* 25:7278–87

Spoelgen R, von Arnim CA, Thomas AV, Peltan ID, Koker M, et al. 2006. Interaction of the cytosolic domains of sorLA/LR11 with the amyloid precursor protein (APP) and beta-secretase beta-site APP-cleaving enzyme. *J. Neurosci.* 26:418–28

Steinerman JR, Irizarry M, Scarmeas N, Raju S, Brandt J, et al. 2008. Distinct pools of beta-amyloid in Alzheimer disease-affected brain: a clinicopathologic study. *Arch. Neurol.* 65:906–12

Takasugi N, Tomita T, Hayashi I, Tsuruoka M, Niimura M, et al. 2003. The role of presenilin cofactors in the gamma-secretase complex. *Nature* 422:438–41

Tiraboschi P, Hansen LA, Masliah E, Alford M, Thal LJ, Corey-Bloom J. 2004. Impact of APOE genotype on neuropathologic and neurochemical markers of Alzheimer disease. *Neurology* 62:1977–83

Tokuda T, Calero M, Matsubara E, Vidal R, Kumar A, et al. 2000. Lipidation of apolipoprotein E influences its isoform-specific interaction with Alzheimer's amyloid beta peptides. *Biochem. J.* 348(Pt. 2):359–65

Verghese J, LeValley A, Derby C, Kuslansky G, Katz M, et al. 2006. Leisure activities and the risk of amnestic mild cognitive impairment in the elderly. *Neurology* 66:821–27

Wang P, Yang G, Mosier DR, Chang P, Zaidi T, et al. 2005. Defective neuromuscular synapses in mice lacking amyloid precursor protein (APP) and APP-Like protein 2. *J. Neurosci.* 25:1219–25

Willem M, Garratt AN, Novak B, Citron M, Kaufmann S, et al. 2006. Control of peripheral nerve myelination by the beta-secretase BACE1. *Science* 314:664–66

Wolfe MS, Xia W, Ostaszewski BL, Diehl TS, Kimberly WT, Selkoe DJ. 1999. Two transmembrane aspartates in presenilin-1 required for presenilin endoproteolysis and gamma-secretase activity. *Nature* 398:513–17

Yang LB, Lindholm K, Yan R, Citron M, Xia W, et al. 2003. Elevated beta-secretase expression and enzymatic activity detected in sporadic Alzheimer disease. *Nat. Med.* 9:3–4

Yang Z, Cool BH, Martin GM, Hu Q. 2006. A dominant role for FE65 (APBB1) in nuclear signaling. *J. Biol. Chem.* 281:4207–14

Yankner BA, Dawes LR, Fisher S, Villa-Komaroff L, Oster-Granite ML, Neve RL. 1989. Neurotoxicity of a fragment of the amyloid precursor associated with Alzheimer's disease. *Science* 245:417–20

Yankner BA, Lu T. 2009. Amyloid beta-protein toxicity and the pathogenesis of Alzheimer disease. *J. Biol. Chem.* 284:4755–59

Young-Pearse TL, Bai J, Chang R, Zheng JB, LoTurco JJ, Selkoe DJ. 2007. A critical function for beta-amyloid precursor protein in neuronal migration revealed by in utero RNA interference. *J. Neurosci.* 27:14459–69

Yuede CM, Zimmerman SD, Dong H, Kling MJ, Bero AW, et al. 2009. Effects of voluntary and forced exercise on plaque deposition, hippocampal volume, and behavior in the Tg2576 mouse model of Alzheimer's disease. *Neurobiol. Dis.* 35:426–32

Zambrano N, Minopoli G, de Candia P, Russo T. 1998. The Fe65 adaptor protein interacts through its PID1 domain with the transcription factor CP2/LSF/LBP1. *J. Biol. Chem.* 273:20128–33

Zuccato C, Cattaneo E. 2009. Brain-derived neurotrophic factor in neurodegenerative diseases. *Nat. Rev. Neurol.* 5:311–22

Motor Functions of the Superior Colliculus

Neeraj J. Gandhi[1-4] and Husam A. Katnani[2,4]

Departments of Otolaryngology,[1] Bioengineering,[2] and Neuroscience;[3] Center for Neural Basis of Cognition,[4] University of Pittsburgh, Pittsburgh, Pennsylvania 15213; email: neg8@pitt.edu

Annu. Rev. Neurosci. 2011. 34:205–31

First published online as a Review in Advance on March 29, 2011

The *Annual Review of Neuroscience* is online at neuro.annualreviews.org

This article's doi: 10.1146/annurev-neuro-061010-113728

0147-006X/11/0721-0205$20.00

Keywords

saccade, fixation, head, gaze shift, reach, vibrissae, pinnae, sonar, vector averaging, vector summation

Abstract

The mammalian superior colliculus (SC) and its nonmammalian homolog, the optic tectum, constitute a major node in processing sensory information, incorporating cognitive factors, and issuing motor commands. The resulting action—to orient toward or away from a stimulus—can be accomplished as an integrated movement across oculomotor, cephalomotor, and skeletomotor effectors. The SC also participates in preserving fixation during intersaccadic intervals. This review highlights the repertoire of movements attributed to SC function and analyzes the significance of results obtained from causality-based experiments (microstimulation and inactivation). The mechanisms potentially used to decode the population activity in the SC into an appropriate movement command are also discussed.

Contents

INTRODUCTION

Visualize a pitcher throwing a fastball to the batter, who swings the bat and makes solid contact with the baseball. It is driven sharply and directly back at the pitcher, who in turn reacts to snatch the ball just before it hits him. To record the out and also to save himself from injury, the pitcher must rely on visual cues about the location and trajectory of the ball after it is hit, auditory information such as the timing and intensity of the contact between the bat and ball, as well as proprioceptive signals and/or internal representations of body positions after releasing the ball. Neural signals representing these events must be integrated and transformed into a coordinated movement of the eyes, head, hands, and body to make the catch. Such a simple and seemingly reflexive orienting behavior is produced by a rather complicated set of neural processes that successfully accomplish the sensory to motor transformation.

The superior colliculus (SC) is a major node for mediating sensorimotor transformations (Hall & Moschovakis 2004, Sparks & Mays 1990, Stein & Meredith 1993). Residing on the roof of the brain stem, this subcortical structure contains seven alternating fibrous and cellular laminae. Neurons in the superficial layers (stratum zonale, stratum griseum superficiale, and stratum opticum) are responsive nearly exclusively to visual stimuli appearing at specific locations in the contralateral hemifield. A subset of cells in the intermediate (stratum griseum intermedium, and stratum album intermedium) and deeper (stratum griseum profundum, and stratum album profundum) layers, collectively referred to as the deep layers, expresses sensitivity to sensory stimuli from several modalities (e.g., vision, audition, somatosensation), and another overlapping group of neurons discharges a vigorous premotor burst during the orienting movement. This review is intended to provide a critical assessment on various topics related to the motor functions of the SC. The repertoire of movements attributed to the SC and its nonmammalian homolog, the optic tectum (OT), is considered first, followed by an evaluation of collicular mechanisms for saccade generation. We specifically address the dynamic features of ensemble activity that permit a balance between fixation and redirection of the visual axis and the computational rules that best explain how population activity in the SC is decoded to produce saccades.

REPERTOIRE OF MOVEMENTS PRODUCED BY THE SUPERIOR COLLICULUS

Eye Movements

When the head is restrained from moving, neurons in the deep layers of the SC discharge a burst prior to contraversive saccades. The elevated activity exists for only a range of saccade vectors that define each cell's movement field (Straschill & Hoffmann 1970, Wurtz &

SC: superior colliculus

OT: optic tectum

Goldberg 1972). The response is maximal for an optimum amplitude and direction, also known as the center or "hot spot" of the movement field, and graded for increasing and decreasing vectors (**Figure 1b**). Neurons in the SC are organized according to their movement field centers (**Figure 1a**) (Sparks et al. 1976). Cells located rostrally within the SC discharge vigorously for small-amplitude saccades, whereas units found at more caudal locations burst optimally for large-size movements. Neurons within the medial and lateral regions are most active during saccades with upward and downward components, respectively. Microstimulation of the SC produces saccade vectors that conform to this organization (McHaffie & Stein 1982, Robinson 1972, Stanford et al. 1996, Straschill & Rieger 1973). The topography of saccade amplitude onto the SC is logarithmic: A disproportionately large amount of SC tissue is allocated for small saccades, corresponding to movements to a parafoveal space, whereas a relatively compressed region is attributed for larger amplitude saccades to peripheral locations. In contrast, the saccade direction map along the mediolateral extent of the SC is fairly linear (Ottes et al. 1986).

Because each SC neuron bursts for a range of saccades, it follows that a population of cells, generally envisioned as a Gaussian mound, emits spikes for every movement (**Figure 1c**). In general, ~28% of the neurons in the deep layers (Munoz & Wurtz 1995b) discharge a premotor burst prior to the generation of a saccade. The size of the active population and the total number of spikes produced in the premotor burst are invariant across all saccades (Anderson et al. 1998, Van Opstal & Goossens 2008), but the mound is centered at the site corresponding to the executed saccade vector. To produce a wide range of saccade amplitudes from a relatively constant output, SC projections to the brainstem burst generator have to be weighted. Consistent with this hypothesis, the number of terminal boutons deployed onto the horizontal component of the burst generator increases monotonically across the rostral-caudal extent of the SC (Moschovakis et al. 1998).

The terminology used to categorize saccade-related neurons in the SC has evolved as investigators probed for specific functionality. It is important to note that these neurons are distinctly different from visual neurons, which reside in the superficial layers of the SC. Visual neurons respond only to sensory (visual) stimulation, and although saccade-related neurons can have sensory responses, they are primarily involved with motor function. The current nomenclature for saccade-related neurons is based primarily on studies performed by Munoz and colleagues (Munoz & Guitton 1991; Munoz & Wurtz 1993a, 1995a). Three major classes exist within the SC:

1. Saccade-related burst neurons emit a high-frequency volley of spikes prior to producing the high-speed eye movement.
2. Buildup neurons discharge a low-level or prelude response that accumulates gradually during the sensorimotor integration period before transitioning into a high-frequency burst to produce the saccade. This low-frequency discharge has been attributed to processes such as motor preparation, target selection, attention, and working memory (see Sparks 1999 for a critical commentary).
3. Fixation neurons in the rostral SC discharge at a tonic rate during visual fixation and pause activity during most, but not all, saccades.

Although used for classification purposes, detailed analyses show that discharge features of SC neurons actually span the continuum across these categories, and in some cases, the labels can offer inaccurate insights into a neuron's function. For example, the so-called fixation neurons actually discharge for a range of saccade amplitudes that include both micro- (Hafed et al. 2009) and small-amplitude macrosaccades (Gandhi & Keller 1997, Krauzlis et al. 1997, Munoz & Wurtz 1993a).

Along with providing saccade vector coordinates, population activity in the SC has also been noted to contribute to the kinematics of

Burst generator: neural elements in the paramedian and mesencephalic reticular formations for the generation of horizontal and vertical components of saccades, respectively

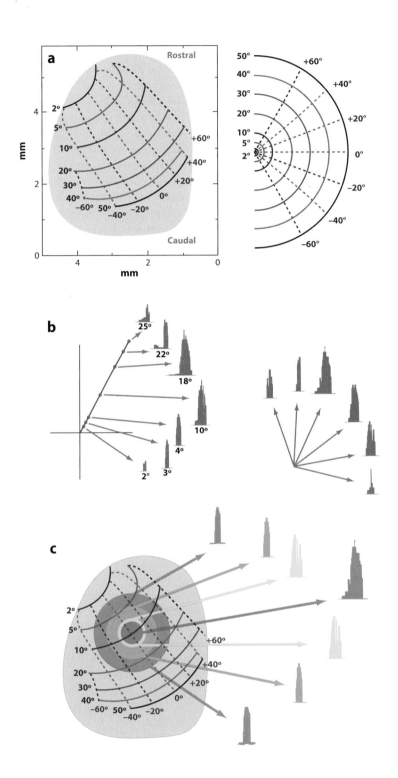

a saccade. The dual-coding hypothesis (Sparks & Mays 1990) states that the firing rate of the high-frequency burst contributes to the speed of the movement, whereas the locus of population activity on the SC map indicates the desired saccade vector. Indeed, amplitude-matched visually guided saccades are faster than memory-guided eye movements (Gnadt et al. 1991, Smit et al. 1987), and the accompanying premotor discharge is more vigorous when the visual target remains illuminated (Edelman & Goldberg 2003). Stimulation-evoked saccades also exhibit a similar relationship with stimulation parameters. Their velocity waveforms scale, up to a saturation limit, with stimulation frequency (Stanford et al. 1996) and intensity (Van Opstal et al. 1990). The amplitude is site specific provided that the stimulation duration is long enough to complete the movement. Prolonged stimulation produces a "staircase" of saccades interrupted with brief intersaccadic intervals (Breznen et al. 1996, Missal et al. 1996, Robinson 1972, Stryker & Schiller 1975).

Eye movements other than saccades have also been associated with SC function (for a review, see Gandhi & Sparks 2004). Briefly, a subset of neurons that discharge during saccades alter their response characteristics during combined saccade-vergence movements (Walton & Mays 2003). Stimulation of the SC can either perturb coordinated saccade-vergence movements (Chaturvedi & van Gisbergen 1999), and for some sites in the

rostral SC, microstimulation can even induce vergence eye movements (Chaturvedi & Van Gisbergen 2000) and lens accommodation (Sawa & Ohtsuka 1994). Activity of neurons in the rostral SC is also correlated with smooth-pursuit eye movements (Krauzlis 2003; Krauzlis et al. 1997, 2000), although microstimulation of the region does not produce such movements (Basso et al. 2000). A recent study also demonstrated that microstimulation of the barn-owl OT evokes pupil dilation (Netser et al. 2010).

Eye and Head Movements

When the head is free to move, large-amplitude changes in the line of sight cannot be produced by a saccadic eye movement alone. Such gaze shifts are generally executed as a coordinated movement of the eyes and head (**Figure 2a**) (also see Freedman 2008 for a review). Typically, the onset of the gaze shift is initiated by a saccadic eye-in-head movement, and the head movement lags behind. The offsets of the gaze shift and ocular saccade often coincide, although some discrepancies have been reported. The time of peak head velocity appears synchronized to the end of the gaze shift (Chen & Tehovnik 2007), and the head movement continues for 100–200 ms after gaze shift has terminated, during which the eyes counter-rotate in the orbits to stabilize gaze. By varying the initial positions of the eyes in orbits (and restraining movements of the torso and lower

Gaze shift: a change in visual axis; small changes are typically completed by a saccade; for large changes, the ocular saccade is nested with a head movement; for even larger changes, other skeletal segments can be integrated with the coordinated eye-head movement

---

Figure 1

Fundamental properties of the superior colliculus (SC) for the generation of saccades. (*a*) A schematic of the topographic organization of contralateral saccade vectors (*left*) is encoded in retinotopic coordinates. Isoradial and isodirectional bands are shown as solid and dashed lines, respectively. The radial and directions bands are identified in green and blue numbers, respectively. Each band is represented in a different color. The mapping of these bands in the contralateral hemifield is shown in the right panel. A disproportionately large amount of SC space is used to produce small amplitude saccades relative to the caudal SC areas that produce larger vectors. (*b*) Neurons in the deep SC layers discharge for a range of saccade amplitudes and directions. Its location on the SC map dictates the optimal vector for which the cell emits its maximal burst. Burst profiles are shown for different amplitude saccades in the optimal direction (*left*) and for several optimal amplitude saccades in various directions (*right*). Adapted from Sparks & Gandhi (2003). (*c*) Population response for the generation of a saccade can be envisioned as a mound of activity across a large portion of the deep SC layers. The amplitude and direction of the executed saccade typically matches with the vector encoded at the locus of maximal activity. Neurons that are active but are located away from the center exhibit a suboptimal burst. Adapted from Sparks & Gandhi (2003).

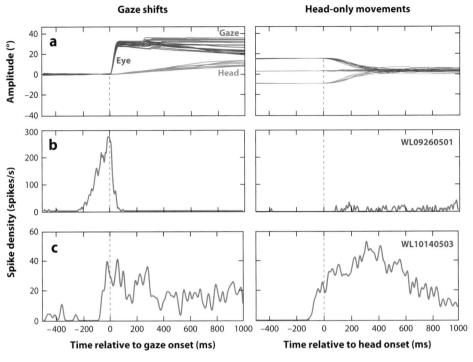

Gaze shifts   **Head-only movements**

Figure 2

Activity of superior colliculus (SC) neurons during coordinated eye-head movements (gaze shifts) and head-only movements. (*a*) Several examples of gaze shifts (*left*) and head-only movements (*right*) are plotted as a function of time. For gaze shifts (*left*), the change in line of sight (equivalently, gaze or eye-in-space) (*green traces*) is produced initially by rapidly moving the eyes within the orbits (eye-in-head) (*blue traces*). The head movement (*orange traces*) typically lags gaze onset, but it can continue for several hundred milliseconds after the termination of the gaze shift, during which the eyes counter-rotate in the orbits. During head-only movements (*right*), gaze remains stable as the eyes counter-rotate in the orbits. (*b*) Average spike density waveform of a SC neuron that resembles a classical gaze-related burst neuron. This cell produced a high-frequency burst for optimal size gaze shifts (*left*), while its activity was negligible for all head-only movements (*right*). (*c*) Average spike density waveform of another SC neuron that responds during head-only movements (*right*). It also discharges for gaze shifts (*left*), but the duration of activity outlasts the duration of the gaze shift and is better correlated with head duration. Note that the firing rate is too low to be a high-frequency burst even when optimal-amplitude head-only movements and gaze shifts were produced. The traces illustrated in the top row are not the specific movements generated during the neural recordings shown in the bottom two rows. Adapted from Walton et al. (2007), with permission.

extremities), the same amplitude gaze shift can be produced by different combinations of eye and head movements.

The subset of SC neurons that discharge a high-frequency premotor burst before head-restrained saccades also exhibit similar neural activity prior to head-unrestrained gaze shifts. The high-frequency component is optimal for a desired change in gaze, not its individual eye or head component (Freedman & Sparks 1997).

Suprathreshold microstimulation of the SC in cats and monkeys produces coordinated eye-head movements with characteristics comparable to visually guided gaze shifts (Freedman et al. 1996; Guillaume & Pélisson 2001, 2006; Harris 1980; Klier et al. 2001; Paré & Guitton 1994; Roucoux et al. 1980). Data suggest that the SC encodes gaze displacement in retinal coordinates (Klier et al. 2001). For a given stimulation site, roughly the same amplitude gaze shift is elicited across variations in the initial eye-in-head position. The amplitudes of the saccadic eye component and the head movement vary inversely as a function of initial eye-in-head position. Thus, as the eyes are initially deviated increasingly in the direction opposite to that of the stimulation-evoked movement, the amplitude of the saccadic eye component increases and that of the accompanying head movement decreases. In species with a negligible oculomotor range, such as the owl and the bat, the change in gaze is produced nearly entirely by the head (du Lac & Knudsen 1990, Valentine et al. 2002).

Changing the frequency of stimulation proportionally modifies the speed of both eye and head components and, therefore, also the speed of the gaze shift (Freedman et al. 1996). Interestingly, prolonged stimulation continues to drive the head movement, albeit at a slower speed, even after the gaze shift comes to an end (du Lac & Knudsen 1990, Freedman et al. 1996). The head-movement amplitude in such cases violates its lawful relationship with gaze amplitude and instead correlates better with stimulation duration. These results suggest that the place code component of the dual-coding hypothesis appears valid only for the line of sight (gaze).

The processing of SC activity by neural elements controlling the neck musculature is not dependent on the generation of a gaze shift. Electromyography (EMG) of neck muscles reveals a transient sensory response linked to the onset of a visual stimulus in the ipsilateral hemifield (Corneil et al. 2004), but perhaps most effectively during reflexive movement tasks (see Pruszynski et al. 2010); the SC could be the primary source of this short-latency response. Low-frequency stimulation of the SC also evokes low-level EMG in the deep neck muscles contralaterally, even when the head is restrained (Corneil et al. 2002a, Roucoux et al. 1980). The EMG response is generally smaller, but not negligible, when stimulation is applied within the rostral SC, and it increases for stimulation delivered at more caudal sites. For any SC site, neck muscle EMG also increases dur-

ing the period leading to saccade onset, reflecting a correlation with movement preparation (Corneil et al. 2007, Rezvani & Corneil 2008). When the head is unrestrained, similar stimulation parameters can evoke head-only movements (the eyes counter-rotate in the orbits) that precede the onset of a gaze shift (Corneil et al. 2002b, Pélisson et al. 2001). When a gaze shift does follow, the EMG response increases significantly prior to the higher velocity head movement associated with the gaze shift. These results are consistent with the hypothesis that the SC output is processed by two separate pathways in the brain stem. The oculomotor pathway produces the saccadic eye component of the gaze shift and the head pathway innervates the neck muscles. A key distinction between the two is that the eye pathway is potently inhibited by the pontine omnipause neurons (OPNs). These neurons prevent the premature execution of eye movements until the SC output reaches a threshold, which is usually associated with the high-frequency burst. An absent or significantly weaker gating mechanism on the head pathway permits the generation of head-only movements that can precede gaze shifts (Corneil et al. 2002b, Gandhi & Sparks 2007, Grantyn et al. 2010, Guitton et al. 1990).

We can generate head movements without changing our line of sight, such as when nodding. Does the SC contribute to the generation of such head-only movements? Or is the relationship between SC output and neck-muscle response constrained only to head movements associated with gaze shifts? Many SC neurons in the deeper layers indeed exhibit various types of modulations when nonhuman primates generate active head-only movements in a controlled, experimental setting (**Figure 2a**) (Walton et al. 2007). Some neurons increase their firing rates (**Figure 2c**), whereas others show modest suppression. The maximal firing rates of these neurons are an order of magnitude lower than the high-frequency bursts observed for head-restrained saccades and head-unrestrained gaze shifts (**Figure 2b**). In general, there are substantial differences between SC neurons that respond during

EMG: electromyography

Omnipause neuron (OPN): neuron located along the midline in the paramedian pontine reticular formation (oculomotor pons), discharging at a tonic rate during fixation and becoming quiescent during saccades

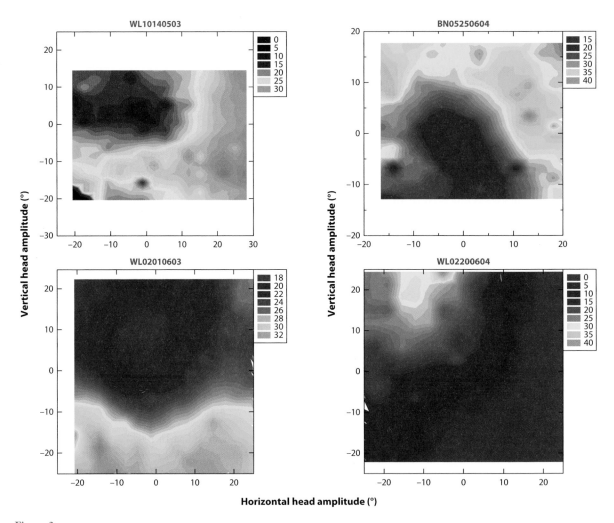

Figure 3

Movement field characterization for head-only movements of four superior colliculus (SC) neurons. Average firing rate during the head-only movement (legend key) is plotted as a function of its horizontal and vertical components. The filled contour plots were constructed from individual trial data points. Adapted from Walton et al. (2007), with permission.

head-only movements and those that are active prior to gaze shifts. For example, no reliable relationship is observed between firing rate and head movement parameters such as velocity, amplitude, or position. High-frequency-burst neurons respond prior to the generation of a limited range of gaze vectors; in contrast, no such circumscribed response fields are noted for head-only movements (**Figure 3**). Although head-movement related neurons are found throughout the deeper layers of the SC, they lack a topographical organization based on head amplitude or direction. Thus, although SC neurons exhibit responsivity to head-only movements, they appear to be different from those generating head movements that accompany gaze shifts.

Pinnae and Whisker Movements

In addition to saccadic movements of the eyes and head (if it is unrestrained), stimulation of the SC can evoke movements of the whiskers and pinnae, more commonly in species that routinely use these mobile sensors for interactions with the environment (Cowie & Robinson 1994, Hemelt & Keller 2008, McHaffie & Stein 1982, Stein & Clamann 1981, Valentine et al. 2002). In the echolocating bat, for example, contralateral pinnae movements can be evoked with low-threshold currents (<25 μA) and with latencies (~20 ms) as short as those for saccades in nonhuman primates (Valentine et al. 2002). Stimulation of rostral sites evokes forward movements of both ears, as if orienting to a target that is straight ahead. Large-amplitude, backward movement of both ears is evoked from posterior sights, conforming to the notion of orienting to a stimulus that is behind the animal. Medial and lateral sites evoke pinnae movements with upward and downward directions, respectively. With the exception of stimulation of rostral sites, the contralateral pinna moves first. Recruitment of the ipsilateral ear as well as the complexity of pinnae movement is often a function of stimulation parameters at posterior sites, suggesting that the entire response is not determined solely by the locus of stimulation. The topography of pinnae movements is similar in cats (Stein & Clamann 1981), although descriptions of SC stimulation-evoked pinnae movements are scarce in other animals. It is generally reported that stimulation of ventral regions in monkeys (Cowie & Robinson 1994) and caudal sites in rodents (McHaffie & Stein 1982) are most effective at producing ear movements.

Hemelt & Keller (2008) recently performed a thorough investigation of SC control of vibrissae movements in rats. Stimulation of effective sites evoked a sustained protraction of the whisker pad that lasted for the duration of the stimulation. A frequency of 333 Hz appeared optimal for producing large-amplitude movements, which scaled with current intensity above threshold (~25 μA). Thus, the protraction magnitude was not site specific, unlike for saccades. The deeper layers of the rodent SC exhibits a dorsoventral topography for the laterality of vibrissae movements: Protractions of the contralateral and ipsilateral whisker pads were evoked from dorsal and ventral aspects, respectively. Bilateral movements were evoked from intermediate regions, but current spread to both dorsal and ventral regions could account for the observation.

What is the role of the SC in producing vibrissae movements? Kinematic properties of sustained vibrissae protraction (amplitude) evoked from the SC are distinct from the rhythmic whisking behavior (frequency) associated with the motor cortex (Cramer & Keller 2006). Hence, the SC may regulate the amplitude and positional control of whisking (Hemelt & Keller 2008). Interestingly, the putative tecto-facial neurons in the region, from which stimulation-evoked vibrissae movements are likely produced, do not respond to trigeminal inputs, precluding them from directly mediating the short-latency reflex loop connecting the trigeminal and facial neurons (Kleinfeld et al. 1999, Hemelt & Keller 2008).

Eye-Head-Body Movements

A combined eye-head movement may not be sufficient to produce very large changes in gaze, such as when looking behind. Coordination across multiple body segments, including the body and feet, is required (e.g., Hollands et al. 2004, McCluskey & Cullen 2007). In such cases, the gaze shift is not necessarily completed in a single movement. Instead, multiple, smaller-amplitude movements with brief intervals of steady gaze are used to fixate the desired location (Anastasopoulos et al. 2009; but also see Degani et al. 2010). To the best of our knowledge, descriptions of extracellular recordings from SC neurons during controlled eye-head-body movements do not exist in literature. A microstimulation approach, however, has been applied and has implicated the SC in controlling body movements. Microstimulation can induce whole-body movements in freely moving cats (Hess et al. 1946, Schaefer 1970, Syka & Radil-Weiss 1971), whole-body and circling behavior in rodents (Dean et al. 1986, Tehovnik & Yeomans 1986), whole-body turns in frogs (Ewert 1984), body and tail movements in goldfish (Herrero et al. 1998), head and body movements in snakes (Dacey & Ulinski 1986), and swimming in lampreys (Saitoh et al. 2007). When permitted by the oculomotor range, stimulation also produces an eye movement, and like the cases described with pinnae, vibrissae, and head movements, the extent of the accompanying body movement varies with stimulation parameters. For example, the frequency of contraversive circling increases monotonically with the frequency and current of stimulation delivered to the caudal SC in rodents (Tehovnik 1989). Tail movements are evoked by stimulation of increasingly posterior sites in the goldfish OT, and both the amplitude and complexity of the eye-body-tail movement increases with stimulation parameters (Herrero et al. 1998). Stimulation duration also plays a critical role in determining the movement evoked by the lamprey OT. A site that evokes just eye movements with short-duration stimulation can also produce complex, swimming patterns with substantially prolonged stimulation duration (Saitoh et al. 2007). These prolonged stimulation results are reminiscent of the complex actions evoked by long-duration stimulation of the precentral cortex in monkeys (Graziano et al. 2002).

The predominant motor function of the SC is to shift gaze toward a stimulus located in the contralateral hemifield. Neural commands for such orienting or approach movements are relayed through the crossed tecto-(reticulo-)spinal tract, also referred to as the predorsal bundle. However, the SC also participates in the generation of ipsiversive eye-head-body movements that resemble aversive actions generated to escape from predators or avoid harmful situations. Commands for movements away from a stimulus are processed by the ipsilateral tecto-(reticulo-)spinal projection that originates laterally and ventrally in the SC (Sparks 1986). Indeed, stimulation of ventral SC sites in the rat protracts the ipsilateral whisker pad (Hemelt & Keller 2008). Small punctate lesions laterally in the frog OT impairs kinematics of escape behavior but keeps prey capture behav-

ior intact (King & Comer 1996). Stimulation of the caudal extent of the rodent SC produces ipsiversive head and body movements, including circling (Sahibzada et al. 1986). Stimulation of posterior sites in the goldfish can induce ipsiversive tail movements that reflect an escape-like swimming response (Herrero et al. 1998). Furthermore, properties of the escape response (number of tail beats, velocity, etc.) depend on stimulation parameters, suggesting that the SC activity both triggers and modulates the kinematics of the escape or avoidance response.

Reach Movements

Interacting with the environment can also require coordinating gaze shifts with arm movements, for example, when extending the arm to catch a ball. Neurons in the deeper layers of the SC and the underlying reticular formation are active prior to such reach movements (Werner 1993). Many show a single or biphasic burst of activity that correlates with simultaneously recorded EMG activity of shoulder, arm, and trunk muscles in nonhuman primates. The strongest correlations are obtained between activity in reticular formation neurons and EMG of proximal limb muscles (Stuphorn et al. 1999, Werner et al. 1997a). The posterior deltoid shoulder muscle in humans can also exhibit a brief EMG response linked to visual target onset (only when a reflexive manual response is required), and it has been speculated that the transient burst could be relayed through the SC (Pruszynski et al. 2010). The majority of SC "reach" neurons are intermingled with but distinct from visuomotor and motor neurons that burst for saccadic eye movements (Stuphorn et al. 2000, Werner et al. 1997b). A subset of reach neurons is modulated by the axis of visual fixation. Reach-related activity in another group of neurons was independent of gaze, and these neurons were found in the deepest portion of the SC and the underlying reticular formation. As observed for head movements (see above), reach-related neurons do not exhibit a topographical organization normally associated with saccade-related responses in the SC.

Results from microstimulation studies have been interpreted in support of a functional role of the SC in arm-movement control. Qualitatively assessed observation of proximal shoulder movement was reported after stimulation of deep SC layers (Cowie & Robinson 1994), but current spread to the underlying reticular formation could have evoked the movement. In addition, the movement could have been a generalized twitch or shrug of the contralateral shoulder, and not necessarily a reach action. Stimulation of the SC during forelimb movements by cats does, however, perturb the ongoing trajectory (Courjon et al. 2004).

Collicular response also exhibits modulation during other aspects of visually guided movements. For example, when reaching for an object, the visual axis often shifts to the target position and maintains a prolonged fixation until the hand makes contact (Neggers & Bekkering 2000, 2001). The extended fixation is reflected in enhanced activity in neurons in the rostral pole of the SC, potentially revealing a neural substrate of "gaze anchoring" during coordinated eye-hand movements (Reyes-Puerta et al. 2010). Another group of neurons in the SC responds specifically when the hand comes in contact with an object (Nagy et al. 2006), and the magnitude of activity increases with the intensity of the push. As these neurons are located in the intermediate and deep layers, it is likely that the activity reflects a premotor signal, although a sustained somatosensory component cannot be ruled out. On average, the neural responses are comparable for both ipsilateral and contralateral arms. As with reach movements, no topographical organization was detected.

Sonar Vocalization

Much like primates inspect the environment by shifting gaze between objects of interest, echolocating bats navigate in darkness by emitting sonar vocalization and processing returning echoes. Thus, sonar vocalization along with head and pinnae movements constitute the major effectors bats use to detect targets, and studies indicate that the SC participates in the movement of all three (Schuller & Radtke-Schuller 1990, Sinha & Moss 2007, Valentine et al. 2002). Microstimulation of the SC evokes sonar vocalizations whose time-frequency characteristics resemble the calls the same bats produced for echolocation for tracking targets (Valentine et al. 2002). The threshold for evoking a vocalization is typically less than 10 μA. The response latency is normally greater than 100 ms, and pinnae and/or head movements typically precede the onset of vocalization. The number of sonar pulses elicited increases with stimulation parameters, particularly current intensity and duration.

Extracellular activity recorded during sonar calls generated by bats during target tracking reveals two bouts of increased activity (Sinha & Moss 2007). The short-lead event is tightly coupled to and leads vocal onset by less than 5 ms, and the long-lead event is more variable and precedes the call by 20–30 ms; the early activity is short lived, as it returns to baseline between the two epochs. Sinha & Moss (2007) propose that the short-lead event triggers or times the vocalization, which in turn gates the neural response to the aural output but preserves sensitivity to its echo. The long-lead event, in contrast, could represent premotor activity, as its interval is correlated with call duration during target tracking.

DIFFERENTIAL CONTROL OF EXTRAOCULAR AND NONEXTRAOCULAR MUSCLES

Assessments based on neural activity offer correlative support for a functional role of the SC in the control of extraocular and skeletomotor actions. Microstimulation-based results suggest causality but indicate that the SC output alone is sufficient. In contrast, inactivation experiments appear best suited to address whether SC signals are necessary to produce gaze shifts coordinated across multiple effectors. Injections of only hundreds of nanoliters of either lidocaine or muscimol produce profound effects on head-restrained saccades encoded by the inactivated region (Aizawa & Wurtz 1998; Hikosaka

& Wurtz 1983, 1985, 1986; Lee et al. 1988; Quaia et al. 1998). Such saccades generally display longer reaction times, attenuated velocity profiles, and prolonged durations. The direct connectivity between the SC and the saccade burst generator in the brainstem (reviewed by Moschovakis et al. 1996) can readily account for the observed deficits.

Gaze shifts generated with the head free to move are also expected to be compromised given the contributions of the saccadic eye component. The more interesting issue then is whether the head component of the combined movement is also attenuated after SC inactivation. Intuitively, a pronounced effect is expected because a weaker SC command is attempting to move a significant inertial load. Reversible inactivation with lidocaine does compromise the latency, velocity, and duration of head-unrestrained gaze shifts, but the effect is manifested nearly entirely by the saccadic ocular component (**Figure 4**) (Walton et al. 2008). The initiation of the head movement is delayed, but the change is modest compared with the increase in reaction time of the gaze shift. All other head-movement features remain unattenuated; somewhat counter-intuitively, the peak head velocity increases slightly.

A negligible or absent effect of small-scale inactivation of the mammalian SC seems to generalize to all skeletomotor actions. For example, neither whisker movements in rodents (A. Keller, personal communication) nor reach movements in primates (K.P. Hoffmann, personal communication) appears compromised after localized SC inactivation. The active avoidance response typically triggered by a fearful conditioned response also remains intact after a lesion in the rodent SC contralateral to the conditioned stimulus (Cohen & Castro-Alamancos 2007). Both latency and kinematics remain unaltered during active head-only movements after inactivation of the monkey SC (Walton et al. 2008).

One possible explanation for the differential effect on extraocular and skeletomotor effectors is grounded in the distinct encoding mechanisms for gaze and other muscles. Motor

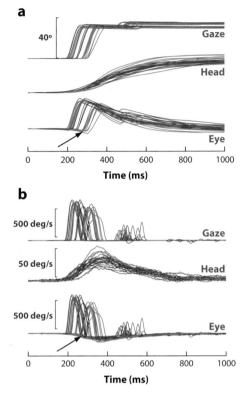

Figure 4

Effects of reversible inactivation with lidocaine on head-unrestrained gaze shifts. (*a*) Position and (*b*) velocity waveforms of several ~40° gaze shifts are plotted before (*blue*) and after (*red*) a microinjection in the caudal superior colliculus. The onsets of both gaze and head components are delayed, but the head movement initiates sooner. Thus, the eyes counter-rotate in the orbits before gaze onset (*arrows*). Peak gaze velocity is also reduced after the inactivation. In contrast, the peak head velocity sometimes increases a modest amount. Adapted from Walton et al. (2008), with permission.

commands for redirecting the line of sight appear to be mediated by a place code, whereas putative signals for skeletomotor actions have a weak, if any, topographical organization. Thus, it is possible that a chemical injection of hundreds of nanoliters of lidocaine or muscimol is not sufficient to dampen the SC drive to nonextraocular muscles. Lidocaine injections roughly an order of magnitude larger in volume also fail to produce deficits on head movements in monkeys (Walton et al. 2008), although increasingly larger inactivations/lesions have yielded deficits in head movements in cats (Isa & Sasaki 2002, Lomber et al. 2001). A drawback of very large inactivations, however, is that the deficits could be evoked by other factors. Consider the example in which head movements and gaze shifts are severely compromised after kainic acid lesions in the SC and the underlying reticular formation in cat (Isa & Sasaki 2002). It is difficult to discount the possibility that the observed effect is largely due to inactivation of the underlying mesencephalic reticular formation, which is known to produce severe deficits in head movements, even with small volume injections (Klier et al. 2002, Waitzman et al. 2000). Also, unilateral cryogenic inactivation of the superficial and deeper layers produces profound deficits on the generation of gaze shifts, including head movements (Lomber et al. 2001). These deficits, however, are absent after bilateral cooling (Lomber & Payne 1996), suggesting that the lack of overt orienting behavior may occur from deficits in nonmotoric function such as spatial hemineglect.

Another interpretation of the inactivation results is that skeletomotor function is also controlled by extracollicular pathway(s) that can fully compensate for an absent or weakened SC signal. One parallel input likely originates in the cortex, which has access to the spinal cord through cortico-spinal and cortico-reticulo-spinal pathways. For head movements, the supplementary eye fields (Chen & Walton 2005, Martinez-Trujillo et al. 2003) and premotor cortical areas may be the main sources of cortical input. Stimulation of the frontal eye fields also induces neck muscle activity (Elsley et al. 2007, Guitton & Mandl 1978) and coordinated eye-head movements (Chen 2006, Knight & Fuchs 2007, Tu & Keating 2000), but a significant component of its output is thought to connect through the SC (Hanes & Wurtz 2001, Komatsu & Suzuki 1985, Stanton et al. 1988). The putative cortical and SC pathways likely merge in the mesencephalic and pontine reticular formations. The convergence, however, is most likely a nonlinear combination of two (or more) streams; if they did combine linearly, inactivation of one pathway is predicted to produce a partial deficit in head movements. The parallel pathways probably express redundant motor commands, and the efficacy of each individual channel can be gated or modulated by cognitive, mechanical (length-tension property), and proprioceptive contributions. Thus, in the case where one pathway is weakened, the intact drive could enhance its throughput to control the compromised effectors. Perhaps simultaneous lesions of more than one pathway may unmask an impaired output. Recall the classic finding that saccades can be executed after a lesion of either the SC or the frontal eye fields, but not after inactivation of both structures (Schiller et al. 1979). A comparable observation also exists for sensory processes that mediate the active avoidance response in rodents (Cohen & Castro-Alamancos 2007): A lesion of either the SC or the somatosensory thalamus has no effect on the active-avoidance response triggered by a conditioned stimulus applied contralaterally, but simultaneous inactivation of both structures abolishes the behavior. If the cortical component is indeed an important parallel pathway, then deficits in skeletomotor actions should be prevalent after tectal lesions in species that lack a neocortex. Indeed, lesions of the OT in the barn owl and frog do appear to compromise a large range of contraversive head (Knudsen et al. 1993, Wagner 1993) and body movements (King & Comer 1996, Kostyk & Grobstein 1987), respectively.

FIXATION CONTROL BY ROSTRAL SUPERIOR COLLICULUS

Robinson's (1972) demonstration of a topographical organization of saccades evoked by microstimulation of the primate SC established the long-standing perspective that the deeper layers of the SC consist of a uniform saccade zone. Conforming to this idea, the rostral pole of the SC is predicted to encode very-small-amplitude eye movements. Nevertheless, the saccade-zone hypothesis was questioned with the discovery of so-called fixation neurons in the rostral pole of the SC. These cells discharge at a tonic rate during fixation and pause during most saccades (Munoz & Guitton 1989, 1991; Munoz & Wurtz 1993a; Peck 1989). Inactivation of the rostral SC reduces the latency of large-amplitude saccades and disrupts the animal's ability to maintain visual fixation during delayed or memory-guided tasks (Munoz & Wurtz 1993b). Conversely, stimulation of this region increases the latency of large-amplitude saccades (Munoz & Wurtz 1993b). Such data led to the view that the rostral pole of the SC serves as a "fixation zone" that stabilizes gaze. The theory implements the idea of a "see-saw"-like lateral interaction network in the SC, in which the rostral portion functions as an inhibitory system that facilitates visual fixation by suppressing activity in the caudal SC and thereby preventing saccade generation (Meredith & Ramoa 1998, Munoz & Istvan 1998). Fixation itself was proposed to be mediated through direct projections from the rostral SC to the OPNs (Büttner-Ennever et al. 1999, Gandhi & Keller 1997, Paré & Guitton 1994). However, several studies have disputed the proposed fixation hypothesis. First, physiological data revealed differences between the discharge properties of OPNs and fixation-related neurons in the rostral SC. For example, the end of pause in the rostral SC neurons lags the resumption of activity in the OPNs as well as saccade offset (Everling et al. 1998). Second, large-amplitude saccades perturbed by stimulation in the rostral SC can be better interpreted as colliding saccades instead of interrupted saccades associated with stimulation of the OPNs (Gandhi & Keller 1999). Finally, neural recordings have revealed that fixation-related neurons within the rostral SC actually discharge a burst for small contraversive saccades (Gandhi & Keller 1997, Krauzlis et al. 1997, Munoz & Wurtz 1993a), and this activity pattern is comparable to the bursts generated for larger saccades by neurons in the caudal SC. An enhanced neural response can be observed for even the involuntary microsaccades (typically <12 min arc) that occur while fixating

(Hafed et al. 2009, Rolfs 2009, Steinman et al. 1973). Furthermore, the burst exhibits selectivity for a specific amplitude and direction and is continuous with the topography of the saccade zone. Thus, SC function has reverted back to the classic view in which it comprises a continuous representation of all saccade amplitudes and directions, with the rostral SC representing some of the smallest eye movements.

It then becomes important to ask, (how) can a continuous saccade vector map that encodes for microsaccades and small-amplitude movements in the rostral SC preserve fixation? One potential mechanism utilizes the observation that many rostral SC neurons discharge a burst for a range of saccades in both ipsiversive and contraversive directions (Hafed et al. 2009, Munoz & Wurtz 1993a), functionally linking the two visual fields to represent central locations. Hafed et al. (2009) proposed a model in which fixation is maintained by balancing the premotor activity across the rostral regions of the two colliculi. The locus of activity in their model is extracted from a stochastic process with a mean of zero amplitude. Microsaccades are triggered when the balance of activity sufficiently shifts the locus of activity from zero. Their model also simulates the effects of inactivation of the rostral SC. When a subset of model neurons is silenced, the spatial distribution of active neurons on the intact side is reduced to maintain the balance of motor activity and thus preserve fixation (**Figure 5**). In addition, the reduction of activity also results in less overall variability, causing a lower probability for stochastic fluctuations in the locus of activity and therefore producing fewer microsaccades.

There is a concrete, conceptual difference between a mechanism based on the balance of ipsiversive and contraversive movement commands (Hafed et al. 2009) and a scheme based on gaze stabilization (Munoz & Guitton 1989; Munoz & Wurtz 1993a, 1993b). Nevertheless, a commonality can exist between the two as demonstrated by the observation that the timing of microsaccades relative to stimulus onset also influences saccade reaction times (Hafed & Krauzlis 2010, Rolfs et al. 2006).

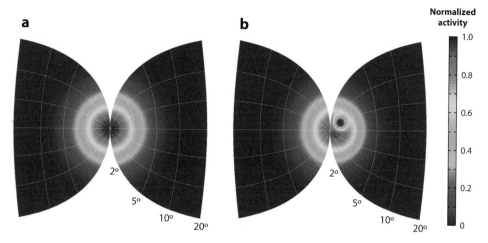

a

b

Normalized activity

1.0

0.8

0.6

0.4

0.2

0

2°

5°

10°

20°

2°

5°

10°

20°

Figure 5

Schematic of distribution of activity in deep superior colliculus (SC) layers during fixation. (*a*) Population activity across the two colliculi is balanced during fixation of a visual target. (*b*) An inactivation of a small region in the rostral SC (*blue circle inside right SC*) induces a compensatory shift in activity in the intact side (*left SC*). Ensemble activity on the lesion side is also redistributed due to inter- and intracollicular interactions. The net activity across the two SC, however, remains balanced to preserve continued fixation on or near the foveal target. Illustration generated based on results by Hafed et al. (2009).

Peripheral visual stimuli presented during fixation but in register with microsaccades were more effective in attenuating the visual bursts in the caudal SC than when shown in absence of microsaccades (Hafed & Krauzlis 2010). Correspondingly, the reaction times of saccades directed to the visual target were greater if the stimulus was presented around the time of a microsaccade. These observations suggest that as the balance of activity shifts more toward one colliculus to trigger a microsaccade, the resulting greater activity in the rostral SC acts through the lateral interaction network to suppress visual and premotor build up in the caudal SC. Therefore, the basic mechanisms of a gaze-stabilization theory could exist within principles based on balancing motor activity across the SC.

MECHANISMS FOR DECODING SUPERIOR COLLICULUS ACTIVITY

When immersed within a visual environment with many potential stimuli, the oculomotor system must first select an object for a saccade goal. The oculomotor system has served as a useful tool to probe mechanisms of target selection (Schall 1995, Schall & Thompson 1999), and the SC, among other structures, plays a major role in this process (Kim & Basso 2008, McPeek & Keller 2002, Shen & Paré 2007). The typical laboratory paradigm involves presentation of multiple visual stimuli, in which a specific feature, such as color, distinguishes the singleton (saccade target) from distractors. Each visual stimulus excites an ensemble of SC neurons in the deeper layers, and the result of the competing populations governs the observed eye movement. Given that the saccade usually ends on one of the visual stimuli, a computational mechanism that computes the vector average across all the active populations is a poor predictor of the observed data. Accurate performance is better associated with the degree of separation between target and distractor activity distributions (Kim & Basso 2008). A winner-takes-all scheme, in which the saccade is driven to the stimulus represented with the highest ensemble activity better utilizes the separation between activity distributions and therefore performs better than an averaging mechanism. A Bayesian model, maximum a posteriori estimate (Kim & Basso 2010), performs slightly better than either an averaging or a winner-takes-all mechanism, suggesting that a read-out mechanism for target and/or motor selection in the SC may be based on a probability framework.

Although these competitive mechanisms can contribute to selecting the neural population that will ultimately guide the eye movement, they do not indicate how the spatially coded contribution of each neuron within the selected (or winner) population of SC neurons is integrated and decoded to produce an eye movement. Two controversial models have dominated the oculomotor field in hypothesizing the proper mechanism for deciphering SC activity: vector averaging and vector summation. Early saccade models utilized static-ensemble-coding schemes, in which SC motor activity specifies only the metric coordinates of the saccade displacement. Dynamical properties such as trajectory and kinematics are assumed to be reflected by the operation of a feedback mechanism downstream of the SC, such as in the pons (Jürgens et al. 1981, Robinson 1975) or cerebellum (Lefèvre et al. 1998, Quaia et al. 1999). The vector-averaging model (**Figure 6a**) hypothesized that an active population in the SC is computed by taking the weighted average of the vector contribution of each neuron (e.g., Lee et al. 1988, Walton et al. 2005). The saccade goal $\vec{S}$ is computed by

$$\vec{S} = \frac{\sum_{n=1}^{N} r_n \cdot \vec{R}_n}{\sum_{n=1}^{N} r_n},$$

where r_n is the mean firing rate of cell n in the motor map and $\vec{R}_n$ is the optimal vector encoded by that neuron. In this format, the level of activity has no direct relation to either saccade trajectory or its kinematics. Early success of the model came from its ability to account accurately for the findings generated from several experiments. Some examples include the following: (*a*) Simultaneous microstimulation at two points within the SC evokes a single saccade

whose amplitude and direction are predicted by the weighted average of the two saccades generated when each site is stimulated independently (Katnani et al. 2009, Robinson 1972). (*b*) Local inactivation within the SC generates saccades with dysmetria patterns that conform to an averaging hypothesis (Lee et al. 1988). (*c*) The timing and initial direction of curved saccades, generated by using a double-step paradigm, are accurately predicted by the computation of an averaging scheme (Port & Wurtz 2003).

Experimental findings that dispute the vector-averaging model also exist. For example, stimulation-evoked saccades can have a sigmoidal dependency with current intensity (Van Opstal et al. 1990). Stimulation frequency may also have a similar effect (see table 3 in Stanford et al. 1996, although the authors did not comment on this observation; also see Groh 2011). This relationship reveals a flaw in the averaging computation, because a strict interpretation of this mechanism indicates that a single spike in the colliculus can generate a maximal amplitude vector. The model, however, can be appended with the addition of a parameter to demonstrate amplitude dependency (Van Opstal & Goossens 2008):

$$\vec{S} = \frac{\sum_{n=1}^{N} r_n \cdot \vec{m}_n}{K + \sum_{n=1}^{N} r_n},$$

where K is a constant that presumably could be an inhibitory threshold. When the total population activity in the SC is low, K can dominate the denominator term and reduce the amplitude of the programmed saccade. If population activity is high, K can be neglected and the computation returns to an original averaging scheme. Another limitation of the averaging mechanism persists in how the computation can be implemented physiologically. Although network architectures that can accomplish normalization have been proposed (Carandini & Heeger 1994, Groh 2001), there is still no substantial anatomical evidence in the oculomotor system to support this structure.

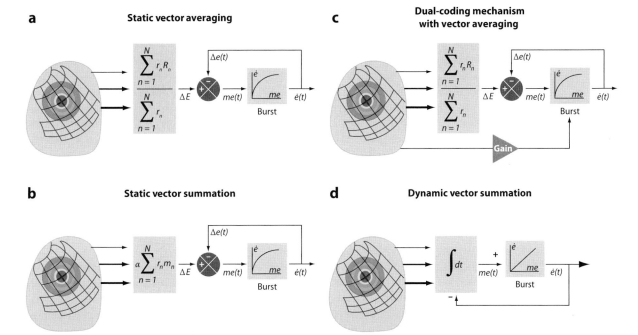

a **Static vector averaging**

$$\dfrac{\displaystyle\sum_{n=1}^{N} r_n R_n}{\displaystyle\sum_{n=1}^{N} r_n}$$

ΔE $\Delta e(t)$ $me(t)$ $\dot e$ me $\dot e(t)$

Burst

c **Dual-coding mechanism with vector averaging**

$$\dfrac{\displaystyle\sum_{n=1}^{N} r_n R_n}{\displaystyle\sum_{n=1}^{N} r_n}$$

ΔE $\Delta e(t)$ $me(t)$ $\dot e$ me $\dot e(t)$

Burst

Gain

b **Static vector summation**

$$\alpha \sum_{n=1}^{N} r_n m_n$$

ΔE $\Delta e(t)$ $me(t)$ $\dot e$ me $\dot e(t)$

Burst

d **Dynamic vector summation**

$$\int dt$$

$me(t)$ $\dot e$ me $\dot e(t)$

Burst

Figure 6

Frameworks of contemporary models for decoding superior colliculus (SC) activity to generate saccadic eye movements. (*a*) Static averaging decoding model that defines desired saccade metrics by using a vector averaging computation (r_n is the mean firing rate of cell n and $\vec R_n$ is the optimal vector encoded by that cell). (*b*) Static summation decoding model uses vector summation to define saccade metrics ($\vec m_n$ is the vector contribution of cell n and α is a fixed scaling constant). For both static averaging and summation models, the trajectory and kinematics are controlled downstream by nonlinear local feedback. (*c*) Dual-coding hypothesis model shares some of the framework of the static vector-averaging model. In addition, the firing rate of the SC across time can modulate the gain of the burst generator. In this manner, SC output now contributes to both metrics and kinematics. (*d*) Dynamic summation model integrates across time the spikes from an active population. The accumulating activity specifies the intended movement trajectory. Each spike from an SC cell adds a fixed, site-specific "mini" vector contribution to the movement command. In contrast to the other frameworks, the movement is controlled downstream by linear feedback. The projections from the SC are weighted (*thickness of lines* and *size of arrows*) according to its origin site along the rostral-caudal dimension. Model parameters: ΔE, desired eye displacement; $\Delta e(t)$, current eye displacement; $me(t)$, dynamic motor error; $\dot e(t)$, current eye velocity; $\int dt$, temporal integration; "burst," brainstem burst generator. Adapted from Goossens & Van Opstal (2006).

Vector summation is recognized as a more physiological mechanism for decoding a motor command (Georgopoulos et al. 1986). For deciphering SC output for the generation of saccades, it hypothesizes that each active SC neuron contributes a vector that is weighted by the mean firing rate of the cell (**Figure 6b**). The resulting sum of these weighted vectors produces the desired saccade $\vec S$ (Van Gisbergen et al. 1987) as

$$\vec S = \alpha \sum_{n=1}^{N} r_n \vec m_n,$$

where $\vec m_n$ is the vector contribution elicited by cell n and α is a fixed scaling constant. As with the averaging scheme, this simple summation model also does not incorporate any means to explain saccade kinematics. The strength of the model is exhibited by its simplicity, intuitive nature, and ability to produce normometric saccades. Its shortcomings, however, become prevalent when tested with more complex motivations (i.e., simultaneous stimulation of two SC sites and inactivation), but they too can be accounted for by incorporating intracollicular connectivity features such as local excitation and distal inhibition (e.g., Behan & Kime 1996, Isa & Hall 2009, Lee et al. 1997, McIlwain 1982, Meredith & Ramoa 1998, Munoz & Istvan 1998, Pettit et al. 1999, Takahashi et al. 2010). Initial implementation of lateral interaction was shown through an inhibitory tuning parameter (Van Opstal & Van Gisbergen 1989). The addition of inhibition provided a cutoff during the summation of two vectors to simulate weighted averaging saccades seen with simultaneous stimulation. Moreover, a later model that incorporated both visual and motor layers of the SC (Arai et al. 1994) demonstrated that vector summation can generate normometric, averaging, and express saccades. Yet another version (Badler & Keller 2002) additionally emphasized that lateral interactions shift the locus of ensemble activity when a subset of model neurons is "inactivated," and the resulting endpoints of simulated saccades match both experimental data and predictions based on an averaging mechanism.

The discussion of both models to this point has focused on computation of only the desired saccade movement (metric). However, accumulating evidence suggests that the level of activity within the SC does influence the saccade kinematics (see the dual-coding hypothesis discussed in the Eye Movements section above), indicating that the changes in collicular activity across time now become significant to saccade programming. A vector-averaging theory of dynamic ensemble coding has been addressed mainly conceptually (**Figure 6c**). In essence, the firing rate of SC activity modulates the gain of the brainstem burst generator (Nichols & Sparks 1996, Sparks & Mays 1990). Hence, an attenuated burst, such as after partial inactivation of the SC or during memory-guided saccades, evokes a slower amplitude-matched saccade. With this implementation, not only the metrics, but also the kinematics of a movement could be explained by an averaging scheme. Furthermore, Van Opstal & Goossens (2008) incorporated the notion of gain modulation into a vector averaging computation that utilized instantaneous firing rate across time. As a result, the vector-averaging model provided

dynamic estimates of the saccade goal; however, simulations revealed that the computation did not capture saccade kinematics effectively and was relatively insensitive to temporal changes in the SC burst profile.

A detailed and quantified vector summation computation has been developed under the theoretical framework of dynamic ensemble coding (**Figure 6d**) (Goossens & Van Opstal 2006, Van Opstal & Goossens 2008). The model proposes that a saccade is computed by the vector summation of all individual cell contributions across time:

$$\vec{S}(t) = \sum_{k=1}^{N_A} \sum_{n=1}^{N_S} \vec{s}_k \cdot \delta(t - \tau_{k,n}),$$

where N_S indicates the number of spikes of cell k (counted from 20 ms before saccade onset to 20 ms before saccade offset), N_A equals the total number of cells in the population, $\delta(x)$ denotes the Dirac impulse function (i.e., an individual spike), $\tau_{k,n}$ represents the time of the nth spike of cell k, and $\vec{s}_k$ is a scaled-eye-displacement vector generated by a single spike of cell k (scaling is determined by the model's cell density). Thus, SC neurons now relate the cumulative number of spikes in the active population to the ongoing eye displacement. With such a scheme, the SC output now specifies the desired saccade trajectory, including its kinematics. Simulations of the model revealed several saccade-related properties that other models cannot incorporate without additional assumptions. First, the decoding computation accounts for stretching of horizontal and vertical saccade components necessary for oblique saccades. Second, SC activity encodes the nonlinear main sequence, in contrast to the long-believed idea that the kinematic nonlinearity originates from a local feedback circuit in the brainstem. Thus, the SC acts as a nonlinear vectorial pulse generator where the spatial temporal activity patterns in the motor SC encode desired saccade kinematics, without having to use nonlinear mechanisms such as normalization of activity. Although the model's mechanism reveals advantageous properties, a shortcoming arises in that the computation will always yield a vector sum when tested with the contribution of two sites. To account for averaging saccades, the model must incorporate an additional saturation criterion that could potentially be introduced through modeled lateral interactions.

The evolution of describing ensemble decoding in the SC has produced two conceptually distinct frameworks that continue to be contrasting equivalents in the field. Predictions of the vector-averaging and vector-summation models have provoked many experiments attempting to validate hypotheses and reveal emerging properties, yet a coherent direction has still not been achieved. Different methods are needed to better differentiate between the two models. One such approach may arise from microstimulation experiments that exploit the dependencies of saccade characteristics on stimulation parameters. By systematically lowering the range of stimulation parameters, one may be able to introduce more variability in the resulting saccade output, and characterization of this variability to the known inputs could reveal insightful properties on the spatiotemporal decoding of the SC output (Brecht et al. 2004, Gandhi & Katnani 2009, Katnani & Gandhi 2010).

NONMOTORIC FUNCTIONS OF THE SUPERIOR COLLICULUS

The intent of this review is to focus primarily on the motoric functions of the SC. This perspective does not diminish a collicular role in mediating other processes, but their inclusion is beyond the scope of this review. Some examples of systems-level phenomena in which the SC have been implicated include multimodal sensory processing (Stein & Meredith 1993, Stein & Stanford 2008), target selection (Krauzlis et al. 2004, McPeek & Keller 2002), attention (Ignashchenkova et al. 2004, Kustov & Robinson 1996), motor preparation (Dorris et al. 1997, 2007), goal representation (Bergeron et al. 2003, Krauzlis & Carello 2003), and reward-related modulation (Ikeda & Hikosaka 2003). In fact, the potentially simultaneous contributions of multiple higher-level functions often confound interpretations of SC activity during complicated behavioral tasks (Sparks 1999). Future studies will need to address systematically if the SC controls a unifying signal (or command) across sensory, motor, and cognitive processes and whether the specific label attributed to its neural activity is task and training dependent.

SUMMARY POINTS

1. Neurons in the deeper layers of the SC are topographically organized for changes in gaze, but spatial organization for other effectors has not been detected.

2. Whereas the locus of ensemble activity dictates the size of gaze shift, the extent and complexity of skeletal segments appear best linked with stimulation duration.

3. Inactivation of the SC produces profound effects on gaze via the saccadic eye component. In contrast, skeletomotor effectors, particularly in mammals, are minimally impaired. These results suggest that the SC is crucial for the control of saccades and that parallel and perhaps redundant pathways can guide skeletal actions.

4. Neurons in the rostral pole of the SC actually burst for microsaccades and small-amplitude movements. They preserve fixation by balancing premotor activity that encodes for small leftward and rightward saccades across the two rostral SC. This theory is conceptually different from the previously proposed notion that fixation is achieved by signals that stabilize gaze.

5. Two prevalent computational models exist to explain the mechanisms for decoding SC motor activity: vector averaging and vector summation. Early models of both computations specified only saccade metrics, relying on downstream mechanisms to control kinematics. However, with experimental evidence suggesting that the SC motor activity reflects dynamic processes, both models have transitioned to account for kinematics. Averaging schemes incorporate a dual-coding hypothesis, whereas a recent summation scheme proposes that SC output encodes desired trajectory and velocity. The evolution of each computation has demonstrated strengths and weaknesses, leaving both models, to this point, as contrasting equivalents.

DISCLOSURE STATEMENT

The authors are not aware of any affiliations, memberships, funding, financial holdings, or any other conflicts of interests that might be perceived as affecting the objectivity of this review.

ACKNOWLEDGMENTS

We thank Udaya Jagadisan for comments on the manuscript. N.J.G. is supported by the National Eye Institute (R01 EY015485, P30 EY008098) and National Institute of Deafness and Communication Disorders (P30 DC0025205). H.A.K. is supported by an institutional training grant from the National Institute of General Medical Science (T32 GM081760).

LITERATURE CITED

Aizawa H, Wurtz RH. 1998. Reversible inactivation of monkey superior colliculus. I. Curvature of saccadic trajectory. *J. Neurophysiol.* 79:2082–96

Anastasopoulos D, Ziavra N, Hollands M, Bronstein A. 2009. Gaze displacement and inter-segmental coordination during large whole body voluntary rotations. *Exp. Brain Res.* 193:323–36

Anderson RW, Keller EL, Gandhi NJ, Das S. 1998. Two-dimensional saccade-related population activity in superior colliculus in monkey. *J. Neurophysiol.* 80:798–817

Arai K, Keller EL, Edelman JA. 1994. Two-dimensional neural network model of the primate saccadic system. *Neural Netw.* 7:1115–35

Badler JB, Keller EL. 2002. Decoding of a motor command vector from distributed activity in superior colliculus. *Biol. Cybern.* 86:179–89

Basso MA, Krauzlis RJ, Wurtz RH. 2000. Activation and inactivation of rostral superior colliculus neurons during smooth-pursuit eye movements in monkeys. *J. Neurophysiol.* 84:892–908

Behan M, Kime NM. 1996. Intrinsic circuitry in the deep layers of the cat superior colliculus. *Vis. Neurosci.* 13:1031–42

Bergeron A, Matsuo S, Guitton D. 2003. Superior colliculus encodes distance to target, not saccade amplitude, in multi-step gaze shifts. *Nat. Neurosci.* 6:404–13

Brecht M, Singer W, Engel AK. 2004. Amplitude and direction of saccadic eye movements depend on the synchronicity of collicular population activity. *J. Neurophysiol.* 92:424–32

Breznen B, Lu SM, Gnadt JW. 1996. Analysis of the step response of the saccadic feedback: system behavior. *Exp. Brain Res.* 111:337–44

Büttner-Ennever JA, Horn AK, Henn V, Cohen B. 1999. Projections from the superior colliculus motor map to omnipause neurons in monkey. *J. Comp. Neurol.* 413:55–67

Carandini M, Heeger DJ. 1994. Summation and division by neurons in primate visual cortex. *Science* 264:1333–36

Chaturvedi V, van Gisbergen JA. 1999. Perturbation of combined saccade-vergence movements by microstimulation in monkey superior colliculus. *J. Neurophysiol.* 81:2279–96

Chaturvedi V, Van Gisbergen JA. 2000. Stimulation in the rostral pole of monkey superior colliculus: effects on vergence eye movements. *Exp. Brain Res.* 132:72–78

Chen LL. 2006. Head movements evoked by electrical stimulation in the frontal eye field of the monkey: evidence for independent eye and head control. *J. Neurophysiol.* 95:3528–42

Chen LL, Tehovnik EJ. 2007. Cortical control of eye and head movements: integration of movements and percepts. *Eur. J. Neurosci.* 25:1253–64

Chen LL, Walton MM. 2005. Head movement evoked by electrical stimulation in the supplementary eye field of the rhesus monkey. *J. Neurophysiol.* 94:4502–19

Cohen JD, Castro-Alamancos MA. 2007. Early sensory pathways for detection of fearful conditioned stimuli: tectal and thalamic relays. *J. Neurosci.* 27:7762–76

Corneil BD, Munoz DP, Olivier E. 2007. Priming of head premotor circuits during oculomotor preparation. *J. Neurophysiol.* 97:701–14

Corneil BD, Olivier E, Munoz DP. 2002a. Neck muscle activity evoked by stimulation of the monkey superior colliculus. I. Topography and manipulation of stimulation parameters. *J. Neurophysiol.* 88:1980–99

Corneil BD, Olivier E, Munoz DP. 2002b. Neck muscle activity evoked by stimulation of the monkey superior colliculus. II. Relationships with gaze shift initiation and comparison to volitional head movements. *J. Neurophysiol.* 88:2000–18

Corneil BD, Olivier E, Munoz DP. 2004. Visual responses on neck muscles reveal selective gating that prevents express saccades. *Neuron* 42:831–41

Courjon JH, Olivier E, Pelisson D. 2004. Direct evidence for the contribution of the superior colliculus in the control of visually guided reaching movements in the cat. *J. Physiol.* 556:675–81

Cowie RJ, Robinson DL. 1994. Subcortical contributions to head movements in macaques. I. Contrasting effects of electrical stimulation of a medial pontomedullary region and the superior colliculus. *J. Neurophysiol.* 72:2648–64

Cramer NP, Keller A. 2006. Cortical control of a whisking central pattern generator. *J. Neurophysiol.* 96:209–17

Dacey DM, Ulinski PS. 1986. Optic tectum of the eastern garter snake, Thamnophis sirtalis. I. Efferent pathways. *J. Comp. Neurol.* 245:1–28

Dean P, Redgrave P, Sahibzada N, Tsuji K. 1986. Head and body movements produced by electrical stimulation of superior colliculus in rats: effects of interruption of crossed tectoreticulospinal pathway. *Neuroscience* 19:367–80

Degani AM, Danna-Dos-Santos A, Robert T, Latash ML. 2010. Kinematic synergies during saccades involving whole-body rotation: a study based on the uncontrolled manifold hypothesis. *Hum. Mov. Sci.* 29:243–58

Dorris MC, Olivier E, Munoz DP. 2007. Competitive integration of visual and preparatory signals in the superior colliculus during saccadic programming. *J. Neurosci.* 27:5053–62

Dorris MC, Paré M, Munoz DP. 1997. Neuronal activity in monkey superior colliculus related to the initiation of saccadic eye movements. *J. Neurosci.* 17:8566–79

du Lac S, Knudsen EI. 1990. Neural maps of head movement vector and speed in the optic tectum of the barn owl. *J. Neurophysiol.* 63:131–46

Edelman JA, Goldberg ME. 2003. Saccade-related activity in the primate superior colliculus depends on the presence of local landmarks at the saccade endpoint. *J. Neurophysiol.* 90:1728–36

Elsley JK, Nagy B, Cushing SL, Corneil BD. 2007. Widespread presaccadic recruitment of neck muscles by stimulation of the primate frontal eye fields. *J. Neurophysiol.* 98:1333–54

Everling S, Paré M, Dorris MC, Munoz DP. 1998. Comparison of the discharge characteristics of brain stem omnipause neurons and superior colliculus fixation neurons in monkey: implications for control of fixation and saccade behavior. *J. Neurophysiol.* 79:511–28

Ewert JP. 1984. Tectal mechanisms that underlie prey-catching and avoidance behavior in toads. In *Comparative Neurology of the Optic Tectum*, ed. H Vañegas, pp. 247–416. New York: Plenum

Freedman EG. 2008. Coordination of the eyes and head during visual orienting. *Exp. Brain Res.* 190:369–87

Freedman EG, Sparks DL. 1997. Activity of cells in the deeper layers of the superior colliculus of the rhesus monkey: evidence for a gaze displacement command. *J. Neurophysiol.* 78:1669–90

Freedman EG, Stanford TR, Sparks DL. 1996. Combined eye-head gaze shifts produced by electrical stimulation of the superior colliculus in rhesus monkeys. *J. Neurophysiol.* 76:927–52

Gandhi NJ, Katnani HA. 2009. Single and dual microstimulation in the superior colliculus: Effects of stimulation intensity and frequency. *Soc. Neurosci. Abstr. Program No. 851.7*

Gandhi NJ, Keller EL. 1997. Spatial distribution and discharge characteristics of superior colliculus neurons antidromically activated from the omnipause region in monkey. *J. Neurophysiol.* 78:2221–25

Gandhi NJ, Keller EL. 1999. Comparison of saccades perturbed by stimulation of the rostral superior colliculus, the caudal superior colliculus, and the omnipause neuron region. *J. Neurophysiol.* 82:3236–53

Gandhi NJ, Sparks DL. 2004. Changing views of the role of the superior colliculus in the control of gaze. In *The Visual Neurosciences*, ed. LM Chalupa, JS Werner, pp. 1449–65. Boston: MIT Press

Gandhi NJ, Sparks DL. 2007. Dissociation of eye and head components of gaze shifts by stimulation of the omnipause neuron region. *J. Neurophysiol.* 98:360–73

Georgopoulos AP, Schwartz AB, Kettner RE. 1986. Neuronal population coding of movement direction. *Science* 233:1416–19

Gnadt JW, Bracewell RM, Andersen RA. 1991. Sensorimotor transformation during eye movements to remembered visual targets. *Vis. Res.* 31:693–715

Goossens HH, Van Opstal AJ. 2006. Dynamic ensemble coding of saccades in the monkey superior colliculus. *J. Neurophysiol.* 95:2326–41

Grantyn A, Kuze B, Brandi AM, Thomas MA, Quenech'du N. 2010. Direct projections of omnipause neurons to reticulospinal neurons: A double-labeling light microscopic study in the cat. *J. Comp. Neurol.* 518:4792–812

Graziano MSA, Taylor CSR, Moore T, Cooke DF. 2002. The cortical control of movement revisited. *Neuron* 36:349–62

Groh JM. 2001. Converting neural signals from place codes to rate codes. *Biol. Cybern.* 85:159–65

Groh JM. 2011. Effects of initial eye position on saccades evoked by microstimulation in the primate superior colliculus: implications for models of the SC read-out process. *Front. Integr. Neurosci.* 4:130:1–16

Guillaume A, Pélisson D. 2001. Gaze shifts evoked by electrical stimulation of the superior colliculus in the head-unrestrained cat. I. Effect of the locus and of the parameters of stimulation. *Eur. J. Neurosci.* 14:1331–44

Guillaume A, Pélisson D. 2006. Kinematics and eye-head coordination of gaze shifts evoked from different sites in the superior colliculus of the cat. *J. Physiol.* 577:779–94

Guitton D, Mandl G. 1978. Frontal 'oculomotor' area in alert cat. II. Unit discharges associated with eye movements and neck muscle activity. *Brain Res.* 149:313–27

Guitton D, Munoz DP, Galiana HL. 1990. Gaze control in the cat: studies and modeling of the coupling between orienting eye and head movements in different behavioral tasks. *J. Neurophysiol.* 64:509–31

Hafed ZM, Goffart L, Krauzlis RJ. 2009. A neural mechanism for microsaccade generation in the primate superior colliculus. *Science* 323:940–43

Hafed ZM, Krauzlis RJ. 2010. Microsaccadic suppression of visual bursts in the primate superior colliculus. *J. Neurosci.* 30:9542–47

Hall WC, Moschovakis A. 2004. *The Superior Colliculus: New Approaches for Studying Sensorimotor Integration.* Boca Raton, FL: CRC Press. 324 pp.

Hanes DP, Wurtz RH. 2001. Interaction of the frontal eye field and superior colliculus for saccade generation. *J. Neurophysiol.* 85:804–15

Harris LR. 1980. The superior colliculus and movements of the head and eyes in cats. *J. Physiol.* 300:367–91

Hemelt ME, Keller A. 2008. Superior colliculus control of vibrissa movements. *J. Neurophysiol.* 100:1245–54

Herrero L, Rodriguez F, Salas C, Torres B. 1998. Tail and eye movements evoked by electrical microstimulation of the optic tectum in goldfish. *Exp. Brain Res.* 120:291–305

Hess WR, Burgi S, Bucher V. 1946. Motorische funktion des tektalund tegmentalgebietes. *Psychiat. Neurol.* 112:1–52

Hikosaka O, Wurtz RH. 1983. Effects on eye movements of a GABA agonist and antagonist injected into monkey superior colliculus. *Brain Res.* 272:368–72

Hikosaka O, Wurtz RH. 1985. Modification of saccadic eye movements by GABA-related substances. I. Effect of muscimol and bicuculline in monkey superior colliculus. *J. Neurophysiol.* 53:266–91

Hikosaka O, Wurtz RH. 1986. Saccadic eye movements following injection of lidocaine into the superior colliculus. *Exp. Brain Res.* 61:531–39

Hollands MA, Ziavra NV, Bronstein AM. 2004. A new paradigm to investigate the roles of head and eye movements in the coordination of whole-body movements. *Exp. Brain Res.* 154:261–66

Ignashchenkova A, Dicke PW, Haarmeier T, Thier P. 2004. Neuron-specific contribution of the superior colliculus to overt and covert shifts of attention. *Nat. Neurosci.* 7:56–64

Ikeda T, Hikosaka O. 2003. Reward-dependent gain and bias of visual responses in primate superior colliculus. *Neuron* 39:693–700

Isa T, Hall WC. 2009. Exploring the superior colliculus in vitro. *J. Neurophysiol.* 102:2581–93

Isa T, Sasaki S. 2002. Brainstem control of head movements during orienting; organization of the premotor circuits. *Prog. Neurobiol.* 66:205–41

Jürgens R, Becker W, Kornhuber HH. 1981. Natural and drug-induced variations of velocity and duration of human saccadic eye movements: evidence for a control of the neural pulse generator by local feedback. *Biol. Cybern.* 39:87–96

Katnani HA, Gandhi NJ. 2010. Analysis of current and frequency stimulation permutations in the superior colliculus. *Soc. Neurosci. Abstr. Program No. 77.9*

Katnani HA, Van Opstal AJ, Gandhi NJ. 2009. Evaluating models of decoding superior colliculus activity for saccade generation. *Soc. Neurosci. Abstr. Program No. 851.8*

Kim B, Basso MA. 2008. Saccade target selection in the superior colliculus: a signal detection theory approach. *J. Neurosci.* 28:2991–3007

Kim B, Basso MA. 2010. A probabilistic strategy for understanding action selection. *J. Neurosci.* 30:2340–55

King JR, Comer CM. 1996. Visually elicited turning behavior in Rana pipiens: Comparative organization and neural control of escape and prey capture. *J. Comp. Physiol. A* 178:293–305

Kleinfeld D, Berg RW, O'Connor SM. 1999. Anatomical loops and their electrical dynamics in relation to whisking by rat. *Somatosens. Mot. Res.* 16:69–88

Klier EM, Wang H, Constantin AG, Crawford JD. 2002. Midbrain control of three-dimensional head orientation. *Science* 295:1314–16

Klier EM, Wang H, Crawford JD. 2001. The superior colliculus encodes gaze commands in retinal coordinates. *Nat. Neurosci.* 4:627–32

Knight TA, Fuchs AF. 2007. Contribution of the frontal eye field to gaze shifts in the head-unrestrained monkey: effects of microstimulation. *J. Neurophysiol.* 97:618–34

Knudsen EI, Knudsen PF, Masino T. 1993. Parallel pathways mediating both sound localization and gaze control in the forebrain and midbrain of the barn owl. *J. Neurosci.* 13:2837–52

Komatsu H, Suzuki H. 1985. Projections from the functional subdivisions of the frontal eye field to the superior colliculus in the monkey. *Brain Res.* 327:324–27

Kostyk SK, Grobstein P. 1987. Neuronal organization underlying visually elicited prey orienting in the frog–I. Effects of various unilateral lesions. *Neuroscience* 21:41–55

Krauzlis RJ. 2003. Neuronal activity in the rostral superior colliculus related to the initiation of pursuit and saccadic eye movements. *J. Neurosci.* 23:4333–44

Krauzlis RJ, Basso MA, Wurtz RH. 1997. Shared motor error for multiple eye movements. *Science* 276:1693–95

Krauzlis RJ, Basso MA, Wurtz RH. 2000. Discharge properties of neurons in the rostral superior colliculus of the monkey during smooth-pursuit eye movements. *J. Neurophysiol.* 84:876–91

Krauzlis RJ, Carello CD. 2003. Going for the goal. *Nat. Neurosci.* 6:332–33

Krauzlis RJ, Liston D, Carello CD. 2004. Target selection and the superior colliculus: goals, choices and hypotheses. *Vis. Res.* 44:1445–51

Kustov AA, Robinson DL. 1996. Shared neural control of attentional shifts and eye movements. *Nature* 384:74–77

Lee C, Rohrer WH, Sparks DL. 1988. Population coding of saccadic eye movements by neurons in the superior colliculus. *Nature* 332:357–60

Lee PH, Helms MC, Augustine GJ, Hall WC. 1997. Role of intrinsic synaptic circuitry in collicular sensorimotor integration. *Proc. Natl. Acad. Sci. USA* 94:13299–304

Lefèvre P, Quaia C, Optican LM. 1998. Distributed model of control of saccades by superior colliculus and cerebellum. *Neural Netw.* 11:1175–90

Lomber SG, Payne BR. 1996. Removal of two halves restores the whole: reversal of visual hemineglect during bilateral cortical or collicular inactivation in the cat. *Vis. Neurosci.* 13:1143–56

Lomber SG, Payne BR, Cornwell P. 2001. Role of the superior colliculus in analyses of space: superficial and intermediate layer contributions to visual orienting, auditory orienting, and visuospatial discriminations during unilateral and bilateral deactivations. *J. Comp. Neurol.* 441:44–57

Martinez-Trujillo JC, Wang H, Crawford DJ. 2003. Electrical stimulation of the supplementary eye fields in the head-free macaque evokes kinematically normal gaze shifts. *J. Neurophysiol.* 89:2961–74

McCluskey MK, Cullen KE. 2007. Eye, head, and body coordination during large gaze shifts in rhesus monkeys: movement kinematics and the influence of posture. *J. Neurophysiol.* 97:2976–91

McHaffie JG, Stein BE. 1982. Eye movements evoked by electrical stimulation in the superior colliculus of rats and hamsters. *Brain Res.* 247:243–53

McIlwain JT. 1982. Lateral spread of neural excitation during microstimulation in intermediate gray layer of cat's superior colliculus. *J. Neurophysiol.* 47:167–78

McPeek RM, Keller EL. 2002. Saccade target selection in the superior colliculus during a visual search task. *J. Neurophysiol.* 88:2019–34

Meredith MA, Ramoa AS. 1998. Intrinsic circuitry of the superior colliculus: pharmacophysiological identification of horizontally oriented inhibitory interneurons. *J. Neurophysiol.* 79:1597–602

Missal M, Lefèvre P, Delinte A, Crommelinck M, Roucoux A. 1996. Smooth eye movements evoked by electrical stimulation of the cat's superior colliculus. *Exp. Brain Res.* 107:382–90

Moschovakis AK, Kitama T, Dalezios Y, Petit J, Brandi AM, Grantyn AA. 1998. An anatomical substrate for the spatiotemporal transformation. *J. Neurosci.* 18:10219–29

Moschovakis AK, Scudder CA, Highstein SM. 1996. The microscopic anatomy and physiology of the mammalian saccadic system. *Prog. Neurobiol.* 50:133–254

Munoz DP, Guitton D. 1989. Fixation and orientation control by the tecto-reticulo-spinal system in the cat whose head is unrestrained. *Rev. Neurol.* 145:567–79

Munoz DP, Guitton D. 1991. Control of orienting gaze shifts by the tectoreticulospinal system in the head-free cat. II. Sustained discharges during motor preparation and fixation. *J. Neurophysiol.* 66:1624–41

Munoz DP, Guitton D, Pélisson D. 1991. Control of orienting gaze shifts by the tectoreticulospinal system in the head-free cat. III. Spatiotemporal characteristics of phasic motor discharges. *J. Neurophysiol.* 66:1642–66

Munoz DP, Istvan PJ. 1998. Lateral inhibitory interactions in the intermediate layers of the monkey superior colliculus. *J. Neurophysiol.* 79:1193–209

Munoz DP, Wurtz RH. 1993a. Fixation cells in monkey superior colliculus. I. Characteristics of cell discharge. *J. Neurophysiol.* 70:559–75

Munoz DP, Wurtz RH. 1993b. Fixation cells in monkey superior colliculus. II. Reversible activation and deactivation. *J. Neurophysiol.* 70:576–89

Munoz DP, Wurtz RH. 1995a. Saccade-related activity in monkey superior colliculus. I. Characteristics of burst and buildup cells. *J. Neurophysiol.* 73:2313–33

Munoz DP, Wurtz RH. 1995b. Saccade-related activity in monkey superior colliculus. II. Spread of activity during saccades. *J. Neurophysiol.* 73:2334–48

Nagy A, Kruse W, Rottmann S, Dannenberg S, Hoffmann KP. 2006. Somatosensory-motor neuronal activity in the superior colliculus of the primate. *Neuron* 52:525–34

Neggers SF, Bekkering H. 2000. Ocular gaze is anchored to the target of an ongoing pointing movement. *J. Neurophysiol.* 83:639–51

Neggers SF, Bekkering H. 2001. Gaze anchoring to a pointing target is present during the entire pointing movement and is driven by a non-visual signal. *J. Neurophysiol.* 86:961–70

Netser S, Ohayon S, Gutfreund Y. 2010. Multiple manifestations of microstimulation in the optic tectum: eye movements, pupil dilations and sensory priming. *J. Neurophysiol.* 104:108–18

Nichols MJ, Sparks DL. 1996. Component stretching during oblique stimulation-evoked saccades: the role of the superior colliculus. *J. Neurophysiol.* 76:582–600

Ottes FP, Van Gisbergen JA, Eggermont JJ. 1986. Visuomotor fields of the superior colliculus: a quantitative model. *Vis. Res.* 26:857–73

Paré M, Guitton D. 1994. The fixation area of the cat superior colliculus: Effects of electrical stimulation and direction connection with brainstem omnipause neurons. *Exp. Brain Res.* 101:109–22

Peck CK. 1989. Visual responses of neurones in cat superior colliculus in relation to fixation of targets. *J. Physiol.* 414:301–15

Pélisson D, Goffart L, Guillaume A, Catz N, Raboyeau G. 2001. Early head movements elicited by visual stimuli or collicular electrical stimulation in the cat. *Vis. Res.* 41:3283–94

Pettit DL, Helms MC, Lee P, Augustine GJ, Hall WC. 1999. Local excitatory circuits in the intermediate gray layer of the superior colliculus. *J. Neurophysiol.* 81:1424–27

Port NL, Wurtz RH. 2003. Sequential activity of simultaneously recorded neurons in the superior colliculus during curved saccades. *J. Neurophysiol.* 90:1887–903

Pruszynski JA, King GL, Boisse L, Scott SH, Flanagan JR, Munoz DP. 2010. Stimulus-locked responses on human arm muscles reveal a rapid neural pathway linking visual input to arm motor output. *Eur. J. Neurosci.* 32:1049–57

Quaia C, Aizawa H, Optican LM, Wurtz RH. 1998. Reversible inactivation of monkey superior colliculus. II. Maps of saccadic deficits. *J. Neurophysiol.* 79:2097–110

Quaia C, Lefèvre P, Optican LM. 1999. Model of the control of saccades by superior colliculus and cerebellum. *J. Neurophysiol.* 82:999–1018

Reyes-Puerta V, Philipp R, Lindner W, Hoffmann KP. 2010. The role of the rostral superior colliculus in gaze anchoring during reach movements. *J. Neurophysiol.* 103:3153–66

Rezvani S, Corneil BD. 2008. Recruitment of a head-turning synergy by low-frequency activity in the primate superior colliculus. *J. Neurophysiol.* 100:397–411

Robinson DA. 1972. Eye movements evoked by collicular stimulation in the alert monkey. *Vis. Res.* 12:1795–808

Robinson DA. 1975. Oculomotor control signals. In *Basic Mechanisms of Ocular Motility and Their Clinical Implications*, ed. P Bach-y-Rita, G Lennerstrand, pp. 337–74. Oxford: Pergamon

Rolfs M. 2009. Microsaccades: small steps on a long way. *Vis. Res.* 49:2415–41

Rolfs M, Laubrock J, Kliegl R. 2006. Shortening and prolongation of saccade latencies following microsaccades. *Exp. Brain Res.* 169:369–76

Roucoux A, Guitton D, Crommelinck M. 1980. Stimulation of the superior colliculus in the alert cat. II. Eye and head movements evoked when the head is unrestrained. *Exp. Brain Res.* 39:75–85

Sahibzada N, Dean P, Redgrave P. 1986. Movements resembling orientation or avoidance elicited by electrical stimulation of the superior colliculus in rats. *J. Neurosci.* 6:723–33

Saitoh K, Menard A, Grillner S. 2007. Tectal control of locomotion, steering, and eye movements in lamprey. *J. Neurophysiol.* 97:3093–108

Sawa M, Ohtsuka K. 1994. Lens accommodation evoked by microstimulation of the superior colliculus in the cat. *Vis. Res.* 34:975–81

Schaefer KP. 1970. Unit analysis and electrical stimulation in the optic tectum of rabbits and cats. *Brain Behav. Evol.* 3:222–40

Schall JD. 1995. Neural basis of saccade target selection. *Rev. Neurosci.* 6:63–85

Schall JD, Thompson KG. 1999. Neural selection and control of visually guided eye movements. *Annu. Rev. Neurosci.* 22:241–59

Schiller PH, True SD, Conway JL. 1979. Effects of frontal eye field and superior colliculus ablations on eye movements. *Science* 206:590–92

Schuller G, Radtke-Schuller S. 1990. Neural control of vocalization in bats: mapping of brainstem areas with electrical microstimulation eliciting species-specific echolocation calls in the rufous horseshoe bat. *Exp. Brain Res.* 79:192–206

Shen K, Paré M. 2007. Neuronal activity in superior colliculus signals both stimulus identity and saccade goals during visual conjunction search. *J. Vis.* 7:15.1–3

Sinha SR, Moss CF. 2007. Vocal premotor activity in the superior colliculus. *J. Neurosci.* 27:98–110

Smit AC, Van Gisbergen JA, Cools AR. 1987. A parametric analysis of human saccades in different experimental paradigms. *Vis. Res.* 27:1745–62

Sparks DL. 1986. Translation of sensory signals into commands for control of saccadic eye movements: role of primate superior colliculus. *Physiol. Rev.* 66:118–71

Sparks DL. 1999. Conceptual issues related to the role of the superior colliculus in the control of gaze. *Curr. Opin. Neurobiol.* 9:698–707

Sparks DL, Gandhi NJ. 2003. Single cell signals: an oculomotor perspective. *Prog. Brain Res.* 142:35–53

Sparks DL, Holland R, Guthrie BL. 1976. Size and distribution of movement fields in the monkey superior colliculus. *Brain Res.* 113:21–34

Sparks DL, Mays LE. 1990. Signal transformations required for the generation of saccadic eye movements. *Annu. Rev. Neurosci.* 13:309–36

Stanford TR, Freedman EG, Sparks DL. 1996. Site and parameters of microstimulation: evidence for independent effects on the properties of saccades evoked from the primate superior colliculus. *J. Neurophysiol.* 76:3360–81

Stanton GB, Goldberg ME, Bruce CJ. 1988. Frontal eye field efferents in the macaque monkey: II. Topography of terminal fields in midbrain and pons. *J. Comp. Neurol.* 271:493–506

Stein BE, Clamann HP. 1981. Control of pinna movements and sensorimotor register in cat superior colliculus. *Brain Behav. Evol.* 19:180–92

Stein BE, Meredith MA. 1993. *The Merging of the Senses*. Cambridge, MA: MIT Press

Stein BE, Stanford TR. 2008. Multisensory integration: current issues from the perspective of the single neuron. *Nat. Rev. Neurosci.* 9:255–66

Steinman RM, Haddad GM, Skavenski AA, Wyman D. 1973. Miniature eye movement. *Science* 181:810–19

Straschill M, Hoffmann KP. 1970. Activity of movement sensitive neurons of the cat's tectum opticum during spontaneous eye movements. *Exp. Brain Res.* 11:318–26

Straschill M, Rieger P. 1973. Eye movements evoked by focal stimulation of the cat's superior colliculus. *Brain Res.* 59:211–27

Stryker MP, Schiller PH. 1975. Eye and head movements evoked by electrical stimulation of monkey superior colliculus. *Exp. Brain Res.* 23:103–12

Stuphorn V, Bauswein E, Hoffmann KP. 2000. Neurons in the primate superior colliculus coding for arm movements in gaze-related coordinates. *J. Neurophysiol.* 83:1283–99

Stuphorn V, Hoffmann KP, Miller LE. 1999. Correlation of primate superior colliculus and reticular formation discharge with proximal limb muscle activity. *J. Neurophysiol.* 81:1978–82

Syka J, Radil-Weiss T. 1971. Electrical stimulation of the tectum in freely moving cats. *Brain Res.* 28:567–72

Takahashi M, Sugiuchi Y, Shinoda Y. 2010. Topographic organization of excitatory and inhibitory commissural connections in the superior colliculi and their functional roles in saccade generation. *J. Neurophysiol.* 104:3146–67

Tehovnik EJ. 1989. Head and body movements evoked electrically from the caudal superior colliculus of rats: pulse frequency effects. *Behav. Brain Res.* 34:71–78

Tehovnik EJ, Yeomans JS. 1986. Two converging brainstem pathways mediating circling behavior. *Brain Res.* 385:329–42

Tu TA, Keating EG. 2000. Electrical stimulation of the frontal eye field in a monkey produces combined eye and head movements. *J. Neurophysiol.* 84:1103–6

Valentine DE, Sinha SR, Moss CF. 2002. Orienting responses and vocalizations produced by microstimulation in the superior colliculus of the echolocating bat, Eptesicus fuscus. *J. Comp. Physiol. A* 188:89–108

Van Gisbergen JA, Van Opstal AJ, Tax AA. 1987. Collicular ensemble coding of saccades based on vector summation. *Neuroscience* 21:541–55

Van Opstal AJ, Goossens H. 2008. Linear ensemble-coding in midbrain superior colliculus specifies the saccade kinematics. *Biol. Cybern.* 98:561–77

Van Opstal AJ, Van Gisbergen JA. 1989. A nonlinear model for collicular spatial interactions underlying the metrical properties of electrically elicited saccades. *Biol. Cybern.* 60:171–83

Van Opstal AJ, Van Gisbergen JA, Smit AC. 1990. Comparison of saccades evoked by visual stimulation and collicular electrical stimulation in the alert monkey. *Exp. Brain Res.* 79:299–312

Wagner H. 1993. Sound-localization deficits induced by lesions in the barn owl's auditory space map. *J. Neurosci.* 13:371–86

Waitzman DM, Silakov VL, DePalma-Bowles S, Ayers AS. 2000. Effects of reversible inactivation of the primate mesencephalic reticular formation. I. Hypermetric goal-directed saccades. *J. Neurophysiol.* 83:2260–84

Walton MMG, Bechara BP, Gandhi NJ. 2007. Role of the primate superior colliculus in the control of head movements. *J. Neurophysiol.* 98:2022–37

Walton MMG, Bechara BP, Gandhi NJ. 2008. Effect of reversible inactivation of superior colliculus on head movements. *J. Neurophysiol.* 99:2479–95

Walton MMG, Sparks DL, Gandhi NJ. 2005. Simulations of saccade curvature by models that place superior colliculus upstream from the local feedback loop. *J. Neurophysiol.* 93:2354–58

Walton MMG, Mays LE. 2003. Discharge of saccade-related superior colliculus neurons during saccades accompanied by vergence. *J. Neurophysiol.* 90:1124–39

Werner W. 1993. Neurons in the primate superior colliculus are active before and during arm movements to visual targets. *Eur. J. Neurosci.* 5:335–40

Werner W, Dannenberg S, Hoffmann KP. 1997a. Arm-movement-related neurons in the primate superior colliculus and underlying reticular formation: comparison of neuronal activity with EMGs of muscles of the shoulder, arm and trunk during reaching. *Exp. Brain Res.* 115:191–205

Werner W, Hoffmann KP, Dannenberg S. 1997b. Anatomical distribution of arm-movement-related neurons in the primate superior colliculus and underlying reticular formation in comparison with visual and saccadic cells. *Exp. Brain Res.* 115:206–16

Wurtz RH, Goldberg ME. 1972. Activity of superior colliculus in behaving monkey. III. Cells discharging before eye movements. *J. Neurophysiol.* 35:575–86

Olfactory Maps in the Brain

Venkatesh N. Murthy

Department of Molecular and Cellular Biology and Center for Brain Science,
Harvard University, Cambridge, Massachusetts 02138; email: vnmurthy@fas.harvard.edu

Annu. Rev. Neurosci. 2011. 34:233–58

The *Annual Review of Neuroscience* is online at
neuro.annualreviews.org

This article's doi:
10.1146/annurev-neuro-061010-113738

Keywords

odor map, chemotopy, neural computation

Abstract

The responses of neural elements in many sensory areas of the brain
vary systematically with their physical position, leading to a topographic
representation of the outside world. Sensory representation in the olfactory system has been harder to decipher, in part because it is difficult to
find appropriate metrics to characterize odor space and to sample this
space densely. Recent experiments have shown that the arrangement
of glomeruli, the elementary units of processing, is relatively invariant
across individuals in a species, yet it is flexible enough to accommodate
new sensors that might be added. Evidence supports the existence of
coarse spatial domains carved out on a genetic or functional basis, but a
systematic organization of odor responses or neural circuits on a local
scale is not evident. Experiments and theory that relate the properties
of odorant receptors to the detailed wiring diagram of the downstream
olfactory circuits and to behaviors they trigger may reveal the design
principles that have emerged during evolution.

Contents

INTRODUCTION: NEURAL MAPS

A topographic map (neuroanatomy) is the ordered projection of a sensory surface, like the retina or the skin, or an effector system, like the musculature, to one or more structures of the central nervous system.

–Wikipedia

A map in the brain refers to an orderly representation of some physical feature of the outside world or a computed value derived from these features (Chklovskii & Koulakov 2004, Kaas 1997, Knudsen et al. 1987, Kohonen & Hari 1999). Such maps are typically found in sensory or motor systems. A common type of sensory map is a topographic map, which is an ordered projection of the sensory surface to higher brain regions; for example, neighboring points in visual space activate neighboring points on the retina, and this relation is preserved through several subsequent stages of processing (Knierim & Van Essen 1992, Talbot & Marshall 1941, Wandell et al. 2007). Another type of neural map is the computational map, in which the value of a computed parameter is mapped systematically across a neural structure—for example, a map of sound source location in the auditory brain stem (Knudsen et al. 1987).

One plausible theory for the existence of such ordered maps is based on the cost of wiring in the brain (Chklovskii & Koulakov 2004, Mitchison 1991). The idea is that it is better to construct a network in which most computation is local, relying on short connections between nearby neurons (Chklovskii & Koulakov

2004). Shorter axons cost less in terms of volume occupied and energy expended, as well as in terms of guidance defects during development (Chklovskii & Koulakov 2004). In many sensory systems, computations are local because important interactions usually occur between neighboring points in the stimulus space. Indeed, the very arrangement of sensory information in a given brain region—the sensory map—can reveal something about which neuronal signals are combined in subsequent circuits.

We have some intuition about which sensory stimuli should be mapped close to each other in the cases of vision, touch, and hearing. The nature of maps is much less obvious for olfaction. Odorant molecules are diverse and vary in size, charge distribution, bond saturation, and three-dimensional structure; indeed, a small molecule can be characterized by hundreds of parameters (Haddad et al. 2008b, Rossiter 1996). Therefore, on the face of it, even a fraction of these parameters cannot be systematically mapped onto a two-dimensional surface (Cleland 2010). Some investigators have argued that some privileged parameters are selected for during evolution and can be mapped systematically over the bulb. However, in the absence of a sensory ethology of olfaction in the commonly studied species, it is difficult to know which odors have special meaning to the animal (other than pheromones, which are not discussed here). This issue is less straightforward than it may appear; the use of natural images to dissect neural circuits is in its infancy even in the visual system, where the stimulus is more readily parameterized (Rieke & Rudd 2009).

It is important to note at the outset that a region of the brain may respond to sensory stimuli in a repeatable, but arbitrary pattern with no smooth or gradual representation of some stimulus attribute. Such a fractured pattern is not really a map in the conventional neurobiological sense, but is instead a sort of look-up table for neural circuits that read it (**Figure 1**). In Knudsen's words (Knudsen et al. 1987), "neurons selective for a particular parameter value or set of parameter values often are clustered to form functional modules in the brain, but there is no systematic variation in their tuning, i.e., there is no map" (pp. 41–42).

In this review, I discuss the organization of olfactory sensory regions in relation to odorant receptors (ORs) and odorant molecules. The focus is decidedly on mammals, but I occasionally invoke insect and fish studies to draw parallels. In addition, the discussion for mammals is restricted to the main olfactory system, with only occasional reference to the pheromone system. There is also a slant toward recent studies, and older work is frequently referenced indirectly through key review articles.

THE SPACE OF ODORS

Which metrics should be used to characterize odor space? Historically, this question has been tackled by asking humans to assign a set of perceptual descriptors to odorants, which can then be compared with chemical descriptors of these odorants (Amoore 1974, Schiffman 1974, Wise et al. 2000). Attempts to obtain a small set of olfactory primaries, a set of basic odors, upon which all odor percepts are built were not satisfactory, and investigators focused on defining some basic axes along which odor perception may lawfully vary (Wise et al. 2000).

This approach has been revived recently, and investigators have found interesting relations between odor quality and physicochemical structure of odors. Sobel and colleagues have found that variation in physicochemical properties along a single principal component can reasonably predict the pleasantness of an odor (Haddad et al. 2008a, Khan et al. 2007). Koulakov and colleagues (unpublished material; **http://arxiv.org/ftp/arxiv/papers/0907/0907.3964.pdf**) have analyzed previously cataloged data and found that psychophysical descriptors vary along a small number of dimensions that could be related to physicochemical parameters. The basic axes (or eigenvectors) that describe the relevant physicochemical descriptors seem to be related to the size of odorants, but not to measures such as carbon chain length or presence of aromatic

OR: odorant receptor

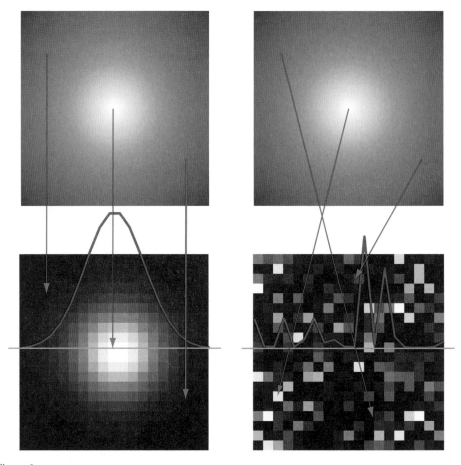

Figure 1

Map versus look-up table. A neural map is present when a variable, intensity of light in space in this example, is smoothly represented by a population of neurons (*bottom left*). By contrast, if a neural layer represents this variable in a fragmented pattern, it is not a map, even if the pattern is predictable.

groups. An important question that remains unanswered is whether odors that lie close together in the perceptual space are mapped systematically in the olfactory epithelium (OE) or the olfactory bulb (OB). A recent survey of the responses of a large number of mouse and human ORs to a large panel of odorants is a valuable starting point (Saito et al. 2009). If the spatial location of the sensory neurons expressing these tested mouse ORs or their glomerular targets can be determined, one can ask whether a specific physicochemical parameter is systematically mapped in physical space.

OE: olfactory epithelium

OB: olfactory bulb

THE SENSORY SURFACE: OLFACTORY EPITHELIUM IN MAMMALS

Odorants are sensed by a family of ORs expressed in olfactory sensory neurons (OSNs) (Buck & Axel 1991). This family has ~1,000 members in mice and rats (Zhang & Firestein 2002), ~50 in adult *Drosophila* (Brochtrup & Hummel 2011), and ~140 in zebrafish (Yoshihara 2009). Other chemosensory epithelia such as the vomeronasal organ and receptors in mammals (Kaupp 2010) are not discussed here. ORs transduce odor binding to electrical activity in OSNs through signaling mechanisms that remain to be fully clarified (Kaupp 2010). In the main olfactory system of the mouse, each sensory neuron expresses only one type of receptor, and neurons expressing a particular receptor type are scattered in a large zone within the OE (Mombaerts 2004, Serizawa et al. 2004). Different receptors are expressed in distinct zones that are continuous and highly overlapping (Miyamichi et al. 2005). In effect, there is a coarse map of receptors in the OE (**Figure 2**), but even in a small region of the epithelium there will be a mosaic of neurons expressing dozens of OR types. What restricts the expression of a given receptor to a subregion of the epithelium is not fully clear (Imai et al. 2010, Sakano 2010).

OSN: olfactory sensory neuron

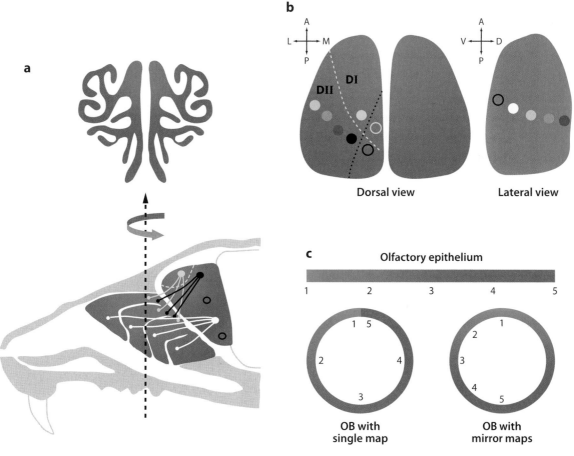

Figure 2

Mapping from the olfactory epithelium (OE) to the olfactory bulb (OB). (*a*) OR expression varies gradually in a rough dorsoventral direction in the OE, schematized by shading that goes from blue to red. A given OR is expressed by olfactory sensory neurons (OSNs) located within a subregion of the OE, which spans the entire OE in the rostrocaudal direction. A coronal section of the turbinates is illustrated; the shading represents the progression of OR expression. Homotypic OSNs (*shown by small circles in the OE*) are shown to converge on glomeruli. The dorsal-most region delineated by the yellow line is the DI domain, where OSNs expressing class I ORs converge. (*b*) Dorsal view of the OB shows the DI and DII domains, as well as the axis of reflection of the mirror-image maps (*dotted black line*). Lateral view is shown at right. (*c*) Illustration of how linear variation of a parameter can be mapped onto a closed contour. If the mapping occurs over all 360°, there will be a singularity (where 1 and 5 meet). If the map takes up 180° and is reflected, then there is no seam. Circles represent coronal sections of the OB.

The expression of several molecular markers carves out two zones—dorsal (D) and ventral (V)—within the OE (Sakano 2010). An additional segregation arises in the expression of the so-called class I and II ORs, whose genes are also segregated in chromosome loci (Zhang et al. 2004). These two classes of OR genes are phylogenetically distinct (Niimura & Nei 2005b), and class II ORs are expressed everywhere in the OE. Class I ORs, however, have restricted expression in the dorsal part of the OE (Miyamichi et al. 2005, Tsuboi et al. 2006, Zhang et al. 2004) and may code for special odorants (Freitag et al. 1995). Class I and II ORs are spatially mixed in the dorsal OE.

Even though each OSN expresses only one type of OR, it can still respond to a range of chemicals with different affinities, and a given odor can activate many different ORs with different efficacies (Araneda et al. 2000, Hallem et al. 2004, Malnic et al. 1999, Saito et al. 2009). Odor recognition, at least in generalist olfactory systems, likely occurs by combinatorial activation of many different types of OSNs.

Is there any functional significance to the preorganized (coarse) spatial map in the epithelium, or is it simply a developmental convenience to help enforce the one neuron/one OR rule (Sakano 2010, Zou et al. 2009)? To shed light on this question, one must know the relation between an OR and its ligand specificity. Unlike in other sensory systems, there is no clear a priori functional meaning for the position of the sensory neuron within the epithelium. Positional information in the OE could become important if the differential solubility of odorants or size segregate them within the olfactory sensory organ (Schoenfeld & Cleland 2005). Intriguingly, class I ORs, which are expressed in the dorsal parts of the OE, may preferentially bind to hydrophilic molecules (Saito et al. 2009). The dorsal OE, however, is also populated by class II ORs. In sum, no clear relation has been determined between the spatial position of an OSN and its odor tuning, and direct measurements from the OE may be necessary.

Olfactory receptors in adult *Drosophila* are found in the antennae and in the maxillary palp (Clyne et al. 1999, Vosshall et al. 1999). The sensilla, structures that house the sensory neurons, fall into distinct morphological classes—basiconic, trichoid, and coeloconic sensilla—that may selectively signal pheromone or general-purpose chemicals (Brochtrup & Hummel 2011, Imai et al. 2010, Luo & Flanagan 2007, Su et al. 2009). The morphological classes of sensilla also have preferential distribution on the sense organs, but a given OR is expressed in a wide region of the antenna or maxillary palp, with an apparently stochastic nearest-neighbor relationship (Couto et al. 2005, Luo & Flanagan 2007).

Zebrafish have two major types of OSNs (ciliated and microvillous) that express different classes of receptors that resemble the main olfactory and the vomeronasal systems of mice (Sato et al. 2005). The morphology, the OR class, the signal transduction machinery, and the projection patterns all point to differences in the function of these two types (Yoshihara 2009). Within each class, no systematic differences in receptor location have been described. OSNs projecting to a single glomerulus (and therefore likely to express the same receptor) are scattered throughout the OE (Baier et al. 1994).

In sum, other than broad expression domains, no obvious topography akin to those in other sensory systems has been identified in the olfactory sensory surface.

WIRING FROM OLFACTORY EPITHELIUM TO OLFACTORY BULB

In vertebrates, OSNs in the nose project their axons to the brain and form anatomical units known as glomeruli, arrayed in a layer on the surface of the OB (Mombaerts 1996, Mombaerts et al. 1996). Each glomerulus receives input from OSNs that express a single OR type. Conversely, OSNs expressing a particular OR project to (usually) two glomeruli in the ipsilateral OB (Mombaerts et al. 1996); not surprisingly, the number of glomeruli is approximately twice the number of receptor genes. Because each glomerulus receives input from only one OR type, its spectrum of odor responses is defined by the ligand-binding properties of the OR. The layout of glomeruli on the bulb forms a two-dimensional map of ORs, and by extension odorants (**Figure 2**).

Three types of topographic projections have been discussed in the literature pertaining to projection from the sensory epithelium to the first brain region: rhinotopy, odotopy, and chemotopy (Johnson & Leon 2007, Xu et al. 2000). Rhinotopy refers to a pattern of projection in which the position in the nose (the OE, to be precise) is mapped on the target area. This mapping can also be translated into a form of "receptoro-topy" if the position in the nasal epithelium dictates which type of receptor is expressed. Odotopy refers to the general idea that individual odors activate distinct parts of the OB. Such an activation pattern can now be rationalized given the one glomerulus/one OR organization. Chemotopy refers to a systematic representation in spatial coordinates of some feature of odorants, for example, the number of carbon chains in an aliphatic molecule. As discussed below, there is evidence for rhinotopy and odotopy in mammalian brains, but chemotopy has been more difficult to understand.

The sorting of like axons (i.e., axons of OSNs expressing the same OR) and their glomerular coalescence is guided by a hierarchical set of guidance mechanisms (Luo & Flanagan 2007, Mombaerts 2006, Sakano 2010). In addition to the identity of the OR itself, the glomerular position is influenced by many contextual factors (Feinstein & Mombaerts 2004). The postsynaptic targets in the OB are largely dispensable for the formation of the glomerular layout, and OSN axons may even be presorted before they arrive in the bulb (Bozza et al. 2009, Kobayakawa et al. 2007).

Broad domains in the OB surface have been defined on the basis of molecular markers, the type of OR expressed, and the position of the OSNs in the epithelium (Bozza et al. 2009,

Kobayakawa et al. 2007, Sakano 2010). The position of the OSN in the OE is mapped in the dorsoventral (D-V) direction in the OB, as appreciated even before the discovery of ORs (Astic & Saucier 1986). We now know that the zone of expression of a particular OR is correlated with the dorsoventral position of the target glomerulus in the OB (Miyamichi et al. 2005, Sakano 2010) (**Figure 2**). Because of the overlapping expression of the class I and II ORs in the dorsal OE, OSNs expressing either class can go to the dorsal surface of the OB. However, these two classes are cleanly separated in the dorsal OB, with an anterodorsal DI domain and a caudoventral DII domain (Sakano 2010) (**Figure 2**). The rest of the class II ORs are represented on the ventral side of the OB. The mapping in the D-V axis of the OB is thought to involve at least two sets of repulsive ligand/receptor pairs: Slits/Robo2 and Sema3F/Nrp2 (Takeuchi et al. 2010). In the proposed mechanism (Sakano 2010), the OR–glomerulus relation is indirect because the OSN position independently controls both.

Unlike for the D-V axis, there is no relation between glomerular position in the anterior-posterior (A-P) axis and the OSN position in the epithelium. Instead, OR-mediated intracellular signaling may provide a rough glomerular address in the A-P direction (Imai et al. 2006). Clear evidence exists for the role of cAMP in glomerular targeting (Col et al. 2007, Imai et al. 2006, Zou et al. 2007). A gradient of cAMP levels is thought somehow to cause transcriptional regulation of classic axon guidance molecules such as Nrp1 and Sema3A (in opposite directions), leading to pretarget axon sorting that then establishes the A-P position (Sakano 2010).

In addition to these global positioning mechanisms, local sorting of axons and the final glomerular coalescence are thought to occur through hemophilic and heterophilic axon-axon interactions mediated by molecules such as the ORs themselves, Kirrel2/3, and ephrin-A/EphA (Mombaerts 2006, Sakano 2010). The final stages of axon segregation and coalescence into glomeruli may also involve neural activity, perhaps through its effect on expression of the different cell surface molecules (Mombaerts 2006, Sakano 2010).

A peculiar feature of the mammalian OB is that each OR has two cognate glomeruli on the OB, leading to two glomerular layouts in each hemisphere (Johnson & Leon 2007, Mori et al. 2006). The two layouts are placed such that they are "reflections" of one another on the bulbar surface (**Figure 2c**). Their function remains obscure, but it may be that mirror layouts arose as a developmental convenience that avoids a severe discontinuity or fracture. An open two-dimensional sheet (the OE) needs to be mapped to a closed two-dimensional sheet (the OB), and a simple linear map in both dimensions will lead to a discontinuity (**Figure 2c**). Having duplicated mirror-symmetric maps may help avoid this. The axis of symmetry of the mirror-image layouts is along the A-P axis, approximately perpendicular to the D-V progression of the glomerular position signal.

The basic wiring from the sensory organ to the first stage of processing is preserved across a wide variety of species. In *Drosophila*, OSNs expressing a particular receptor converge on an anatomically identifiable glomerulus in the antennal lobe (Gao et al. 2000, Vosshall et al. 2000). The three types of sensilla that house OSNs have differential distribution on the sense organs, and they are also loosely mapped on the antennal lobe (Couto et al. 2005, Imai et al. 2010, Luo & Flanagan 2007). OSNs belonging to the same sensillum (therefore, located physically close to each other) can project to glomeruli that are close together or far apart (Couto et al. 2005, Endo et al. 2007). A detailed analysis of responses of ORs to a large set of odorants concluded that the cognate glomeruli had no neighborhood relation in functional space—that is, adjacent glomeruli are functionally no more similar than distant ones (Hallem & Carlson 2006). In the zebrafish, OSNs that converge to a single glomerulus are randomly scattered over the OE, and no known connectional topography exists (Baier et al. 1994). The one glomerulus/one receptor rule may be largely preserved, but there is no obvious order in the projection from OE to OB (Sato et al. 2007).

REPRODUCIBLE ARRANGEMENT OF ODORANT RECEPTORS IN THE GLOMERULAR LAYER

The elaborate sequence of events that lead to the convergence of axons to glomeruli and their proper segregation may suggest that the final spatial layout of glomeruli is variable across hemispheres or individuals because of accumulating errors. However, genetic studies using tagged olfactory receptors and visual observation in relation to anatomical landmarks have made it clear that the position of a glomerulus is not random and lies within a region of ~30 glomeruli (Mombaerts 2006, Mombaerts et al. 1996).

Functional studies have also noted that roughly similar glomerular activation patterns are recorded across animals when specific odorants are used (Johnson & Leon 2007, Rubin & Katz 1999, Uchida et al. 2000, Wachowiak & Cohen 2001), again suggesting some degree of repeatability of glomerular position. A global layout of activated regions in the glomerular layer using 2-deoxyglucose uptake has been painstakingly constructed by Leon and colleagues for a large number of odorants (Johnson & Leon 2007), which points to relative stereotypy of activation at least on a coarse level. A recent study used a large set of odorants to image ~200 glomeruli at higher resolution in mice and rats and determined that the average error in the relative position of dorsal glomeruli was on the order of 1 glomerular diameter (Soucy et al. 2009), both across two hemispheres of an individual and across different individuals.

The precision in relative glomerular position, with only some local jitter, supports a model of glomerular targeting that involves some degree of stochasticity in the final step. Glomeruli are placed more precisely in the mediolateral axis than in the A-P axis (Oka et al. 2006, Soucy et al. 2009). Because the mediolateral axis is essentially collinear with the

D-V direction, the data suggest that the placement guided by cues related to the position of the OSNs (Slits/Robo2 and Sema3F/Nrp2) may have better accuracy than those driving the A-P position (Nrp1/Sema3A). Glomerular positions are invariant across individuals in flies, and each glomerulus can be assigned a specific label (Imai et al. 2010). Glomerular layout is thought to be precise in fish, although the exact positional precision has not been quantified.

The available data indicate that for a given repertoire of ORs, the corresponding glomeruli are arranged in a predictable and precise pattern on the surface of the OB.

RELATION BETWEEN ODORS AND GLOMERULAR POSITION

Does the fact that glomeruli have specific spatial addresses lead to something more than a look-up table for downstream neural circuits, for example, a meaningful map in the space of odorants?

The idea that some metric of odorant structure will be smoothly mapped on the OB surface has inspired many studies. Early work in zebrafish noted that different classes of odorants activate distinct regions of the OB (Friedrich & Korsching 1998). This idea has been pursued in rats and mice using different odorants from several chemical classes such as aldehydes, ketones, alcohols, and acids, and a vast catalog of glomerular activity patterns has been obtained (Johnson & Leon 2007, Mori et al. 2006, Xu et al. 2000). As noted by Johnson & Leon (2007), three general features of odor responses in the rodent OB require explanation, and they are discussed in sequence.

Clustered Responses

Responses to a particular class of odorants appear in localized clusters, and each cluster may represent a particular odor quality (Mori et al. 2006). The fact that glomeruli responding to a particular odorant often cluster together is likely to be the consequence of the similarity of the corresponding ORs (see below).

Because glomerular responses are a function of the ORs expressed by the corresponding OSNs, an important consideration is how the coarse organization of odor responses is related to the domains carved out by ORs themselves, for example, class I and II ORs (Zhang et al. 2004). Two recent studies reported that at least some odors activate the DI and DII domains, specifically (Bozza et al. 2009, Matsumoto et al. 2010). More extensive sampling of the odor space is necessary before firm conclusions can be drawn. The positional mapping of OSNs along the D-V axis in the bulb is not accompanied by any obvious progression of odor responses, other than a tendency for heavier molecules to activate the ventral OB in rats.

Activation domains have also been identified in the fish OB (Friedrich & Korsching 1998, Nikonov & Caprio 2001). Amino acids activate glomeruli in the anterior-lateral subregion of the bulb, and bile acids activate a medial subregion (Friedrich & Korsching 1998). Odorants were chosen on the basis of behavioral responses and educated guesses about natural stimuli that fish might encounter (Hara 1994), and it is interesting to consider whether adding other chemically diverse odorants will redistrict or preserve these domains.

Fine-Scale Organization of Responding Glomeruli

In addition to clustering of responses in spatial domains, several studies have emphasized within-domain ordering of responses (Johnson & Leon 2007, Mori et al. 2006). A frequently studied "chemotopic" progression involves straight-chain aldehydes or alcohols, with varying carbon chain lengths (Johnson & Leon 2000, Uchida et al. 2000). Each class of these odorants activates a cluster of glomeruli, and within a cluster response foci seem to shift gradually as chain length is altered. Such a progression is not observed for all odor domains, and when present it heads in a D-V direction (Johnson & Leon 2007). Because a rough D-V positional map exists between the OE and the

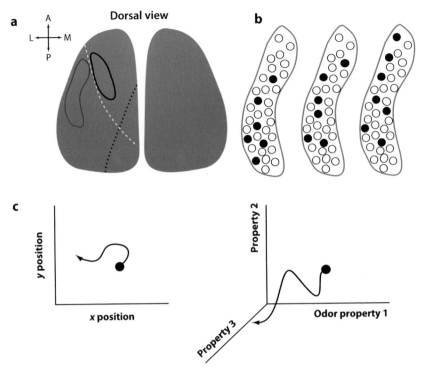

Figure 3

Coarse and fine-scale structure of glomerular responses. (*a*) Some classes of odors such as ketones (*red line*) and aliphatic acids (*black*) activate clusters of glomeruli. In some cases, these may even respect boundaries demarcated by molecular markers (*yellow line*). (*b*) Within such a domain, responding glomeruli (*filled circles*) can be intermingled with inactive glomeruli (*open circles*). When different related odorants are presented, the overall center of mass of the responding glomeruli may shift gradually, but often different glomeruli respond to different odors. (*c*) Illustration of how the glomerular position (*left*) may move in two dimensions as the OR sequence is altered. At the same time, the odor sensitivity may be altered; but this occurs in high dimensions, so its evolution may be independent of the glomerular position.

OB, it will be of interest to determine whether there is a systematic progression of response tuning in the OE position.

The gradual spatial procession of glomerular responses to odorants is not readily seen in higher-resolution imaging studies. In particular, active glomeruli that are activated by the domain-specific class of odorants (aliphatic aldehydes, for example) are interrupted by many inactive glomeruli (Bozza et al. 2004, Soucy et al. 2009), leading to a rather discontinuous progression of responses (**Figure 3**). These inactive glomeruli, however, can respond readily to other odorants not obviously related to aldehydes (Soucy et al. 2009). Indeed, the use of a large set of chemically diverse odor-

ants to activate dorsal glomeruli in the same animal allowed investigators to compare response tuning of different glomeruli. Several interesting observations were made (Soucy et al. 2009). First, even though different odorants of a given chemical class activate clusters of glomeruli within a domain, they are not always the same glomeruli. Coarse-scale analysis will be unable to differentiate distinct responses of adjacent glomeruli. Second, odors of a particular class activate only a subset of the glomeruli present within a domain or module. Third, neighboring glomeruli were no more similar to each other than were glomeruli separated by greater distances, when all odors were considered (and not just a few classes of odorants).

Neighboring glomeruli in flies can have diverse odor responses, and odors of a given class can activate distinct glomeruli (Hallem & Carlson 2006). Although a direct analysis of glomerular response similarity and spatial separation has not been reported in fish, inspection of published response patterns indicates a strong neighborhood relation (Friedrich & Korsching 1998).

The existence of local diversity of glomerular responses is not necessarily at odds with a domain organization for a set of odorants. The coarse-scale order in odor responses may arise from some overall similarity of ORs that innervate a given region of the OB (Feinstein & Mombaerts 2004, Miyamichi et al. 2005). The fine-scale diversity may arise because it is possible for ORs with similar sequences to bind to different ligands (Araneda et al. 2000, Saito et al. 2009). Therefore, although a degree of similarity at a coarse, domain level is possible owing to overall similarity of ORs, fine-scale organization may be difficult, given the disconnect between mechanisms that select for ligand binding and those that determine glomerular location (see below).

Similar Glomerular Activity Patterns for Chemically Similar Odorants

A final aspect of the spatial organization of functional odor responses is that the patterns of glomerular activity for chemically related odorants can be very similar. This similarity, however, is inevitable because (*a*) a given OR often responds to chemically similar odorants (Araneda et al. 2000, Saito et al. 2009), (*b*) a given glomerulus receives inputs from OSNs expressing the same OR (Mombaerts 1996), and (*c*) the position of a glomerulus is relatively precise (Soucy et al. 2009). If a given odorant activates a set of ORs, then a related odorant (how one defines that may be tricky) would likely activate similar ORs. But this similarity does not automatically imply the existence of a map.

Quantitative and unbiased clustering of odorants based on their OSN response patterns in *Drosophila* indicated that odors of a

particular chemical class fall into groups, but there are many outliers (Hallem & Carlson 2006). A global analysis in the rat and mouse OB using lower-resolution methods revealed that the overall activation pattern is indeed similar for odorants within a particular class (Johnson & Leon 2007). Metrics derived from an exhaustive set of physicochemical properties of odorants can be used to predict the pattern of neural responses in several species, including mice (Haddad et al. 2008a, Saito et al. 2009), but these metrics are not simple measures such as carbon chain length.

Data from different species have failed to identify general principles of functional organization of glomeruli in OB analogs. Within a given species, there may be some large domains of functionally similar glomeruli, but fine-scale diversity seems to be the norm.

POTENTIAL INDEPENDENCE OF LIGAND SPECIFICITY AND GLOMERULAR POSITION

Although there may not (yet) be a way of predicting the location of a glomerulus just by its sequence, the overall glomerular layout is predictable for a given repertoire of ORs and genomic loci. For example, although the layouts for mice and rats are different from each other, they are individually predictable (Soucy et al. 2009). Why is it that the glomerular layout is repeatable, but the neighborhood relation among glomeruli in functional terms is difficult to predict? The major reason is likely the complex relation among OR sequence, its ligand specificity, and the location on the glomerular surface.

There is a coarse relation between the sequence of an OR and the location of its corresponding glomerulus; similar sequences may have a tendency to go to neighboring bulbar locations, but the displacement can sometimes be substantial for even small differences in sequence (Feinstein & Mombaerts 2004, Miyamichi et al. 2005, Wang et al. 1998). It became clear from these heroic gene-targeting experiments that the receptor sequence does

JG: juxtaglomerular

M: mitral

T: tufted

SA: short axon

PG: periglomerular

not provide a unique address on the bulb, and other factors strongly influence the position. One clear determinant of glomerular position is the chromosome locus from which the OR gene is expressed; the same OR gene expressed from different loci can target very different glomerular positions (Bozza et al. 2009, Feinstein & Mombaerts 2004, Miyamichi et al. 2005). The large OR gene repertoire in mice seems to have arisen mainly from tandem gene duplication (Niimura & Nei 2005a), so novel receptors appearing during evolution may be somewhat constrained in their target choice within the glomerular layout. Another, potentially related, factor that determines the glomerular target is the identity of the OSN itself, independent of its OR choice (Bozza et al. 2009).

Glomerular position, therefore, arises from developmental constraints, with neighborhood relation established by similarity of sequence, proximity on chromosome, transcriptional time, enhancer elements, etc. A newly evolved receptor may, by virtue of sequence similarity, go to a bulbar position that is close to another receptor. However, even small mutations can alter odor selectivity unpredictably. In particular, because odor binding depends on many residues, such small changes may leave intact binding affinity for many odorants but may introduce new odor-binding ability. Displacement in the space of sequence may not correspond to commensurate displacements in functional space (**Figure 3**).

In summary, it seems most useful to consider the glomerular layout on the OB surface as an arbitrary, but relatively stable look-up table for any particular species and examine how the higher brain areas interpret these.

POSTSYNAPTIC TRANSFORMATION IN THE GLOMERULAR LAYER

OSN axons make synaptic connections with several targets in the OB, including juxtaglomerular (JG), mitral (M), and tufted (T) cells (**Figure 4**). I discuss how odor representation is transformed by bulbar circuits, mainly in relation to the spatial layout, ignoring temporal modulations (Laurent et al. 2001).

A vast majority of neurons that send their dendrites to the glomerular layer elaborate their dendritic arbors within a single glomerulus (Shepherd et al. 2004), which makes a compelling argument for the independence of glomeruli and no special relation between neighboring glomeruli. In addition to the principal cells (M/T cells), several JG neurons ramify their dendrites within the glomerulus. These cells are diverse and variously release GABA, glutamate, or dopamine (Shepherd et al. 2004, Wachowiak & Shipley 2006). One class of JG neuron has its processes in many glomeruli; these are the short axon (SA) cells, which are dopaminergic as well as GABAergic (Kiyokage et al. 2010).

Intraglomerular interactions can alter signal strength in postsynaptic neurons and alter the spatial pattern of odor-induced activity. Excitatory connections within the glomerulus, in the form of both gap junctions and synaptic connections (Wachowiak & Shipley 2006), can lead to signal amplification. Recurrent and feedforward inhibition arises from periglomerular (PG) cells excited by OSN inputs directly or through recurrent excitation from M/T cells (De Saint Jan et al. 2009, Hayar et al. 2005). Intraglomerular inhibition may serve as a gain-control mechanism acting presynaptically on OSN synapses or through postsynaptic inhibition of principal cells (Gire & Schoppa 2009, Shao et al. 2009, Wachowiak et al. 2005). Weak sensory inputs may be shunted out by PG cells (Gire & Schoppa 2009), helping sharpen odor-tuning curves in principal cells (Cleland 2010).

More elaborate changes in the spatial pattern of active regions can be achieved through interglomerular interactions. Interglomerular interactions in the glomerular layer are likely mediated by SA cells, which are expected to be inhibitory because of their GABAergic/dopaminergic identity. There is also some evidence for excitatory interactions (Aungst et al. 2003), which remains to be further explored. SA cells project to dozens of glomeruli (Kiyokage et al. 2010), but it is not clear whether the long projections are axons or dendrites and whether their connection strength is distance dependent.

If lateral interactions are not specific, interglomerular inhibition could mediate a form of normalization that averages activity across many glomeruli and leads to a general suppression of responses, similar to that seen in insects (Olsen et al. 2010). If, however, lateral inhibition is specific (genetically predefined?), the spatial pattern of odor responses could be altered in more complex ways (**Figure 4b**). A recent study found that PG cells had very broad responses compared with their genetically tagged parent glomerulus (Tan et al. 2010), presenting direct functional evidence

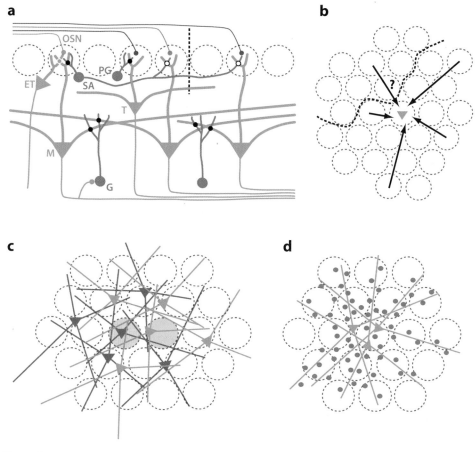

Figure 4

Circuits in the OB. (*a*) Schematic of the major neuronal types in the olfactory bulb (OB) and their connections. Filled circles on the intersection of overlapping processes denote bidirectional synaptic connections, and filled circles that abut neurons indicate polarized axo-dendritic synapses. (*b*) Overhead view of the OB showing glomeruli as a two-dimensional array (*dashed circles*). Mitral cell activity can be influenced by a number of scattered glomeruli, as illustrated by arrows. Domain boundary is shown by a thick dashed line, and whether there are interactions across domain boundaries is not known (*question mark*). (*c*) Mitral cells affiliated with neighboring glomeruli (*separated by green and gray colors*) are likely to be intermingled. Lateral dendrites are sketched as radiating lines from the cell bodies. (*d*) Granule cells within reach of lateral dendrites of mitral cells affiliated with one glomerulus are shown by blue dots. The geometry of mitral cell dendrites dictates that their density falls off with distance and that the density of connected granule cells will also drop off with distance. Abbreviations: G, granule; ET, external tufted; M, mitral; OSN, olfactory sensory neuron; PG, periglomerular; SA, short axon; T, tufted.

that PG cells integrate information from multiple glomeruli.

Glutamatergic external tufted (ET) cells, the principal cells in the glomerular region, receive direct inputs from OSNs (Hayar et al. 2005) and are involved in integrating information between homologous, mirror-image glomeruli with the bulb (Belluscio et al. 2002). A recent study noted that all T cells (presumably including some ET cells) had much sparser responses than did their inputs (Tan et al. 2010), suggesting intra- or interglomerular GABAergic inhibition. It is important for future experiments to obtain cellular-resolution spatial layout of identified JG cell responses in relation to the glomerular responses. Multiphoton imaging may offer a way forward.

No clear distinctions exist between glomerular and deeper-layer circuits in flies and fish, and they are discussed below in the section, Spatial Organization of Functionally Related Granule Cells.

SPATIAL PATTERNS IN OUTPUT NEURON LAYER

The major relay cells in the OB, the M/T cells, are thought to receive direct inputs from OSNs and lateral excitation through dendro-dendritic contacts within glomeruli. Inhibitory inputs arise from PG cells in the glomerular layer and granule cells in the external plexiform layer (EPL). Each M/T cell has its primary dendrite in a single glomerulus, and the 20 or so M/T cells associated with a single glomerulus are located below the glomerulus within a radius of 2–5 glomeruli (Buonviso et al. 1991). On the basis of geometric considerations, a small region in the M cell layer should contain M cells from multiple glomeruli (**Figure 4c**). Indeed, pairs of M cells adjacent to each other have a probability of 0.3 or less of innervating the single glomerulus (Dhawale et al. 2010). Whether there is a preferential arrangement of M cells of a particular kind (functionally or glomerular-identity based) in a local neighborhood remains to be determined. Given the local diversity of glomerular odor tuning, it would not be entirely surprising if M cells within a local region have diverse odor tuning; this author's unpublished data support this supposition (Albeanu et al. 2009).

The circuitry and the synaptic properties of connections between M/T cells and granule cells have been extensively investigated in vitro (Egger & Urban 2006, Schoppa & Urban 2003). Similarly, odor responses of M/T cells have been studied for several decades, and much is known about the odor selectivity and dynamics of odor responses (recent examples include Cury & Uchida 2010, Fantana et al. 2008, Tan et al. 2010). However, much less is known about the spatial organization and to what extent it reflects the inputs. Intrabulbar interactions are likely to help sharpen the tuning curves of M/T cells, perhaps without changing the overall spatial pattern of responses (Mori et al. 1999, Tan et al. 2010).

A direct comparison of the responses of the input and output neurons associated with a genetically identified glomerulus recently revealed that M cells responded more sparsely to a large set of odorants than did the cognate input neurons (Tan et al. 2010). If these findings are generalized to other glomeruli, one would predict that the spatial pattern of M cell responses would be a sparser version of the glomerular patterns. More complex transformation of information in the OB is certainly possible because of the extensive connectivity and the feedback projections from higher brain areas (Restrepo et al. 2009). In addition, the temporal evolution of responses in M/T cells (Cury & Uchida 2010) could lead to systematic changes in spatial patterns of activity, as seen in zebrafish.

An important question raised by the presence of coarse-scale domain organization is how M/T cells across the domain boundaries interact anatomically and functionally. If indeed the class of receptors and their domains are important for segregating functional responses in the input layer, one might expect that M/T cells situated close to the boundaries have asymmetric or at least specialized interactions. More generally, it seems imperative to understand how many of the lateral interactions among M/T cells are hard-wired and invariant, and how many of them vary across hemispheres or individuals (**Figure 4b**). Exciting developments in the field of optogenetics promise to help uncover the connection logic in the OB. In particular, the ability to activate individual glomeruli using light will allow a detailed mapping of the effective connections between a large number of glomeruli and an individual M/T cell (Arenkiel et al. 2007, Dhawale et al. 2010). These techniques will inform us about any asymmetry or anisotropy in M/T cell connections, for example, across domain boundaries.

The spatial organization of M cell responses has been studied in zebrafish using multiphoton imaging (Yaksi et al. 2007, 2009). Most M cells in fish have their apical tufts in single glomeruli just as they do in mammals and insects (Fuller et al. 2006, Miyasaka et al. 2009). Odor responses of M/T cells appear to follow the spatial pattern of inputs soon after odor presentation (Yaksi et al. 2007). Neighboring M cells are more similar to each other in odor tuning than are those with greater separation (Yaksi et al. 2007). The spatial patterns of responses become sparse and the chemotopic organization largely disappears with increasing time after odor exposure (Yaksi et al. 2007). Odor responses in principal neurons in the antennal lobe of flies have also been characterized using calcium imaging as well as electrophysiology, and the emerging view is that principal cells are more broadly tuned than are their OSN inputs (Masse et al. 2009). No specific organization by spatial location is evident, perhaps not surprising given the lack of positional information noted above.

SPATIAL ORGANIZATION OF FUNCTIONALLY RELATED GRANULE CELLS

Granule cells in the mammalian OB, which outnumber principal cells by a large margin, receive input via lateral dendrites and axon collaterals of M cells, as well as axons from cortical areas (Shepherd et al. 2004). The connectivity patterns of granule cells are not fully understood. A given M cell extends lateral dendrites to distances exceeding 800 microns in rodents, but whether they are capable of exciting granule cells uniformly over this distance is unknown. Similarly, whether a given granule cell receives inputs from M/T cells spread over a radius of 800 microns is also unknown. Granule cells are heterogeneous in their soma location, the principal cells they target, and their age, raising the distinct possibility that they are part of distinct microcircuits in the OB (Kelsch et al. 2010, Lledo et al. 2006).

Granule cell responses to odorants can be subthreshold or can involve action potentials (Cang & Isaacson 2003, Wellis & Scott 1990). Recordings from granule cells have been sparse, in part because they are technically challenging. Although earlier studies hinted that granule cell responses may be sparse, a recent study reported denser responses to a large panel of odorants (Tan et al. 2010). The density of responses increased with odor concentration, indicating that lateral inhibition may be more prominent at higher concentrations. However, none of these studies sheds light on any spatial pattern of responses. Lower-resolution, "one-shot" experiments using immediate-early genes or 2-deoxyglucose to report activity indicate that there might be a columnar organization in granule cell responses (Guthrie et al. 1993), perhaps owing to geometric reasons (**Figure 4d**). Real-time imaging studies in the future may shed light on whether granule cells can mediate long-distance interactions between M/T cells and whether there are functional correlates to the patchy organization hinted at by viral tracing studies (Willhite et al. 2006).

Imaging studies in zebrafish indicate that interneurons equivalent to granule cells responded more densely to odorants (Yaksi et al. 2007). The coarse spatial segregation observed at the level of M cells was much less obvious, and interneurons could respond to multiple classes of odorants that normally do not activate the same M cells. All these indicate that interneurons integrate input from a disparate set of M cells. Lateral interactions mediated by interneurons in *Drosophila* are complicated and include excitation and inhibition (Masse et al. 2009), with no noteworthy spatial aspect.

STEREOTYPY OF CIRCUITRY WITHIN THE OLFACTORY BULB

How much of the computation in the OB is predicated on precise or prespecified circuitry? For example, if specific glomeruli are placed close together by design to allow neighbors to interact, then there would be preferential connections between neurons getting inputs from these glomeruli. If such interactions are mediated by lateral inhibition, which has been suggested to have a center-surround organization (Yokoi et al. 1995), connections could be somewhat nonspecific, but distance dependent. However, recent data suggest that only a few scattered glomeruli are likely to influence the firing of a given M/T cell (Fantana et al. 2008). We do not know if these surround interactions are specific and repeatable across animals.

A specific type of intrabulbar connection—the one linking homologous glomeruli within a bulb—is indeed rather precise and plastic (Lodovichi et al. 2003, Marks et al. 2006). ET cells from one glomerulus send their axons to a small region of the internal plexiform layer below the homologous glomerulus and make synapses on granule cells. Whether these granule cells affect only M/T cells that are associated with the homologous glomerulus is unknown, but seems likely. The cellular and functional specificity of this circuit will help formulate its precise role.

MAPS IN HIGHER BRAIN REGIONS

M/T cells of the OB project through the lateral olfactory tract to many different brain regions, including the anterior olfactory nucleus (AON), the anterior and posterior piriform cortices (aPC and pPC), the olfactory tubercle (OT), the cortical amygdala (coA), and the entorhinal cortex (EC) (Shipley & Ennis 1996). Classic anatomical studies using anterograde and retrograde tracers have found very little evidence for any sort of positional topography in projections, with the exception of a subregion of the AON (Shipley & Ennis 1996).

The AON, which is considered a cortical area, has strong reciprocal connections with the OB and piriform cortex (PC) (Brunjes et al. 2005). Broadly, the AON can be divided into pars principalis (the core of the AON) and the pars externa, a thin ring of cells surrounding the core region. The anterior olfactory nucleus, pars externa (AONpE), has strong back projections to the OB but does not seem to project to downstream cortical areas such as the PC (Brunjes et al. 2005).

The lateral olfactory tract (LOT) maintains a broad organization in the D-V axis to match the source M/T cells in the OB (Walz et al. 2006). The projections of M/T cells to the AONpE maintain a rough topography, with M/T cells from the medial side of the OB projecting to medial AONpE and similarly for the lateral side (Brunjes et al. 2005). There is also recent evidence for topographic projections in the A-P axis, as well (Yan et al. 2008). Projections from the OB to other parts of the AON show no topography (Brunjes et al. 2005).

A very interesting feature of the AON is its backprojections to the OB (Brunjes et al. 2005). The AONpE again exhibits topography in the backprojection; for example, dorsal and ventral AONpE project to dorsal and ventral OB, respectively (Brunjes et al. 2005). A recent study extended these findings to the A-P axis of the OB and found exquisite topography in the backprojections, even to the contralateral bulb (Yan et al. 2008). Similar to the intrabulbar projections between homologous glomeruli, the excitatory projections from AONpE terminate in the granule cell layer and presumably activate those granule cells that are closest to the somata of M/T cells with apical dendrites in the appropriate glomerulus. These precise bilateral connections must have a role in coordinating information from the two nostrils (Yan et al. 2008).

The PC is the largest target area of the OB in mammals and has been divided into subregions—the aPC and the pPC—with the aPC divided further into dorsal and ventral parts (Haberly 2001). The three-layered PC is phylogenetically older than the more widespread six-layer cortex. Inputs from the OB arrive through the lateral olfactory tract, with synapses made in layer 1a of the PC (Haberly 2001, Isaacson 2010). Cortico-cortical association synapses are made in layer 1b (Haberly 2001, Isaacson 2010). Layers 2 and 3 contain cell bodies of pyramidal cells as well as some interneurons (Isaacson 2010).

M/T cell axons innervating the PC have widespread but patchy arbors, with no specific topography (Ojima et al. 1984). M cells project throughout the PC, but T cells do not reach much of the pPC (Haberly 2001, Nagayama et al. 2010). Single M cell axons make collaterals throughout the PC (Ojima et al. 1984). Indirect functional studies using immediate-early gene expression to assay neuronal activity (Illig & Haberly 2003), as well as extracellular spike recordings, have found that cells activated by a particular odor are distributed widely in the PC (Litaudon et al. 2003, Rennaker et al. 2007). Whole cell patch clamp recordings of PC cells indicated that they responded to only a few of the odorants tested, and the fraction of cells responding to a given odor was estimated to be ~10% (Poo & Isaacson 2009). More heterogeneous responses were reported in awake, head-fixed mice using cell-attached recordings and post hoc cell identification (Zhan & Luo 2010). In addition to finding sparse responders, investigators also saw cells with broad receptive fields and cells exhibiting inhibitory responses (Zhan & Luo 2010). Using multiphoton microscopy, Stettler & Axel (2009) directly demonstrated spatially distributed PC cell responses to odors. This method allows researchers to examine responses from a large population of neurons to many odors at the same time. Any given odor activated ~10% of the imaged neurons, widely distributed across the cortex. Neighboring cells had diverse odor receptive fields, with no evidence of clustering or patchiness.

The anatomical and physiological evidence converges on a model of connectivity from the bulb to the cortex that is spatially non-specific and integrating inputs from multiple glomeruli (Haberly 2001). Exactly how many glomeruli project to a single pyramidal cell and whether there are specific connections preferentially from the different broad domains in the OB remain unknown. The cellular mechanisms of sparse odor responses are beginning to be elucidated and may involve widespread inhibition as well as recurrent corticocortical connections (Isaacson 2010). Lest one become too enamored with sparse responses, it is worth recalling theoretical arguments indicating that an optimal tuning curve width (narrow versus wide) depends on noise covariance, the number of variables being coded, and their impact on downstream circuits (Salinas 2006). A better understanding of these properties in the olfactory system may shed light on the coding strategy and the type of feature extraction.

In addition to direct input from the OB, the PC also receives profuse corticocortical connections arising locally from the ipsilateral PC, arriving from both ipsi- and contralateral anterior olfactory nucleus as well as from the contralateral PC (Haberly 2001). These connections have been invoked in models of the PC as an associative memory network (Haberly 2001). None of the regions giving rise to these corticocortical projections is known to have topographic projections from the OB; AONpE, the one region that has a map of the bulb, does not send projections to other cortical areas (Brunjes et al. 2005). A given PC cell is likely to receive inputs from both hemispheres and respond to odors from both nostrils, but the details of binasal interactions are not fully clear (Wilson 1997). An intriguing question is how association inputs modify the response tuning of PC cells. At first blush, it might seem that recurrent excitation should serve to broaden responses unless they are made very selectively. Distinct subregions within the PC are anatomically and functionally different, but whether they differ in the spatial arrangement of odor responses is not known (Haberly 2001).

The OT is considered part of the ventral striatum, a high-level cognitive region, but it receives direct monosynaptic inputs from the OB, perhaps preferentially from T cells (Nagayama et al. 2010, Shipley & Ennis 1996). Odor responses have been recorded in the OT (Wesson & Wilson 2010). To date, there is no evidence for organized projections from the OB or PC to the OT. There is also no known systematic organization of response properties in either odor space or brain space in OB targets such as the olfactory amygdala, the entorhinal cortex, and the tenia tecta (Shipley & Ennis 1996).

In the zebrafish, as in mammals, M cells send their projections to many higher areas (Miyasaka et al. 2009). Individual M cell axons innervate large regions of the telencephalon, but MCs from different domains of the bulb may have distinct patterns of innervations (Miyasaka et al. 2009). Functionally, there are reports of odotopic maps in higher brain areas in the catfish (Nikonov et al. 2005). Recent imaging studies in the zebrafish shed light on odor responses in two telencephalic target regions, Vv and Dp (Yaksi et al. 2009). Neurons in the Vv, a subpallial region, pool information from M cells and respond to more odors than do individual M cells. Neurons in Dp, a region homologous to the PC, have sparser responses with widespread inhibitory synaptic inputs (Yaksi et al. 2009). There is, however, no topographic map of responses, and cells responding to specific chemical classes (amino acids, for example) are distributed across the entire region.

Projections from the antennal lobe to higher regions in insects exhibit varying degrees of stereotypy. Of the two major projection regions, the lateral horn is thought to be sufficient for basic olfactory behaviors, and accordingly, the spatial pattern of axonal input from antennal lobe is highly stereotyped across animals (Jefferis et al. 2007, Marin et al. 2002, Tanaka et al. 2004, Wong et al. 2002). Furthermore, principal neurons associated with glomeruli that sense fruity odors appear to have projection areas distinct from those associated with pheromone-sensing glomeruli

(Jefferis et al. 2007). Whether these correspond to functional segregation remains to be seen. The other target of the antennal lobe is the mushroom body, which is important for associative learning (Keene & Waddell 2007). Projection neurons from the antennal lobe have variable axonal arbors in the mushroom body, and Kenyon cells are thought to integrate information from multiple glomeruli (Masse et al. 2009). At the present level of understanding, synaptic integration and response properties in the mushroom body are quite similar to those in the mammalian PC.

HOW CAN MAPS AID COMPUTATION IN THE OLFACTORY SYSTEM?

In the end, we must return to the question of what maps might actually do for the olfactory system. I consider some of the potential reasons for the existence of and the potential function of olfactory maps.

Wiring Length

As noted in the introduction, sensory maps could reduce wiring cost. In the OB, very few of the identified connections involve preferential nearest-neighbor interactions. In the glomerular layer, most dendrites are localized to a single glomerulus. Therefore, considerations of wiring optimization do not, at present, shed light on why there should be odor maps. More information on whether there are preferential connections that are invariant across individuals may catalyze further thinking about wiring optimization—for example, do granule cells link preferred pairs of M/T cells?

Decorrelation

Local circuits with lateral inhibition are thought to sharpen response tuning to stimuli, to decorrelate stimuli, and to reduce redundancy. In the OB, because neighboring glomeruli are not particularly similar to each other functionally (Soucy et al. 2009), lateral interactions cannot be confined to nearest neighbors. In fact, placing a diverse set of glomeruli within a small region may facilitate greater sampling of the odor space with local interactions. Such spatially unstructured lateral inhibition can be used to decorrelate M cell activity (Arevian et al. 2008, Cleland 2010). Even M/T cells targeting the same glomerulus can become decorrelated in their responses to odors, presumably through lateral inhibition (Dhawale et al. 2010).

Clustering of responses to a particular class of odorants may facilitate local processing, for example to discriminate among similar odorants, sometimes referred to as contrast enhancement (Mori et al. 1999). Although the nature of contrast enhancement in olfaction is not clear, discussions have typically focused on discriminating among molecules within a chemical class (Johnson & Leon 2007, Yokoi et al. 1995). The main evidence for contrast enhancement is the increase in the strength of M cell responses to nonoptimal stimuli when GABAergic inhibition is blocked (Tan et al. 2010, Yokoi et al. 1995). Lateral inhibitory interactions are usually assumed to occur in the deeper layers but may also occur in the glomerular layer (Aungst et al. 2003, McGann et al. 2005, Vucinic et al. 2006). However, because local representations are rather fragmented, lateral interactions may need to be specific (Fantana et al. 2008). An alternate hypothesis to explain contrast enhancement discounts lateral interactions altogether and invokes intraglomerular processing (Cleland 2010). Additional experiments blocking inhibition caused by a specific population of cells (using optogenetics, for example) are necessary to test these hypotheses.

Beyond decorrelating the responses of different neurons, Friedrich and colleagues have proposed that the OB performs pattern decorrelation (Niessing & Friedrich 2010, Wiechert et al. 2010). In this process, the overall population response is orthogonalized to allow easier discrimination of the activity patterns by downstream circuits. Whether pattern decorrelation is necessary in mammalian OB, with its fragmented map and greater local diversity in glomerular responses (Soucy et al. 2009), is not clear.

Domain Organization

What could be the function of broad domains within the OB, identified either by chemical classes of odors activating various glomeruli or by classes of odorant receptors innervating the glomeruli? These coarse maps may be helpful for genetically defining connectivity to different downstream areas. For example, the dorsal and ventral parts of the OB may project to different brain regions. Indeed, recent studies note that although both dorsal and ventral regions may contain glomeruli responding to specific odorants, abolishing dorsal glomeruli genetically affects innate fear responses but not learned fear responses to predator odors (Kobayakawa et al. 2007). This intriguing study suggests that odor maps cannot be interpreted without additional knowledge of projection patterns. Studies in fish have also offered tantalizing hints of this sort of domain-specific connectivity (Koide et al. 2009).

Another important question to be addressed is whether there are privileged lateral interactions between M/T cells within a domain or those located between domains (**Figure 4b**). Fantana et al. (2008) could predict M/T cell odor tuning by taking into consideration only ~200 visualized glomeruli (only about one-tenth of all glomeruli). Whether this is simply a consequence of the total length of the lateral dendrites of M/T cells or whether this hints at some domain organization remains to be seen.

Gain Control

Another computation thought to happen in the OB is gain control, which is essential for sensory systems to deal with a large range of stimulus intensities. In the olfactory system, one major form of gain control is localized to the presynaptic terminals of OSNs (Kazama & Wilson 2008, Murphy et al. 2004, Wachowiak et al.

2005, Wachowiak & Shipley 2006), which together reduce the activation of principal cells. A precise and organized layout of glomeruli is presumably not necessary to carry out gain control independently in each glomerulus. A different sort of gain control can also be achieved by interglomerular inhibition, in which the overall activity levels of glomeruli may be used to reduce the activity of all responding glomeruli proportionately (Cleland et al. 2007, Olsen et al. 2010). Such an interglomerular normalization, which can be divisive or subtractive, will preserve the relative activity levels of glomeruli and may be important for any coding strategy that does not involve simple labeled lines. Even interglomerular inhibition may not require maps in the traditional sense because there is not much evidence for preferential nearest-neighbor interactions. If lateral interactions in the OB are, in fact, sparse and distance independent (Fantana et al. 2008), an important question is whether the interactions are selective and precise. Gain control operations, as proposed to date, do not seem to need ordered topographic maps.

Multiple Maps for Different Scenes

In the retina, it has been apparent for some time that multiple views of the visual scene are broadcast to the brain by different classes of ganglion cells (Wassle 2004). Is there a similar strategy in the olfactory system, by which multiple odor images are relayed from the OB? For example, M and T cells may be relaying different aspects of the odor scene—M cells carrying more processed, fine features and T cells carrying more broad categories—to different target areas. Because there are two dozen or so M cells associated with an individual glomerulus, it is also interesting to speculate whether there are systematic distinctions among them; anatomical evidence indicates such differences (Orona et al. 1984). Combining molecular genetics and targeted recordings from identified cells has led to dramatic advances in retinal neurobiology in recent years, and this approach is likely to bear fruit in the olfactory system, as well.

Mirror-image representation in the mammalian OB is also an interesting issue. Are the two layouts essentially redundant and serve to increase signal to noise? Or do they have distinct functions and targets? Mirror-image layouts may have arisen as a developmental convenience and were subsequently harnessed for some yet-unknown function (**Figure 2c**).

SUMMARY

Solid evidence now supports the existence of a coarse topographic map from the receptor sheet to the first stage of processing at least in some animals. If and how such a map translates to a functional olfactory map continues to be difficult to resolve. Any functional organization in terms of local neighborhood similarity has to arise, in part, in the receptor sheet itself. An alternate hypothesis for positional mapping in the periphery is that it is essential for a predictable arrangement of glomeruli. A broad mapping from OE to OB is achieved with chemical gradients and positional cues, but local diversity may be inevitable. The OB circuit will have to be built to cope with this fractured layout, relying less on neighborhood relation and more on spatially distributed lateral interactions.

Future studies aimed at neural circuit analysis will likely bear the most fruit. These might include (*a*) analysis of the projection patterns of M/T cells from different bulbar domains, (*b*) examination of M/T cell lateral dendrites to determine any rules they may obey at domain boundaries, (*c*) analysis of any reproducibility and precision in the lateral interaction circuits within the OB, (*d*) analysis of the glomerular receptive fields of various cell types in the OB and its targets, (*e*) a detailed comparison of OR sequence and spatial position of the cognate glomerulus, and (*f*) analysis of potential scaling or morphological rules governing M/T and granule cell dendritic arbors.

DISCLOSURE STATEMENT

The author is not aware of any affiliations, memberships, funding, or financial holdings that might be perceived as affecting the objectivity of this review.

ACKNOWLEDGMENTS

I thank the members of my laboratory for stimulating discussions. Work in my laboratory related to this review was supported by an anonymous foundation and funds from Harvard University.

LITERATURE CITED

Albeanu DF, Knopfel T, Murthy VN. 2009. Imaging mitral cell population activity in vivo in the mouse olfactory bulb. *Soc. Neurosci. Annu. Meet.* Abstr. 68.8

Amoore JE. 1974. Evidence for the chemical olfactory code in man. *Ann. NY Acad. Sci.* 237:137–43

Araneda RC, Kini AD, Firestein S. 2000. The molecular receptive range of an odorant receptor. *Nat. Neurosci.* 3:1248–55

Arenkiel BR, Peca J, Davison IG, Feliciano C, Deisseroth K, et al. 2007. In vivo light-induced activation of neural circuitry in transgenic mice expressing channelrhodopsin-2. *Neuron* 54:205–18

Arevian AC, Kapoor V, Urban NN. 2008. Activity-dependent gating of lateral inhibition in the mouse olfactory bulb. *Nat. Neurosci.* 11:80–87

Astic L, Saucier D. 1986. Anatomical mapping of the neuroepithelial projection to the olfactory bulb in the rat. *Brain Res. Bull.* 16:445–54

Aungst JL, Heyward PM, Puche AC, Karnup SV, Hayar A, et al. 2003. Centre-surround inhibition among olfactory bulb glomeruli. *Nature* 426:623–29

Baier H, Rotter S, Korsching S. 1994. Connectional topography in the zebrafish olfactory system: random positions but regular spacing of sensory neurons projecting to an individual glomerulus. *Proc. Natl. Acad. Sci. USA* 91:11646–50

Belluscio L, Lodovichi C, Feinstein P, Mombaerts P, Katz LC. 2002. Odorant receptors instruct functional circuitry in the mouse olfactory bulb. *Nature* 419:296–300

Bozza T, McGann JP, Mombaerts P, Wachowiak M. 2004. In vivo imaging of neuronal activity by targeted expression of a genetically encoded probe in the mouse. *Neuron* 42:9–21

Bozza T, Vassalli A, Fuss S, Zhang JJ, Weiland B, et al. 2009. Mapping of class I and class II odorant receptors to glomerular domains by two distinct types of olfactory sensory neurons in the mouse. *Neuron* 61:220–33

Brochtrup A, Hummel T. 2011. Olfactory map formation in the *Drosophila* brain: genetic specificity and neuronal variability. *Curr. Opin. Neurobiol.* 21:85–92

Brunjes PC, Illig KR, Meyer EA. 2005. A field guide to the anterior olfactory nucleus (cortex). *Brain Res. Brain Res. Rev.* 50:305–35

Buck L, Axel R. 1991. A novel multigene family may encode odorant receptors: a molecular basis for odor recognition. *Cell* 65:175–87

Buonviso N, Chaput MA, Scott JW. 1991. Mitral cell-to-glomerulus connectivity: an HRP study of the orientation of mitral cell apical dendrites. *J. Comp. Neurol.* 307:57–64

Cang J, Isaacson JS. 2003. In vivo whole-cell recording of odor-evoked synaptic transmission in the rat olfactory bulb. *J. Neurosci.* 23:4108–16

Chklovskii DB, Koulakov AA. 2004. Maps in the brain: What can we learn from them? *Annu. Rev. Neurosci.* 27:369–92

Cleland TA. 2010. Early transformations in odor representation. *Trends Neurosci.* 33:130–39

Cleland TA, Johnson BA, Leon M, Linster C. 2007. Relational representation in the olfactory system. *Proc. Natl. Acad. Sci. USA* 104:1953–58

Clyne PJ, Warr CG, Freeman MR, Lessing D, Kim J, Carlson JR. 1999. A novel family of divergent seven-transmembrane proteins: candidate odorant receptors in *Drosophila*. *Neuron* 22:327–38

Col JA, Matsuo T, Storm DR, Rodriguez I. 2007. Adenylyl cyclase-dependent axonal targeting in the olfactory system. *Development* 134:2481–89

Couto A, Alenius M, Dickson BJ. 2005. Molecular, anatomical, and functional organization of the *Drosophila* olfactory system. *Curr. Biol.* 15:1535–47

Cury KM, Uchida N. 2010. Robust odor coding via inhalation-coupled transient activity in the mammalian olfactory bulb. *Neuron* 68:570–85

De Saint Jan D, Hirnet D, Westbrook GL, Charpak S. 2009. External tufted cells drive the output of olfactory bulb glomeruli. *J. Neurosci.* 29:2043–52

Dhawale AK, Hagiwara A, Bhalla US, Murthy VN, Albeanu DF. 2010. Non-redundant odor coding by sister mitral cells revealed by light addressable glomeruli in the mouse. *Nat. Neurosci.* 13:1404–12

Egger V, Urban NN. 2006. Dynamic connectivity in the mitral cell-granule cell microcircuit. *Semin. Cell Dev. Biol.* 17:424–32

Endo K, Aoki T, Yoda Y, Kimura K, Hama C. 2007. Notch signal organizes the *Drosophila* olfactory circuitry by diversifying the sensory neuronal lineages. *Nat. Neurosci.* 10:153–60

Fantana AL, Soucy ER, Meister M. 2008. Rat olfactory bulb mitral cells receive sparse glomerular inputs. *Neuron* 59:802–14

Feinstein P, Mombaerts P. 2004. A contextual model for axonal sorting into glomeruli in the mouse olfactory system. *Cell* 117:817–31

Freitag J, Krieger J, Strotmann J, Breer H. 1995. Two classes of olfactory receptors in *Xenopus laevis*. *Neuron* 15:1383–92

Friedrich RW, Korsching SI. 1998. Chemotopic, combinatorial, and noncombinatorial odorant representations in the olfactory bulb revealed using a voltage-sensitive axon tracer. *J. Neurosci.* 18:9977–88

Fuller CL, Yettaw HK, Byrd CA. 2006. Mitral cells in the olfactory bulb of adult zebrafish (*Danio rerio*): morphology and distribution. *J. Comp. Neurol.* 499:218–30

Gao Q, Yuan B, Chess A. 2000. Convergent projections of *Drosophila* olfactory neurons to specific glomeruli in the antennal lobe. *Nat. Neurosci.* 3:780–85

Gire DH, Schoppa NE. 2009. Control of on/off glomerular signaling by a local GABAergic microcircuit in the olfactory bulb. *J. Neurosci.* 29:13454–64

Guthrie KM, Anderson AJ, Leon M, Gall C. 1993. Odor-induced increases in c-fos mRNA expression reveal an anatomical "unit" for odor processing in olfactory bulb. *Proc. Natl. Acad. Sci. USA* 90:3329–33

Haberly LB. 2001. Parallel-distributed processing in olfactory cortex: new insights from morphological and physiological analysis of neuronal circuitry. *Chem. Senses* 26:551–76

Haddad R, Khan R, Takahashi YK, Mori K, Harel D, Sobel N. 2008a. A metric for odorant comparison. *Nat. Methods* 5:425–29

Haddad R, Lapid H, Harel D, Sobel N. 2008b. Measuring smells. *Curr. Opin. Neurobiol.* 18:438–44

Hallem EA, Carlson JR. 2006. Coding of odors by a receptor repertoire. *Cell* 125:143–60

Hallem EA, Ho MG, Carlson JR. 2004. The molecular basis of odor coding in the *Drosophila* antenna. *Cell* 117:965–79

Hara TJ. 1994. Olfaction and gustation in fish: an overview. *Acta Physiol. Scand.* 152:207–17

Hayar A, Shipley MT, Ennis M. 2005. Olfactory bulb external tufted cells are synchronized by multiple intraglomerular mechanisms. *J. Neurosci.* 25:8197–208

Illig KR, Haberly LB. 2003. Odor-evoked activity is spatially distributed in piriform cortex. *J. Comp. Neurol.* 457:361–73

Imai T, Sakano H, Vosshall LB. 2010. Topographic mapping—the olfactory system. *Cold Spring Harb. Perspect. Biol.* 2:a001776

Imai T, Suzuki M, Sakano H. 2006. Odorant receptor-derived cAMP signals direct axonal targeting. *Science* 314:657–61

Isaacson JS. 2010. Odor representations in mammalian cortical circuits. *Curr. Opin. Neurobiol.* 20:328–31

Jefferis GS, Potter CJ, Chan AM, Marin EC, Rohlfing T, et al. 2007. Comprehensive maps of *Drosophila* higher olfactory centers: spatially segregated fruit and pheromone representation. *Cell* 128:1187–203

Johnson BA, Leon M. 2000. Odorant molecular length: one aspect of the olfactory code. *J. Comp. Neurol.* 426:330–38

Johnson BA, Leon M. 2007. Chemotopic odorant coding in a mammalian olfactory system. *J. Comp. Neurol.* 503:1–34

Kaas JH. 1997. Topographic maps are fundamental to sensory processing. *Brain Res. Bull.* 44:107–12

Kaupp UB. 2010. Olfactory signalling in vertebrates and insects: differences and commonalities. *Nat. Rev. Neurosci.* 11:188–200

Kazama H, Wilson RI. 2008. Homeostatic matching and nonlinear amplification at identified central synapses. *Neuron* 58:401–13

Keene AC, Waddell S. 2007. *Drosophila* olfactory memory: single genes to complex neural circuits. *Nat. Rev. Neurosci.* 8:341–54

Kelsch W, Sim S, Lois C. 2010. Watching synaptogenesis in the adult brain. *Annu. Rev. Neurosci.* 33:131–49

Khan RM, Luk CH, Flinker A, Aggarwal A, Lapid H, et al. 2007. Predicting odor pleasantness from odorant structure: pleasantness as a reflection of the physical world. *J. Neurosci.* 27:10015–23

Kiyokage E, Pan YZ, Shao Z, Kobayashi K, Szabo G, et al. 2010. Molecular identity of periglomerular and short axon cells. *J. Neurosci.* 30:1185–96

Knierim JJ, Van Essen DC. 1992. Visual cortex: cartography, connectivity, and concurrent processing. *Curr. Opin. Neurobiol.* 2:150–55

Knudsen EI, du Lac S, Esterly SD. 1987. Computational maps in the brain. *Annu. Rev. Neurosci.* 10:41–65

Kobayakawa K, Kobayakawa R, Matsumoto H, Oka Y, Imai T, et al. 2007. Innate versus learned odour processing in the mouse olfactory bulb. *Nature* 450:503–8

Kohonen T, Hari R. 1999. Where the abstract feature maps of the brain might come from. *Trends Neurosci.* 22:135–39

Koide T, Miyasaka N, Morimoto K, Asakawa K, Urasaki A, et al. 2009. Olfactory neural circuitry for attraction to amino acids revealed by transposon-mediated gene trap approach in zebrafish. *Proc. Natl. Acad. Sci. USA* 106:9884–89

Laurent G, Stopfer M, Friedrich RW, Rabinovich MI, Volkovskii A, Abarbanel HD. 2001. Odor encoding as an active, dynamical process: experiments, computation, and theory. *Annu. Rev. Neurosci.* 24:263–97

Litaudon P, Amat C, Bertrand B, Vigouroux M, Buonviso N. 2003. Piriform cortex functional heterogeneity revealed by cellular responses to odours. *Eur. J. Neurosci.* 17:2457–61

Lledo PM, Alonso M, Grubb MS. 2006. Adult neurogenesis and functional plasticity in neuronal circuits. *Nat. Rev. Neurosci.* 7:179–93

Lodovichi C, Belluscio L, Katz LC. 2003. Functional topography of connections linking mirror-symmetric maps in the mouse olfactory bulb. *Neuron* 38:265–76

Luo L, Flanagan JG. 2007. Development of continuous and discrete neural maps. *Neuron* 56:284–300

Malnic B, Hirono J, Sato T, Buck LB. 1999. Combinatorial receptor codes for odors. *Cell* 96:713–23

Marin EC, Jefferis GS, Komiyama T, Zhu H, Luo L. 2002. Representation of the glomerular olfactory map in the *Drosophila* brain. *Cell* 109:243–55

Marks CA, Cheng K, Cummings DM, Belluscio L. 2006. Activity-dependent plasticity in the olfactory intrabulbar map. *J. Neurosci.* 26:11257–66

Masse NY, Turner GC, Jefferis GS. 2009. Olfactory information processing in *Drosophila*. *Curr. Biol.* 19:R700–13

Matsumoto H, Kobayakawa K, Kobayakawa R, Tashiro T, Mori K, Sakano H. 2010. Spatial arrangement of glomerular molecular-feature clusters in the odorant-receptor class domains of the mouse olfactory bulb. *J. Neurophysiol.* 103:3490–500

McGann JP, Pirez N, Gainey MA, Muratore C, Elias AS, Wachowiak M. 2005. Odorant representations are modulated by intra- but not interglomerular presynaptic inhibition of olfactory sensory neurons. *Neuron* 48:1039–53

Mitchison G. 1991. Neuronal branching patterns and the economy of cortical wiring. *Proc. Biol. Sci.* 245:151–58

Miyamichi K, Serizawa S, Kimura HM, Sakano H. 2005. Continuous and overlapping expression domains of odorant receptor genes in the olfactory epithelium determine the dorsal/ventral positioning of glomeruli in the olfactory bulb. *J. Neurosci.* 25:3586–92

Miyasaka N, Morimoto K, Tsubokawa T, Higashijima S, Okamoto H, Yoshihara Y. 2009. From the olfactory bulb to higher brain centers: genetic visualization of secondary olfactory pathways in zebrafish. *J. Neurosci.* 29:4756–67

Mombaerts P. 1996. Targeting olfaction. *Curr. Opin. Neurobiol.* 6:481–86

Mombaerts P. 2004. Odorant receptor gene choice in olfactory sensory neurons: the one receptor-one neuron hypothesis revisited. *Curr. Opin. Neurobiol.* 14:31–36

Mombaerts P. 2006. Axonal wiring in the mouse olfactory system. *Annu. Rev. Cell Dev. Biol.* 22:713–37

Mombaerts P, Wang F, Dulac C, Chao SK, Nemes A, et al. 1996. Visualizing an olfactory sensory map. *Cell* 87:675–86

Mori K, Nagao H, Yoshihara Y. 1999. The olfactory bulb: coding and processing of odor molecule information. *Science* 286:711–15

Mori K, Takahashi YK, Igarashi KM, Yamaguchi M. 2006. Maps of odorant molecular features in the mammalian olfactory bulb. *Physiol. Rev.* 86:409–33

Murphy GJ, Glickfeld LL, Balsen Z, Isaacson JS. 2004. Sensory neuron signaling to the brain: properties of transmitter release from olfactory nerve terminals. *J. Neurosci.* 24:3023–30

Nagayama S, Enerva A, Fletcher ML, Masurkar AV, Igarashi KM, et al. 2010. Differential axonal projection of mitral and tufted cells in the mouse main olfactory system. *Front. Neural Circuits* 4:120

Niessing J, Friedrich RW. 2010. Olfactory pattern classification by discrete neuronal network states. *Nature* 465:47–52

Niimura Y, Nei M. 2005a. Comparative evolutionary analysis of olfactory receptor gene clusters between humans and mice. *Gene* 346:13–21

Niimura Y, Nei M. 2005b. Evolutionary dynamics of olfactory receptor genes in fishes and tetrapods. *Proc. Natl. Acad. Sci. USA* 102:6039–44

Nikonov AA, Caprio J. 2001. Electrophysiological evidence for a chemotopy of biologically relevant odors in the olfactory bulb of the channel catfish. *J. Neurophysiol.* 86:1869–76

Nikonov AA, Finger TE, Caprio J. 2005. Beyond the olfactory bulb: an odotopic map in the forebrain. *Proc. Natl. Acad. Sci. USA* 102:18688–93

Ojima H, Mori K, Kishi K. 1984. The trajectory of mitral cell axons in the rabbit olfactory cortex revealed by intracellular HRP injection. *J. Comp. Neurol.* 230:77–87

Oka Y, Katada S, Omura M, Suwa M, Yoshihara Y, Touhara K. 2006. Odorant receptor map in the mouse olfactory bulb: in vivo sensitivity and specificity of receptor-defined glomeruli. *Neuron* 52:857–69

Olsen SR, Bhandawat V, Wilson RI. 2010. Divisive normalization in olfactory population codes. *Neuron* 66:287–99

Orona E, Rainer EC, Scott JW. 1984. Dendritic and axonal organization of mitral and tufted cells in the rat olfactory bulb. *J. Comp. Neurol.* 226:346–56

Poo C, Isaacson JS. 2009. Odor representations in olfactory cortex: "sparse" coding, global inhibition, and oscillations. *Neuron* 62:850–61

Rennaker RL, Chen CF, Ruyle AM, Sloan AM, Wilson DA. 2007. Spatial and temporal distribution of odorant-evoked activity in the piriform cortex. *J. Neurosci.* 27:1534–42

Restrepo D, Doucette W, Whitesell JD, McTavish TS, Salcedo E. 2009. From the top down: flexible reading of a fragmented odor map. *Trends Neurosci.* 32:525–31

Rieke F, Rudd ME. 2009. The challenges natural images pose for visual adaptation. *Neuron* 64:605–16

Rossiter KJ. 1996. Structure-odor relationships. *Chem. Rev.* 96:3201–40

Rubin BD, Katz LC. 1999. Optical imaging of odorant representations in the mammalian olfactory bulb. *Neuron* 23:499–511

Saito H, Chi Q, Zhuang H, Matsunami H, Mainland JD. 2009. Odor coding by a mammalian receptor repertoire. *Sci. Signal.* 2:ra9

Sakano H. 2010. Neural map formation in the mouse olfactory system. *Neuron* 67:530–42

Salinas E. 2006. How behavioral constraints may determine optimal sensory representations. *PLoS Biol.* 4:e387

Sato Y, Miyasaka N, Yoshihara Y. 2005. Mutually exclusive glomerular innervation by two distinct types of olfactory sensory neurons revealed in transgenic zebrafish. *J. Neurosci.* 25:4889–97

Sato Y, Miyasaka N, Yoshihara Y. 2007. Hierarchical regulation of odorant receptor gene choice and subsequent axonal projection of olfactory sensory neurons in zebrafish. *J. Neurosci.* 27:1606–15

Schiffman SS. 1974. Physicochemical correlates of olfactory quality. *Science* 185:112–17

Schoenfeld TA, Cleland TA. 2005. The anatomical logic of smell. *Trends Neurosci.* 28:620–27

Schoppa NE, Urban NN. 2003. Dendritic processing within olfactory bulb circuits. *Trends Neurosci.* 26:501–6

Serizawa S, Miyamichi K, Sakano H. 2004. One neuron-one receptor rule in the mouse olfactory system. *Trends Genet.* 20:648–53

Shao Z, Puche AC, Kiyokage E, Szabo G, Shipley MT. 2009. Two GABAergic intraglomerular circuits differentially regulate tonic and phasic presynaptic inhibition of olfactory nerve terminals. *J. Neurophysiol.* 101:1988–2001

Shepherd GM, Chen WR, Greer CA. 2004. Olfactory bulb. In *The Synaptic Organization of the Brain*, ed. GM Shepherd, pp. 165–216. New York: Oxford Univ. Press

Shipley MT, Ennis M. 1996. Functional organization of olfactory system. *J. Neurobiol.* 30:123–76

Soucy ER, Albeanu DF, Fantana AL, Murthy VN, Meister M. 2009. Precision and diversity in an odor map on the olfactory bulb. *Nat. Neurosci.* 12:210–20

Stettler DD, Axel R. 2009. Representations of odor in the piriform cortex. *Neuron* 63:854–64

Su CY, Menuz K, Carlson JR. 2009. Olfactory perception: receptors, cells, and circuits. *Cell* 139:45–59

Takeuchi H, Inokuchi K, Aoki M, Suto F, Tsuboi A, et al. 2010. Sequential arrival and graded secretion of Sema3F by olfactory neuron axons specify map topography at the bulb. *Cell* 141:1056–67

Talbot S, Marshall W. 1941. Physiological studies on neural mechanisms of visual localization and discrimination. *Am. J. Opthalmol.* 24:1255–63

Tan J, Savigner A, Ma M, Luo M. 2010. Odor information processing by the olfactory bulb analyzed in gene-targeted mice. *Neuron* 65:912–26

Tanaka NK, Awasaki T, Shimada T, Ito K. 2004. Integration of chemosensory pathways in the *Drosophila* second-order olfactory centers. *Curr. Biol.* 14:449–57

Tsuboi A, Miyazaki T, Imai T, Sakano H. 2006. Olfactory sensory neurons expressing class I odorant receptors converge their axons on an antero-dorsal domain of the olfactory bulb in the mouse. *Eur. J. Neurosci.* 23:1436–44

Uchida N, Takahashi YK, Tanifuji M, Mori K. 2000. Odor maps in the mammalian olfactory bulb: domain organization and odorant structural features. *Nat. Neurosci.* 3:1035–43

Vosshall LB, Amrein H, Morozov PS, Rzhetsky A, Axel R. 1999. A spatial map of olfactory receptor expression in the *Drosophila* antenna. *Cell* 96:725–36

Vosshall LB, Wong AM, Axel R. 2000. An olfactory sensory map in the fly brain. *Cell* 102:147–59

Vucinic D, Cohen LB, Kosmidis EK. 2006. Interglomerular center-surround inhibition shapes odorant-evoked input to the mouse olfactory bulb in vivo. *J. Neurophysiol.* 95:1881–87

Wachowiak M, Cohen LB. 2001. Representation of odorants by receptor neuron input to the mouse olfactory bulb. *Neuron* 32:723–35

Wachowiak M, McGann JP, Heyward PM, Shao Z, Puche AC, Shipley MT. 2005. Inhibition [corrected] of olfactory receptor neuron input to olfactory bulb glomeruli mediated by suppression of presynaptic calcium influx. *J. Neurophysiol.* 94:2700–12

Wachowiak M, Shipley MT. 2006. Coding and synaptic processing of sensory information in the glomerular layer of the olfactory bulb. *Semin. Cell Dev. Biol.* 17:411–23

Walz A, Omura M, Mombaerts P. 2006. Development and topography of the lateral olfactory tract in the mouse: imaging by genetically encoded and injected fluorescent markers. *J. Neurobiol.* 66:835–46

Wandell BA, Dumoulin SO, Brewer AA. 2007. Visual field maps in human cortex. *Neuron* 56:366–83

Wang F, Nemes A, Mendelsohn M, Axel R. 1998. Odorant receptors govern the formation of a precise topographic map. *Cell* 93:47–60

Wassle H. 2004. Parallel processing in the mammalian retina. *Nat. Rev. Neurosci.* 5:747–57

Wellis DP, Scott JW. 1990. Intracellular responses of identified rat olfactory bulb interneurons to electrical and odor stimulation. *J. Neurophysiol.* 64:932–47

Wesson DW, Wilson DA. 2010. Smelling sounds: olfactory-auditory sensory convergence in the olfactory tubercle. *J. Neurosci.* 30:3013–21

Wiechert MT, Judkewitz B, Riecke H, Friedrich RW. 2010. Mechanisms of pattern decorrelation by recurrent neuronal circuits. *Nat. Neurosci.* 13:1003–10

Wikipedia. *Topographic map (neuroanatomy)*. **http://en.wikipedia.org/wiki/Topographic_map_%28neuroanatomy%29**

Willhite DC, Nguyen KT, Masurkar AV, Greer CA, Shepherd GM, Chen WR. 2006. Viral tracing identifies distributed columnar organization in the olfactory bulb. *Proc. Natl. Acad. Sci. USA* 103:12592–97

Wilson DA. 1997. Binaral interactions in the rat piriform cortex. *J. Neurophysiol.* 78:160–69

Wise PM, Olsson MJ, Cain WS. 2000. Quantification of odor quality. *Chem. Senses* 25:429–43

Wong AM, Wang JW, Axel R. 2002. Spatial representation of the glomerular map in the *Drosophila* protocerebrum. *Cell* 109:229–41

Xu F, Greer CA, Shepherd GM. 2000. Odor maps in the olfactory bulb. *J. Comp. Neurol.* 422:489–95

Yaksi E, Judkewitz B, Friedrich RW. 2007. Topological reorganization of odor representations in the olfactory bulb. *PLoS Biol.* 5:e178

Yaksi E, von Saint Paul F, Niessing J, Bundschuh ST, Friedrich RW. 2009. Transformation of odor representations in target areas of the olfactory bulb. *Nat. Neurosci.* 12:474–82

Yan Z, Tan J, Qin C, Lu Y, Ding C, Luo M. 2008. Precise circuitry links bilaterally symmetric olfactory maps. *Neuron* 58:613–24

Yokoi M, Mori K, Nakanishi S. 1995. Refinement of odor molecule tuning by dendrodendritic synaptic inhibition in the olfactory bulb. *Proc. Natl. Acad. Sci. USA* 92:3371–75

Yoshihara Y. 2009. Molecular genetic dissection of the zebrafish olfactory system. *Results Probl. Cell Differ.* 47:97–120

Zhan C, Luo M. 2010. Diverse patterns of odor representation by neurons in the anterior piriform cortex of awake mice. *J. Neurosci.* 30:16662–72

Zhang X, Firestein S. 2002. The olfactory receptor gene superfamily of the mouse. *Nat. Neurosci.* 5:124–33

Zhang X, Rogers M, Tian H, Zou DJ, Liu J, et al. 2004. High-throughput microarray detection of olfactory receptor gene expression in the mouse. *Proc. Natl. Acad. Sci. USA* 101:14168–73

Zou DJ, Chesler A, Firestein S. 2009. How the olfactory bulb got its glomeruli: a just so story? *Nat. Rev. Neurosci.* 10:611–18

Zou DJ, Chesler AT, Le Pichon CE, Kuznetsov A, Pei X, et al. 2007. Absence of adenylyl cyclase 3 perturbs peripheral olfactory projections in mice. *J. Neurosci.* 27:6675–83

The Cognitive Neuroscience of Human Memory Since H.M.

Larry R. Squire[1,2,3,4] and John T. Wixted[4]

[1]Veterans Affairs Healthcare System, San Diego, California 92161

[2]Department of Psychiatry, [3]Department of Neurosciences, and [4]Department of Psychology, University of California, San Diego, La Jolla, California 92093; email: lsquire@ucsd.edu, jwixted@ucsd.edu

Annu. Rev. Neurosci. 2011. 34:259–88

First published online as a Review in Advance on March 29, 2011

The *Annual Review of Neuroscience* is online at neuro.annualreviews.org

This article's doi: 10.1146/annurev-neuro-061010-113720

0147-006X/11/0721-0259$20.00

Keywords

medial temporal lobe, hippocampus, neocortex, anterograde amnesia, retrograde amnesia

Abstract

Work with patient H.M., beginning in the 1950s, established key principles about the organization of memory that inspired decades of experimental work. Since H.M., the study of human memory and its disorders has continued to yield new insights and to improve understanding of the structure and organization of memory. Here we review this work with emphasis on the neuroanatomy of medial temporal lobe and diencephalic structures important for memory, multiple memory systems, visual perception, immediate memory, memory consolidation, the locus of long-term memory storage, the concepts of recollection and familiarity, and the question of how different medial temporal lobe structures may contribute differently to memory functions.

Contents

INTRODUCTION

In the earliest systematic writings about human memory, it was already appreciated that the study of memory impairment can provide valuable insights into the structure and organization of normal function (Ribot 1881, Winslow 1861). This tradition of research has continued to prove fruitful and has yielded a broad range of fundamental information about the structure and organization of memory. What is memory? Is it one thing or many? What are the concepts and categories that guide our current understanding of how memory works and that underlie the classification of its disorders? It is sometimes not appreciated that the concepts and categories used in current discussions

of memory are not fixed and were not easily established. Even the question of which cognitive operations reflect memory and which depend on other faculties has a long history of empirical work and discussion.

One needs only to sample nineteenth-century writings to recognize how differently memory was viewed then and now. For example, in his classic treatment of memory disorders, Ribot (1881) considered amnesias due to neurological injury together with amnesias due to psychological trauma. And he viewed aphasia and agnosia as disorders of memory, wherein (in aphasia, for example) patients have lost their memory for words or memory for the movements needed to produce words. Today, aphasia is considered a deficit of language, and agnosia a deficit of visual perception. Memory is affected but only as part of a more fundamental defect in a specific kind of information processing.

The notion that the study of brain injury can elucidate the organization of memory was itself a matter for empirical inquiry. If brain regions were highly interconnected, and the brain's functions distributed and integrated one with another, then damage to any one area would produce a global impairment, blurred across multiple faculties and affecting all of mental life. But the fact of the matter is different. The brain is highly specialized and modular, with different regions dedicated to specific operations. As a result, localized damage can produce strikingly specific effects, including a selective and circumscribed impairment of memory.

The idea that functions of the nervous system can be localized was already well accepted by the end of the nineteenth century. This localizationist view had its roots in the writings of Gall (1825) and was supported by the experimental work of Broca (1861), Ferrier (1876), Fritsch & Hitzig (1870), and others (see Finger 1994). Yet, these ideas centered mainly around sensory functions, motor control, and language and did not usefully address the topic of memory. Then, in the early twentieth century, an influential program of experimental work in rodents investigated directly the localization of memory with the conclusion that memory is distributed throughout the cortex and that the contribution to memory is equivalent across regions (Lashley 1929). This idea was strongly challenged (Hebb 1949, Hunter 1930) by the alternative, and more modern, interpretation that memory storage is indeed distributed but that different areas store different features of the whole. Still, as the midpoint of the twentieth century approached, memory functions, while distributed, were thought to be well integrated with perceptual and intellectual functions, and no region of the brain was believed to be disproportionately dedicated to memory. All that was about to change.

In 1957, Brenda Milner reported the profound effect on memory of bilateral medial temporal lobe resection, carried out to relieve epilepsy in a patient who became known as H.M. (1926–2008) (Scoville & Milner 1957, Squire 2009) (**Figure 1**). Remarkably, H.M. exhibited profound forgetfulness but in the absence of any general intellectual loss or perceptual disorders. He could not form new memories (anterograde amnesia) and also could not access some memories acquired before his surgery (retrograde amnesia). His impairment extended to both verbal and non-verbal material, and it involved information acquired through all sensory modalities. These findings established the fundamental principle that memory is a distinct cerebral function, separable from other perceptual and cognitive abilities, and also identified the medial aspect of the temporal lobe as important for memory. The early descriptions of H.M. can be said to have inaugurated the modern era of memory research, and the findings from H.M. enormously influenced the direction of subsequent work.

ANATOMY OF MEMORY

The work with H.M. is sometimes cited incorrectly as evidence of the importance of the hippocampus for memory, but this particular point could not of course be established by a large lesion that included not only the hippocampus but also the amygdala together with the adja-

cent parahippocampal gyrus. Which structures within H.M.'s lesion are important for memory became understood only gradually during the 1980s following the successful development of an animal model of human amnesia in the nonhuman primate (Mishkin 1978). Cumulative studies in the monkey (Murray 1992, Squire & Zola-Morgan 1991, Zola-Morgan et al. 1994) considerably clarified this issue. The important structures proved to be the hippocampus and the adjacent entorhinal, perirhinal, and parahippocampal cortices, which make up much of the parahippocampal gyrus (**Figure 2**).

One particularly instructive case of human memory impairment became available during this same time period (Zola-Morgan et al. 1986). R.B. developed a moderately severe, enduring impairment following an ischemic episode in 1978. During the five years until his death, his memory deficit was well documented with formal tests. Detailed histological examination of his brain revealed a circumscribed bilateral lesion involving the entire CA1 field of the hippocampus. Note that a lesion confined to the CA1 field must substantially disrupt hippocampal function because the CA1 field is a bottleneck in the unidirectional chain of processing that begins at the dentate gyrus and ends in the subiculum and entorhinal cortex. R.B. was the first case of memory impairment following a lesion limited to the hippocampus that was supported by extensive neuropsychological testing as well as neuropathological analysis.

The findings from R.B., considered together with the much more severe impairment in H.M., made two useful points. First, damage to the hippocampus itself is sufficient to produce a clinically significant and readily detectable memory impairment. Second, additional damage to the adjacent cortical regions along the parahippocampal gyrus (as in H.M.) greatly exacerbates the memory impairment. These same conclusions about the neuroanatomy of modest and severe memory impairment were also established in the monkey (Zola-Morgan et al. 1994).

Another case was subsequently described (patient G.D.) with a histologically confirmed bilateral lesion confined to the CA1 field and with a memory impairment very similar to R.B. (Rempel-Clower et al. 1996). Two other patients were also of interest. L.M. and W.H. had somewhat more severe memory impairment than did R.B. and G.D., but the impairment was still moderate in comparison to H.M. (Rempel-Clower et al. 1996). Histological examination

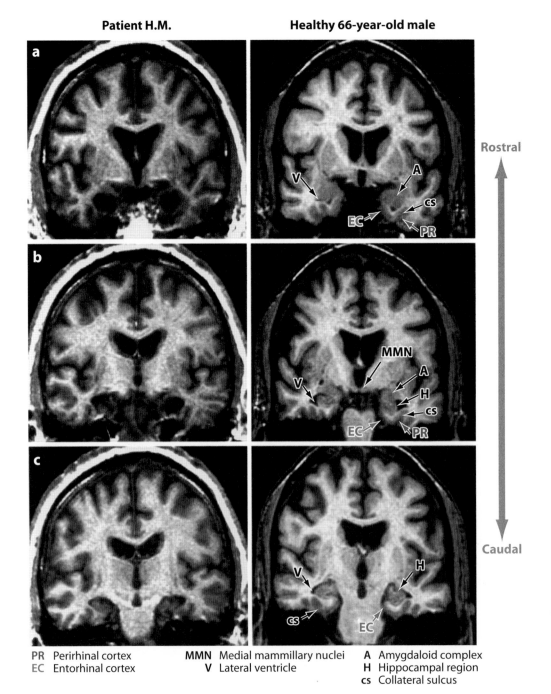

Patient H.M.   **Healthy 66-year-old male**

Rostral

Caudal

PR	Perirhinal cortex	MMN	Medial mammillary nuclei	A	Amygdaloid complex
EC	Entorhinal cortex	V	Lateral ventricle	H	Hippocampal region
				cs	Collateral sulcus

revealed extensive bilateral lesions of the hippocampal region, involving all the CA fields and the dentate gyrus. There was also some cell loss in entorhinal cortex and, for W.H., cell loss in the subiculum, as well. The more severe memory impairment in these two cases, in comparison to R.B. and G.D., could be due to the additional damage within the hippocampus or to the cell loss in entorhinal cortex.

There are only a small number of cases where detailed neuropsychological testing and thorough neurohistological analysis have combined to demonstrate memory impairment after limited hippocampal damage or larger medial temporal lobe lesions (see also Victor & Agamanolis 1990). Yet, neuroanatomical information is essential because it lays the groundwork for classifying memory disorders, for understanding qualitative and quantitative differences between patients, and for addressing questions about how specific structures may contribute differently to memory functions. Nonetheless, in the absence of histological data, valuable information can be obtained from structural imaging. Methods for high-resolution imaging of hippocampal damage were developed some time ago (Press et al. 1989), and quantitative data can now be obtained that provide reliable estimates of tissue volume (Gold & Squire 2005). These estimates are based on guidelines defined histologically and use landmarks in the medial temporal lobe that are visible on MRI (Insausti et al. 1998a,b).

An interesting observation has emerged from calculations of hippocampal volume in memory-impaired patients, usually patients who have sustained an anoxic episode. Across a number of reports, hippocampal volume (or area in the coronal plane) is typically reduced by ~40% [41%, $n = 10$ (Isaacs et al. 2003); 44%, $n = 5$ (Shrager et al. 2008); 43%, $n = 4$ (Squire et al. 1990); 45%, $n = 1$ (Cipolotti et al. 2001); 46%, $n = 1$ (Mayes et al. 2002)]. Neurohistological data from two of these patients (L.M. and W.H.) suggest an explanation for this striking consistency. As described above, these two patients had extensive cell loss in the hippocampus as well as in the dentate gyrus. Accordingly, a reduction in hippocampal volume of 40%, as estimated by MRI, may indicate a nearly complete loss of hippocampal neurons. The tissue collapses, but it does not disappear entirely. A volume loss in the hippocampus of ~40% may represent a maximum value for some etiologies of memory impairment.

While medial temporal lobe structures have received the most attention in studies of memory and memory impairment, it is notable that damage to the diencephalic midline also impairs memory. The deficit has essentially the same features as in medial temporal lobe amnesia. The best-known cause of diencephalic amnesia is alcoholic Korsakoff's syndrome. Here, damage to the medial dorsal thalamic nucleus (alone or perhaps in combination with damage to the mammillary nuclei) has been associated with memory impairment (Victor et al. 1989). Another survey of Korsakoff's syndrome documented damage to these two structures and, in addition, identified a role for the anterior thalamic nuclei (Harding et al. 2000). Six cases that were studied both neuropsychologically and neurohistologically (Gold & Squire 2006, Mair

Figure 1

Left column. Magnetic resonance images arranged from rostral (*a*) to caudal (*c*) through the temporal lobe of patient H.M. (in 1993 at age 67) and a 66-year-old healthy male (*right*). The comparison brain illustrates the structures that appear to have been removed during H.M.'s surgery in 1953. The lesion was bilaterally symmetrical, extending caudally 5.4 cm on the left side and 5.1 cm on the right. The full caudal extent of abnormal tissue is not illustrated. The damage included medial temporal polar cortex, most of the amygdaloid complex, virtually all the entorhinal cortex, and approximately the rostral half of the hippocampal region (dentate gyrus, hippocampus, and subicular complex). The perirhinal cortex was substantially damaged except for its ventrocaudal aspect. The more posterior parahippocampal cortex (areas TF and TH, not shown here) was largely intact. Adapted from Corkin et al. (1997) with permission from the Society for Neuroscience.

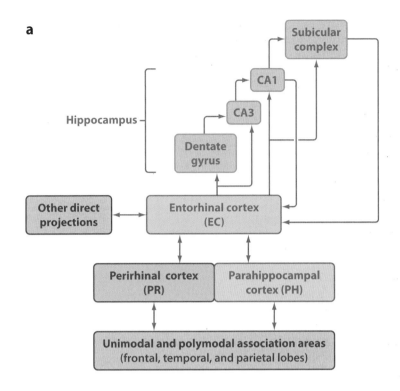

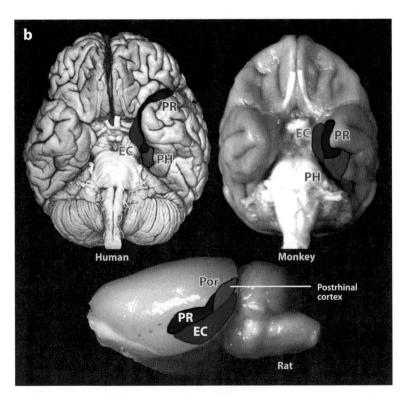

et al. 1979, Mayes et al. 1988) consistently identified damage in the medial thalamus (as well as in the mammillary nuclei for the five cases with Korsakoff's syndrome). Two regions of thalamus were implicated by these cases and by two neuroimaging studies of diencephalic amnesia (Squire et al. 1989, von Cramon et al. 1985): first, the medial dorsal nucleus and the adjacent internal medullary lamina; and second, the mammillothalamic tract and its target, the anterior thalamic nuclei. Damage to either of these regions can cause memory impairment. These diencephalic nuclei and tracts are anatomically related to the medial temporal lobe. The perirhinal cortex originates projections to the medial dorsal nucleus that enter through the internal medullary lamina, and the hippocampal formation projects both to the rostrally adjacent anterior nuclei and to the mammillary nuclei. These anatomical connections likely explain why patients with medial temporal or diencephalic lesions exhibit the same core deficit.

PRINCIPLES OF ORGANIZATION SUGGESTED BY H.M.'S FINDINGS

The early descriptions of H.M suggested four principles about how memory is organized in the brain. First, despite his debilitating and pervasive memory impairment, H.M. successfully acquired a motor skill. This finding raised the possibility that memory is not a single thing. Second, because his memory impairment appeared to be well circumscribed, the structures damaged in memory-impaired patients were thought not to be involved in intellectual and perceptual functions. Third, H.M. had a considerable capacity for sustained attention, including the ability to retain information for a period of time after it was first encountered. This finding suggested that medial temporal lobe structures are not needed for immediate memory or for the rehearsal and maintenance of material in what would now be termed working memory. Fourth, H.M. appeared to have good access to facts and events from time periods remote to his surgery. This observation suggested that the medial temporal lobe cannot be the ultimate storage site for long-term memory. Permanent memory must be stored elsewhere, presumably in neocortex. In the years since H.M. was described, each of these ideas has been the topic of extensive experimental work.

During the 1960s and 1970s, when human memory impairment began to be systematically studied, there was considerable debate about whether medial temporal and diencephalic structures were concerned more with storage or with retrieval. The findings from H.M. led to the view that these structures are needed for memory storage, that is, for the establishment of new representations in long-term memory. If these structures are unable to participate in forming long-term memory, then representations established in immediate memory are presumably lost or perhaps achieve some disorganized state. Consider the case of transient amnesic episodes (transient global amnesia or the memory impairment associated with electroconvulsive therapy). Here, the events that occur during the period of anterograde amnesia are not subsequently remembered

---

Figure 2

(*a*) Schematic view of the medial temporal lobe memory system for declarative memory, which is composed of the hippocampus and the perirhinal, entorhinal, and parahippocampal cortices. In addition to the connections shown here, there are also weak projections from the perirhinal and parahippocampal cortices to the CA1-subiculum border. (*b*) Ventral view of a human brain (*upper left*), monkey brain (*upper right*), and a lateral view of a rat brain (*lower center*). The major cortical components of the medial temporal lobe are highlighted and outlined. The hippocampus is not visible from the surface and in the human lies beneath the cortex of the medial temporal lobe. Its anterior extent lies below the posterior entorhinal (*red*) and perirhinal (*purple*) cortices, and the main body of the hippocampus lies beneath the parahippocampal cortex. In the rat, the parahippocampal cortex is termed postrhinal cortex. Abbreviations: EC, entorhinal cortex; PH, parahippocampal cortex (*dark yellow*); Por, postrhinal cortex; PR, perirhinal cortex.

after recovery from the amnesic condition. New learning again becomes possible, but events from the amnesic episode do not return to memory. Thus, if medial temporal lobe or diencephalic structures are not functional at the time of learning, memory is not established in a usable way and does not become available at a later time. More direct investigations of this issue using single-cell recording in monkeys have reached similar conclusions (Higuchi & Miyashita 1996; see Squire 2006). The idea is that the synaptic changes that would ordinarily represent acquired information in long-term memory either are lost altogether or fail to develop into a stable, coherent ensemble.

MULTIPLE MEMORY SYSTEMS

The memory impairment in H.M. and other patients is narrower than once thought in that not all kinds of learning and memory are affected. The first hint of this idea came when H.M. was found capable of learning a hand-eye coordination skill (mirror drawing) over a period of days, despite having no recollection of practicing the task before (Milner 1962). Although this finding showed that memory was not unitary, for some time it was thought that motor skill learning was a special case and that all the rest of memory is of one piece and is impaired in amnesia. Subsequently, it was discovered that motor-skill learning is but one example of a large domain of learning and memory abilities, all of which are intact in H.M. and other patients. H.M.'s motor skill learning marked the beginning of a body of experimental work that would eventually establish the biological reality of two major forms of memory.

An early insight was that perceptual skills and cognitive skills, not just motor skills, are preserved in amnesia. Specifically, amnesic patients acquired at a normal rate the perceptual skill of reading mirror-reversed words, despite poor memory for the task itself and for the words that were read (Cohen & Squire 1980). This finding was the basis for the formulation of a brain-based distinction between two major forms of memory, which afford either declarative or procedural knowledge. Declarative knowledge referred to knowledge available as conscious recollections about facts and events. Procedural knowledge referred primarily to skill-based information, where what has been learned is embedded in acquired procedures.

Subsequently, memory-impaired patients were found to exhibit intact priming effects (see Tulving & Schacter 1990). For example, patients (like healthy volunteers) could name pictures of objects 100 ms faster when the pictures had been presented previously than when they were presented for the first time and independently of whether patients could recognize the pictures as familiar (Cave & Squire 1992).

Another important insight was the idea that the neostriatum (not the medial temporal lobe) is important for the sort of gradual, feedback-guided learning that results in habit memory (Mishkin et al. 1984). Thus, memory-impaired patients learned at a normal rate when explicit memorization was not useful (for example, when the outcome of each trial was determined probabilistically and performance needed to be based on a gut feeling) (Knowlton et al. 1996). Furthermore, tasks that healthy volunteers could learn rapidly by memorization (such as the concurrent learning of eight different, two-choice object discriminations) could also be learned successfully by profoundly amnesic patients, albeit very gradually (healthy volunteers required fewer than 80 trials; patients required more than 1000 trials). Although memory became robust in the patients after extended training (>90% accuracy), it differed from the memory acquired by healthy volunteers in that what was learned was outside of awareness and was rigidly organized (performance collapsed when the task format was modified) (Bayley et al. 2005a).

Given the wide variety of learning and memory phenomena that could be demonstrated in patients (for example, priming and habit learning), the perspective eventually shifted to a framework that accommodated multiple memory systems, not just two kinds of memory.

Indeed, one could ask what the various kinds of memory that were preserved in patients had in common aside from the fact that they were not declarative. Accordingly, the term nondeclarative was introduced with the idea that declarative memory refers to one kind of memory system and that nondeclarative memory is an umbrella term referring to several additional memory systems (Squire & Zola-Morgan 1988). Nondeclarative memory includes skills and habits, simple forms of conditioning, emotional learning, priming, and perceptual learning, as well as phylogenetically early forms of behavioral plasticity such as habituation and sensitization.

Declarative memory is the kind of memory that is referred to when the term memory is used in everyday language. Declarative memory allows remembered material to be compared and contrasted. The stored representations are flexible, accessible to awareness, and can guide performance in a variety of contexts. Declarative memory is representational. It provides a way of modeling the external world, and it is either true or false. Nondeclarative memory is neither true nor false. It is dispositional and is expressed through performance rather than recollection. These forms of memory provide for myriad unconscious ways of responding to the world. In no small part, by virtue of the unconscious status of the nondeclarative forms of memory, they create some of the mystery of human experience. Here arise the dispositions, habits, and preferences that are inaccessible to conscious recollection but that nevertheless are shaped by past events, influence our behavior and mental life, and are an important part of who we are.

VISUAL PERCEPTION

Formal testing of patient H.M. over the years documented his good performance on intelligence tests and on other tests of perceptual function and lexical knowledge (Kensinger et al. 2001, Milner et al. 1968). He could detect the anomalous features of cartoon drawings, and he performed above the control mean on the Mooney "Closure" task, which requires participants to find a face in a chaotic black and white pattern with incomplete contour (Milner et al. 1968). This perspective, that visual perception is intact after large medial temporal lobe lesions, was eventually challenged, first by work in monkeys (Eacott et al. 1994) and later by studies in humans (Lee et al. 2005a,b). These studies proposed that the perirhinal cortex, one of the structures damaged in H.M., is important for complex visual perceptual tasks involving stimuli with substantial feature overlap. It was also proposed that the hippocampus is needed when spatial processing is required, as in visual discriminations involving scenes.

Although some subsequent studies appeared to provide additional support for this perspective (Barense et al. 2007, Lee & Rudebeck 2010), attempts to replicate some of the key early work and to find impairments with new tests were unsuccessful (Shrager et al. 2006). Comprehensive reviews of this topic (Suzuki 2009, 2010) raised three important issues. First, a consideration of neuroanatomic and neurophysiological data emphasizes that the perirhinal cortex has unique characteristics that distinguish it from the laterally adjacent, unimodal visual area TE. The perirhinal cortex is a polymodal association area with strong connections to the hippocampus and entorhinal cortex, and it is difficult to view the perirhinal cortex as a visual area and as a continuation of the ventral visual pathway (Suzuki 2010).

Second, many of the studies designed to test visual perception, particularly studies in monkeys, involve a significant memory requirement. Thus, impaired associative learning or impaired long-term memory for the stimulus material could have contributed to many of the deficits reported after perirhinal lesions in monkeys. Even in studies of humans, impaired associative learning could result in deficient performance when different test stimuli need to be judged against the same two comparison stimuli on every trial (Graham et al. 2006). Indeed, in a new study that explored this issue, patients with hippocampal lesions were impaired when the same comparison stimuli were used on every trial but were fully intact when the stimuli were unique to every trial (Kim et al. 2011). Using fixed comparison stimuli gives an advantage to those who can remember because one can learn what to look for in the test stimuli to decide which comparison stimulus it most closely resembles.

Third, patients who exhibit impaired performance on tasks of visual perception may have significant damage to lateral temporal cortex in addition to medial temporal lobe damage. This idea merits consideration, given that two of the three patients with medial temporal lobe damage who were impaired were reported to have damage lateral to the medial lobe (Barense et al. 2007; Lee et al. 2005a,b; Lee & Rudebeck 2010). Also, estimates of damage in most of the patients who were impaired were based on ratings of single sections through the lateral temporal cortex, not on quantitative measures of the entire region, thus leaving large amounts of tissue unexamined.

The importance of thorough neuroanatomical measurement in neuropsychological studies of memory cannot be overstated. Many current disagreements about the facts and ideas emerging from neuropsychological research on human memory can be traced to concerns about the locus and extent of lesions. If a deficit is expected but not found, perhaps the damage is less extensive than believed. If a deficit is not expected but is found, perhaps the damage is more extensive than has been detected. There is no substitute for thorough, quantitative descriptions of damage based on magnetic resonance imaging, as well as (where possible) detailed neurohistological description of the postmortem brain.

The possible role of perirhinal cortex in certain kinds of visual perception remains a topic of discussion and will benefit from detailed analysis of the lesions in the cases under study. At the present time, the weight of evidence from experimental lesion studies in monkeys, neurophysiological studies, and human neuropsychological studies continues to support the view that medial temporal lobe structures are important for declarative memory and not for perceptual functions (see also Clark et al. 2011).

IMMEDIATE MEMORY AND WORKING MEMORY

The early descriptions of H.M. emphasized how capable he was at focusing his attention and at retaining information for short periods of time (Milner et al. 1968). For example, he could retain a three-digit number for 15 minutes by continuous rehearsal, using what would now be termed working memory (Baddeley 2003). Yet when his attention was diverted, he forgot the whole event. In one dramatic demonstration, participants heard digit strings of increasing length (Drachman & Arbit 1966) (**Figure 3a**). Each string was presented as many times as needed until it was reported back correctly. Then, a new digit string was presented that was one digit longer than the previous one. Controls made their first errors with strings of eight digits and were eventually able to repeat strings as long as 20 digits (with no more than 25 repetitions at any one string length). In contrast, H.M. exhibited a marked discontinuity in performance as the string length increased. He repeated up to six digits correctly on his first try (six was his preoperative digit span), but he never succeeded at seven digits, even though he was given 25 repetitions of the same string. The interpretation was that at short string lengths H.M. could rely on his intact immediate memory and that he failed when the material to be remembered was more than could be held in mind. That is, he failed when the material exceeded his immediate memory capacity.

Time is not the key factor that determines how long information can be retained by patients like H.M. The relevant factors are immediate memory capacity and how successfully material can be maintained in working memory through rehearsal. Maintenance of information is difficult when material is difficult to rehearse (e.g., faces and designs). Moreover, working memory capacity can be quite limited, and typically only three or four simple visual objects can be maintained (Cowan 2001, Fukuda et al. 2010). With these considerations in mind, it is perhaps not surprising that impaired performance after medial temporal lobe lesions has sometimes been reported at short retention intervals, usually when the task requires learning complex material or learning the relations between items (e.g., object-location associations) (Finke et al. 2008, Hannula et al. 2006, Kan et al. 2007, Olson et al. 2006). In these cases, the important question is whether working memory capacity has been exceeded and performance must rely on long-term memory, or whether working memory sometimes depends on the medial temporal lobe. Methods that are independent of the particular task that is used are needed to decide this question.

One approach to this issue seems promising in cases where the retention interval is long enough (about 8 seconds) to allow a manipulation to be introduced during the interval (Shrager et al. 2008). Controls (but not patients) were given either distraction or no distraction between study and test. Across experiments involving names, faces, or object-location associations, patient performance was related to how distraction affected controls. The patients were impaired when distraction had no effect on control performance, and the patients were intact when distraction disrupted control performance. These results suggested that the patients were impaired when the task

depended minimally on working memory (as indicated by the ineffectiveness of distraction on control performance), and they performed well when the task depended substantially on working memory (as indicated by the disruptive effect of distraction on controls). Thus, for the kinds of material studied here, including relational information for objects and locations, working memory appears to be intact after medial temporal lobe damage.

A possible approach in cases where the retention interval is very short (1–3 seconds) is based on the early study of digit span, described above. Participants saw different numbers of objects (1 to 7) arranged on a tabletop and then immediately tried to reproduce the array on an adjacent table (Jeneson et al. 2010) (**Figure 3b**). The same study-test sequence was repeated (up to a maximum of ten times) until participants correctly placed each object within a specified distance of its original location. The finding was that performance was intact when only a few object locations needed to be remembered. However, just as was found for digit strings, there was an abrupt discontinuity in performance with larger numbers of object locations. For example, patient G.P. (who has large medial temporal lobe lesions similar to H.M.'s lesions) learned 1, 2, or 3 object locations as quickly as

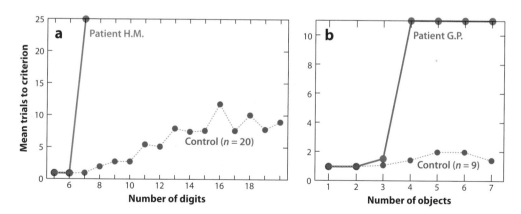

Figure 3

Intact working memory and impaired long-term memory. (*a*) The number of trials needed to succeed at each string length for patient H.M. and controls. H.M. could not succeed at repeating back 7 digits even after 25 attempts with the same string. (*b*) The number of trials needed to learn the locations of different numbers of objects for patient G.P. and controls. G.P. could not reproduce the locations of four objects, even after 10 attempts with the same display (panel *a* adapted from Drachman & Arbit 1966, with permission from the American Medical Association, and panel *b* adapted from Jeneson et al. 2010).

did controls, needing no more than one or two tries at each stage. However, when four object locations needed to be remembered, he could not succeed even in 10 attempts with the same array. These findings suggest that the maintenance of relational information (in this case, object-location associations) can proceed normally, even in patients with large medial temporal lobe lesions. An impairment is evident only when a capacity limit is reached, at which point performance must depend, at least in part, on long-term memory.

These observations support the view that patients with medial temporal lobe lesions can succeed at remembering whatever they have encountered, so long as the material to be remembered can be supported by a limited-capacity, short-term memory system (see also Jeneson et al. 2011). This formulation touches on a large and fundamental issue: whether there is any ability at all that depends on the hippocampus and related structures, even when a task can be managed within working memory. That is, do these structures perform any online computations for which the distinction between working memory and long-term memory is irrelevant?

This is a question of considerable current interest. It runs through discussions of perceptual functions and discussions of relational memory (as considered in this section and the preceding section). The issue is especially prominent in discussions of spatial cognition. For example, the ability to path integrate (i.e., the ability to use self-motion cues to keep track of a reference location as one moves though space) has been proposed to have a fundamental dependency on the hippocampus and entorhinal cortex. That is, these structures are proposed to carry out computations essential for path integration, regardless of the memory load or the retention interval (Whitlock et al. 2008). Furthermore, the hippocampus is proposed to be necessary for constructing a spatially correct mental image of either a remembered scene or an imagined scene (Bird et al. 2010, Bird & Burgess 2008), a task that need not involve recollection at all.

In the case of path integration, humans can succeed at simple paths in the absence of hippocampus and entorhinal cortex so long as the task can be managed within 30–40 seconds (presumably supported by working memory) (Shrager et al. 2008). In the case of spatial imagining, patients with severe memory impairment can describe routes around their childhood neighborhoods, including when main routes are blocked and alternative routes must be found (Rosenbaum et al. 2000, Teng & Squire 1999). Furthermore, in one study, patients with hippocampal damage successfully imagined future events and provided a normal number of spatial referents (Squire et al. 2010; see Hassabis et al. 2007 for a deficit on a similar task). These demonstrations appear straightforward and would seem to raise doubts about the idea that the hippocampus performs online computations. Yet there is an alternate perspective. Specifically, it has been suggested that spatial representations can be established outside the hippocampus, and in parallel with hippocampal representations, but using somewhat different computations (Bird & Burgess 2008, Whitlock et al. 2008). By this account, some spatial tasks that are accomplished successfully after hippocampal damage are in fact being accomplished using different structures and different computations than are used by healthy individuals.

The idea is that, despite intact performance in patients, some tasks are still hippocampus-dependent and could be shown to be so if one could devise tasks that can only be done with computations unique to the hippocampus. This is an interesting perspective and one that, in principle, could be applied to any example of intact performance in patients. It will be difficult to resolve issues like these without understanding which strategies are used in any particular case and without gaining experimental control over them. In addition, tasks that can be solved by different structures and using different strategies may be associated with inconsistent deficits after hippocampal lesions. In contrast, there are some tasks that depend on the medial temporal lobe, where performance deficits are invariably pronounced, and where performance cannot be made to appear normal by recruiting other brain structures or by using different strategies. These are tasks that assess the ability to form conscious long-term memory of facts and events, and the inability to carry out this function appears to be the central deficit in H.M. and other patients with medial temporal lobe lesions.

REMOTE MEMORY AND MEMORY CONSOLIDATION

A key insight about the organization of memory came with early observations of H.M.'s capacity to remember information that he acquired before his surgery in 1953. Initially, he was described as having a loss of memory (retrograde amnesia) covering the three years immediately preceding surgery and with earlier memories "seemingly normal" (Scoville & Milner 1957, p. 17). About ten years later, the impression was similar as there did not appear to have been any change in H.M.'s capacity to recall remote events antedating his operation, such as incidents from his early school years, a high school attachment, or jobs he had held in his late teens and early twenties (Milner et al. 1968, p. 216).

The first study of this issue with formal tests asked H.M. to recognize faces of persons who had become famous in the decades 1920–1970 (Marslen-Wilson & Teuber 1975). As expected, he performed poorly in the postmorbid period (the 1950s and 1960s) but did as well as or better than age-matched controls at recognizing faces from the premorbid period (the 1920s–1940s). This important finding implied that medial temporal lobe structures are not the ultimate storage sites for acquired memories. Memories that initially require the integrity of medial temporal lobe structures must be reorganized as time passes after learning so as to gradually become independent of these structures. The extent of retrograde amnesia provides an indication of how long this process takes.

Retrograde amnesia can be either temporally limited, covering a few years, or prolonged, depending on the locus and extent of the damage. Patients with damage thought to be restricted to the hippocampus had retrograde amnesia for past news events that extended only a few years into the premorbid period (Manns et al. 2003b). By contrast, patients with large medial temporal lobe lesions (damage to hippocampus plus parahippocampal gyrus) exhibited extended retrograde amnesia that covered several decades, albeit sparing memories acquired in early life (patients E.P. and G.P.; Bayley et al. 2006, Bright et al. 2006). The possibility that some amount of more lateral damage (e.g., in the fusiform gyrus) contributed to the extended retrograde impairment in E.P. and G.P. cannot be excluded.

There has been particular interest in the status of autobiographical memories for unique events following medial temporal lobe damage, and in recent years methods have been developed to assess the detail with which such recollections can be reproduced. In the earliest formal assessments of H.M. (Sagar et al. 1985), he produced well-formed autobiographical memories from age 16 and younger (his surgery occurred at age 27). However, the situation seemed to change as H.M. aged. In a later update (Corkin 2002), H.M. (now 76 years old) was reported to have memories of childhood, but the memories appeared fact-like and lacked detail. It was stated that he could not reproduce a single event that was specific to time and place. In a formal study reported a few years later (Steinvorth et al. 2005), he was also impaired in recollecting events from his early life. It was concluded that autobiographical memories remain dependent on the medial temporal lobe so long as the memories persist.

This conclusion about H.M. is complicated by the findings from MRI scans obtained in 2002 and 2003 (Salat et al. 2006). These scans documented a number of significant changes since his first MRI scans from 1992–1993 (Corkin et al. 1997) (**Figure 1**). Specifically, the scans showed cortical thinning, subcortical atrophy, large amounts of abnormal white matter, and subcortical infarcts. All these features were thought to have developed during the past decade, and they complicate the interpretation of neuropsychological data collected during and after this period. Considering the earlier reports that he could successfully retrieve past autobiographical memories (Milner et al. 1968, Scoville & Milner 1957), it is possible that remote autobiographical memories were in fact intact during the early years after surgery but were later compromised by neurological change. It is also possible that the available memories faded with time because they could not be strengthened through rehearsal and relearning.

Other work has supported the earlier descriptions of H.M. For example, methods similar to those used to assess H.M. have also been used to evaluate autobiographical memory in other patients with hippocampal damage or larger medial temporal lobe lesions (Bright et al. 2006, Kirwan et al. 2008). These patients had intact autobiographical memory from their early lives. The following example illustrates a well-formed autobiographical memory produced by E.P. about his early life, one of 18 that he produced. In this case, he was asked for a specific recollection in response to the cue word "fire." Like most recollections, his narrative contains both fact-based and event-specific information. Note the several repetitions in the narrative, which reflect his severe anterograde amnesia.

> Dad had 3 1/2 acres of property in Castro Valley and the back property would just grow and would be dry and for some reason, I didn't do it, but somehow or other the next thing we knew is that it was starting to burn. I told dad and he called the Castro Valley fire department. They came up and they got it out real quick. However it started I don't know. He had 3 1/2 acres of property and he just let it grow. It would be grass or whatever. Who knows how it started, but it started to burn. Dad called the Castro Valley fire department and they came up and all the volunteers came in and they got it out in a matter of 10–15 minutes. They stamped it out. They don't know how it started. I was 16–17, in that bracket. Dad had 3 1/2 acres of property. It was summer time, 1938. Those sort of things I think you remember. (Bayley et al. 2003, p. 139)

The same finding of intact early memories was reported in 10 patients with medial temporal lobe lesions in a study of emotional (and remote) autobiographical memories (Buchanan et al. 2005), and in two other patients (M.R. and P.D.), using a simpler assessment device (Eslinger 1998). In another study of four patients with medial temporal lobe damage and variable damage to anterior and posterior temporal neocortex (Rosenbaum et al. 2008), one patient (S.J.) was reported to have extended retrograde amnesia for autobiographical memory. The other three patients were less impaired, performing poorly in time periods closer to the onset of their amnesia. The impairment in S.J. was attributed to hippocampal damage. Alternatively, it is difficult to rule out a substantial contribution from the damage that was identified in neocortex.

It is noteworthy that, not infrequently, patients have been described as having extensive and ungraded retrograde amnesia (i.e., unrelated to how long ago the memory was formed) (for examples, see Bright et al. 2006, Cipolotti et al. 2001, Noulhiane et al. 2007, Rosenbaum et al. 2008, Sanders & Warrington 1971). This pattern of impairment has sometimes been taken to mean that the hippocampus (or related structures) is required as long as a memory persists. Yet, in many cases testing did not cover early adulthood and adolescence, so it is possible that the amnesia was not as ungraded as it appeared to be. In other cases, the damage was known to extend substantially into lateral temporal neocortex (see Bright et al. 2006 and Squire & Bayley 2007 for consideration of several cases). In one report of patients with unilateral temporal lobe resections, autobiographical memory was impaired across all past time periods (Noulhiane et al. 2007). In these patients, damage was recorded in the medial temporal lobe as well as in the temporal pole and in the anterior aspect of the superior, middle, and inferior temporal gyri. It is difficult to know to what extent this damage outside the medial temporal lobe might have contributed to the impairment. Significant damage to lateral temporal or frontal cortex can severely impair performance on tests of remote memory, including tests of autobiographical memory about early life [7 cases, Bright et al. (2006); patients H.C., P.H., and G.T., Bayley et al. (2005b); patient E.K., Eslinger (1998)]. If lateral temporal cortex, for example, is a site of long-term memory storage (Mishkin 1982, Miyashita 1993), then lateral temporal damage would be expected to cause severe and extended retrograde amnesia. The difficulty is knowing in any particular case to what extent such damage is responsible for impaired remote memory.

Among several single-case studies reporting impaired memory for early-life events (see Squire & Bayley 2007 for discussion), patient V.C. has been the most carefully documented. The volume of his lateral temporal lobes was reported as normal. Yet, it is striking that V.C.'s 1/9 score on the childhood portion of the autobiographical memory interview differs sharply from the good scores (and sometimes maximum scores of 9) obtained on the same test by as many as 12 patients with MRI documentation of limited medial temporal lobe damage [$n = 8$, Bayley et al. (2006); $n = 2$, Eslinger (1998); $n = 1$, Kapur & Brooks (1999); $n = 1$, Schnider et al. (1995). With the possible exception of V.C., we are unaware of memory-impaired patients who have damage limited to the medial temporal lobe (as documented by neurohistology or thorough MRI) and who do so poorly at recollecting remote autobiographical memories (**Figure 4**).

The finding that retrograde amnesia is temporally limited after damage to the medial temporal lobe implies a process of reorganization whereby over time memories become less dependent on medial temporal lobe structures. As time passes after learning, the role of medial temporal lobe structures diminishes and a more permanent memory gradually develops, presumably in neocortex. According to a different perspective, only fact-based memories (not autobiographical memories) make this transition (Winocur et al. 2010). This view discounts the possible importance of neocortical damage in patients with impaired autobiographical remembering of remote events and attributes the impairment specifically to hippocampal damage.

Some studies in experimental animals have directly tracked neural activity and structural changes in the hippocampus and neocortex after learning. Expression patterns of c-Fos described gradually decreasing activity in the mouse hippocampus after learning and parallel increases in a number of cortical regions (Frankland & Bontempi 2005). These findings and others (Restivo et al. 2009) reflect the increasing importance of distributed cortical regions for the representation of memory as time passes. The idea is not that memory is literally transferred from the hippocampus to neocortex but that gradual changes in the neocortex increase the complexity, distribution, and connectivity among multiple cortical regions. The next section considers what the study of patients has contributed to understanding the organization and storage of long-term memory.

MEMORY IN THE NEOCORTEX

The view that emerged from the study of H.M. and other patients is that medial temporal lobe structures are uniquely specialized to establish and maintain declarative memories. Other structures support the initial perception and processing of an experience, and these other structures are also critical for the long-term storage of the experience. A long-standing view is that the cortical processing of a multisensory experience leaves a distributed record in the same multiple regions that initially performed the processing. For example, neurons in visual areas store the visual aspect of a multisensory experience, neurons in auditory areas store the auditory aspect of the experience, other areas store the spatial aspects, and so on. According to this view, any act of remembering consists of the coordinated reactivation of the distributed neocortical regions that were engaged at the time of encoding (Damasio 1989, De Renzi 1982, Mishkin 1982, Squire 1987). When a memory is first formed, this reactivation depends on the hippocampus and related structures, but once

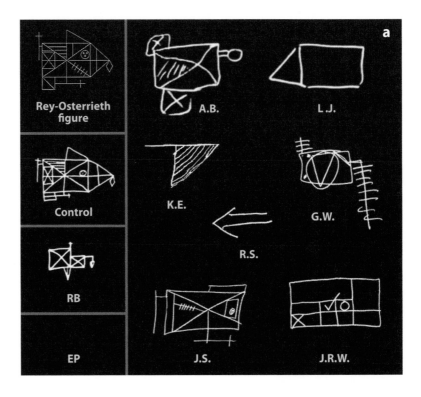

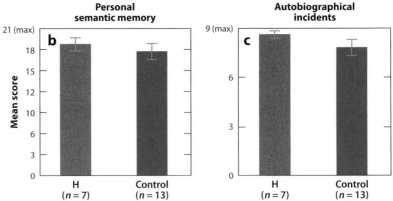

Figure 4

(*a*) Participants copied the Rey-Osterrieth figure illustrated in the small box in the upper left and 10–15 min later, without forewarning, tried to reproduce it from memory. The reproduction by a representative control is shown below the target figure. The left panel also shows the reproduction by patient R.B., who had histologically identified lesions of the CA1 field of the hippocampus (Zola-Morgan et al. 1986). Patient E.P., who had large medial temporal lobe lesions, did not recall copying a figure and declined to guess. The right section shows reproductions by seven patients with circumscribed damage to the hippocampus. Panels *b* and *c* show scores for the same seven patients (H) and 13 controls on the autobiographical memory interview, childhood portion (Kopelman et al. 1989). These findings suggest that patients who fail to produce any of the complex figure (like E.P.) or who are deficient at producing either remote semantic memories (A, maximum score, 21) or remote autobiographical events (B, maximum score, 9) will prove to have damage beyond the hippocampus. Indeed, even E.P. with his large lesions limited mainly to the medial temporal lobes, obtained maximal scores on these two tests (21/21 and 9/9).

memory is fully consolidated, reactivation can occur independently in neocortex.

A considerable body of evidence supporting the reactivation view has come from studies using fMRI (see Buckner & Wheeler 2001, Danker & Anderson 2010 for reviews). For example, several studies have found that the modality-specific or category-specific processes engaged at encoding tend to be re-engaged at retrieval (e.g., Polyn et al. 2005, Wheeler et al. 2000, Woodruff et al. 2005). This perspective of remembering implies that the dedicated processing areas of the neocortex can also be viewed as memory areas. However, rather than broadly encoding and consolidating memories, like the structures of the medial temporal lobe, each neocortical region operates within a very specific domain, and each region stores only specific features of an experience. It follows then that the same neocortical lesions that selectively impair processing in one particular domain should also cause correspondingly specific anterograde and retrograde memory impairments within the same domain. Although an extensive literature documents the selective information-processing deficits that are associated with different cortical lesions, the effects of those lesions on new learning and past remembering are only rarely considered. Here, we consider the cognitive effects of selective processing deficits with a view toward also identifying the effects on memory.

Achromatopsia

Finding selective anterograde memory impairment in association with a selective perceptual processing deficit would not be surprising. That is, if a perceptual deficit is present in one modality (e.g., visual perception), it should also be difficult to learn new material presented in the same modality. In addition, there should be consequences for remembering the past. Specifically, a selective deficit in processing particular features of visual material should selectively compromise the ability to recollect the same features in a previous memory, while leaving other aspects of the memory intact. This

idea is illustrated by "The Case of the Color-blind Painter" (Sacks 1995). An accomplished painter was involved in an automobile accident at the age of 65, which rendered him completely color blind. Although the anatomical basis of his disability was not identified, it was thought to have been caused by damage to regions dedicated to the perception of color (possibly including area V4). The disability itself was striking. The patient could discriminate between wavelengths of light, even though the different wavelengths no longer gave rise to the perception of different colors. Instead, different wavelengths gave rise to the perception of different shades of gray. Because this was a case of acquired cerebral achromatopsia (i.e., cortical color blindness), it was possible to ask about the status of previously established memories that had once included the subjective experience of color. If color in early memories depends on the same cortical structures that support the perception of color, then previously intact memories that were once retrieved in color should now be retrieved in black and white. Indeed, the case description leaves little doubt that the patient's experience—both going forward and looking back—was now completely (and selectively) devoid of color. Although he retained abstract semantic knowledge of color, he could neither perceive nor later remember the color of objects presented to him (anterograde impairment). In addition, he could not subjectively experience color in his earlier (and once chromatic) memories (retrograde impairment). For example, he knew that his lawn was green, but he reported that he could no longer visualize it in green when he tried to remember what it once looked like.

Prosopagnosia

Similar effects have been documented by formal testing in cases of acquired prosopagnosia (impaired recognition of faces, or face blindness). The cardinal complaint of patients diagnosed with prosopagnosia is that they have a selective retrograde memory deficit. That is, once-recognizable faces no longer yield a memory signal, even though other aspects of one's memory for the same individuals are preserved. For example, a patient who could not recognize his mother's face might continue to recognize the sound of her voice and still be able to recall his prior experiences with her.

Patient L.H., a 37-year-old man, sustained a severe closed-head injury in an automobile accident at the age of 18 (Farah et al. 1995a,b). His brain damage involved bilateral inferior temporo-occipital regions, as well as the right inferior frontal lobe and right anterior temporal lobe. Although general intellectual and elementary visual capabilities were preserved following the accident, L.H. became profoundly impaired at recognizing previously familiar faces. Along with this retrograde memory deficit, L.H. also exhibited a perceptual processing deficit that was selective for upright faces. For example, on a same/different face discrimination task, L.H. performed worse than controls at discriminating upright faces (consistent with a face perception deficit), but he performed unexpectedly better than controls at discriminating inverted faces (indicating that general perceptual abilities were preserved). Patient L.H. also exhibited anterograde amnesia for new faces. For example, L.H. and controls were presented with black and white photographs of both faces and common objects and asked to memorize them (Farah et al. 1995a). On a later recognition test, control subjects performed at the same level for faces and nonface objects. L.H's ability to remember faces was selectively impaired.

The retrograde memory deficit associated with acquired prosopagnosia is not confined to recognition memory but applies as well to recalling and imaging the past. In one study, (Barton & Cherkasova 2003), seven patients with adult-onset prosopagnosia performed comparative judgments about the configuration of famous faces that they tried to retrieve from memory (e.g., "Who has the more angular face: George Washington or Abraham Lincoln?"). The famous faces used in this test were presumably familiar before the onset of prosopagnosia. Even so, the patients were severely impaired on the face imagery task. Together, the findings from acquired prosopagnosia—a modular perceptual processing deficit associated with selective anterograde and retrograde amnesia—suggest that the same areas that support the perception of faces also support the long-term memory of faces.

Amusia

This same set of findings, whereby an acquired and relatively modular processing deficit is associated with corresponding memory deficits (both anterograde and retrograde), has also been reported in a patient who lost the ability to recognize familiar music while retaining other perceptual and intellectual functions (amusia). Patient I.R. suffered bilateral brain damage at the age of 28 after undergoing a series of operations to clip aneurysms on the left and right middle cerebral arteries (Peretz et al. 1998, Peretz & Gagnon 1999). At the time she was tested (in her early 40s), CT scans indicated that the superior temporal gyrus was severely damaged bilaterally, and the lesion also extended to involve structures in the frontal cortex and anterior inferior parietal lobule.

I.R. was of normal intelligence, and her overall memory ability was normal as well. In addition, she exhibited no evidence of a hearing impairment according to standard audiometric tests, and except for music she had no difficulty recognizing familiar environmental sounds. However, tunes that were once familiar to her were now unrecognizable, and she could no longer sing music from memory (which she had previously been able to do). Her selective retrograde amnesia for previously familiar music was also accompanied by a selective perceptual deficit for music. Musical perception was tested using a same/different format in which two short excerpts were presented in succession (e.g., Mozart's piano concerto #27 followed by Mozart's piano concerto #23). Controls found this task so easy that they made no errors even when the interstimulus interval was long (20 s) and filled with conversation, but I.R.'s performance was no better than 80% correct even when the interstimulus interval was short (4 s).

She also exhibited anterograde amnesia for new music. A list of 15 briefly presented melodies was presented for study. On a subsequent old/new recognition test involving the 15 old melodies intermixed with 15 new ones, her memory performance was no better than chance (whereas control performance exceeded 85% correct). Thus, as with the cases of acquired achromatopsia and acquired prosopagnosia discussed earlier, impairments associated with acquired amusia imply a close connection between information processing and storage. The specificity of her anterograde and retrograde memory deficits corresponded directly to the specificity of her perceptual deficit.

Knowledge Systems

The findings considered here are consistent with the idea that memory storage in the neocortex reflects the outcome of the perceptual processing and analysis that occurred at the time of learning. A related literature concerns the status of stored semantic knowledge and its relation to information processing. These studies do not document a deficit in specific perceptual processing modules. Instead, they document the effects of cortical lesions (e.g., to posterior temporal cortex) on previously acquired knowledge within specific semantic categories, and they relate these deficits to the kinds of processing involved when the knowledge was first acquired.

The idea that knowledge systems may be organized by semantic categories was discussed by Warrington & Shallice (1984). They described four patients with widespread bilateral lesions (following herpes simplex encephalitis) that included the medial and lateral temporal lobes. In addition to having global amnesia, all four patients exhibited an asymmetry in their ability to identify animate and inanimate objects. They had a selective impairment in the ability to name or describe pictures of animate objects (e.g., animals and plants). By contrast, their ability to name or describe pictures of inanimate objects (e.g., broom, pencil, umbrella) appeared to be preserved. Assuming that all the objects were previously familiar to the patients, the findings describe a category-specific retrograde memory impairment.

Other patients exhibited the opposite impairment. For example, patient Y.O.T., who had damage to the left temporoparietal region (thought to have resulted from a thromboembolism), showed relatively preserved knowledge of living things and poor knowledge of inanimate objects (Warrington & McCarthy 1987). However, her comprehension of body parts and fabrics was anomalous in that she exhibited knowledge about fabric names (nonliving things) and poor knowledge about body parts (living things). In addition, Warrington & McCarthy (1987) noted that patient J.B.R. [one of the four patients previously described by Warrington & Shallice (1984)], who had exhibited a selective loss of knowledge about living things, nevertheless had preserved knowledge about body parts (living things) and poor knowledge about fabrics (nonliving things). These findings suggested that the principle by which knowledge is organized in the brain concerns whether objects are identified mainly by their physical features (form, color, texture, etc.) or by their function and how they are used. Generally, the animate/inanimate distinction fits this principle, but the exceptions are telling. Most animals are identified by their physical attributes, not by what can be done with them. By contrast, small inanimate objects are usually identified by their functions and how they are used (e.g., sweep with a broom, write with a pencil). However, some living things (such as body parts) are identified largely by their function, and some nonliving things (such as fabrics) are identified largely by their texture and shape. A recent comprehensive review of neuroimaging evidence strongly supports this account of stored semantic knowledge (Martin 2007).

If these category-specific retrograde memory deficits reflect the loss of knowledge that was initially acquired through category-specific processing, then a corresponding anterograde memory deficit would be expected, as well. Thus, for example, a patient who exhibits a selective deficit in naming or describing objects that are defined by how they are used should also exhibit a selective deficit in learning novel objects that are defined by how they are used. To our knowledge, this prediction has not been tested.

RECOLLECTION AND FAMILIARITY

In recent years, there has been extended investigation of the idea that the different medial temporal lobe structures (hippocampus, entorhinal cortex, perirhinal cortex, and parahippocampal cortex) may support different memory functions. The study of H.M. could not address this issue because his bilateral lesions included most of these structures. However, other patients, especially patients with limited hippocampal lesions, have been useful in this regard.

One issue that has commanded considerable attention concerns the roles played by the hippocampus and perirhinal cortex in recognition memory. Recognition memory is thought to be supported by two processes, recollection and familiarity (Atkinson & Juola 1974,

Mandler 1980). Recollection involves remembering specific contextual details about a prior learning episode; familiarity involves simply knowing that an item was presented without having available any additional information about the learning episode. According to one view, both the hippocampus and the perirhinal cortex contribute to recollection and familiarity (Squire et al. 2007, Wixted & Squire 2010). According to a different view, the hippocampus and perirhinal cortex selectively support recollection and familiarity, respectively (Brown & Aggleton 2001, Eichenbaum et al. 2007).

Recall versus Recognition

One approach to investigating this issue has been to compare performance on an old/new recognition task, which is widely thought to be supported by both recollection and familiarity, with performance on a task of free recall, which is thought to depend mainly on recollection. (In a free recall task, subjects are presented with a list of items to memorize and are later asked to recall those items in any order they wish.) Because old/new recognition can be partially supported by familiarity, the question of interest is whether the performance of patients with hippocampal lesions is disproportionately better on an old/new recognition task in comparison to free recall.

Several case studies and group studies have asked this question of patients with adult-onset bilateral lesions that, according to quantitative MRI, are limited to the hippocampus. The case studies differ in their findings about the status of old/new recognition memory (Aggleton et al. 2005, Cipolotti et al. 2006, Mayes et al. 2002). Because the differing results may reflect individual differences, group studies are more informative. Two group studies have shown that the degree of impairment is similar when old/new recognition and free recall are compared (Kopelman et al. 2007, Manns et al. 2003a) (**Figure 5**). Another group study involved 56 hypoxic patients with damage believed to be limited to the hippocampus (no radiological information was available)

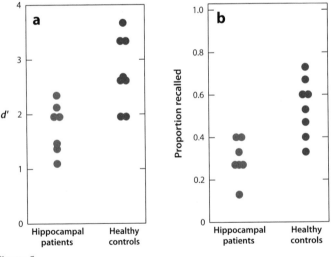

Figure 5

Individual recognition (*a*) and recall scores (*b*) for hippocampal patients (*n* = 7) and healthy controls (*n* = 8) from Manns et al. (2003a). When the patient scores for recognition and recall are converted to z-scores based on the mean and standard deviation of the corresponding control scores, the recognition deficit (−1.59) is statistically indistinguishable from the recall deficit (−1.81), *p* > 0.60. *d'* = discriminability.

(Yonelinas et al. 2002). The patients were less impaired on old/new recognition than on free recall. However, this conclusion was later shown to result from the remarkably aberrant recognition performance of a single 1 of the 55 control subjects (Wixted & Squire 2004). With that one outlier removed from the analysis, the patients and controls exhibited similar levels of impairment on recall and recognition. The recognition z-score for the patients was −0.59 (before removal of the outlier, z = −0.39), and the recall z-score was a statistically indistinguishable −0.68. Thus, the available group studies are consistent in showing that the degree of memory impairment in patients with lesions limited to the hippocampus is similar for old/new recognition (which is substantially supported by familiarity) and for free recall (which is fully dependent on recollection). These findings suggest that the hippocampus is important for both recollection and familiarity.

Remember/Know Procedure

Another method that has been used to investigate the role of the medial temporal lobe in recollection and familiarity is the Remember/Know procedure, which is based on subjective reports of whether recollection is available when an item is declared old. Participants report Remember when they can recollect something about the original encounter with the item (e.g., its context, what thoughts they had), and they report Know when they judge the item to be familiar but cannot recollect anything about its presentation. The Remember/Know judgments made by patients and controls are often converted into quantitative estimates of recollection and familiarity based on a widely used but controversial model of recognition memory (Yonelinas 1994). Using this method, some studies have reported that recollection is selectively impaired in patients with hippocampal lesions (Yonelinas et al. 2002), whereas other studies have found impairments in both recollection and familiarity (Manns et al. 2003a). A difficulty with deriving quantitative estimates of recollection and familiarity from

Remember/Know judgments is that the assumptions of the model that is used to derive estimates have generally not been supported by empirical test (e.g., Heathcote 2003, Rotello et al. 2005, Slotnick 2010, Slotnick & Dodson 2005). In particular, Know judgments reflect weaker memory than do Remember judgments, as measured by both confidence and accuracy (e.g., Dunn 2004, Squire et al. 2007). Thus, a supposed impairment in recollection (Remembering) after hippocampal lesions could simply mean that the patients have few strong memories (and that what would have been strong memories are now weak memories), not that recollection is selectively affected. The Remember/Know procedure could be used to study recollection and familiarity effectively if Remember and Know judgments were first equated for confidence and accuracy, but this approach has not been used in patient studies to date.

Analysis of the Receiver Operating Characteristic

Still another method that has been used to estimate recollection and familiarity has been to fit the Yonelinas (1994) dual-process model to receiver operating characteristic (ROC) data. This is the same model that has often been used to estimate recollection and familiarity using the Remember/Know procedure. An ROC is a plot of the hit rate versus the false alarm rate across different decision criteria. Typically, multiple pairs of hit and false alarm rates are obtained by asking subjects to provide confidence ratings for their old/new recognition decisions. A pair of hit and false alarm rates is then computed for each level of confidence, and the paired values are plotted across the confidence levels. The points of an ROC typically trace out a curvilinear path that can be characterized in terms of its symmetry relative to the negative diagonal (**Figure 6**). The dual-process model proposed by Yonelinas (1994) holds that the degree of asymmetry in an ROC directly reflects the degree to which the recollection process is involved in recognition decisions. Accordingly, a symmetrical ROC indicates that

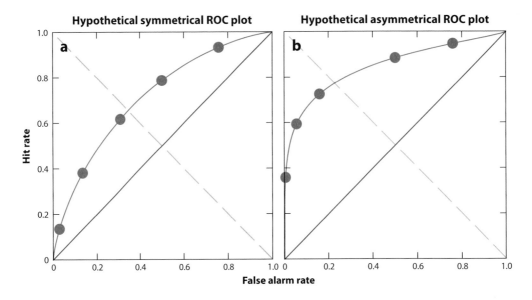

Figure 6

Symmetrical (*a*) and asymmetrical (*b*) receiver operating characteristic (ROC) plots with hypothetical data shown as filled red circles. The axis of symmetry is the negative diagonal (*dashed gray line*), and chance performance is indicated by the positive diagonal (*solid blue line*). The symmetrical ROC (*a*) reflects relatively weak memory (the data fall close to the positive diagonal), and the asymmetrical ROC (*b*) reflects stronger memory (the data fall farther from the positive diagonal).

recognition decisions were based solely on familiarity, and an asymmetrical ROC indicates that recollection occurred for some of the items, as well.

The finding that memory-impaired patients produce symmetrically curvilinear ROCs, whereas controls produce asymmetrical curvilinear ROCs, has been interpreted to mean that the recollection process is selectively impaired by hippocampal lesions (Yonelinas et al. 1998, 2000). However, once again, this is a model-dependent interpretation, and much evidence that has accumulated against this model in recent years instead supports an alternative signal-detection model (e.g., Dunn 2004, 2008; Wixted 2007; Wixted & Mickes 2010). According to the signal-detection model, a symmetrical ROC does not indicate familiarity-based responding but simply reflects weaker memory. Because patients have weaker memory than do controls, the fact that patients tend to exhibit symmetrical ROCs is not surprising.

The question is whether patients can exhibit asymmetrical ROCs (like controls) once

the strength of memory is equated. In one study, patients with lesions limited to the hippocampus were studied under two conditions (weak and strong memory) (Wais et al. 2006). In the weak condition, patients studied 50-item word lists, as did matched controls. As expected, the controls performed better than the patients did. In addition (again as expected), the control ROC was asymmetrical, and the patient ROC was symmetrical. To equate for overall memory strength, patients also studied lists of 10 items, which improved their memory performance to a level similar to that of the controls who had studied 50-item lists. In this condition, the patient ROC and the control ROC were similarly asymmetrical. These results show that patients can exhibit asymmetric ROCs, which have been taken to denote performance based on recollection. The results further suggest that the typical finding of asymmetrical ROCs for controls and symmetrical ROCs for patients does not necessarily indicate a selective deficit in recollection but can reflect a difference in overall memory strength.

Newer (Model-Free) Methods

The effects of hippocampal lesions on recollection and familiarity can also be studied in a way that does not depend on the assumptions of any specific psychological model. If hippocampal lesions selectively impair recollection, and preserve familiarity, then patients with hippocampal lesions should commonly experience strong, familiarity-based recognition that is unaccompanied by recollection. Furthermore, this experience should occur even more frequently in patients than controls because ordinarily when strong, familiarity-based recognition occurs (e.g., seeing a familiar face), details about prior encounters are remembered, as well.

In a formal test of this prediction, five patients with circumscribed hippocampal damage studied 25 words in one of two contexts (source A or source B) (Kirwan et al. 2010). Old/new recognition memory for the words was then tested using a six-point confidence scale (1 = sure new, 6 = sure old). For items endorsed as old, participants were also asked to make a source recollection decision (was the item learned in context A or B?). Old decisions made with high confidence but in the absence of successful source recollection would thus correspond to strong, familiarity-based recognition without recollection. The results were that there was no increased tendency for this experience to occur in patients relative to controls. If anything, the experience was less frequent in the patients. The simplest explanation for this result is that hippocampal damage impairs familiarity as well as recollection.

In summary, a large body of evidence based on the Remember/Know procedure and ROC analysis has been interpreted to mean that the hippocampus subserves recollection and plays no role in familiarity. It is often not appreciated that this interpretation is based on a specific model that equates weak memory with familiarity and strong memory with recollection [the model proposed by Yonelinas (1994)]. However, familiarity can sometimes be strong, and recollection can sometimes be weak (Wixted & Mickes 2010). In studies that do not depend on

this model, the results suggest that hippocampal lesions impair both recollection and familiarity (Kirwan et al. 2010, Wais et al. 2006).

The fact that a memory strength confound can explain why earlier studies have failed to detect impaired familiarity in hippocampal patients should not be taken to mean that "memory strength" is a concept that usefully informs the functional organization of medial temporal lobe structures. Consideration of how these structures contribute differently to memory properly begins with neuroanatomy. Information from neocortex enters the medial temporal lobe at different points (Suzuki & Amaral 1994). Perirhinal cortex receives strong input from unimodal visual areas, and the parahippocampal cortex receives prominent projections from areas important for spatial cognition, including posterior parietal cortex. This anatomical specialization suggests that perirhinal cortex may be especially important for visual memory (regardless whether a task requires recollection), and the parahippocampal cortex may be important for spatial memory. The finding of severe impairment in monkeys in visual associative tasks after perirhinal lesions (Murray et al. 1993) and in spatial tasks after parahippocampal cortex lesions (Malkova & Mishkin 2003) conforms to this suggestion. The hippocampus itself receives input from the adjacent cortex and is thus in a position to combine the operations of memory formation that are carried out by the more specialized structures that project to it. As expected, hippocampal lesions impair both visual memories and spatial memories. The impairment in memory formation is only modestly severe because many memory functions can be carried out by the adjacent cortex [for additional discussion of differences in the function of medial temporal lobe structures, see Squire et al. (2007), Wixted & Squire (2011)].

GROUP STUDIES AND MULTIPLE METHODS

The study of patients with medial temporal lobe lesions (especially the severely impaired patient H.M.) has led to dramatic advances in understanding the structure and organization of memory. As work progressed, many studies came to focus on smaller lesions and less severe impairments. Furthermore, many of these studies were based on single cases and investigated specific questions about particular aspects of the impairment (for example, recall versus recognition or recollection versus familiarity). Such questions are difficult to settle from individual case studies because the expected effects are relatively small. Under such conditions, a deficit may reflect (unmeasured) premorbid individual differences rather than the effect of a focal brain lesion. Accordingly, for many questions single case studies are more suggestive than conclusive, and group studies are needed to answer experimental questions in a compelling way.

The advantage of group studies is that individual variability tends to be averaged out. However, group studies are useful only to the extent that the lesions can be documented and quantified with MRI. Some group studies have studied patients with assumed lesions, such as patients with modest memory impairments due to hypoxia who are studied on the untested assumption that their lesions are limited to the hippocampus. Given techniques currently available for quantifying the locus and extent of lesions, the use of such techniques in both single-case and group studies should become standard practice.

It is important to emphasize that studies of patients with lesions provide only one of many experimental approaches to investigating the organization of memory. The same issues have been usefully investigated in experimental animals with lesions (e.g., rats and monkeys), in single-unit recording studies of animals and humans, in studies using functional neuroimaging or transcranial magnetic stimulation (TMS), and in studies of genetically modified mice. Each approach has its own advantages and disadvantages, such that one can expect that their combined application will provide the best opportunity for further discovery.

CONCLUSIONS

The early descriptions of H.M. changed how human memory was understood. What became clear as a result of work with H.M.—and what remains clear today—is that the structures of the medial temporal lobe are essential for normal memory function. Specifically, these structures are thought to be important for the formation of memory and for the maintenance of memory for a period of time after learning. Although active lines of research investigate the possibility that these structures also contribute to other domains of cognitive function (e.g., visual perception, working memory, and online computations supporting spatial cognition), the half-century of research that began with H.M. has shown that profound impairment after medial temporal lobe damage occurs in only one domain, specifically, in what is now termed declarative memory.

The elements of long-term memory are stored in the neocortex (not in the medial temporal lobe) as products of the distributed, domain-specific processing that occurred in different regions of neocortex at the time of learning. Thus, long-term memory for whole events is widely represented, but the multiple areas that are involved each store distinct components of information. In addition, acts of remembering involve the reactivation of the same neocortical regions that initially processed and stored what was learned. The role of the medial temporal lobe is to consolidate the distributed elements of memory into a coherent and stable ensemble (a process that can take years). Many questions remain about how consolidation occurs, as well as about memory storage, memory retrieval, and the specific functions of the different medial temporal lobe structures and the different areas of neocortex. These topics encompass what has become a substantial and fruitful tradition of research within systems and cognitive neuroscience—a tradition that began with the study of H.M.

DISCLOSURE STATEMENT

The authors are not aware of any affiliations, memberships, funding, or financial holdings that might be perceived as affecting the objectivity of this review.

ACKNOWLEDGMENTS

This work was supported by the Medical Research Service of the Department of Veterans Affairs, NIMH grant MH24600 to L.R.S., and NIMH grant MH082892 to J.T.W.

LITERATURE CITED

Aggleton JP, Vann SD, Denby C, Dix S, Mayes AR, et al. 2005. Sparing of the familiarity component of recognition memory in a patient with hippocampal pathology. *Neuropsychologia* 43:810–23

Atkinson RC, Juola JF. 1974. Search and decision processes in recognition memory. In *Contemporary Developments in Mathematical Psychology*, ed. DH Krantz, RC Atkinson, P Suppes, pp. 243–90. San Francisco: Freeman

Baddeley A. 2003. Working memory: looking back and looking forward. *Nat. Rev. Neurosci.* 4:829–39

Barense MD, Gaffan D, Graham KS. 2007. The human medial temporal lobe processes online representations of complex objects. *Neuropsychologia* 45:2963–74

Barton JJS, Cherkasova M. 2003. Face imagery and its relation to perception and covert recognition in prosopagnosia. *Neurology* 61:220–25

Bayley PJ, Frascino JC, Squire LR. 2005a. Robust habit learning in the absence of awareness and independent of the medial temporal lobe. *Nature* 436:550–53

Bayley PJ, Gold JJ, Hopkins RO, Squire LR. 2005b. The neuroanatomy of remote memory. *Neuron* 46:799–810

Bayley PJ, Hopkins RO, Squire LR. 2003. Successful recollection of remote autobiographical memories by amnesic patients with medial temporal lobe lesions. *Neuron* 37:135–44

Bayley PJ, Hopkins RO, Squire LR. 2006. The fate of old memories after medial temporal lobe damage. *J. Neurosci.* 26:13311–17

Bird CM, Burgess N. 2008. The hippocampus and memory: insights from spatial processing. *Nat. Rev. Neurosci.* 9:182–94

Bird CM, Capponi C, King JA, Doeller CF, Burgess N. 2010. Establishing the boundaries: the hippocampal contribution to imagining scenes. *J. Neurosci.* 30:11688–95

Bright P, Buckman JR, Fradera A, Yoshimasu H, Colchester ACF, Kopelman MD. 2006. Retrograde amnesia in patients with hippocampal, medial temporal, temporal lobe, or frontal pathology. *Learn. Mem.* 13:545–57

Broca PP. 1861. Remarks on the seat of the faculty of articulate language, followed by an observation of aphemia. In *Some Papers on the Cerebral Cortex*, transl. G von Bonin, pp. 199–220. Springfield, IL: Thomas

Brown MW, Aggleton JP. 2001. Recognition memory: What are the roles of the perirhinal cortex and hippocampus? *Nat. Rev. Neurosci.* 2:51–61

Buchanan TW, Tranel D, Adolphs R. 2005. Emotional autobiographical memories in amnesic patients with medial temporal lobe damage. *J. Neurosci.* 25:3151–60

Buckner RL, Wheeler ME. 2001. The cognitive neuroscience of remembering. *Nat. Rev. Neurosci.* 2:624–34

Cave C, Squire LR. 1992. Intact and long-lasting repetition priming in amnesia. *J. Exp. Psychol. Learn. Mem. Cog.* 18:509–20

Cipolotti L, Bird C, Good T, Macmanus D, Rudge P, Shallice T. 2006. Recollection and familiarity in dense hippocampal amnesia: a case study. *Neuropsychologia* 44:489–506

Cipolotti L, Shallice T, Chan D, Fox N, Scahill R, et al. 2001. Long-term retrograde amnesia...the crucial role of the hippocampus. *Neuropsychologia* 2:151–72

Clark RE, Reinagel P, Broadbent N, Flister E, Squire LR. 2011. Intact performance on feature ambiguous discriminations in rats with lesions of the perirhinal cortex. *Neuron.* In press

Cohen NJ, Squire LR. 1980. Preserved learning and retention of pattern analyzing skill in amnesia: dissociation of knowing how and knowing that. *Science* 210:207–9

Corkin S. 2002. What's new with the amnesic patient H.M.? *Nat. Rev. Neurosci.* 3:153–60

Corkin S, Amaral DG, Gonzalez RG, Johnson KA, Hyman BT. 1997. H.M.'s medial temporal lobe lesion: findings from magnetic resonance imaging. *J. Neurosci.* 17:3964–80

Cowan N. 2001. The magical number 4 in short-term memory: a reconsideration of mental storage capacity. *Behav. Brain Sci.* 24:87–185

Damasio AR. 1989. Time-locked multiregional retroactivation: a systems-level proposal for the neural substrates of recall and recognition. *Cognition* 33:25–62

Danker JF, Anderson JR. 2010. The ghosts of brain states past: remembering reactivates the brain regions engaged during encoding. *Psychol. Bull.* 136:87–102

De Renzi E. 1982. Memory disorders following focal neocortical damage. *Philos. R. Soc. London [Biol.]* 298:73–83

Drachman DA, Arbit J. 1966. Memory and the hippocampal complex. II. Is memory a multiple process? *Arch. Neurol.* 15:52–61

Dunn JC. 2004. Remember-Know: a matter of confidence. *Psychol. Rev.* 111:524–42

Dunn JC. 2008. The dimensionality of the remember–know task: a state-trace analysis. *Psychol. Rev.* 115:426–46t

Eacott MJ, Gaffan D, Murray EA. 1994. Preserved recognition memory for small sets, and impaired stimulus identification for large sets, following rhinal cortex ablations in monkeys. *Euro. J. Neurosci.* 6:1466–78

Eichenbaum H, Yonelinas AR, Ranganath C. 2007. The medial temporal lobe and recognition memory. *Annu. Rev. Neurosci.* 30:123–52

Eslinger PJ. 1998. Autobiographical memory after temporal lobe lesions. *Neurocase* 4:481–95

Farah MJ, Levinson KL, Klein KL. 1995a. Face perception and within-category discrimination in prosopagnosia. *Neuropsychologia* 33:661–74

Farah MJ, Wilson KD, Drain HM, Tanaka J. 1995b. The inverted face inversion effect in prosopagnosia: evidence for mandatory, face-specific perceptual mechanisms. *Vision Res.* 35:2089–93

Ferrier D. 1876. *The Functions of the Brain*. London: Smith, Elder and Company

Finger S. 1994. *Origins of Neuroscience: A History of Explorations into Brain Function*. New York: Oxford Univ. Press

Finke C, Braun M, Ostendorf F, Lehmann TN, Hoffmann KT, et al. 2008. The human hippocampal formation mediates short-term memory of colour-location associations. *Neuropsychologia* 46:614–23

Frankland PW, Bontempi B. 2005. The organization of recent and remote memories. *Nat. Rev. Neurosci.* 6:119–30

Fritsch G, Hitzig E. 1870. On the electrical excitability of the cerebrum. In *Some Papers on the Cerebral Cortex*, transl. G von Bonin. pp. 73–86. Springfield, IL: Thomas

Fukuda K, Edward A, Vogel EK. 2010. Discrete capacity limits in visual working memory. *Curr. Opin. Neurobiol.* 20:177–82

Gall FJ. 1825. *Sur les Fonctions du Cerveau et sur Celles de Chacune des Ses Parties*, 6 Vol. Paris: Bailliere

Gold JJ, Squire LR. 2005. Quantifying medial temporal lobe damage in memory-impaired patients. *Hippocampus* 15:79–85

Gold JJ, Squire LR. 2006. The anatomy of amnesia: neurohistological analysis of three new cases. *Learn. Mem.* 13:699–710

Graham KS, Scahill VL, Hornberger M, Barense MD, Lee AC, et al. 2006. Abnormal categorization and perceptual learning in patients with hippocampal damage. *J. Neurosci.* 26:7547–54

Hannula DE, Tranel D, Cohen NJ. 2006. The long and short of it: relational memory impairments in amnesia, even at short lags. *J. Neurosci.* 26:8352–59

Harding A, Halliday G, Caine D, Kril J. 2000. Degeneraton of anterior thalamic nuclei differentiates alcoholics with amnesia. *Brain* 123:141–54

Hassabis D, Kumaran D, Vann SD, Maguire EA. 2007. Patients with hippocampal amnesia cannot imagine new experiences. *Proc. Natl. Acad. Sci. USA* 104:1726–31

Heathcote A. 2003. Item recognition memory and the ROC. *J. Exp. Psych. Learn. Mem. Cogn.* 29:1210–30

Hebb DO. 1949. *The Organization of Behavior.* New York: Wiley

Higuchi S, Miyashita Y. 1996. Formation of mnemonic neuronal responses to visual paired associates in inferotemporal cortex is impaired by perirhinal and entorhinal lesions. *Proc. Natl. Acad. Sci. USA* 93:739–43

Hunter WS. 1930. A consideration of Lashley's theory of equipotentiality of cerebral action. *J. Gen. Psych.* 3:444–68

Insausti R, Insausti AM, Sobreviela M, Salinas A, Martinez-Penuela J. 1998a. Human medial temporal lobe in aging: anatomical basis of memory preservation. *Microsc. Res. Techn.* 43:8–15

Insausti R, Juottonen K, Soininen H, Insausti AM, Partanen K, et al. 1998b. MR volumetric analysis of the human entorhinal, perirhinal, and temporopolar cortices. *Am. J. Neuroradiol.* 19:659–71

Isaacs EB, Vargha-Khadem F, Watkins KE, Lucas A, Mishkin M, Gadian DG. 2003. Developmental amnesia and its relationship to degree of hippocampal atrophy. *Proc. Natl. Acad. Sci. USA* 100:13060–63

Jeneson A, Mauldin KN, Hopkins RO, Squire LR. 2011. The role of the hippocampus in retaining relational information across short delays: the importance of memory load. *Learn. Mem.* In press

Jeneson A, Mauldin KN, Squire LR. 2010. Intact working memory for relational information after medial temporal lobe damage. *J. Neurosci.* 30:13624–29

Kan IP, Giovanello KS, Schnyer DM, Makris N, Verfaellie M. 2007. Role of the medial temporal lobes in relational memory: neuropsychological evidence from a cued recognition paradigm. *Neuropsychologia* 45:2589–97

Kapur N, Brooks DJ. 1999. Temporally-specific retrograde amnesia in two cases of discrete bilateral hippocampal pathology. *Hippocampus* 9:247–54

Kensinger E, Ullman MT, Corkin S. 2001. Bilateral medial temporal lobe damage does not affect lexical or grammatical processing: evidence from amnesic patient H.M. *Hippocampus* 11:347–60

Kim S, Jeneson A, van der Horst AS, Frascino JC, Hopkins RO, Squire LR. 2011. Memory, visual discrimination performance, and the human hippocampus. *J. Neurosci.* 31:2624–29

Kirwan CB, Bayley PJ, Galvan VV, Squire LR. 2008. Detailed recollection of remote autobiographical memory after damage to the medial temporal lobe. *Proc. Natl. Acad. Sci. USA* 105:2676–80

Kirwan CB, Wixted JT, Squire LR. 2010. A demonstration that the hippocampus supports both recollection and familiarity. *Proc. Natl. Acad. Sci. USA* 107:344–48

Knowlton BJ, Mangels JA, Squire LR. 1996. A neostriatal habit learning system in humans. *Science* 273:1399–402

Kopelman MD, Bright P, Buckman J, Fradera A, Yoshimasu H, et al. 2007. Recall and recognition memory in amnesia: patients with hippocampal, medial temporal, temporal lobe or frontal pathology. *Neuropsychologia* 45:1232–46

Kopelman MD, Wilson BA, Baddeley AD. 1989. The autobiographical memory interview: a new assessment of autobiographical and personal semantic memory in amnesic patients. *J. Clin. Exp. Neuropsy.* 5:724–44

Lashley KS. 1929. *Brain Mechanisms and Intelligence: A Quantitative Study of Injuries to the Brain.* Chicago: Chicago Univ. Press

Lee AC, Buckley MJ, Pegman SJ, Spiers H, Scahill VL, et al. 2005a. Specialization in the medial temporal lobe for processing of objects and scenes. *Hippocampus* 15:782–97

Lee AC, Bussey TJ, Murray EA, Saksida LM, Epstein RA, et al. 2005b. Perceptual deficits in amnesia: challenging the medial temporal lobe 'mnemonic' view. *Neuropsychologia* 43:1–11

Lee AC, Rudebeck SR. 2010. Human medial temporal lobe damage can disrupt the perception of single objects. *J. Neurosci.* 30:6588–94

Mair WGP, Warrington EK, Weiskrantz L. 1979. Memory disorder in Korsakoff psychosis. A neuropathological and neuropsychological investigation of two cases. *Brain* 102:749–83

Malkova L, Mishkin M. 2003. One-trial memory for object-place associations after separate lesions of hippocampus and posterior parahippocampal region in the monkey. *J. Neurosci.* 23:1956–65

Mandler G. 1980. Recognizing: the judgment of previous occurrence. *Psychol. Rev.* 87:252–71

Manns JR, Hopkins RO, Reed JM, Kitchener EG, Squire LR. 2003a. Recognition memory and the human hippocampus. *Neuron* 37:171–80

Manns JR, Hopkins RO, Squire LR. 2003b. Semantic memory and the human hippocampus. *Neuron* 37:127–33

Marslen-Wilson WD, Teuber HL. 1975. Memory for remote events in anterograde amnesia: recognition of public figures from news photographs. *Neuropsychologia* 13:353–64

Martin A. 2007. The representation of object concepts in the brain. *Ann. Review. Psychol.* 58:25–45

Mayes AR, Holdstock JS, Isaac CL, Hunkin NM, Roberts N. 2002. Relative sparing of item recognition memory in a patient with adult-onset damage limited to the hippocampus. *Hippocampus* 12:325–40

Mayes AR, Meudell PR, Mann D, Pickering A. 1988. Location of lesions in Korsakoff's syndrome: neuropsychological and neuropathological data on two patients. *Cortex* 3:367–88

Milner B. 1962. Les troubles de la memoire accompagnant des lesions hippocampiques bilaterales. In *Physiologie de l'hippocampe*, pp. 257–72. Paris: Cent. Natl. Rech. Sci.

Milner B, Corkin S, Teuber HL. 1968. Further analysis of the hippocampal amnesic syndrome: 14 year follow-up study of H.M. *Neuropsychologia* 6:215–34

Mishkin M. 1978. Memory in monkeys severely impaired by combined but not by separate removal of amygdala and hippocampus. *Nature* 273:297–98

Mishkin M. 1982. A memory system in the monkey. *Philos. Trans. R. Soc. London Ser. B* 1089:83–95

Mishkin M, Malamut B, Bachevalier J. 1984. Memories and habits: two neural systems. In *Neurobiology of Learning and Memory*, ed. G Lynch, JL McGaugh, NM Weinberger, pp. 65–77. New York: Guilford

Miyashita Y. 1993. Inferior temporal cortex: where visual perception meets memory. *Annu. Rev. Neurosci.* 16:245–63

Murray EA. 1992. Medial temporal lobe structures contributing to recognition memory: the amygdaloid complex versus rhinal cortex. In *The Amygdala: Neurobiological Aspects of Emotion, Memory, and Mental Dysfunction*, ed. JP Aggleton, pp. 453–70. London: Wiley-Liss

Murray EA, Gaffan D, Mishkin M. 1993. Neural substrates of visual stimulus-stimulus association in rhesus monkeys. *J. Neurosci.* 13:4549–61

Noulhiane M, Piolino P, Hasboun D, Clemenceau S, Baulac M, Samson S. 2007. Autobiographical memory after temporal lobe resection: neuropsychological and MRI volumetric findings. *Brain* 130:3184–99

Olson IR, Page K, Moore KS, Chatterjee A, Verfaellie M. 2006. Working memory for conjunctions relies on the medial temporal lobe. *J. Neurosci.* 26:4596–601

Peretz I, Gagnon L. 1999. Dissociation between recognition and emotional judgments for melodies. *Neurocase* 5:21–30

Peretz I, Gagnon L, Bouchard B. 1998. Music and emotion: perceptual determinants, immediacy, and isolation after brain damage. *Cognition* 68:111–41

Polyn SM, Natu VS, Cohen JD, Norman KA. 2005. Category-specific cortical activity precedes retrieval during memory search. *Science* 310:1963–66

Press GA, Amaral DG, Squire LR. 1989. Hippocampal abnormalities in amnesic patients revealed by high-resolution magnetic resonance imaging. *Nature* 341:54–57

Rempel-Clower N, Zola SM, Squire LR, Amaral DG. 1996. Three cases of enduring memory impairment following bilateral damage limited to the hippocampal formation. *J. Neurosci.* 16:5233–55

Restivo L, Vetere G, Bontempi B, Ammassari-Teule M. 2009. The formation of recent and remote memory is associated with time-dependent formation of dendritic spines in the hippocampus and anterior cingulate cortex. *J. Neurosci.* 29:8206–14

Ribot T. 1881. *Les Maladies de la Memoire* [English translation: Diseases of Memory]. New York: Appleton-Century-Crofts

Rosenbaum RS, Moscovitch M, Foster JK, Schnyer DM, Gao F, et al. 2008. Patterns of autobiographical memory loss in medial-temporal lobe amnesic patients. *J. Cogn. Neurosci.* 20:1490–506

Rosenbaum RS, Priselac S, Kohler S, Black SE, Gao F, et al. 2000. Remote spatial memory in an amnesic person with extensive bilateral hippocampal lesions. *Nat. Neurosci.* 3:1044–48

Rotello CM, Macmillan NA, Reeder JA, Wong M. 2005. The remember response: subject to bias, graded, and not a process-pure indicator of recollection. *Psychon. Bull. Rev.* 12:865–73

Sacks O. 1995. The case of the colorblind painter. In *An Anthropologist on Mars*, pp. 3–41. New York: Random House

Sagar HH, Cohen NJ, Corkin S, Growdon JM. 1985. Dissociations among processes in remote memory. In *Memory Dysfunctions*, ed. DS Olton, E Gamzu, S Corkin, pp. 533–35. New York: Ann. NY Acad. Sci.

Salat DH, van der Kouwe AJW, Tuch DS, Quinn BT, Fischl B, et al. 2006. Neuroimaging H.M.: a 10-year follow-up examination. *Hippocampus* 16:936–45

Sanders HI, Warrington DK. 1971. Memory for remote events in amnesic patients. *Brain* 94:661–68

Schnider A, Bassetti C, Schnider A, Gutbrod K, Ozdoba C. 1995. Very severe amnesia with acute onset after isolated hippocampal damage due to systemic lupus erythematosus. *J. Neurol. Neurosurg. Psychiatry* 59:644–46

Scoville WB, Milner B. 1957. Loss of recent memory after bilateral hippocampal lesions. *J. Neurol. Neurosurg. Psychiatry* 20:11–21

Shrager Y, Gold JJ, Hopkins RO, Squire LR. 2006. Intact visual perception in memory-impaired patients with medial temporal lobe lesions. *J. Neurosci.* 26:2235–40

Shrager Y, Levy DA, Hopkins RO, Squire LR. 2008. Working memory and the organization of brain systems. *J. Neurosci.* 28:4818–22

Slotnick SD. 2010. "Remember" source memory ROCs indicate recollection is a continuous process. *Memory* 18:27–39

Slotnick SD, Dodson CS. 2005. Support for a continuous (single-process) model of recognition memory and source memory. *Mem. Cogn.* 33:151–70

Squire LR. 1987. *Memory and Brain*. New York: Oxford Univ. Press

Squire LR. 2006. Lost forever or temporarily misplaced? The long debate about the nature of memory impairment. *Learn. Mem.* 13:522–29

Squire LR. 2009. The legacy of patient H.M. for neuroscience. *Neuron* 61:6–9

Squire LR, Amaral DG, Press GA. 1990. Magnetic resonance imaging of the hippocampal formation and mammillary nuclei distinguish medial temporal lobe and diencephalic amnesia. *J. Neurosci.* 10:3106–17

Squire LR, Amaral DG, Zola-Morgan S, Kritchevsky M, Press GA. 1989. Description of brain injury in the amnesic patient N.A. based on magnetic resonance imaging. *Exp. Neurol.* 105:23–25

Squire LR, Bayley PJ. 2007. The neuroscience of remote memory. *Curr. Opin. Neurobiol.* 17:185–96

Squire LR, van der Horst AS, McDuff SGR, Frascino JC, Hopkins RO, Mauldin KN. 2010. Role of the hippocampus in remembering the past and imagining the future. *Proc. Natl. Acad. Sci. USA* 107:19044–48

Squire LR, Wixted JT, Clark RE. 2007. Recognition memory and the medial temporal lobe: a new perspective. *Nat. Rev. Neurosci.* 8:872–83

Squire LR, Zola-Morgan S. 1988. Memory: brain systems and behavior. *Trends Neurosci.* 11:170–75

Squire LR, Zola-Morgan S. 1991. The medial temporal lobe memory system. *Science* 253:1380–86

Steinvorth S, Levine B, Corkin S. 2005. Medial temporal lobe structures are needed to re-experience remote autobiographical memories: evidence from H.M. and W.R. *Neuropsychologia* 43:479–96

Suzuki WA. 2009. Perception and the medial temporal lobe: evaluating the current evidence. *Neuron* 61:657–66

Suzuki WA. 2010. Untangling memory from perception in the medial temporal lobe. *Trends Cogn. Sci.* 1:195–200

Suzuki WA, Amaral DG. 1994. Topographic organization of the reciprocal connections between the monkey entorhinal cortex and the perirhinal and parahippocampal cortices. *J. Neurosci.* 14:1856–77

Teng E, Squire LR. 1999. Memory for places learned long ago is intact after hippocampal damage. *Nature* 400:675–77

Tulving E, Schacter DL. 1990. Priming and human memory systems. *Science* 247:301–6

Victor M, Adams RD, Collins GH, eds. 1989. *The Wernicke-Korsakoff Syndrome and Related Neurological Disorders due to Alcoholism and Malnutrition*. Philadelphia: F.A. Davis

Victor M, Agamanolis J. 1990. Amnesia due to lesions confined to the hippocampus: a clinical-pathological study. *J. Cogn. Neurosci.* 2:246–57

von Cramon DY, Hebel N, Schuri U. 1985. A contribution to the anatomical basis of thalamic amnesia. *Brain* 108:993–1008

Wais P, Wixted JT, Hopkins RO, Squire LR. 2006. The hippocampus supports both the recollection and the familiarity components of recognition memory. *Neuron* 49:459–68

Warrington EK, McCarthy RA. 1987. Categories of knowledge. Further fractionations and an attempted integration. *Brain* 110:1273–96

Warrington EK, Shallice T. 1984. Category specific semantic impairments. *Brain* 107:829–53

Wheeler ME, Petersen SE, Buckner RL. 2000. Memory's echo: vivid remembering activates sensory-specific cortex. *Proc. Natl. Acad. Sci. USA* 97:11125–29

Whitlock JR, Sutherland RJ, Witter MP, Moser MB, Moser EI. 2008. Navigating from hippocampus to parietal cortex. *Proc. Natl. Acad. Sci. USA* 105:14755–62

Winocur G, Moscovitch M, Bontempi JB. 2010. Memory formation and long-term retention in humans and animals: convergence towards a transformation account of hippocampal-neocortical interactions. *Neuropsychologia* 48:2339–56

Winslow F. 1861. *On Obscure Diseases of the Brain and Disorders of the Mind*. London: John W. Davies

Wixted JT. 2007. Dual-process theory and signal-detection theory of recognition memory. *Psychol. Rev.* 114:152–76

Wixted JT, Mickes L. 2010. A continuous dual-process model of remember/know judgments. *Psychol. Rev.* 117:1025–54

Wixted JT, Squire LR. 2004. Recall and recognition are equally impaired in patients with selective hippocampal damage. *Cogn. Aff. Behav. Neurosci.* 4:58–66

Wixted JT, Squire LR. 2010. The role of the human hippocampus in familiarity-based recognition memory. *Behav. Brain Res.* 215:197–208

Wixted JT, Squire LR. 2011. The medial temporal lobe and the attributes of memory. *Trends Cogn. Sci.* In press

Woodruff CC, Johnson JD, Uncapher MR, Rugg MD. 2005. Content-specificity of the neural correlates of recollection. *Neuropsychologia* 43:1022–32

Yonelinas AP. 1994. Receiver-operating characteristics in recognition memory: evidence for a dual-process model. *J. Exp. Psychol. Learn. Mem. Cogn.* 20:1341–54

Yonelinas AP, Kroll NEA, Dobbins IG, Lazzara MM, Knight RT. 1998. Recollection and familiarity deficits in amnesia: convergence of remember/know, process dissociation, and receiver operating characteristic data. *Neuropsychology* 12:1–17

Yonelinas AP, Kroll NEA, Quamme JR, Lazzara MM, Sauve MJ, et al. 2002. Effects of extensive temporal lobe damage or mild hypoxia on recollection and familiarity. *Nat. Neurosci.* 5:1236–41

Zola-Morgan S, Squire LR, Amaral DG. 1986. Human amnesia and the medial temporal region: enduring memory impairment following a bilateral lesion limited to field CA1 of the hippocampus. *J. Neurosci.* 6:2950–67

Zola-Morgan S, Squire LR, Ramus SJ. 1994. Severity of memory impairment in monkeys as a function of locus and extent of damage within the medial temporal lobe memory system. *Hippocampus* 4:483–95

Deep Brain Stimulation for Psychiatric Disorders

Paul E. Holtzheimer and Helen S. Mayberg

Department of Psychiatry and Behavioral Sciences, Emory University School of Medicine, Atlanta, Georgia 30322; email: pholtzh@emory.edu, hmayber@emory.edu

Annu. Rev. Neurosci. 2011. 34:289–307

The *Annual Review of Neuroscience* is online at neuro.annualreviews.org

This article's doi:
10.1146/annurev-neuro-061010-113638

0147-006X/11/0721-0289$20.00

Keywords

major depression, treatment-resistant depression, obsessive-compulsive disorder, tourette syndrome

Abstract

Medications, psychotherapy, and other treatments are effective for many patients with psychiatric disorders. However, with currently available interventions, a substantial number of patients experience incomplete resolution of symptoms, and relapse rates are high. In the search for better treatments, increasing interest has focused on focal neuromodulation. This focus has been driven by improved neuroanatomical models of mood, thought, and behavior regulation, as well as by more advanced strategies for directly and focally altering neural activity. Deep brain stimulation (DBS) is one of the most invasive focal neuromodulation techniques available; data have supported its safety and efficacy in a number of movement disorders. Investigators have produced preliminary data on the safety and efficacy of DBS for several psychiatric disorders, as well. In this review, we describe the development and justification for testing DBS for various psychiatric disorders, carefully consider the available clinical data, and briefly discuss potential mechanisms of action.

Contents

INTRODUCTION

Diseases desperate grown,

By desperate alliances are relieved,

Or not at all.

William Shakespeare, *Hamlet*, Act 4, Scene 3

Psychiatric disorders are among the most prevalent and costly ailments worldwide. Nearly 50% of all Americans will meet criteria for a mental disorder diagnosis in their lifetime, more than 28% will meet criteria for an anxiety disorder, and more than 20% will meet criteria for a depressive disorder (Kessler et al. 2005a). Depressive disorders alone are the highest contributor to the burden of disease in middle- and high-income countries and are the third highest contributor worldwide (WHO 2008). Suicide, which is almost always associated with the presence of a mental disorder, is the eleventh leading cause of death in the United States, third among persons aged 15–24 years, and fourth among persons aged 18–65 years (**http://www.cdc.gov/violenceprevention/ suicide**).

Established treatments for many psychiatric disorders include medications and psychotherapy. Various other interventions have demonstrated efficacy for specific disorders [e.g., electroconvulsive therapy (ECT) for depression, mania, and catatonia; light therapy for seasonal affective disorder; and vagus nerve stimulation (VNS) and transcranial magnetic stimulation (TMS) for medication-resistant depression]. However, despite the proven effectiveness of available treatments, a substantial number of patients fail to fully remit (i.e., become symptom free) or maintain symptomatic improvement over time. This lack of effectiveness for some has led to the search for novel strategies, with more invasive approaches being considered and tested for the most severe and treatment-resistant patients.

Deep brain stimulation (DBS) is an invasive neurosurgical intervention being investigated for several psychiatric disorders, most notably treatment-resistant depression (TRD) and treatment-refractory obsessive-compulsive disorder (OCD), but also Tourette's Syndrome (TS), Alzheimer's dementia (AD), and addiction. The rationale for using DBS in the treatment of psychiatric disorders is based on its effectiveness in several movement disorders and the development of detailed neuroanatomical models for regulating emotion, cognition, and behavior.

DBS is achieved by implanting one or more electrode arrays (leads) into a specific region of the brain via burr holes in the skull using neuroimaging-guided stereotactic neurosurgical techniques (where the target for implantation is calculated within a three-dimensional coordinate system based on external landmarks) (**Figure 1**). Each array generally contains several (typically four) electrode contacts spanning 10–20 mm. DBS leads are connected via subcutaneous extension wires to one or more subcutaneously implanted pulse generators (IPGs) containing the system battery and the computer that drives stimulation; therefore, the system is completely internalized within the patient's body. Parameters can be noninvasively set and adjusted via a handheld computer interface. Stimulation parameters can vary widely, e.g., in terms of frequency, pulse width, and voltage/amplitude. In the clinical setting, common DBS parameters are 60–130 Hz, 60–200 µs

DBS: deep brain stimulation

TRD: treatment-resistant depression

OCD: obsessive-compulsive disorder

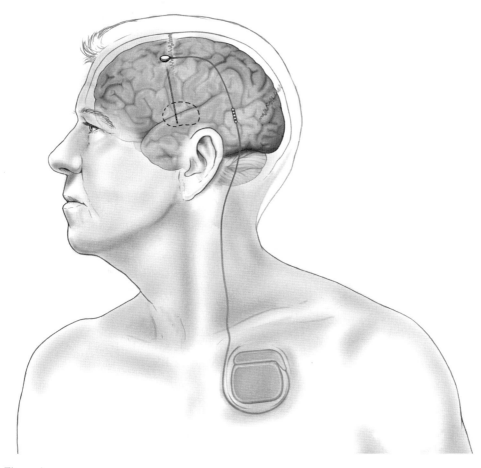

Figure 1

Diagram of a deep brain stimulation (DBS) system. Electrodes are implanted via stereotactic neurosurgery and attached to an implanted pulse generator via subcutaneous extension wires. Image courtesy of David Peace, Department of Neurosurgery, University of Florida School of Medicine.

pulse width, and 2–10 volts (V). Stimulation can be either monopolar (with one or more DBS contacts serving as the cathode and the IPG serving as the anode) or bipolar (with individual DBS contacts serving as the cathode and anode). In some systems, a duty cycle can be used such that stimulation is on and off for specified periods of time. Serious risks associated with DBS surgery include intracranial hemorrhage, infection, and complications associated with anesthesia. The DBS system could malfunction or break, making replacement of one or more components necessary. Finally, repeated minor surgery is needed to replace the

IPG. Acute and chronic effects of stimulation depend on the site of stimulation.

In this review, we briefly describe the history of neurosurgery for psychiatric disorders to emphasize that this approach is not new, but has been previously limited by the neuroanatomical models used to select targets and by available neurosurgical techniques. We then describe, with a focus on movement disorders, how the refinement of neuroanatomical models and neurosurgical techniques led to the establishment of ablative neurosurgery and DBS as reasonable approaches for severe, treatment-refractory brain disorders. Next, the available

PD: Parkinson's
disease

GPi: internal globus
pallidus

data on the safety and efficacy of DBS for psychiatric disorders are presented and critically evaluated, with a brief discussion of the small but growing database on potential mechanisms of action. We conclude with recommendations for a research agenda going forward.

NEUROSURGERY FOR PSYCHIATRIC DISORDERS

The development of neurosurgical techniques for focal ablation and stimulation have historically been driven by attempts to treat intractable psychiatric disease [for an overview, please see the comprehensive historical review by Hariz and colleagues (2010)]. Prior to the 1950s, no specific medications existed for the treatment of severe psychiatric disorders. Given the disabling and often lethal nature of these illnesses, aggressive and often invasive treatments were pursued, including malarial pyrotherapy (Epstein 1936), hypoglycemic coma (Sakel 1937), electroconvulsive therapy (Bini 1938), and neurosurgery. The first surgical attempts to treat severely psychotic patients occurred as early as 1891 (Burckhardt 1891) with limited success. With the emergence of relatively crude models of the functional and structural neuroanatomy of mood and behavior regulation [e.g., Papez' circuit (Papez 1937)], researchers hypothesized that abnormal mood and behavioral regulation derived from dysfunctional thalamo-cortical communication (Moniz 1937). Neurosurgical treatments thus shifted to disrupting the white matter tracts connecting these regions, culminating in the prefrontal leucotomy (Moniz 1937) which was later referred to as the prefrontal lobotomy. The surgical methods available for severing these tracts were limited; procedures were often done blindly and resulted in relatively large lesions.

The mid-twentieth-century discovery of medications with antimanic (Cade 1949, Schou et al. 1954), antipsychotic (Bower 1954, Winkelman 1954), and antidepressant effects (Bailey et al. 1959, Kiloh et al. 1960, Kuhn 1958) essentially ended the lobotomy era. However, using novel stereotactic neurosurgical

techniques allowing more focal ablation (initially viewed as a potential substitute for prefrontal leucotomy in psychiatric patients) (Hariz et al. 2010, Spiegel et al. 1947), surgery for severe, intractable psychiatric disorders has continued in a limited fashion. Current approaches include subcaudate tractotomy, dorsal anterior cingulotomy, anterior capsulotomy, and limbic leucotomy (combing a subcaudate tractotomy with a cingulotomy) (**Figure 2**) (Cosgrove 2000). Naturalistic and nondisease-specific response rates of 22%–75% for these procedures have been reported for patients with severe, treatment-resistant psychiatric illness, including treatment-resistant OCD and depression (Cosgrove 2000, Cosgrove & Rauch 2003, Sachdev & Sacher 2005, Shields et al. 2008). Risks include weight gain, cognitive impairments, personality change, and epilepsy (Moreines et al. 2011, Sachdev & Sacher 2005).

DBS FOR MOVEMENT DISORDERS

As neurosurgery for psychiatric disorders diminished in popularity, stereotactic neurosurgery continued as a strategy to treat severe movement disorders. On the basis of early motor regulation models and preliminary non-stereotactic surgical results, Spiegel and colleagues attempted subcortical lesions for the treatment of Huntington's chorea (Spiegel & Wycis 1950). A variety of subcortical targets were then proposed for treating tremor in Parkinson's disease (PD) (Andy et al. 1963). The refinement of neuroanatomical models (Alexander et al. 1986, Bergman et al. 1990) further established lesions of the internal globus pallidus (GPi), ventrolateral thalamus, and/or subthalamus as accepted treatments for a number of medication-refractory movement disorders, including PD (Burchiel 1995, Walter & Vitek 2004), essential tremor (Burchiel 1995, Pahwa et al. 2000), and dystonia (Tasker 1990, Vitek et al. 1998). With the recognition that high-frequency (50–200 Hz) stimulation of the thalamus and subthalamus achieved beneficial effects similar to thalamotomy (Benabid et al.

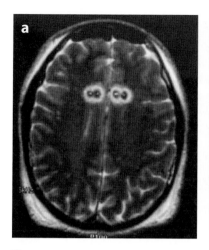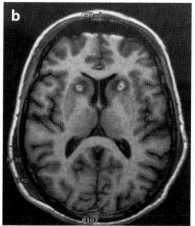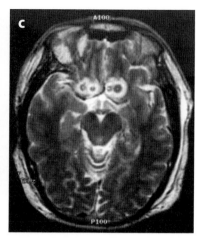

Figure 2

Postoperative MRI scans of three ablative neurosurgical procedures currently available for psychiatric disorders: (*a*) anterior cingulotomy, (*b*) anterior capsulotomy, and (*c*) subcaudate tractotomy. A limbic leucotomy is achieved by combining an anterior cingulotomy (*a*) with a subcaudate tractotomy (*c*). Images courtesy of G. Rees Cosgrove, Department of Neurosurgery, Brown University.

1991, Benabid et al. 1987, Hariz et al. 2010), DBS became a realistic alternative to ablation. Currently, DBS has largely replaced ablative surgery for the treatment of severe, refractory PD, essential tremor, and dystonia (Wichmann & Delong 2006). Targets include the thalamus for tremor and the subthalamic nucleus (STN) and GPi for PD (Moro et al. 2010b) and dystonia (Hung et al. 2007, Krause et al. 2004, Lozano & Abosch 2004, Sun et al. 2007). On the basis of more detailed models of motor regulation, investigators have proposed additional DBS targets for specific PD symptoms, such as the pedunculopontine nucleus for postural instability (Moro et al. 2010a, Plaha & Gill 2005). This historical progression highlights how the joint development of improved neuroanatomical models and neurosurgical techniques has been critical to establishing DBS as a valuable treatment option.

DBS FOR PSYCHIATRIC DISORDERS

Investigators began exploring the effects of intracranial stimulation in psychiatric patients as early as the 1950s, although reports are largely anecdotal (Hariz et al. 2010, Heath 1963, Heath et al. 1955). As such, DBS was not viewed as a viable treatment alternative until 1999 when Nuttin et al. (1999) published results of chronic, high-frequency stimulation of the bilateral anterior limbs of the internal capsule in four patients with severe, treatment-refractory OCD. Over the past decade, several reports have emerged that described effects of DBS at various targets and for various psychiatric indications (**Figure 3**). In general, these data derive from small, open-label studies without a comparator group and must therefore be viewed as encouraging but highly preliminary.

Obsessive-Compulsive Disorder

OCD is defined by the presence of one or more markedly distressing or functionally impairing obsessions (intrusive, ruminative thoughts, images, or impulses) and/or compulsions (highly repetitive mental or physical acts such as hand washing, checking locks, counting, praying, etc.); compulsions are often performed to diminish the anxiety/distress associated with obsessions. OCD affects about 1% of the U.S. population (Kessler et al. 2005b) and

STN: subthalamic nucleus

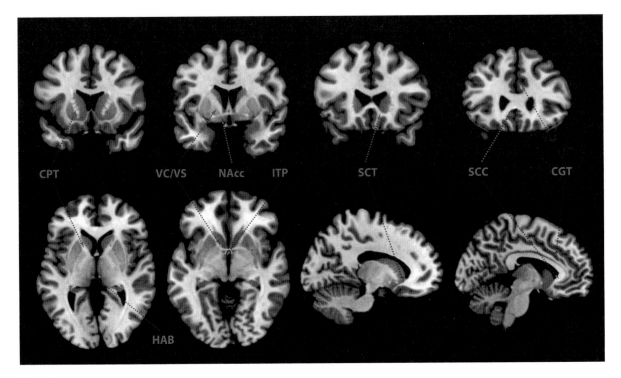

Figure 3

Location of various targets for ablation or DBS for psychiatric disorders. Abbreviations: CGT, cingulotomy; CPT, capsulotomy; HAB, habenula; ITP, inferior thalamic peduncle; SCC, subcallosal cingulate; SCT, subcaudate tractotomy; VC/VS, ventral capsule/ventral striatum. Note that the current VC/VS target is more posterior and inferior to the classic CPT target (Greenberg et al. 2010); the VC/VS and NAcc targets span a very similar area (Greenberg et al. 2010, Schlaepfer et al. 2008); the SCC target is more medial and anterior to the classic SCT target; and the classic CGT target is in the dorsal anterior cingulate (not rostral or subcallosal).

is associated with significant distress and disability (Hollander et al. 2010, Huppert et al. 2009). Evidence-based treatments include medications (primarily antidepressant and antipsychotic medications that modulate monoaminergic neurotransmission) (Goodman et al. 2000, Kellner 2010, Soomro et al. 2008) and cognitive-behavioral psychotherapy (Gava et al. 2007). However, up to 60% of patients do not respond adequately to available treatments (Goodman et al. 2000, Kellner 2010, Pallanti & Quercioli 2006, Simpson et al. 2008), making treatment-refractory OCD a serious public health problem.

Ventral capsule/ventral striatum. As described above, several ablative procedures, including capsulotomy, had shown moderate efficacy in severe, treatment-refractory OCD.

However, side effects, when they occurred, were generally irreversible, leading to an interest in DBS as a possible alternative. In the first report of bilateral DBS of the anterior internal capsule for OCD (Nuttin et al. 1999), four patients were implanted and three were reported to have beneficial effects (although the definition of improvement is not provided). More detailed data are presented for one subject: This patient had a 90% reduction in compulsive behavior per her family, and behavioral testing suggested acute positive effects with stimulation (Nuttin et al. 1999). A number of case reports/series followed that suggested beneficial effects of DBS of the ventral capsule (VC) and/or ventral striatum (VS) in treatment-refractory OCD (Abelson et al. 2005, Anderson & Ahmed 2003, Aouizerate et al. 2004).

VC: ventral capsule

VS: ventral striatum

An international, multicenter group implanted 26 patients over 8 years using similar but not identical protocols (Greenberg et al. 2010). All but three patients had a comorbid mood disorder. Response was defined as a decrease in the Yale-Brown Obsessive Compulsive Scale (YBOCS) of $\geq 35\%$ from baseline. At 3 months ($N = 26$), 50% of patients were responders; at 6 months ($N = 24$), 46% were responders; at 1 year ($N = 21$), 48% were responders; at 2 years ($N = 17$), 65% were responders; and at 3 years ($N = 12$), 58% were responders. Significant decreases in depression severity and increases in global functioning were also observed. Over the course of this study, location of the DBS electrodes was moved posterior and inferior from the classical capsulotomy target on the basis of clinical results; this adjustment was associated with lower stimulation intensity and better response rates in the latter patients. Adverse events related to surgery, the device, or stimulation included intracranial hemorrhage ($N = 2$), seizure ($N = 1$), infection ($N = 1$), lead or extension wire break ($N = 2$), hypomania ($N = 1$), and increased suicidal ideation (in at least four patients, although it appears this result was associated with a planned stimulation off phase (i.e., no stimulation was provided during this phase) at one of the sites in at least some patients. Acute effects of stimulation included hypomania-like episodes, mood worsening, improved anxiety, worsened anxiety (including one panic attack), irritability, cognitive abnormalities, and sensorimotor effects; in general, these symptoms resolved with either continued stimulation or a parameter change. Battery life ranged from 6–18 months, and battery depletion (or cessation of stimulation due to other causes) was noted to be mostly associated with a return of depressive rather than OCD symptoms.

Goodman et al. (2010) recently reported a blinded, staggered-onset, sham-controlled study of DBS of the VC/VS for OCD. Six patients with severe, treatment-refractory OCD were included. At one month postsurgery (without stimulation), subjects were randomized to one month of active (on) versus sham (off) DBS. After another month, the sham patients received active stimulation and were followed for one month. No patient showed significant improvement during the one month postsurgery or during the one-month sham period; however, in the first few months of the study, there were no statistically significant differences between patients receiving early versus delayed stimulation. Following 12 months of chronic DBS, 4 of 6 patients were responders ($\geq 35\%$ decrease in the YBOCS score from baseline). Side effects were similar to those previously reported with DBS at this target, and battery depletion/stimulation cessation were similarly associated with a return of depressive rather than OCD symptoms. Most patients showed no cognitive changes with chronic stimulation, although some showed improvements and others showed decrements.

Nucleus accumbens. The VC/VS target for DBS includes the ventral portion of the anterior internal capsule and the VS. The nucleus accumbens (a major component of the VS) has also been explored as a discrete target for DBS treating OCD. In a sham-controlled, crossover study, Huff and colleagues (2010) implanted 10 patients with OCD with a DBS electrode in the right nucleus accumbens. Patients were randomized to three months of active versus sham DBS then crossed over to the other condition for three months; all patients received active stimulation after these six months. There were no statistically significant differences between active and sham DBS. After 12 months of active stimulation, one patient (10%) achieved a response ($\geq 35\%$ decrease in the YBOCS score from baseline) and five patients (50%) achieved a partial response ($\geq 25\%$ decrease in the YBOCS score from baseline). Adverse events were similar to those seen previously with VC/VS stimulation. Although preliminary, these data suggest that DBS of the internal capsule may be necessary to treat OCD. This finding may relate to a proposed mechanism of action of DBS involving activation of white matter fibers.

Subthalamic nucleus. On the basis of observations of decreased compulsive behaviors in PD patients following STN DBS, Mallet et al. (2008) conducted a double-blind, crossover, sham-controlled trial of STN DBS for treatment-refractory OCD. At three months postsurgery, patients were randomized to three months of active versus sham DBS, followed by a one-month washout (no stimulation) phase, then crossed over to the other condition for an additional three months. Active DBS was statistically superior to sham stimulation in reducing OCD symptoms. Following three months of active stimulation, 75% of subjects met response criteria ($\geq$25% decrease in the YBOCS score from baseline) versus 38% of subjects meeting response criteria following three months of sham stimulation.

Inferior thalamic peduncle. The inferior thalamic peduncle (ITP) is a white matter bundle that includes fibers connecting thalamic nuclei with ventral prefrontal cortex. Presumably based on earlier models suggesting that disruption of thalamic-prefrontal connections could have beneficial effects in psychiatric disorders, one group has tested ITP DBS for treatment-refractory OCD in a small case series (Jimenez-Ponce et al. 2009). Following 12 months of open-label DBS, the authors report a 100% response rate ($\geq$35% decrease in the YBOCS score from baseline). Acute stimulation was associated with anxiety and autonomic symptoms in all subjects and confusion/disorientation in one subject; however, these effects occurred at contacts that were not optimal for long-term stimulation. No other adverse events were described.

TRD

Major depressive disorder (MDD) is a syndrome characterized by depressed mood and/or anhedonia (decreased ability to experience pleasure) and a combination of additional symptoms, including sleep and appetite abnormalities, low energy, psychomotor retardation (or agitation in some patients), decreased attention/concentration, low self-esteem and feeling of guilt, and suicidal ideation. MDD has a one-year U.S. prevalence of 7% (Kessler et al. 2005b), is the leading cause of years lost due to disability worldwide, and represents the largest contributor to overall disease burden in middle- and high-income countries (WHO 2008). Treatments for depression include medications, psychotherapy, and a number of other somatic interventions (such as ECT and light therapy). Despite the availability of treatments, two-thirds of patients do not remit (i.e., become symptom free) with an initial medication, one-third do not remit with multiple treatments, and at least 20% remain significantly symptomatic despite multiple, often aggressive interventions (Holtzheimer & Mayberg 2011, Rush et al. 2006). For those patients that respond well to treatment, depressive relapse is quite common, occurring in 60% to nearly 100% of patients depending on prior history of relapse and level of treatment resistance (APA 2000, Rush et al. 2006, Sackeim et al. 2001). Treatment-resistant depression (TRD), therefore, has a prevalence of ~1%–3%, and better treatments for achieving and maintaining remission are clearly needed.

Subcallosal cingulate white matter. As neuroimaging techniques have greatly improved over the past three decades, an impressive literature has accumulated on the neuroanatomical bases of depression and the antidepressant treatment response. On the basis of a converging dataset (Mayberg 2009, Price & Drevets 2010), Mayberg hypothesized that DBS modulation of the subcallosal cingulate (SCC) white matter would help bring about the functional neuroanatomical changes previously associated with an antidepressant response (across a number of treatments) and lead to symptomatic improvement in severe TRD patients. Critically, this proposed target was not a prior site for ablative neurosurgery in psychiatric disorders: The standard target for a cingulotomy was located in the mid-cingulate cortex dorsal to the corpus callosum, and the target for a subcaudate tractotomy was several millimeters more lateral.

An initial, open-label pilot study demonstrated clear antidepressant effects following chronic (6 months) SCC DBS in 4 of 6 TRD patients, with 3 patients achieving full remission (Mayberg et al. 2005). A number of these patients experienced acute, positive effects with intraoperative stimulation (such as a sense of calmness, increased alertness, and enhanced connectedness with others). No adverse effects of acute or chronic stimulation were described. Three patients developed infections requiring antibiotics; the DBS system was removed in 2 patients (both nonresponders) because of infection. This study was expanded to include a total of 20 patients followed for one year (Lozano et al. 2008). A 60% antidepressant response rate (defined as a ≥50% decrease in depression severity from baseline) was seen after 6 months of chronic DBS, and 55% of patients were responders at one year. Seventy-two percent of responders at 6 months were responders at one year, with 3 additional patients (all nonresponders at 6 months) achieving an antidepressant response at one year. Long-term follow-up, using an intent-to-treat analysis, found a 45% response rate and 15% remission rate after two years and a 60% response rate and 50% remission after three years of chronic stimulation (Kennedy et al. 2011). Adverse events included infection, 1 perioperative seizure (of unclear etiology), worsening mood/irritability unrelated to stimulation, and perioperative headache/incision pain. No adverse cognitive effects were identified, and improvements across several domains were observed (McNeely et al. 2008). Two probable suicides occurred in patients who had previously achieved response and/or remission and some degree of functional recovery. One occurred after three years of stimulation, and the other occurred after six years of stimulation. Both probable suicides were deemed unrelated to chronic stimulation or any change in stimulation parameters or other treatments.

Ventral capsule/ventral striatum. Earlier reports of VC/VS DBS for OCD consistently described significant antidepressant effects, independent of changes in OCD symptoms. Thus investigators hypothesized that VC/VS DBS may serve as an antidepressant treatment even in patients without comorbid OCD. A three-site pilot study found a 40% response rate following 6 months of open-label VC/VS DBS in 15 TRD patients, with a 53% response rate at last follow-up (an average of 24 ± 15 months after onset of stimulation, with a range of 6–51 months) (Malone et al. 2009). Acute stimulation was associated with both positive effects (improved mood, spontaneous smiling, decreased anxiety, and increased awareness/energy) and negative effects [autonomic symptoms (tachycardia, flushing), increased anxiety, perseverative speech, and involuntary facial movements]; the number of patients experiencing each was not described, though it is stated that these effects could be attenuated with a stimulation parameter change. Adverse events included 2 cases of hypomania/mixed-bipolar state, 5 cases of increased depression/suicidality, 1 case of perioperative pain, 1 device failure (DBS lead fracture), and 2 cases of syncope. Maintenance of response and recurrence rates are not described in the report.

Nucleus accumbens. The nucleus accumbens (NAcc) has been implicated in neural pathways related to reward processing and motivation (Humphries & Prescott 2010, Knutson et al. 2008, Sesack & Grace 2010). As such, Schlaepfer et al. (2008) hypothesized that DBS modulation of the NAcc may have specific effects on anhedonia, a core symptom of the depressive symptom that could effectively lead to an antidepressant treatment response in TRD. An initial report of three TRD patients receiving NAcc DBS described improvements in depression ratings and hedonic response with active stimulation that were reversed when stimulation was turned off (Schlaepfer et al. 2008). This cohort was expanded to 10 patients, and a 50% antidepressant response rate was seen following 12 months of chronic, active NAcc DBS (Bewernick et al. 2010). Acute positive effects were briefly described but reported not to be predictive of a longer-term outcome. Adverse effects were generally similar to those seen with VC/VS DBS; one completed suicide occurred but was judged to be unrelated to chronic stimulation.

Inferior thalamic peduncle. A single case report described effects of 24 months of ITP DBS in a woman with TRD, borderline personality disorder, and bulimia (Jimenez et al. 2005). In general, notable and mostly sustained antidepressant effects were observed with chronic stimulation. A slight return of depressive symptoms was seen following double-blind discontinuation following eight months of active stimulation. Anxiety, fear, and significant adverse autonomic effects were noted with acute stimulation of certain contacts and resolved when stimulation ceased. A "pleasant sensation" (p. 589) with decreased anxiety was found with acute stimulation of other contacts. No adverse effects of chronic stimulation were described.

Stria medullaris thalami (habenula). The habenula is a midbrain structure composed of several small nuclei (also called the habenular complex) that lies dorsal and caudal to the main body of the dorsal thalamus. The habenula is involved in the coordination of monoaminergic neurotransmission (primarily serotonergic and noradrenergic) and receives input from cortical and subcortical structures via the stria medullaris thalami. Given its neuroanatomy, and the clear role for monoaminergic neurotransmission in the pathophysiology of depression, Sartorius & Henn (2007) proposed that the habenula is as a potential target for DBS for TRD. A single case report described effects of DBS of the stria medullaris thalami in a woman with TRD who required biweekly ECT to prevent severe depressive relapse (although relapses occasionally occurred even with ECT at this frequency) (Sartorius et al. 2010). She was implanted during a period of full remission following ECT, treated with DBS at 5V, and experienced a severe relapse within 3 weeks following implantation. Stimulation intensity was progressively increased to 10.5V, and remission

was again achieved. No acute stimulation effects were observed, and no adverse events were reported.

Other Disorders

Addiction. Human and animal studies have helped define the neural circuitry of addiction (Koob & Volkow 2010), indicating that focal neuromodulation within these systems may have clinical benefit. Case reports have associated nucleus accumbens DBS with benefits for nicotine addiction (Kuhn et al. 2009, Mantione et al. 2010) and alcoholism (Muller et al. 2009). It is not clear whether DBS acts by reducing the rewarding quality of the abused substance, perhaps by replacing the desire for the substance with stimulation [e.g., analogous to intracranial self-stimulation (Jacques 1979, Olds & Fobes 1981)] or by decreasing cue-associated reinstatement of addictive behavior, thereby reducing likelihood of relapse (Muller et al. 2009, Vassoler et al. 2008).

Alzheimer's disease. Alzheimer's disease (AD) is a neurodegenerative/neuropsychiatric disorder that affects 1%–2% of the U.S. population (Hebert et al. 2003). More than 10% of those over age 65 (Evans et al. 1989) suffer from AD. AD treatments are notably limited to behavioral interventions and acetylcholinesterase inhibitors and/or N-methyl-D-aspartic acid receptor antagonists that may slow but do not prevent further cognitive decline (Burns & Iliffe 2009). On the basis of a case report of memory enhancement with DBS of the fornix in a patient with obesity (Hamani et al. 2008), Laxton et al. (2010) performed a small open-label study of fornical/hypothalamic DBS in AD patients, which showed potential efficacy in patients stable on anticholinesterase medication(s) for at least six months. It is unknown whether this approach is superior to currently available medications. The potential mechanism of action of this approach is unclear, although the authors propose that activation of fornical axons leads to downstream activation of other brain regions involved in memory (Laxton et al. 2010).

Tourette's syndrome. Tourette's syndrome (TS) is a neuropsychiatric disorder characterized by potentially disabling vocal and motor tics, with a prevalence of ~1% (Robertson 2008). TS is associated with a number of psychiatric comorbidities, including OCD (Swain et al. 2007). A subset of TS patients may have disabling symptoms despite standard medical treatment (including medications and behavioral intervention) (Mink 2009, Temel & Visser-Vandewalle 2004). Investigators have tried ablative neurosurgery in such patients with mixed success (Temel & Visser-Vandewalle 2004); among the various targets attempted, lesions of the centromedian-parafascicular complex of the thalamus have been most effective (Mink 2009). On the basis of results in related conditions (e.g., OCD and movement disorders), researchers have hypothesized that DBS of the GPi or VC/VS is also efficacious for treatment-refractory TS.

A number of case reports for thalamic DBS for TS have been published (Mink 2009). An open-label study of 18 patients showed initial positive results with follow-up of 3–18 months (Servello et al. 2008), and a 2-year assessment of this cohort identified significant decreases in tic severity, obsessive-compulsive symptoms, anxiety, and depression (Porta et al. 2009). Case reports also suggest potential efficacy of GPi and VC/VS DBS for severe TS patients (Mink 2009).

Mechanism of Action

Neuronal effects. The development of DBS as an alternative for ablative neurosurgery posited that high-frequency DBS served as a "reversible lesion" by inhibiting the activity in the stimulated gray matter (Benabid et al. 1991, Benazzouz et al. 1995). Further research has clarified that the mechanisms of action for DBS are more complex, including elements of inhibition as well as axonal excitation (Iremonger et al. 2006, McIntyre et al. 2004). Thus, the neuronal effects of DBS will depend greatly on location and the mix of cell bodies and passing white matter fibers in the field of stimulation (and whether these are inhibitory, excitatory, or modulatory). The various parameters of stimulation will also impact the effects of DBS. For example, suppression of tremor with thalamic stimulation is generally not achieved with stimulation frequencies below 50 Hz, and an optimal effect is seen with frequencies greater than 100 Hz (Benabid et al. 1991). However, 60–130 Hz GPi stimulation may be similarly effective for dystonia (Alterman et al. 2007, Isaias et al. 2009, Moro et al. 2009). The efficacy of DBS may depend on the activity state of the region being impacted, as well as the ability of DBS to adequately modulate this. For example, presumably abnormal oscillatory activity in the STN and GPi has been implicated in the pathophysiology of PD (Brown et al. 2001, 2004; Hammond et al. 2007, Levy et al. 2002, Weinberger et al. 2006), and the therapeutic effects of DBS may depend on modulation of this activity (Brown et al. 2004, Kuhn et al. 2008). We do not know whether this represents a general mechanism of action for DBS in other disorders.

Animal studies are beginning to address the direct neuronal effects of DBS at targets used for psychiatric disorders. Hamani and colleagues (2010a,b) carried out a series of experiments using high-frequency stimulation of the prelimbic (PL)/infralimbic (IL) cortex in rats—a region homologous to the SCC target for DBS for TRD. Stimulation was associated with an antidepressant-like effect in the forced swim test (FST), an animal model for testing potential antidepressant treatments; 130 Hz stimulation was more effective than 20 Hz, and stimulation in the PL/IL boundary region was more effective than pure IL stimulation (Hamani et al. 2010a). However, lesions of this region did not show the same effects, and lesions of local gray matter (preserving passing white matter fibers) did not prevent the antidepressant-like effects of stimulation (Hamani et al. 2010b); these authors further showed that the antidepressant-like effects of PL/IL stimulation depended on an intact serotonergic but not noradrenergic system (Hamani et al. 2010b). These results suggest that the effects of SCC DBS may be due to excitation of passing white matter fibers (possibly from brain stem monoaminergic nuclei) rather than local inhibitory effects. However, two other groups have described antidepressant-like effects of transient inactivation of rat IL cortex using the FST (Scopinho et al. 2010, Slattery et al. 2010).

Using the quinpirole model of OCD, high-frequency stimulation of the rat nucleus accumbens (both shell and core components) showed anticompulsive efficacy (Mundt et al. 2009). High-frequency but not low-frequency stimulation of the entopeduncular nucleus and globus pallidus (homologues of the human external and internal globus pallidus) showed anticompulsive effects in the signal attenuation model (Klavir et al. 2010). Both lesions and high-frequency stimulation of the rat STN have shown anticompulsive effects in both the quinpirole and signal attenuation OCD models (Klavir et al. 2009, Winter et al. 2008), suggesting that stimulation and lesions are functionally equivalent at this target.

System-level effects. Beyond the immediate neural effects of DBS, we presume that behavioral effects result from neural network activity modulation. In PD, overly synchronized oscillations within thalamo-cortico-striatal loops are associated with motor symptoms (Hammond et al. 2007), and DBS may work by modulating communication between these regions (Gradinaru et al. 2009, Hammond et al. 2007, Iremonger et al. 2006). Similarly, MDD and OCD (and most other psychiatric disorders) are viewed as network-level disorders in which behavioral abnormalities arise from dysfunction within a discrete network of interconnected brain regions (Freyer et al. 2010, Harrison et al. 2009, Mayberg 2009).

Animal and human studies have helped begin to elucidate the specific networks impacted by focal neuromodulation. Nucleus accumbens high-frequency stimulation in rats altered function and oscillatory power and coherence within a network involving the medial frontal cortex, the orbitofrontal cortex (OFC), and the mediodorsal thalamus (McCracken & Grace 2007, 2009). As above, the antidepressant-like effects of PL/IL DBS may depend on the integrity of passing white matter fibers, as well as an intact serotonergic system (Hamani et al. 2010b). A positron emission tomography (PET) study of six OCD patients showed that acute VC/VS DBS was associated with increased blood flow in the OFC, the anterior cingulate cortex, the basal ganglia, and the thalamus (Rauch et al. 2006); another PET study showed decreased OFC metabolism with chronic (3–6 weeks) VC/VS DBS for OCD in 2 of 3 patients (Abelson et al. 2005). Following 3 and 6 months of effective SCC DBS for TRD, investigators saw decreased blood flow at the SCC target, the medial prefrontal cortex, the OFC, the insula, and the midbrain; they saw increased blood flow in the mid-cingulate cortex and the dorsolateral prefrontal cortex. Lozano et al. (2008) saw similar changes in metabolism in the expanded cohort. The brain regions showing changes with DBS had been previously implicated in the neurobiology of depression and/or antidepressant treatment response (Mayberg 2009). With chronic NAcc DBS for TRD, decreased metabolism was seen in a number of cortical and subcortical brain regions implicated in the pathophysiology of depression (no areas of increased metabolism were identified) (Bewernick et al. 2010). A structural imaging study using diffusion tensor imaging tractography—an MRI imaging method and analytic approach that can delineate white matter tracts in vivo (Behrens et al. 2003, Johansen-Berg et al. 2005)—found that projections from the two major white matter targets for DBS for TRD (SCC and VC/VS) were largely divergent, but overlapped at several potentially important regions, including the medial prefrontal cortex, the amygdala/hippocampus, the nucleus accumbens, and the thalamus (Gutman et al. 2009). These regions may represent common remote regions impacted by DBS and necessary for its antidepressant effects.

SUMMARY AND FUTURE DIRECTIONS

Neurosurgical intervention for psychiatric disorders has a long history. However, earlier attempts were profoundly limited by the relative crudeness of available techniques and the neuroanatomical models employed. As both technical approaches and neuroanatomical models of mood, thought, and behavior have advanced, DBS has emerged as an intervention with the potential to ameliorate symptoms and restore function in patients with the most severe and treatment-refractory psychiatric disorders. Preliminary data on safety and efficacy of DBS at various targets for OCD and TRD are highly encouraging. However, these data are, by and large, open label and lack any control comparison. Although these patients are among the most difficult to treat, the possibility of a placebo effect cannot be discounted (Goetz et al. 2008). Therefore, these early data should be replicated in larger, placebo-controlled trials prior to widespread clinical acceptance.

As research progresses, a number of important issues will need to be addressed. First, the specific effects of various stimulation parameters at each target for each disorder should be carefully considered and tested. Current stimulation parameters in psychiatric disorders have been modeled after those used in movement disorders, which emerged from the earlier (mostly incorrect) understanding that DBS was simply a reversible lesion. The specific effects of frequency, pulse width, duty cycle (i.e., the ration of time on to time off), and stimulation intensity should be investigated at each putative target. The effects of unilateral versus bilateral stimulation should also be considered (Guinjoan et al. 2010, Hamani et al. 2010a). This work will be advanced by modeling approaches that incorporate anatomical data (Butson et al. 2006) as well as by animal studies. Second, the comparative safety and efficacy of the various targets for each disorder should be established, similar to efforts in movement disorders (Moro et al. 2010b). This research should also incorporate mechanism of action studies to help identify whether each approach is more appropriate for a specific subgroup of patients; such efforts will be greatly supported by academic, federal, and industry collaboration. Third, a careful consideration of the value of DBS versus ablation is needed. For some disorders, targets, and patient populations, ablation may be preferred over chronic stimulation (Gross 2008). Fourth, better animal models of psychiatric disorders are needed. Animal studies will be critical to refining stimulation targets and parameters. Studies of healthy and behaviorally normal animals may be useful in some cases, although studies in animal models of disease will be necessary, as well. However, there are currently no accepted models of treatment-resistant OCD or depression. To develop better models, it may be necessary to reframe more drastically how a particular psychiatric disorder is defined, especially in the case of treatment-resistant conditions (Holtzheimer & Mayberg 2011). Also, the interpretation of animal studies will need to consider the state of the animal being tested, e.g., awake and behaving versus anesthetized.

DISCLOSURE STATEMENT

Dr. Holtzheimer has received grant funding from the Greenwall Foundation, NARSAD, National Institutes of Health Loan Repayment Program, NIMH, and Northstar, Inc.; he has received consulting fees from AvaCat Consulting, St. Jude Medical Neuromodulation, and Oppenheimer & Co. Dr. Mayberg has a consulting agreement with St. Jude Medical Neuromodulation, which has licensed her intellectual property to develop SCC DBS for the treatment of severe depression (US 2005/0033379A1). The terms of these arrangements have been reviewed and approved by Emory University in accordance with their conflict-of-interest policies. Dr. Mayberg has current grant funding from the Dana Foundation, NARSAD, National Institute of Mental Health (NIMH), Stanley Medical Research Institute, and Woodruff Foundation.

LITERATURE CITED

Abelson JL, Curtis GC, Sagher O, Albucher RC, Harrigan M, et al. 2005. Deep brain stimulation for refractory obsessive-compulsive disorder. *Biol. Psychiatry* 57:510–16

Alexander GE, DeLong MR, Strick PL. 1986. Parallel organization of functionally segregated circuits linking basal ganglia and cortex. *Annu. Rev. Neurosci.* 9:357–81

Alterman RL, Shils JL, Miravite J, Tagliati M. 2007. Lower stimulation frequency can enhance tolerability and efficacy of pallidal deep brain stimulation for dystonia. *Mov. Disord.* 22:366–68

Anderson D, Ahmed A. 2003. Treatment of patients with intractable obsessive-compulsive disorder with anterior capsular stimulation. Case report. *J. Neurosurg.* 98:1104–8

Andy OJ, Jurko MF, Sias FR Jr. 1963. Subthalamotomy in treatment of parkinsonian tremor. *J. Neurosurg.* 20:860–70

Aouizerate B, Cuny E, Martin-Guehl C, Guehl D, Amieva H, et al. 2004. Deep brain stimulation of the ventral caudate nucleus in the treatment of obsessive-compulsive disorder and major depression. Case report. *J. Neurosurg.* 101:682–86

APA. 2000. *Diagnostic and Statistical Manual of Mental Disorders–Text Revision (DSM-IV-TR)*. Washington, DC: Am. Psychiatr. Assoc.

Bailey SD, Bucci L, Gosline E, Kline NS, Park IH, et al. 1959. Comparison of iproniazid with other amine oxidase inhibitors, including W-1544, JB-516, RO 4-1018, and RO 5-0700. *Ann. N. Y. Acad. Sci.* 80:652–68

Behrens TE, Johansen-Berg H, Woolrich MW, Smith SM, Wheeler-Kingshott CA, et al. 2003. Non-invasive mapping of connections between human thalamus and cortex using diffusion imaging. *Nat. Neurosci.* 6:750–57

Benabid AL, Pollak P, Gervason C, Hoffmann D, Gao DM, et al. 1991. Long-term suppression of tremor by chronic stimulation of the ventral intermediate thalamic nucleus. *Lancet* 337:403–6

Benabid AL, Pollak P, Louveau A, Henry S, de Rougemont J. 1987. Combined (thalamotomy and stimulation) stereotactic surgery of the VIM thalamic nucleus for bilateral Parkinson disease. *Appl. Neurophysiol.* 50:344–46

Benazzouz A, Piallat B, Pollak P, Benabid AL. 1995. Responses of substantia nigra pars reticulata and globus pallidus complex to high frequency stimulation of the subthalamic nucleus in rats: electrophysiological data. *Neurosci. Lett.* 189:77–80

Bergman H, Wichmann T, DeLong MR. 1990. Reversal of experimental parkinsonism by lesions of the subthalamic nucleus. *Science* 249:1436–38

Bewernick BH, Hurlemann R, Matusch A, Kayser S, Grubert C, et al. 2010. Nucleus accumbens deep brain stimulation decreases ratings of depression and anxiety in treatment-resistant depression. *Biol. Psychiatry* 67:110–16

Bini L. 1938. Experimental researches on epileptic attacks induced by the electric current. *Am. J. Psychiatry* 94:172–74

Bower WH. 1954. Chlorpromazine in psychiatric illness. *N. Engl. J. Med.* 251:689–92

Brown P, Mazzone P, Oliviero A, Altibrandi MG, Pilato F, et al. 2004. Effects of stimulation of the subthalamic area on oscillatory pallidal activity in Parkinson's disease. *Exp. Neurol.* 188:480–90

Brown P, Oliviero A, Mazzone P, Insola A, Tonali P, Di Lazzaro V. 2001. Dopamine dependency of oscillations between subthalamic nucleus and pallidum in Parkinson's disease. *J. Neurosci.* 21:1033–38

Burchiel KJ. 1995. Thalamotomy for movement disorders. *Neurosurg. Clin. N. Am.* 6:55–71

Burckhardt G. 1891. On cortical resection as a contribution to the operative treatment of psychosis. *Psychiatrie psychischgerichtliche Medizin* 47:463–548

Burns A, Iliffe S. 2009. Alzheimer's disease. *BMJ* 338:b158

Butson CR, Cooper SE, Henderson JM, McIntyre CC. 2006. Predicting the effects of deep brain stimulation with diffusion tensor based electric field models. *Med. Image Comput. Comput. Assist. Interv.* 9:429–37

Cade JFJ. 1949. Lithium salts in the treatment of psychotic excitement. *Med. J. Austr.* 2:349–52

Cosgrove GR. 2000. Surgery for psychiatric disorders. *CNS Spectr.* 5:43–52

Cosgrove GR, Rauch SL. 2003. Stereotactic cingulotomy. *Neurosurg. Clin. N. Am.* 14:225–35

Epstein NN. 1936. Artificial fever as a therapeutic procedure. *Cal West Med.* 44:357–58

Evans DA, Funkenstein HH, Albert MS, Scherr PA, Cook NR, et al. 1989. Prevalence of Alzheimer's disease in a community population of older persons. Higher than previously reported. *JAMA* 262:2551–56

Freyer T, Kloppel S, Tuscher O, Kordon A, Zurowski B, et al. 2011. Frontostriatal activation in patients with obsessive-compulsive disorder before and after cognitive behavioral therapy. *Psychol. Med.* 41:207–16

Gava I, Barbui C, Aguglia E, Carlino D, Churchill R, et al. 2007. Psychological treatments versus treatment as usual for obsessive compulsive disorder (OCD). *Cochrane Database Syst. Rev.*: CD005333

Goetz CG, Wuu J, McDermott MP, Adler CH, Fahn S, et al. 2008. Placebo response in Parkinson's disease: comparisons among 11 trials covering medical and surgical interventions. *Mov. Disord.* 23:690–99

Goodman WK, Foote KD, Greenberg BD, Ricciuti N, Bauer R, et al. 2010. Deep brain stimulation for intractable obsessive compulsive disorder: pilot study using a blinded, staggered-onset design. *Biol. Psychiatry* 67:535–42

Goodman WK, Ward HE, Kablinger AS, Murphy TK. 2000. Biological approaches to treatment-resistant obsessive compulsive disorder. In *Obsessive Compulsive Disorders: Contemporary Issues in Treatment*, ed. WK Goodman, MV Rudorfer, J Maser, pp. 333–70. New York: Erlbaum

Gradinaru V, Mogri M, Thompson KR, Henderson JM, Deisseroth K. 2009. Optical deconstruction of parkinsonian neural circuitry. *Science* 324:354–59

Greenberg BD, Gabriels LA, Malone DA Jr, Rezai AR, Friehs GM, et al. 2010. Deep brain stimulation of the ventral internal capsule/ventral striatum for obsessive-compulsive disorder: worldwide experience. *Mol. Psychiatry* 15:64–79

Gross RE. 2008. What happened to posteroventral pallidotomy for Parkinson's disease and dystonia? *Neurotherapeutics* 5:281–93

Guinjoan SM, Mayberg HS, Costanzo EY, Fahrer RD, Tenca E, et al. 2010. Asymmetrical contribution of brain structures to treatment-resistant depression as illustrated by effects of right subgenual cingulum stimulation. *J. Neuropsychiatry Clin. Neurosci.* 22:265–77

Gutman DA, Holtzheimer PE, Behrens TE, Johansen-Berg H, Mayberg HS. 2009. A tractography analysis of two deep brain stimulation white matter targets for depression. *Biol. Psychiatry* 65:276–82

Hamani C, Diwan M, Isabella S, Lozano AM, Nobrega JN. 2010a. Effects of different stimulation parameters on the antidepressant-like response of medial prefrontal cortex deep brain stimulation in rats. *J. Psychiatr. Res.* 44:683–87

Hamani C, Diwan M, Macedo CE, Brandao ML, Shumake J, et al. 2010b. Antidepressant-like effects of medial prefrontal cortex deep brain stimulation in rats. *Biol. Psychiatry* 67:117–24

Hamani C, McAndrews MP, Cohn M, Oh M, Zumsteg D, et al. 2008. Memory enhancement induced by hypothalamic/fornix deep brain stimulation. *Ann. Neurol.* 63:119–23

Hammond C, Bergman H, Brown P. 2007. Pathological synchronization in Parkinson's disease: networks, models and treatments. *Trends Neurosci.* 30:357–64

Hariz MI, Blomstedt P, Zrinzo L. 2010. Deep brain stimulation between 1947 and 1987: the untold story. *Neurosurg. Focus* 29:E1

Harrison BJ, Soriano-Mas C, Pujol J, Ortiz H, Lopez-Sola M, et al. 2009. Altered corticostriatal functional connectivity in obsessive-compulsive disorder. *Arch. Gen. Psychiatry* 66:1189–200

Heath RG. 1963. Electrical self-stimulation of the brain in man. *Am. J. Psychiatry* 120:571–77

Heath RG, Monroe RR, Mickle WA. 1955. Stimulation of the amygdaloid nucleus in a schizophrenic patient. *Am. J. Psychiatry* 111:862–63

Hebert LE, Scherr PA, Bienias JL, Bennett DA, Evans DA. 2003. Alzheimer disease in the US population: prevalence estimates using the 2000 census. *Arch. Neurol.* 60:1119–22

Hollander E, Stein DJ, Fineberg NA, Marteau F, Legault M. 2010. Quality of life outcomes in patients with obsessive-compulsive disorder: relationship to treatment response and symptom relapse. *J. Clin. Psychiatry* 71:784–92

Holtzheimer PE, Mayberg HS. 2011. Stuck in a rut: rethinking depression and its treatment. *Trends Neurosci.* 34:1–9

Huff W, Lenartz D, Schormann M, Lee SH, Kuhn J, et al. 2010. Unilateral deep brain stimulation of the nucleus accumbens in patients with treatment-resistant obsessive-compulsive disorder: outcomes after one year. *Clin. Neurol. Neurosurg.* 112:137–43

Humphries MD, Prescott TJ. 2010. The ventral basal ganglia, a selection mechanism at the crossroads of space, strategy, and reward. *Prog. Neurobiol.* 90:385–417

Hung SW, Hamani C, Lozano AM, Poon YY, Piboolnurak P, et al. 2007. Long-term outcome of bilateral pallidal deep brain stimulation for primary cervical dystonia. *Neurology* 68:457–59

Huppert JD, Simpson HB, Nissenson KJ, Liebowitz MR, Foa EB. 2009. Quality of life and functional impairment in obsessive-compulsive disorder: a comparison of patients with and without comorbidity, patients in remission, and healthy controls. *Depress. Anxiety* 26:39–45

Iremonger KJ, Anderson TR, Hu B, Kiss ZH. 2006. Cellular mechanisms preventing sustained activation of cortex during subcortical high-frequency stimulation. *J. Neurophysiol.* 96:613–21

Isaias IU, Alterman RL, Tagliati M. 2009. Deep brain stimulation for primary generalized dystonia: long-term outcomes. *Arch. Neurol.* 66:465–70

Jacques S. 1979. Brain stimulation and reward: "pleasure centers" after twenty-five years. *Neurosurgery* 5:277–83

Jimenez F, Velasco F, Salin-Pascual R, Hernandez JA, Velasco M, et al. 2005. A patient with a resistant major depression disorder treated with deep brain stimulation in the inferior thalamic peduncle. *Neurosurgery* 57:585–93; discussion 85–93

Jimenez-Ponce F, Velasco-Campos F, Castro-Farfan G, Nicolini H, Velasco AL, et al. 2009. Preliminary study in patients with obsessive-compulsive disorder treated with electrical stimulation in the inferior thalamic peduncle. *Neurosurgery* 65:203–9; discussion 9

Johansen-Berg H, Behrens TE, Sillery E, Ciccarelli O, Thompson AJ, et al. 2005. Functional-anatomical validation and individual variation of diffusion tractography-based segmentation of the human thalamus. *Cereb. Cortex* 15:31–39

Kellner M. 2010. Drug treatment of obsessive-compulsive disorder. *Dialogues Clin. Neurosci.* 12:187–97

Kennedy SH, Giacobbe P, Rizvi SJ, Placenza FM, Nishikawa Y, et al. 2011. Deep brain stimulation for treatment-resistant depression: follow-up after 3 to 6 years. *Am. J. Psychiatry.* **http://ajp.psychiatryonline.org/cgi/content/abstract/appi.ajp.2010.10081187v1**

Kessler RC, Berglund P, Demler O, Jin R, Merikangas KR, Walters EE. 2005a. Lifetime prevalence and age-of-onset distributions of DSM-IV disorders in the National Comorbidity Survey Replication. *Arch. Gen. Psychiatry* 62:593–602

Kessler RC, Chiu WT, Demler O, Merikangas KR, Walters EE. 2005b. Prevalence, severity, and comorbidity of 12-month DSM-IV disorders in the National Comorbidity Survey Replication. *Arch. Gen. Psychiatry* 62:617–27

Kiloh LG, Child JP, Latner G. 1960. A controlled trial of iproniazid in the treatment of endogenous depression. *J. Mental Sci.* 106:1139–44

Klavir O, Flash S, Winter C, Joel D. 2009. High frequency stimulation and pharmacological inactivation of the subthalamic nucleus reduces 'compulsive' lever-pressing in rats. *Exp. Neurol.* 215:101–9

Klavir O, Winter C, Joel D. 2011. High but not low frequency stimulation of both the globus pallidus and the entopeduncular nucleus reduces 'compulsive' lever-pressing in rats. *Behav. Brain Res.* 216:84–93

Knutson B, Bhanji JP, Cooney RE, Atlas LY, Gotlib IH. 2008. Neural responses to monetary incentives in major depression. *Biol. Psychiatry* 63:686–92

Koob GF, Volkow ND. 2010. Neurocircuitry of addiction. *Neuropsychopharmacology* 35:217–38

Krause M, Fogel W, Kloss M, Rasche D, Volkmann J, Tronnier V. 2004. Pallidal stimulation for dystonia. *Neurosurgery* 55:1361–68; discussion 68–70

Kuhn AA, Kempf F, Brucke C, Gaynor Doyle L, Martinez-Torres I, et al. 2008. High-frequency stimulation of the subthalamic nucleus suppresses oscillatory beta activity in patients with Parkinson's disease in parallel with improvement in motor performance. *J. Neurosci.* 28:6165–73

Kuhn J, Bauer R, Pohl S, Lenartz D, Huff W, et al. 2009. Observations on unaided smoking cessation after deep brain stimulation of the nucleus accumbens. *Eur. Addict. Res.* 15:196–201

Kuhn R. 1958. The treatment of depressive states with G 22355 (imipramine hydrochloride). *Am. J. Psychiatry* 115:459–64

Laxton AW, Tang-Wai DF, McAndrews MP, Zumsteg D, Wennberg R, et al. 2010. A phase I trial of deep brain stimulation of memory circuits in Alzheimer's disease. *Ann. Neurol.* 68:521–34

Levy R, Ashby P, Hutchison WD, Lang AE, Lozano AM, Dostrovsky JO. 2002. Dependence of subthalamic nucleus oscillations on movement and dopamine in Parkinson's disease. *Brain* 125:1196–209

Lozano AM, Abosch A. 2004. Pallidal stimulation for dystonia. *Adv. Neurol.* 94:301–8

Lozano AM, Mayberg HS, Giacobbe P, Hamani C, Craddock RC, Kennedy SH. 2008. Subcallosal cingulate gyrus deep brain stimulation for treatment-resistant depression. *Biol. Psychiatry* 64:461–67

Mallet L, Polosan M, Jaafari N, Baup N, Welter ML, et al. 2008. Subthalamic nucleus stimulation in severe obsessive-compulsive disorder. *N. Engl. J. Med.* 359:2121–34

Malone DA Jr, Dougherty DD, Rezai AR, Carpenter LL, Friehs GM, et al. 2009. Deep brain stimulation of the ventral capsule/ventral striatum for treatment-resistant depression. *Biol. Psychiatry* 65:267–75

Mantione M, van de Brink W, Schuurman PR, Denys D. 2010. Smoking cessation and weight loss after chronic deep brain stimulation of the nucleus accumbens: therapeutic and research implications: case report. *Neurosurgery* 66:E218; discussion E18

Mayberg HS. 2009. Targeted electrode-based modulation of neural circuits for depression. *J. Clin. Invest.* 119:717–25

Mayberg HS, Lozano AM, Voon V, McNeely HE, Seminowicz D, et al. 2005. Deep brain stimulation for treatment-resistant depression. *Neuron* 45:651–60

McCracken CB, Grace AA. 2007. High-frequency deep brain stimulation of the nucleus accumbens region suppresses neuronal activity and selectively modulates afferent drive in rat orbitofrontal cortex in vivo. *J. Neurosci.* 27:12601–10

McCracken CB, Grace AA. 2009. Nucleus accumbens deep brain stimulation produces region-specific alterations in local field potential oscillations and evoked responses in vivo. *J. Neurosci.* 29:5354–63

McIntyre CC, Savasta M, Kerkerian-Le Goff L, Vitek JL. 2004. Uncovering the mechanism(s) of action of deep brain stimulation: activation, inhibition, or both. *Clin. Neurophysiol.* 115:1239–48

McNeely HE, Mayberg HS, Lozano AM, Kennedy SH. 2008. Neuropsychological impact of Cg25 deep brain stimulation for treatment-resistant depression: preliminary results over 12 months. *J. Nerv. Ment. Dis.* 196:405–10

Mink JW. 2009. Clinical review of DBS for Tourette Syndrome. *Front Biosci. (Elite Ed.)* 1:72–76

Moniz E. 1937. Prefrontal leucotomy in the treatment of mental disorders. *Am. J. Psychiatry* 93:1379–85

Moreines JL, McClintock SM, Holtzheimer PE. 2011. Neuropsychologic effects of neuromodulation techniques for treatment-resistant depression: a review. *Brain Stimul.* 4:17–27

Moro E, Hamani C, Poon YY, Al-Khairallah T, Dostrovsky JO, et al. 2010a. Unilateral pedunculopontine stimulation improves falls in Parkinson's disease. *Brain* 133:215–24

Moro E, Lozano AM, Pollak P, Agid Y, Rehncrona S, et al. 2010b. Long-term results of a multicenter study on subthalamic and pallidal stimulation in Parkinson's disease. *Mov. Disord.* 25:578–86

Moro E, Piboolnurak P, Arenovich T, Hung SW, Poon YY, Lozano AM. 2009. Pallidal stimulation in cervical dystonia: clinical implications of acute changes in stimulation parameters. *Eur. J. Neurol.* 16:506–12

Muller UJ, Sturm V, Voges J, Heinze HJ, Galazky I, et al. 2009. Successful treatment of chronic resistant alcoholism by deep brain stimulation of nucleus accumbens: first experience with three cases. *Pharmacopsychiatry* 42:288–91

Mundt A, Klein J, Joel D, Heinz A, Djodari-Irani A, et al. 2009. High-frequency stimulation of the nucleus accumbens core and shell reduces quinpirole-induced compulsive checking in rats. *Eur. J. Neurosci.* 29:2401–12

Nuttin B, Cosyns P, Demeulemeester H, Gybels J, Meyerson B. 1999. Electrical stimulation in anterior limbs of internal capsules in patients with obsessive-compulsive disorder. *Lancet* 354:1526

Olds ME, Fobes JL. 1981. The central basis of motivation: intracranial self-stimulation studies. *Annu. Rev. Psychol.* 32:523–74

Pahwa R, Lyons K, Koller WC. 2000. Surgical treatment of essential tremor. *Neurology* 54:S39–44

Pallanti S, Quercioli L. 2006. Treatment-refractory obsessive-compulsive disorder: methodological issues, operational definitions and therapeutic lines. *Prog. Neuropsychopharmacol. Biol. Psychiatry* 30:400–12

Papez JW. 1937. A proposed mechanism of emotion. *Arch. Neurol. Psychiatry* 38:725–43

Plaha P, Gill SS. 2005. Bilateral deep brain stimulation of the pedunculopontine nucleus for Parkinson's disease. *Neuroreport* 16:1883–87

Porta M, Brambilla A, Cavanna AE, Servello D, Sassi M, et al. 2009. Thalamic deep brain stimulation for treatment-refractory Tourette syndrome: two-year outcome. *Neurology* 73:1375–80

Price JL, Drevets WC. 2010. Neurocircuitry of mood disorders. *Neuropsychopharmacology* 35:192–216

Rauch SL, Dougherty DD, Malone D, Rezai A, Friehs G, et al. 2006. A functional neuroimaging investigation of deep brain stimulation in patients with obsessive-compulsive disorder. *J. Neurosurg.* 104:558–65

Robertson MM. 2008. The prevalence and epidemiology of Gilles de la Tourette syndrome. Part 1: the epidemiological and prevalence studies. *J. Psychosom. Res.* 65:461–72

Rush AJ, Trivedi MH, Wisniewski SR, Nierenberg AA, Stewart JW, et al. 2006. Acute and longer-term outcomes in depressed outpatients requiring one or several treatment steps: a STAR*D report. *Am. J. Psychiatry* 163:1905–17

Sachdev P, Sacher J. 2005. Long-term outcome of neurosurgery for treatment of resistant depression. *J. Neuropsychiatry Clin. Neurosci.* 17:478–85

Sackeim HA, Haskett RF, Mulsant BH, Thase ME, Mann JJ, et al. 2001. Continuation pharmacotherapy in the prevention of relapse following electroconvulsive therapy: a randomized controlled trial. *JAMA* 285:1299–307

Sakel M. 1937. The origin and nature of the hypoglycemic therapy of the psychoses. *Bull. N. Y. Acad. Med.* 13:97–109

Sartorius A, Henn FA. 2007. Deep brain stimulation of the lateral habenula in treatment resistant major depression. *Med. Hypotheses* 69:1305–8

Sartorius A, Kiening KL, Kirsch P, von Gall CC, Haberkorn U, et al. 2010. Remission of major depression under deep brain stimulation of the lateral habenula in a therapy-refractory patient. *Biol. Psychiatry* 67:e9–11

Schlaepfer TE, Cohen MX, Frick C, Kosel M, Brodesser D, et al. 2008. Deep brain stimulation to reward circuitry alleviates anhedonia in refractory major depression. *Neuropsychopharmacology* 33:368–77

Schou M, Juel-Nielsen N, Stromgren E, Voldby H. 1954. The treatment of manic psychoses by the administration of lithium salts. *J. Neurol. Neurosurg. Psychiatry* 17:250–60

Scopinho AA, Scopinho M, Lisboa SF, Correa FM, Guimaraes FS, Joca SR. 2010. Acute reversible inactivation of the ventral medial prefrontal cortex induces antidepressant-like effects in rats. *Behav. Brain Res.* 214:437–42

Servello D, Porta M, Sassi M, Brambilla A, Robertson MM. 2008. Deep brain stimulation in 18 patients with severe Gilles de la Tourette syndrome refractory to treatment: the surgery and stimulation. *J. Neurol. Neurosurg. Psychiatry* 79:136–42

Sesack SR, Grace AA. 2010. Cortico-basal ganglia reward network: microcircuitry. *Neuropsychopharmacology* 35:27–47

Shields DC, Asaad W, Eskandar EN, Jain FA, Cosgrove GR, et al. 2008. Prospective assessment of stereotactic ablative surgery for intractable major depression. *Biol. Psychiatry* 64:449–54

Simpson HB, Foa EB, Liebowitz MR, Ledley DR, Huppert JD, et al. 2008. A randomized, controlled trial of cognitive-behavioral therapy for augmenting pharmacotherapy in obsessive-compulsive disorder. *Am. J. Psychiatry* 165:621–30

Slattery DA, Neumann I, Cryan JF. 2011. Transient inactivation of the infralimbic cortex induces antidepressant-like effects in the rat. *J. Psychopharmacol.* In Press

Soomro GM, Altman D, Rajagopal S, Oakley-Browne M. 2008. Selective serotonin re-uptake inhibitors (SSRIs) versus placebo for obsessive compulsive disorder (OCD). *Cochrane Database Syst. Rev:* CD001765

Spiegel EA, Wycis HT. 1950. Pallidothalamotomy in chorea. *Arch. Neurol. Psychiatry* 64:295–96

Spiegel EA, Wycis HT, Marks M, Lee AJ. 1947. Stereotaxic apparatus for operations on the human brain. *Science* 106:349–50

Sun B, Chen S, Zhan S, Le W, Krahl SE. 2007. Subthalamic nucleus stimulation for primary dystonia and tardive dystonia. *Acta Neurochir. Suppl.* 97:207–14

Swain JE, Scahill L, Lombroso PJ, King RA, Leckman JF. 2007. Tourette syndrome and tic disorders: a decade of progress. *J. Am. Acad. Child. Adolesc. Psychiatry* 46:947–68

Tasker RR. 1990. Thalamotomy. *Neurosurg. Clin. N. Am.* 1:841–64

Temel Y, Visser-Vandewalle V. 2004. Surgery in Tourette syndrome. *Mov. Disord.* 19:3–14

Vassoler FM, Schmidt HD, Gerard ME, Famous KR, Ciraulo DA, et al. 2008. Deep brain stimulation of the nucleus accumbens shell attenuates cocaine priming-induced reinstatement of drug seeking in rats. *J. Neurosci.* 28:8735–39

Vitek JL, Zhang J, Evatt M, Mewes K, DeLong MR, et al. 1998. GPi pallidotomy for dystonia: clinical outcome and neuronal activity. *Adv. Neurol.* 78:211–19

Walter BL, Vitek JL. 2004. Surgical treatment for Parkinson's disease. *Lancet Neurol.* 3:719–28

Weinberger M, Mahant N, Hutchison WD, Lozano AM, Moro E, et al. 2006. Beta oscillatory activity in the subthalamic nucleus and its relation to dopaminergic response in Parkinson's disease. *J. Neurophysiol.* 96:3248–56

WHO. 2008. *The Global Burden of Disease: 2004 update.* Geneva: World Health Organ.

Wichmann T, Delong MR. 2006. Deep brain stimulation for neurologic and neuropsychiatric disorders. *Neuron* 52:197–204

Winkelman NW Jr. 1954. Chlorpromazine in the treatment of neuropsychiatric disorders. *J. Am. Med. Assoc.* 155:18–21

Winter C, Mundt A, Jalali R, Joel D, Harnack D, et al. 2008. High frequency stimulation and temporary inactivation of the subthalamic nucleus reduce quinpirole-induced compulsive checking behavior in rats. *Exp. Neurol.* 210:217–28

Three-Dimensional Transformations for Goal-Directed Action

J. Douglas Crawford,[1] Denise Y.P. Henriques,[2] and W. Pieter Medendorp[3]

[1]York Centre for Vision Research, Canadian Action and Perception Network, and Departments of Psychology and [2]Kinesiology and Health Sciences, York University, Toronto, Ontario, Canada, M3J 1P3; email: jdc@yorku.ca, deniseh@yorku.ca

[3]Radboud University Nijmegen, Donders Institute for Brain, Cognition and Behaviour, 6525 HR, Nijmegen, The Netherlands; email: p.medendorp@donders.ru.nl

Annu. Rev. Neurosci. 2011. 34:309–31

First published online as a Review in Advance on March 29, 2011

The *Annual Review of Neuroscience* is online at neuro.annualreviews.org

This article's doi:
10.1146/annurev-neuro-061010-113749

0147-006X/11/0721-0309$20.00

Keywords

vision, saccades, reach, reference frames, parietal cortex, brainstem

Abstract

Much of the central nervous system is involved in visuomotor transformations for goal-directed gaze and reach movements. These transformations are often described in terms of stimulus location, gaze fixation, and reach endpoints, as viewed through the lens of translational geometry. Here, we argue that the intrinsic (primarily rotational) 3-D geometry of the eye-head-reach systems determines the spatial relationship between extrinsic goals and effector commands, and therefore the required transformations. This approach provides a common theoretical framework for understanding both gaze and reach control. Combined with an assessment of the behavioral, neurophysiological, imaging, and neuropsychological literature, this framework leads us to conclude that (*a*) the internal representation and updating of visual goals are dominated by gaze-centered mechanisms, but (*b*) these representations must then be transformed as a function of eye and head orientation signals into effector-specific 3-D movement commands.

Contents

INTRODUCTION

3-D: three-
dimensional

Reference position:
the zero location
and/or orientation,
from which other
locations and/or
orientations are
measured

Day-to-day life can be described as a series of
goal-directed behaviors, sometimes relatively
simple and direct, such as pressing a door-
bell, and sometimes highly abstract through
space and time, such as planning a university
education. Here, we focus on the spatial transformations for two simple and well-studied
behaviors: gaze and hand movements made
immediately, or after a short delay, toward a

visual goal. We consider how these systems
represent spatial goals, how these systems
update spatial goals during self-motion, and
finally how they transform goals into action.

The major theme of this review is that internal representations and transformations, even
for extrinsic goals, cannot be divorced from the
underlying three-dimensional (3-D) geometry
that links the sensors to the effectors. This geometry affects not only how stimuli project onto
the sensory apparatus, but also how visual activation maps onto the correct pattern of effector commands. These mappings, or transformations, must account for the translational,
and especially rotational, geometry of the eyes,
head, and shoulder. It may seem tempting to ig-
nore some of these details, but the brain has no
such luxury. Here, we focus on how these details
are incorporated into the feed-forward (open-
loop) transformations for movement. Viewed
from this perspective, the early spatial transformations for visually guided gaze and reach
movements show several common principles.

Unless stated otherwise, the behavioral data
referenced below pertain to observations that
hold for both the human and the monkey. Animal models continue to advance the boundaries
of known physiology in this field, but wherever
possible, we emphasize recent advances in human systems neuroscience. But first, we provide
the necessary background of mathematical and
geometric concepts.

GEOMETRIC FOUNDATIONS

The Vocabulary of Spatial Transformations

Positions and movements are normally represented as vectors, which for our purposes can
be loosely defined as 3-D arrows with a certain
length and direction. The point at the tail of
this arrow coincides with the zero vector, often called the reference position. To be meaningful, these must all be defined within some
coordinate frame. The latter incorporates two
concepts: A reference frame is some rigid body,
useful for describing the relative location or

orientation of the body we want to represent. In neuroscience, reference frames are typically divided into two categories: egocentric reference frames, where a location is represented relative to some part of the body, such as the retina, eye, head, or torso, and allocentric reference frames, where a location is represented relative to an external object. In the case of motor control, one generally chooses the more stable insertion point of a set of muscles as the egocentric frame of reference. For example, the head is the logical frame for eye movement, and the torso is the logical frame for head and arm movement. One can then fix a set of coordinate axes within this frame and use some arbitrary unit along these axes to specify the components of the vector. These topics have been reviewed previously (Soechting & Flanders 1992), and rigorous definitions can be found in any linear algebra text. Sometimes there is confusion about the meaning of an "eye-centered" reference frame. A frame of reference could be both eye-centered in the sense that its directional coordinates are fixed with the rotating eye and head-centered in the sense that the ego center of these coordinates is located at some fixed point in the head. Therefore, we use the term gaze-centered to denote a directional coordinate system that rotates with the eye.

When discussing position/movement, it is important to distinguish between location/translation and orientation/rotation. These two types of position/motion have very different mathematical properties. Vectors representing the former commute (they add in any order), but the latter do not: The math of rotations is highly nonlinear and generally is influenced by the initial orientation. As we see in the next section, the early geometry of visuomotor control is dominated by orientation/rotation. And yet the vast majority of models that deal with this system use translational math that only approximates rotations over a small range. This principle was first noted in the context of oculomotor control (Tweed & Vilis 1987), but it has implications for nearly every process described below.

The 3-D Geometry of Visual-Motor Transformations

Gaze direction determines the two-dimensional (2-D) direction of the visual stimulus that falls on the fovea. However, the spatial correspondence of proximal stimuli on other points on the retina to the locations and the orientations of distal stimuli is determined by the complete 3-D orientation of the eye, including torsion. (Conversely, one can only infer the correct plan for a goal-directed movement from knowledge of both the proximal visual stimulus and 3-D eye orientation; see 3-D Reference Frame Transformations, below.) Here we define torsion as rotation about a head-fixed axis aligned with the primary gaze direction. Defined thus, Listing's law (**Figure 1a**) states that torsion is held at zero (in practice, within $\pm 1°$). Mechanical factors likely play a role in implementing some aspects of Listing's law (Demer 2006a, Ghasia & Angelaki 2005, Klier et al. 2006), but we know that they do not constrain torsion because Listing's law is obeyed only for smooth pursuit and saccades with the head fixed (Ferman et al. 1987, Haslwanter et al. 1991, Tweed & Vilis 1987) and for gaze fixations during head translation (Angelaki et al. 2003). Other types of eye movement abandon or modify Listing's law to optimize different factors such as retinal stabilization and binocular vision (Misslisch & Tweed 2001, Tweed 1997).

During natural gaze behaviors, subjects use a Fick strategy to move the head. This strategy implies that the head assumes orientations near zero torsion in Fick coordinates (**Figure 1b**), i.e., orientations that can be reached by a horizontal rotation about a body-fixed vertical axis and a vertical rotation about a head-fixed horizontal axis, with only minor, random variations about the third fronto-caudal torsional axis (Glenn & Vilis 1992, Klier et al. 2003, Medendorp et al. 1999).

In natural behavior, both the eyes and the head contribute to gaze direction, where the former contributes more vertical and the latter more horizontal (Freedman & Sparks 1997),

Reference frame: a rigid body in which coordinate axes are embedded, thereby used to define the directions of rotation and/or translation for some other mobile rigid body

2-D: two-dimensional

Listing's law: the kinematic rule that describes 3-D eye orientation during eye movements when the head is motionless

Fick strategy: a kinematic strategy in which the head rotates about a body-fixed vertical axis and head-fixed horizontal axis, as in Fick coordinates

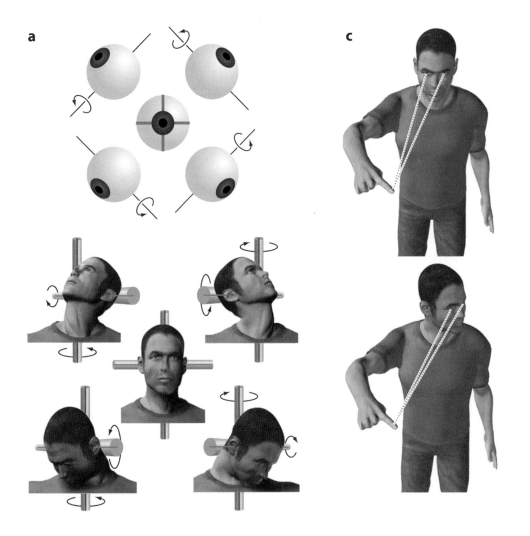

a

c

Figure 1

Geometric constraints on the visuomotor system. (*a*) Listing's law states that the eye assumes only orientations (e.g., peripheral panels) that can be reached from a central primary eye position (*center panel*) through fixed-axis rotations about axes within a head-fixed plane—here, the plane of the page. Curving arrows show direction of rotation about four example axes (right-hand rule applies). Torsion is defined as rotation about the axis aligned with gaze at primary eye position—here, orthogonal to the page. (*b*) The Fick strategy states that the head assumes only orientations that can be reached through rotations about a body fixed vertical axis (*black lines embedded in gray cylinders*) and a head-fixed horizontal axis (*green lines and cylinders*). (*c*) The geometry of reach is influenced by 3-D constraints on eye, head, and arm orientation and also by translations of the eye during head rotation.

which results in a Fick-like constraint on eye-in-space orientation, i.e., with torsion minimized about the visual axis (Glenn & Vilis 1992, Klier et al. 2003, Radau et al. 1994). Because both the eye and the head each show random biological errors in torsional control during gaze, these errors sum to produce torsional variability of up to $\pm 10°$, a factor rarely accounted for in visual or motor experiments.

The eyes and shoulder joint are essentially capable of rotation only, but head motion and its visual consequences are more complex (**Figure 1c**). Because the spine attaches near the back of the head and the eyes are near the front, any head rotation causes the two eyes to translate in different directions relative to space (Crane et al. 1997, Crane & Demer 1997, Medendorp et al. 2000) and the shoulder (Henriques et al. 2003, Henriques & Crawford 2002). Separation of the eyes is crucial for stereoscopic vision, but it also provides two different head-centered reference locations for visual direction. Psychophysical experiments suggest that visual direction is aligned to each eye independently (Erkelens & van de Grind 1994), to a central cyclopean eye (Ono et al. 2002), or to a dominant eye (Porac & Coren 1976), likely depending on the task. In visuomotor tasks that encourage monocular alignment, dominance may switch, depending on the field of view (Banks et al. 2004, Khan & Crawford 2001).

SPATIAL CODING AND UPDATING OF THE GOAL

Figure 2a provides an overview of the human brain structures that will be referred to in the remainder of this review, as well as their functional connectivity for the saccade and reach systems. The functional anatomy and effector-specificity of the human brain are not yet as clear as those of the monkey, but there appear to be many homologs between the two species (Amiez & Petrides 2009, Beurze et al. 2009, Culham & Valyear 2006, Filimon et al. 2009, Picard & Strick 2001). For example, in posterior parietal cortex (PPC), the saccade and reach areas located in monkey lateral (LIP) and medial (MIP) intraparietal cortex (Andersen & Buneo 2002) appear to correspond to mIPS in the human (Van Der Werf et al. 2010, Vesia et al. 2010). **Figure 2b** provides a flow diagram of the major transformations that we discuss.

Goal Coding versus Sensory and Motor Coding

High-level goal representations are closely associated with working memory and the dissociation of future intentions from current sensorimotor events (di Pellegrino & Wise 1993, Goldman-Rakic 1992). Here, we restrict this notion to entail early visuomotor representations of desired gaze and hand positions. If these encode spatial goals, one should be able to discriminate this activity from both sensory and motor events. Anti-saccade (or anti-reach) tasks dissociate the direction of the visual stimulus from the direction of the internal goal for movement (Guitton et al. 1985, Munoz & Everling 2004). Subjects are trained or asked to move in the direction opposite of the stimulus (pro-saccades/pro-reaches refer to movements made directly to the target). Recordings from monkey LIP and MIP during anti tasks suggest that most neurons are tuned for the movement direction, some encode the visual stimulus direction, and some switch from the latter to the former during the trial (Gail & Andersen 2006, Hallett 1978, Kusunoki et al. 2000, Zhang & Barash 2000). Similarly, human PPC is spatially selective for direction in prosaccades/reaches and remaps this activity to tune to the opposite direction during anti tasks (Medendorp et al. 2005, Van Der Werf et al. 2008). However, when subjects were instead trained to point while looking through left-right reversing prisms, the spatially selective activity in most PPC areas [superior parieto-occipital cortex (SPOC), mIPS, visual areas V3, 7] remained tied to the visual direction of the goal, not the movement direction (Fernandez-Ruiz et al. 2007). Only one PPC region—the angular gyrus (**Figure 2**)—showed the opposite effect. Taken together, these experiments suggest that visuomotor areas such as SPOC primarily code the spatial goal for movements.

Egocentric versus Allocentric Coding

The dorsal stream of vision (terminating in parietal-frontal movement areas) is, by default,

PPC: posterior parietal cortex

LIP: lateral intraparietal cortex (monkey)

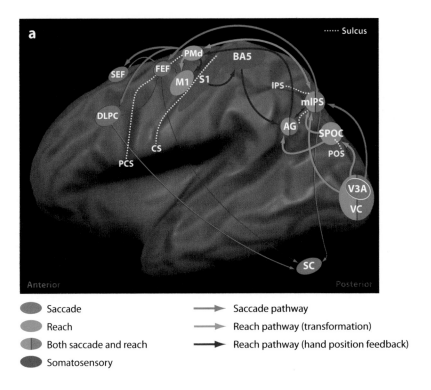

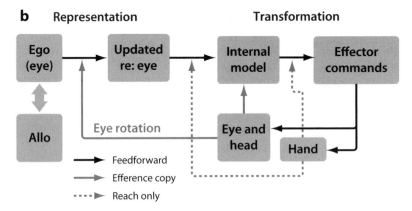

b   Representation   Transformation

Ego (eye) — Updated re: eye — Internal model — Effector commands

Allo

Eye rotation

Eye and head

Hand

Feedforward
Efference copy
Reach only

Figure 2

Overview of visuomotor brain areas and transformations. (*a*) Schematic representation of human brain (lateral view) regions involved in processing of visuomotor transformations and eye-hand coordination: VC, visual cortex (V3A); AG, angular gyrus; mIPS, mid-posterior intraparietal sulcus; and SPOC, superior parieto-occipital cortex; S1, primary somatosensory area for arm movements (proprioception); BA5, Brodmann area 5; M1, primary motor cortex; PMd, dorsal premotor cortex; FEF, frontal eye fields; SEF, supplementary eye fields; DLPC, dorsolateral prefrontal cortex; SC, superior colliculus; PCS, precentral sulcus; CS, central sulcus; IPS, intraparietal sulcus; POS, parieto-occipital sulcus. (*b*) Primarily eye-centered ego(centric) goal representations interact with allo(centric) representations and are updated as a function of eye rotation. These signals are then put through an inverse internal model of the eye-head-torso system to compute motor effector commands for limb and gaze control. Efference copies derived from the latter provide position and movement signals for the internal model and updating, respectively, whereas hand position signals derived from multiple sources are used in computations of the reach command (see text for details).

involved in egocentric coding, i.e., relative to some part of the body (Goodale & Milner 1992, Schenk 2006). Neurophysiological studies have attempted to determine the frame of reference by dissociating the candidate frames (most often the eye and head) while recording sensorimotor receptive fields. With some exceptions (e.g., Avillac et al. 2005, Mullette-Gillman et al. 2009), most studies of goal-related activity in PPC, frontal cortex saccade areas, and superior colliculus (SC) suggest a gaze-centered, eye-fixed frame of reference (Andersen & Buneo 2002, Colby & Goldberg 1999). Consistent with this, fMRI recordings show egocentric directional tuning over human parietal and frontal visuomotor areas (Kastner et al. 2007, Levy et al. 2007, Medendorp et al. 2006, Schluppeck et al. 2005, Sereno et al. 2001), with gaze-centered coding in PPC and dorsal premotor cortex and body-centered coding for reaching near motor cortex (Beurze et al. 2010). However, this scheme may depend on the sensory modality used to aim the action: When the goal stimulus is somatosensory, PPC seems capable of switching from gaze-centered to body-centered coordinates (Bernier & Grafton 2010).

In contrast, the ventral visual stream (including occipital-temporal areas involved in object recognition) is more closely associated with allocentric coding, i.e., relative to some stable external visual cue (Goodale & Milner 1992, Schenk 2006). The brain likely relies more on these mechanisms when memory delays increase (Glover & Dixon 2004, Goodale & Haffenden 1998, Obhi & Goodale 2005), perhaps because allocentric codes are more stable over time (Carrozzo et al. 2002, Lemay et al. 2004, McIntyre et al. 1998). However, to influence behavior, allocentric signals must somehow enter the action stream (**Figure 2b**). Consistent with this notion, egocentric codes appear in visual area 7a before allocentric codes do (Crowe et al. 2008). Monkeys trained to saccade toward a particular end of an object show object-centered spatial tuning in supplementary eye fields (SEF), area 7a and LIP (Olson & Gettner 1996, Olson & Tremblay

2000, Sabes et al. 2002, Tremblay & Tremblay 2002). However, these areas may use objects as a reference position, whereas the underlying reference frame may still be egocentric. For example, Deneve & Pouget (2003) showed, with the use of neural network models, that object-centered spatial tuning can arise from neurons with gaze-centered receptive fields that show object-modulated firing rates.

When both egocentric and allocentric cues are available, the brain uses both (Battaglia-Mayer et al. 2003, Diedrichsen et al. 2004, Sheth & Shimojo 2004), incorporating allocentric information at least until movement begins (Hay & Redon 2006, Krigolson & Heath 2004). Allocentric and egocentric cues are combined on the basis of both actual reliability and subjective judgments of their relative reliability (Byrne & Crawford 2010).

Spatial Updating: Behavioral Aspects

Animals are not always motionless when planning goal-directed movements. Often self motion invalidates the spatial relationship between extrinsic stimuli and the intrinsic sensory representations they produced. One option would be to wait for new sensory feedback, but this would introduce processing delays (e.g., the duration of a saccade + the latency for visual feedback) that could at times mean the difference between life and death. Moreover, this combined latency (~200 ms) multiplied by 3–4 saccades/second would render us functionally blind during most of our waking lives. To avoid such delays and blind periods, the brain must derive a predictive representation of visual space from brief visual glimpses and copies of motor commands (Ariff et al. 2002, Desmurget & Grafton 2000, Mehta & Schaal 2002, Wolpert & Ghahramani 2000). The process that updates spatial presentations during self-generated or passively induced motion is called spatial updating.

Spatial updating is often studied in the double-step task, in which subjects view a target, produce an intervening eye movement, and then move toward the first target. Saccades can

SC: superior colliculus

SEF: supplementary eye fields

Spatial updating: updating the representation of an external goal within some intrinsic frame to compensate for self-generated or passively induced motion of that frame

be aimed with reasonable accuracy toward remembered targets after an intervening saccade (Hallett & Lightstone 1976, Mays & Sparks 1980), smooth-pursuit eye movement (Baker et al. 2003, Blohm et al. 2005, Daye et al. 2010, Schlag et al. 1990), eye-head gaze shift (Herter & Guitton 1998, Vliegen et al. 2005), full body rotation and translation (Klier et al. 2005, 2007; Klier et al. 2008; Medendorp et al. 2003b), and torsional rotation of the eyes, head, and body (Klier et al. 2005, Medendorp et al. 2002, Van Pelt et al. 2005). Likewise, humans and monkeys can reach or point toward remembered targets after an intervening eye movement or full body motion (Henriques et al. 1998, Poljac & van den Berg 2003, Pouget et al. 2002, Sorrento & Henriques 2008, Thompson & Henriques 2008, Van Pelt & Medendorp 2007).

The studies cited above were performed in dark conditions, forcing subjects to rely on their own egocentric sense of target direction. Visuomotor systems may use different strategies when visual feedback is available (Flanagan et al. 2008). However, even when visual feedback is available, humans are more accurate at aiming movements when spatially updated memory of the goal is also available (Vaziri et al. 2006).

Theoretical Mechanisms for Spatial Updating

As we have already seen, the early spatial representations for visual goals, from the retina to the PPC and some areas of frontal cortex, utilize primarily an eye-fixed, gaze-centered code. This code could be used in two general ways to provide spatial updating. First, it could be compared with eye, head, and even body position. For example, many of these same areas contain subtle eye-position modulations called gain fields (Andersen & Buneo 2002, Boussaoud & Bremmer 1999, Sahani & Dayan 2003) that could, in theory, transform gaze-centered signals into successively more stable frames such as the head or body (Zipser & Andersen 1988). The problem with this scheme is that (although motor commands are eventually encoded in effector-specific, muscle-based coordinates) there is little evidence for visuospatial representation in such intermediate spatial maps.

The alternative is to use the internal sense of self-motion to remap the goal representation within gaze-centered coordinates, so that after the eye movement it corresponds to the correct retinal location at final eye position (Colby & Goldberg 1999). This model was originally simulated by subtracting a vector representing the intervening eye movement from another vector representing the goal to obtain a third vector representing the final saccade direction in retinal coordinates (Moschovakis & Highstein 1994, Waitzman et al. 1991). This does not quite work in real-world conditions because (*a*) spatial updating of saccades is noncommutative (Klier et al. 2007, Smith & Crawford 2001), and (*b*) during torsional eye rotations, goals on the opposite side of gaze need to be updated in opposite directions (Crawford & Guitton 1997, Medendorp et al. 2002, Smith & Crawford 2001). However, the remapping model does work when the correct 3-D math is used. Neural network simulations show that these noncommutative operations can be performed through a combination of physiologically realistic eye orientation and movement commands (Keith & Crawford 2008).

Experimental Evidence for Remapping

Remapping occurs in virtually every area of the monkey brain associated with saccade and reach goal coding, including early visual areas (Nakamura & Colby 2002), LIP (Duhamel et al. 1992a, Gnadt & Andersen 1988, Heiser & Colby 2006), SEF (Russo & Bruce 2000), frontal eye fields (FEF) (Sommer & Wurtz 2008, Umeno & Goldberg 1997), MIP (Batista et al. 1999, Buneo et al. 2002), and the SC (Walker et al. 1995). Many neurons in these areas show peri-saccadic changes consistent with a recalculation of future saccade goals with respect to the new eye position, sometimes beginning even before the saccade (Duhamel et al. 1992a, Umeno & Goldberg 1997, Walker et al. 1995). Recent experiments suggest that this is accomplished in part through signals routed from the brainstem, via the thalamus, to the cortex (Sommer & Wurtz 2008).

Gaze-centered remapping was first demonstrated in the human using a psychophysical paradigm (Henriques et al. 1998). This experiment relies on the control finding that humans overestimate the angle between a remembered peripheral pointing target and gaze direction (Bock 1986, McGuire & Sabes 2009). When subjects were additionally required to make a saccade between seeing, remembering, and pointing toward a central target (**Figure 3***a*), the resulting pointing errors matched the final (updated) target-gaze angle, not the angle at the time of viewing. The same result occurred for pointing to targets at different distances (Medendorp & Crawford 2002), after body translations (Van Pelt & Medendorp 2007), after smooth-pursuit eye movements (Thompson & Henriques 2008), for pointing to goals inferred from expanding motion patterns (Poljac & van den Berg 2003), or proprioceptive and auditory targets (Jones & Henriques 2010, Pouget et al. 2002), and for repeated pointing movements to the same remembered target (Sorrento & Henriques 2008).

Human cortical remapping has been confirmed using several different approaches. fMRI recordings demonstrated that both remembered movement goals (Medendorp et al. 2003a) and passively remembered stimuli (Merriam et al. 2003, 2007) remap between the intraparietal sulci on opposite hemispheres during saccades (**Figure 3***b*). Application of transcranial magnetic stimulation (TMS) pulses to the same cortical area disrupts remapping (Chang & Ro 2007, Morris et al. 2007). Unilateral optic ataxia patients show gaze-centered reach deficits that remap across saccades—from the "good" to "bad" hemifield, and vice versa (Khan et al. 2005b). Bidirectional saccadic updating was present in a patient with just one hemisphere (Herter & Guitton 1998) and recovered after anterior commissurotomy (Berman et al. 2005). Damage to frontal-parietal cortex can also produce deficits consistent with an impairment to the

signal that drives updating (Duhamel et al. 1992b, Heide et al. 1995).

These findings do not show that gaze-centered remapping is the only mechanism for spatial updating. For example, patients with bilateral parietal-occipital damage appear to retain a different, nonretinal mechanism (Khan et al. 2005a). However, gaze-centered updating is likely the dominant mechanism for updating visual saccade and reach goals.

Encoding and Updating in Depth

In the previous section, we consider the encoding and updating of visual direction for action, but this leaves out an essential component: depth. Distance is a significant variable in the programming of vergence eye movements and reaching movement. It is generally assumed that target depth and direction are processed in functionally distinct visuomotor channels (Cumming & DeAngelis 2001, Vindras et al. 2005).

Depth perception is typically associated with binocular disparity. If one can correctly match, point-by-point, the images on the two retinas, then geometry dictates that they will be slightly deviated on the basis of the difference in distance of the target relative to the individual eyes, the interocular distance, and the 3-D orientations of the eye and head (Wei et al. 2003). The binocular version of Listing's law partially reduces the degrees of freedom of this comparison (Tweed 1997), but in the absence of other visual cues, knowledge of 3-D eye and head orientation is required (Blohm et al. 2008). But these egocentric mechanisms are normally supplemented by allocentric cues based on object features and pictorial information, such as relative size, perspective, occlusion, and convergence of lines (Howard & Rogers 1995, Wei et al. 2003).

Spatial updating also occurs in depth, i.e., humans partially compensate for changes in vergence angle that occur between sensation and action (Krommenhoek & Van Gisbergen 1994). A recent study by Van Pelt & Medendorp (2008) used a variation of the

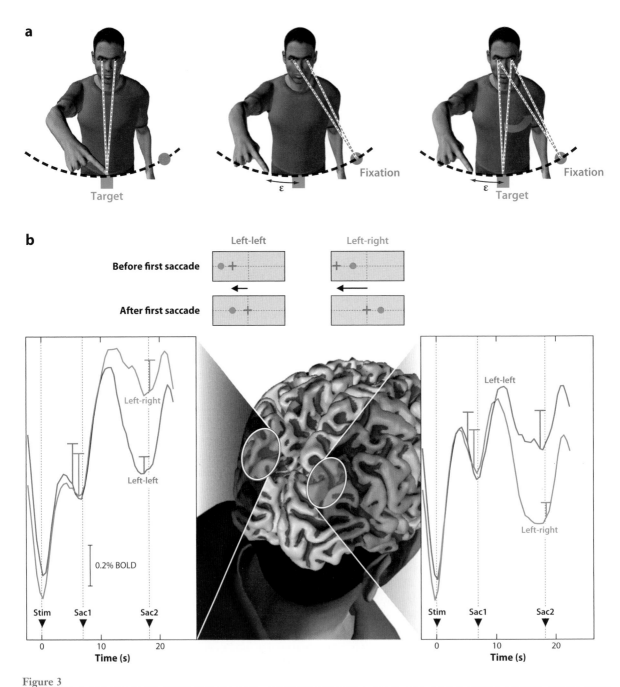

Figure 3

Psychophysical and neuroimaging evidence for gaze-centered spatial updating in the human. (*a*) Reaches to memorized visual targets, presented on the fovea, are relatively accurate (control trial), whereas reaches to peripheral targets show a clear directional bias (fixation trial). Reaches to a foveally presented target, but shifted to the periphery by an intervening saccade (saccade trial), show the same bias as do fixation trials, suggesting that the target is updated relative to gaze (Henriques et al. 1998). (*b*) A bilateral region in the PPC (*red*) shows gaze-centered spatial updating during the intervening saccade task. When eye movements reversed the side of the remembered target location relative to fixation, the region exchanged activity across the two cortical lobules (left-right trial). Modified from Medendorp et al. (2003a)

paradigm by Henriques et al. (1998) to show that similar principles hold for depth updating. Considering the binocular fixation point as the 3-D depth equivalent to the gaze point, these authors measured errors for reaching to targets relative to that depth. These results suggest that targets of reaching movements are updated in both direction and depth relative to the binocular fixation point (Van Pelt & Medendorp 2008).

In more complex motion conditions, direction and depth cannot be regarded as independent variables in the neural computations of spatial updating. In motion parallax, the change of visual direction depends on target depth during head translation. Psychophysical evidence shows that the brain takes this translation-depth geometry into account when programming the direction of saccades after an intervening translation, even compensating for eye translations produced by head rotation (Klier et al. 2008, Li et al. 2005, Li & Angelaki 2005, Medendorp et al. 2003b, Van Pelt & Medendorp 2007). To control such behavior in gaze-centered coordinates, the updater circuit must synthesize information about self motion with object depth information to remap each target by a different amount (Medendorp et al. 2003b).

TRANSFORMATION OF THE GOAL INTO A MOVEMENT COMMAND

Once a goal has been selected (Schall & Thompson 1999), and a desired action chosen (Cisek & Kalaska 2010), the representations described in the previous sections must be transformed into commands suitable for action (**Figure 2b**). In real-world circumstances, this transformation would be combined with visual feedback (Gomi 2008), but here we focus on the feed-forward mechanisms required for rapid, accurate action.

Computing the Displacement Vector

In theory, motor systems could function by specifying desired postural patterns and letting the effector drift to that position

(Bizzi et al. 1984, Feldman 1986). However, physiological experiments suggest that early saccade and reach areas are concerned primarily with developing a plan to displace gaze and/or hand position.

Retinal stimulation defines a desired gaze displacement, implicitly relative to current gaze direction, in eye-fixed coordinates. Subsequent oculomotor codes maintain this gaze-centered organization, computing eye velocity and orientation commands only at the final premotor stage before motoneurons (Robinson 1975). The exception occurs for depth saccades, in which current and desired binocular fixation must be compared to program a disconjugate saccade component. Saccade-related neurons in LIP show modulations related to both initial and desired depth (e.g., Genovesio et al. 2007).

A fixed relationship rarely exists between initial hand position, the goal, and gaze direction. The only way to compute the reach vector is to compare initial and desired hand position. For translational motion of the hand, it is sufficient to subtract a vector representing initial hand position from a vector representing desired hand position in the same frame. Investigators have historically assumed that this was done either entirely in visual coordinates or by transforming the visual goal into proprioceptive coordinates.

Sober & Sabes (2005) showed that when vision is available, humans compare the target to both visual and proprioceptive sensation of hand position and optimally integrate these signals depending on the stage of motor planning; however, they tend to rely more on vision especially in the early stages of motor planning. Other psychophysical experiments in healthy and brain-damaged humans have supported the notion that the reach vector is calculated either in gaze-centered coordinates (Chang et al. 2009, Khan et al. 2005b, Pisella et al. 2009, Pisella & Mattingley 2004) or in a mix of gaze and somatosensory coordinates (Beurze et al. 2006; Blangero et al. 2007; Khan et al. 2005a,b).

Several regions within PPC (**Figure 2**) play a role in both visual and proprioceptive calculations of the 3-D reach vector. Parietal area 5 is modulated both by target depth signals and by initial hand position (Ferraina et al. 2009b). Human angular gyrus appears to play a special role for incorporating the somatosensory sense of hand position into the reach vector (Vesia et al. 2010). Moreover, the PPC appears to possess the signals necessary for computation of movements in depth (see Ferraina et al. 2009a for review). Many neurons in areas such as LIP and the parietal reach region (PRR) are sensitive to both visual direction and retinal disparity (Bhattacharyya et al. 2009, Genovesio & Ferraina 2004, Gnadt & Mays 1995). Activity in most of these neurons is also modulated by vergence angle (Bhattacharyya et al. 2009, Genovesio & Ferraina 2004, Sakata et al. 1980). Consistent with these findings, damage to PPC produces deficits in both reach direction and depth (Baylis & Baylis 2001; Khan et al. 2005a,b; Striemer et al. 2009).

Buneo et al. (2002) showed that neurons in monkey PPC (area 5 and PRR) can show gaze-centered responses with hand-position modulations, consistent with calculation of the movement vector in visual coordinates. These responses persisted even when hand was not visible, suggesting that proprioceptively derived estimates had been transformed into gaze-centered coordinates. Recently, Beurze et al. (2010) reported similar findings in the human brain using fMRI. Other experiments suggest that dorsal premotor cortex (PMd) and PRR neurons show a relative-position code for target, gaze, and hand position (Pesaran et al. 2006, 2010) and/or encode target position in gaze coordinates with opposing gain modulations for gaze and hand position (Chang et al. 2009).

3-D Reference Frame Transformations: Behavioral Aspects

As we have already seen, retinal codes are predominant throughout the visuomotor system, at least at the explicit level revealed by receptive field mapping and fMRI. How are these gaze-centered codes converted into commands for eye movement relative to the head and arm movements relative to the torso?

Reference frame transformations have historically been considered from the viewpoint of position coding, where retinal position is compared with eye position to compute target position relative to the head, and this is compared to head position to compute target position relative to the body (Flanders et al. 1992). For relative position/displacement codes, the need for such comparisons disappears in frames that only translate with respect to each other (Andersen & Buneo 2002, Goldberg & Colby 1992). However, the frames of reference for visuomotor transformations (eye, head, torso) primarily rotate with respect to each other (**Figure 1**). The mathematics of rotations dictates that the representation of a movement or position in one of these frames corresponds to different representations in the other frames as a complex, nonlinear product of their relative orientations (Blohm & Crawford 2007, Crawford & Guitton 1997). Small movement and position vectors restricted to a frontal plane (like a laboratory stimulus screen) are relatively immune to the effects, but this result does not hold in general real-world conditions. For example, if gaze is simply directed 90° to the left, a forward reach in body coordinates is now a rightward reach in eye coordinates. In most circumstances, these reference frame projections produce more complex distortions in gaze (**Figure 4a**) and reach (**Figure 4b**) space. In a system that relies on relative position/displacement codes, these nonlinearities become the central problem in reference frame transformations, and this can be solved only by a transformation that includes a model of eye/head orientations and rotational geometry.

A brain that does not account for this geometry would produce predictable errors in generating rapid movements (Crawford & Guitton 1997). Behavioral studies in humans have shown that saccades to visual targets partially account for torsional eye orientations and fully account for eye positions in Listing's plane (Klier & Crawford 1998). A recent study has also shown that smooth-pursuit eye movements compensate for these factors (Blohm et al. 2006, Daye et al. 2010).

Pointing movements toward horizontally displaced targets also compensate for geometric relationships related to vertical eye position and the way it distorts the retinal projection (Crawford et al. 2000). Moreover, the internal models for reach and pointing movements also account for the translational linkage geometry (**Figure 1c**) between the centers of rotation of the eye, head, and shoulder (Henriques et al. 2003, Henriques & Crawford 2002).

A recent study (Blohm & Crawford 2007) combined all these features, modeling visually guided reach with the use of a direct transformation from visual coordinates to shoulder coordinates, accounting for only the translational geometry of the system, versus a system with a full internal model of eye-head-shoulder linkage (**Figure 1**) and nonlinear reference frame transformations (**Figure 4b**). As expected, the former model predicted errors in both reach direction and depth as a function of initial eye orientation, whereas the latter model predicted perfect reach. Tested the same way, real reaches showed various unrelated offsets and noise in the absence of visual feedback, but they did not show any of the errors predicted by the direct transformation model, even in the initial stages before proprioceptive feedback could occur.

3-D Reference Frame Transformations—Neural Mechanisms

The best theoretical candidate for reference frame transformations in the brain arises from studies of gain fields and their variants (Blohm et al. 2009, Pouget et al. 2002, Zipser & Andersen 1988). As mentioned above, these describe postural modulations (such as eye position) on visual-motor receptive fields. Eye and head position gain fields have been identified in essentially every area of the brain implicated in visuomotor transformations, from occipital cortex (Galletti & Battaglini 1989, Weyand & Malpeli 1993), to parietal eye and reach fields (Andersen & Mountcastle 1983, Brotchie et al. 1995, Chang & Snyder 2010, Galletti

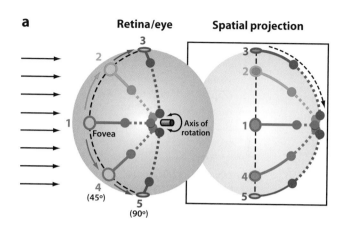

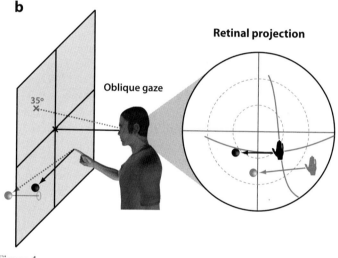

Figure 4

Influence of 3-D gaze orientation on the spatial relationship between visual input and motor output. (*a*) Projection of retinal coordinates (*middle panel*) onto a space-fixed reference frame (*right panel*). Imagine two horizontal vectors, painted onto the retina so that they project rightward from the fovea (*green empty circle*) by 40° (*solid green line*) and 80° (*discontinuous green line*) at position 1. Imagine that this eye-fixed assembly is now rotated up and down to positions 2–5 (*color coded for each eye orientation*). Although remaining horizontal in eye coordinates, these vectors are no longer horizontal in space coordinates. For example, an imaginary light source (*rightward arrows to the left*) casts a shadow on the right with a converging pattern, becoming more convergent with increasing eye orientation and vector length. Similar patterns of gaze shifts were observed during stimulation of the SC. Adapted from Klier et al. (2001). (*b*) Converse case of space coordinates (*left*) mapping onto retinal coordinates (*right*) during reach. A desired leftward trajectory in space coordinates (*black arrow*) is distorted on the retina by eye orientation (here an oblique gaze direction). If not taken into account, this would result in directional and depth errors (*red arrow*). Adapted from Blohm et al. (2009).

et al. 1995), to frontal cortex gaze and reach control centers (Boussaoud & Bremmer 1999, Boussaoud et al. 1998), and even subcortical structures (Groh & Sparks 1996a,b; Van Opstal et al. 1995). The original account of gain fields assumed the use of visual goal-in-space code (Zipser & Andersen 1988), which has since been questioned (Colby & Goldberg 1999), but the nonlinear geometry described in the last section gives new significance to this theory.

Artificial neural networks can be trained to transform visual targets into saccades (Smith & Crawford 2005) or reach movements (Blohm et al. 2009) using the correct 3-D geometry (**Figures 1** and **4**). These networks develop intermediate units that show gain fields similar to those seen in real physiology. Moreover, when probed with simulated receptive field mapping and microstimulation, individual units can show both a sensor-fixed frame of reference for the former and effector-fixed frames for the latter. This shows that (*a*) unit recording and stimulation reveal different neuron properties, and (*b*) individual units should show a fixed input-output relation when they perform a transformation.

These modeling studies suggest that electrical microstimulation reveals the reference frame to which a neural structure projects and should differ from the input code (derived from receptive field mapping) when a transformation is occurring. A 3-D reference frame analysis on gaze shifts evoked from the SC (Klier et al. 2001) and LIP (Constantin et al. 2007) showed that their position dependencies simply arise from the projection of light on an eye-fixed spherical frame. But these results also suggest that the 3-D reference frame transformation for gaze saccades occurs only as late as at the level of the brain stem/cerebellum. SEF stimulation evoked gaze shifts toward intermediate, eye-, head-, and body-fixed frames (Martinez-Trujillo et al. 2003b, Park et al. 2006), suggesting a capacity for more complex and arbitrary reference frame transformations in the frontal cortex.

The analogous 3-D analysis has not been done for reach, but as we have seen, supe-rior parietal structures appear to encode primarily visual targets with a gaze-centered code (Batista et al. 1999, Bhattacharyya et al. 2009), intermediate structures such as angular gyrus (Fernandez-Ruiz et al. 2007, Vesia et al. 2010) and ventral premotor cortex (Beurze et al. 2010, Kakei et al. 2001) employ progressively more extrinsic reach codes, and structures closer to the motor output for reach employ successively more effector-related spatial codes (Beurze et al. 2010, Hoshi & Tanji 2004, Scott 2003). Moreover, the latter structures continue to encode gaze-fixed signals (Boussaoud & Bremmer 1999, Cisek & Kalaska 2002) and yet produce complex and coordinated movements when stimulated (Graziano et al. 2002a,b). This seeming paradox could reflect a transition from sensory to motor codes such as that seen in 3-D network models (Blohm et al. 2009).

The 2-D to 3-D Transformation

Finally, the lower-dimensional neural codes discussed in the previous sections must be converted into the commands that implement the higher-dimensional behavioral geometry shown in **Figure 1**. The mechanisms that convert 2-D gaze commands into 3-D eye rotations and implement Listing's law have been the subject of intense theoretical debate (e.g., Quaia et al. 1998, Quaia & Optican 1998, Raphan 1998, Tweed & Vilis 1987). The analogous transformations for reach have also been modeled (Lieberman et al. 2006).

High-level gaze-control centers (SC, FEF, SEF) appear to encode the desired 2-D direction of gaze, leaving 3-D eye and head control downstream (Monteon et al. 2010; Hepp et al. 1993; Klier et al. 2003; Martinez-Trujillo et al. 2003a,b; van Opstal et al. 1991). In contrast, the reticular formation saccade generator (Henn et al. 1989, Luschei & Fuchs 1972) and the neural integrator that holds eye and head orientation (Crawford et al. 1991, Fukushima 1991, Helmchen et al. 1996, Klier et al. 2002) utilize a 3-D coordinate system. Thus, a 2-D to 3-D transformation must occur between these stages.

The default 2-D to 3-D transformation co-operates with mechanical factors to maintain eye and head orientation within the Listing and Fick ranges (**Figure 1**). The brainstem co-ordinates for 3-D eye control are organized such that they effectively collapse into 2-D axes in Listing's plane during symmetric bilateral activation of the midbrain (Crawford 1994, Crawford & Vilis 1992). The position-dependent torsional saccade axis tilts required (counterintuitively) to keep eye position in Listing's plane (Tweed et al. 1990, Tweed & Vilis 1987) are then implemented mechanically (Ghasia & Angelaki 2005; Klier et al. 2006, 2011), possibly by pulley-like actions of tissues surrounding the eye muscles (Demer 2006a,b). Similar neuro-muscular principles hold for head control: the 3-D brainstem coordinates for head control align with Fick coordinates (Klier et al. 2007), but neck anatomy also facilitates head rotations in Fick coordinates (Graf et al. 1995).

Different neural mechanisms are required to generate torsional movements toward or away from these 2-D ranges, for example during head-unimmobilized gaze shifts (Klier et al. 2003). A bilateral imbalance of input to the midbrain coordinate system is required to produce torsional components. What chooses the correct level of torsion? The cerebellum may influence torsional control in both the vestibular system, via outputs to vestibular eye-head cells with 3-D properties (Ghasia et al. 2008), and the saccade system, via inputs from the paramedian pontine reticular formation (Van Opstal et al. 1995). Consistent with this, it has been observed that Listing's plane is degraded in patients with damage to the cerebellum (Straumann et al. 2000).

Analogous neural mechanisms may come into play for 3-D reach constraints, but these are less understood at this time.

CONCLUSIONS

Neural recordings from the human and monkey suggest that gaze and reach movements toward visual goals are controlled by separate, but overlapping neural control systems. When considered from the perspective of the 3-D geometry of the spatial relationship between goal representations and effector commands, and the associated computational problems that must be solved, these two systems show several common principles (**Figure 2b**). First, their early representational phases are dominated by gaze-centered mechanisms (although these co-exist with other mechanisms, both egocentric and allocentric). Second, these gaze-centered signals are remapped during self motion. Third, upon selection for potential action, these representations are put through a series of transformations, involving computation of the movement vector (for depth saccades and reach), a successive series of reference frame transformations, and finally elaboration of these higher-level/low-dimensional plans into multidimensional motor commands. The role of some of these stages and their corresponding physiology—such as the prevalence of eye-position signals throughout the visuomotor system—becomes fully clear only when one takes the complete 3-D geometry of the system into account. Given the commonalities that emerge in these two systems, one would expect similar physiological solutions to arise whenever other sensorimotor systems encounter similar computational problems.

DISCLOSURE STATEMENT

The authors are not aware of any affiliations, memberships, funding, or financial holdings that might be perceived as affecting the objectivity of this review.

ACKNOWLEDGMENTS

The authors thank Dr. Luc Selen for help with preparing **Figures 1b** and **1c, 3,** and **4b,** and Dr. Michael Vesia for help preparing **Figure 2a**. Dr. Crawford's work was supported by a Canadian Institutes of Health Research (CIHR) Canada Research Chair and grants from CIHR, the National Science and Engineering Research Council (NSERC) and the Canada Foundation for Innovation (CFI). Dr. Henriques's work was supported by grants from NSERC, CFI, the Ministry of Research and Innovation (Early Researcher Award), and the J.P. Bickell Foundation. Dr. Henriques is an Alfred P. Sloan fellow. Dr. Medendorp's work was supported by grants from the Human Frontier Science Program (HFSP), the Netherlands Organisation for Scientific Research (NWO), and the Donders Center for Cognition.

LITERATURE CITED

Amiez C, Petrides M. 2009. Anatomical organization of the eye fields in the human and non-human primate frontal cortex. *Prog. Neurobiol.* 89:220–30

Andersen RA, Buneo CA. 2002. Intentional maps in posterior parietal cortex. *Annu. Rev. Neurosci.* 25:189–220

Andersen RA, Mountcastle VB. 1983. The influence of the angle of gaze upon the excitability of the light-sensitive neurons of the posterior parietal cortex. *J. Neurosci.* 3:532–48

Angelaki DE, Zhou HH, Wei M. 2003. Foveal versus full-field visual stabilization strategies for translational and rotational head movements. *J. Neurosci.* 23:1104–8

Ariff G, Donchin O, Nanayakkara T, Shadmehr R. 2002. A real-time state predictor in motor control: study of saccadic eye movements during unseen reaching movements. *J. Neurosci.* 22:7721–29

Avillac M, Deneve S, Olivier E, Pouget A, Duhamel JR. 2005. Reference frames for representing visual and tactile locations in parietal cortex. *Nat. Neurosci.* 8:941–49

Baker JT, Harper TM, Snyder LH. 2003. Spatial memory following shifts of gaze. I. Saccades to memorized world-fixed and gaze-fixed targets. *J. Neurophysiol.* 89:2564–76

Banks MS, Ghose T, Hillis JM. 2004. Relative image size, not eye position, determines eye dominance switches. *Vis. Res.* 44:229–34

Batista AP, Buneo CA, Snyder LH, Andersen RA. 1999. Reach plans in eye-centered coordinates. *Science* 285:257–60

Battaglia-Mayer A, Caminiti R, Lacquaniti F, Zago M. 2003. Multiple levels of representation of reaching in the parieto-frontal network. *Cereb. Cortex* 13:1009–22

Baylis GC, Baylis LL. 2001. Visually misguided reaching in Balint's syndrome. *Neuropsychologia* 39:865–75

Berman RA, Heiser LM, Saunders RC, Colby CL. 2005. Dynamic circuitry for updating spatial representations. I. Behavioral evidence for interhemispheric transfer in the split-brain macaque. *J. Neurophysiol.* 94:3228–48

Bernier PM, Grafton ST. 2010. Human posterior parietal cortex flexibly determines reference frames for reaching based on sensory context. *Neuron* 68:776–88

Beurze SM, de Lange FP, Toni I, Medendorp WP. 2009. Spatial and effector processing in the human parietofrontal network for reaches and saccades. *J. Neurophysiol.* 101:3053–62

Beurze SM, Toni I, Pisella L, Medendorp WP. 2010. Reference frames for reach planning in human parietofrontal cortex. *J. Neurophysiol.* 104:1736–45

Beurze SM, Van Pelt S, Medendorp WP. 2006. Behavioral reference frames for planning human reaching movements. *J. Neurophysiol.* 96:352–62

Bhattacharyya R, Musallam S, Andersen RA. 2009. Parietal reach region encodes reach depth using retinal disparity and vergence angle signals. *J. Neurophysiol.* 102:805–16

Bizzi E, Accornero N, Chapple W, Hogan N. 1984. Posture control and trajectory formation during arm movement. *J. Neurosci.* 4:2738–44

Blangero A, Ota H, Delporte L, Revol P, Vindras P, et al. 2007. Optic ataxia is not only 'optic': impaired spatial integration of proprioceptive information. *Neuroimage* 36(Suppl. 2):T61–68

Blohm G, Crawford JD. 2007. Computations for geometrically accurate visually guided reaching in 3-D space. *J. Vis.* 7:4, 1–22

Blohm G, Keith GP, Crawford JD. 2009. Decoding the cortical transformations for visually guided reaching in 3D space. *Cereb. Cortex* 19:1372–93

Blohm G, Khan AZ, Ren L, Schreiber KM, Crawford JD. 2008. Depth estimation from retinal disparity requires eye and head orientation signals. *J. Vis.* 8:3, 1–23

Blohm G, Missal M, Lefevre P. 2005. Processing of retinal and extraretinal signals for memory-guided saccades during smooth pursuit. *J. Neurophysiol.* 93:1510–22

Blohm G, Optican LM, Lefevre P. 2006. A model that integrates eye velocity commands to keep track of smooth eye displacements. *J. Comput. Neurosci.* 21:51–70

Bock O. 1986. Contribution of retinal versus extraretinal signals towards visual localization in goal-directed movements. *Exp. Brain Res.* 64:476–82

Boussaoud D, Bremmer F. 1999. Gaze effects in the cerebral cortex: reference frames for space coding and action. *Exp. Brain Res.* 128:170–80

Boussaoud D, Jouffrais C, Bremmer F. 1998. Eye position effects on the neuronal activity of dorsal premotor cortex in the macaque monkey. *J. Neurophysiol.* 80:1132–50

Brotchie PR, Andersen RA, Snyder LH, Goodman SJ. 1995. Head position signals used by parietal neurons to encode locations of visual stimuli. *Nature* 375:232–35

Buneo CA, Jarvis MR, Batista AP, Andersen RA. 2002. Direct visuomotor transformations for reaching. *Nature* 416:632–36

Byrne PA, Crawford JD. 2010. Cue reliability and a landmark stability heuristic determine relative weighting between egocentric and allocentric visual information in memory-guided reach. *J. Neurophysiol.* 103:3054–69

Carrozzo M, Stratta F, McIntyre J, Lacquaniti F. 2002. Cognitive allocentric representations of visual space shape pointing errors. *Exp. Brain Res.* 147:426–36

Chang E, Ro T. 2007. Maintenance of visual stability in the human posterior parietal cortex. *J. Cogn. Neurosci.* 19:266–74

Chang SW, Papadimitriou C, Snyder LH. 2009. Using a compound gain field to compute a reach plan. *Neuron* 64:744–55

Chang SW, Snyder LH. 2010. Idiosyncratic and systematic aspects of spatial representations in the macaque parietal cortex. *Proc. Natl. Acad. Sci. USA* 107:7951–56

Cisek P, Kalaska JF. 2002. Modest gaze-related discharge modulation in monkey dorsal premotor cortex during a reaching task performed with free fixation. *J. Neurophysiol.* 88:1064–72

Cisek P, Kalaska JF. 2010. Neural mechanisms for interacting with a world full of action choices. *Annu. Rev. Neurosci.* 33:269–98

Colby CL, Goldberg ME. 1999. Space and attention in parietal cortex. *Annu. Rev. Neurosci.* 22:319–49

Constantin AG, Wang H, Martinez-Trujillo JC, Crawford JD. 2007. Frames of reference for gaze saccades evoked during stimulation of lateral intraparietal cortex. *J. Neurophysiol.* 98:696–709

Crane BT, Demer JL. 1997. Human gaze stabilization during natural activities: translation, rotation, magnification, and target distance effects. *J. Neurophysiol.* 78:2129–44

Crane BT, Viirre ES, Demer JL. 1997. The human horizontal vestibulo-ocular reflex during combined linear and angular acceleration. *Exp. Brain Res.* 114:304–20

Crawford JD. 1994. The oculomotor neural integrator uses a behavior-related coordinate system. *J. Neurosci.* 14:6911–23

Crawford JD, Cadera W, Vilis T. 1991. Generation of torsional and vertical eye position signals by the interstitial nucleus of Cajal. *Science* 252:1551–53

Crawford JD, Guitton D. 1997. Visual-motor transformations required for accurate and kinematically correct saccades. *J. Neurophysiol.* 78:1447–67

Crawford JD, Henriques DY, Vilis T. 2000. Curvature of visual space under vertical eye rotation: implications for spatial vision and visuomotor control. *J. Neurosci.* 20:2360–68

Crawford JD, Vilis T. 1992. Symmetry of oculomotor burst neuron coordinates about Listing's plane. *J. Neurophysiol.* 68:432–48

Crowe DA, Averbeck BB, Chafee MV. 2008. Neural ensemble decoding reveals a correlate of viewer- to object-centered spatial transformation in monkey parietal cortex. *J. Neurosci.* 28:5218–28

Culham JC, Valyear KF. 2006. Human parietal cortex in action. *Curr. Opin. Neurobiol.* 16:205–12

Cumming BG, DeAngelis GC. 2001. The physiology of stereopsis. *Annu. Rev. Neurosci.* 24:203–38

Daye PM, Blohm G, Lefevre P. 2010. Saccadic compensation for smooth eye and head movements during head-unrestrained two-dimensional tracking. *J. Neurophysiol.* 103:543–56

Demer JL. 2006a. Current concepts of mechanical and neural factors in ocular motility. *Curr. Opin. Neurol.* 19:4–13

Demer JL. 2006b. Evidence supporting extraocular muscle pulleys: refuting the platygean view of extraocular muscle mechanics. *J. Pediatr. Ophthalmol. Strabismus.* 43:296–305

Deneve S, Pouget A. 2003. Basis functions for object-centered representations. *Neuron* 37:347–59

Desmurget M, Grafton S. 2000. Forward modeling allows feedback control for fast reaching movements. *Trends Cogn. Sci.* 4:423–31

Diedrichsen J, Werner S, Schmidt T, Trommershauser J. 2004. Immediate spatial distortions of pointing movements induced by visual landmarks. *Percept. Psychophys.* 66:89–103

di Pellegrino G, Wise SP. 1993. Visuospatial versus visuomotor activity in the premotor and prefrontal cortex of a primate. *J. Neurosci.* 13:1227–43

Duhamel JR, Colby CL, Goldberg ME. 1992a. The updating of the representation of visual space in parietal cortex by intended eye movements. *Science* 255:90–92

Duhamel JR, Goldberg ME, Fitzgibbon EJ, Sirigu A, Grafman J. 1992b. Saccadic dysmetria in a patient with a right frontoparietal lesion. The importance of corollary discharge for accurate spatial behaviour. *Brain* 115(Pt. 5):1387–402

Erkelens CJ, van de Grind WA. 1994. Binocular visual direction. *Vis. Res.* 34:2963–69

Feldman AG. 1986. Once more on the equilibrium-point hypothesis (lambda model) for motor control. *J. Mot. Behav.* 18:17–54

Ferman L, Collewijn H, Van den Berg AV. 1987. A direct test of Listing's law—I. Human ocular torsion measured in static tertiary positions. *Vis. Res.* 27:929–38

Fernandez-Ruiz J, Goltz HC, DeSouza JF, Vilis T, Crawford JD. 2007. Human parietal "reach region" primarily encodes intrinsic visual direction, not extrinsic movement direction, in a visual motor dissociation task. *Cereb. Cortex* 17:2283–92

Ferraina S, Battaglia-Mayer A, Genovesio A, Archambault P, Caminiti R. 2009a. Parietal encoding of action in depth. *Neuropsychologia* 47:1409–20

Ferraina S, Brunamonti E, Giusti MA, Costa S, Genovesio A, Caminiti R. 2009b. Reaching in depth: hand position dominates over binocular eye position in the rostral superior parietal lobule. *J. Neurosci.* 29:11461–70

Filimon F, Nelson JD, Huang RS, Sereno MI. 2009. Multiple parietal reach regions in humans: cortical representations for visual and proprioceptive feedback during on-line reaching. *J. Neurosci.* 29:2961–71

Flanagan JR, Terao Y, Johansson RS. 2008. Gaze behavior when reaching to remembered targets. *J. Neurophysiol.* 100:1533–43

Flanders M, Tillery SI, Soechting JF. 1992. Early stages in a sensorimotor transformation. *Behav. Brain Sci.* 15:309–62

Freedman EG, Sparks DL. 1997. Eye-head coordination during head-unrestrained gaze shifts in rhesus monkeys. *J. Neurophysiol.* 77:2328–48

Fukushima K. 1991. The interstitial nucleus of Cajal in the midbrain reticular formation and vertical eye movement. *Neurosci. Res.* 10:159–87

Gail A, Andersen RA. 2006. Neural dynamics in monkey parietal reach region reflect context-specific sensorimotor transformations. *J. Neurosci.* 26:9376–84

Galletti C, Battaglini PP. 1989. Gaze-dependent visual neurons in area V3A of monkey prestriate cortex. *J. Neurosci.* 9:1112–25

Galletti C, Battaglini PP, Fattori P. 1995. Eye position influence on the parieto-occipital area PO (V6) of the macaque monkey. *Eur. J. Neurosci.* 7:2486–501

Genovesio A, Brunamonti E, Giusti MA, Ferraina S. 2007. Postsaccadic activities in the posterior parietal cortex of primates are influenced by both eye movement vectors and eye position. *J. Neurosci.* 27:3268–73

Genovesio A, Ferraina S. 2004. Integration of retinal disparity and fixation-distance related signals toward an egocentric coding of distance in the posterior parietal cortex of primates. *J. Neurophysiol.* 91:2670–84

Ghasia FF, Angelaki DE. 2005. Do motoneurons encode the noncommutativity of ocular rotations? *Neuron* 47:281–93

Ghasia FF, Meng H, Angelaki DE. 2008. Neural correlates of forward and inverse models for eye movements: evidence from three-dimensional kinematics. *J. Neurosci.* 28:5082–87

Glenn B, Vilis T. 1992. Violations of Listing's law after large eye and head gaze shifts. *J. Neurophysiol.* 68:309–18

Glover S, Dixon P. 2004. A step and a hop on the Muller-Lyer: illusion effects on lower-limb movements. *Exp. Brain Res.* 154:504–12

Gnadt JW, Andersen RA. 1988. Memory related motor planning activity in posterior parietal cortex of macaque. *Exp. Brain Res.* 70:216–20

Gnadt JW, Mays LE. 1995. Neurons in monkey parietal area LIP are tuned for eye-movement parameters in three-dimensional space. *J. Neurophysiol.* 73:280–97

Goldberg ME, Colby CL. 1992. Oculomotor control and spatial processing. *Curr. Opin. Neurobiol.* 2:198–202

Goldman-Rakic PS. 1992. Working memory and the mind. *Sci. Am.* 267:110–17

Gomi H. 2008. Implicit online corrections of reaching movements. *Curr. Opin. Neurobiol.* 18:558–64

Goodale MA, Haffenden A. 1998. Frames of reference for perception and action in the human visual system. *Neurosci. Biobehav. Rev.* 22:161–72

Goodale MA, Milner AD. 1992. Separate visual pathways for perception and action. *Trends Neurosci.* 15:20–25

Graf W, de Waele C, Vidal PP. 1995. Functional anatomy of the head-neck movement system of quadrupedal and bipedal mammals. *J. Anatomy* 186(Pt. 1):55–74

Graziano MS, Taylor CS, Moore T. 2002a. Complex movements evoked by microstimulation of precentral cortex. *Neuron* 34:841–51

Graziano MS, Taylor CS, Moore T. 2002b. Probing cortical function with electrical stimulation. *Nat. Neurosci.* 5:921

Groh JM, Sparks DL. 1996a. Saccades to somatosensory targets. II. Motor convergence in primate superior colliculus. *J. Neurophysiol.* 75:428–38

Groh JM, Sparks DL. 1996b. Saccades to somatosensory targets. III. Eye-position-dependent somatosensory activity in primate superior colliculus. *J. Neurophysiol.* 75:439–53

Guitton D, Buchtel HA, Douglas RM. 1985. Frontal lobe lesions in man cause difficulties in suppressing reflexive glances and in generating goal-directed saccades. *Exp. Brain Res.* 58:455–72

Hallett PE. 1978. Primary and secondary saccades to goals defined by instructions. *Vis. Res.* 18:1279–96

Hallett PE, Lightstone AD. 1976. Saccadic eye movements towards stimuli triggered by prior saccades. *Vis. Res.* 16:99–106

Haslwanter T, Straumann D, Hepp K, Hess BJ, Henn V. 1991. Smooth pursuit eye movements obey Listing's law in the monkey. *Exp. Brain Res.* 87:470–72

Hay L, Redon C. 2006. Response delay and spatial representation in pointing movements. *Neurosci. Lett.* 408:194–98

Heide W, Blankenburg M, Zimmermann E, Kompf D. 1995. Cortical control of double-step saccades: implications for spatial orientation. *Ann. Neurol.* 38:739–48

Heiser LM, Colby CL. 2006. Spatial updating in area LIP is independent of saccade direction. *J. Neurophysiol.* 95:2751–67

Helmchen C, Rambold H, Buttner U. 1996. Saccade-related burst neurons with torsional and vertical on-directions in the interstitial nucleus of Cajal of the alert monkey. *Exp. Brain Res.* 112:63–78

Henn V, Hepp K, Vilis T. 1989. Rapid eye movement generation in the primate. Physiology, pathophysiology, and clinical implications. *Rev. Neurol.* 145:540–45

Henriques DY, Crawford JD. 2002. Role of eye, head, and shoulder geometry in the planning of accurate arm movements. *J. Neurophysiol.* 87:1677–85

Henriques DY, Klier EM, Smith MA, Lowy D, Crawford JD. 1998. Gaze-centered remapping of remembered visual space in an open-loop pointing task. *J. Neurosci.* 18:1583–94

Henriques DY, Medendorp WP, Gielen CC, Crawford JD. 2003. Geometric computations underlying eye-hand coordination: orientations of the two eyes and the head. *Exp. Brain Res.* 152:70–78

Hepp K, Van Opstal AJ, Straumann D, Hess BJ, Henn V. 1993. Monkey superior colliculus represents rapid eye movements in a two-dimensional motor map. *J. Neurophysiol.* 69:965–79

Herter TM, Guitton D. 1998. Human head-free gaze saccades to targets flashed before gaze-pursuit are spatially accurate. *J. Neurophysiol.* 80:2785–89

Hoshi E, Tanji J. 2004. Differential roles of neuronal activity in the supplementary and presupplementary motor areas: from information retrieval to motor planning and execution. *J. Neurophysiol.* 92:3482–99

Howard IP, Rogers BJ. 1995. *Binocular Vision and Stereopsis*. New York: Oxford Univ. Press. 736 pp.

Jones SA, Henriques DY. 2010. Memory for proprioceptive and multisensory targets is partially coded relative to gaze. *Neuropsychologia* 48:3782–92

Kakei S, Hoffman DS, Strick PL. 2001. Direction of action is represented in the ventral premotor cortex. *Nat. Neurosci.* 4:1020–25

Kastner S, DeSimone K, Konen CS, Szczepanski SM, Weiner KS, Schneider KA. 2007. Topographic maps in human frontal cortex revealed in memory-guided saccade and spatial working-memory tasks. *J. Neurophysiol.* 97:3494–507

Keith GP, Crawford JD. 2008. Saccade-related remapping of target representations between topographic maps: a neural network study. *J. Comput. Neurosci.* 24:157–78

Khan AZ, Crawford JD. 2001. Ocular dominance reverses as a function of horizontal gaze angle. *Vis. Res.* 41:1743–48

Khan AZ, Pisella L, Rossetti Y, Vighetto A, Crawford JD. 2005a. Impairment of gaze-centered updating of reach targets in bilateral parietal-occipital damaged patients. *Cereb. Cortex* 15:1547–60

Khan AZ, Pisella L, Vighetto A, Cotton F, Luaute J, et al. 2005b. Optic ataxia errors depend on remapped, not viewed, target location. *Nat. Neurosci.* 8:418–20

Klier EM, Angelaki DE, Hess BJ. 2005. Roles of gravitational cues and efference copy signals in the rotational updating of memory saccades. *J. Neurophysiol.* 94:468–78

Klier EM, Angelaki DE, Hess BJ. 2007. Human visuospatial updating after noncommutative rotations. *J. Neurophysiol.* 98:537–44

Klier EM, Crawford JD. 1998. Human oculomotor system accounts for 3-D eye orientation in the visual-motor transformation for saccades. *J. Neurophysiol.* 80:2274–94

Klier EM, Hess BJ, Angelaki DE. 2008. Human visuospatial updating after passive translations in three-dimensional space. *J. Neurophysiol.* 99:1799–809

Klier EM, Meng H, Angelaki DE. 2006. Three-dimensional kinematics at the level of the oculomotor plant. *J. Neurosci.* 26:2732–37

Klier EM, Meng H, Angelaki DE. 2011. Revealing the kinematics of the oculomotor plant with tertiary eye positions and ocular counterroll. *J. Neurophysiol.* 105:640–49

Klier EM, Wang H, Constantin AG, Crawford JD. 2002. Midbrain control of three-dimensional head orientation. *Science* 295:1314–16

Klier EM, Wang H, Crawford JD. 2001. The superior colliculus encodes gaze commands in retinal coordinates. *Nat. Neurosci.* 4:627–32

Klier EM, Wang H, Crawford JD. 2003. Three-dimensional eye-head coordination is implemented downstream from the superior colliculus. *J. Neurophysiol.* 89:2839–53

Krigolson O, Heath M. 2004. Background visual cues and memory-guided reaching. *Hum. Mov. Sci.* 23:861–77

Krommenhoek KP, Van Gisbergen JA. 1994. Evidence for nonretinal feedback in combined version-vergence eye movements. *Exp. Brain Res.* 102:95–109

Kusunoki M, Gottlieb J, Goldberg ME. 2000. The lateral intraparietal area as a salience map: the representation of abrupt onset, stimulus motion, and task relevance. *Vis. Res.* 40:1459–68

Lemay M, Bertram CP, Stelmach GE. 2004. Pointing to an allocentric and egocentric remembered target. *Mot. Control* 8:16–32

Levy I, Schluppeck D, Heeger DJ, Glimcher PW. 2007. Specificity of human cortical areas for reaches and saccades. *J. Neurosci.* 27:4687–96

Li N, Angelaki DE. 2005. Updating visual space during motion in depth. *Neuron* 48:149–58

Li N, Wei M, Angelaki DE. 2005. Primate memory saccade amplitude after intervened motion depends on target distance. *J. Neurophysiol.* 94:722–33

Liebermann DG, Biess A, Gielen CC, Flash T. 2006. Intrinsic joint kinematic planning. II: hand-path predictions based on a Listing's plane constraint. *Exp. Brain Res.* 171:155–73

Luschei ES, Fuchs AF. 1972. Activity of brain stem neurons during eye movements of alert monkeys. *J. Neurophysiol.* 35:445–61

Martinez-Trujillo JC, Klier EM, Wang H, Crawford JD. 2003a. Contribution of head movement to gaze command coding in monkey frontal cortex and superior colliculus. *J. Neurophysiol.* 90:2770–76

Martinez-Trujillo JC, Wang H, Crawford JD. 2003b. Electrical stimulation of the supplementary eye fields in the head-free macaque evokes kinematically normal gaze shifts. *J. Neurophysiol.* 89:2961–74

Mays LE, Sparks DL. 1980. Saccades are spatially, not retinocentrically, coded. *Science* 208:1163–65

McGuire LM, Sabes PN. 2009. Sensory transformations and the use of multiple reference frames for reach planning. *Nat. Neurosci.* 12:1056–61

McIntyre J, Stratta F, Lacquaniti F. 1998. Short-term memory for reaching to visual targets: psychophysical evidence for body-centered reference frames. *J. Neurosci.* 18:8423–35

Medendorp WP, Crawford JD. 2002. Visuospatial updating of reaching targets in near and far space. *Neuroreport* 13:633–36

Medendorp WP, Goltz HC, Vilis T. 2005. Remapping the remembered target location for anti-saccades in human posterior parietal cortex. *J. Neurophysiol.* 94:734–40

Medendorp WP, Goltz HC, Vilis T. 2006. Directional selectivity of BOLD activity in human posterior parietal cortex for memory-guided double-step saccades. *J. Neurophysiol.* 95:1645–55

Medendorp WP, Goltz HC, Vilis T, Crawford JD. 2003a. Gaze-centered updating of visual space in human parietal cortex. *J. Neurosci.* 23:6209–14

Medendorp WP, Smith MA, Tweed DB, Crawford JD. 2002. Rotational remapping in human spatial memory during eye and head motion. *J. Neurosci.* 22:RC196

Medendorp WP, Tweed DB, Crawford JD. 2003b. Motion parallax is computed in the updating of human spatial memory. *J. Neurosci.* 23:8135–42

Medendorp WP, van Gisbergen JA, Horstink MW, Gielen CC. 1999. Donders' law in torticollis. *J. Neurophysiol.* 82:2833–38

Medendorp WP, van Gisbergen JA, Van Pelt S, Gielen CC. 2000. Context compensation in the vestibuloocular reflex during active head rotations. *J. Neurophysiol.* 84:2904–17

Mehta B, Schaal S. 2002. Forward models in visuomotor control. *J. Neurophysiol.* 88:942–53

Merriam EP, Genovese CR, Colby CL. 2003. Spatial updating in human parietal cortex. *Neuron* 39:361–73

Merriam EP, Genovese CR, Colby CL. 2007. Remapping in human visual cortex. *J. Neurophysiol.* 97:1738–55

Misslisch H, Tweed D. 2001. Neural and mechanical factors in eye control. *J. Neurophysiol.* 86:1877–83

Monteon JA, Constantin AG, Wang H, Martinez-Trujillo J, Crawford JD. 2010. Electrical stimulation of the frontal eye fields in the head-free macaque evokes kinematically normal 3D gaze shifts. *J. Neurophysiol.* 104:3462–75

Morris AP, Chambers CD, Mattingley JB. 2007. Parietal stimulation destabilizes spatial updating across saccadic eye movements. *Proc. Natl. Acad. Sci. USA* 104:9069–74

Moschovakis AK, Highstein SM. 1994. The anatomy and physiology of primate neurons that control rapid eye movements. *Annu. Rev. Neurosci.* 17:465–88

Mullette-Gillman OA, Cohen YE, Groh JM. 2009. Motor-related signals in the intraparietal cortex encode locations in a hybrid, rather than eye-centered reference frame. *Cereb. Cortex* 19:1761–75

Munoz DP, Everling S. 2004. Look away: the anti-saccade task and the voluntary control of eye movement. *Nat. Rev. Neurosci.* 5:218–28

Nakamura K, Colby CL. 2002. Updating of the visual representation in monkey striate and extrastriate cortex during saccades. *Proc. Natl. Acad. Sci. USA* 99:4026–31

Obhi SS, Goodale MA. 2005. The effects of landmarks on the performance of delayed and real-time pointing movements. *Exp. Brain Res.* 167:335–44

Olson CR, Gettner SN. 1996. Brain representation of object-centered space. *Curr. Opin. Neurobiol.* 6:165–70

Olson CR, Tremblay L. 2000. Macaque supplementary eye field neurons encode object-centered locations relative to both continuous and discontinuous objects. *J. Neurophysiol.* 83:2392–411

Ono H, Mapp AP, Howard IP. 2002. The cyclopean eye in vision: The new and old data continue to hit you right between the eyes. *Vis. Res.* 42:1307–24

Park J, Schlag-Rey M, Schlag J. 2006. Frames of reference for saccadic command tested by saccade collision in the supplementary eye field. *J. Neurophysiol.* 95:159–70

Pesaran B, Nelson MJ, Andersen RA. 2006. Dorsal premotor neurons encode the relative position of the hand, eye, and goal during reach planning. *Neuron* 51:125–34

Pesaran B, Nelson MJ, Andersen RA. 2010. A relative position code for saccades in dorsal premotor cortex. *J. Neurosci.* 30:6527–37

Picard N, Strick PL. 2001. Imaging the premotor areas. *Curr. Opin. Neurobiol.* 11:663–72

Pisella L, Mattingley JB. 2004. The contribution of spatial remapping impairments to unilateral visual neglect. *Neurosci. Biobehav. Rev.* 28:181–200

Pisella L, Sergio L, Blangero A, Torchin H, Vighetto A, Rossetti Y. 2009. Optic ataxia and the function of the dorsal stream: contributions to perception and action. *Neuropsychologia* 47:3033–44

Poljac E, van den Berg AV. 2003. Representation of heading direction in far and near head space. *Exp. Brain Res.* 151:501–13

Porac C, Coren S. 1976. The dominant eye. *Psychol. Bull.* 83:880–97

Pouget A, Deneve S, Duhamel JR. 2002. A computational perspective on the neural basis of multisensory spatial representations. *Nat. Rev. Neurosci.* 3:741–47

Quaia C, Optican LM. 1998. Commutative saccadic generator is sufficient to control a 3-D ocular plant with pulleys. *J. Neurophysiol.* 79:3197–215

Quaia C, Optican LM, Goldberg ME. 1998. The maintenance of spatial accuracy by the perisaccadic remapping of visual receptive fields. *Neural Netw.* 11:1229–40

Radau P, Tweed D, Vilis T. 1994. Three-dimensional eye, head, and chest orientations after large gaze shifts and the underlying neural strategies. *J. Neurophysiol.* 72:2840–52

Raphan T. 1998. Modeling control of eye orientation in three dimensions. I. Role of muscle pulleys in determining saccadic trajectory. *J. Neurophysiol.* 79:2653–67

Robinson DA. 1975. Oculomotor control signals. In *Basic Mechanisms of Ocular Motility and Their Clinical Implications*, ed. G Iennerstrand, P Bach-y-Rita, pp. 337–74: Oxford, UK: Pergamon

Russo GS, Bruce CJ. 2000. Supplementary eye field: representation of saccades and relationship between neural response fields and elicited eye movements. *J. Neurophysiol.* 84:2605–21

Sabes PN, Breznen B, Andersen RA. 2002. Parietal representation of object-based saccades. *J. Neurophysiol.* 88:1815–29

Sahani M, Dayan P. 2003. Doubly distributional population codes: simultaneous representation of uncertainty and multiplicity. *Neural Comput.* 15:2255–79

Sakata H, Shibutani H, Kawano K. 1980. Spatial properties of visual fixation neurons in posterior parietal association cortex of the monkey. *J. Neurophysiol.* 43:1654–72

Schall JD, Thompson KG. 1999. Neural selection and control of visually guided eye movements. *Annu. Rev. Neurosci.* 22:241–59

Schenk T. 2006. An allocentric rather than perceptual deficit in patient D.F. *Nat. Neurosci.* 9:1369–70

Schlag J, Schlag-Rey M, Dassonville P. 1990. Saccades can be aimed at the spatial location of targets flashed during pursuit. *J. Neurophysiol.* 64:575–81

Schluppeck D, Glimcher P, Heeger DJ. 2005. Topographic organization for delayed saccades in human posterior parietal cortex. *J. Neurophysiol.* 94:1372–84

Scott SH. 2003. The role of primary motor cortex in goal-directed movements: insights from neurophysiological studies on non-human primates. *Curr. Opin. Neurobiol.* 13:671–77

Sereno MI, Pitzalis S, Martinez A. 2001. Mapping of contralateral space in retinotopic coordinates by a parietal cortical area in humans. *Science* 294:1350–54

Sheth BR, Shimojo S. 2004. Extrinsic cues suppress the encoding of intrinsic cues. *J. Cogn. Neurosci.* 16:339–50

Smith MA, Crawford JD. 2001. Implications of ocular kinematics for the internal updating of visual space. *J. Neurophysiol.* 86:2112–17

Smith MA, Crawford JD. 2005. Distributed population mechanism for the 3-D oculomotor reference frame transformation. *J. Neurophysiol.* 93:1742–61

Sober SJ, Sabes PN. 2005. Flexible strategies for sensory integration during motor planning. *Nat. Neurosci.* 8:490–97

Soechting JF, Flanders M. 1992. Moving in three-dimensional space: frames of reference, vectors, and coordinate systems. *Annu. Rev. Neurosci.* 15:167–91

Sommer MA, Wurtz RH. 2008. Brain circuits for the internal monitoring of movements. *Annu. Rev. Neurosci.* 31:317–38

Sorrento GU, Henriques DY. 2008. Reference frame conversions for repeated arm movements. *J. Neurophysiol.* 99:2968–84

Straumann D, Zee DS, Solomon D. 2000. Three-dimensional kinematics of ocular drift in humans with cerebellar atrophy. *J. Neurophysiol.* 83:1125–40

Striemer C, Locklin J, Blangero A, Rossetti Y, Pisella L, Danckert J. 2009. Attention for action? Examining the link between attention and visuomotor control deficits in a patient with optic ataxia. *Neuropsychologia* 47:1491–99

Thompson AA, Henriques DY. 2008. Updating visual memory across eye movements for ocular and arm motor control. *J. Neurophysiol.* 100:2507–14

Tremblay F, Tremblay LE. 2002. Cortico-motor excitability of the lower limb motor representation: a comparative study in Parkinson's disease and healthy controls. *Clin. Neurophysiol.* 113:2006–12

Tweed D. 1997. Visual-motor optimization in binocular control. *Vis. Res.* 37:1939–51

Tweed D, Cadera W, Vilis T. 1990. Computing three-dimensional eye position quaternions and eye velocity from search coil signals. *Vis. Res.* 30:97–110

Tweed D, Vilis T. 1987. Implications of rotational kinematics for the oculomotor system in three dimensions. *J. Neurophysiol.* 58:832–49

Umeno MM, Goldberg ME. 1997. Spatial processing in the monkey frontal eye field. I. Predictive visual responses. *J. Neurophysiol.* 78:1373–83

Van Der Werf J, Jensen O, Fries P, Medendorp WP. 2008. Gamma-band activity in human posterior parietal cortex encodes the motor goal during delayed prosaccades and antisaccades. *J. Neurosci.* 28:8397–405

Van Der Werf J, Jensen O, Fries P, Medendorp WP. 2010. Neuronal synchronization in human posterior parietal cortex during reach planning. *J. Neurosci.* 30:1402–12

van Opstal AJ, Hepp K, Hess BJ, Straumann D, Henn V. 1991. Two- rather than three-dimensional representation of saccades in monkey superior colliculus. *Science* 252:1313–15

Van Opstal AJ, Hepp K, Suzuki Y, Henn V. 1995. Influence of eye position on activity in monkey superior colliculus. *J. Neurophysiol.* 74:1593–610

Van Pelt S, Medendorp WP. 2007. Gaze-centered updating of remembered visual space during active whole-body translations. *J. Neurophysiol.* 97:1209–20

Van Pelt S, Medendorp WP. 2008. Updating target distance across eye movements in depth. *J. Neurophysiol.* 99:2281–90

Van Pelt S, Van Gisbergen JA, Medendorp WP. 2005. Visuospatial memory computations during whole-body rotations in roll. *J. Neurophysiol.* 94:1432–42

Vaziri S, Diedrichsen J, Shadmehr R. 2006. Why does the brain predict sensory consequences of oculomotor commands? Optimal integration of the predicted and the actual sensory feedback. *J. Neurosci.* 26:4188–97

Vesia M, Prime SL, Yan X, Sergio LE, Crawford JD. 2010. Specificity of human parietal saccade and reach regions during transcranial magnetic stimulation. *J. Neurosci.* 30:13053–65

Vindras P, Desmurget M, Viviani P. 2005. Error parsing in visuomotor pointing reveals independent processing of amplitude and direction. *J. Neurophysiol.* 94:1212–24

Vliegen J, Van Grootel TJ, Van Opstal AJ. 2005. Gaze orienting in dynamic visual double steps. *J. Neurophysiol.* 94:4300–13

Waitzman DM, Ma TP, Optican LM, Wurtz RH. 1991. Superior colliculus neurons mediate the dynamic characteristics of saccades. *J. Neurophysiol.* 66:1716–37

Walker MF, Fitzgibbon EJ, Goldberg ME. 1995. Neurons in the monkey superior colliculus predict the visual result of impending saccadic eye movements. *J. Neurophysiol.* 73:1988–2003

Wei M, DeAngelis GC, Angelaki DE. 2003. Do visual cues contribute to the neural estimate of viewing distance used by the oculomotor system? *J. Neurosci.* 23:8340–50

Weyand TG, Malpeli JG. 1993. Responses of neurons in primary visual cortex are modulated by eye position. *J. Neurophysiol.* 69:2258–60

Wolpert DM, Ghahramani Z. 2000. Computational principles of movement neuroscience. *Nat. Neurosci.* 3(Suppl.):1212–17

Zhang M, Barash S. 2000. Neuronal switching of sensorimotor transformations for antisaccades. *Nature* 408:971–75

Zipser D, Andersen RA. 1988. A back-propagation programmed network that simulates response properties of a subset of posterior parietal neurons. *Nature* 331:679–84

Neurobiology of Economic Choice: A Good-Based Model

Camillo Padoa-Schioppa

Department of Anatomy and Neurobiology, Washington University in St. Louis, St. Louis, Missouri 63110; email: camillo@wustl.edu

Annu. Rev. Neurosci. 2011. 34:333–59

First published online as a Review in Advance on March 29, 2011

The *Annual Review of Neuroscience* is online at neuro.annualreviews.org

This article's doi: 10.1146/annurev-neuro-061010-113648

0147-006X/11/0721-0333$20.00

Keywords

neuroeconomics, subjective value, action value, orbitofrontal cortex, transitivity, adaptation, abstract representation

Abstract

Traditionally the object of economic theory and experimental psychology, economic choice recently became a lively research focus in systems neuroscience. Here I summarize the emerging results and propose a unifying model of how economic choice might function at the neural level. Economic choice entails comparing options that vary on multiple dimensions. Hence, while choosing, individuals integrate different determinants into a subjective value; decisions are then made by comparing values. According to the good-based model, the values of different goods are computed independently of one another, which implies transitivity. Values are not learned as such, but rather computed at the time of choice. Most importantly, values are compared within the space of goods, independent of the sensorimotor contingencies of choice. Evidence from neurophysiology, imaging, and lesion studies indicates that abstract representations of value exist in the orbitofrontal and ventromedial prefrontal cortices. The computation and comparison of values may thus take place within these regions.

Contents

INTRODUCTION

Economic choice can be defined as the behavior observed when individuals make choices based solely on subjective preferences. Since at least the seventeenth century, this behavior has been the central interest of economic theory (which justifies the term economic choice) and also a frequent area of research in experimental psychology. In the past decade, however, economic choice has attracted substantial interest in neuroscience for at least three reasons. First, economic choice is an intrinsically fascinating topic, intimately related to deep philosophical questions such as free will and moral behavior. Second, over many generations, economists and psychologists accumulated a rich body of knowledge, identifying concepts and quantitative relationships that describe economic choice. In fact, economic choice is a rare case of high cognitive function for which such a formal and established behavioral description exists. This rich "psychophysics" can now be used to both guide and constrain research in neuroscience. Third, economic choice is directly relevant to a constellation of mental and neurological disorders, including frontotemporal dementia, obsessive-compulsive disorder, and drug addiction. These reasons explain the blossoming of an area of research referred to as neuroeconomics (Glimcher et al. 2008).

In a nutshell, research in neuroeconomics aspires to describe the neurobiological processes and cognitive mechanisms that underlie economic choices. Although the field is still in its infancy, significant progress has been made already. Examples of economic choice include the choice between different ice cream flavors in a gelateria, the choice between different houses for sale, and the choice between different financial investments in a retirement plan. Notably, options available for choice in different situations can vary on a multitude of dimensions. For example, different flavors of ice cream evoke different sensory sensations and may be consumed immediately; different houses may vary for their price, their size, the school district, and the distance from work; different financial investments may carry different degrees of risk, with returns available in a distant, or not-so-distant, future. How does the brain generate choices in the face of this enormous variability? Economic and psychological theories of choice behavior have a cornerstone in the concept of value. While choosing, individuals assign values to the available options; a decision is then made by comparing these values. Hence, while options can vary on multiple dimensions, value represents a common unit of measure with which to make a comparison. From this perspective, understanding the neural mechanisms of economic choice amounts to describing how values are computed and compared in the brain.

Much research in recent years thus focused on the neural representation of economic value. As detailed in this review, a wealth of results obtained with a variety of techniques—single-cell recordings in primates and rodents, functional imaging in humans, lesion studies in multiple species, etc.—indicates that neural representations of value exist in several brain areas and that lesions in some of these areas—most notably the orbitofrontal cortex (OFC) and ventromedial prefrontal cortex (vmPFC)—specifically impair choice behavior. In essence, the brain actually computes values when subjects make economic choices.

To appreciate the significance of this proposition, it is helpful to step back and consider a historical and theoretical perspective. Neoclassic economic theory can be thought of as a rigorous mathematical construct founded on a limited set of axioms (Kreps 1990). In this framework, the concept of value is roughly as follows. Under few and reasonable assumptions, any large set of choices can be accounted for as if the choosing subject maximized an internal value function. Thus values are central to the economist's description of choice behavior. Note, however, that the concept of value in economics is behavioral and analytical, not psychological. The fact that choices are effectively described in terms of values does not imply that subjects actually assign values while choosing. Thus by taking an "as if" stance, economic theory explicitly avoids stating what mental processes actually underlie choice behavior. The distinction between an "as if" theory and a psychological theory may seem subtle if not evanescent. However, this distinction is critical in economics, and it helps us to appreciate the contribution of recent research in neuroscience. The "as if" stance captures a fundamental limit: On the basis of behavior alone, values cannot be measured independently of choice. Consequently, the assertion that choices maximize values is intrinsically circular. The observation that values are actually computed in the brain essentially breaks this circularity. Indeed, once the correspondence between a neural signal and a behavioral measure of value has been established, that neural signal provides an independent measure of value, in principle dissociable from choices. In other words, the assertion that choices maximize values becomes potentially falsifiable and thus truly scientific (Popper 1963). For this reason, I view the discovery that

values are indeed encoded at the neural level as a major conceptual advance and perhaps the most important result of neuroeconomics to date.

With this perspective, the purposes of this article are threefold. First, I review the main experimental results on the neural mechanisms of value encoding and economic choice. Second, I place the current knowledge in a unifying framework, proposing a model of how economic choice may function at the neural level. And third, I indicate areas of current debate and suggest directions for future research. The article is organized as follows. The next section introduces basic concepts and outlines a good-based model of economic choice. The third section describes the standard neuroeconomic method used to assess the neural encoding of subjective value. The fourth section summarizes a large body of work from animal neurophysiology, human imaging, and lesion studies, which provides evidence for an abstract representation of value. The fifth section discusses the neural encoding of action values and their possible relevance to economic choice. The final section highlights open issues that require further experimental work. Overall, I hope to provide a comprehensive, though necessarily not exhaustive, overview of this field.

ECONOMIC CHOICE: A GOOD-BASED MODEL

What cognitive and neural computations take place when individuals make economic choices? In broad strokes, my proposal is as follows. I embrace the view that economic choice is a distinct mental function (Padoa-Schioppa 2007) and that it entails assigning values to the available options. The central proposition of the model is that the brain maintains an abstract representation of "goods" and that the choice process—the computation and comparison of values—takes place within this space of goods. Thus I refer to this proposal as a good-based model of economic choice. I define a commodity as a unitary amount of a specified good independent of the circumstances in which it is available (e.g., quantity, cost, delay). The value

OFC: orbitofrontal cortex

vmPFC: ventromedial prefrontal cortex

Good: a commodity and a collection of determinants

Good-based models: values are computed and compared independent of the sensorimotor contingencies of choice (goods space). The choice outcome subsequently guides an action plan

Action value: a neuron encodes an action value if it is preferentially active when a particular action is planned and if it is modulated by the value associated with that action

of each good is computed at the time of choice on the basis of multiple determinants, which include the specific commodity, its quantity, the current motivational state, the cost, the behavioral context of choice, etc. The collection of these determinants thus defines the good. While choosing, individuals compute the values of different options independently of one another. This computation does not depend on the sensorimotor contingencies of choice (the spatial configuration of the offers or the specific action that will implement the choice outcome). These contingencies may, however, affect values in the form of action costs (the actions necessary to obtain different goods often bear different costs). The model proposed here assumes that the action cost (i.e., the physical effort) is computed, represented in a nonspatial way, and integrated with other determinants in the computation of subjective value. According to the good-based model, computation and comparison of values take place within prefrontal regions, including the OFC, the vmPFC, and possibly other areas. The choice outcome—the chosen good and/or the chosen value—then guides the selection of a suitable action (good-to-action transformation). The good-based model, depicted in **Figure 1**, is thus defined by the following propositions:

1. Economic choice is a distinct mental function, qualitatively different from other overt behaviors that can be construed as involving a choice (e.g., perceptual decisions, associative learning). Economic choice entails assigning values to the available options.
2. A good is defined by a commodity and a collection of determinants that characterize the conditions under which the commodity is offered. Determinants can be either external (e.g., cost, time delay, risk, ambiguity) or internal to the subject (e.g., motivational state, (im)patience, risk attitude, ambiguity attitude).
3. The brain maintains an abstract representation of goods. More specifically, when a subject makes a choice, different sets of neurons represent the identities and

values of different goods. The ensemble of these sets of neurons provides a space of goods. This representation is abstract in the sense that the encoding of values does not depend on the sensorimotor contingencies of choice. Choices take place within this representation; values are computed and compared in the space of goods.

4. Some determinants may be learned through experience (e.g., the cost of a particular good), whereas other determinants may not be learned (e.g., the motivational state, the behavioral context). The process of value assignment implies an integration of different determinants. Thus the value of each good is computed "online" at the time of choice.
5. While choosing, individuals normally compute the values of different goods independently of one another. Such menu invariance implies transitive preferences.
6. Values computed in different behavioral conditions can vary by orders of magnitude. The encoding of value adapts to the range of values available in any given condition and thus maintains high sensitivity.
7. With respect to brain structures, the computation and comparison of values take place within prefrontal regions, including OFC, vmPFC, and possibly other regions. The choice outcome then guides a good-to-action transformation that originates in prefrontal regions and culminates in premotor regions, including parietal, precentral, and subcortical regions.
8. In addition to providing the bases for economic choices, subjective values inform a variety of neural systems, including sensory and motor systems (through attention and attention-like mechanisms), learning (e.g., through mechanisms of reinforcement learning), emotion (including autonomic functions), etc.

As illustrated below, this good-based model accounts for a large body of experimental results. It also makes several predictions that need to be tested in future work. In this respect, the

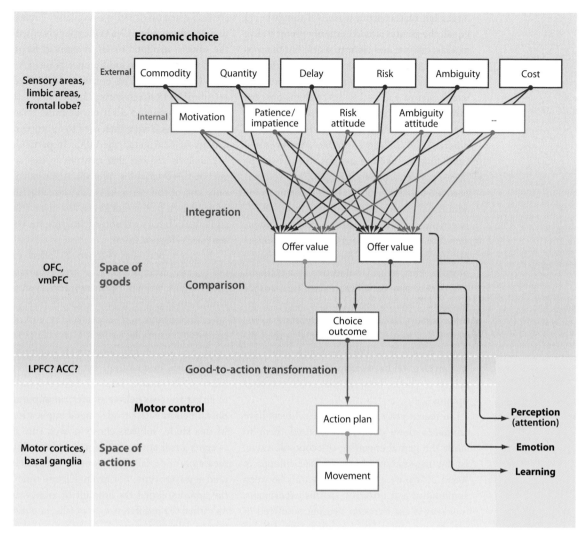

Figure 1

Good-based model. The value of each good is computed integrating multiple determinants, of which some are external (commodity, quantity, etc.) and others are internal (motivation, (im)patience, etc.). Offer values of different goods are computed independently of one another and then compared to make a decision. This comparison takes place within the space of goods. The choice outcome (chosen good, chosen value) then guides an action plan through a good-to-action transformation. Values and choice outcomes also inform other brain systems, including sensory and motor systems (through attention and attention-like mechanisms), associative learning (e.g., through mechanisms of reinforcement learning), emotion (including autonomic functions), etc. Abbreviations: orbitofrontal cortex (OFC), ventromedial prefrontal cortex (vmPFC), lateral prefrontal cortex (LPFC), anterior cingulate cortex (ACC).

good-based model proposed here should be regarded as a working hypothesis. Notably, my proposal differs from other models of economic choice previously discussed by other authors. Below, I highlight these differences and suggest possible approaches to assess the merits of different proposals.

MEASURING ECONOMIC VALUE AND ITS NEURAL REPRESENTATION

Consider a person choosing between two houses for sale at the same price: One house is smaller but closer to work, and the other is larger but further from work. All things being equal, the person would certainly prefer to live in a large house and close to work, but that option is beyond her budget. Thus while comparing houses to make a choice, the person must weigh against each other two dimensions, the distance from work and the square footage of the house. Physically, these two dimensions are different and incommensurable. However, the value that the chooser assigns to the two options provides a common scale, a way to compare the two dimensions. Thus intrinsic to the concept of value is the notion of a trade-off between physically distinct and competing dimensions (i.e., different determinants). This example also highlights two fundamental attributes of value. First, value is subjective. For example, one person may be willing to live in a smaller house to avoid a long commute, whereas another person may accept a long commute to enjoy a larger house. Second, measuring the subjective value assigned by a particular individual to a given good necessarily requires asking the subject to choose between that good and other options.

In recent years, neuroscience scholars have embraced these concepts and used them to study the neural encoding of economic value. In the first study to do so (Padoa-Schioppa & Assad 2006), we examined trade-offs between commodity and quantity. In this experiment, monkeys chose between two juices offered in variable amounts. The two juices were labeled A and B, with A preferred. When offered one drop of juice A versus one drop of juice B (offer 1B:1A), the animals chose juice A. However, if the animals were thirsty; they generally preferred larger amounts of juice to smaller amounts of juice. The amounts of the two juices offered against each other varied from trial to trial, which induced a commodity-quantity trade-off in the choice pattern. For example, in one session (**Figure 2a,b**), offer types included 0B:1A, 1B:2A, 1B:1A, 2B:1A, 3B:1A, 4B:1A, 6B:1A, 10B:1A, and 3B:0A. The monkey generally chose 1A when 1B, 2B, or 3A were available as an alternative; it was roughly indifferent between the two juices when offered 4B:1A; and it

chose B when 6B or 10B was available. Thus the monkey assigned to 1A a value roughly equal to the value it assigned to 4B. A sigmoid fit provided a more precise indifference point: 1A = 4.1B (**Figure 2b**). This equation established a relationship between juices A and B. On this basis, we computed a variety of value-related variables, which were then used to interpret the activity of neurons in the OFC. In particular, the analysis showed that neurons in this area encode three variables: offer value (the value of only one of the two juices), chosen value (the value chosen by the monkey in any given trial), and taste (a binary variable identifying the chosen juice) (**Figure 2c–e**).

In our experiment (**Figure 2**), offers varied on two dimensions: juice type (commodity) and juice amount (quantity). However, the same method can be applied when offers vary on other dimensions, such as probability, cost, delay, etc. For example, Kable & Glimcher (2007) conducted on human subjects an experiment on temporal discounting. People and animals often prefer smaller rewards delivered earlier to larger rewards delivered later, an important phenomenon with broad societal implications. In this study, subjects chose in each trial between a small amount of money delivered immediately and a larger amount of money delivered at a later time. For given delivery time T, the authors varied the amount of money and identified the indifference point (the amount of money delivered at time T such that the subject would be indifferent between the two options). The authors repeated this procedure for different delivery times T. Indifference points—fitted with a hyperbolic function—provided a measure of the subjective value choosers assigned to time-discounted money. During the experiment, the authors recorded the blood-oxygen-level-dependent (BOLD) signal. In the analyses, they used the measure of subjective value obtained from the indifference points as a regressor for the neural activity. Their results showed that the vmPFC encodes time-discounted values (see also Kim et al. 2008, Kobayashi & Schultz 2008, Louie & Glimcher 2010).

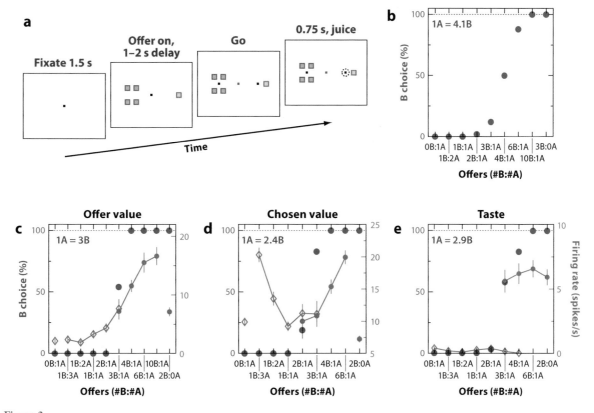

Figure 2

Measuring subjective values: value encoding in the orbitofrontal cortex (OFC). (*a*) Economic choice task. In this experiment, monkeys chose between different juices offered in variable amounts. Different colors indicated different juice types, and the number of squares indicated different amounts. In the trial depicted here, the animal was offered 4 drops of peppermint tea (juice B) versus 1 drop of grape juice (juice A). The monkey indicated its choice with an eye movement. (*b*) Choice pattern. The x-axis represents different offer types ranked by the ratio #B:#A. The y-axis represents the percent of trials in which the animal chose juice B. The monkey was roughly indifferent between 1A and 4B. A sigmoid fit indicated, more precisely, that 1A = 4.1B. The relative value (4.1 here) is a subjective measure in multiple senses. First, it depends on the two juices. Second, given two juices, it varies for different individuals. Third, for any individual and two given juices, it varies depending, for example, on the motivational state of the animal (thirst). Thus to examine the neural encoding of economic values, it is necessary to examine neural activity in relation to the subjective values measured concurrently. (*c*) OFC neuron encoding the offer value. Blue circles indicate the behavioral choice pattern (*relative value in the upper left*), and red symbols indicate the neuronal firing rate. Hollow red diamonds and filled red circles refer, respectively, to trials in which the animal chose juice A and juice B. A linear relationship exists between the activity of the cell and the quantity of juice B offered to the monkey. (*d*) OFC neuron encoding the chosen value. A linear relationship exists between the activity of the cell and the value chosen by the monkey in each trial. For this session, 1A = 2.4B. The activity of the cell was low when the monkey chose 1A or 2B, higher when the monkey chose 2A or 4B, and highest when the monkey chose 1A or 6B. Neurons encoding the chosen value are thus identified on the basis of the relative value of the two juices. (*e*) OFC neuron encoding the taste. The activity of the cell is binary depending on the chosen juice but independent of its quantity (panels *d–e*, same conventions as in panel *c*). Adapted from Padoa-Schioppa & Assad (2006), *Nature* (Nature Publishing Group), and from Padoa-Schioppa (2009), *Journal of Neuroscience* (used with permission from the Society for Neuroscience).

An interesting procedure to measure indifference points is to perform a "second price auction." For example, in a study by Plassmann et al. (2007), hungry human subjects were asked to declare the highest price they would be willing to pay for a given food (i.e., their indifference point, also called "reservation price"). Normally, people would try to save money and declare a price lower than their true reservation price. However, second price auctions discourage them from doing so by randomly generating a second price after the subjects have declared their own price. If the second price is lower than the declared price, subjects get to buy the food and pay the second price; if the second price is higher than the declared price, subjects do not get to buy the food at all. In these conditions, the optimal strategy for subjects is to declare their true reservation price. This procedure thus measures for each subject the indifference point between food and money. Using this measure, Plassmann et al. confirmed that the BOLD signal in the OFC encodes the value subjects assigned to different foods (see also De Martino et al. 2009).

In summary, to measure the neural representation of subjective value, it is necessary to let the subject choose between alternative offers, infer values from the indifference point, and use that measure to interpret neural signals. This experimental method—used widely in primate neurophysiology (Kim et al. 2008, Kimmel et al. 2010, Klein et al. 2008, Kobayashi & Schultz 2008, Louie & Glimcher 2010, O'Neill & Schultz 2010, Sloan et al. 2010, Watson & Platt 2008) and human imaging (Brooks et al. 2010, Christopoulos et al. 2009, De Martino et al. 2009, FitzGerald et al. 2009, Gregorios-Pippas et al. 2009, Hsu et al. 2009, Levy et al. 2010, Peters & Buchel 2009, Pine et al. 2009, Shenhav & Greene 2010)—is now standard in neuroeconomics.

AN ABSTRACT REPRESENTATION OF ECONOMIC VALUE

In this section, I review the evidence from neural recordings and lesion studies indicating that the representation of value in OFC and vmPFC is abstract and causally linked to economic choices. I then describe how this representation of value is affected by the behavioral context choice, and I discuss the evidence suggesting that values are computed online.

Evidence From Neural Recordings

A neuronal representation of value can be said to be abstract (i.e., in the space of goods) if two conditions are met. First, the encoding should be independent of the sensorimotor contingencies of choice. In particular, the activity representing the value of any given good should not depend on the action executed to obtain that good. Second, the encoding should be domain general. In other words, the activity should represent the value of the good affected by all the relevant determinants (commodity, quantity, risk, cost, etc.). Current evidence for such an abstract representation is most convincing for two brain areas: OFC and vmPFC. In this subsection and the next, I review the main experimental results from, respectively, neural recordings and lesion studies.

In our original study (**Figure 2**), we examined a large number of variables that OFC neurons might possibly encode, including offer value, chosen value, other value (the value of the unchosen good), total value, value difference (chosen value minus unchosen value), taste, etc. Several statistical procedures were used to identify a small set of variables that would best account for the neuronal population. The results can be summarized as follows. First, offer value, chosen value, and taste accounted for the activity of neurons in the OFC significantly better than any other variable examined in the study. Any additional variable explained less than 5% of responses. Second, the encoding of value in OFC was independent of the sensorimotor contingencies of the task. Indeed, less than 5% of OFC neurons were significantly modulated by the spatial configuration of the offers on the monitor or by the direction of the eye movement. Third, each neuronal response encoded only one variable, and the encoding was linear. Indeed, a linear regression of the firing rate onto the encoded variables generally provided a very good fit, and adding terms to the regression (quadratic terms or additional variables) usually failed to improve the fit significantly. Fourth, the timing of the encoding appeared to match the mental processes monkeys presumably undertook during each trial. In particular, neurons encoding the offer value—the variable on which choices were presumably based—were the most prominent immediately after the offers were presented to the animal (Padoa-Schioppa & Assad 2006).

With respect to the first condition—independence from sensorimotor contingencies—the evidence for an abstract representation of values thus seems robust. Indeed, consistent results were obtained in several other single-cell studies in primates (Grattan & Glimcher 2010, Kennerley & Wallis 2009, Roesch & Olson 2005).

With respect to the second condition—domain generality—current evidence for an abstract representation of value is clearly supportive. Indeed, domain generality has been examined extensively using functional imaging in humans. For example, Peters & Büchel (2009) let subjects choose between different money offers that could vary on two dimensions: delivery time and probability. Using the method described above, they found that neural activity in the OFC and ventral striatum encoded subjective values as affected by either delay or risk. In another study, Levy et al. (2010) let subjects choose between money offers that varied either for risk or for ambiguity. Using the same method, they found that the BOLD signal in vmPFC and ventral striatum encoded subjective values under both conditions. [More recent evidence suggests that the ventral striatum is not involved in choice per se (Cai et al. 2011).] De Martino et al. (2009) compared the encoding of subjective value when individuals gain or lose money—an important distinction because behavioral measures of value are typically reference-dependent (Kahneman & Tversky 1979). They found that OFC activity encoded the subjective value under either gains or losses. Taken together, these results consistently support a domain general representation of subjective value in OFC and vmPFC. As a caveat, I note that because of the low spatial resolution, functional imaging data cannot rule out that different determinants

of value might be encoded by distinct, but anatomically nearby, neuronal populations.

Several determinants of choice have also been examined at the level of single neurons. For example, Roesch & Olson (2005) delivered to monkeys different quantities of juice with variable delays. They found that OFC neurons were modulated by both variables and that neurons that increased their firing rates for increasing juice quantities generally decreased their firing rate for increasing time delays. Although the study did not provide a measure of subjective value, the results do suggest an integrated representation of value. In related work, Morrison & Salzman (2009) delivered to monkeys positive or negative stimuli (juice drops or air puffs). Consistent with domain generality, neuronal responses in the OFC had opposite signs. In another study, Kennerley et al. (2009) found a sizable population of OFC neurons modulated by three variables: the juice quantity, the action cost, and the probability of receiving the juice at the end of the trial. Notably, the firing rate generally increased as a function of the juice quantity and of the probability and decreased as a function of the action cost (or the other way around). Thus the modulation across determinants was congruent. Although these experiments did not measure subjective value, the results clearly support the notion of a domain-general representation.

In conclusion, a wealth of empirical evidence is consistent with the notion that OFC and vmPFC harbor an abstract representation of value, although the issue of domain generality needs confirmation at the level of single cells and for determinants not yet tested. Interestingly, insofar as a representation of value exists in rodents (Schoenbaum et al. 2009, van Duuren et al. 2007), it does not appear to meet the conditions for abstraction defined here. Indeed, several groups found that neurons in the rodent OFC are spatially selective (Feierstein et al. 2006, Roesch et al. 2006). Furthermore, experiments that manipulated two determinants of value found that different neuronal populations in the rat OFC represent reward magnitude and time delay—a striking difference with primates (Roesch et al. 2006, Roesch & Olson 2005). The reasons for this discrepancy are not clear (Zald 2006). However, Wise (2008) noted that the architecture of the orbital cortex in rodents and primates is qualitatively different, which suggests that an abstract representation of value emerged late in evolution in parallel with the expansion of the frontal lobe. At the same time, it cannot be excluded that domain-general value signals exist in other regions of the rodent brain.

Evidence From Lesion Studies

While establishing a link between OFC and vmPFC and the encoding of value, the evidence reviewed so far does not demonstrate a causal relationship between neural activity in these areas and economic choices. Such relationship emerges from lesion studies. In this respect, one of the most successful experimental paradigms is that of reinforcer devaluation. In these experiments, animals choose between two different foods. During training sessions, animals reveal their normal preferences. Before test sessions, however, animals are given free access to their preferred food. Following such selective satiation, control animals switch their preferences and choose their usually-less-preferred food. In contrast, in animals with OFC lesions, this satiation effect disappears. After OFC lesions, animals continue to choose the same food and thus seem incapable of computing values. This result has been replicated by several groups in both rodents (Gallagher et al. 1999, Pickens et al. 2003) and monkeys (Izquierdo et al. 2004; Kazama & Bachevalier 2009; Machado & Bachevalier 2007a,b). Notably, OFC lesions specifically affect value-based decisions as distinguished, for example, from strategic (i.e., rule-based) decisions (Baxter et al. 2009) or from perceptual judgments (Fellows & Farah 2007).

In the scheme of **Figure 1**, selective satiation alters subjective values by manipulating the motivational state of the animal. However, OFC lesions disrupt choice behavior also when trade-offs involve other determinants of value. For example, with respect to risk, several groups reported that patients with OFC lesions present atypical risk-seeking behavior (Damasio 1994, Rahman et al. 1999). Along similar lines, Hsu et al. (2005) found that OFC patients are much less adverse to ambiguity compared with normal subjects. OFC lesions affect choices also when the trade-off involves a social determinant such as fairness, as observed in the ultimatum game (Koenigs & Tranel 2007). With respect to time delays, OFC patients are sometimes described as impulsive (Berlin et al. 2004). However, animal studies on the effects of OFC lesions on intertemporal choices provide diverse results. Specifically, Winstanley et al. (2004) found that rats with OFC lesions are more patient than control animals, whereas Mobini et al. (2002) found the opposite effect. Notably, Winstanley et al. trained animals before the lesion, whereas Mobini et al. trained animals after the lesion. Moreover, in another study, Rudebeck et al. (2006) found that intertemporal preferences following OFC lesions are rather malleable; lesioned animals that initially seemed more impulsive than controls became indistinguishable from controls after performing in a forced-delay version of the task. In the scheme of **Figure 1**, these results may be explained as follows. Choices are normally based on values integrated in the OFC. Absent the OFC, animals choose in a not-value-based fashion, with one determinant taking over. Training affects what option animals default to when OFC is ablated.

One determinant of choice for which current evidence is arguably more controversial is action cost. Arguments against domain generality have been based in particular on two sets of experiments conducted by Rushworth and colleagues. In a first experiment (Rudebeck et al. 2006, Walton et al. 2002), rats could choose between two possible options, one of which was more effortful but more rewarding. The authors found that the propensity to choose the effortful option was reduced after lesions to the anterior cingulate cortex (ACC) but was not significantly altered after OFC lesions. In another study (Rudebeck et al. 2008), the authors tested monkeys with ACC or OFC lesions in two variants of a matching task, in which the correct response was identified either by a particular object (object-based) or by a particular action (action-based). Both sets of lesions reduced performance in both tasks. However, ACC lesions had a comparatively higher effect on the action-based than on the object-based variant, whereas the contrary was true for OFC lesions. On this basis, several investigators proposed that stimulus values (i.e., good values defined disregarding action costs) and action costs are computed separately, in OFC and ACC respectively (Rangel & Hare 2010, Rushworth et al. 2009). Although this proposal deserves further examination, it can be noted that the results of Rushworth and colleagues do not actually rule out a domain-general representation of value in the OFC. Indeed, as illustrated above for intertemporal choices, ablating a valuation center does not necessarily lead to a consistent bias for or against one determinant of value. Thus the results of the first experiment (Rudebeck et al. 2006)—which, in fact, have not been replicated in primates (Kennerley et al. 2006)—do implicate the ACC in some computaton related to action costs but are not conclusive on the OFC. Conversely, the second study (Rudebeck et al. 2008) is less obviously relevant to the issue of value encoding because matching tasks do not necessarily require an economic choice in the sense defined here. Indeed, in matching tasks, there is always a correct answer; subjects are required to infer it from previous trials, not to state a subjective preference (Padoa-Schioppa 2007). Even assuming that animals undertake in matching tasks the same cognitive and neural processes underlying economic choice, it is difficult to establish whether impairments observed after selective brain lesions are due to deficits in learning or in choosing. Finally, in the study of Rudebeck et al. (2008), the action-based variant of the task was much more difficult than the object-based variant of the task (many more errors), and OFC lesions disrupted performance in both variants. Hence, it is possible that OFC lesions selectively interfered with the choice component of the task (and thus

affected both variants equally), whereas ACC lesions affected only the action-based variant. In conclusion, current evidence on choices in the presence of action cost can certainly be reconciled with the hypothesis that OFC harbors an abstract and domain-general representation of subjective value.

To summarize, OFC and vmPFC lesions disrupt choices as defined by a variety of different determinants. Although lesion studies typically lack fine spatial resolution, the results are generally consistent with a domain-general representation of subjective value. Most important, the disruptive effect of OFC and vmPFC lesions on choice behavior establishes a causal link between the neuronal representation of subjective value found by neural recordings in these areas and economic choices.

Choosing in Different Contexts: Menu Invariance and Range Adaptation

The results reviewed in the previous sections justify the hypothesis that choices are based on values computed in OFC and vmPFC. Notably, different neurons in the OFC encode different variables (**Figure 2**). In a computational sense, the valuation stage underlying the choice is captured by neurons encoding the offer value. Thus according to the current hypothesis, choices are based on the activity of these neurons. In this respect, a critical question is whether and how the encoding of value depends on the behavioral context of choice. There are at least two aspects to this issue.

First, for any given offer, a variety of different goods may be available as an alternative. For example, in a gelateria, a person might choose between nocciola and pistacchio or, alternatively, between nocciola and chocolate. A critical question is whether the value a subject assigns to a given good depends on what other good is available as an alternative (i.e., on the menu). Notably, this question is closely related to another critical question: whether preferences are transitive. Given three goods A, B, and C, transitivity holds true if A > B and B > C imply A > C (where > stands for "is preferred to"). Preference transitivity is a hallmark of rational choice behavior and one of the most fundamental assumptions of economic theory (Kreps 1990). Transitivity and menu invariance are closely related because preferences may violate transitivity only if values depend on the menu (Grace 1993, Tversky & Simonson 1993). Although transitivity violations can sometimes be observed (Shafir 2002, Tversky 1969), in most circumstances human and animal choices indeed satisfy transitivity. In a second study, we showed that the representation of value in the OFC is invariant for changes of menu (Padoa-Schioppa & Assad 2008). In this experiment, the monkey chose among 3 juices labeled A, B, and C, in decreasing order of preference. Juices were offered pairwise, and trials with the 3 juice pairs (A:B, B:C, and C:A) were interleaved. Neuronal responses encoding the offer value of one particular juice typically did not depend on the juice offered as an alternative (**Figure 3**), and similar results were obtained for chosen value neurons and taste neurons. If choices are indeed based on values encoded in the OFC, menu invariance may thus be the neurobiological origin of preference transitivity. Corroborating this hypothesis, Fellows & Farah (2007) found that patients with OFC lesions asked to express preference judgments for different foods violate transitivity significantly more often than do both control subjects and patients with dorsal prefrontal lesions—an effect not observed with perceptual judgments (e.g., in the assessment of different colors).

Second, values computed in different behavioral conditions can vary substantially. For example, the same individual may choose sometimes between goods worth a few dollars (e.g., when choosing between different ice cream flavors in a gelateria) and other times between goods worth many thousands of dollars (e.g., when choosing between different houses for sale). At the same time, any representation of value is ultimately limited to a finite range of neuronal firing rates. Moreover, given a range of possible values, an optimal (i.e., maximally sensitive) representation of value would fully

exploit the range of possible firing rates. These considerations suggest that the neuronal encoding of value may adapt to the range of values available in any given condition—a hypothesis I recently confirmed (Padoa-Schioppa 2009). The basic result is illustrated in **Figure 4**, which depicts the activity of 937 offer value neurons from the OFC. Different neurons were recorded in different sessions, and the range of values offered to the monkey varied from session to session. Yet, the distribution of activity ranges measured for the population did not depend on the range of values offered to the monkey. OFC neurons adapted their gain (i.e., the slope of the linear encoding) in such a way that a given range of firing rates described different ranges of values in different behavioral conditions. Corroborating results of Kobayashi et al. (2010) indicate that this adaptation can take place within 15 trials. Interestingly, neuronal firing rates in OFC do not depend on whether the encoded juice is preferred or nonpreferred in that particular session (Padoa-Schioppa 2009).

It has often been discussed whether the brain represents values as relative or absolute (Seymour & McClure 2008). This question can be rephrased by asking which parameters of the behavioral context do or do not affect the encoding of value. The results illustrated here indicate that the encoding of value in the OFC is menu invariant and range adapting. Importantly, although menu invariance and range adaptation hold in normal circumstances, when preferences are stable and transitive, these neural properties may be violated in the presence of choice fallacies (Camerer 2003, Frederick et al. 2002, Kahneman & Tversky 2000, Tversky & Shafir 2004)—a promising topic for future research (Kalenscher et al. 2010).

Online Computation of Economic Values

Although they indicate that an abstract representation of good values is encoded in prefrontal areas, the results discussed so far do not address how this representation is formed. In this respect, two broad hypotheses can be

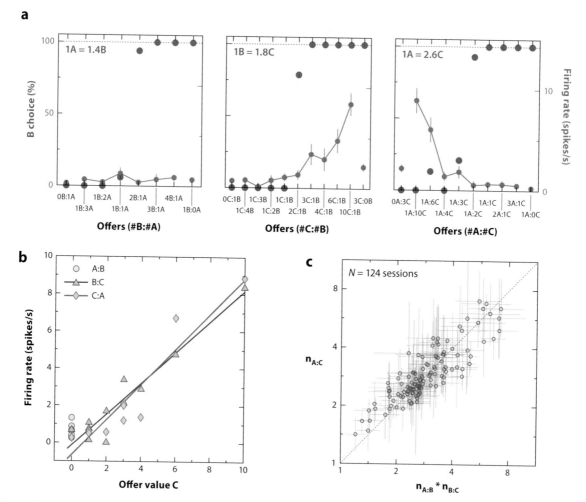

Figure 3

Menu invariance and preference transitivity. (*a*) One neuron encoding the offer value. In this experiment, monkeys chose between 3 juices (A, B, and C) offered pairwise. The three panels refer, respectively, to trials A:B, B:C, and C:A. In each panel, the x-axis represents different offer types, blue circles indicate the behavioral choice pattern, and red symbols indicate the neuronal firing rate. This neuron encodes the variable offer value C independently of whether juice C is offered against juice B or juice A. In trials A:B, the cell activity is low and not modulated. (*b*) Linear encoding. Same neuron as in panel *a*, with the firing rate (y-axis) plotted against the encoded variable (x-axis) separately for different juice pairs (indicated by different symbols, see legend). (*c*) Value transitivity. For each juice pair X:Y, the relative value n_{XY} is measured from the indifference point. The three relative values satisfy transitivity if (in a statistical sense) $n_{AB} * n_{BC} = n_{AC}$. In this scatter plot, each circle indicates one session ($\pm$ SD) and the two axes indicate, respectively, $n_{AB} * n_{BC}$ and n_{AC}. Data lie along the identity line, indicating that subjective values measured in this experiment satisfy transitivity. Choices based on a representation of value that is menu invariant are necessarily transitive. Adapted from Padoa-Schioppa & Assad (2008), *Nature Neuroscience* (Nature Publishing Group).

entertained. One possibility is that values are learned through experience and retrieved from memory at the time of choice. Alternatively, values could be computed online at the time of choice. In observance with a long tradition

in experimental psychology (Skinner 1953, Sutton & Barto 1998), referred to as behaviorism, economic choice is often discussed within the framework of, or as intertwined with, associative learning (Glimcher 2008, Montague

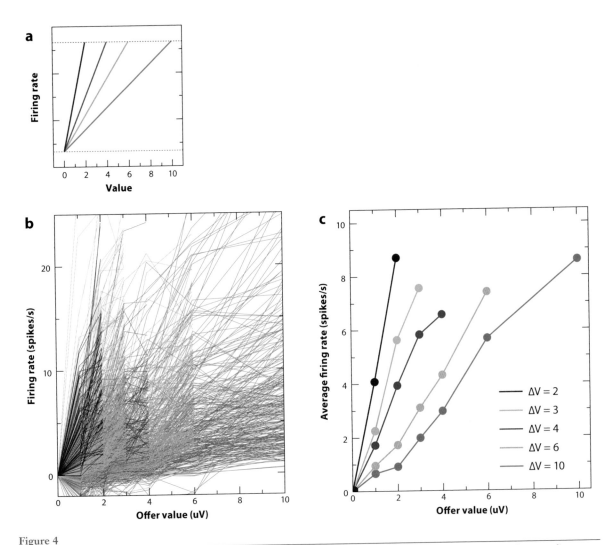

Figure 4

Range adaptation in the valuation system. (*a*) Model of neuronal adaptation. The cartoon depicts the activity of a value-encoding neuron adapting to the range of values available in different conditions. The x-axis represents value, the y-axis represents the firing rate, and different colors refer to different value ranges. In different conditions, the same range of firing rates encodes different value ranges. (*b*) Neuronal adaptation in the orbitofrontal cortex. The figure illustrates the activity of 937 offer value responses. Each line represents the activity of one neuron (y-axis) plotted against the offer value (x-axis). Different responses were recorded with different value ranges (see color labels). Although activity ranges vary widely across the population, the distribution of activity ranges does not depend on the value range. (*c*) Population averages. Each line represents the average obtained from neuronal responses in panel *b*. Adaptation can be observed for any value because average responses are separated throughout the value spectrum. Similar results were obtained for neurons encoding the chosen value. Adaptation was also observed for individual cells recorded with different value ranges (not shown). Adapted from Padoa-Schioppa (2009), *Journal of Neuroscience* (used with permission from the Society for Neuroscience).

et al. 2006, Rangel et al. 2008). Indeed, it is often assumed that subjective values are learned and retrieved from memory. Several considerations suggest, however, that values are more likely not learned and retrieved, but rather computed online at the time of choice. Intuitively, this proposition follows from the fact that people and animals choose often and effectively between novel goods and/or in novel situations. Consider, for example, a person choosing between two possible cocktails in a bar. The person might be familiar with both drinks. Yet, her choice will likely depend on unlearned determinants such as the motivational state (e.g., does she "feel like" a dry or sweet drink at this time), the behavioral context (e.g., what cocktail did her friend order), etc. Thus describing her choice on the basis of learned-and-retrieved values seems difficult.

Experimental evidence for values being computed online comes from an elegant series of studies conducted by Dickinson, Rescorla, Balleine, and their colleagues on reinforcer devaluation in rats (Adams & Dickinson 1981, Balleine & Ostlund 2007, Colwill & Rescorla 1986). In the simplest version of the experiment, animals were trained to perform a task (e.g., pressing a lever) to receive a given food. Subsequently, the animals were selectively satiated with that food and tested in the task. Critically, animals were tested "in extinction" (the food was not actually delivered upon successful execution of the task). Thus the performance of the animals gradually degraded over trials during the test phase. Most important, however, the performance of satiated animals was significantly lower than that of control animals throughout the test phase (**Figure 5**). In other words, satiated animals assigned to the food a lower value compared with that assigned by controls—an interpretation confirmed by a variety of control studies and in a free-choice version of the experiment (Balleine & Dickinson 1998). To my understanding, this result is at odds with the hypothesis that values are learned during training, stored in memory, and simply retrieved at the time of choice. Indeed, if this were the case, rats would retrieve in the test phase the value learned in the training phase, which is the same for experimental animals and control animals. In contrast, this result suggests that animals compute values online on the basis of both current motivation and previously acquired knowledge (see sidebar, Further Discussion on Reinforcer Devaluation). Adams (1982) also found that overtraining, which presumably induced a habit, made animals insensitive to devaluation.

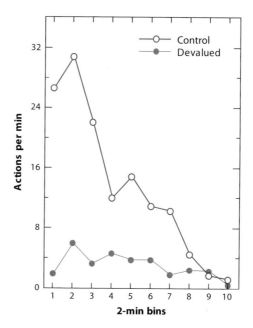

Figure 5

Effects of selective devaluation. In the training phase of this study, rats learned to perform a task (lever press or chain pull) to obtain a reward (food pellet or starch, in a counterbalanced design). Before testing, animals were selectively satiated with one of the two foods (devaluation). They were then tested in extinction. Thus their performance, measured in actions per minute (y-axis), dropped over time (x-axis) for either food reward. Critically, the performance for the devalued food (*red filled symbols*) was consistently below that for the control food (*blue empty symbols*). Adapted from Balleine & Dickinson (1998), *Neuropharmacology* (used with permission from Elsevier).

In summary, intuition and empirical evidence suggest that subjective values are computed online at the time of choice, not learned and retrieved from memory. However, more work is necessary to understand how the neural systems of valuation and associative learning interact and inform each other. Most important for the present purposes, the neural mechanisms by which different determinants—including learned and unlearned determinants—are integrated in the computation of values remain unknown. Although these mechanisms likely involve a variety of sensory, limbic, and association

FURTHER DISCUSSION ON REINFORCER DEVALUATION

If values were learned and retrieved from memory, the results obtained in studies of reinforcer devaluation (**Figure 5**) would have to be interpreted with the assumption that during the devaluation phase the brain automatically updates stored values to reflect the new motivational state. However, this hypothesis seems hardly credible if one considers the fact that the motivational appeal of different goods is in perpetual evolution. For example, the value an individual would assign to any given food changes many times a day, during and after every meal, every time the individual exercises, or simply over time as sugar levels in the bloodstream get lower. Thus the hypothesis that values are learned and retrieved implies that the brain holds and constantly updates a large look-up table of values—a rather expensive design. The hypothesis put forth here—that values are computed only when needed—appears more parsimonious.

Action-based models: economic decisions are made by comparing action values

areas, further research is necessary to shed light on this critical aspect of choice behavior.

ACTION VALUES AND THEIR POSSIBLE RELEVANCE TO ECONOMIC CHOICE

As reviewed in the previous section, a defining trait of the representation of value found in the OFC and the vmPFC is that values are encoded independently of the sensorimotor contingencies of choice. In contrast, in other brain areas, values modulate neuronal activity that is primarily sensory and/or motor. Such "nonabstract" representations have been found in numerous regions, including the dorsolateral prefrontal cortex (Kim et al. 2008, Leon & Shadlen 1999), the anterior cingulate cortex (Matsumoto et al. 2003, Seo & Lee 2007, Shidara & Richmond 2002), the posterior cingulate cortex (McCoy et al. 2003), the lateral intraparietal area (Louie & Glimcher 2010, Sugrue et al. 2004), the dorsal premotor area, the supplementary motor area, the frontal eye fields (Roesch & Olson 2003), the supplementary eye fields (Amador et al. 2000), the superior colliculus (Ikeda &

Hikosaka 2003, Thevarajah et al. 2010), the striatum (Kawagoe et al. 1998, Kim et al. 2009, Lau & Glimcher 2008, Samejima et al. 2005), and the centromedian nucleus of the thalamus (Minamimoto et al. 2005). A comprehensive review of all the relevant work is beyond my current purpose. However, I discuss here the possible significance of these value representations for economic choice.

Nonabstract value modulations are often interpreted in the "space of actions." In other words, the spatially selective component of the neural activity is interpreted as encoding a potential action, and the value modulation is interpreted as a bias contributing to the process of action selection. Thus many experimental results have been or can be described in terms of action values. In broad terms, a neuron can be said to encode an action value if it is preferentially active when a particular action is planned and if it is modulated by the value associated with that action. Influential theoretical accounts posit that decisions are made ultimately on the basis of action values (Kable & Glimcher 2009, Rangel & Hare 2010). According to these action-based models, values are attached to different possible actions in the form of action values, and the decision—the comparison between values—unfolds as a process of action selection. This view of economic choice is clearly in contrast with the good-based proposal. Thus it is important to discuss whether current evidence for the neuronal encoding of action values can be reconciled with the good-based model proposed here. In this respect, a few considerations are needed.

First, in some cases, spatially selective signals modulated by value may be better interpreted as sensory rather than motor. In perceptual domains, value modulates activity by the way of attention—a more valuable visual stimulus inevitably draws higher attention. Thus such value signals may be best described in terms of spatial attention (Maunsell 2004). For example, neurons in the lateral intraparietal area (LIP) activate both in response to visual stimuli placed in their response field and in anticipation of an eye movement. Value modulations recorded in economic choice tasks are strong during presentation of the visual stimulus and significantly lower before the saccade, when movement-related activity dominates (Louie & Glimcher 2010). This observation suggests that value modulates activity in this area by way of attention, a view bolstered by the fact that value modulations in LIP are normalized as predicted by psychophysical theories of attention (Bundesen 1990, Dorris & Glimcher 2004). Similar arguments may apply to other brain areas where neural activity interpreted in terms of action values is most likely not genuinely motor.

Second, action values possibly relevant to economic choice should be distinguished from action values defined in the context of reinforcement learning (RL) (Sutton & Barto 1998). Models of RL typically describe an agent facing a problem with multiple possible actions, one of which is objectively correct. In this context, an action value is an estimate of future rewards for a given action, and the agent learns action values by trial and error. According to behaviorism, any behavior, including economic choice, results from stimulus-response associations. Thus the behaviorist equates action values defined in RL to action values possibly relevant to economic choice. As noted above, a general problem with the behaviorist account is that people and animals can and often do choose effectively between novel goods. The RL variant of this account has the additional problem that choosing a particular good may require different actions at different times. For these reasons, action values possibly relevant to economic choice cannot be equated to action values defined in RL. Consequently, evidence for neuronal encoding of action values gathered using tasks that include a major learning component—instrumental conditioning (Samejima et al. 2005), dynamic matching tasks (Lau & Glimcher 2008, Sugrue et al. 2004), or n-armed bandit tasks—and obtained inferring values from RL models must be considered with caution. This issue is particularly relevant for brain regions, such as the dorsal striatum, that have been clearly linked to associative learning as distinguished from

action selection (Kim et al. 2009, Williams & Eskandar 2006).

Third and most important, value signals can modulate physiological processes downstream of and unrelated to the decision. A compelling example is provided by Roesch & Olson (2003), who trained three monkeys in a variant of the memory saccade task. At the beginning of each trial, a cue indicated whether the amount of juice delivered for a correct response would be large or small. The authors found neuronal modulations consistent with action values in the frontal eye fields, the supplementary eye fields, the premotor cortex, and the supplementary motor area. Strikingly, modulations consistent with action values were also found in the electromyographic (EMG) activity of neck and jaw muscles (**Figure 6**), which suggests that value modulations recorded in cortical motor areas in this experiment—and possibly in other experiments—may be downstream of and unrelated to any decision in the sense defined here.

Taken together, these considerations suggest that evidence for the neural encoding of action values and their possible relevance to economic choices should be vetted against alternative hypotheses. With this premise, what evidence is necessary to hypothesize that an action value signal contributes to economic choice in the sense postulated by action-based models? It is reasonable to require three minimal conditions: (*a*) Neural activity must be genuinely motor, (*b*) neural activity must be modulated by subjective value, and (*c*) neural activity must not be downstream of the decision. These three conditions provide a more restrictive definition of action value. To my knowledge, evidence of neuronal activity satisfying these three conditions has never been reported. In fact, even relaxing condition *b*, I am not aware of any result that satisfies both conditions *a* and *c*. In particular, for activity encoding action values recorded in genuinely motor regions (which presumably satisfied condition *a*), it is generally difficult to rule out that responses were computationally downstream of the decision process (see sidebar, Separation Between Decision

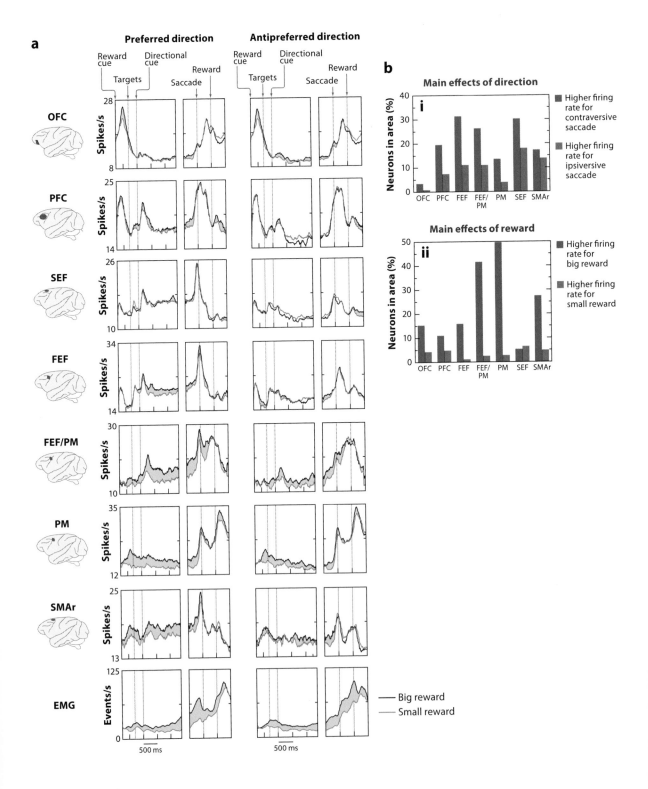

a

Preferred direction

Antipreferred direction

Reward cue • Directional cue • Targets • Saccade • Reward

Reward cue • Directional cue • Targets • Saccade • Reward

OFC — Spikes/s 28 / 8

PFC — Spikes/s 25 / 14

SEF — Spikes/s 26 / 10

FEF — Spikes/s 34 / 14

FEF/PM — Spikes/s 30 / 10

PM — Spikes/s 35 / 12

SMAr — Spikes/s 25 / 13

EMG — Events/s 125 / 0

500 ms — 500 ms

— Big reward
— Small reward

b

i **Main effects of direction**

Neurons in area (%)

■ Higher firing rate for contraversive saccade

■ Higher firing rate for ipsiversive saccade

OFC PFC FEF FEF/PM PM SEF SMAr

ii **Main effects of reward**

Neurons in area (%)

■ Higher firing rate for big reward

■ Higher firing rate for small reward

OFC PFC FEF FEF/PM PM SEF SMAr

Making and Action Planning in Other Mental Functions).

In summary, neural activity encoding action values can contribute to a decision if it encodes action, it encodes value, and it does not follow the decision. Of course, the current lack of evidence for such neural activity, per se, does not falsify action-based models of economic choice. At the same time, current evidence on the encoding of action values can certainly be reconciled with the good-based model and thus does not challenge the present proposal.

OPEN QUESTIONS AND DEVELOPMENTS

As illustrated in the previous sections, the good-based model explains a wealth of experimental results in the literature. At the same time, many aspects of this model remain to be tested. In this section, I briefly discuss two issues that seem particularly urgent.

Perhaps the most distinctive trait of the good-based model is the proposal that values are compared in the space of goods, independent of the actions necessary to implement choices. In this view, action values do not contribute to economic choice per se. Thus the good-based model is in contrast with action-based models, according to which choices are ultimately made by comparing the value of different action plans (Glimcher et al. 2005,

Rangel & Hare 2010). Ultimately, assessing between the two models requires tasks that can (in principle) dissociate in time economic choice from action planning. Consistent with the current proposal, recent work suggests that choices can be made independent of action planning (Cai & Padoa-Schioppa 2010, Wunderlich et al. 2010). Many aspects of this issue, however, remain to be clarified. For example, in many situations, goods available for choice require courses of action associated with different costs. The hypothesis put forth here—that action costs are integrated with other determinants of value in a nonspatial representation—remains to be tested. Also, in most circumstances, a choice ultimately leads to an action. Thus if choices indeed take place in the space of goods, a fundamental question is how choice outcomes are transformed into action plans. The good-to-action transformation, or series of transformations, is poorly understood and should be investigated in future work.

Another important issue is the relative role of OFC and vmPFC in economic choice and value-guided behavior. These two regions roughly correspond to two anatomically defined networks named, respectively, the orbital network (OFC) and the medial network (vmPFC) (Ongur & Price 2000). In an elegant series of studies, Price and colleagues showed that these two networks have distinct and largely segregated anatomical connections

Action plan: specification of an intended movement. Its neural representation reflects the spatial nature of the action

---

Figure 6

Action value signals downstream of the decision. (*a*) Activity profiles from the orbitofrontal cortex (OFC), the lateral prefrontal cortex (PFC), the supplementary eye fields (SEF), the frontal eye fields (FEF), the premotor cortex (PM), the supplementary motor area (SMA), and the muscle's electromyographic activity (EMG). For each brain region, dark green and light green refer, respectively, to trials with high and low value. Left and right panels refer to saccades toward, respectively, the preferred and antipreferred directions. For each area, the overall difference between the activity observed in the left and right panels (highlighted in *bi*) can be interpreted as encoding the action. The difference between the dark green and light green traces (light green area, highlighted in *bii*) is a value modulation. (*b*) Summary of action value signals. The top panel (*bi*) highlights the encoding of possible actions (*contraversive and ipsiversive for blue and red bars, respectively*). The bottom panel (*bii*) highlights value modulations (*positive and negative encoding for blue and red bars, respectively*). Action encoding is minimal in the OFC but significant in all motor areas. In contrast, value modulation is significant both in the OFC and in motor areas. Notably. there is a strong value modulation also in the EMG (bottom panels in panel *a*). Muscles certainly do not contribute to economic choice—a clear example of action value unrelated to the decision. Thus value modulations in the motor areas—which ultimately control the motor output—are most likely related to value modulations in the EMG not to the decision process per se. Adapted from Roesch & Olson (2003, 2005), *Journal of Neurophysiology* (used with permission from the American Physiology Society).

SEPARATION BETWEEN DECISION MAKING AND ACTION PLANNING IN OTHER MENTAL FUNCTIONS

The observation that decision making is separable from action planning appears to remain valid beyond the domain of economic choice. Indeed, neural activity recorded during different tasks that involve a decision generally violates condition *a* (the activity should be genuinely motor) and/or condition *c* (the activity should not be downstream of the decision). Consider, for example, condition *c*. To demonstrate that a neural process satisfies it, it is necessary to design experiments that can—at least in principle—dissociate in time between decision making and action planning (Bennur & Gold 2011, Cai & Padoa-Schioppa 2010, Cisek & Kalaska 2002, Gold & Shadlen 2003, Horwitz et al. 2004, Wunderlich et al. 2010). Evidence that decisions cannot be made in the absence of action planning would support the action-based hypothesis (i.e., that decision making is embedded in motor or premotor systems). However, we are not aware of any such evidence. In fact, several studies found evidence to the contrary. For example, recent results by Bennur & Gold (2011) demonstrate that perceptual decisions can occur in the absence of any action planning. In another study, Cisek & Kalaska (2002) explicitly designed a task to satisfy condition *c* and obtained results that strikingly violate condition *a*. In their experiment, monkeys were first shown two potential targets for a reaching movement. Subsequently, the ambiguity was resolved in favor of one of the two targets. Insofar as this task requires a "decision," neurons encoding potential movements prior to the final instruction would be consistent with the decision unfolding as a process of action selection. Remarkably, the authors did not find any evidence for such neurons. Indeed, cells in motor and premotor cortices (areas F1 and F2) did not activate before the final instruction. Conversely, neurons that activated prior to the final instruction were from prefrontal cortex (area F7) and thus most likely not motor (Picard & Strick 2001).

(Price & Drevets 2010). The orbital network receives inputs from nearly all sensory modalities and from limbic regions, consistent with a role in integrating different determinants into a value signal. In contrast, the medial network is strongly interconnected with the hypothalamus and brain stem, suggesting a role in the control of autonomic functions and visceromotor responses (Price 1999). Indeed, neural activity in this region is known to correlate with heart rate and skin conductance (Critchley 2005, Fredrikson et al. 1998, Ziegler et al. 2009). The relationship between decision making, emotion, and autonomic functions, while often discussed, remains substantially unclear. One possibility is that autonomic responses play a direct role in decision making (Damasio 1994). Another possibility is that values and decisions, made independently, inform emotion and autonomic responses. A third possibility is that decisions emerge from the interplay of multiple decision systems (McClure et al. 2004). The scheme of **Figure 1** is somewhat intermediate. Indeed, I posit the existence of a unitary representation of value, which integrates sensory stimuli and motivational states. In turn, values inform emotional and autonomic responses. However, more work is necessary to clarify the relation between motivation, emotion, and autonomic responses.

CONCLUSIONS

I have reviewed current knowledge on the neural mechanisms of economic choice and, more specifically, on how values are computed, represented, and compared when individuals make a choice. I have also presented a good-based model that provides a unifying framework and accounts for current results. Finally, I have discussed open issues that should be examined in the future.

Much work in the past few years was designed to test the hypothesis that, while making choices, individuals indeed assign subjective values to the available goods. This proposition has now been successfully tested with respect to a variety of determinants: commodity, quantity, risk, delay, effort, and others. Although other determinants remain to be examined, current evidence affords the provisional conclusion that economic values are indeed represented at the neuronal level. This conclusion might appear deceptively foreknown. In fact, a concept of value rooted in neural evidence is a paradigmatic step forward compared with how values have been conceptualized in the past century. Indeed, both behaviorism and neoclassical economics—arguably the dominant theories of choice in psychology and economics since the 1930s—explicitly state that values are purely descriptive entities, not mental states. For this reason, the demonstration that economic values are neurally and thus psychologically real entities may be regarded as a major success for the emerging field of neuroeconomics.

SUMMARY POINTS

1. Different types of decision (perceptual decisions, economic choice, action selection, etc.) involve different mental operations and different brain mechanisms. Economic choice involves assigning values to different goods and comparing these values.

2. Measuring the neural representation of economic value requires letting subjects choose between different options, inferring subjective values from the indifference point, and using that measure to analyze neural activity.

3. A representation of economic value is abstract if neural activity does not depend on the sensorimotor contingencies of choice and if the representation is domain-general. Such an abstract representation exists in the OFC and vmPFC. Lesions to these areas specifically disrupt economic choice behavior.

4. The representation of value in the OFC is menu invariant: Values assigned to different goods are independent of one another. Menu invariance implies preference transitivity.

5. Values computed in different behavioral conditions may vary substantially. The representation of value in the OFC is range adapting: A given range of neural activity represents different value ranges in different behavioral conditions.

6. While computing the value of a given good, subjects integrate a variety of determinants. Some determinants may be learned, whereas other determinants may not be learned. Thus values are computed online at the time of choice.

7. A neural representation of action values may contribute to economic choice if three conditions are met: Neural activity must be genuinely motor, neural activity must be modulated by subjective value, and neural activity must not be downstream of the decision.

8. In addition to guiding an action, values and choice outcomes inform a variety of cognitive and neural systems, including sensory and motor systems (through perceptual attention and attention-like mechanisms), learning (e.g., through mechanisms of reinforcement learning), and emotion (including autonomic functions).

FUTURE ISSUES

1. Where in the brain are different determinants of value (e.g., risk, cost, delay) computed, and how are they represented?

2. The process of integrating multiple determinants into a value signal can be thought of as analogous to computing a nonlinear function with many arguments. How is this computation implemented at the neuronal level? Can it be captured with a computational model?

3. Through which neuronal mechanisms are different values compared to make a decision? Are the underlying algorithms similar to those observed in other brain systems?

4. Assuming that choices indeed take place in goods space, through which neuronal mechanisms is a choice outcome transformed into an action plan?

5. In the OFC and other areas, neurons may encode values in a positive or negative way (i.e., the encoding slope may be positive or negative). Do these two neuronal populations play different roles in choice behavior?

6. Abstract representations of value appear to exist in the primate OFC and vmPFC, but the relative contributions of these two brain regions to choice behavior are not clear. In fact, the anatomical connectivity of the orbital network and medial network is markedly different. How do OFC and vmPFC contribute to economic choices?

7. No abstract representation of value has yet been found in the rodent OFC—a striking difference with primates. Possible reasons for this discrepancy include a poor homology between "OFC" as defined in different species, the hypothesis that an abstract representation of value may have emerged late in evolution, and differences in experimental procedures. How can differences among species be explained best?

8. Choice traits such as temporal discounting, risk aversion, and loss aversion ultimately affect subjective values. Thus their neuronal correlates may be and have been observed by measuring neural activity encoding subjective value. However, these measures generally do not explain the neurobiological origin of these choice traits. Can temporal discounting and other choice traits be explained as the result of specific neuronal properties?

DISCLOSURE STATEMENT

The author is not aware of any affiliations, memberships, funding, or financial holdings that might be perceived as affecting the objectivity of this review.

ACKNOWLEDGMENTS

This article is dedicated to my father Tommaso Padoa-Schioppa, in memory. I am grateful to John Assad and Xinying Cai for helpful discussions and comments on the manuscript. My research is supported by the NIMH (grant MH080852) and the Whitehall Foundation (grant 2010-12-13).

LITERATURE CITED

Adams CD. 1982. Variations in the sensitivity of instrumental responding to reinforcer devaluation. *Q. J. Exp. Psych.* 34B:77–98

Adams CD, Dickinson A. 1981. Instrumental responding following reinforcer devaluation. *Q. J. Exp. Psych.* 33B:109–21

Amador N, Schlag-Rey M, Schlag J. 2000. Reward-predicting and reward-detecting neuronal activity in the primate supplementary eye field. *J. Neurophysiol.* 84:2166–70

Balleine BW, Dickinson A. 1998. Goal-directed instrumental action: contingency and incentive learning and their cortical substrates. *Neuropharmacology* 37:407–19

Balleine BW, Ostlund SB. 2007. Still at the choice-point: action selection and initiation in instrumental conditioning. *Ann. N.Y. Acad. Sci.* 1104:147–71

Baxter MG, Gaffan D, Kyriazis DA, Mitchell AS. 2009. Ventrolateral prefrontal cortex is required for performance of a strategy implementation task but not reinforcer devaluation effects in rhesus monkeys. *Eur. J. Neurosci.* 29:2049–59

Bennur S, Gold JI. 2011. Distinct representations of a perceptual decision and the associated oculomotor plan in the monkey lateral intraparietal area. *J. Neurosci.* 31:913–21

Berlin HA, Rolls ET, Kischka U. 2004. Impulsivity, time perception, emotion and reinforcement sensitivity in patients with orbitofrontal cortex lesions. *Brain* 127:1108–26

Brooks AM, Pammi VS, Noussair C, Capra CM, Engelmann JB, Berns GS. 2010. From bad to worse: striatal coding of the relative value of painful decisions. *Front. Neurosci.* 4:176

Bundesen C. 1990. A theory of visual attention. *Psychol. Rev.* 97:523–47

Cai X, Kim S, Lee D. 2011. Heterogeneous coding of temporally discounted values in the dorsal and ventral striatum during intertemporal choice. *Neuron* 69:170–82

Cai X, Padoa-Schioppa C. 2010. Dissociating economic choice from action selection. *Soc. Neurosci. Meet. Plann.* Abstr. 813.1

Camerer C. 2003. *Behavioral Game Theory: Experiments in Strategic Interaction*. Princeton, NJ: Russell Sage Found., Princeton Univ. Press

Christopoulos GI, Tobler PN, Bossaerts P, Dolan RJ, Schultz W. 2009. Neural correlates of value, risk, and risk aversion contributing to decision making under risk. *J. Neurosci.* 29:12574–83

Cisek P, Kalaska JF. 2002. Simultaneous encoding of multiple potential reach directions in dorsal premotor cortex. *J. Neurophysiol.* 87:1149–54

Colwill RM, Rescorla RA. 1986. Associative structures in instrumental conditioning. In *The Psychology of Learning and Motivation*, ed. GH Bower, pp. 55–104. Orlando, FL: Academic

Critchley HD. 2005. Neural mechanisms of autonomic, affective, and cognitive integration. *J. Comp. Neurol.* 493:154–66

Damasio AR. 1994. *Descartes' Error: Emotion, Reason, and the Human Brain*. New York: Putnam. xix, 312 pp.

De Martino B, Kumaran D, Holt B, Dolan RJ. 2009. The neurobiology of reference-dependent value computation. *J. Neurosci.* 29:3833–42

Dorris MC, Glimcher PW. 2004. Activity in posterior parietal cortex is correlated with the relative subjective desirability of action. *Neuron* 44:365–78

Feierstein CE, Quirk MC, Uchida N, Sosulski DL, Mainen ZF. 2006. Representation of spatial goals in rat orbitofrontal cortex. *Neuron* 51:495–507

Fellows LK, Farah MJ. 2007. The role of ventromedial prefrontal cortex in decision making: judgment under uncertainty or judgment per se? *Cereb. Cortex* 17:2669–74

FitzGerald TH, Seymour B, Dolan RJ. 2009. The role of human orbitofrontal cortex in value comparison for incommensurable objects. *J. Neurosci.* 29:8388–95

Frederick S, Loewenstein G, O'Donogue T. 2002. Time discounting and time preference: a critical review. *J. Econ. Lit.* 40:351–401

Fredrikson M, Furmark T, Olsson MT, Fischer H, Andersson J, Langstrom B. 1998. Functional neuroanatomical correlates of electrodermal activity: a positron emission tomographic study. *Psychophysiology* 35:179–85

Gallagher M, McMahan RW, Schoenbaum G. 1999. Orbitofrontal cortex and representation of incentive value in associative learning. *J. Neurosci.* 19:6610–14

Glimcher PW. 2008. Choice: towards a standard back-pocket model. See Glimcher et al. 2008, pp. 501–19

Glimcher PW, Camerer CF, Fehr E, Poldrack RA. 2008. *Neuroeconomics: Decision Making and the Brain*. Amsterdam: Elsevier. 538 pp.

Glimcher PW, Dorris MC, Bayer HM. 2005. Physiological utility theory and the neuroeconomics of choice. *Games Econ. Behav.* 52:213–56

Gold JI, Shadlen MN. 2003. The influence of behavioral context on the representation of a perceptual decision in developing oculomotor commands. *J. Neurosci.* 23:632–51

Grace RC. 1993. Violations of transitivity: implications for a theory of contextual choice. *J. Exp. Anal. Behav.* 60:185–201

Grattan LE, Glimcher PW. 2010. No evidence for spatial tuning in orbitofrontal cortex. *Soc. Neurosci. Meet. Plann.* Abstr. 102.23

Gregorios-Pippas L, Tobler PN, Schultz W. 2009. Short-term temporal discounting of reward value in human ventral striatum. *J. Neurophysiol.* 101:1507–23

Horwitz GD, Batista AP, Newsome WT. 2004. Representation of an abstract perceptual decision in macaque superior colliculus. *J. Neurophysiol.* 91:2281–96

Hsu M, Bhatt M, Adolphs R, Tranel D, Camerer CF. 2005. Neural systems responding to degrees of uncertainty in human decision-making. *Science* 310:1680–83

Hsu M, Krajbich I, Zhao C, Camerer CF. 2009. Neural response to reward anticipation under risk is nonlinear in probabilities. *J. Neurosci.* 29:2231–37

Ikeda T, Hikosaka O. 2003. Reward-dependent gain and bias of visual responses in primate superior colliculus. *Neuron* 39:693–700

Izquierdo A, Suda RK, Murray EA. 2004. Bilateral orbital prefrontal cortex lesions in rhesus monkeys disrupt choices guided by both reward value and reward contingency. *J. Neurosci.* 24:7540–48

Kable JW, Glimcher PW. 2007. The neural correlates of subjective value during intertemporal choice. *Nat. Neurosci.* 10:1625–33

Kable JW, Glimcher PW. 2009. The neurobiology of decision: consensus and controversy. *Neuron* 63:733–45

Kahneman D, Tversky A. 1979. Prospect theory: an analysis of decision under risk. *Econometrica* 47:263–91

Kahneman D, Tversky A, eds. 2000. *Choices, Values and Frames.* Cambridge, UK/New York: Russell Sage Found./Cambridge Univ. Press. xx, 840 pp.

Kalenscher T, Tobler PN, Huijbers W, Daselaar SM, Pennartz CM. 2010. Neural signatures of intransitive preferences. *Front. Hum. Neurosci.* 4:49

Kawagoe R, Takikawa Y, Hikosaka O. 1998. Expectation of reward modulates cognitive signals in the basal ganglia. *Nat. Neurosci.* 1:411–16

Kazama A, Bachevalier J. 2009. Selective aspiration or neurotoxic lesions of orbital frontal areas 11 and 13 spared monkeys' performance on the object discrimination reversal task. *J. Neurosci.* 29:2794–804

Kennerley SW, Dahmubed AF, Lara AH, Wallis JD. 2009. Neurons in the frontal lobe encode the value of multiple decision variables. *J. Cogn. Neurosci.* 21:1162–78

Kennerley SW, Wallis JD. 2009. Encoding of reward and space during a working memory task in the orbitofrontal cortex and anterior cingulate sulcus. *J. Neurophysiol.* 102:3352–64

Kennerley SW, Walton ME, Behrens TE, Buckley MJ, Rushworth MF. 2006. Optimal decision making and the anterior cingulate cortex. *Nat. Neurosci.* 9:940–47

Kim H, Sul JH, Huh N, Lee D, Jung MW. 2009. Role of striatum in updating values of chosen actions. *J. Neurosci.* 29:14701–12

Kim S, Hwang J, Lee D. 2008. Prefrontal coding of temporally discounted values during intertemporal choice. *Neuron* 59:161–72

Kimmel DL, Rangel A, Newsome WT. 2010. Value representations in the primate orbitofrontal cortex during cost-benefit decision making. *Soc. Neurosci. Meet. Plann.* Abstr. 129.14

Klein JT, Deaner RO, Platt ML. 2008. Neural correlates of social target value in macaque parietal cortex. *Curr. Biol.* 18:419–24

Kobayashi S, Pinto de Carvalho O, Schultz W. 2010. Adaptation of reward sensitivity in orbitofrontal neurons. *J. Neurosci.* 30:534–44

Kobayashi S, Schultz W. 2008. Influence of reward delays on responses of dopamine neurons. *J. Neurosci.* 28:7837–46

Koenigs M, Tranel D. 2007. Irrational economic decision-making after ventromedial prefrontal damage: evidence from the Ultimatum Game. *J. Neurosci.* 27:951–56

Kreps DM. 1990. *A Course in Microeconomic Theory.* Princeton, NJ: Princeton Univ. Press. 850 pp.

Lau B, Glimcher PW. 2008. Value representations in the primate striatum during matching behavior. *Neuron* 58:451–63

Leon MI, Shadlen MN. 1999. Effect of expected reward magnitude on the response of neurons in the dorsolateral prefrontal cortex of the macaque. *Neuron* 24:415–25

Levy I, Snell J, Nelson AJ, Rustichini A, Glimcher PW. 2010. Neural representation of subjective value under risk and ambiguity. *J. Neurophysiol.* 103:1036–47

Louie K, Glimcher PW. 2010. Separating value from choice: delay discounting activity in the lateral intraparietal area. *J. Neurosci.* 30:5498–507

Machado CJ, Bachevalier J. 2007a. The effects of selective amygdala, orbital frontal cortex or hippocampal formation lesions on reward assessment in nonhuman primates. *Eur. J. Neurosci.* 25:2885–904

Machado CJ, Bachevalier J. 2007b. Measuring reward assessment in a semi-naturalistic context: the effects of selective amygdala, orbital frontal or hippocampal lesions. *Neuroscience* 148:599–611

Matsumoto K, Suzuki W, Tanaka K. 2003. Neuronal correlates of goal-based motor selection in the prefrontal cortex. *Science* 301:229–32

Maunsell JH. 2004. Neuronal representations of cognitive state: reward or attention? *Trends Cogn. Sci.* 8:261–65

McClure SM, Laibson DI, Loewenstein G, Cohen JD. 2004. Separate neural systems value immediate and delayed monetary rewards. *Science* 306:503–7

McCoy AN, Crowley JC, Haghighian G, Dean HL, Platt ML. 2003. Saccade reward signals in posterior cingulate cortex. *Neuron* 40:1031–40

Minamimoto T, Hori Y, Kimura M. 2005. Complementary process to response bias in the centromedian nucleus of the thalamus. *Science* 308:1798–801

Mobini S, Body S, Ho MY, Bradshaw CM, Szabadi E, et al. 2002. Effects of lesions of the orbitofrontal cortex on sensitivity to delayed and probabilistic reinforcement. *Psychopharmacology (Berl.)* 160:290–98

Montague PR, King-Casas B, Cohen JD. 2006. Imaging valuation models in human choice. *Annu. Rev. Neurosci.* 29:417–48

Morrison SE, Salzman CD. 2009. The convergence of information about rewarding and aversive stimuli in single neurons. *J. Neurosci.* 29:11471–83

O'Neill M, Schultz W. 2010. Coding of reward risk by orbitofrontal neurons is mostly distinct from coding of reward value. *Neuron* 68:789–800

Ongur D, Price JL. 2000. The organization of networks within the orbital and medial prefrontal cortex of rats, monkeys and humans. *Cereb. Cortex* 10:206–19

Padoa-Schioppa C. 2007. Orbitofrontal cortex and the computation of economic value. *Ann. N.Y. Acad. Sci.* 1121:232–53

Padoa-Schioppa C. 2009. Range-adapting representation of economic value in the orbitofrontal cortex. *J. Neurosci.* 29:14004–14

Padoa-Schioppa C, Assad JA. 2006. Neurons in orbitofrontal cortex encode economic value. *Nature* 441:223–26

Padoa-Schioppa C, Assad JA. 2008. The representation of economic value in the orbitofrontal cortex is invariant for changes of menu. *Nat. Neurosci.* 11:95–102

Peters J, Buchel C. 2009. Overlapping and distinct neural systems code for subjective value during intertemporal and risky decision making. *J. Neurosci.* 29:15727–34

Picard N, Strick PL. 2001. Imaging the premotor areas. *Curr. Opin. Neurobiol.* 11:663–72

Pickens CL, Saddoris MP, Setlow B, Gallagher M, Holland PC, Schoenbaum G. 2003. Different roles for orbitofrontal cortex and basolateral amygdala in a reinforcer devaluation task. *J. Neurosci.* 23:11078–84

Pine A, Seymour B, Roiser JP, Bossaerts P, Friston KJ, et al. 2009. Encoding of marginal utility across time in the human brain. *J. Neurosci.* 29:9575–81

Plassmann H, O'Doherty J, Rangel A. 2007. Orbitofrontal cortex encodes willingness to pay in everyday economic transactions. *J. Neurosci.* 27:9984–88

Popper K. 1963. Science: conjectures and refutations. In *Conjectures and Refutations*, pp. 43–86. London/New York: Routledge Classics

Price JL. 1999. Prefrontal cortical networks related to visceral function and mood. *Ann. N.Y. Acad. Sci.* 877:383–96

Price JL, Drevets WC. 2010. Neurocircuitry of mood disorders. *Neuropsychopharmacology* 35:192–216

Rahman S, Sahakian BJ, Hodges JR, Rogers RD, Robbins TW. 1999. Specific cognitive deficits in mild frontal variant of frontotemporal dementia. *Brain* 122(Pt. 8):1469–93

Rangel A, Camerer C, Montague PR. 2008. A framework for studying the neurobiology of value-based decision making. *Nat. Rev. Neurosci.* 9:545–56

Rangel A, Hare T. 2010. Neural computations associated with goal-directed choice. *Curr. Opin. Neurobiol.* 20:262–70

Roesch MR, Olson CR. 2003. Impact of expected reward on neuronal activity in prefrontal cortex, frontal and supplementary eye fields and premotor cortex. *J. Neurophysiol.* 90:1766–89

Roesch MR, Olson CR. 2005. Neuronal activity in primate orbitofrontal cortex reflects the value of time. *J. Neurophysiol.* 94:2457–71

Roesch MR, Taylor AR, Schoenbaum G. 2006. Encoding of time-discounted rewards in orbitofrontal cortex is independent of value representation. *Neuron* 51:509–20

Rudebeck PH, Behrens TE, Kennerley SW, Baxter MG, Buckley MJ, et al. 2008. Frontal cortex subregions play distinct roles in choices between actions and stimuli. *J. Neurosci.* 28:13775–85

Rudebeck PH, Walton ME, Smyth AN, Bannerman DM, Rushworth MF. 2006. Separate neural pathways process different decision costs. *Nat. Neurosci.* 9:1161–68

Rushworth MF, Mars RB, Summerfield C. 2009. General mechanisms for making decisions? *Curr. Opin. Neurobiol.* 19:75–83

Samejima K, Ueda Y, Doya K, Kimura M. 2005. Representation of action-specific reward values in the striatum. *Science* 310:1337–40

Schoenbaum G, Roesch MR, Stalnaker TA, Takahashi YK. 2009. A new perspective on the role of the orbitofrontal cortex in adaptive behaviour. *Nat. Rev. Neurosci.* 10:885–92

Seo H, Lee D. 2007. Temporal filtering of reward signals in the dorsal anterior cingulate cortex during a mixed-strategy game. *J. Neurosci.* 27:8366–77

Seymour B, McClure SM. 2008. Anchors, scales and the relative coding of value in the brain. *Curr. Opin. Neurobiol.* 18:173–78

Shafir S. 2002. Context-dependent violations of rational choice in honeybees (*Apis mellifera*) and gray jays (*Perisoreus canadensis*). *Behav. Ecol. Sociobiol.* 51:180–87

Shenhav A, Greene JD. 2010. Moral judgments recruit domain-general valuation mechanisms to integrate representations of probability and magnitude. *Neuron* 67:667–77

Shidara M, Richmond BJ. 2002. Anterior cingulate: single neuronal signals related to degree of reward expectancy. *Science* 296:1709–11

Skinner BF. 1953. *Science and Human Behavior*. New York: Macmillan. 461 pp.

Sloan J, Kennerley SW, Wallis JD. 2010. Neural encoding of cost-based decisions: effort versus delay. *Soc. Neurosci. Meet. Plann.* Abstr. 805.16

Sugrue LP, Corrado GS, Newsome WT. 2004. Matching behavior and the representation of value in the parietal cortex. *Science* 304:1782–87

Sutton RS, Barto AG. 1998. *Reinforcement Learning: An Introduction*. Cambridge, MA: MIT Press. xviii, 322 pp.

Thevarajah D, Webb R, Ferrall C, Dorris MC. 2010. Modeling the value of strategic actions in the superior colliculus. *Front. Behav. Neurosci.* 3:57

Tversky A. 1969. The intransitivity of preferences. *Psychol. Rev.* 76:31–48

Tversky A, Shafir E. 2004. *Preference, Belief, and Similarity: Selected Writings*. Cambridge, MA: MIT Press. xvi, 1023 pp.

Tversky A, Simonson I. 1993. Context-dependent preferences. *Manag. Sci.* 39:117–85

van Duuren E, Escamez FA, Joosten RN, Visser R, Mulder AB, Pennartz CM. 2007. Neural coding of reward magnitude in the orbitofrontal cortex of the rat during a five-odor olfactory discrimination task. *Learn. Mem.* 14:446–56

Walton ME, Bannerman DM, Rushworth MF. 2002. The role of rat medial frontal cortex in effort-based decision making. *J. Neurosci.* 22:10996–1003

Watson KK, Platt ML. 2008. Orbitofrontal neurons encode both social and nonsocial rewards. *Soc. Neurosci. Meet. Plann.* Abstr. 691.11

Williams ZM, Eskandar EN. 2006. Selective enhancement of associative learning by microstimulation of the anterior caudate. *Nat. Neurosci.* 9:562–68

Winstanley CA, Theobald DE, Cardinal RN, Robbins TW. 2004. Contrasting roles of basolateral amygdala and orbitofrontal cortex in impulsive choice. *J. Neurosci.* 24:4718–22

Wise SP. 2008. Forward frontal fields: phylogeny and fundamental function. *Trends Neurosci.* 31:599–608

Wunderlich K, Rangel A, O'Doherty JP. 2010. Economic choices can be made using only stimulus values. *Proc. Natl. Acad. Sci. USA* 107:15005–10

Zald DH. 2006. The rodent orbitofrontal cortex gets time and direction. *Neuron* 51:395–97

Ziegler G, Dahnke R, Yeragani VK, Bar KJ. 2009. The relation of ventromedial prefrontal cortex activity and heart rate fluctuations at rest. *Eur. J. Neurosci.* 30:2205–10

RELATED RESOURCES

Heyman GM. 2009. *Addiction: A Disorder of Choice*. Cambridge, MA: Harvard Univ. Press. 200 pp.

Kagel JH, Battalio RC, Green L. 1995. *Economic Choice Theory: An Experimental Analysis of Animal Behavior*. Cambridge, UK: Cambridge Univ. Press. 230 pp.

Niehans J. 1990. *A History of Economic Theory: Classic Contributions, 1720–1980*. Baltimore, MD: Johns Hopkins Univ. Press. 578 pp.

Padoa-Schioppa C. 2008. The syllogism of neuro-economics. *Econ. Philos.* 24:449–57

Ross D. 2005. *Economic Theory and Cognitive Science: Microexplanation*. Cambridge, MA: MIT Press. 444 pp.

The Extraction of 3D Shape in the Visual System of Human and Nonhuman Primates

Guy A. Orban

Laboratorium voor Neuro-en Psychofysiologie, KU Leuven Medical School, Leuven, Belgium; email: guy.orban@med.kuleuven.be

Annu. Rev. Neurosci. 2011. 34:361–88

First published online as a Review in Advance on March 29, 2011

The *Annual Review of Neuroscience* is online at neuro.annualreviews.org

This article's doi: 10.1146/annurev-neuro-061010-113819

0147-006X/11/0721-0361$20.00

Keywords

cerebral cortex, functional imaging, action potential, depth, motion

Abstract

Depth structure, the third dimension of object shape, is extracted from disparity, motion, texture, and shading in the optic array. Gradient-selective neurons play a key role in this process. Such neurons occur in CIP, AIP, TEs, and F5 (for first- or second-order disparity gradients), in MT/V5, in FST (for speed gradients), and in CIP and TEs (for texture gradients). Most of these regions are activated during magnetic resonance scanning in alert monkeys by comparing 3D conditions with the 2D controls for the different cues. Similarities in activation patterns of monkeys and humans tested with identical paradigms suggest that like gradient-selective neurons are found in corresponding human cortical areas. This view gains credence as the homologies between such areas become more evident. Furthermore, 3D shape-processing networks are similar in the two species, with the exception of the greater involvement of human posterior parietal cortex in the extraction of 3D shape from motion. Thus we can begin to understand how depth structure is extracted from motion, disparity, and texture in the primate brain, but the extraction of depth structure from shading and that of wire-like objects requires further scrutiny.

Contents

INTRODUCTION

Gradient: a derivative of a quantitative property, such as speed, in the retinal image(s) along an axis in the image

Depth structure: shape in the third dimension

Although we live in a three-dimensional (3D) world, surprisingly little is known about the processing of 3D shape in the primate brain. We move within a 3D environment, we interact with 3D objects, including animals and conspecifics, and the processing of visual 3D information can be of vital importance, if one is, say, walking along a cliff or fighting off a tiger. My interest in 3D shape arose from interactions, during the mid-1990s, with two eminent scientists, Jan Koenderink and Olivier Faugeras, in the framework of interdisciplinary EU projects (the Insight series). In this article, I review the progress made these 15 years in understanding the first steps of 3D shape processing. Results show that the brain extracts representations from the visual array that map directly onto a real-world object property, here 3D shape, as predicted by Gibson (1950). The quantities extracted from the optic array are primarily the gradients of visual parameters that covary with changes in depth along 3D surfaces, most notably disparity and speed gradients for the stereo and motion cues, respectively. The timeliness of this review is further underscored by the recent developments in the movie and television industry, introducing 3D movies such as Avatar.

3D Shape and Depth

Three-dimensional shape is defined by the boundary edges and surfaces of a 3D object. Because projection onto the retina loses explicit depth information that must be recovered by subsequent cerebral processing, it is customary to subdivide visual 3D shape into a fronto-parallel component, the 2D shape, and a depth component, the depth structure (Durand et al. 2007). Because the processing of 2D shape is ubiquitous throughout the dorsal and ventral visual streams (see Kourtzi & Connor 2011, this issue), this article addresses the extraction of depth structure from the visual array. Depth structure is nothing more than the variation of depth along the surface or edge of an object. Hence, many of the depth cues also provide information about depth structure.

The most sensitive depth cue is stereopsis, which exploits the small differences in relative horizontal positions of retinal images in the two eyes, a consequence of the slightly different viewpoints of the two eyes. Monocular depth cues include motion, texture, and shading. The motion cue may be compounded because retinal speed depends on distance from the eye, and direction of motion on the retina reverses at the fixation point (Roy et al. 1992). Other depth cues, such as blur or occlusion, provide less information about depth structure. This review concentrates on the four cues pertaining to the depth structure of surfaces, stereopsis, motion, texture, and shading, which we refer to as 3D shape from disparity (3D SFD; see **Supplemental Figures 1** and **2**. Follow the **Supplemental Material link** from the Annual Reviews home page at **http://www.annualreviews.org**), from motion (3D SFM; **Supplemental Videos 1–4**), from texture (3D SFT; **Supplemental Figure 3**), and from shading (3D SFS; **Supplemental Figure 3**). Notice that the first two cues also apply to thin elongated 3D objects (e.g., a wire) or edges of a surface, supplemented by perspective as a third cue. Closure constitutes a possible fourth cue for edge-like structures, which may be processed in early extrastriate areas (S. Sunaert, J. Todd, and G. Orban, unpublished observations).

Depth Structure of Surfaces and Depth Orders

The depth structure of a surface is basically defined by the variation of depth along the surface and hence is intimately related to nonzero depth orders (**Figure 1a–c**). Zero-order depth is simply range or distance relative to the observer or the fixation point. It includes both a qualitative aspect, whether one surface is in front of another, and a quantitative aspect, i.e., the distance between two surfaces, i.e., depth intervals (Anzai & DeAngelis 2010). First-order depth is the derivative of depth along an axis in the front-parallel plane. It reflects 3D orientation, i.e., orientation in depth, specified by two parameters: tilt and slant, respectively indicating the direction and the degree by which the planar surface deviates from the fronto-parallel plane. Second-order depth, or depth curvature, is the derivative of 3D orientation along an axis of the fronto-parallel plane. This derivative specifies a surface curved in depth, with the orientation, sign, and degree of curvature as parameters. Such a singly curved surface is but one member

of a larger family of second-order surfaces defined by curvature in two orthogonal directions, the so-called quadratic surfaces. Koenderink (1990) proposed shape index and curvedness as the two parameters of quadratic or second-order surfaces. The shape index ranges from -1 to $+1$, with values of -0.5 or 0.5 corresponding to ridges, values between -0.5 and 0.5 to saddles, and values smaller than -0.5 or larger than 0.5 to dimples or bumps (**Figure 1d**). These surfaces need orientation as an additional parameter to be fully described because ridges and saddles can have different orientations in the fronto-parallel plane. Higher-order depth corresponds to variations in the curvature in depth.

First-order depth is ambiguous with respect to depth structure insofar as it may signal the 3D orientation of a bounding surface of a 3D object, such as a pyramid, or simply of a planar object, such as a disc. Second-order depth is necessarily related to depth structure. Moreover, for stereo, it is invariant with distance from the eye, enhancing its attractiveness as a descriptor of depth structure. The importance of second-order depth for 3D shape was not well appreciated by Marr (1982), who considered mainly 3D surface orientation and its discontinuities. It is worth noting that discontinuities in either zero- or first-order depth correspond to edges of objects, not to surfaces.

Depth orders specify only local depth structure of surfaces. Hence this information needs to be combined with size information to define the depth structure of an object. Small-to-medium-sized surfaces (up to $10°$–$15°$) most likely belong to 3D objects; larger surfaces describe the 3D layout of the environment. At the other end of the size scale, 3D texture is defined by nonzero-depth orders on a very small spatial scale. Notice that in first approximation the extraction of depth structure assumes that the 3D object or the 3D surface is rigid. So far, very little work has addressed deforming surfaces. Notice also that specifying the layout of 3D surfaces within the 2D shape defined by the occluding edges does not exhaust the volumetric perception of 3D objects, which may also include the processing of the 3D

3D SFD: 3D shape from disparity

3D SFM: 3D shape from motion

3D SFT: 3D shape from texture

3D SFS: 3D shape from shading

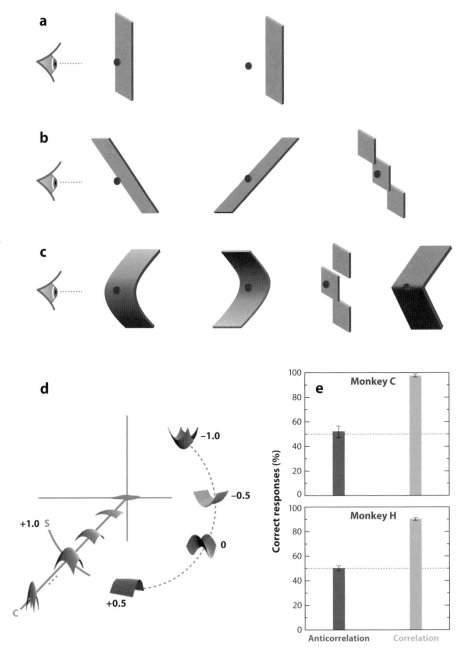

Figure 1

Stimulus definitions for 3D shape and behavioral evidence in the monkey. (*a–c*): orders of depth: zero- (*a*), first- (*b*), and second- (*c*) order surfaces seen from the left, red dot: fixation point, on the right in panels *b* and *c*, zero-order and first-order approximations. (*d*) Shape index (S, polar angle) and curvedness (C, radial dimension) of quadratic surfaces as defined by Koenderink (1990). (*e*) Behavior:% correct for discrimination of convex and concave disparity-defined surfaces in correlated and anticorrelated stereograms (Janssen et al. 2000b).

orientation and its derivative at their occluding edges.

Stereopsis is the most sensitive cue for depth structure, and it also applies to all depth orders and recovers the sign of depth curvature. However, its distance from the subject over which it operates is restricted, a limitation far less relevant for motion. Motion is almost as sensitive as stereopsis but provides ambiguous information about the sign of curvature. Texture provides excellent first-order depth information about surfaces, which shading does not do. Shading, on the other hand, provides clear second-order depth information, although its sign depends strongly on the position of the light source (Hanazawa & Komatsu 2001).

PERCEPTION OF 3D SHAPE

Numerous psychophysical studies have shown that humans perceive 3D shape well (for review, see Todd 2004) using any of the four cues (Norman et al. 2004). Several tasks have been used to study 3D shape perception, but one that stands out is a 3D shape adjustment task (Georgieva et al. 2009, Koenderink et al. 2001). Subjects view a randomly deformed sphere (a potato; see **Supplemental Figure 3**) onto which a horizontal line is superimposed. The subject's task is to manipulate a comparison horizontal line until its profile matches as precisely as possible the perceived relief along the horizontal line on the potato. Line height generally correlates very closely with the actual relief, but the slope of a line fitted to the data is typically less than one, indicating that subjects underestimate the amount of relief (Todd 2004). Three-dimensional shape from motion and disparity develops as early as two and four months after birth, respectively (Arterberry & Yonas 2000, Yonas et al. 1987). Sensitivity for pictorial cues develops a few months later (Tsuruhara et al. 2009).

Because we use the macaque as an animal model, it is important to verify that this species perceives 3D shape. This is difficult to establish, but several psychophysical tests indicate that monkeys achieve thresholds for detecting or discriminating 3D shapes similar to those of humans. Janssen et al. (2003) showed that monkeys can discriminate convex and concave singly curved disparity-defined surfaces, data that have been confirmed in subsequent experiments using singly (**Figure 2** and **Supplemental Figure 4**) or doubly curved (Verhoef et al. 2010) surfaces. Moreover, anticorrelated stereograms produced no 3D shape perception (**Figure 1e**) (Janssen et al. 2003), as in humans (Cumming et al. 1998). Monkeys also perceive 3D structure from motion and detect the coherence of a 3D rotating cylinder even more quickly than do humans (Siegel & Andersen 1990). Zhang & Schiller (2008) confirmed these findings using another task, in which the monkey had to indicate the point protruding furthest from the background. Tsutsui et al. (2002) demonstrated that monkeys can cross-match the 3D orientations of texture-defined and disparity-defined surfaces and can generalize to new orientations and texture patterns. **Figure 2c** and **Supplemental Figure 4** confirm that monkeys generalize from disparity-defined to texture-defined singly curved surfaces, when discriminating convex from concave surfaces, even though performance levels were relatively low. The responsiveness of monkeys to the texture cue has also been demonstrated in infant monkeys (Gunderson et al. 1993). Finally, Zhang et al. (2007), using an oddity task, have provided evidence that monkeys can also perceive 3D shape from shading, although performance was poorer in monkeys than in humans. Thus the available evidence suggests macaque monkeys process 3D shape, particularly from motion or disparity, and combine cues (Schiller et al. 2011) in ways similar to humans.

EXTRACTION AND PROCESSING OF DEPTH STRUCTURE IN THE MONKEY MODEL

A 3D Shape-Processing Network in the Monkey

I tentatively propose the following network (**Figure 3**) for processing 3D shape in the

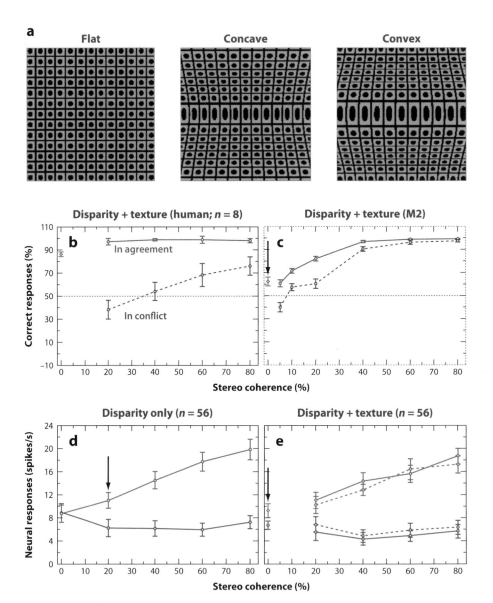

Figure 2

Mixing disparity and texture for second-order depth in behavior and TEs (stereo part of TE) neurons. (*a*) Stimuli: example of flat, single-curved concave and convex texture-defined surfaces; (*b, c*) % correct in discrimination of convex and concave surfaces defined by texture and disparity in agreement or conflict for different % disparity coherences (0% disparity coherence indicates only texture cue, *red*) for eight human subjects (*b*) and one monkey subject (M2, *c*); (*d, e*) average response of 56 TEs neurons, recorded in monkeys M1 and M2 (see here and in **Supplemental Figure 4**. Follow the **Supplemental Material link** from the Annual Reviews home page at **http://www.annualreviews.org**), to disparity-defined surfaces for preferred (*blue*) and nonpreferred (*purple*) curvature as a function of stereo coherence for surfaces defined only by disparity (*d*) and surfaces defined by both disparity and texture (*e*) in agreement (*solid lines*) and in conflict (*dashed lines*). Vertical bars indicate standard errors. Red and dark yellow symbols in panel *e* indicate responses to texture-only preferred and nonpreferred curvature, respectively; arrow in *d* points to the coherence threshold for discrimination (see **Supplemental Figure 4**), arrows in *c* and *e* point to texture cue only conditions. Unpublished data from Y Liu, R Vogels, and GA Orban.

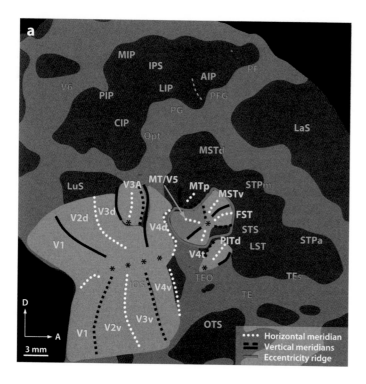

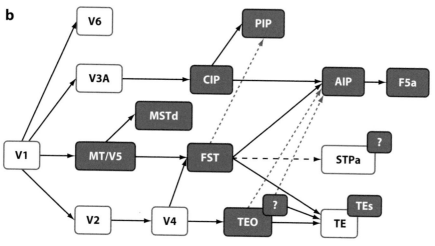

Figure 3

Monkey 3D shape areas: location and proposed network. (*a*) Flat map showing the visual system of the monkey, including indications of retinotopic areas (*yellow*) with borders defined by meridians, as indicated, and other cortical regions (*orange*). Abbreviations: IPS, intraparietal sulcus; LaS, lateral sulcus; LuS, lunate sulcus; OTS, occipito-temporal sulcus; STS, superior temporal sulcus. (*b*) Network of visual areas showing main connections. Blue-shaded boxes indicate areas processing 3D shape; question marks indicate a putative area near TEO extracting disparity gradients and a part of STPa (anterior part of superior temporal polysensory region) that may process speed gradients. Dashed arrows indicate weak or uncertain functional connections. Whereas texture and disparity gradients seem to be extracted by the same neurons in the caudal intraparietal area (CIP), segregated populations may extract depth structure from disparity, texture, and shading in or near TEO.

CIP: caudal intraparietal area

AIP: anterior intraparietal area

MT/V5: middle temporal area or V5

FST: the floor of the temporal sulcus area

IPS: intraparietal sulcus

STS: superior temporal sulcus

TEO and TE: two neighboring architectonic fields, which together constitute the infero-temporal cortex (IT) of monkeys; they likely contain multiple cortical areas

ITG: inferior temporal gyrus

macaque brain, hypothesizing that depth structure is extracted at three points in the visual system. First, depth structure is extracted from disparity and texture in caudal intraparietal area (CIP) (Sakata et al. 2005), an area dealing with the depth structure of both objects and the environment, hence the wide range of surface sizes it handles. The CIP may not use shading as a cue because slanted surfaces are important for the layout of the environment and shading provides no first-order depth information. Depth-structure information related to objects is further relayed to the anterior intraparietal area (AIP), then to F5a, to which the AIP projects (Borra et al. 2008). In this process, quantitative information gains importance for guiding prehension, explaining the dominance of disparity, the more sensitive cue, over texture. Second, depth structure is extracted from motion in two steps involving gradient-selective neurons in the middle temporal area (MT)/V5 and the floor of the temporal sulcus area (FST) (Mysore et al. 2010b). The motion information, which is relatively precise, is projected to the AIP in the intraparietal sulcus (IPS), to the anterior part of the superior temporal polysensory region (STPa) in the superior temporal sulcus (STS), and probably to parts of the TE. Finally, depth structure is extracted from texture and shading in ventral and dorsal TEO. This extraction may not be based on a gradient mechanism, as in the CIP or MT/V5-FST, but on an analysis of the 2D orientation distribution, represented in V4. Indeed Fleming et al. (2004) have shown that the locations and sizes of those distributions' peaks provide information about the minimum slant in texture and the minimum curvature in shading. This also may help explain why shading is apparently processed only in the ventral pathway. The FST and TEO are very close to one another, so perhaps some yet-unidentified region in this intermediate stage of the ventral stream (see ? in **Figure 3**) extracts depth structure from the remaining cue, stereo, using a gradient-based mechanism. This region located near TEO would then send signals into the lower bank of the STS and particularly to TEs, where depth

structure from stereo is represented. It may also correspond to the caudal inferior temporal gyrus (ITG) region in humans, where all three static cues are processed (see below).

Experimental Strategy

How does one address a problem as complex as the extraction of depth structure from the optic array? Single-cell studies and functional imaging are quite complementary because they operate at different spatio-temporal scales. Indeed, single-cell recordings are well suited to demonstrate the analytical power of cortical neurons by documenting their selectivities for stimulus parameters or features. However, this technique samples neuronal activity in a very local and often biased manner. Functional magnetic resonance imaging (fMRI) is generally used to localize perceptual or cognitive functions across the entire brain or large portions of it by testing for significant differences between experimental and control conditions, either in clusters of single voxels or in patterns of activity over many voxels. fMRI is very sensitive and capable of detecting small activity differences between conditions, but it does not reflect neuronal preferences in the sense of parameter selectivity. Indeed, selective neurons within an area will prefer all parameter values, and pooling over many neurons yields an MR signal largely independent of parameter value. By using these techniques in tandem in the monkey, where both techniques are now well-established (Logothetis et al. 1999, Vanduffel et al. 2001), one takes maximum advantage of these complementarities. We are ultimately interested in the human brain, in which direct single-cell recordings are difficult to obtain (Fried et al. 1997). Functional imaging is used routinely in humans. Its findings are difficult to relate to single-cell recordings in monkeys, however, because it amounts to solving an equation with two unknowns: technique and species. By adding functional imaging in the monkey, we reduce the problem to two equations, each with a single unknown (**Figure 4**): technique, comparing single cells and fMRI in the monkey, and species, comparing fMRI in awake humans and monkeys (see sidebar, Parallel Functional Imaging). Thus fMRI in the alert monkey not only complements single-cell studies, but also serves to establish the link with the human work.

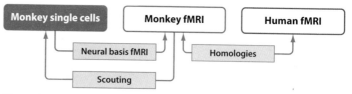

Figure 4

Experimental strategy: the triadic approach integrating monkey and human studies through fMRI in the alert monkey. Note that the link between single-cell activity and monkey fMRI refers to the relationship between magnetic resonance paradigms and neuronal selectivity.

Disparity-Gradient Selective Neurons in the Intraparietal Sulcus

Neurons selective for first-order disparity gradients were first described in the CIP by Shikata et al. (1996), but proof of higher-order selectivity had to wait until Taira et al. (2000) showed that selectivity was maintained despite changes in fixation distance. These neurons were subsequently observed in a delayed match to sample (DMS) task with 2-s intervals between test and sample. Muscimol injections in the CIP impaired performance on this task in 50% of the cases (Tsutsui et al. 2001). It is unclear whether this finding indicates that stereoscopic analysis in the CIP is critical for surface orientation perception or that the CIP is critical for bridging the interval between the test and the sample. Indeed, in human imaging (Cornette et al. 2002), ventral or parietal regions were activated in a same-different task for short (300 ms) or long intervals (6 s), respectively. In support of the role of the CIP in short-term memory, a subsequent study (Tsutsui et al. 2003) showed that CIP neurons remain selective for 3D surface orientation during the 2-s delay between sample and test stimuli.

CIP neurons are also selective for first-order texture gradients (Tsutsui et al. 2002), with matched preferences for texture and disparity (**Figure 5a**). In this study, investigators used large textured surfaces (30° diameter), much larger than stimuli used initially (Taira et al. 2000) to investigate disparity selectivity (6.3° squares). A similar convergence between first-order selectivities for texture- and disparity-defined surfaces was observed by Liu et al. (2004) in TEs. The small stimuli (5° by 5°), having various 2D outlines used to accommodate TEs neurons' selectivity for 2D shape, still portrayed large 3D surfaces because these were strongly slanted. Thus, the texture stimuli used in the CIP and TEs suggest that neurons are selective for texture element size gradients, in which the dimension that varies is orthogonal to the gradient (Todd & Thaler 2010). In

PARALLEL FUNCTIONAL IMAGING

Even with the same stimulus parameters and paradigms in monkey and human functional magnetic resonance imaging (fMRI), a number of differences, other than the species difference, remain.

1. In our studies, macaque fMRI, unlike human fMRI, uses a contrast agent (MION or equivalent) and is more sensitive and specific. Signal signs are opposite and hemodynamic response functions are different, but these are minor issues.
2. Scanning parameters differ, especially voxel size. Typically human and monkey imaging have used 3-mm and 2-mm isotropic voxels, respectively. Sizes have recently been reduced to 2-mm and 1-mm isotropic voxels, respectively, resulting in sampling ratios close to the tenfold ratio between human and monkey cortical surfaces.
3. The number of subjects is smaller (2–4) in monkey versus human studies (12–20), owing to the lesser variability between macaques. Hence, human fMRI more frequently uses random rather than fixed effects with more smoothing.
4. Monkey and human subjects are both highly motivated but by different means: control of water supply and monetary rewards, respectively.
5. To control for possible differences in attentiveness during passive viewing, subjects perform a high-acuity task requiring them to pay acute attention to a central stimulus. This control has invariably confirmed the results obtained with passive viewing.

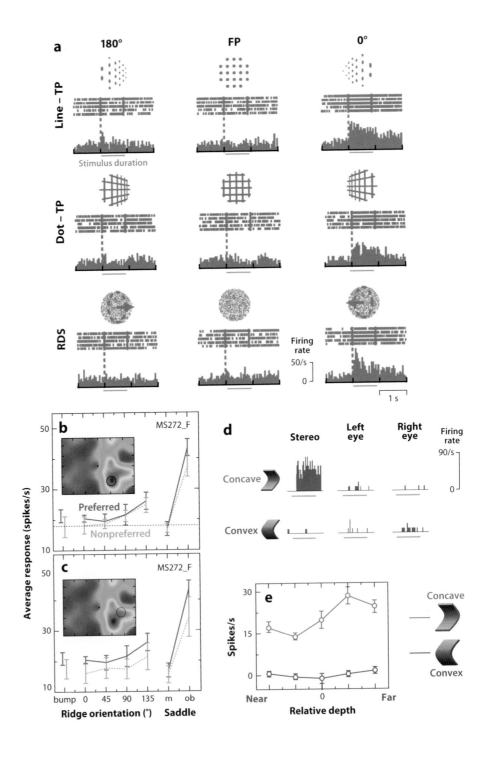

a

180°   FP   0°

Line – TP

Stimulus duration

Dot – TP

RDS

Firing
rate
50/s
0

1 s

b   MS272_F

50

Preferred

Nonpreferred

c   MS272_F

50

30

Average response (spikes/s)

10

bump 0 45 90 135   m ob
Ridge orientation (°)   Saddle

d   Stereo   Left
eye   Right
eye   Firing
rate
90/s
0

Concave

Convex

e

30

Spikes/s

15

0

Near   0   Far
Relative depth

Concave

Convex

contrast, gradients used in the 3D shape perception of textured potatoes are supposedly based on gradients of the texture element dimension parallel to the gradient direction (Todd & Thaler 2010). Finally, CIP neurons are also selective for perspective gradients created by deforming the stimulus outline (Tsutsui et al. 2001), and the optimal tilt for perspective matches that for disparity. Thus the selectivity of CIP neurons for first-order gradients is well established. Katsuyama et al. (2010) recently reported that CIP neurons in one animal were selective for second-order stereo surfaces defined by the shape index.

The Sakata group has proposed that disparity-gradient selectivity arises in the CIP and is then transmitted to the AIP (Nakamura et al. 2001, Sakata et al. 2005, Tsutsui et al. 2005). Second-order disparity selectivity for vertically curved surfaces has indeed been observed in the AIP (Srivastava et al. 2009). Compared with their TEs counterparts, AIP second-order disparity-selective neurons respond much earlier, their selectivity for variations in disparity is coarser, and it emphasizes the metric changes more than sign reversals in curvature. Neurons selective for first-order disparity gradients have also been reported in the AIP (Srivastava et al. 2009). The short response latencies of gradient-selective neurons in the AIP, much shorter than their ventral counterparts, certainly support the scheme proposed by the Sakata group. Second-order

disparity-selective neurons have also been observed in ventral premotor cortex area F5a (Theys et al. 2009), which receives direct input from the AIP (Borra et al. 2008).

The involvement of two cortical regions, the CIP and the AIP, located in the lateral bank of the IPS, in the extraction and processing of depth structure from stereo has been amply confirmed by fMRI studies. Two types of paradigms have been used. One, inspired by the perceptual function, contrasts experimental conditions in which stimuli appeared as 3D, with control conditions in which only 2D shape was perceived. Other paradigms start with the properties of known gradient-selective neurons and attempt to devise a contrast that selectively activates regions housing such neurons.

The extraction of 3D shape from disparity was investigated with a random line (RL) perceptual paradigm (**Supplemental Figure 2**) in which three conditions were compared: RLs slanted in depth depicting a 3D shape (3D shape), RLs in multiple fronto-parallel planes (zero-order depth), and RLs in the fixation plane (no depth). Although regions outside the IPS, e.g., in the caudal part of the STS, posterior to MT/V5, were activated in the contrast 3D shape minus zero-order depth, we concentrate here on the three activation sites in the IPS: one in the CIP, another in the anterior lateral intraparietal area (LIP) and the AIP, and a third one in the posterior intraparietal area (PIP) and possibly the medial intraparietal area

RL: random lines

LIP: lateral intraparietal area

PIP: posterior intraparietal area

Figure 5

Gradient-selective neurons: a first-order disparity- and texture-gradient selective caudal intraparietal area (CIP) neuron (*a*); a second-order speed gradient selective floor of the temporal sulcus (FST) area neuron (*b,c*) and a second-order disparity-gradient selective TEs neuron (*d,e*). (*a*) Responses of a CIP neuron to three orientations of dot texture (dot-TP), line texture (line-TP), and random dot stereogram (RDS) surfaces; (*b,c*) responses of an FST neuron to speed-defined quadratic surfaces for preferred (*solid dark yellow line*) and nonpreferred (*dotted orange line*) mean speed and two positions in the RF (*black circles in insets showing the RF map*); (*d*) responses of a TEs neuron to singly curved disparity-defined concave and convex surfaces for binocular and monocular presentations; (*e*) response for the same neuron to concave and convex surfaces for different positions in depth of the stimuli. In *a*, horizontal lines indicate stimulus presentation, numbers indicate tilt, FP stands for fronto-parallel, dashed lines and arrows in the RDS indicate the surface orientation and the surface normal, and vertical calibration bar indicates firing rate; in *d* vertical line indicates 90 spikes/s; in *a, d, e* horizontal lines indicate stimulus duration. Data from Tsutsui et al. (2002), Mysore et al. (2010a), and Janssen et al. (2000a).

(MIP) (Durand et al. 2007). Activity in both regions in the lateral bank, and to a lesser degree in those in the medial bank, was reduced by a scrambling of object images (Kourtzi & Kanwisher 2000), indicating that these regions also process 2D shape.

The second MR paradigm was inspired by the properties of TEs neurons (neuronal paradigm), which are disparity selective, responding better to binocular than monocular stimuli (Janssen et al. 2000a), and selective for second (first)-order disparity gradients, responding better to curved (or slanted) than flat stereo surfaces (Janssen et al. 2000b). Therefore, a factorial design was used with stereo (present or absent) and nature of surfaces (curved or slanted versus flat) as factors, and the interactions between these factors should correspond to regions housing second (first)-order gradient-selective neurons. The stimuli, identical to those of the single-cell studies (**Supplemental Figure 1**), were presented in the stereo conditions at different positions in depth to emulate the position in depth invariance of the higher-order TEs neurons. Finally, the distributions of disparities were matched as well as possible in the curved (slanted) and flat conditions. For curved surfaces, interaction between the stereo and surface-type factors, indicating stronger response, relative to monocular controls, to curved than to flat stereo surfaces, reached significance in the anterior IPS (in the AIP and anterior LIP) and F5a (Durand et al. 2007, Joly et al. 2009). For slanted surfaces, the interactions occurred mainly in the AIP, consistent with the observation (Srivastava et al. 2009) that first-order disparity-selective neurons were more numerous in the AIP than in the TEs. The absence of interactions in the CIP was ascribed to the small stimulus size better suited to the TEs than to CIP neurons. In subsequent studies using a more sensitive MR technique (3T instead of 1.5T and several coils in parallel), interactions were also observed in the CIP (Van Dromme et al. 2010).

The importance of stimulus size for driving CIP was demonstrated in one animal in which linear texture gradients were tested (O. Joly, J. Todd, W. Vanduffel, and G. Orban, unpublished observations), using a factorial design, with surface type (slanted versus front-parallel) and retinal size (small versus large) as factors. The interaction of size and type yielded an activation site in the CIP, consistent with the large stimuli used by Tsutsui et al. (2002). The involvement of CIP in 3D SFT was demonstrated by Nelissen et al. (2009), using large, second-order textured stimuli (see below). So far, little evidence for processing of depth structure from shading has been obtained in the IPS, although the AIP was activated in one animal tested by Nelissen et al. (2009).

Speed-Gradient Selective Neurons in the Posterior Superior Temporal Sulcus

Selectivity for first-order speed gradients was demonstrated in MT/V5 neurons by Xiao et al. (1997a). The speed invariance of the MT/V5 gradient selectivity was demonstrated in alert animals by Nguyenkim & DeAngelis (2004) and Mysore et al. (2010b). These authors compared MT/V5 and the FST in the same animals and observed more first-order selective neurons in FST than in MT/V5. Nguyenkim & DeAngelis (2003) reported MT/V5 neurons selective for surface orientation defined by disparity, but such selectivity was relatively modest, compared with that in the CIP or TEs. In a subsequent study in which surface orientations defined by speed, disparity, or texture gradients were compared, MT/V5 neurons were more selective for surface orientation defined by speed than by disparity gradients, texture being rather ineffective (Nguyenkim & DeAngelis 2004). A number of studies have examined the responses of MT/V5 neurons while monkeys made judgments about the (ambiguous) direction of motion of a rotating transparent cylinder (Bradley et al. 1998, Dodd et al. 2001). This task reveals mechanisms related to the monkey's perception of order in depth, i.e., zero-order depth, but requires no judgment about 3D shape (Born & Bradley 2005).

Second-order selectivity based on motion was investigated in the FST using a representative set of quadratic surfaces, including a bump, ridges, and saddles, corresponding to different values of the shape index (**Supplemental Videos 1–2**) and several orientations of these ridges and saddles (Mysore et al. 2010b). Sixty percent of the responsive FST neurons were selective for second-order SFM. **Figure 5b** illustrates a neuron selective for a saddle. This selectivity was invariant for changes in speed and position (**Figure 5b,c**), demonstrating the higher-order nature of the selectivity. Additional invariance was observed for changes in stimulus size, monocular or binocular presentation, and the nature of the motion cue (speed gradient, direction reversal, or both). This second-order selectivity was already present in 45% of the MT/V5 neurons, which represent a processing stage earlier than FST neurons (Ungerleider & Desimone 1986). Compared with their MT/V5 counterparts, FST neurons were more selective, and selective in a more invariant manner, for a wider range of surfaces and throughout the stimulation interval (Mysore et al. 2010b). Hence, these authors proposed that the extraction of speed gradients involves two stages: At the level of MT/V5, a surround-based mechanism extracts linear gradients because (*a*) the antagonistic surrounds are asymmetric and restricted to a single suppressive zone in 50% of the surround neurons (Xiao et al. 1997b) and (*b*) the speed sensitivity of this surround zone is relative (Xiao et al. 1997a), in agreement with computational models (Buracas & Albright 1996). A similar mechanism but one with a double-symmetric suppressive zone in the surround (Xiao et al. 1995) may extract second-order gradients. This idea is consistent with the weak position invariance of the selectivity, with the relationship between preferred orientations for first- and second-order selectivity and with the restriction of selectivity to tilted planes and ridges in MT/V5 (Mysore et al. 2010b). The convergence of several MT/V5 neurons having offset centers, and possibly selective for orthogonal ridges, combined with inputs from

V4 selective for kinetic contours created by opposing directions of motion (Mysore et al. 2006), could explain the full range of properties of FST speed-gradient selective neurons.

MT/V5 and FST send the speed gradient signals further along the STS, most notably in MSTd. Duffy & Wurtz (1997) reported that MSTd neurons were sensitive to the inversion of the speed gradient present in expansion patterns. MSTd neurons selective for first-order speed gradients superimposed onto optic flow components were reported by Sugihara et al. (2002), indicating that motion patterns other than translation can carry speed gradients. It remains to be seen whether selectivity for 3D SFM, reported for STP neurons tested with coherent and incoherent rotating spheres (Andersen & Siegel 2005), survives more sophisticated testing with a wider range of speeds, positions, and 3D shapes.

Several fMRI experiments have confirmed the role of MT/V5 and the FST in 3D SFM. In one experiment, two conditions with nine connected RLs undergoing either rotation around a vertical axis or translation at uniform speed were contrasted in a perceptual paradigm (**Supplemental Videos 3** and **4**). Indeed, the first display evokes the perception of a 3D wireframe figure rotating in depth, and the second, that of a flat figure moving in the fronto-parallel plane. This contrast yielded a significant activation of MT/V5 and the FST (Vanduffel et al. 2002a). The activation was marginally higher in the FST than in MT/V5 (Nelissen et al. 2006, figure 10*c*), which is consistent with the larger proportion of gradient-selective neurons in FST. Mysore et al. (2010b) compared FST neuronal responses to rotating and translating lines identical to those used in these MR studies. FST neurons responded significantly more strongly to rotating lines than to translating lines, and the percent difference was similar to that of MR signals in the FST documented by Nelissen et al. (2006). Three-dimensional shape and 2D motion control conditions were also compared by Sereno et al. (2002), who observed MR activation of MT/V5 and the FST, consistent with the neuronal properties in these areas, but also saw a widespread activation in anterior STS and in the posterior two-thirds of the IPS not observed by Vanduffel et al. (2002a). There are many differences between the two studies: Sereno et al. (2002) used textured surfaces, not randomly connected lines, an anaesthetized preparation rather than awake monkeys, and lower statistical thresholds. Investigators recently confirmed the involvement of MT/V5 and the FST in a third MR study contrasting second- and zero-order speed distributions (Mysore et al. 2010a) using the surface stimuli of Mysore et al. (2010b).

The latter fMRI study also provided evidence for further processing in the lower bank of the STS and the convexity of the TE. The initial study (Vanduffel et al. 2002a) reported that activation by 3D SFM did not reach significance in the IPS, but a later study using a somewhat different group of monkeys revealed weak but significant activations in the AIP and the PIP (Durand et al. 2007). One animal with a clear SFM activation in the AIP showed a weak activation in the stereo experiments, but it is too early to draw conclusions from this observation.

Extraction of Depth Structure in the Ventral Stream

Second-order disparity-gradient selectivity has been documented in the ventral stream by Janssen et al. (1999), who tested several second-order vertical configurations of disparity variation (some corresponding to a horizontal ridge; see **Supplemental Figure 1**), which are more easily produced (avoiding spurious texture density changes). They verified, however, that neurons can be selective for vertical or horizontal ridges or both (Janssen et al. 2001). This disparity selectivity was initially discovered in a small region in the rostral part of the lower bank of the STS, labeled TEs (Janssen et al. 2000a). Higher-order selectivity was demonstrated by showing invariance of selectivity for changes in position in depth (**Figure 5d,e**) (Janssen et al. 1999, 2000a). TEs neurons (see Orban et al. 2006b for review) give a fine description of the

disparity variations, emphasize the change in sign from convex to concave, and are not selective for anticorrelated stereograms (Janssen et al. 2000b, 2003). Neurons selective for first-order disparity gradients have also been reported in the TEs (Janssen et al. 2000b, Liu et al. 2004). It remains unclear whether the TEs and the region behind it, where Yamane et al. (2008) reported selectivity for combinations of depth curvatures (see Kourtzi & Connor 2011, this volume), constitute a single area or two. AIP gradient-selective neurons have much shorter latencies than do their TEs counterparts; hence, we cannot deny that TEs neurons inherit their selectivity via AIP input (Borra et al. 2008). It is more likely that disparity gradients are extracted earlier in the ventral pathway, but it is unclear where. Little selectivity for disparity gradients has been observed in V4 for surface stimuli (Hegde & Van Essen 2005). But the TEO, which has not yet been explored with single-unit recordings, remains a possibility, as does the region described by Yamane et al. (2008).

Selectivity for second-order vertical variations in texture has been tested in the TEs (Liu et al. 2002). Because the 3D texture stimuli also have a 2D interpretation, the standard strategy has been to test the convergence of disparity and texture signals onto single neurons that have similar preferences for the two cues. Horizontal ridges were portrayed by a combination of disparity and texture (**Figure 2**), either in agreement or in conflict, and the strength of the disparity was manipulated by reducing stereo coherence. When stereo was strong, selectivity depended entirely on this cue. At 20% stereo coherence—where the monkey can still use the stereo cue to distinguish convex from concave (**Supplemental Figure 4**)—selectivity was significantly lower when cues conflicted than when they agreed. At zero coherence, i.e., with only the texture cue, the selectivity for curvature sign was weak but matched that for disparity (**Figure 2e**). Thus the disparity cue was stronger than the texture cue for second-order gradients in the TEs. However, convergence between first-order selectivities for texture- and disparity-defined surfaces was observed in the TEs by Liu et al. (2004).

The selectivity for shading has been investigated using randomly deformed spheres (potatoes) in V4 (Arcizet et al. 2009) and in the infero-temporal (IT) cortex (Vangeneugden et al. 2006). Selectivity for 3D shape was tested for four directions of illumination and compared with that for 2D shape controls. These studies were relatively disappointing because selectivity depended on illumination direction, and the small response differences between 3D and 2D shape barely reached significance in the TEO and were not significant in V4 or the TE.

In contrast with the results obtained in the IPS and the STS, the fMRI results did not match the single-cell data particularly well in the ventral stream. For the pictorial cues, Nelissen et al. (2009) used the perceptual strategy, again comparing conditions in which 3D shape, or only 2D shape, was perceived. Because a single control condition could not control for all low-level 2D aspects present in the 3D shape conditions, they followed the human experiment (Georgieva et al. 2008), using a battery of control conditions that collectively included all low-level properties of the 3D displays and required activation in the conjunction of subtractions comparing the 3D condition to each of the control conditions. The 3D SFT activation in the ventral stream was restricted to the ventral TEO (Nelissen et al. 2009). No activation of the TEs was observed, reflecting either the low number of first-order selective neurons in the TEs or technical limitations, because the signals in the STS were relatively weak in the 1.5T. These results only partially replicate those of Sereno et al. (2002), who reported widespread activation in the occipito-temporal cortex, including in MT/V5, but not in the TEO. Activation by 3D SFS was also restricted (Nelissen et al. 2009) and limited to the dorsal TEO and the adjacent portion of V4. These very restricted MR activations may explain the largely negative single-cell results obtained so far for shading.

Interactions between the stereo and surface type factors, indicating stronger activation, relative to monocular controls, for curved rather than flat surfaces, yielded activation sites in the lower STS bank in a region near the TEs (Joly et al. 2009). Such interactions were not observed for slanted surfaces. A subsequent study at 3T (Van Dromme et al. 2010) has revealed more extensive activation of the lower bank of the STS, yet posterior to the TEs, as well as an activation near the TEO.

Overview of Monkey Studies

Gradient-selective neurons form a family of higher-order neurons that occur in at least seven predominantly middle- or high-level visual areas: MT/V5, FST, MSTd, CIP, AIP, TEs, and F5a (**Table 1**). In most of these areas, zero-, first-, and second-order selective neurons co-occur. For example, MT/V5 neurons are selective for depth specified by motion parallax (Nadler et al. 2008, 2009), which can be considered zero-order selectivity for motion, more precisely depth intervals, complementing the selectivity for qualitative order in depth documented with the rotating cylinder (Bradley et al. 1998).

In general, a good match can be found between the MR activation sites obtained

Table 1   Areas housing gradient selective neurons[a]

Order/cue	Motion	Stereo	Texture
First order	MT/V5, MSTd, FST	CIP, TEs, AIP, MT/V5?	CIP, TEs
Second order	MT/V5, FST, STP?	TEs, AIP, F5a, CIP?	TEs?

[a]Abbreviations: AIP, anterior intraparietal area; CIP, caudal intraparietal area; FST, floor of the temporal sulcus area; MSTd, dorsal part of the medial temporal sulcus area; MT/V5, middle temporal area or V5; STP, superior temporal polysensory region; TEs, stereo part of TE.

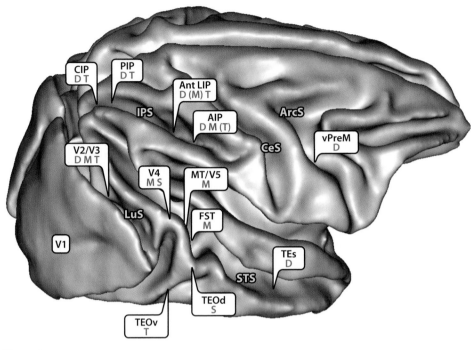

Figure 6

Overview of monkey functional magnetic resonance imaging (fMRI): activation sites for the four cues indicated by capital letters: D, disparity (*red text*); M, motion (*green text*); T, texture (*blue text*); and S, shading (*dark yellow text*). Abbreviations: ArcS, arcuate sulcus; CeS, central sulcus; IPS, intraparietal sulcus; LuS, lunate sulcus; STS, superior temporal sulcus. Note that the texture and motion cue activations were weak in PIP.

(**Figure 6**) and the cortical areas that house gradient-selective neurons (**Table 1**), especially if one notes that the activations in early cortex (V2 and V3) occur at the edges of the stimulus representation and may reflect depth discontinuities or attentional effects. Such effects occur in V1–V3 of both monkeys and humans, typically at the edge of stimuli (Kolster et al. 2010, Saygin & Sereno 2008). Beyond V2–V3, fMRI produces, if anything, more widespread activation than expected from the locations of gradient-selective neurons, e.g., in V4 for 3D SFM and V3 for 3D SFD. The V3 activation could either be an edge effect or represent sensitivity for relative disparity, i.e., a nonspecific effect owing to a lower-order confound. More recent experiments in the 3T using steps in zero-order disparity (**Figure 1a**) support the latter view (Van Dromme et al. 2010). The interpretation of the dorsal V4 activation is

unclear because no relevant single-cell data are available. It may represent a region unknown to house gradient-selective neurons, as may be the case for the PIP or the anterior LIP in 3D SFM or 3D SFD, respectively. Further studies combining fMRI and single-cell recording in the same animal are needed to clarify these issues.

HUMAN FUNCTIONAL IMAGING STUDIES OF 3D SHAPE PROCESSING

Parallel Human and Monkey fMRI Studies

The ramifications of parallel human and monkey fMRI studies extend beyond the comparison of functional neuroanatomy into the homologies between cortical areas in the two species, which remain largely uncharted. In

Cortical area: a functional cortical entity defined by four criteria: archi- and myelo-architectonics, anatomical connections, topographic organization, and functional properties

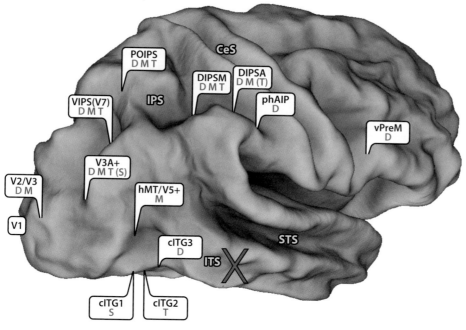

Figure 7

Overview of human functional magnetic resonance imaging (fMRI): activation sites for the four cues indicated by capital letters: D, disparity (*red text*); M, motion (*green text*); T, texture (*blue text*); and S, shading (*dark yellow text*). Abbreviations: see list in **Figure 6**; ITS, inferior temporal sulcus; X, lack of magnetic resonance signals. VIPS may correspond to several retinotopic areas including V7.

addition, combining these two approaches with single-cell recordings (**Figure 4**) allows one to extrapolate from single-cell results to humans by devising MR paradigms that activate the monkey areas in which a given selectivity occurs, porting those paradigms to human imaging, and obtaining activations in homologous areas. This method circumvents certain problems plaguing human fMRI when addressing neuronal selectivity because the underlying assumptions were either questionable, as for repetition suppression (Sawamura et al. 2006), or inapplicable to certain areas, as with multivoxel analysis of orientation selectivity in V4 (Vanduffel et al. 2002b). Hence, this section starts with the human fMRI experiments using exactly the same stimuli and paradigms as those used in the monkey (Durand et al. 2009; Georgieva et al. 2008, 2009; Orban et al. 1999; Vanduffel et al. 2002a). These studies included many control experiments [up to

eight additional experiments in Georgieva et al. (2008)], and the reader is referred to the original publications for descriptions of these controls (see also **Figure 8**).

A comparison of **Figures 6** and **7**, summarizing the activation sites obtained with the four 3D shape cues in the two species, reveals several striking similarities. The agreement between monkey and human activation patterns is even more remarkable if one considers homologies. There is clear support for the homology of the DIPSM (dorsal IPS medial region) and the DIPSA (dorsal IPS anterior region) with the monkey anterior LIP and posterior AIP, respectively (Durand et al. 2009), and of the phAIP (putative human homolog of AIP) with the anterior AIP (Georgieva et al. 2009). The following homologies are only putative: the VIPS (ventral IPS region) and the POIPS (parieto-occipito IPS region) with the CIP and the PIP, respectively (Durand et al. 2009), and

Activation pattern: distribution of differential MR activity between experimental and control conditions over the brain or cortical surface

Homology: cross-species similarities of a cortical area because it derives from that of a common ancestor, here of primates

hMT/V5+: the human middle temporal/V5 complex corresponding to monkey MT/V5 and a number of its satellites; for human MT/V5, see Kolster et al. (2010)

the caudal ITG with the TEO (Nelissen et al. 2009). Finally, firm evidence now indicates that hMT/V5+ includes homologues of MT/V5 and the FST (Kolster et al. 2009, 2010). The first similarity between the two species, shown in **Figures 6** and **7**, is the absence of activation in the striate cortex. Second, the extraction of 3D shape involves both the dorsal and the ventral (Ungerleider & Mishkin 1982) visual pathways. More precisely, three cues, disparity, motion, and texture, are processed in the two pathways, whereas shading is processed predominantly in the ventral stream. Third, in the dorsal pathway, activation is observed at even the most anterior level, the AIP in the monkey and the phAIP in humans, whereas this was not the case in the ventral stream. Ventral activation sites occur at the middle level, with the exception of the site identified as the TEs, the human counterpart of which is not yet known. Caution is needed regarding this difference between dorsal and ventral streams because susceptibility artifacts may limit visualization in the temporal regions of human cortex (X in **Figure 7**) (Georgieva et al. 2009). Fourth, in the parietal cortex, multiple cues are processed within the same areas. On the other hand, in the occipito-temporal cortex, including in MT/V5, cues are processed largely in segregated areas: in MT/V5 and the FST for motion and in the dorsal and ventral TEO and corresponding human sites in the caudal ITG for shading and texture. The main species difference lies in the activation of the human V3A complex, but it is likely that its stereo activation reflects relative depth rather than depth structure (Preston et al. 2008, Tsao et al. 2003b).

Other Human fMRI Studies

The human fMRI results, summarized in **Figure 7**, are in general agreement with those obtained in other laboratories with different stimuli. Several studies (Beer et al. 2009, Klaver et al. 2008, Martinez-Trujillo et al. 2005, Murray et al. 2003, Paradis et al. 2000, Yamamoto et al. 2008) have confirmed our 3D SFM observations, with the exception that surface stimuli activate more ventral areas than do RLs (Beer et al. 2009, Kriegeskorte et al. 2003, Orban et al. 2006a). Beer et al. (2009) also observed a 3D SFM activation in the human STS. The timing of the activation in the various regions has been revealed by magnetoencephalography (MEG) experiments (Iwaki et al. 2007, Jiang et al. 2008). To control for motion in 3D space present in the rotating RL stimuli, we performed an additional experiment whereby subjects attended to either the 3D shape or the 3D motion, thereby confirming the involvement of hMT/V5+ in the processing of 3D shape (Peuskens et al. 2004).

The 3D SFD experiments agree with the recent studies of Preston et al. (2008, 2009). Georgieva et al. (2009) have correlated MR activity with the strength of the 3D shape percept, itself derived from the regression provided by the 3D shape adjustment task. Significant correlation was restricted to the V3A complex and the adjacent area V7/VIPS. Similar regions were implicated in 3D shape estimation experiments by Preston et al. (2009) using after-effects produced in ambiguous stimuli by binocular and monocular depth cues. Preston et al. (2008) investigated only far and near distinctions but found that dorsal parietal areas represented the disparity structure metrically while the lateral occipital complex (LOC) did so categorically, a finding reminiscent of the difference between higher-order disparity-selective neurons in the AIP and the TEs (Srivastava et al. 2009). Other studies in which subjects made judgments about the stereo-defined 3D shape have emphasized the role of hMT/V5+ and the LOC in such tasks (Chandrasekaran et al. 2007, Welchman et al. 2005), but these regions were defined by localizer tests and included multiple cortical areas (Kolster et al. 2010), making difficult any links with the voxel-based approach of Georgieva et al. (2009). One of the studies (Chandrasekaran et al. 2007) observed the parietal activations described by Georgieva et al. (2009) but found that their activity decreased with accuracy of judgments.

Several studies (Gerardin et al. 2010; Humphrey et al. 1997; Konen & Kastner 2008;

Kourtzi et al. 2003; Moore & Engel 2001; Shikata et al. 2001, 2008; Taira et al. 2001) have attempted to locate regions involved in extracting 3D shape from texture and shading but have reported activation sites only partially matching our results. The two main inconsistencies concern the involvement of the parietal cortex in 3D SFS and the parts of the LOC involved in this process. The discrepancies probably reflect the many differences between these experiments and ours: differences in stimuli or control conditions or in task, exploration restricted to posterior visual regions or to regions defined by localizer tests rather than whole brain, or the use of repetitive suppression, which can yield questionable results (Sawamura et al. 2006).

3D Shape Processing in Human and Monkey

The similarities of activation patterns in humans and monkeys tested with identical paradigms (**Figures 6** and **7**), and the homologies outlined above, suggest that the 3D shape network in humans is very similar to that in monkeys. The major exception is related to the capacity of human, but not monkey, V3A to process motion information (Tootell et al. 1997, Vanduffel et al. 2001). This capacity endows the VIPS, the putative homolog of the CIP, with the additional ability to extract depth structure from motion (Orban et al. 2006a), explaining the stronger activation of human compared with monkey parietal cortex by 3D SFM (Vanduffel et al. 2002a).

Similarities in activation patterns for 3D shape processing in the two species also suggest that the human areas activated by 3D shape house gradient-selective neurons similar to those recorded in monkeys. As noted above, the MR paradigms may err on the nonspecific side and thus yield false positives. Therefore, one should derive such conclusions only for areas whose homology is at least putative. For example, the DIPSA likely houses higher-order disparity-selective neurons. Because the DIPSA may not be a single cortical area, there is room for alternative interpretations. On the other hand, considerable evidence suggests that recently identified human MT/V5 (Kolster et al. 2010) and monkey MT/V5 (Kolster et al. 2009) are homologous. Hence, this human cortical area likely houses speed-gradient selective neurons (**Figure 8**).

TWO TYPES OF 3D SHAPE EXTRACTION

Even 3D surfaces have boundaries, and Janssen et al. (2001) have shown that TEs neurons can independently process the depth structures of the boundaries and of the surface itself. This dissociation is reminiscent of the distinction between axis and plate neurons in the CIP (Sakata et al. 2005). Computationally, the extraction of depth curvature of edges and surfaces requires different operations (Li & Zucker 2006a,b), which suggests the involvement of different areas. Far less is known about the areas involved in processing the depth structure of boundaries. It has recently been observed that boundary stimuli are more efficient at driving higher-order disparity-selective neurons in the AIP (Srivastava et al. 2008), whereas in the TEs, boundaries and surfaces were equally effective (Janssen et al. 2001). Furthermore, investigators have reported that V4 neurons are selective for the 3D orientations of line stimuli (Hinkle & Connor 2002), which is not the case for surface stimuli (Hegde & Van Essen 2005). Also, line stimuli are more effective for 3D SFM in parietal cortex, and conversely, surface stimuli are more effective in occipito-temporal regions (Orban et al. 2006a, figure 2). Hence, further work is needed to clarify the networks extracting and processing these two types of 3D shape information.

BEYOND THE EXTRACTION OF 3D SHAPE

So far I have considered the extraction of depth structure from the four cues separately, although single-cell studies using monocular static cues have combined disparity with texture to circumvent ambiguity present in pictorial-cue stimuli. The MR activation patterns

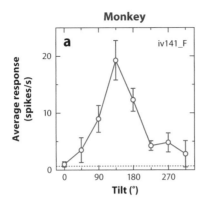

Monkey

a iv141_F

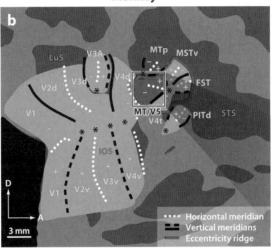

Monkey

b

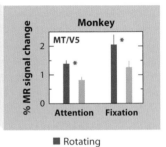

Monkey

MT/V5

Attention   Fixation

■ Rotating
■ Translating

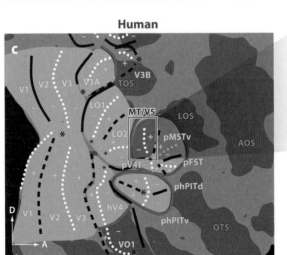

Human

c

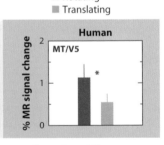

Human

MT/V5

* Significant differences

•••• Horizontal meridian
▬▬ Vertical meridians
▬▬ Eccentricity ridge

(**Figures 6** and **7**) suggest that the different cues are processed in the same parietal areas, which does not necessarily guarantee convergence at the single-cell level. Up to now, neuronal convergence has been observed only in the CIP (Tsutsui et al. 2002), although it likely occurs in the AIP also. Cortical areas in which multiple depth structure cues converge consistently at the neuronal level are likely to be final representations where depth structure can be used for behavioral purposes. The representation in the CIP probably lays out the 3D structure of the environment, part of the metric description of space, necessary for visual control of action in space. Consider a winding mountain path along a cliff. The representation in the AIP lays out the 3D shape of objects for visual control of grasping (Sakata et al. 2005). Grasping a disc requires a different grip than does grasping an orange.

Cue convergence has also been documented in the ventral stream. First-order gradients defined by disparity and texture converge consistently onto TEs neurons (Liu et al. 2004). Yamane et al. (2008) reported that disparity, shading, and texture converged onto neurons located mainly in the lower bank of the STS, stretching broadly speaking from the TEO into the TEs. Because responses were sought using complex curved surfaces defined by all three cues, the demonstration that selectivity for surface fragments was retained for isolated cues leaves open the question of cue consistency in

this region. I propose that cues converge not only in the lower bank of the STS, as documented by Yamane et al. (2008), but also in overlapping and neighboring regions involved in the extraction of the 3D shape of the static (Tsao et al. 2003a) and the moving human body (Perrett et al. 1985, Singer and Sheinberg 2010) and face, as discussed by Mysore et al. (2010b). Indeed, perceptual judgments about actions improve in 3D conditions (Jackson & Blake 2010).

Haptics also provides information about the 3D shape of objects. In humans, 3D shape from tactile information (Amedi et al. 2002) activates regions close to the caudal ITG site activated by 3D shape-from-disparity and monocular cues (Georgieva et al. 2009). Thus, somatosensory information and visual information may converge relatively early in the ventral stream. Developmentally, haptic 3D shape may be important for calibrating visual 3D shape, which in turn may guide the development of 2D shape selectivity. This developmental sequence may reconcile cases of developmental prosopagnosia (Laeng & Caviness 2001) with the predominantly 2D selectivity of face-selective neurons (Freiwald et al. 2009). At the level of V4, orientation information provided by both vision and touch converges (Maunsell et al. 1991). Hence, similar analyses of orientation distributions in both modalities may give rise to a multimodal depth structure representation in the neighborhood of the TEO. Thus neurons selective for first- and second-order depth in or near the

---

Figure 8

Integration of human and monkey 3D shape studies: the example of area MT/V5. (*a*) First-order speed-gradient selective neuron recorded in monkey MT/V5; (*b*) flat map showing the location of MT/V5 in the monkey; (*c*) flat map with location of MT/V5 in the human. Insets in *b* and *c* plot the magnetic resonance (MR) signal change (relative to static control) for rotating (*dark blue*) and translating (*light blue*) random lines (RL) in three monkeys (both when fixating and when attention was distracted by a high-acuity bar task) and six humans (when fixating; vertical bars indicate standard errors (SE). Asterisks indicate significant differences between conditions. Considerable evidence indicates that human and monkey MT/V5 are homologous (Kolster et al. 2010). Both show similar MR response differences to rotating and translating RL. In the monkey, this difference corresponds with the presence of speed-gradient selective neurons, hence human MT/V5 in all likelihood also houses speed-gradient selective neurons. Data from Vanduffel et al. (2002a), Kolster et al. (2009, 2010) and Mysore et al. (2010b). Note that the high-acuity bar task (*b*) was one of the typical control experiments in the imaging studies.

middle ventral pathway may provide the building blocks for describing the depth structures of objects or object parts, just as orientation- and curvature-selective neurons in V4 (Patsupathy & Connor 2002) do for such descriptions in 2D. The same may hold true for building visual representations of actions (Jackson & Blake 2010) and gestures, at least those based on temporal derivatives of shape (Giese & Poggio 2003, Singer & Sheinberg 2010). The importance of the 3D descriptions of actions and objects may decrease with development to the point that depth structure may not be required for adults to recognize objects (Pizlo et al. 2010), although further studies using meaningful images of objects and actions to test 3D selectivity are warranted. At present, regions such as the TEs or that described by Yamane et al. (2008) seem to play a supplemental role in object representation in IT during adulthood, very much as color does (Matsumora et al. 2008). The representations can be called on when details about 3D shape are needed to better identify visual objects or to better describe their physiological or emotional state.

Finally, most of the experiments reviewed here were performed in passively fixating subjects, although subjects in some human fMRI experiments performed a task. As noted by Shikata et al. (2008), using a task implies that task differences unrelated to stimulus depth structure may determine the MR activation pattern. This ambiguity may also apply to the studies of Tsutsui et al. (2001, 2002) in which monkeys performed a DMS task (see above). Given task-dependent processing in sensory systems (Colombo et al. 1990, Orban & Vogels 1998), all studies using tasks will face similar problems, except where the task is unrelated to the stimuli studied (**Figure 8b**). Hence, it is difficult to interpret results obtained in subjects performing 3D shape tasks without preliminary information provided by passive experiments. The present review thus paves the way for experiments using tasks (Tsutsui et al. 2001, Verhoef et al. 2010) to demonstrate causal links between cortical areas and behavior. Even in these experiments, results will depend on the properties of the stimuli (depth structure) and the task components.

FUTURE ISSUES

1. Further work is needed on the mechanisms extracting depth structure from texture and shading, as well as the processes for extracting depth structure from boundaries and surfaces.

2. Once these mechanisms have been fully identified, their corresponding wiring diagrams will have to be worked out.

3. How does 3D shape information derived from different visual cues converge? To what extent is this process generalized to other senses such as the somato-sensory system?

4. How is 3D shape information used in different tasks, whether related to guidance of motor behavior or for perceptual purposes?

5. How is the human 3D shape network affected by brain damage from trauma or disease? How does this compare to experimental lesions or inactivation studies?

DISCLOSURE STATEMENT

The author is not aware of any affiliations, memberships, funding, or financial holdings that might be perceived as affecting the objectivity of this review.

ACKNOWLEDGMENTS

The author thanks J. Todd for the excellent collaboration spanning 15 years and for comments on the manuscript. The author is indebted to S. Mysore for help with the figures and references and to H. Kolster and O. Joly for designing figures. He thanks S. Raiguel, P. Janssen, W. Vanduffel, and R. Vogels for comments on earlier versions of the manuscript. The author has enjoyed the multiple collaborations underlying the work reported in this article with his co-PIs, P. Janssen, W. Vanduffel, and R. Vogels, and with all past and present members and associates of the Laboratorium voor Neuro-en Pyschofysiologie. Supported by Geconcerteerde Onderzoeks Acties (GOA), Fonds voor Wetenschappelijk Onderzoek (FWO), Inter-University Attraction poles (IAP), Excellentie Financiering (EF), and European Union (EU) projects from Information Society Technologies (IST) Future and Emerging Technologies (FET).

LITERATURE CITED

Amedi A, Jacobson G, Hendler T, Malach R, Zohary E. 2002. Convergence of visual and tactile shape processing in the human lateral occipital complex. *Cereb. Cortex* 12(11):1202–12

Anderson KC, Siegel RM. 2005. Three-dimensional structure-from-motion selectivity in the anterior superior temporal polysensory area, STPa, of the behaving monkey. *Cereb. Cortex* 15(9):1299–307

Anzai A, DeAngelis GC. 2010. Neural computations underlying depth perception. *Curr. Opin. Neurobiol.* 20:367–75

Arcizet F, Jouffrais C, Girard P. 2009. Coding of shape from shading in area V4 of the macaque monkey. *BMC Neurosci.* 10:140

Arterberry ME, Yonas A. 2000. Perception of three-dimensional shape specified by optic flow by 8-week old infants. *Percept. Psychophys.* 62(3):550–56

Beer AL, Watanabe T, Ni R, Sasaki Y, Andersen GJ. 2009. 3D surface perception from motion involves a temporal-parietal network. *Eur. J. Neurosci.* 30(4):703–13

Born RT, Bradley DC. 2005. Structure and function of visual area MT. *Annu. Rev. Neurosci.* 28:157–89

Borra E, Belmalih A, Calzavara R, Gerbella M, Murata A, et al. 2008. Cortical connections of the macaque anterior intraparietal (AIP) area. *Cereb. Cortex* 18(5):1094–111

Bradley DC, Chang GC, Andersen RA. 1998. Encoding of three-dimensional structure-from-motion by primate MT neurons. *Nature* 392:714–17

Buracas GT, Albright TD. 1996. Contribution of area MT to perception of three-dimensional shape: a computational study. *Vis. Res.* 36(6):869–87

Chandrasekaran C, Canon V, Dahmen JC, Kourtzi Z, Welchman AE. 2007. Neural correlates of disparity-defined shape discrimination in the human brain. *J. Neurophysiol.* 97(2):1553–65

Colombo M, D'Amato MR, Rodman HR, Gross CG. 1990. Auditory association cortex lesions impair auditory short-term memory in monkeys. *Science* 247(4940):336–38

Cornette L, Dupont P, Orban GA. 2002. The neural substrate of orientation short-term memory and resistance to distractor items. *Eur. J. Neurosci.* 15(1):165–75

Cumming BG, Shapiro SE, Parker AJ. 1998. Disparity detection in anticorrelated stereograms. *Perception* 27(11):1367–77

DeYoe EA, Carman GJ, Bandettini P, Glickman S, Wieser J, et al. 1996. Mapping striate and extrastriate visual areas in human cerebral cortex. *Proc. Natl. Acad. Sci. USA* 93(6):2382–86

Dodd JV, Krug K, Cumming BG, Parker AJ. 2001. Perceptually bistable three-dimensional figures evoke high choice probabilities in cortical area MT. *J. Neurosci.* 21(13):4809–21

Duffy CJ, Wurtz RH. 1997. Medial superior temporal area neurons respond to speed patterns in optic flow. *J. Neurosci.* 17(8):2839–51

Durand JB, Nelissen K, Joly O, Wardak C, Todd JT, et al. 2007. Anterior regions of monkey parietal cortex process visual 3D shape. *Neuron* 55(3):493–505

Durand JB, Peeters R, Norman JF, Todd JT, Orban GA. 2009. Parietal regions processing visual 3D shape extracted from disparity. *Neuroimage* 46(4):1114–26

Fleming RW, Torralba A, Adelson EH. 2004. Specular reflections and the perception of shape. *J. Vis.* 4(9):798–820

Freiwald WA, Tsao DY, Livingstone MS. 2009. A face feature space in the macaque temporal lobe. *Nat. Neurosci.* 12(9):1187–96

Fried I, MacDonald KA, Wilson CL. 1997. Single neuron activity in human hippocampus and amygdala during recognition of faces and objects. *Neuron* 18:753–65

Georgieva S, Peeters R, Kolster H, Todd JT, Orban GA. 2009. The processing of three-dimensional shape from disparity in the human brain. *J. Neurosci.* 29(3):727–42

Georgieva S, Todd JT, Peeters R, Orban GA. 2008. The extraction of 3D shape from texture and shading in the human brain. *Cereb. Cortex* 18(10):2416–38

Gerardin P, Kourtzi Z, Mamassian P. 2010. Prior knowledge of illumination for 3D perception in the human brain. *Proc. Natl. Acad. Sci. USA* 107(37):16309–14

Gibson JJ. 1950. *The Perception of the Visual World*. Boston: Houghton Mifflin

Giese MA, Poggio T. 2003. Neural mechanisms for the recognition of biological movements. *Nat. Rev. Neurosci.* 4(3):179–92

Gunderson VM, Yonas A, Sargent PL, Grant-Webster KS. 1993. Infant macaque monkeys respond to pictorial depth. *Psychol. Sci.* 4:93–98

Hanazawa A, Komatsu H. 2001. Influence of the direction of elemental luminance gradients on the responses of V4 cells to textured surfaces. *J. Neurosci.* 21(12):4490–97

Hegde J, Van Essen DC. 2005. Role of primate visual area V4 in the processing of 3-D shape characteristics defined by disparity. *J. Neurophysiol.* 94(4):2856–66

Hinkle DA, Connor CE. 2002. Three-dimensional orientation tuning in macaque area V4. *Nat. Neurosci.* 5(7):665–70

Humphrey GK, Goodale MA, Bowen CV, Gati JS, Vilis T, et al. 1997. Differences in perceived shape from shading correlate with activity in early visual areas. *Curr. Biol.* 7:144–47

Iwaki S, Bonmassar G, Belliveau JW. 2007. Neuromagnetic brain responses during 3D object perception from 2D optic flow. *Int. Congr. Ser.* 1300:543–46

Jackson S, Blake R. 2010. Neural integration of information specifying human structure from form, motion and depth. *J. Neurosci.* 30(3):838–48

Janssen P, Vogels R, Liu Y, Orban GA. 2001. Macaque inferior temporal neurons are selective for three-dimensional boundaries and surfaces. *J. Neurosci.* 21(23):9419–29

Janssen P, Vogels R, Liu Y, Orban GA. 2003. At least at the level of inferior temporal cortex, the stereo correspondence problem is solved. *Neuron* 37(4):693–701

Janssen P, Vogels R, Orban GA. 1999. Macaque inferior temporal neurons are selective for disparity-defined three-dimensional shapes. *Proc. Natl. Acad. Sci. USA* 96(14):8217–22

Janssen P, Vogels R, Orban GA. 2000a. Selectivity for 3D shape that reveals distinct areas within macaque inferior temporal cortex. *Science* 288(5473):2054–56

Janssen P, Vogels R, Orban GA. 2000b. Three-dimensional shape coding in inferior temporal cortex. *Neuron* 27(2):385–97

Jiang Y, Boehler CN, Nonnig N, Duzel E, Hopf JM, et al. 2008. Binding 3-D object perception in the human visual cortex. *J. Cogn. Neurosci.* 20(4):553–62

Joly O, Vanduffel W, Orban GA. 2009. The monkey ventral premotor cortex processes 3D shape from disparity. *Neuroimage* 47(1):262–72

Katsuyama N, Yamashita A, Sawada K, Naganuma T, Sakata H, Taira M. 2010. Functional and histological properties of caudal intraparietal area of macaque monkey. *Neuroscience* 167(1):1–10

Klaver P, Lichtensteiger J, Bucher K, Dietrich T, Loenneker T, Martin E. 2008. Dorsal stream development in motion and structure-from-motion. *Neuroimage* 39:1815–23

Koenderink JJ. 1990. *Solid Shape*. Cambridge, MA: MIT Press

Koenderink JJ, van Doorn AJ, Kappers AM, Todd JT. 2001. Ambiguity and the 'mental eye' in pictorial relief. *Perception* 30(4):431–48

Kolster H, Mandeville JB, Arsenault JT, Ekstrom LB, Wald LL, Vanduffel W. 2009. Visual field map clusters in macaque extrastriate cortex. *J. Neurosci.* 29(21):7031–39

Kolster H, Peeters R, Orban GA. 2010. The retinotopic organization of the human middle temporal area MT/V5 and its cortical neighbors. *J. Neurosci.* 30(29):9801–20

Konen CS, Kastner S. 2008. Two hierarchically organized neural systems for object information in human visual cortex. *Nat. Neurosci.* 11(2):224–31

Kourtzi Z, Connor CE. 2011. Neural representations for object perception: structure, category, and adaptive coding. *Annu. Rev. Neurosci.* 34:45–67

Kourtzi Z, Erb M, Grodd W, Bulthoff HH. 2003. Representation of the perceived 3-D object shape in the human lateral occipital complex. *Cereb. Cortex* 13(9):911–20

Kourtzi Z, Kanwisher N. 2000. Cortical regions involved in perceiving object shape. *J. Neurosci.* 20(9):3310–18

Kriegeskorte N, Sorger B, Naumer M, Schwarzbach J, van den BE, et al. 2003. Human cortical object recognition from a visual motion flowfield. *J. Neurosci.* 23(4):1451–63

Laeng B, Caviness VS. 2001. Prosopagnosia as a deficit in encoding curved surface. *J. Cogn. Neurosci.* 13(5):556–76

Li G, Zucker SW. 2006a. Contextual inference in contour-based stereo correspondence. *Int. J. Comput. Vis.* 69:59–75

Li G, Zucker SW. 2006b. Differential geometric consistency extends stereo to curved surfaces. *Proc. Eur. Conf. Comp. Vis. (ECCV'06)*, ed. A Leonardis, H Bischof, A Pinz, pp. 44–57. New York: Springer Verlag

Liu Y, Vogels R, Orban GA. 2002. The effect of texture depth cue on the disparity selectivity of macaque inferior temporal neurons. *Soc. Neurosci. Abstr.* 56.13

Liu Y, Vogels R, Orban GA. 2004. Convergence of depth from texture and depth from disparity in macaque inferior temporal cortex. *J. Neurosci.* 24(15):3795–800

Logothetis NK, Guggenberger H, Peled S, Pauls J. 1999. Functional imaging of the monkey brain. *Nat. Neurosci.* 2:555–62

Marr D. 1982. *Vision.* San Francisco: Freeman

Martinez-Trujillo JC, Tsotsos JK, Simine E, Pomplun M, Wildes R, et al. 2005. Selectivity for speed gradients in human area MT/V5. *NeuroReport* 16(5):435–38

Matsumora T, Koida K, Komatsu H. 2008. Relationship between color discrimination and neural responses in the inferior temporal cortex of the monkey. *J. Neurophysiol.* 100(6):3361–74

Maunsell JH, Sclar G, Nealey TA, DePriest DD. 1991. Extraretinal representations in area V4 in the macaque monkey. *Vis. Neurosci.* 7(6):561–73

Moore C, Engel SA. 2001. Neural response to perception of volume in the lateral occipital complex. *Neuron* 29(1):277–86

Murray SO, Olshausen BA, Woods DL. 2003. Processing shape, motion and three-dimensional shape-from-motion in the human cortex. *Cereb. Cortex* 13(5):508–16

Mysore SG, Vogels R, Kolster H, Vanduffel W, Orban GA. 2010a. Cortical network of 3D-structure from motion (3D-SFM) in the macaque: a functional imaging study. *Soc. Neurosci. Abstr.* 776.9

Mysore SG, Vogels R, Raiguel SE, Orban GA. 2006. Processing of kinetic boundaries in macaque V4. *J. Neurophysiol.* 95(3):1864–80

Mysore SG, Vogels R, Raiguel SE, Todd JT, Orban GA. 2010b. The selectivity of FST neurons for 3D structure from motion. *J. Neurosci.* 30(46):15491–508

Nadler JW, Angelaki DE, DeAngelis GC. 2008. A neural representation of depth from motion parallax in macaque visual cortex. *Nature* 452(7187):642–45

Nadler JW, Nawrot M, Angelaki DE, DeAngelis GC. 2009. MT neurons combine visual motion with smooth eye movement signal to code depth-sign from motion parallax. *Neuron* 63:523–32

Nakamura H, Kuroda T, Wakita M, Kusunoki M, Kato A, et al. 2001. From three-dimensional space vision to prehensile hand movements: the lateral intraparietal area links the area V3A and the anterior intraparietal area in macaques. *J. Neurosci.* 21(20):8174–87

Nelissen K, Joly O, Durand JB, Todd JT, Vanduffel W, Orban GA. 2009. The extraction of depth structure from shading and texture in the macaque brain. *PLoS One* 4(12):e8306

Nelissen K, Vanduffel W, Orban GA. 2006. Charting the lower superior temporal region, a new motion-sensitive region in monkey superior temporal sulcus. *J. Neurosci.* 26(22):5929–47

Nguyenkim JD, DeAngelis GC. 2003. Disparity-based coding of three-dimensional surface orientation by macaque middle temporal neurons. *J. Neurosci.* 23(18):7117–28

Nguyenkim JD, DeAngelis GC. 2004. Macaque MT neurons are selective for 3D surface orientation defined by multiple cues. *Soc. Neurosci. Abstr.* 368.12

Norman JF, Todd JT, Orban GA. 2004. Perception of three-dimensional shape from specular highlights, deformations of shading, and other types of visual information. *Psychol. Sci.* 15(8):565–70

Orban GA, Claeys K, Nelissen K, Smans R, Sunaert S, et al. 2006a. Mapping the parietal cortex of human and non-human primates. *Neuropsychologia* 44(13):2647–67

Orban GA, Janssen P, Vogels R. 2006b. Extracting 3D structure from disparity. *Trends Neurosci.* 29(8):466–73

Orban GA, Sunaert S, Todd JT, Van HP, Marchal G. 1999. Human cortical regions involved in extracting depth from motion. *Neuron* 24(4):929–40

Orban GA, Van Essen DC, Vanduffel W. 2004. Comparative mapping of higher visual areas in monkeys and humans. *Trends Cogn. Sci.* 8(7):315–24

Orban GA, Vogels R. 1998. The neuronal machinery involved in successive orientation discrimination. *Prog. Neurobiol.* 55(2):117–47

Paradis AL, Cornilleau-Peres V, Droulez J, van-de-Moortele PF, Lobel E, et al. 2000. Visual perception of motion and 3-D structure from motion: an fMRI study. *Cereb. Cortex* 10:772–83

Patsupathy A, Connor CE. 2002. Population coding of shape in area V4. *Nat. Neurosci.* 5(12):1332–38

Perrett DI, Smith PA, Mistlin AJ, Chitty AJ, Head AS, et al. 1985. Visual analysis of body movements by neurones in the temporal cortex of the macaque monkey: a preliminary report. *Behav. Brain Res.* 16(2–3):153–70

Peuskens H, Claeys KG, Todd JT, Norman JF, Van HP, Orban GA. 2004. Attention to 3-D shape, 3-D motion, and texture in 3-D structure from motion displays. *J. Cogn. Neurosci.* 16(4):665–82

Pizlo Z, Sawada T, Li Y, Kropatsch WG, Steinman RM. 2010. New approach to the perception of 3D shape based on veridicality, complexity, symmetry and volume. *Vis. Res.* 50:1–11

Preston TJ, Kourtzi Z, Welchman AE. 2009. Adaptive estimation of three-dimensional structure in the human brain. *J. Neurosci.* 29(6):1688–98

Preston TJ, Li S, Kourtzi Z, Welchman AE. 2008. Multivoxel pattern selectivity for perceptually relevant binocular disparities in the human brain. *J. Neurosci.* 28(44):11315–27

Roy JP, Komatsu H, Wurtz RH. 1992. Disparity sensitivity of neurons in monkey extrastriate area MST. *J. Neurosci.* 12(7):2478–92

Sakata H, Tsutsui K, Taira M. 2005. Toward an understanding of the neural processing for 3D shape perception. *Neuropsychologia* 43(2):151–61

Sawamura H, Orban GA, Vogels R. 2006. Selectivity of neuronal adaptation does not match response selectivity: a single-cell study of the FMRI adaptation paradigm. *Neuron* 49(2):307–18

Saygin AP, Sereno MI. 2008. Retinotopy and attention in human occipital, temporal, parietal, and frontal cortex. *Cereb. Cortex* 18(9):2158–68

Schiller PH, Slocum WM, Jao B, Weiner VS. 2011. The integration of disparity, shading and motion parallax cues for depth perception in humans and monkeys. *Brain Res.* 1377:67–77

Sereno ME, Trinath T, Augath M, Logothetis NK. 2002. Three-dimensional shape representation in monkey cortex. *Neuron* 33:635–52

Shikata E, Hamzei F, Glauche V, Knab R, Dettmers C, et al. 2001. Surface orientation discrimination activates caudal and anterior intraparietal sulcus in humans: an event-related fMRI study. *J. Neurophysiol.* 85(3):1309–14

Shikata E, McNamara A, Sprenger A, Hamzei F, Glauche V, et al. 2008. Localization of human intraparietal areas AIP, CIP, and LIP using surface orientation and saccadic eye movement tasks. *Hum. Brain Mapp.* 29(4):411–21

Shikata E, Tanaka Y, Nakamura H, Taira M, Sakata H. 1996. Selectivity of the parietal visual neurons in 3D orientation of surface of stereoscopic stimuli. *NeuroReport* 7(14):2389–94

Siegel RM, Andersen RA. 1990. The perception of structure from visual motion in monkey and man. *J. Cogn. Neurosci.* 2:306–19

Singer JM, Sheinberg DL. 2010. Temporal cortex neurons encode articulated actions as slow sequences of integrated poses. *J. Neurosci.* 30(8):3133–45

Srivastava S, Orban GA, De Maziere PA, Janssen P. 2009. A distinct representation of three-dimensional shape in macaque anterior intraparietal area: fast, metric, and coarse. *J. Neurosci.* 29(34):10613–26

Srivastava S, Orban GA, Janssen P. 2008. Disparity-defined 3D boundary and surface selectivity in macaque area AIP. *Soc. Neurosci. Abstr.* 462.4

Sugihara H, Murakami I, Shenoy KV, Andersen RA, Komatsu H. 2002. Response of MSTd neurons to simulated 3D orientation of rotating planes. *J. Neurophysiol.* 87(1):273–85

Taira M, Nose I, Inoue K, Tsutsui K. 2001. Cortical areas related to attention to 3D surface structures based on shading: an fMRI study. *Neuroimage* 14:959–66

Taira M, Tsutsui KI, Jiang M, Yara K, Sakata H. 2000. Parietal neurons represent surface orientation from the gradient of binocular disparity. *J. Neurophysiol.* 83(5):3140–46

Theys T, Van Loon J, Goffin J, Janssen P. 2009. Selectivity for disparity-defined three-dimensional shape in macaque premotor cortex. *Soc. Neurosci. Abstr.* 852.9

Todd JT. 2004. The visual perception of 3D shape. *Trends Cogn. Sci.* 8(3):115–21

Todd JT, Thaler L. 2010. The perception of 3D shape from texture based on directional width gradients. *J. Vis.* 10(5):17:1–13

Tootell RB, Mendola JD, Hadjikhani NK, Ledden PJ, Liu AK, et al. 1997. Functional analysis of V3A and related areas in human visual cortex. *J. Neurosci.* 17(18):7060–78

Tsao DY, Freiwald WA, Knutsen TA, Mandeville JB, Tootell RB. 2003a. Faces and objects in macaque cerebral cortex. *Nat. Neurosci.* 6(9):989–95

Tsao DY, Vanduffel W, Sasaki Y, Fize D, Knutsen TA, et al. 2003b. Stereopsis activates V3A and caudal intraparietal areas in macaques and humans. *Neuron* 39(3):555–68

Tsuruhara A, Sawada T, Kanazawa S, Yamaguchi MK, Yonas A. 2009. Infant's ability to form a common representation of an object's shape from different pictorial cues: a transfer-across-cues study. *Infant Behav. Dev.* 32(4):468–75

Tsutsui K, Jiang M, Sakata H, Taira M. 2003. Short-term memory and perceptual decision for three-dimensional visual features in the caudal intraparietal sulcus (area CIP). *J. Neurosci.* 23(13):5486–95

Tsutsui K, Jiang M, Yara K, Sakata H, Taira M. 2001. Integration of perspective and disparity cues in surface-orientation-selective neurons of area CIP. *J. Neurophysiol.* 86(6):2856–67

Tsutsui K, Sakata H, Naganuma T, Taira M. 2002. Neural correlates for perception of 3D surface orientation from texture gradient. *Science* 298(5592):409–12

Tsutsui K, Taira M, Sakata H. 2005. Neural mechanisms of three-dimensional vision. *Neurosci. Res.* 51(3):221–29

Ungerleider LG, Desimone R. 1986. Cortical connections of visual area MT in the macaque. *J. Comp Neurol.* 248(2):190–222

Ungerleider LG, Mishkin M. 1982. Two cortical visual systems. In *Analysis of Visual Behavior*, ed. DJ Ingle, MA Goodale, RJW Mansfield, pp. 549–86. Cambridge: MIT Press

Van Dromme IC, Janssen P, Kolster H, Vanduffel W. 2010. Mapping 3D-shape processing in the macaque monkey. *Soc. Neurosci. Abstr.* 19.7

Vanduffel W, Fize D, Mandeville JB, Nelissen K, Van HP, et al. 2001. Visual motion processing investigated using contrast agent-enhanced fMRI in awake behaving monkeys. *Neuron* 32(4):565–77

Vanduffel W, Fize D, Peuskens H, Denys K, Sunaert S, et al. 2002a. Extracting 3D from motion: differences in human and monkey intraparietal cortex. *Science* 298(5592):413–15

Vanduffel W, Tootell RB, Schoups AA, Orban GA. 2002b. The organization of orientation selectivity throughout macaque visual cortex. *Cereb. Cortex* 12(6):647–62

Vangeneugden J, Koteles K, Orban GA, Vogels R. 2006. The coding of 3-D shape from shading in macaque areas TE and TEO. *Perception* 35(Suppl.):34

Verhoef BE, Vogels R, Janssen P. 2010. Contribution of inferior temporal and posterior parietal activity to three-dimensional shape perception. *Curr. Biol.* 20(10):909–13

Welchman AE, Deubelius A, Conrad V, Bulthoff HH, Kourtzi Z. 2005. 3D shape perception from combined depth cues in human visual cortex. *Nat. Neurosci.* 8(6):820–27

Xiao DK, Raiguel S, Marcar V, Koenderink J, Orban GA. 1995. Spatial heterogeneity of inhibitory surrounds in the middle temporal visual area. *Proc. Natl. Acad. Sci. USA* 92(24):11303–6

Xiao DK, Marcar VL, Raiguel SE, Orban GA. 1997a. Selectivity of macaque MT/V5 neurons for surface orientation in depth specified by motion. *Eur. J. Neurosci.* 9(5):956–64

Xiao DK, Raiguel S, Marcar V, Orban GA. 1997b. The spatial distribution of the antagonistic surround of MT/V5 neurons. *Cereb. Cortex* 7(7):662–77

Yamamoto T, Takahashi S, Hankawa T, Urayama S-I, Aso T, et al. 2008. Neural correlates of the stereokinetic effect revealed by functional magnetic resonance imaging. *J. Vis.* 8(10):1–17

Yamane Y, Carlson ET, Bowman KC, Wang Z, Connor CE. 2008. A neural code for three-dimensional object shape in macaque inferotemporal cortex. *Nat. Neurosci.* 11(11):1352–60

Yonas A, Arterberry ME, Granrud CE. 1987. Four-month-old infants' sensitivity to binocular and kinetic information for three-dimensional-object shape. *Child Dev.* 58(4):910–17

Zhang Y, Schiller PH. 2008. The effect of overall stimulus velocity on motion parallax. *Vis. Neurosci.* 25(1):3–15

Zhang Y, Weiner VS, Slocum WM, Schiller PH. 2007. Depth from shading and disparity in humans and monkeys. *Vis. Neurosci.* 24(2):207–15

RELATED RESOURCES

Orban GA. 2008. Higher order visual processing in macaque extrastriate cortex. *Physiol. Rev.* 88(1):59–89

Orban GA. 2008. Three-dimensional shape: cortical mechanisms of shape extraction. In *The Senses: A Comprehensive Reference*, ed. AI Basbaum, A Kaneko, GM Shepherd, G Westheimer, pp. 245–74. San Diego: Academic

The Development and Application of Optogenetics

Lief Fenno,[1,2] Ofer Yizhar,[1] and Karl Deisseroth[1,3,4]

[1]Department of Bioengineering, [2]Neuroscience Program, [3]Departments of Psychiatry and Behavioral Sciences, [4]Howard Hughes Medical Institute, Stanford University, Stanford, California 94305; email: deissero@stanford.edu

Annu. Rev. Neurosci. 2011. 34:389–412

The *Annual Review of Neuroscience* is online at neuro.annualreviews.org

This article's doi: 10.1146/annurev-neuro-061010-113817

Keywords

channelrhodopsin, halorhodopsin, bacteriorhodopsin, electrophysiology

Abstract

Genetically encoded, single-component optogenetic tools have made a significant impact on neuroscience, enabling specific modulation of selected cells within complex neural tissues. As the optogenetic toolbox contents grow and diversify, the opportunities for neuroscience continue to grow. In this review, we outline the development of currently available single-component optogenetic tools and summarize the application of various optogenetic tools in diverse model organisms.

Contents

INTRODUCTION

In describing unrealized prerequisites for assembling a general theory of the mind, Francis Crick observed that the ability to manipulate individual components of the brain would be needed, requiring "a method by which all neurons of just one type could be inactivated, leaving the others more or less unaltered" (Crick 1979, p. 222). Extracellular electrical manipulation does not readily achieve true inactivation, and even electrical excitation, while allowing for temporal precision in stimulating within a given volume, lacks specificity for cell type. However, pharmacological and genetic manipulations can be specific to cells with certain expression profiles (in the best case) but lack temporal precision on the timescale of neural coding and signaling.

Because no prior technique has achieved both high-temporal and cellular precision within intact mammalian neural tissue, there has been strong pressure to develop a new class of technology. As a result of these efforts, neurons now may be controlled with optogenetics for fast, specific excitation or inhibition within systems as complex as freely moving mammals [for example, with microbial opsin methods, light-induced inward cation currents may be used to depolarize the neuronal membrane and positively modulate firing of action potentials, while optical pumping of chloride ions can induce outward currents and membrane hyperpolarization, thereby inhibiting spiking (**Figure 1**)]. These optogenetic tools of microbial origin (**Figure 1**) may be readily targeted to subpopulations of neurons within heterogeneous tissue and function on a temporal scale commensurate with physiological rates of spiking or critical moments in behavioral tests, with fast deactivation upon cessation of light. With these properties, microbe-derived optogenetic tools fulfill the criterion set forth by Crick in 1979 (Deisseroth 2010, 2011).

EARLY EFFORTS TOWARD OPTICAL CONTROL

The microbial opsin approach is heir to a long tradition of using light as an intervention in biology. With chromophore-assisted laser inactivation, light can be used to inhibit targeted proteins by destroying them [what a geneticist would call "loss of function" (Schmucker et al. 1994)]; conversely, lasers can be used to stimulate neurons directly in a way that could be adapted (in principle) to control fluorescently labeled, genetically targeted cells [what a geneticist would call "gain of function" (Fork 1971, Hirase et al. 2002)]. Next, various cascades of genes, and combinations of genes with chemicals, were tested as multicomponent

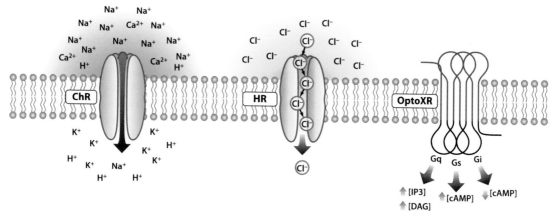

Figure 1

Optogenetic tool families. Channelrhodopsins conduct cations and depolarize neurons upon illumination (*left*). Halorhodopsins conduct chloride ions into the cytoplasm upon yellow light illumination (*center*). OptoXRs are rhodopsin-GPCR (G protein–coupled receptor) chimeras that respond to green (500 nm) light with activation of the biological functions dictated by the intracellular loops used in the hybrid (*right*).

strategies for optical control; rhodopsin and arrestin genes from *Drosophila* photoreceptors were combined to light-sensitize neurons (Zemelman et al. 2002); ligand-gated channels, combined with ultraviolet (UV)-light photolysis of caged agonists, were developed for *Drosophila* experiments (Lima & Miesenbock 2005, Zemelman et al. 2003); and UV light–isomerizable chemicals linked to genetically encoded channels were employed in cultured cells and in zebrafish (Banghart et al. 2004, Szobota et al. 2007, Volgraf et al. 2006). These efforts have been reviewed (Gorostiza & Isacoff 2008, Miesenbock & Kevrekidis 2005) and while elegant, have thus far been found to be limited to various extents in speed, targeting, tissue penetration, and/or applicability because of their multicomponent nature. Here, we review development and application efforts focused on the distinct single-component optogenetic tools, such as microbial opsins, over the past six years since they were first implemented.

MICROBIAL OPSINS

Species from multiple branches of the animal kingdom have evolved mechanisms to sense electromagnetic radiation in their

environments. Likewise many microbes, in the absence of complex eye structures employed by metazoans, have developed light-activated proteins for a variety of purposes. For some, this serves as a mechanism of homeostasis to remain at a certain depth in the ocean (Beja et al. 2000, 2001); for others, this helps maintain osmotic balance in a highly saline environment (Stoeckenius 1985). These and other diverse roles are, in many cases, fulfilled by a family of seven-transmembrane, light-responsive proteins encoded by opsin genes.

Opsin genes are divided into two distinct superfamilies: microbial opsins (type I) and animal opsins (type II). Opsin proteins from both families require retinal, a vitamin A–related organic cofactor that serves as the antenna for photons; when retinal is bound, the functional opsin proteins are termed rhodopsins. Retinal covalently attaches to a conserved lysine residue of helix 7 by forming a protonated retinal Schiff base (RSBH+). The ionic environment of the RSB, defined by the residues of the binding pocket, dictates the spectral and kinetic characteristics of each individual protein. Upon absorption of a photon, retinal isomerizes and triggers a sequence of conformational changes within the opsin partner.

The photoisomerized retinal is the trigger for subsequent structural rearrangements and activities performed by these proteins.

Although both opsin families encode seven-transmembrane structures, sequence homology between the two families is extremely low; homology within families, however, is high (25%–80% residue similarity) (Man et al. 2003). Whereas type I opsin genes are found in prokaryotes, algae, and fungi (Spudich 2006), type II opsin genes are present only in higher eukaryotes and are responsible mainly for vision (but also play roles in circadian rhythm and pigment regulation) (Sakmar 2002, Shichida & Yamashita 2003). Type II opsin genes encode G protein–coupled receptors (GPCRs) and, in the dark, bind retinal in the 11-*cis* configuration. Upon illumination, retinal isomerizes to the all-*trans* configuration and initiates the reactions that underlie the visual phototransduction second messenger cascade. After photoisomerization, the retinal-protein linkage is hydrolyzed; free all-*trans* retinal then diffuses out of the protein and is replaced by a fresh 11-*cis* retinal molecule for another round of signaling (Hofmann et al. 2009).

In contrast, type I opsins more typically encode proteins that utilize retinal in the all-*trans* configuration, which photoisomerizes upon photon absorption to the 13-*cis* configuration. Unlike the situation with type II rhodopsins, the activated retinal molecule in type I rhodopsins does not dissociate from its opsin protein but thermally reverts to the all-*trans* state while maintaining a covalent bond to its protein partner (Haupts et al. 1997). Type I opsins encode several distinct subfamilies of protein, discussed in more detail below. The central operating principle of these elegant molecular machines [established for this broad family of opsins since bacteriorhodopsin (BR) in 1971 (Oesterhelt & Stoeckenius 1971) and now including halorhodopsins and channelrhodopsins (**Figure 1**)] is their unitary nature. They combine the two tasks of light sensation and ion flux into a single protein (with bound small organic cofactor), encoded by a single gene. In 2005, one of these micro-

bial opsins was brought to neuroscience as the first single-component optogenetic tool (Boyden et al. 2005), and the other microbial opsin subfamilies followed close behind. For example, channelrhodopsin-1 (ChR1) (Nagel et al. 2002) and channelrhodopsin-2 (ChR2) (Nagel et al. 2003) from *Chlamydomonas reinhardtii* are blue-light-activated nonspecific cation channels. In common with other type I opsins, these proteins require retinal as the photon-sensing cofactor to function. In response to light stimulation, the channel shuttles from the dark-adapted state through a stereotyped progression of functional and conformational states, eventually (in the absence of further light stimulation) reaching the dark-adapted state once again. These different states, which all have unique spectroscopic signatures, are collectively referred to as the photocycle, which (as for BR and halorhodopsin in earlier work) has been extensively investigated (Bamann et al. 2010, Berndt et al. 2010, Ernst et al. 2008, Hegemann et al. 2005, Ritter et al. 2008, Stehfest et al. 2010, Stehfest & Hegemann 2010).

The size, kinetic properties, and wavelength sensitivity of photocurrents resulting from activation of an individual protein are a direct result of its photocycle topology, ion selectivity, and activation/deactivation/inactivation time constants (Ernst et al. 2008, Hegemann et al. 2005, Ritter et al. 2008). Typically, a transient peak photocurrent, evoked at the onset of light stimulation, decays modestly to a steady-state photocurrent even in the presence of continuous light, owing in part to the desensitization of a certain population of channels (Nagel et al. 2003). The desensitized population can recover in the dark with a characteristic time constant on the order of 5 seconds, giving rise to a similar peak photocurrent if a second light pulse is applied after sufficient time has elapsed (Nagel et al. 2003). The fraction of desensitized proteins is crucial for determining the reliability of light stimulation during prolonged experiments (e.g., behavioral or long-term physiological experiments) where light is applied for extended periods. This issue is addressed in more detail below.

OPTOGENETIC TOOLS FOR NEURONAL EXCITATION

Many years passed after the discovery of BRs, channelrhodopsins, and halorhodopsins, prior to the development of optogenetics. As noted above, optogenetics with microbial opsins began with a channelrhodopsin, introduced into hippocampal neurons in 2005 (Boyden et al. 2005) where it was found to confer millisecond-precision control of neuronal spiking. A number of additional reports followed over the next year (Bi et al. 2006, Ishizuka et al. 2006, Li et al. 2005, Nagel et al. 2005). Moreover, this turned out to be a single-component system; through a remarkable twist of nature, investigators found that sufficient retinal is present in mammalian brains (and as later established, in all vertebrate tissues tested thus far) to enable functional expression of these optogenetic tools as single components, in the absence of any added chemical or other gene (Deisseroth et al. 2006, Zhang et al. 2006). The optogenetic toolbox has been vastly expanded since the original 2005 discovery to include dozens of single-component proteins activated by various wavelengths of light, with various ion conductance regulation properties that operate in neurons over a wide range of speeds (from milliseconds to tens of minutes), enabling broad experimental configurations and opportunities.

Initial improvements in ChR2 were carried out in an incremental fashion and focused on improving expression and photocurrent in mammalian systems. Human codon optimization for improved expression of ChR2 (Boyden et al. 2005, Gradinaru et al. 2007) was combined with substitution of histidine for arginine at position 134 to increase steady-state current size, although the mechanism of the latter effect (delayed channel closure) also significantly impaired temporal precision and high-speed spiking (ChR2-H134R; Nagel et al. 2005, Gradinaru et al. 2007). Better solutions were needed, and further diversification of the optogenetic toolbox via mutational engineering has proven to be challenging but highly productive (**Figure 2**).

First, because many microbial tools do not express well in mammalian neurons, general strategies for enabling mammalian expression were required. Membrane-trafficking modifications turned out to be crucial, beginning with the observation in 2008 that endoplasmic reticulum (ER)-export motifs were helpful for achieving high, safe expression of halorhodopsins (Gradinaru et al. 2008, Zhao et al. 2008), a principle that turned out to be extendable [to other classes of membrane-trafficking motif (Gradinaru et al. 2010)] and generalizable [to most microbial opsins tested (Gradinaru et al. 2010)]. Next, chimera strategies (using hybrids of ChR1 and ChR2) were found to be helpful in giving rise to improved expression and spiking properties for channelrhodopsins (Lin et al. 2009, Wang et al. 2009).

Finally, much subsequent opsin engineering for optogenetics was carried out on the basis of hypothesized ChR structures. To date, there has been no reported crystal structure of any excitatory optogenetic tools, although structures exist for halorhodopsin and for the proton pump BR (Luecke et al. 1999a,b), which can function to inhibit neurons when heterologously expressed (Gradinaru et al. 2010); proton conductance is a property shared by ChR2, whether through channel or possible pumping mechanisms (Feldbauer et al. 2009). Capitalizing upon homology between BR and ChRs, we have introduced a number of mutations that modify various properties of the opsins (discussed below) and have described a framework by which these improvements can be applied to other novel opsins [indeed, this direction has come full circle, with insights from ChRs now used to improve the optogenetic function of BR itself (Gradinaru et al. 2010)].

These structural hypothesis-based opsin engineering efforts were spurred by inherent limitations of the channelrhodopsin system (**Figure 2**). Specifically, the deactivation time constant of ChR2 upon cessation of light (~10–12 ms in neurons) imposed a limit on temporal precision, leading in some cases to artifactual multiplets of spikes after even single brief light pulses (as well as a plateau potential

Blue (τ_off/peak activation wavelength)

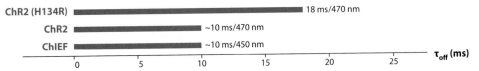

ChR2 (H134R) — 18 ms/470 nm
ChR2 — ~10 ms/470 nm
ChIEF — ~10 ms/450 nm

τ_{off} (ms)
0   5   10   15   20   25

Red-shifted (τ_off/peak activation wavelength)

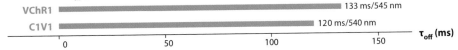

VChR1 — 133 ms/545 nm
C1V1 — 120 ms/540 nm

τ_{off} (ms)
0   50   100   150

Blue ChETA (τ_off/peak activation wavelength)

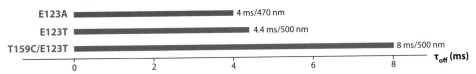

E123A — 4 ms/470 nm
E123T — 4.4 ms/500 nm
T159C/E123T — 8 ms/500 nm

τ_{off} (ms)
0   2   4   6   8

Red-shifted ChETA (τ_off/peak activation wavelength)

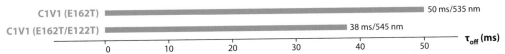

C1V1 (E162T) — 50 ms/535 nm
C1V1 (E162T/E122T) — 38 ms/545 nm

τ_{off} (ms)
0   10   20   30   40   50

Bistable (τ_off/peak activation wavelength/peak inactivation wavelength)

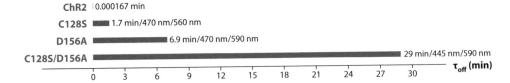

ChR2 — 0.000167 min
C128S — 1.7 min/470 nm/560 nm
D156A — 6.9 min/470 nm/590 nm
C128S/D156A — 29 min/445 nm/590 nm

τ_{off} (min)
0   3   6   9   12   15   18   21   24   27   30

Biochemical signaling (τ_off/peak activation wavelength)

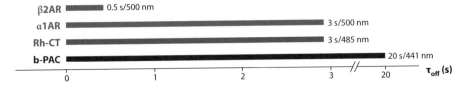

β2AR — 0.5 s/500 nm
α1AR — 3 s/500 nm
Rh-CT — 3 s/485 nm
b-PAC — 20 s/441 nm

τ_{off} (s)
0   1   2   3 // 20

Inhibitory (τ_off/peak activation wavelength)

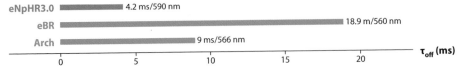

eNpHR3.0 — 4.2 ms/590 nm
eBR — 18.9 m/560 nm
Arch — 9 ms/566 nm

τ_{off} (ms)
0   5   10   15   20

when pulsing blue light at frequencies above 40 hz; the next pulse of light would occur before all the ChRs had deactivated, leading eventually to failed deinactivation of native voltage-gated sodium channels and thus missed spikes and further loss of fidelity). The latter problem was compounded by the desensitization of ChR itself even in the presence of light, leading to further missed spikes late in trains. Addressing part of this challenge, the chimeric opsins noted above (e.g., ChIEF; Lin et al. 2009), demonstrated reduced desensitization in cultured neurons, allowing more robust spiking over trains in culture as well as stronger currents. In another approach addressing both desensitization and deactivation, considering the crystal structure of BR led to modification of the counterion residue E123 of ChR2 to threonine or alanine; the resulting faster opsin is referred to as ChETA (**Figure 2**) (Gunaydin et al. 2010). This substitution introduced two advantages over wild-type ChR2. First, it reduced desensitization during light exposure, with the result that light pulses late in high-frequency trains became as likely as early light pulses to drive spikes (a very important property referred to as temporal stationarity). Second, it destabilized the active conformation of retinal, speeding spontaneous isomerization to the inactive state after light-off and thus closing the channel much more quickly after cessation of light than wild-type or improved ChR2 variants. The resulting functional consequences of ChETA mutations are temporal stationarity, reduced extra spikes, reduced plateau potentials, and improved high-frequency spike following at 200 Hz or more over sustained trains, even within intact mammalian brain tissue (Gunaydin et al. 2010).

Whereas many experimental designs employ optogenetic tools to initiate precise spiking, alternative paradigms may instead require the researcher to simply alter the excitability of a target neuronal population. Indeed, it is often important to bias the activity of a particular neuronal population without specifically driving action potentials or synchronous activity. To facilitate experiments examining altered excitability, Berndt et al. (2009) developed the step-function opsins (SFOs). SFOs are a family of ChR mutants that display bistable behavior—orders-of-magnitude prolonged activity after termination of the light stimulus—first instantiated as mutations in position 128 in ChR2 (cysteine to serine, alanine, or threonine). Again based on homology between ChR2 and BR (Peralvarez-Marin et al. 2004), we mutated this residue to manipulate the interaction between the opsin backbone and the covalently bound retinal photon sensor. In contrast to ChETA, the SFO mutations are designed to stabilize the active retinal isomer, the functional consequence of which is prolonging the active state of the channel even after light-off. SFOs have inactivation time constants on the order of tens of seconds or more instead of milliseconds (**Figure 2**) and can be activated by a single 10-ms pulse of blue light (Berndt et al. 2009). The SFOs can also be deactivated by a pulse of yellow light; the yellow pulse drives isomerization of retinal back to the nonconducting state. A subsequently engineered SFO, the ChR2(D156A) opsin (Bamann et al. 2010), displays an even longer inactivation time constant, which in our hands approaches eight minutes. One potential noted use of opsins with extended time constants could be for scanning two-photon stimulation paradigms (Rickgauer

Figure 2

Spectral and kinetic diversification of optogenetic tools. Deactivation time constants (τ_{off}) and approximate peak activation/inactivation wavelengths are shown for blue light–activated opsins, blue ChETAs (including E123T/T159C) (Berndt et al. 2011, Mattis et al. 2011), red-shifted opsins, red-shifted ChETAs, bistable (SFO) opsins, and modulatory/inhibitory tools in a compact look-up table. ChR2 is listed among the bistable group for scale purposes only. Where precise values are not available, decay kinetics were measured (courtesy of J. Mattis, personal communication) or estimated from published traces; all values were recorded at room temperature (except for optoXRs measured at 37°C), with substantial acceleration in kinetics (~50%) expected for all at 37°C.

& Tank 2009), during which it would be helpful to have persistent accumulating activity as a small two-photon spot scans a cell or tissue of interest.

We have now engineered a third class of SFO by combining the D156 and C128 mutations to produce a ChR2 variant that has a spontaneous deactivation time constant approaching 30 min; this stabilized SFO (SSFO) (O. Yizhar, L. Fenno, M. Prigge, K. Stehfest, J. Paz, F. Schneider, S. Tsunoda, R. Fudim, C. Ramakrishnan, J. Huguenard, P. Hegemann & K. Deisseroth, submitted) induces peak currents of >200 pA. An advantage of having an opsin with such a long time constant is the ability to conduct behavioral protocols in the absence of tethered external light delivery devices (e.g., optical fibers). Because a single pulse of blue light is sufficient to induce activity for a time period extending beyond that of most behavioral paradigms, the fiber may be removed before commencing the experiment. Just as with the original SFO proteins, SSFO may be inactivated by yellow light, allowing for precise control of network dynamics. A final advantage of the SFOs (which scales with their kinetic stability) is orders-of-magnitude greater light sensitivity of transduced cells, particularly for long light pulses, a direct result of the photon integration bestowed by their prolonged deactivation time constant. This property renders SFOs especially attractive as minimally invasive tools for stimulating large brain regions (for example, in primate studies) and deep within tissue.

Separate from, but synergistic with, molecular engineering is the systematic genomic identification of novel opsins. Adapting novel opsins activated by red-shifted wavelengths could enable control of two separate populations of circuit elements within the same physical volume. To this end, we launched genomic search strategies that led to identification of an opsin from *Volvox carteri* (VChR1) (Zhang et al. 2008), which shares homology with ChR2 and similarly functions as a cation channel. In contrast to ChR2, VChR1 is activated by red-shifted light. However, the relatively small currents due to low expression in mammalian neurons, as described in the initial report, have hampered in vivo adaptation of VChR1, even after codon optimization. To this end, we have engineered a chimeric opsin, C1V1 (O. Yizhar, L. Fenno, M. Prigge, K. Stehfest, J. Paz, F. Schneider, S. Tsunoda, R. Fudim, C. Ramakrishnan, J. Huguenard, P. Hegemann & K. Deisseroth, submitted), composed of the first two and one-half helices of ChR1 (a poorly expressing relative of ChR2 from the same organism) (Nagel et al. 2002) and the last four and one-half helices of VChR1. The resulting tool retains the red-shifted activation spectrum of VChR1, but with nanoampere-scale currents that exceed those of ChR2. The large current of C1V1 allows use in vivo and also use of off-peak (redder) light wavelengths, together enabling truly separable control of multiple populations of neurons when used in conjunction with ChR2 (O. Yizhar, L. Fenno, M. Prigge, K. Stehfest, J. Paz, F. Schneider, S. Tsunoda, R. Fudim, C. Ramakrishnan, J. Huguenard, P. Hegemann & K. Deisseroth, submitted). In addition to combinatorial experiments with ChR2, tools with red-shifted activation wavelengths such as C1V1 are also more amenable to use with simultaneous imaging of genetically encoded calcium indicators, such as GCaMP variants (Hires et al. 2008, Zhang & Oertner 2007).

OPTOGENETIC TOOLS FOR NEURONAL INHIBITION

Complementing these tools for precise control of neural excitation, certain light-activated ion pumps may be used for inhibition, although thus far only one ion pump has shown efficacy at modulating behavior in mammals (Tye et al. 2011, Witten et al. 2010, Zhang et al. 2007a): the ER trafficking-enhanced version of a halorhodopsin called NpHR (Gradinaru et al. 2008, 2010; Zhang et al. 2007a) derived from the halobacterium *Natronomonas pharaonis*. In the context of optogenetic application, this yellow light–activated electrogenic chloride pump acts to hyperpolarize the targeted neuron upon activation (**Figure 1**) (Zhang et al. 2007a).

Unlike the excitatory channelrhodopsins, NpHR is a true pump and requires constant light to move through its photocycle. Although optogenetic inhibition with NpHR was shown to operate well in freely moving worms (Zhang et al. 2007a), mammalian brain slices (Zhang et al. 2007a), and cultured neurons (Han & Boyden 2007, Zhang et al. 2007a), several years passed before final validation of this (or any) inhibitory optogenetic tool was obtained by successful application to intact mammals (Tye et al. 2011, Witten et al. 2010) because of membrane trafficking problems that required additional engineering (Gradinaru et al. 2008, 2010; Zhao et al. 2008). Indeed, a number of modifications to NpHR were required to improve its function, initially codon-optimizing the sequence followed by enhancement of its subcellular trafficking (eNpHR2.0 and eNpHR3.0) (Gradinaru et al. 2008, 2010), which resulted in improved membrane targeting and higher currents suitable even for use in human tissue (Busskamp et al. 2010), as well as activation with red-shifted wavelengths at the infrared border (680 nm). Kouyama et al. (2010) published the 2.0A crystal structure of halorhodopsin and illustrate that this protein has a high degree of homology within the retinal binding pocket with the proton pump BR. The two are proposed to distribute charge and store energy from absorbed photons in similar ways.

The diversification of the inhibitory opsin toolbox has been guided by bioinformatics approaches to screen nature for novel inhibitory ion pumps with desirable properties, just as had been successful previously for excitatory opsins (Zhang et al. 2008). In 2010, we and others explored the use of proton pumps (eBR, Mac, and Arch) as optogenetic tools (Gradinaru et al. 2010, Chow et al. 2010), finding evidence for robust efficacy but leaving open questions of long-term tolerability and functionality of proton-motive pumps in mammalian neurons. A major concern is the extent to which pumping of protons into the extracellular space (especially in juxtamembranous compartments difficult to visualize or measure) could have deleterious or noncell-type-specific effects on local tissue, which could show up as light-induced inhibition affecting all recorded units (more units than expected for a given transduction efficiency) or with a slightly slower time course than expected for the fast proton pumps.

Although the proton pumps must be treated with caution until these issues are addressed, many opportunities exist; indeed, we have improved the ability of proton pumps to hyperpolarize neurons following a methodology similar to that used for improving NpHR (Gradinaru et al. 2010), although none of the proton pumps yet described is as kinetically stable or as potent as eNpHR3.0, especially at the safe light levels (<20 mW/mm^2 at the target cell) required for in vivo use (Gradinaru et al. 2010). Indeed, eNpHR3.0 has now delivered the loss-of-function side of the optogenetic coin for freely moving mammals (Tye et al. 2011, Witten et al. 2010), complementing the channelrhodopsins, which delivered gain of function. Witten and colleagues (2010) used eNpHR3.0 to inhibit the cholinergic neurons of the nucleus accumbens and identified a causal role for these rare cells in implementing cocaine conditioning in freely moving mice by enhancing inhibition of inhibitory striatal medium spiny neurons. Tye and colleagues (2011) used eNpHR3.0 to inhibit a specific intra-amygdala projection in freely moving mice, creating a switchable model of anxiety in mammals and implicating a highly defined neural pathway as a native circuit involved in anxiolysis.

OPTOGENETIC TOOLS FOR BIOCHEMICAL CONTROL

The microbial (type I) opsins described up to this point serve strictly as ion flow modulators and control the excitability of a neuron by manipulating its membrane potential: either bringing the voltage nearer or above the threshold for generating an action potential or hyperpolarizing the potential and thereby inhibiting spiking. Several classes of biochemical control have now been achieved, beginning with the use of eukaryotic (type II) opsins for precise control of well-defined GPCR signaling pathways.

Rhodopsin, the light-sensing protein in the mammalian eye, is both an opsin, in that it is covalently bound to retinal and its function is modulated by the absorption of photons, and a GPCR, in that it is coupled on the intracellular side of the membrane to a G-protein, transducin. Virtually all neurons can communicate via GPCRs, which of course respond not only to neuromodulators from dopaminergic, serotonergic, and adrenergic pathways, but also to "fast" neurotransmitters such as glutamate and GABA. Building both on our finding that adequate retinal is present within mammalian brain tissue (Deisseroth et al. 2006, Zhang et al. 2006) and on a long history of elegant GPCR structure-function work from Khorana, Kobilka, Caron, Lefkowitz, and others, we have determined that GPCRs can be converted into light-activated regulators of well-defined biochemical signaling pathways that function within freely moving mammals. These proteins are referred to as optoXRs (Airan et al. 2009). OptoXRs allow for receptor-mediated intracellular signaling with temporal resolution suitable for modulation of behavior in freely moving mice (Airan et al. 2009). OptoXRs are modulated by 500-nm light and now include the alpha-1 and beta-2 adrenergic receptors (Airan et al. 2009), which are coupled to Gq and Gs signaling pathways, respectively, and the 5-HT1a receptor, which is Gi/o coupled (Oh et al. 2010).

Optical control over small GTPases was next achieved in cultured cells by several different laboratories (Levskaya et al. 2009, Wu et al. 2009, Yazawa et al. 2009) using optically modulated protein-protein interactions; although these have not yet been shown to express or display single-component functionality in freely moving mammals, such capability is plausible where flavin or biliverdin chromophores are required and present. Finally, investigators have recently described microbial adenylyl cyclases with lower dark activity than earlier microbial cyclases, and because they employ a flavin chromophore, these tools appear suitable for single-component optogenetic control (Ryu et al. 2010, Stierl et al. 2010). Together, these experiments have extended optogenetic

capability to essentially every cell type, even in nonexcitable tissues, in biology.

DELIVERING OPTOGENETIC TOOLS INTO NEURONAL SYSTEMS

Viral expression systems have the dual advantages of fast/versatile implementation and high infectivity/copy number for robust expression levels. Cellular specificity can be obtained with viruses by specific promoters (if small, specific, and strong enough), by spatial targeting of virus injection, and by restriction of opsin activation to particular cells (or projections of specific cells) via targeted light delivery (Zhang et al. 2010, Diester et al. 2011). Lenti and adeno-associated (AAV) viral vectors have been used successfully to introduce opsins into the mouse, rat, and primate brain (Zhang et al. 2010). Additionally, these have been well tolerated and highly expressed over long periods of time with no reported adverse effects. Lentivirus is easily produced using standard tissue culture techniques and an ultracentrifuge (see Zhang et al. 2010 for protocol). AAV may be produced either by individual laboratories or through core viral facilities. Neither AAV nor lentivirus were found to be highly expressed in the zebrafish, for which Sindbis and rabies are more effective (Zhu et al. 2009). Viruses have been used to target (among other cells) hypocretin neurons (Adamantidis et al. 2007), excitatory pyramidal neurons (Lee et al. 2010, Sohal et al. 2009, Zhang et al. 2007a), and astroglia (Gourine et al. 2010, Gradinaru et al. 2009). For example, one group (Gourine et al. 2010) recently described the use of AAV-delivered ChR2 to control astroglial activity in the brain stem of mice and to dissect a mechanism by which astroglia can transfer systemic information from the blood to neurons underlying homeostasis, in this case directly modulating neurons that manipulate the rate of respiration. However, a major downside of viral expression systems is a maximum genetic payload length; only promoter fragments that are small (less than ~4 kb), specific, and strong may be used, and these are rare. This limitation may be skirted using Cre-driver animals and Cre-dependent viruses, discussed below.

TRANSGENIC ANIMALS

The use of transgenic or knock-in animals obviates viral payload limitations and allows for tighter control of transgene expression using larger promoter fragments or indeed the endogenous genome in full via knock-in. The first transgenic opsin-expressing mouse line was generated using the Thy1 promoter (Arenkiel et al. 2007, Zhao et al. 2008), with widespread expression throughout neocortical layer 5 projection neurons as well as in some subcortical structures (Arenkiel et al. 2007). This mouse line has been widely used, for example, to examine the roles of inhibitory neurons on cortical information processing (Sohal et al. 2009) and the mechanism of action of deep brain stimulation for Parkinson's disease (Gradinaru et al. 2009). Several other groups have subsequently also generated transgenic mouse lines directly expressing opsin genes (Hagglund et al. 2010, Katzel et al. 2010, Thyagarajan et al. 2010).

Caveats to using transgenic mouse lines to directly express optogenetic tools include the time, effort, and cost associated with their production, validation, and maintenance. To enable widespread use of the latest optogenetic tools, investigators have designed opsin-delivering viruses for which opsin expression is dependent on the coexpression of Cre recombinase (**Figure 3**). This doublefloxed inverted open-reading-frame (DIO; reviewed in Zhang et al. 2010) strategy (Atasoy et al. 2008, Sohal et al. 2009, Tsai et al. 2009) situates the opsin gene (inhibitory or excitatory) in the inverted (meaningless) orientation, but the gene is flanked by two sets of incompatible Cre recombinase recognition sequences (Sohal et al. 2009, Tsai et al. 2009). The recombinase recognition sequences are placed such that in the presence of Cre, the ORF is inverted instead of being excised. Reversing the sequence then allows one of the Cre recognition sites to be excised (**Figure 3**), locking the reading frame into

the correct direction and allowing for strong expression of the opsin with (for example) the elongation factor 1-alpha (EF1α) promoter. The specificity of this gene expression can then come (for example) from the targeted expression of Cre in driver rodent lines in which Cre is controlled with high specificity in the context of very large chromosomal promoter-enhancer regions; the DIO strategy thus enables versatile and widespread use of optogenetics with the many (and growing number of) experimental systems selectively expressing Cre recombinase (Geschwind 2004; Gong et al. 2003, 2007; Heintz 2004). This strategy has been used recently in many systems, for example to target dopamine-1 (D1) or dopamine-2 (D2) receptor–expressing neurons of the striatum via transgenic D1-Cre and D2-Cre mouse lines, to examine the effects of their stimulation in the classic direct/indirect-movement pathways (Kravitz et al. 2010). With these same Cre lines, Lobo and colleagues (2010) examined the roles of nucleus accumbens D1 and D2 neurons in modulating cocaine reward. The Cre-dependent optogenetic system allowed for the first time a direct examination of the relationship between neuronal activity of specific neuronal populations and animal behavior, thus paving the way for a deeper understanding of diseases such as Parkinson's disease, depression, and substance abuse.

DEVELOPMENTAL AND LAYER-SPECIFIC TARGETING

The ability to target specific neocortical layers has been a long-sought goal of neuroscience; this can now be achieved either with layer-specific Cre driver lines or with developmental targeting strategies such as in utero electroporation (IUE). As a result, multiple laboratories have now successfully teased apart the role of layer-specific neurons in behavioral paradigms and network dynamics. Optogenetic tools have been well tolerated when electroporated in utero into mouse embryos (Adesnik & Scanziani 2010, Gradinaru et al. 2007, Lewis et al. 2009, Petreanu et al. 2007); IUE may

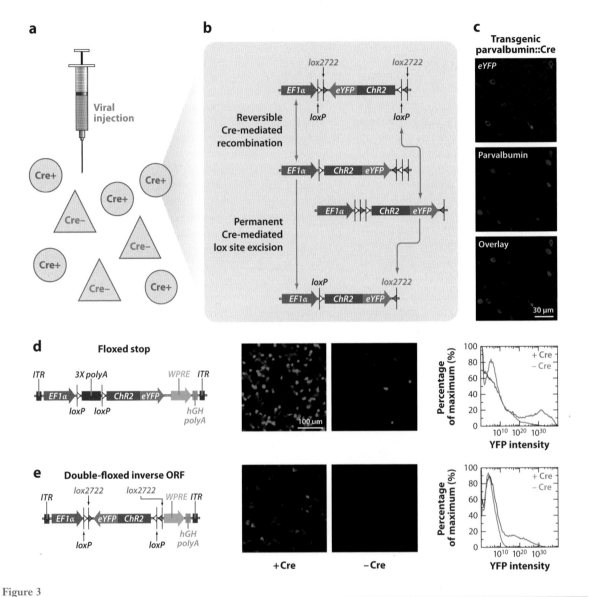

Figure 3

Low-leak Cre-dependent expression using the doublefloxed inverted open-reading-frame (DIO) strategy. The combination of a transgenic mouse expressing Cre recombinase in specific neuronal subtypes and the injection of a virally encoded DIO opsin (*a*) results in the physical inversion of the open reading frame (ORF) in only that population (*b,c*), which may be transient and revert back to the original state or undergo further recombination to be permanently anchored in the sense direction, resulting in functional expression of the opsin (*c*). The DIO strategy may be contrasted with the lox-stop-lox ("floxed STOP") strategy (*d,e*). In the absence of Cre recombinase, lox-stop-lox (*d*) allows for some level of expression leak as assayed by both enhanced yellow fluorescent protein (eYFP) expression and fluorescence-activated cell sorting (FACS) analysis. Because the ORF of an opsin in DIO configuration encodes nonsense (*e*), there is no functional expression in the absence of Cre recombinase. Adapted with permission from Sohal et al. (2009) and F. Zhang and K. Deisseroth.

be used to target specific layers of cortex by incorporating the DNA (with no promoter size limit) into neurons being born during a specific embryonic stage. A major advantage of using IUE or transgenic mice over viral infection is that opsins are expressed at the time of birth, allowing electrophysiological researchers to harvest acute slices at a younger stage. Counteracting this advantage is the fact that transgenic animals typically express lower levels of opsins, likely owing to the reduced gene copy number.

CIRCUIT TARGETING

Another generalizeable strategy for targeting is referred to as projection targeting, which capitalizes on the efficient membrane trafficking of engineered opsin gene products, especially down axons to axon terminals. Light can be delivered not to somata but to axons, thereby recruiting cells defined by virtue of their wiring without any genetic information about the downstream target required (Gradinaru et al. 2007, 2009; Lee et al. 2010; Petreanu et al. 2007). Neurons may also be targeted by projection using viruses that transduce axon terminals, such as herpes simplex virus (HSV) family viruses, certain serotypes of AAV, or pseudotyped lentiviruses. Trans-synaptic targeting may be achieved by exploiting the transcellular trafficking of, for example, the wheat germ agglutinin (WGA) peptide sequence (Gradinaru et al. 2010), which can deliver Cre recombinase to the site of a second Cre-dependent virus injection. Combination strategies are also possible; by crossing transgenic Drd2-GFP mice with mice expressing Cre under the control of Emx1, then injecting DIO-ChR2-mCherry into cortex, Higley & Sabatini (2010) were able to localize synapses originating from cortex (red projections) onto neurons expressing the dopamine-2 receptor (D2) in striatum (green cell bodies) and use brief pulses of blue light to elicit synaptic activity onto the D2 neurons. Combining optogenetic manipulation of the synapse with pharmacology and two-photon glutamate uncaging allowed the investigators to elaborate precisely upon the role of D2 receptors in glutamatergic synaptic transmission.

A noteworthy method pioneered by the Callaway group (Wickersham et al. 2007a,b) using a glycoprotein-deficient pseudotyped rabies virus is yet another technique for monosynaptic circuit tracing. Rabies virus is well known to travel trans-synaptically from neuron to neuron; the virus used in this technique is not able to produce viable packaged copies of itself after moving trans-synaptically and thus will be stopped after one synapse jump. Applying this method enables tracing of all neurons synaptically connected to a single neuron of interest. Two plasmids, one containing the glycoprotein-deficient rabies payload and another containing the glycoprotein, are coelectroporated into a single neuron in vivo. This neuron is then able to produce competent rabies virus; the payload, however, does not encode the coat protein. In this case, the virus is stuck after moving one synapse. By using red and green fluorophores in the two components, the targeted neuron and its synaptic partners may be identified. This system has not yet been integrated with optogenetics, however, and the extremely high levels of expression resulting from rabies virus expression may result in toxicity incompatible with typical optogenetic experiments.

LIGHT DELIVERY AND READOUT HARDWARE FOR OPTOGENETICS

Optogenetics fundamentally relies on light-delivery technology, the development of which has led to improved precision of modulation both in vitro and in vivo. In vitro, optogenetic tools are typically activated with filtered light from mercury arc lamps (e.g., Berndt et al. 2009, Boyden et al. 2005, Gunaydin et al. 2010), lasers (Cardin et al. 2010, Cruikshank et al. 2010, Kravitz et al. 2010, Petreanu et al. 2009), light-emitting diodes (LEDs) (e.g., Adesnik & Scanziani 2010; Grubb & Burrone 2010a,b; Wang et al. 2009), or LED arrays for multisite stimulation (Grossman et al. 2010). In vivo, stimulation of behaving animals has been conducted mostly with laser light delivered to the transduced tissue via optical fibers inserted through chronically implanted cannulas (Adamantidis et al. 2007, Aravanis et al. 2007, Zhang et al. 2010) or with fiber-coupled high-power LEDs (Wang et al. 2010). The chronic delivery of light using implanted infrared-triggered LEDs (Iwai et al. 2011) is in the early developmental stages but promises to open a new direction in optogenetic research.

To achieve rich readouts from optogenetically controlled tissue, major effort has been directed toward generating electrophysiological systems that combine high-density single-unit recordings with optogenetic stimulation in mice and other organisms. The first readouts from in vivo optogenetic modulation were obtained in anesthetized animals using a device composed of a fiberoptic cable integrated with a tungsten electrode (Gradinaru et al. 2007), called an "optrode" (Cardin et al. 2009; Gradinaru et al. 2007, 2009; Sohal et al. 2009; Zhang et al. 2009). More advanced strategies have emerged recently, employing silicone multisite electrodes (Royer et al. 2010) and movable tetrode arrays combined with optical fibers for more flexible interrogation of neural activity in vivo (Lima et al. 2009). Two-photon imaging is another avenue with which optogenetics may be integrated to stimulate and record neural activity simultaneously. Several studies have made progress toward this type of experiment (Andrasfalvy et al. 2010, Mohanty et al. 2008, Papagiakoumou et al. 2010, Rickgauer & Tank 2009, Zhu et al. 2009); it seems that the major limitation that hampers optogenetic activation with two-photon approaches is the combination of rapidly decaying opsin-mediated photocurrents in the setting of typical slow 2P raster-scanning techniques. Modifying the raster scan paradigm (Rickgauer & Tank 2009) or modulating the laser light such that fast activation is possible across wider regions (i.e., an entire cell soma; Papagiakoumou et al. 2010) can address this problem (Shoham 2010). Appropriately rich readouts, when combined with optogenetic inputs, will powerfully facilitate fundamental studies regarding the organization and function of intact, complex neural networks.

OPTOGENETICS IN DIVERSE ANIMAL MODELS

Caenorhabditis elegans

In transgenic nematodes harboring the channelrhodopsin gene, it is possible to control muscle wall motor neuron and mechanosensory neuron activity (Nagel et al. 2005). Zhang and colleagues (2007a) controlled body wall muscle contraction bidirectionally with ChR2 and NpHR, demonstrating the power of combinatorial optogenetics. This concept has been built on with the description of three-color LCD-based multimodal light delivery (Stirman et al. 2011) and digital micromirror device (DMD)/laser light delivery (Leifer et al. 2011), each coupled with tracking software for use with the behaving specimen. The facility of quantifying body-wall contraction and elongation in C. elegans has enabled large-scale investigation of various mutant strains for synaptic protein defects (Liewald et al. 2008, Stirman et al. 2010) and nicotinic acetylcholine receptor function (Almedom et al. 2009). Finally, C. elegans was also used for combined light-based stimulation and readout of neural activity (Guo et al. 2009, Tian et al. 2009), fulfilling the promise of all-optical physiological experiments using optogenetic tools and genetically encoded activity sensors (Scanziani & Hausser 2009).

Fly

Fly lines expressing upstream activation sequence (UAS):ChR2 (Zhang et al. 2007b) have been used to investigate the neuronal basis of the nociceptive response (Hwang et al. 2007) and appetitive/aversive odorant learning at the receptor (Bellmann et al. 2010) or neurotransmitter (Schroll et al. 2006) level and to rescue photosensory mutants (Xiang et al. 2010). Additionally, Hortstein et al. (2009) demonstrated Gal4/UAS targeting of ChR2 to the larval neuromuscular junction system. Creative uses of optogenetic tools in *Drosophila* include validating neurons identified in a screen to probe the proboscis extension reflex (Gordon & Scott

2009), driving monoamine release to validate fast-scanning cyclic voltammetry detection of serotonin and dopamine (Borue et al. 2009), and investigating the innate escape response (Zimmermann et al. 2009). Special considerations are required for this model organism (Pulver et al. 2009). Unlike mammals, flies and worms do not possess levels of endogenous retinal sufficient for the function of optogenetic tools, but food supplement can provide sufficient retinal to drive ChR2 function (Xiang et al. 2010).

Flies also possess innate behavioral responses to blue light that are developmentally dependent (Bellmann et al. 2010, Pulver et al. 2009, Suh et al. 2007, Xiang et al. 2010), complicating behavioral studies using opsins with blue activation spectra. This confound may be partially rectified using fly lines without vision, such as those lacking the *norpA* gene (Bellmann et al. 2010), although norpA-deficient lines remain sensitive to blue light (Xiang et al. 2010). These issues could, in principle, be circumvented with red-shifted optogenetic tools for excitation and inhibition. Complementing eNpHR3.0 for red-shifted inhibition, we have developed an optogenetic toolset for potent red-shifted excitation (O. Yizhar, L. Fenno, M. Prigge, K. Stehfest, J. Paz, F. Schneider, S. Tsunoda, R. Fudim, C. Ramakrishnan, J. Huguenard, P. Hegemann & K. Deisseroth, submitted). These have yet to be tested in *Drosophila*, but their green peak activation wavelength is outside the range of key *Drosophila* photosensory proteins (Xiang et al. 2010); moreover, spiking may be driven with up to 630 nm light, improving the potential for deep-penetrating excitation (Pulver et al. 2009).

Zebrafish

The short generational time and easy integration of foreign DNA into zebrafish are complemented by ease of optogenetic manipulation owing to transparency of the organism (McLean & Fetcho 2011, White et al. 2008). The first use of optogenetic tools in zebrafish (Douglass et al. 2008) appeared in a study examining the role of somatosensory control of escape behavior. The use of ChR2 to drive single spikes in a genetically defined population during the course of movement took advantage of a number of properties of optogenetic tools not available with traditional pharmacological or electrophysiological methods. Neurons were stimulated with a simple setup combining a dissecting scope and epifluorescence source, with light restricted by the microscope aperture. Recent advances (Arrenberg et al. 2009) reported zebrafish lines with eNpHR- enhanced yellow fluorescent protein (eYFP) and ChR2-eYFP expression controlled by the Gal4/UAS system to allow the tools to be easily targeted to specific neuronal subtypes via genetic crosses with zebrafish expressing Gal4 in various cell populations. Of note, Arrenberg et al. (2009) compared various iterations of NpHR and fluorophore and concluded that eNpHR2.0-eYFP had the most reliable and efficient expression. Photoconverting proteins Kaede and Dendra were used to approximate the upper bound of light spread with low numerical aperture, small-diameter (50 um) fibers in place of a microscope aperture, observing that the combination of small-fiber and neuronal targeting allowed for the stimulation of an approximately 30-um-diameter spot.

Other groups have reported using Gal4/UAS systems driving optogenetic tools to examine cardiac function and development (Arrenberg et al. 2010), transduction of sensory neuron mechanoreception (Low et al. 2010), command of swim behavior (Arrenberg et al. 2009, Douglass et al. 2008), and saccade generation (Schoonheim et al. 2010). As an alternative to producing stable transgenic lines, Zhu et al. (2009) undertook a systematic examination of viral infection in zebrafish and found successful ChR2 delivery by the Sindbis and rabies viruses. Of note, they also modulated ChR2 expression using a Tet-inducible expression strategy. A specific technical consideration of implementing optogenetic tools in studies using zebrafish is the stimulation of neurons that express endogenous light-activated proteins; Arrenberg et al. (2009) found that 26% of control neurons in zebrafish had a firing rate modulated by yellow light. This percentage was reduced to 14% in congenitally blind zebrafish lines. The remaining response was postulated to be due to either expression of other optically activated proteins or a thermal response, but it was not investigated further.

Mouse

By far, the most widely published optogenetic model organism to date has been the mouse. Mice represent the majority of transgenic animals, including a vast selection of transgenic lines expressing Cre recombinase in specific subpopulations of neurons (Gong et al. 2007). Mouse embryonic stem cells have also been amenable to expression and interrogation with optogenetic technologies (Stroh et al. 2010). The first report to use channelrhodopsin in behaving mammals examined the contribution of hypothalamic hypocretin (orexin) neurons to sleep and wakefulness (Adamantidis et al. 2007).

Optogenetic modulation in mouse has also yielded control of monoaminergic systems. Recently, in a study that used ChR2, eNpHR2.0, and TH::Cre transgenic mice to modulate the locus coeruleus neurons bidirectionally, these noradrenergic neurons strongly modulated sleep and arousal states (Carter et al. 2010). Using the same TH::Cre transgenic mice, causal relationships were identified between activity patterns in VTA dopamine neurons and reward behavior in mice, showing that phasic dopamine release is more effective than tonic release in driving reward behavior (Tsai et al. 2009). Using optogenetic stimulation of axonal terminals in the nucleus accumbens, investigators recently discovered that dopamine neurons corelease glutamate (Stuber et al. 2010, Tecuapetla et al. 2010). DAT-Cre mice have been used in conjunction with Cre-dependent ChR2 to examine mechanisms underlying dopamine-modulated addiction (Brown et al. 2010). And ChAT-Cre transgenic mice were used in combination with Cre-dependent ChR2 and NpHR viruses

to show that cholinergic interneurons of the nucleus accumbens are key regulators of medium spiny neuron activity and can modulate cocaine-based place preference (Witten et al. 2010). The connectivity of striatal medium spiny neurons themselves has been described using a tetracycline-based ChR2 transgenic system (Chuhma et al. 2011).

Direct optogenetic modulation of principal and local-circuit inhibitory neurons in mouse cortex and hippocampus has also enabled contributions to understanding the complexity of mammalian neural circuit dynamics. Reports on the functions of parvalbumin-expressing fast-spiking interneurons demonstrated directly their involvement in gamma oscillations and information processing in mouse prefrontal (Sohal et al. 2009) and somatosensory (Cardin et al. 2009, 2010) cortex. Focal stimulation of pyramidal neurons in Thy1::ChR2 mice has enabled rapid, functional mapping of motor control across the motor cortex (Ayling et al. 2009, Hira et al. 2009), and axonal stimulation in regions contralateral to injected cortical areas has enabled the mapping of projection patterns in callosal cortical projections (Petreanu et al. 2007). Within local cortical microcircuits, ChR2 has been used to characterize the spatial receptive fields of various neuron types (Katzel et al. 2010, Petreanu et al. 2009, Wang et al. 2007) and to study the basic properties of cortical disynaptic inhibition (Hull et al. 2009). Optogenetics is also being used to discern the possible therapeutic mechanism of cortical intervention in mouse models of depression (Covington et al. 2010) and to develop novel strategies for control of peripheral nerves (Llewellyn et al. 2010).

Mice have also been used to study amygdala circuits involved in fear and anxiety. Johansen and colleagues (2010) used ChR2 to demonstrate the sufficiency of lateral amygdala pyramidal neurons in auditory cued fear conditioning. In two recent reports, functional circuits within the central amygdala were further delineated, demonstrating that distinct subpopulations of inhibitory central amygdala neurons separately gate the acquisition and expression of conditioned fear (Ciocchi et al. 2010, Haubensak et al. 2010). Finally, Tye and colleagues (2011) described the differential effects of activating lateral amygdala projections onto central amygdala neurons in regulating anxiety behaviors. These studies shed new light on fear and anxiety behaviors and demonstrate the utility of optogenetic techniques in dissecting complex local neuronal circuits.

Rat

Rats are important for neuroscience research because of their ability to perform complex behavioral tasks, the relative simplicity of their brains (compared with human and nonhuman primates), and the ability to perform high-density recordings of neural ensembles during free behavior. Recently, virally delivered optogenetic tools were used in rats to examine blood oxygen level–dependent (BOLD) responses in functional magnetic resonance imaging (fMRI) (Lee et al. 2010). Driving ChR2 in excitatory neuronal populations was sufficient to elicit a BOLD response not only in local cortical targets (where both the virus and light delivery optical fiber were targeted) but also in downstream thalamic regions, allowing global maps of activity causally driven by defined cell populations to be obtained within intact living mammals. Optogenetic work in rats has been limited by the availability of viral promoters that are capable of driving specific expression in the absence of transgenic targeting, but the advent of transgenic rat lines expressing Cre recombinase in specific neuronal subtypes (in addition to projection targeting) will greatly expand the potential for using rat models of neural circuit function in health and disease. Outside the central nervous system, optogenetic manipulation in rodents is providing insights into diverse physiological functions. ChR2 was used to modulate rhythmic beating activity in rodent cardiomyocytes, demonstrating the potential for future applications in this field (Bruegmann et al. 2010), and several groups have used opto-genetics to modulate cardiovascular function, breathing, and blood pressure (Abbott et al. 2009a,b; Alilain et al. 2008; Kanbar et al. 2010) in both anesthetized and awake rats.

Primate

Optogenetic modulation of primate neurons (Han et al. 2009, Diester et al. 2011) has been explored by ChR2 delivery to cortical neurons of macaques via lentiviral transduction, but behavioral responses have not yet been observed. eNpHR2.0 has been delivered to human neural tissue in the form of ex vivo human retinas and has shown optogenetic efficacy on physiological measures (Busskamp et al. 2010) with possible relevance to retinitis pigmentosa (RP), a disease in which light-sensing cells degenerate in the retina. By expressing eNpHR2.0 in light-insensitive cone cells, normal phototransduction was restored, as well as center/surround computational features, directional sensitivity, and light-guided behavior. Additionally, Weick et al. (2010) demonstrated the functionality of ChR2 in human embryonic stem cell–derived neurons.

OUTLOOK

The optogenetic toolbox has broadly expanded to include proteins that are powerful and diverse in their ionic selectivity, spectral sensitivity, and temporal resolution. Combined with powerful molecular techniques for transgenic and viral expression in rodents, zebrafish, and flies, the current generation of optogenetic tools may be adapted to an extensive landscape of questions within neuroscience. The current generation of optogenetic tools has been optimized for stronger expression, higher currents, and spectral shifts to allow combinatorial control within the same volume of space. Ongoing improvements to the toolbox will yield molecular tools targeted to subcellular compartments [such as dendrites or axons (Lewis et al. 2009)], tools for two-photon activation, and tools that further expand the optical control of biochemistry. At this moment in time, single-component optogenetics has become a staple in neuroscience laboratories, even as many opportunities remain yet untapped.

DISCLOSURE STATEMENT

The authors are not aware of any affiliations, memberships, funding, or financial holdings that might be perceived as affecting the objectivity of this review.

LITERATURE CITED

Abbott SB, Stornetta RL, Fortuna MG, Depuy SD, West GH, et al. 2009a. Photostimulation of retrotrapezoid nucleus phox2b-expressing neurons in vivo produces long-lasting activation of breathing in rats. *J. Neurosci.* 29:5806–19

Abbott SB, Stornetta RL, Socolovsky CS, West GH, Guyenet PG. 2009b. Photostimulation of channelrhodopsin-2 expressing ventrolateral medullary neurons increases sympathetic nerve activity and blood pressure in rats. *J. Physiol.* 587:5613–31

Adamantidis AR, Zhang F, Aravanis AM, Deisseroth K, de Lecea L. 2007. Neural substrates of awakening probed with optogenetic control of hypocretin neurons. *Nature* 450:420–24

Adesnik H, Scanziani M. 2010. Lateral competition for cortical space by layer-specific horizontal circuits. *Nature* 464:1155–60

Airan RD, Thompson KR, Fenno LE, Bernstein H, Deisseroth K. 2009. Temporally precise in vivo control of intracellular signalling. *Nature* 458:1025–29

Alilain WJ, Li X, Horn KP, Dhingra R, Dick TE, et al. 2008. Light-induced rescue of breathing after spinal cord injury. *J. Neurosci.* 28:11862–70

Almedom RB, Liewald JF, Hernando G, Schultheis C, Rayes D, et al. 2009. An ER-resident membrane protein complex regulates nicotinic acetylcholine receptor subunit composition at the synapse. *EMBO J.* 28:2636–49

Andrasfalvy BK, Zemelman BV, Tang J, Vaziri A. 2010. Two-photon single-cell optogenetic control of neuronal activity by sculpted light. *Proc. Natl. Acad. Sci. USA* 107:11981–86

Aravanis AM, Wang LP, Zhang F, Meltzer LA, Mogri MZ, et al. 2007. An optical neural interface: in vivo control of rodent motor cortex with integrated fiberoptic and optogenetic technology. *J. Neural Eng.* 4:S143–56

Arenkiel BR, Peca J, Davison IG, Feliciano C, Deisseroth K, et al. 2007. In vivo light-induced activation of neural circuitry in transgenic mice expressing channelrhodopsin-2. *Neuron* 54:205–18

Arrenberg AB, Del Bene F, Baier H. 2009. Optical control of zebrafish behavior with halorhodopsin. *Proc. Natl. Acad. Sci. USA* 106:17968–73

Arrenberg AB, Stainier DY, Baier H, Huisken J. 2010. Optogenetic control of cardiac function. *Science* 330:971–74

Atasoy D, Aponte Y, Su HH, Sternson SM. 2008. A FLEX switch targets channelrhodopsin-2 to multiple cell types for imaging and long-range circuit mapping. *J. Neurosci.* 28:7025–30

Ayling OG, Harrison TC, Boyd JD, Goroshkov A, Murphy TH. 2009. Automated light-based mapping of motor cortex by photoactivation of channelrhodopsin-2 transgenic mice. *Nat. Methods* 6:219–24

Bamann C, Gueta R, Kleinlogel S, Nagel G, Bamberg E. 2010. Structural guidance of the photocycle of channelrhodopsin-2 by an interhelical hydrogen bond. *Biochemistry* 49:267–78

Banghart M, Borges K, Isacoff E, Trauner D, Kramer RH. 2004. Light-activated ion channels for remote control of neuronal firing. *Nat. Neurosci.* 7:1381–86

Beja O, Aravind L, Koonin EV, Suzuki MT, Hadd A, et al. 2000. Bacterial rhodopsin: evidence for a new type of phototrophy in the sea. *Science* 289:1902–6

Beja O, Spudich EN, Spudich JL, Leclerc M, DeLong EF. 2001. Proteorhodopsin phototrophy in the ocean. *Nature* 411:786–89

Bellmann D, Richardt A, Freyberger R, Nuwal N, Schwarzel M, et al. 2010. Optogenetically induced olfactory stimulation in *Drosophila* larvae reveals the neuronal basis of odor-aversion behavior. *Front. Behav. Neurosci.* 4:27

Berndt A, Prigge M, Gradmann D, Hegemann P. 2010. Two open states with progressive proton selectivities in the branched channelrhodopsin-2 photocycle. *Biophys. J.* 98:753–61

Berndt A, Schoenenberger P, Mattis J, Tye KM, Deisseroth K, et al. 2011. High-efficiency channelrhodopsins for fast neuronal stimulation at low light levels. *Proc. Natl. Acad. Sci. USA* Submitted

Berndt A, Yizhar O, Gunaydin LA, Hegemann P, Deisseroth K. 2009. Bi-stable neural state switches. *Nat. Neurosci.* 12:229–34

Bi A, Cui J, Ma YP, Olshevskaya E, Pu M, et al. 2006. Ectopic expression of a microbial-type rhodopsin restores visual responses in mice with photoreceptor degeneration. *Neuron* 50:23–33

Borue X, Cooper S, Hirsh J, Condron B, Venton BJ. 2009. Quantitative evaluation of serotonin release and clearance in *Drosophila*. *J. Neurosci. Methods* 179:300–8

Boyden ES, Zhang F, Bamberg E, Nagel G, Deisseroth K. 2005. Millisecond-timescale, genetically targeted optical control of neural activity. *Nat. Neurosci.* 8:1263–68

Brown MT, Bellone C, Mameli M, Labouebe G, Bocklisch C, et al. 2010. Drug-driven AMPA receptor redistribution mimicked by selective dopamine neuron stimulation. *PLoS One* 5:e15870

Bruegmann T, Malan D, Hesse M, Beiert T, Fuegemann CJ, et al. 2010. Optogenetic control of heart muscle in vitro and in vivo. *Nat. Methods* 7:897–900

Busskamp V, Duebel J, Balya D, Fradot M, Viney TJ, et al. 2010. Genetic reactivation of cone photoreceptors restores visual responses in retinitis pigmentosa. *Science* 329:413–17

Cardin JA, Carlen M, Meletis K, Knoblich U, Zhang F, et al. 2009. Driving fast-spiking cells induces gamma rhythm and controls sensory responses. *Nature* 459:663–67

Cardin JA, Carlén M, Meletis K, Knoblich U, Zhang F, et al. 2010. Targeted optogenetic stimulation and recording of neurons in vivo using cell-type-specific expression of channelrhodopsin-2. *Nat. Protoc.* 5:247–54

Carter ME, Yizhar O, Chikahisa S, Nguyen H, Adamantidis A, et al. 2010. Tuning arousal with optogenetic modulation of locus coeruleus neurons. *Nat. Neurosci.* 13:1526–33

Chow BY, Han X, Dobry AS, Qian X, Chuong AS, et al. 2010. High-performance genetically targetable optical neural silencing by light-driven proton pumps. *Nature* 463:98–102

Chuhma N, Tanaka KF, Hen R, Rayport S. 2011. Functional connectome of the striatal medium spiny neuron. *J. Neurosci.* 31:1183–92

Ciocchi S, Herry C, Grenier F, Wolff SB, Letzkus JJ, et al. 2010. Encoding of conditioned fear in central amygdala inhibitory circuits. *Nature* 468:277–82

Covington HE 3rd, Lobo MK, Maze I, Vialou V, Hyman JM, et al. 2010. Antidepressant effect of optogenetic stimulation of the medial prefrontal cortex. *J. Neurosci.* 30:16082–90

Crick FH. 1979. Thinking about the brain. *Sci. Am.* 241:219–32

Cruikshank SJ, Urabe H, Nurmikko AV, Connors BW. 2010. Pathway-specific feedforward circuits between thalamus and neocortex revealed by selective optical stimulation of axons. *Neuron* 65:230–45

Deisseroth K. 2010. Controlling the brain with light. *Sci. Am.* 303:48–55

Deisseroth K. 2011. Optogenetics. *Nat. Methods* 8:26–29

Deisseroth K, Feng G, Majewska AK, Miesenbock G, Ting A, Schnitzer MJ. 2006. Next-generation optical technologies for illuminating genetically targeted brain circuits. *J. Neurosci.* 26:10380–86

Diester I, Kaufman MT, Mogri M, Pashaie R, Goo W, et al. 2011. An optogenetic toolbox designed for primates. *Nat. Neurosci.* 14:387–97

Douglass AD, Kraves S, Deisseroth K, Schier AF, Engert F. 2008. Escape behavior elicited by single, channelrhodopsin-2-evoked spikes in zebrafish somatosensory neurons. *Curr. Biol.* 18:1133–37

Ernst OP, Sánchez Murcia PA, Daldrop P, Tsunoda SP, Kateriya S, Hegemann P. 2008. Photoactivation of channelrhodopsin. *J. Biol. Chem.* 283:1637–43

Feldbauer K, Zimmermann D, Pintschovius V, Spitz J, Bamann C, Bamberg E. 2009. Channelrhodopsin-2 is a leaky proton pump. *Proc. Natl. Acad. Sci. USA* 106:12317–22

Fork RL. 1971. Laser stimulation of nerve cells in aplysia. *Science* 171:907–8

Geschwind D. 2004. GENSAT: a genomic resource for neuroscience research. *Lancet Neurol.* 3:82

Gong S, Doughty M, Harbaugh CR, Cummins A, Hatten ME, et al. 2007. Targeting Cre recombinase to specific neuron populations with bacterial artificial chromosome constructs. *J. Neurosci.* 27:9817–23

Gong S, Zheng C, Doughty ML, Losos K, Didkovsky N, et al. 2003. A gene expression atlas of the central nervous system based on bacterial artificial chromosomes. *Nature* 425:917–25

Gordon MD, Scott K. 2009. Motor control in a *Drosophila* taste circuit. *Neuron* 61:373–84

Gorostiza P, Isacoff EY. 2008. Optical switches for remote and noninvasive control of cell signaling. *Science* 322:395–99

Gourine AV, Kasymov V, Marina N, Tang F, Figueiredo MF, et al. 2010. Astrocytes control breathing through pH-dependent release of ATP. *Science* 329:571–75

Gradinaru V, Mogri M, Thompson KR, Henderson JM, Deisseroth K. 2009. Optical deconstruction of parkinsonian neural circuitry. *Science* 324:354–59

Gradinaru V, Thompson KR, Deisseroth K. 2008. eNpHR: a natronomonas halorhodopsin enhanced for optogenetic applications. *Brain Cell Biol.* 36:129–39

Gradinaru V, Thompson KR, Zhang F, Mogri M, Kay K, et al. 2007. Targeting and readout strategies for fast optical neural control in vitro and in vivo. *J. Neurosci.* 27:14231–38

Gradinaru V, Zhang F, Ramakrishnan C, Mattis J, Prakash R, et al. 2010. Molecular and cellular approaches for diversifying and extending optogenetics. *Cell* 141:154–65

Grossman N, Poher V, Grubb MS, Kennedy GT, Nikolic K, et al. 2010. Multi-site optical excitation using ChR2 and micro-LED array. *J. Neural Eng.* 7:16004

Grubb MS, Burrone J. 2010a. Activity-dependent relocation of the axon initial segment fine-tunes neuronal excitability. *Nature* 465:1070–74

Grubb MS, Burrone J. 2010b. Channelrhodopsin-2 localised to the axon initial segment. *PloS One* 5:e13761

Gunaydin LA, Yizhar O, Berndt A, Sohal VS, Deisseroth K, Hegemann P. 2010. Ultrafast optogenetic control. *Nat. Neurosci.* 13:387–92

Guo ZV, Hart AC, Ramanathan S. 2009. Optical interrogation of neural circuits in *Caenorhabditis elegans*. *Nat. Methods* 6:891–96

Hagglund M, Borgius L, Dougherty KJ, Kiehn O. 2010. Activation of groups of excitatory neurons in the mammalian spinal cord or hindbrain evokes locomotion. *Nat. Neurosci.* 13:246–52

Han X, Boyden ES. 2007. Multiple-color optical activation, silencing, and desynchronization of neural activity, with single-spike temporal resolution. *PloS One* 2:e299

Han X, Qian X, Bernstein JG, Zhou HH, Franzesi GT, et al. 2009. Millisecond-timescale optical control of neural dynamics in the nonhuman primate brain. *Neuron* 62:191–98

Haubensak W, Kunwar PS, Cai H, Ciocchi S, Wall NR, et al. 2010. Genetic dissection of an amygdala microcircuit that gates conditioned fear. *Nature* 468:270–76

Haupts U, Tittor J, Bamberg E, Oesterhelt D. 1997. General concept for ion translocation by halobacterial retinal proteins: the isomerization/switch/transfer (IST) model. *Biochemistry* 36:2–7

Hegemann P, Ehlenbeck S, Gradmann D. 2005. Multiple photocycles of channelrhodopsin. *Biophys. J.* 89:3911–18

Heintz N. 2004. Gene expression nervous system atlas (GENSAT). *Nat. Neurosci.* 7:483

Higley MJ, Sabatini BL. 2010. Competitive regulation of synaptic Ca2+ influx by D2 dopamine and A2A adenosine receptors. *Nat. Neurosci.* 13:958–66

Hira R, Honkura N, Noguchi J, Maruyama Y, Augustine GJ, et al. 2009. Transcranial optogenetic stimulation for functional mapping of the motor cortex. *J. Neurosci. Methods* 179:258–63

Hirase H, Nikolenko V, Goldberg JH, Yuste R. 2002. Multiphoton stimulation of neurons. *J. Neurobiol.* 51:237–47

Hires SA, Tian L, Looger LL. 2008. Reporting neural activity with genetically encoded calcium indicators. *Brain Cell Biol.* 36:69–86

Hofmann KP, Scheerer P, Hildebrand PW, Choe HW, Park JH, et al. 2009. A G protein-coupled receptor at work: the rhodopsin model. *Trends Biochem. Sci.* 34:540–52

Hornstein NJ, Pulver SR, Griffith LC. 2009. Channelrhodopsin2 mediated stimulation of synaptic potentials at *Drosophila* neuromuscular junctions. *J. Vis. Exp.* 25: doi:10.3791/1133. **http://www.jove.com/details.stp?id=1133**

Hull C, Adesnik H, Scanziani M. 2009. Neocortical disynaptic inhibition requires somatodendritic integration in interneurons. *J. Neurosci.* 29:8991–95

Hwang RY, Zhong L, Xu Y, Johnson T, Zhang F, et al. 2007. Nociceptive neurons protect *Drosophila* larvae from parasitoid wasps. *Curr. Biol.* 17:2105–16

Ishizuka T, Kakuda M, Araki R, Yawo H. 2006. Kinetic evaluation of photosensitivity in genetically engineered neurons expressing green algae light-gated channels. *Neurosci. Res.* 54:85–94

Iwai Y, Honda S, Ozeki H, Hashimoto M, Hirase H. 2011. A simple head-mountable LED device for chronic stimulation of optogenetic molecules in freely moving mice. *Neurosci. Res.* In press

Johansen JP, Hamanaka H, Monfils MH, Behnia R, Deisseroth K, et al. 2010. Optical activation of lateral amygdala pyramidal cells instructs associative fear learning. *Proc. Natl. Acad. Sci. USA* 107:12692–97

Kanbar R, Stornetta RL, Cash DR, Lewis SJ, Guyenet PG. 2010. Photostimulation of Phox2b medullary neurons activates cardiorespiratory function in conscious rats. *Am. J. Respir. Crit. Care Med.* 182:1184–94

Katzel D, Zemelman BV, Buetfering C, Wolfel M, Miesenbock G. 2010. The columnar and laminar organization of inhibitory connections to neocortical excitatory cells. *Nat. Neurosci.* 14:100–7

Kouyama T, Kanada S, Takeguchi Y, Narusawa A, Murakami M, Ihara K. 2010. Crystal structure of the light-driven chloride pump halorhodopsin from Natronomonas pharaonis. *J. Mol. Biol.* 396:564–79

Kravitz AV, Freeze BS, Parker PR, Kay K, Thwin MT, et al. 2010. Regulation of parkinsonian motor behaviours by optogenetic control of basal ganglia circuitry. *Nature* 466:622–26

Lee JH, Durand R, Gradinaru V, Zhang F, Goshen I, et al. 2010. Global and local fMRI signals driven by neurons defined optogenetically by type and wiring. *Nature* 465:788–92

Leifer AM, Fang-Yen C, Gershow M, Alkema MJ, Samuel AD. 2011. Optogenetic manipulation of neural activity in freely moving *Caenorhabditis elegans*. *Nat. Methods* 8:147–52

Levskaya A, Weiner OD, Lim WA, Voigt CA. 2009. Spatiotemporal control of cell signalling using a light-switchable protein interaction. *Nature* 461:997–1001

Lewis TL Jr, Mao T, Svoboda K, Arnold DB. 2009. Myosin-dependent targeting of transmembrane proteins to neuronal dendrites. *Nat. Neurosci.* 12:568–76

Li X, Gutierrez DV, Hanson MG, Han J, Mark MD, et al. 2005. Fast noninvasive activation and inhibition of neural and network activity by vertebrate rhodopsin and green algae channelrhodopsin. *Proc. Natl. Acad. Sci. USA* 102:17816–21

Liewald JF, Brauner M, Stephens GJ, Bouhours M, Schultheis C, et al. 2008. Optogenetic analysis of synaptic function. *Nat. Methods* 5:895–902

Lima SQ, Hromadka T, Znamenskiy P, Zador AM. 2009. PINP: a new method of tagging neuronal populations for identification during in vivo electrophysiological recording. *PloS One* 4:e6099

Lima SQ, Miesenbock G. 2005. Remote control of behavior through genetically targeted photostimulation of neurons. *Cell* 121:141–52

Lin JY, Lin MZ, Steinbach P, Tsien RY. 2009. Characterization of engineered channelrhodopsin variants with improved properties and kinetics. *Biophys. J.* 96:1803–14

Llewellyn ME, Thompson KR, Deisseroth K, Delp SL. 2010. Orderly recruitment of motor units under optical control in vivo. *Nat. Med.* 16:1161–65

Lobo MK, Covington HE 3rd, Chaudhury D, Friedman AK, Sun H, et al. 2010. Cell type-specific loss of BDNF signaling mimics optogenetic control of cocaine reward. *Science* 330:385–90

Low SE, Ryan J, Sprague SM, Hirata H, Cui WW, et al. 2010. touche is required for touch-evoked generator potentials within vertebrate sensory neurons. *J. Neurosci.* 30:9359–67

Luecke H, Schobert B, Richter HT, Cartailler JP, Lanyi JK. 1999a. Structural changes in bacteriorhodopsin during ion transport at 2 angstrom resolution. *Science* 286:255–61

Luecke H, Schobert B, Richter HT, Cartailler JP, Lanyi JK. 1999b. Structure of bacteriorhodopsin at 1.55 A resolution. *J. Mol. Biol.* 291:899–911

Man D, Wang W, Sabehi G, Aravind L, Post AF, et al. 2003. Diversification and spectral tuning in marine proteorhodopsins. *EMBO J.* 22:1725–31

Mattis J, Tye KM, Ramakrishnan C, O'Shea D, Gunaydin LA, et al. 2011. Clarifying the optogenetics toolbox: a comprehensive analysis of new and existing opsins for scientific application. *Nat. Meth.* Submitted

McLean DL, Fetcho JR. 2011. Movement, technology and discovery in the zebrafish. *Curr. Opin. Neurobiol.* 21:110–15

Miesenbock G, Kevrekidis IG. 2005. Optical imaging and control of genetically designated neurons in functioning circuits. *Annu. Rev. Neurosci.* 28:533–63

Mohanty SK, Reinscheid RK, Liu X, Okamura N, Krasieva TB, Berns MW. 2008. In-depth activation of channelrhodopsin 2-sensitized excitable cells with high spatial resolution using two-photon excitation with a near-infrared laser microbeam. *Biophys. J.* 95:3916–26

Nagel G, Brauner M, Liewald JF, Adeishvili N, Bamberg E, Gottschalk A. 2005. Light activation of channelrhodopsin-2 in excitable cells of *Caenorhabditis elegans* triggers rapid behavioral responses. *Curr. Biol.* 15:2279–84

Nagel G, Ollig D, Fuhrmann M, Kateriya S, Musti AM, et al. 2002. Channelrhodopsin-1: a light-gated proton channel in green algae. *Science* 296:2395–98

Nagel G, Szellas T, Huhn W, Kateriya S, Adeishvili N, et al. 2003. Channelrhodopsin-2, a directly light-gated cation-selective membrane channel. *Proc. Natl. Acad. Sci. USA* 100:13940–45

Oesterhelt D, Stoeckenius W. 1971. Rhodopsin-like protein from the purple membrane of Halobacterium halobium. *Nat. New Biol.* 233:149–52

Oh E, Maejima T, Liu C, Deneris E, Herlitze S. 2010. Substitution of 5-HT1A receptor signaling by a light-activated G protein-coupled receptor. *J. Biol. Chem.* 285:30825–36

Papagiakoumou E, Anselmi F, Begue A, de Sars V, Gluckstad J, et al. 2010. Scanless two-photon excitation of channelrhodopsin-2. *Nat. Methods* 7:848–54

Perálvarez-Marín A, Márquez M, Bourdelande JL, Querol E, Padrós E. 2004. Thr-90 plays a vital role in the structure and function of bacteriorhodopsin. *J. Biol. Chem.* 279:16403–9

Petreanu L, Huber D, Sobczyk A, Svoboda K. 2007. Channelrhodopsin-2-assisted circuit mapping of long-range callosal projections. *Nat. Neurosci.* 10:663–68

Petreanu L, Mao T, Sternson SM, Svoboda K. 2009. The subcellular organization of neocortical excitatory connections. *Nature* 457:1142–45

Pulver SR, Pashkovski SL, Hornstein NJ, Garrity PA, Griffith LC. 2009. Temporal dynamics of neuronal activation by channelrhodopsin-2 and TRPA1 determine behavioral output in *Drosophila* larvae. *J. Neurophysiol.* 101:3075–88

Rickgauer JP, Tank DW. 2009. Two-photon excitation of channelrhodopsin-2 at saturation. *Proc. Natl. Acad. Sci. USA* 106:15025–30

Ritter E, Stehfest K, Berndt A, Hegemann P, Bartl FJ. 2008. Monitoring light-induced structural changes of channelrhodopsin-2 by UV-visible and Fourier transform infrared spectroscopy. *J. Biol. Chem.* 283:35033–41

Royer S, Zemelman BV, Barbic M, Losonczy A, Buzsaki G, Magee JC. 2010. Multi-array silicon probes with integrated optical fibers: light-assisted perturbation and recording of local neural circuits in the behaving animal. *Eur. J. Neurosci.* 31:2279–91

Ryu MH, Moskvin OV, Siltberg-Liberles J, Gomelsky M. 2010. Natural and engineered photoactivated nucleotidyl cyclases for optogenetic applications. *J. Biol. Chem.* 285:41501–8

Sakmar TP. 2002. Structure of rhodopsin and the superfamily of seven-helical receptors: the same and not the same. *Curr. Opin. Cell Biol.* 14:189–95

Scanziani M, Hausser M. 2009. Electrophysiology in the age of light. *Nature* 461:930–39

Schmucker D, Su AL, Beermann A, Jackle H, Jay DG. 1994. Chromophore-assisted laser inactivation of patched protein switches cell fate in the larval visual system of *Drosophila*. *Proc. Natl. Acad. Sci. USA* 91:2664–68

Schoonheim PJ, Arrenberg AB, Del Bene F, Baier H. 2010. Optogenetic localization and genetic perturbation of saccade-generating neurons in zebrafish. *J. Neurosci.* 30:7111–20

Schroll C, Riemensperger T, Bucher D, Ehmer J, Voller T, et al. 2006. Light-induced activation of distinct modulatory neurons triggers appetitive or aversive learning in *Drosophila* larvae. *Curr. Biol.* 16:1741–47

Shichida Y, Yamashita T. 2003. Diversity of visual pigments from the viewpoint of G protein activation—comparison with other G protein-coupled receptors. *Photochem. Photobiol. Sci.* 2:1237–46

Shoham S. 2010. Optogenetics meets optical wavefront shaping. *Nat. Methods* 7:798–99

Sohal VS, Zhang F, Yizhar O, Deisseroth K. 2009. Parvalbumin neurons and gamma rhythms enhance cortical circuit performance. *Nature* 459:698–702

Spudich JL. 2006. The multitalented microbial sensory rhodopsins. *Trends Microbiol.* 14:480–87

Stehfest K, Hegemann P. 2010. Evolution of the channelrhodopsin photocycle model. *ChemPhysChem* 11:1120–26

Stehfest K, Ritter E, Berndt A, Bartl F, Hegemann P. 2010. The branched photocycle of the slow-cycling channelrhodopsin-2 mutant C128T. *J. Mol. Biol.* 398:690–702

Stierl M, Stumpf P, Udwari D, Gueta R, Hagedorn R, et al. 2010. Light modulation of cellular cAMP by a small bacterial photoactivated adenylyl cyclase, bPAC, of the soil bacterium Beggiatoa. *J. Biol. Chem.* 286:1181–88

Stirman JN, Brauner M, Gottschalk A, Lu H. 2010. High-throughput study of synaptic transmission at the neuromuscular junction enabled by optogenetics and microfluidics. *J. Neurosci. Methods* 191:90–93

Stirman JN, Crane MM, Husson SJ, Wabnig S, Schultheis C, et al. 2011. Real-time multimodal optical control of neurons and muscles in freely behaving *Caenorhabditis elegans*. *Nat. Methods* 8:152–58

Stoeckenius W. 1985. The rhodopsin-like pigments of halobacteria: light-energy and signal transducers in an archaebacterium. *Trends Biochem. Sci.* 10:483–86

Stroh A, Tsai HC, Ping Wang L, Zhang F, Kressel J, et al. 2010. Tracking stem cell differentiation in the setting of automated optogenetic stimulation. *Stem Cells* 29(1):78–88

Stuber GD, Hnasko TS, Britt JP, Edwards RH, Bonci A. 2010. Dopaminergic terminals in the nucleus accumbens but not the dorsal striatum corelease glutamate. *J. Neurosci.* 30:8229–33

Suh GS, Ben-Tabou de Leon S, Tanimoto H, Fiala A, Benzer S, Anderson DJ. 2007. Light activation of an innate olfactory avoidance response in *Drosophila*. *Curr. Biol.* 17:905–8

Szobota S, Gorostiza P, Del Bene F, Wyart C, Fortin DL, et al. 2007. Remote control of neuronal activity with a light-gated glutamate receptor. *Neuron* 54:535–45

Tecuapetla F, Patel JC, Xenias H, English D, Tadros I, et al. 2010. Glutamatergic signaling by mesolimbic dopamine neurons in the nucleus accumbens. *J. Neurosci.* 30:7105–10

Thyagarajan S, van Wyk M, Lehmann K, Lowel S, Feng G, Wassle H. 2010. Visual function in mice with photoreceptor degeneration and transgenic expression of channelrhodopsin 2 in ganglion cells. *J. Neurosci.* 30:8745–58

Tian L, Hires SA, Mao T, Huber D, Chiappe ME, et al. 2009. Imaging neural activity in worms, flies and mice with improved GCaMP calcium indicators. *Nat. Methods* 6:875–81

Tsai HC, Zhang F, Adamantidis A, Stuber GD, Bonci A, et al. 2009. Phasic firing in dopaminergic neurons is sufficient for behavioral conditioning. *Science* 324:1080–84

Tye K, Prakash R, Kim S-Y, Fenno LE, Grosenick L, et al. 2011. Amygdala circuitry mediating reversible and bidirectional control of anxiety. *Nature*. In press

Volgraf M, Gorostiza P, Numano R, Kramer RH, Isacoff EY, Trauner D. 2006. Allosteric control of an ionotropic glutamate receptor with an optical switch. *Nat. Chem. Biol.* 2:47–52

Wang H, Peca J, Matsuzaki M, Matsuzaki K, Noguchi J, et al. 2007. High-speed mapping of synaptic connectivity using photostimulation in channelrhodopsin-2 transgenic mice. *Proc. Natl. Acad. Sci. USA* 104:8143–48

Wang H, Sugiyama Y, Hikima T, Sugano E, Tomita H, et al. 2009. Molecular determinants differentiating photocurrent properties of two channelrhodopsins from chlamydomonas. *J. Biol. Chem.* 284:5685–96

Wang J, Borton DA, Zhang J, Burwell RD, Nurmikko AV. 2010. A neurophotonic device for stimulation and recording of neural microcircuits. *Conf. Proc. IEEE Eng. Med. Biol. Soc.* 1:2935–38

Weick JP, Johnson MA, Skroch SP, Williams JC, Deisseroth K, Zhang SC. 2010. Functional control of transplantable human ESC-derived neurons via optogenetic targeting. *Stem Cells* 28:2008–16

White RM, Sessa A, Burke C, Bowman T, LeBlanc J, et al. 2008. Transparent adult zebrafish as a tool for in vivo transplantation analysis. *Cell Stem Cell* 2:183–89

Wickersham IR, Finke S, Conzelmann KK, Callaway EM. 2007a. Retrograde neuronal tracing with a deletion-mutant rabies virus. *Nat. Methods* 4:47–49

Wickersham IR, Lyon DC, Barnard RJ, Mori T, Finke S, et al. 2007b. Monosynaptic restriction of transsynaptic tracing from single, genetically targeted neurons. *Neuron* 53:639–47

Witten IB, Lin SC, Brodsky M, Prakash R, Diester I, et al. 2010. Cholinergic interneurons control local circuit activity and cocaine conditioning. *Science* 330:1677–81

Wu YI, Frey D, Lungu OI, Jaehrig A, Schlichting I, et al. 2009. A genetically encoded photoactivatable Rac controls the motility of living cells. *Nature* 461:104–8

Xiang Y, Yuan Q, Vogt N, Looger LL, Jan LY, Jan YN. 2010. Light-avoidance-mediating photoreceptors tile the *Drosophila* larval body wall. *Nature* 468:921–26

Yazawa M, Sadaghiani AM, Hsueh B, Dolmetsch RE. 2009. Induction of protein-protein interactions in live cells using light. *Nat. Biotechnol.* 27:941–45

Zemelman BV, Lee GA, Ng M, Miesenbock G. 2002. Selective photostimulation of genetically chARGed neurons. *Neuron* 33:15–22

Zemelman BV, Nesnas N, Lee GA, Miesenbock G. 2003. Photochemical gating of heterologous ion channels: remote control over genetically designated populations of neurons. *Proc. Natl. Acad. Sci. USA* 100:1352–57

Zhang F, Gradinaru V, Adamantidis AR, Durand R, Airan RD, et al. 2010. Optogenetic interrogation of neural circuits: technology for probing mammalian brain structures. *Nat. Protoc.* 5:439–56

Zhang F, Prigge M, Beyriere F, Tsunoda SP, Mattis J, et al. 2008. Red-shifted optogenetic excitation: a tool for fast neural control derived from Volvox carteri. *Nat. Neurosci.* 11:631–33

Zhang F, Wang LP, Boyden ES, Deisseroth K. 2006. Channelrhodopsin-2 and optical control of excitable cells. *Nat. Methods* 3:785–92

Zhang F, Wang LP, Brauner M, Liewald JF, Kay K, et al. 2007a. Multimodal fast optical interrogation of neural circuitry. *Nature* 446:633–39

Zhang J, Laiwalla F, Kim JA, Urabe H, Van Wagenen R, et al. 2009. Integrated device for optical stimulation and spatiotemporal electrical recording of neural activity in light-sensitized brain tissue. *J. Neural Eng.* 6:055007

Zhang W, Ge W, Wang Z. 2007b. A toolbox for light control of *Drosophila* behaviors through channel-rhodopsin 2-mediated photoactivation of targeted neurons. *Eur. J. Neurosci.* 26:2405–16

Zhang YP, Oertner TG. 2007. Optical induction of synaptic plasticity using a light-sensitive channel. *Nat. Methods* 4:139–41

Zhao S, Cunha C, Zhang F, Liu Q, Gloss B, et al. 2008. Improved expression of halorhodopsin for light-induced silencing of neuronal activity. *Brain Cell Biol.* 36:141–54

Zhu P, Narita Y, Bundschuh ST, Fajardo O, Schärer YP, et al. 2009. Optogenetic dissection of neuronal circuits in zebrafish using viral gene transfer and the Tet system. *Front. Neural Circuits* 3:21

Zimmermann G, Wang LP, Vaughan AG, Manoli DS, Zhang F, et al. 2009. Manipulation of an innate escape response in *Drosophila*: photoexcitation of acj6 neurons induces the escape response. *PloS One* 4:e5100

Recovery of Locomotion After Spinal Cord Injury: Some Facts and Mechanisms

Serge Rossignol[1] and Alain Frigon[1,2]

[1]Groupe de Recherche sur le Système Nerveux Central (FRSQ), Department of Physiology, and Multidisciplinary Team in Locomotor Rehabilitation of the Canadian Institutes for Health Research, Université de Montréal, Montreal H3C 3J7, Canada; email: Serge.Rossignol@umontreal.ca

[2]Department of Physiology and Biophysics, Université de Sherbrooke, Sherbrooke J1H 5N4, Canada

Annu. Rev. Neurosci. 2011. 34:413–40

First published online as a Review in Advance on April 4, 2011

The *Annual Review of Neuroscience* is online at neuro.annualreviews.org

This article's doi:
10.1146/annurev-neuro-061010-113746

Keywords

spinalization, central pattern generator, electromyography, kinematics

Abstract

After spinal cord injury (SCI), various sensorimotor functions can recover, ranging from simple spinal reflexes to more elaborate motor patterns, such as locomotion. Locomotor recovery after complete spinalization (complete SCI) must depend on the presence of spinal circuitry capable of generating the complex sequential activation of various leg muscles. This is achieved by an intrinsic spinal circuitry, termed the central pattern generator (CPG), working in conjunction with sensory feedback from the legs. After SCI, different changes in cellular and circuit properties occur spontaneously and can be promoted by pharmacological, electrical, or rehabilitation strategies. After partial SCI, hindlimb locomotor recovery can result from regeneration or sprouting of spared pathways, but also from mechanisms observed after complete SCI, namely changes within the intrinsic spinal circuitry and sensory inputs.

Contents

Spinal cord injury (SCI) models in animals provide an opportunity to study function in the central nervous system (CNS). The spinal cord possesses well-characterized descending and ascending tracts and well-defined inputs of diverse sensory modalities, as well as autonomous circuits that produce a rich variety of quantifiable motor outputs, ranging from simple reflexes to more complex motor patterns, such as scratching, fast paw shake, and locomotion. Multiple aspects of spinal function can be investigated by inactivating or stimulating specific inputs and measuring outputs using behavioral, electrophysiological, genetic, and pharmacological tools. Lesions of spinal tracts or peripheral nerves have been key in demonstrating fundamental neurobiological concepts, some of which are gaining importance in the design of more effective and targeted rehabilitation therapies in humans with SCI.

The present review focuses on locomotor recovery after SCI by first describing some basic concepts, such as pattern generation by intrinsic spinal circuits and its control by descending and sensory inputs. Adaptation within these circuits and how their outputs

SCI: spinal cord injury

Central pattern generator (CPG): a circuitry of interneurons in the spinal cord capable of generating the basic rhythmic alternating activity between multiple groups of muscles with the proper temporal pattern of activation

NE: norepinephrine

5-HT: 5 hydroxytryptamine (serotonin)

T13: thoracic level 13

are modified by SCI to regain functions, such as locomotion, are then discussed. We propose that plasticity within intrinsic spinal circuits is a critical component of hindlimb locomotor recovery after SCI and that inputs from descending and peripheral sources that undergo functional and anatomical changes contribute to recovery mainly by accessing and modulating the modified spinal network.

Figure 1 illustrates some key features of locomotor control, consisting of (*a*) a specialized spinal circuitry, identified as the central pattern generator (CPG); (*b*) several descending pathways; (*c*) neurochemically defined pathways that release neuromodulators, such as norepinephrine (NE) or serotonin (5-HT); and (*d*) afferents from peripheral receptors of different sensory modalities. The roles of these components are largely derived from lesion studies and are discussed below. All these components interact, and after SCI, these interactions are disrupted as new ones become prominent. We first describe the effects of a spinalization, which leaves only the spinal circuits and sensory afferents to organize hindlimb locomotion. We then discuss the effects of a partial spinal lesion in which a combination of spinal, supraspinal, and sensory mechanisms remains to establish new interactions for the generation of hindlimb locomotion.

LOCOMOTOR RECOVERY AFTER COMPLETE SPINAL TRANSECTION

Cats with a complete SCI (i.e., spinalization) at the last thoracic segment (T13) gradually recover hindlimb locomotion on a treadmill following a few weeks of locomotor training. Although initially well studied in kittens (Grillner 1973; Forssberg et al. 1980a,b), hindlimb locomotor recovery after spinalization also occurs in adult cats (Barbeau & Rossignol 1987; Belanger et al. 1996; de Leon et al. 1998a,b; Rossignol et al. 2000, 2002) as well as several other species (Rossignol et al. 1996). In the adult cat, a few days after spinalization, with the forelimbs standing on

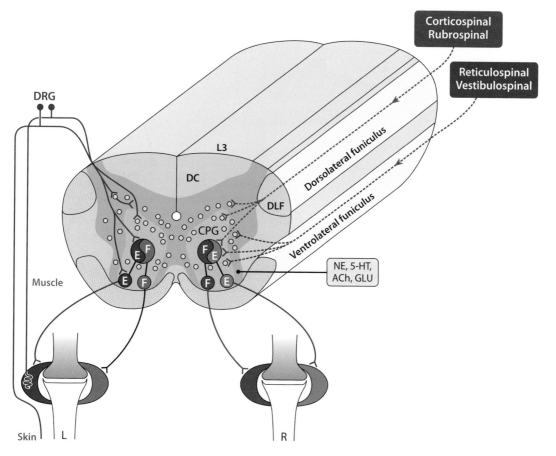

Figure 1

General framework of locomotor control. This scheme shows a lumbar spinal segment (L3-L4) with gray and white matter. In the gray matter, a specialized network of interneurons termed the central pattern generator (CPG) produces alternating activity between flexors and extensors on one side and is coupled with the CPG on the other side. The specialized interneurons activate flexor (F) and extensor (E) motoneurons. Some interneurons are under the influence of the CPG and respond to various inputs in a phase-dependent manner. Other interneurons are outside this zone of influence but receive various inputs as well. Descending inputs and sensory afferents can reach different types of interneurons. The main descending pathways coursing through the dorsolateral funiculus (DLF) are the cortico- and rubrospinal tracts. The reticulo- and vestibulospinal tracts are found in the ventrolateral funiculus (VLF). Other pathways are defined mainly by the type of neurotransmitters released, such as norepinephrine (NE), serotonin (5-HT), glutamate (GLU), or local circuits releasing GABA or acetylcholine (Ach). Sensory afferents originate from muscle or skin and project to the spinal cord through mono-, di-, and polysynaptic pathways. Abbreviations: DC, dorsal columns; DRG, dorsal root ganglion.

a fixed platform (see **Figure 2**) and manual stimulation of the perineum, small alternating steps can be evoked in the hindlimbs, although plantar foot placement is absent and a prominent foot drag is observed. After 2–3 weeks of daily treadmill training, cats can walk with alternate hindlimb movements, hindquarter weight support, and plantar foot contact.

The activity of hindlimb muscles [i.e., electromyography (EMG)] during treadmill walking is remarkably similar before and after spinalization in the same cat (Belanger et al. 1996, figure 5a,c). However, EMG amplitudes in extensor muscles can be of smaller amplitude, probably because of the loss of vestibulospinal or reticulospinal pathways. When present,

EMG:
electromyography

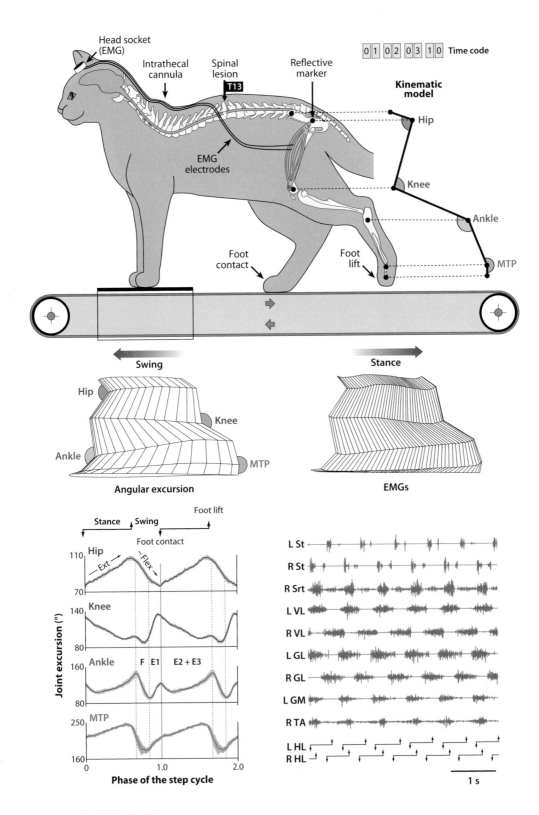

changes in EMG timing are most often observed between flexor muscles operating at different joints, such as hip and knee flexors, which might explain deficits such as foot drag during early swing.

In animal preparations with complete SCI, only the intrinsic spinal circuitry (i.e., CPG) and sensory inputs from peripheral receptors remain to initiate and organize hindlimb locomotion. The spinal locomotor CPG, except under strict experimental conditions, never works in isolation because it dynamically interacts with inputs from the periphery and from descending pathways (comprehensively reviewed in Rossignol et al. 2006). After complete SCI, access to the spinal locomotor circuitry by these inputs is altered because of the obvious loss of descending pathways controlling these circuits and also in regulating inputs from peripheral afferents. The sections below discuss important aspects of the CPG and peripheral sensory inputs in relation to recovery of function after complete SCI.

The Inescapable Central Spinal Pattern Generator

From the extensive work on spinalized animals, a circuitry within the spinal cord capable of generating the basic locomotor rhythm must exist (Grillner 1981, Rossignol 1996, Rossignol et al. 2006). The expression spinal pattern generator refers to the locomotor circuitry remaining after spinalization, which can still interact with tonic and phasic sensory inputs from the hindlimbs. The term central pattern generator should only be used to refer to spinal circuits capable of generating an organized bilateral rhythmic pattern in the absence of descending and sensory inputs. Indeed, in acutely spinalized and paralyzed cats, a well-organized rhythmic output (fictive locomotion) can be recorded directly from flexor or extensor nerves with pharmacological stimulation (L-DOPA) in the complete absence of overt movement (Grillner & Zangger 1979). In chronic spinal cats, fictive locomotion can occur spontaneously (i.e., without drugs), indicating that functional changes have occurred within the spinal locomotor circuitry enabling the spontaneous expression of this endogenous pattern (Pearson & Rossignol 1991).

Functional organization of locomotor-generating circuits. Although the spinal locomotor CPG is better characterized in swimming vertebrates, such as the lamprey (Buchanan 2001, Grillner 2003) and *Xenopus* tadpole (Roberts et al. 1998), a considerable body of work in adult cats and neonatal

Spinal pattern generator: a spinal circuit capable of generating locomotion after complete spinal transection, which is subjected to tonic or phasic sensory afferents from structures located caudal to the transection

Central pattern generator: such a pattern can be generated centrally in absence of phasic sensory afferent inputs after curarization

---

Figure 2

General methodology for the study of locomotion in cats with spinal lesions. (*Top panel*) With complete lesion of the spinal cord at T13, cats are positioned with their forelimbs standing on a platform fixed above the belt while the hindquarters are free to walk on the moving treadmill. With a partial lesion, the cat is free to walk with all four limbs on the treadmill. Pairs of electromyographic (EMG) electrodes are implanted into various muscles (only one pair represented here). The multipin EMG connector is cemented to the skull. Reflective markers are placed at different joints on the hindlimb to reconstruct angular excursions. The swing and stance phases of each cycle can also be determined. To prevent overlap of the stick figures, each one is displaced by an amount equal to the displacement of the foot along the horizontal axis. The bottom panels illustrate normal locomotion at 0.8 m s^{-1}. (*Left bottom panel*) Mean angular excursions of the hip, knee, ankle, and metatarsophalangeal (MTP) joints of 10 cycles of the left hindlimb. Angular values decrease in flexion and increase in extension for all joints. Subphases of the cycle are defined as flexion (F), first extension (E1) during swing, and second extension (E2) and third extension (E3) during stance, as proposed by Philippson (1905). (*Right panel*) The EMGs recorded in both hindlimbs and presented with stance phase duty cycles (*black bars*) in the four limbs. The downward arrows represent touchdown defining the onset of stance, whereas upward arrows represent the onset of liftoff defining the onset of swing (L HL, left hindlimb; R HL, right hindlimb). Muscles are abbreviated as follows: St, semitendinosus (knee flexor/hip extensor); Srt, sartorius (knee extensor, hip flexor); VL, vastus lateralis (knee extensor); GL and GM, gastrocnemius lateralis and medialis (ankle extensors and knee flexors); and TA, tibialis anterior (ankle flexor). The digital (SMPTE) time code (*upper right*) is used to synchronize video and EMG recordings. The spinal lesions are made at T13, unless otherwise specified. Modified with permission from Rossignol et al. 2009.

rodents has identified some key features of the mammalian spinal locomotor CPG (recently reviewed in Kiehn 2006, Goulding 2009), although overall, details of the mammalian CPG remain elusive. In quadrupedal mammals, the four limbs are controlled by distinct, but interconnected, CPGs. The CPGs controlling the hindlimbs can be uncoupled using a hemicord preparation, in which the lumbosacral spinal cord is split longitudinally along the mid-line (Kjaerulff & Kiehn 1997). The CPGs controlling fore- and hindlimb locomotion can be uncoupled by blocking transmission in propriospinal pathways that connect the cervical and lumbar enlargements (Viala & Vidal 1978, Juvin et al. 2005). Humans are thought to employ a coordination similar to quadrupeds during bipedal walking (Dietz & Michel 2009, Zehr et al. 2009). After incomplete SCI, homologous (i.e., between limbs of the same girdle) and homolateral (i.e., between fore- and hindlimbs on the same side) coordination during quadrupedal locomotion can be disrupted to various extents, giving rise to independent walking frequencies in the fore- and hindlimbs or to asymmetrical coupling between limbs in adult cats (Kato et al. 1984, Brustein & Rossignol 1998, Barrière et al. 2010).

The mammalian locomotor CPG is thought to be composed of interconnected modules that coordinate activity around specific joints (Grillner 1981). This ensemble of modules may or may not be dissociated from the rhythm-generating circuitry. A multilayered spinal locomotor CPG, in which rhythm-generation and pattern formation are functionally separated, has been proposed to account for some experimental findings (reviewed in McCrea & Rybak 2008). In this two-level CPG, the rhythm generator controls features of the rhythm (i.e., cycle period, phase durations/transitions) and projects to the pattern-formation level, which coordinates and distributes activity to motor pools. Inputs from peripheral mechanoreceptors or supraspinal structures can regulate activity at either level, including spinal motoneurons. Inputs producing an abrupt phase transition, or prolonging the ongoing phase, thereby disrupting the post-stimulus rhythm, indicate a more direct access to the rhythm-generating circuitry (Hultborn et al. 1998, Pearson et al. 1998).

Spinal localization of locomotor-generating circuits. In limbless vertebrates, such as fish, all segments are interconnected and are more or less equivalent in terms of rhythm-generating capacity to produce a smooth progressive traveling wave that exerts force on the water to move forward or backward, depending on the order of intersegmental coupling (Grillner & Wallen 2002). Although remnants of such rostrocaudal traveling waves exist in mammals (de Seze et al. 2008, Cuellar et al. 2009, Perez et al. 2009), not all lumbosacral segments are equivalent in their capacity to generate the locomotor pattern. For instance, although rhythmogenic properties within the lumbosacral spinal cord are somewhat distributed over several segments, the L3-L4 segments in cats (Marcoux & Rossignol 2000, Langlet et al. 2005, Delivet-Mongrain et al. 2008) and L1-L2 segments in rodents (Cazalets et al. 1995, Kiehn 2006) are critical for rhythm generation. Inactivating or damaging these segments will generally abolish or severely impair locomotion, although other rhythms such as fast paw shake can still be evoked. This segmental heterogeneity has important implications for the recovery of walking after SCI.

A balance between excitation and inhibition. Function within the spinal locomotor network is governed by excitatory and inhibitory connections. During locomotion, motoneurons receive rhythmic alternating push-pull patterns of glutamatergic excitation and glycinergic inhibition during the active and inactive phases, respectively (Shefchyk & Jordan 1985, Cazalets et al. 1996, Grillner 2003). Excitatory connections are sufficient to drive rhythmic bursting because blocking inhibitory transmission, through $GABA_A$ (i.e., bicuculline) and glycine (i.e., strychnine) receptor antagonists, does not abolish oscillatory activity (Kjaerulff & Kiehn 1997, Grillner & Jessell

2009). However, inhibition is necessary to produce appropriate flexion/extension (Cowley & Schmidt 1995) and left-right (Cowley & Schmidt 1995, Kremer & Lev-Tov 1997, Hinckley et al. 2005) alternations (Grillner & Jessell 2009). The flexibility of the system also permits synchronous gaits, such as galloping and hopping, which probably involves reconfiguring inhibitory and excitatory connections within the spinal locomotor CPG (Cowley & Schmidt 1995). After SCI, synchronous rhythmic bursting in flexor and extensor muscles in one limb or between homologous hindlimb muscles can appear (Norreel et al. 2003, Courtine et al. 2008), and this could be related to deficient control in inhibitory pathways (see below).

Intrinsic properties of central pattern generator neurons. The locomotor rhythm is generated by spinal neurons with self-generating oscillatory properties (i.e., pacemaker cells) and through synaptically or electrically coupled excitatory interneurons with synchronized discharges (recently reviewed in Brocard et al. 2010). Intrinsic properties of spinal neurons are thought to facilitate the emergence of the locomotor rhythm. For instance, voltage-dependent persistent inward currents (PICs) that amplify excitatory synaptic inputs and sustain neuronal firing are thought to facilitate rhythmogenesis by timing and shaping locomotor output (Brownstone et al. 1994, Kiehn et al. 1996, Tazerart et al. 2008). PICs could also be involved in initiating pacemaker-like activity in CPG-related interneurons (Tazerart et al. 2008). Monoamines released from brainstem pathways, primarily 5-HT and NE, strongly modulate PIC amplitude and threshold (Heckman et al. 2003, Hultborn et al. 2004). The advent of transgenic mouse lines has provided genetic tools for identifying interneuronal populations involved in generating locomotor output in the mammalian spinal cord (reviewed in Goulding 2009, Grillner & Jessell 2009), which should eventually facilitate identifying key elements of the CPG and how these may change after SCI.

Cellular changes in spinal locomotor-generating circuits. After SCI, changes at the cellular level will directly impact how inputs are processed and integrated, thereby influencing function at all levels of organization, from spinal reflexes to the more complex CPG circuitry. Immediately after SCI, the excitability of spinal interneurons and motoneurons is depressed because of the loss of excitatory neuromodulatory inputs from brainstem-derived pathways, particularly the monoamines 5-HT and NA, which strongly regulate intrinsic neuronal properties, such as PICs (Heckman et al. 2003, Hultborn 2003). The depression of neuronal excitability can be prolonged, and its duration can vary widely between species (Eken et al. 1989, Hultborn 2003). The return of neuronal excitability is required for functional recovery and can be mediated by several factors. In one study, alpha-1 and alpha-2 NE receptors and 5-HT1 receptors were upregulated for 3 months following spinalization in adult cats followed by a return to normal levels, as determined by autoradiography (Giroux 1999). However, locomotor recovery did not result from this upregulation, at least for NE, because blocking NE receptors with yohimbine did not impair spinal locomotion. The role of 5-HT receptors was not addressed more fully because, in cats, 5-HT agonists do not induce locomotion after spinalization, although they modulate locomotor activity.

An elegant recent study showed that some 5-HT receptors became constitutively active following SCI in adult rats, indicating that intracellular signaling occurred without normal ligand binding. More specifically, the 5-HT$_{2C}$ receptor, which regulates a Ca^{2+}-dependent PIC, became constitutively active after complete SCI in adult rats, despite, and perhaps because of, the absence of brainstem-derived 5-HT (Murray et al. 2010). In rodents, such as rats and mice, 5-HT agonists have a role in initiating locomotion after SCI (Antri et al. 2003). Some receptors can also become supersensitive to remaining endogenous sources of neurotransmitters. For example, the amount of 5-HT or NE required to activate motoneuron

Persistent inward current (PIC): dendritic current (calcium, sodium), studied mainly in motoneurons, that can amplify synaptic inputs; influenced by various neurotransmitters

PICs is considerably reduced following chronic complete SCI in adult rats (Harvey et al. 2006, Li et al. 2007, Rank et al. 2007) and cats (Bédard et al. 1979; Chau et al. 1998a,b).

Changes in inhibitory circuits could also play a part in modifying neuronal excitability following SCI. For instance, levels of a GABA-synthesizing enzyme increased within the spinal cord of adult cats after complete SCI (Tillakaratne et al. 2000). Increased levels of inhibitory neurotransmitters (i.e., more inhibition) could depress neuronal excitability and impair specific spinal circuits. PICs are extremely sensitive to postsynaptic inhibition (Heckman et al. 2003) and increased inhibition could prevent or prematurely terminate PICs. Interestingly, levels of the GABA-synthesizing enzyme were downregulated by locomotor treadmill training (i.e., less inhibition) but not by stand training (Tillakaratne et al. 2002). Moreover, administering glycinergic or GABAergic antagonists (i.e., less inhibition) improved hindlimb locomotion in chronic spinal cats that were not responding to treadmill training (Robinson & Goldberger 1985, de Leon et al. 1999). It thus appears that modifying the level of inhibition within spinal circuits through locomotor training or pharmacology facilitates functional recovery.

Although the return of neuronal excitability is required for functional recovery, impaired regulatory mechanisms also produce maladaptive changes, such as spasticity. Spasticity is a complex multifactorial phenomenon, with varying definitions, and symptoms that can include hyperreflexia, hypertonus, clonus, and muscle spasms (reviewed in Nielsen et al. 2007). A major underlying mechanism of spasticity appears to be inappropriate regulation of spinal neuronal excitability. The tail of adult rats has proven an effective experimental model to study neuronal mechanisms involved in spastic-like behavior. In the tail of SCI rats, spastic-like muscle spasms involve a Ca^{2+}-dependent motoneuron PIC (Bennett et al. 2001, Li et al. 2004). With little to no regulation from brainstem pathways, this PIC is not easily turned off, producing muscle spasms and long-lasting

reflexes. Constitutive $5\text{-}HT_{2C}$ receptor activity might contribute to these muscle spasms (Murray et al. 2010). A PIC-like phenomenon has also been linked to spasticity in leg muscles of human SCI subjects (Gorassini et al. 2004).

The emergence of spasticity could also involve impaired regulation of inhibitory circuits between agonist/antagonist pairs. In humans (Crone et al. 2003, Xia & Rymer 2005) and adult rats (Boulenguez et al. 2010), group Ia afferent-mediated reciprocal inhibition between agonist/antagonist pairs can switch to facilitation. The switch from inhibition to facilitation in adult rats was partly attributed to downregulation of potassium-chloride cotransporter 2 (KCC2) in lumbar neurons (Boulenguez et al. 2010, Boulenguez & Vinay 2009). KCC2 expression progressively decreased within the ventral horn following complete or incomplete SCI, and increased levels of intracellular Cl^- diminished the efficacy of synaptic inhibition.

The Key Role Played by Sensory Inputs

Sensory inputs play a key role in the regulation of normal locomotion, which can be altered after SCI. Some of these changes can facilitate or hamper locomotor recovery and even the expression of locomotion.

Spinal reflexes during locomotion. Although sensory inputs are not required to produce the basic rhythm, they are critical in adapting and modulating locomotion to meet the demands of the environment, by selecting behaviorally relevant motor patterns and performing rapid postural corrections. The spinal locomotor CPG circuitry interacts dynamically with afferent inputs from receptors located in muscles, joints, and skin, which shapes the locomotor output (comprehensively reviewed in Rossignol et al. 2006). Reflex responses are state- and phase-dependent, indicating that sensory processing is regulated by context or, in other words, the current configuration of the spinal circuitry. Locomotor-generating circuits are active in this process to ensure pattern stability and appropriate state- and phase-dependent modulation of responses to various converging inputs. Afferent inputs most relevant to walking primarily arise from stretch- and load-sensitive mechanoreceptors located in muscles and skin. Inputs arising from muscles that flex the hip and extend the ankle are particularly important for adjusting phase transitions or reinforcing ongoing muscle activity (reviewed in Rossignol et al. 2006; Pearson 2008).

Sensory inputs from skin mechanoreceptors of the foot are involved in positioning the feet and become increasingly important when performing difficult locomotor tasks in which precise foot placement is critical, such as ladder walking in the cat (Bouyer & Rossignol 2003a). Activating receptors located in the skin of the dorsal foot during the swing phase of walking produces a coordinated reflex response to overcome obstacles or perturbations in cats (Forssberg 1979), human adults (Eng et al. 1994, Schillings et al. 2000), and human infants (Lam et al. 2003). Cutaneous inputs also appear to be involved in scaling the magnitude of other sensory cues when performing postural corrections during locomotion (Bolton & Misiaszek 2009). Cutaneous inputs from the plantar surface of the paw can also reinforce extensor activity in decerebrate cats walking on a treadmill (Duysens & Pearson 1976) or during fictive locomotion (Guertin et al. 1995).

The importance of reflex pathways from group II hip flexor afferents, group I ankle extensor afferents, and low-threshold cutaneous afferents from the foot likely stems from their direct access to the rhythm-generating circuitry of the spinal locomotor CPG, inferred by their ability to reset or entrain the fictive locomotor rhythm in adult cats (reviewed in Rossignol et al. 2006, Frigon et al. 2010). The relative contribution of these various inputs in modulating locomotion most likely changes with behavioral context and following SCI (Frigon & Rossignol 2006a, Pearson 2008). Changes at the cellular level following SCI will directly impact the regulation of peripheral sensory inputs

and their interaction with the spinal locomotor CPG.

Changes in spinal reflexes after spinal lesion. The general role of sensory inputs after complete SCI is illustrated by the fact that deafferentation of one hindlimb drastically perturbs air stepping or the locomotor pattern (Giuliani & Smith 1987). Moreover, experiments by Kato (1989) demonstrated that the isolated hemispinal cord (hemisection and longitudinal myelotomy) depends on afferent feedback to recover locomotion. Locomotor training after SCI is based on the principle that sensory inputs reactivate and reorganize the spinal locomotor circuitry (Rossignol et al. 2006). After SCI, some cutaneous reflexes, which normally have an inhibitory effect on extensor muscles, can exert an excitatory effect, thus increasing the activity of these muscles during the stance phase (Forssberg et al. 1975, Frigon & Rossignol 2008).

If a small portion of sensory feedback is reduced by lesioning specific peripheral nerves before a complete SCI in adult cats, the recovery of hindlimb locomotion is severely impaired, with reduced weight-bearing capability, improper foot placement, nerve-specific deficits, and in some cats an inability to express spinal locomotion (Carrier et al. 1997, Bouyer & Rossignol 2003b, Frigon & Rossignol 2009). If the same denervation is performed after complete SCI, once spinal locomotion has been established, only transient locomotor deficits are observed, indicating that the underlying processes of locomotor recovery after complete SCI require intact sources of peripheral sensory feedback. Indeed, the effectiveness of therapeutic interventions, such as epidural stimulation of the spinal cord in adult rats (Lavrov et al. 2008), is greatly reduced if sensory feedback is diminished. In contrast, providing phasic sensory feedback can facilitate locomotor recovery after SCI in chicks or adult rats (Muir & Steeves 1995, Smith et al. 2006). To summarize, after complete SCI, intrinsic changes at the cellular level of the CPG promote the return of hindlimb locomotion

Mesencephalic locomotor region: a brain stem nucleus below the inferior colliculus that evokes locomotion in the decerebrate cat when electrically stimulated

through interactions with peripheral sensory inputs.

LOCOMOTOR RECOVERY AFTER PARTIAL SPINAL LESIONS

The previous sections summarized observations after complete severance of the spinal cord, and the following sections describe mechanisms implicated in the recovery of locomotion after partial spinal lesions. Damage to specific spinal tracts or lesions that affect several pathways concomitantly produces characteristic deficits and directly influences locomotor recovery. After incomplete SCI, spared pathways originating from supraspinal and propriospinal structures can play an active role in the recovery process, and also in restoring some voluntary control. However, intrinsic spinal circuits and peripheral afferents still remain to initiate and organize hindlimb locomotion.

Accessing the Locomotor Circuitry by Descending Inputs

The role of descending pathways on motor control can be viewed from different perspectives depending on whether it is studied by recording cells of origin, by stimulating specific tracts, or by observing the behavioral consequences of inactivation. **Figure 3** summarizes the location of the main descending pathways in the cat spinal cord (Petras 1967, Holstege & Kuypers 1987) that may be damaged after spinal lesions.

Ventral and ventrolateral lesions (reticulospinal and vestibulospinal pathways). Reticulospinal and vestibulospinal pathways may be critical in initiating locomotion and postural control. Electrical stimulation of the mesencephalic locomotor region elicits locomotion in decerebrate cats through the activation of reticulospinal cells (Shik et al. 1966, Orlovsky & Shik 1976). Various diencephalic and telencephalic structures in turn project directly to the mesencephalic locomotor region and to the reticular formation (Grillner et al. 1997, Jordan 1998). Electrically stimulating reticulospinal (Perreault et al. 1994) or vestibulospinal (Russel & Zajac 1979, Gossard et al. 1996) pathways can reset the hindlimb rhythm during fictive locomotion in the adult cat, indicating

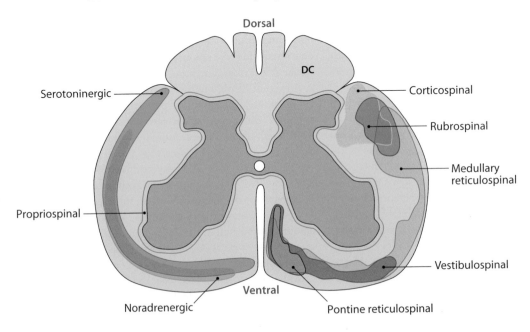

Figure 3

Descending pathways in the cat, based mainly on work by Petras (1967) and Holstege & Kuypers (1987). Modified from Rossignol et al. 2009. Abbreviation: DC, dorsal columns.

direct access to the spinal rhythm-generating circuitry.

It was suggested that sparing a small part of a ventrolateral quadrant is required to initiate hindlimb locomotion after SCI (Afelt 1974; Eidelberg et al. 1981a,b; Contamin 1983). However, it was later shown that cats (Gorska et al. 1990, 1993a,b; Zmyslowski et al. 1993; Bem et al. 1995; Brustein & Rossignol 1999; Rossignol et al. 1999) and monkeys (Vilensky et al. 1992) could walk with their hindlimbs even after large bilateral lesions of ventral pathways at the last thoracic segment (T13). Similarly, humans with a surgical section of ventral pathways for pain relief can retain walking ability (Nathan 1994).

Dorsal/dorsolateral lesions (corticospinal and rubrospinal pathways). Corticospinal and rubrospinal pathways are important for volitional and goal-directed aspects of locomotion, as well as fine control of the distal musculature. They also appear to play a critical role in more difficult locomotor tasks, such as obstacle avoidance and ladder walking (Liddell & Phillips 1944, Drew et al. 2004). Electrically stimulating the pyramidal tract during fictive locomotion in the adult cat can reset the hindlimb rhythm, indicating that the corticospinal tract has direct access to the rhythm-generating circuitry (Leblond et al. 2001). In contrast, the rubrospinal pathway does not reset the hindlimb locomotor rhythm. Thus, although it has similar functions to the corticospinal tract, it may not directly access the spinal rhythm-generating circuitry (Rho et al. 1999).

Cats can walk over ground after large lesions of the dorsolateral white matter, which contains corticospinal and rubrospinal pathways (Gorska et al. 1993b, Zmyslowski et al. 1993, Bem et al. 1995). Studies have shown that lesioning the dorsal/dorsolateral spinal cord produced only transient deficits in overground or treadmill locomotion (Eidelberg & Yu 1981, Gorska et al. 1993b). However, some deficits can persist (e.g., foot drag) even during undemanding locomotion, such as treadmill walking

(Jiang & Drew 1996, Muir et al. 2007, Kanagal & Muir 2008). Coordination between the fore- and hindlimbs is impaired, but the coupling between homologous limbs is not (Jiang & Drew 1996, English 1980, Gorska et al. 2007). Corticospinal pathways are mainly involved in skilled movements and precision movements (Metz et al. 2000b, Ghosh et al. 2009), as well as in anticipatory controls (Yakovenko & Drew 2009), which are affected by SCI.

Other pathways. Aside from the long classical pathways damaged by lesions, others that are less well defined or characterized anatomically may also be damaged.

Propriospinal pathways. Spinal segments are interconnected by short and long intraspinal pathways (i.e., propriospinal) that run close to the gray matter bilaterally (see **Figure 3**). For example, the tract of Lissauer is located at the entry of dorsal roots and distributes afferent inputs to different spinal segments. Other connections may also be established by spinal interneurons that project their axons through short or long pathways. The original work by Sherrington & Laslett (1903) established quite clearly the importance of propriospinal pathways.

It was demonstrated that after corticospinal tract lesions in rats, new connections could be established with the lumbosacral cord through cervical propriospinal pathways (Bareyre et al. 2004). Propriospinal pathways appear to be of considerable importance for volitional aspects of locomotor recovery after an incomplete SCI in adult mice (Courtine et al. 2008) and cats (Kato et al. 1984) through the formation of new functional circuits.

Noradrenergic and serotonergic pathways. Important neuromodulators, such as 5-HT and NE, are synthesized in the brainstem. The serotoninergic pathway originates from subdivisions of the raphe nucleus, whereas the noradrenergic pathway originates from the locus coeruleus. Projections to the spinal cord are widespread, and neurotransmitters may be released at specific synapses or in the environment by a mechanism known as volumic transmission, whereby several neurons can be influenced (**Figure 1**). These pathways release their neurotransmitters throughout the spinal cord, and receptor subtypes are distributed differentially at various segments (Schmidt & Jordan 2000). As such, drugs acting on receptors at different segments may affect different functions. Lesions of the spinal cord will deprive the spinal cord of an important source of neuromodulators, which will vary depending on lesion extent. The loss of neurotransmitters will in turn have important consequences on the membrane properties of target neurons, as shown below. Activating or blocking 5-HT or NE receptors after SCI is thoroughly reviewed elsewhere (Rossignol et al. 2001, 2006; Rossignol 2006).

Multiple pathways severed by contusions or hemisections. Whereas previous paragraphs dealt with rather specific lesions, in contusion or hemisection models, several pathways are severed simultaneously, a situation approaching the clinical situation in which spinal impact due to accidents or falls damages several pathways.

Contusions. Contusions are achieved by various impact devices or by compression clips and damage several pathways simultaneously. Contusive lesions are diffuse (Choo et al. 2007) and result in a central cavitation surrounded by spared white matter (Metz et al. 2000a, Poon et al. 2007). The majority of SCIs in humans are contusions due to an impact to the vertebral column. Contusion models in animals have improved our understanding of biological mechanisms involved in the secondary injuries that follow the initial SCI (Siegal 1995, Young 2002). The extent of spinal white matter damage is strongly related to locomotor deficits (Shields et al. 2005, Li et al. 2006, Majczynski et al. 2007, Poon et al. 2007). The level of contusion is also important because injury at T13-L2 produces greater loss of locomotor function compared with the same injury at L3-L4 (Magnuson et al. 2005). Contusion at

T13-L2 could damage key elements of the spinal CPG for hindlimb locomotion thought to be localized at L1-L2 in the rat (Cazalets et al. 1995, Magnuson & Trinder 1997, Bertrand & Cazalets 2002). Mild contusion SCIs mainly affect the central spinal cord, whereas more severe contusion SCIs also damage pathways located more peripherally. Based on such comparisons, it was found that locomotor recovery did not depend on the sparing of corticospinal or long propriospinal pathways (Basso et al. 1996), indicating a role for short intraspinal circuits.

Hemisections. Unilateral hemisections damage ventral and dorsal tracts primarily on one side. After such lesions, treadmill and overground locomotion recovers in most species (Kato et al. 1984, 1985; Helgren & Goldberger 1993; Bem et al. 1995; Kuhtz-Buschbeck et al. 1996, Saruhashi et al. 1996, Suresh et al. 2000, Courtine et al. 2005, 2008; Gulino et al. 2007; Barrière et al. 2008).

In the first few days after a thoracic (T10–T11) hemisection in cats, the ipsi-lesional hindlimb is flaccid, and the cat walks with a tripod gait, requiring assistance to maintain weight support (Helgren & Goldberger 1993, Barrière et al. 2008). Within 2 weeks, cats recover hindquarter support, with reasonable joint excursions, but a foot drag during swing persists. The coordination between the fore- and hindlimbs is also affected (Helgren & Goldberger 1993, Bem et al. 1995, Kuhtz-Buschbeck et al. 1996, Barrière et al. 2008). With smaller lesions, cats maintain a one-to-one ratio between the fore- and hindlimb cycle durations (Barrière et al. 2010). With large lesions, the fore- and hindlimbs can walk at different rhythms, as is the case for ventral lesions (Brustein & Rossignol 1998). Skilled locomotion (ladder and grid walking) is also impaired in cats, monkeys, and rats following hemisection (Helgren & Goldberger 1993, Suresh et al. 2000, Gulino et al. 2007).

Therefore, although substantial recovery of hindlimb locomotion is observed following lateral hemisection of the spinal cord, some deficits, mostly observed on the side of the lesion, can persist. Deficits observed after hemisection resemble those associated with ventral (transiently impaired body equilibrium, interlimb coupling) and dorsal (impaired skilled locomotion) lesions.

Mechanisms of Locomotor Recovery After Partial Spinal Cord Injury

As illustrated in **Figure 1**, locomotion is controlled at multiple levels of the CNS, and after various types of spinal lesions, the recovery of functions involves optimizing interactions between remaining structures. Although models of complete SCI are important in determining intrinsic spinal mechanisms involved in locomotor recovery, most SCIs in humans are incomplete, and spared descending pathways can still access the spinal circuitry. However, as shown below, intrinsic spinal mechanisms and afferent mechanisms are still critical in locomotor recovery after an incomplete SCI. In turn, new interactions can modify spared structures throughout the CNS, not just the spinal cord. Therefore, the end result is a combination of phenomena with short- and longer-time courses owing to the immediate loss of excitability and connectivity within spinal neurons followed by the subsequent development of compensatory physiological and morphological mechanisms, which optimize basic locomotor function.

The underlying model of research on functional recovery after partial SCI has been more implicit than explicit. Indeed, in partial SCI models, functional recovery has often been attributed to the growth of new axons (regenerating or sprouting) from remaining pathways, but it is not easy to establish the extent to which this recovery depends on these new fibers. Therefore, functional recovery is often thought to result from a combination of regeneration, sprouting, or other ill-defined plastic changes in descending pathways (Cafferty et al. 2008). The review by Bradbury & McMahon (2006) clearly states the problem: "An implicit assumption of much SCI research, at least

until recently, has been that the major goal is to induce damaged axons to regrow, to reconnect to appropriate targets and thereby restore function (as is indeed possible in the peripheral nervous system)." In the following sections we present evidence that changes in sensory afferents, descending pathways, and intrinsic spinal circuitry participate in the recovery of hindlimb locomotion after partial SCI.

Compensation by sensory afferents. After chronic spinal hemisection, sprouting of sensory afferents on the lesioned side is prominent and could partly account for the functional recovery of various motor patterns, including locomotion (Goldberger 1977, Helgren & Goldberger 1993). The sparing of only one dorsal root (L6) can be sufficient to regain locomotion. Although sensory axons from primary afferents can regenerate within the CNS after SCI, there is evidence that afferents invading the dorsal columns remain in a pathophysiological state (Tan et al. 2007) and that sprouting may not be as abundant (Wilson & Kitchener 1996). Deafferenting a hindlimb followed by a hemisection of the spinal cord at L1 can render the limb unusable (Goldberger 1977), indicating that sensory feedback is of crucial importance in the recovery process.

After SCI, the role of sensory feedback from the periphery is heightened because of the loss of descending inputs from supraspinal structures. Moreover, the processing of sensory feedback within spinal circuits at rest and during locomotion is altered after SCI, as revealed by several reflex studies in rodents, cats, and humans (comprehensively reviewed in Frigon & Rossignol 2006b). Following a period of reduced reflex activity, primarily due to the loss of motoneuron or interneuron intrinsic excitability, considerable changes in reflex pathways develop over time, as a consequence of the injury, spontaneous phenomena, and activity-dependent processes, such as locomotor training (Côté et al. 2003, Côté & Gossard 2004). More noticeable changes include the appearance of reflex responses not normally present in the intact state and modified regulation of reflex pathways. Evaluating reflex responses before and after spinal lesions provides indications of how the spinal circuitry is reconfigured by SCI.

Recent studies using recordings in the same cats, before and after a complete or incomplete SCI, showed the appearance of short-latency excitatory responses, evoked by cutaneous nerve stimulation, during the stance phase of locomotion after the spinal lesion, which replaced the more common short-latency inhibition (Forssberg et al. 1975, Frigon & Rossignol 2008, Frigon et al. 2009). A dual spinal lesion paradigm (i.e., partial lesion followed by complete transection) showed that some reflex changes after incomplete SCI resulted from intrinsic spinal changes because they were retained after complete SCI (Frigon et al. 2009).

At rest, longer-latency responses can also appear following SCI in adult rats, evoked by epidural stimulation of the spinal cord (Lavrov et al. 2006), and in humans, elicited by cutaneous nerve stimulation (Roby-Brami & Bussel 1987, Dietz et al. 2009). The appearance of long-latency responses was associated with locomotor recovery (Roby-Brami & Bussel 1987) because they are reminiscent of locomotor-like late discharges in acute spinal cats following administration of L-DOPA (Anden et al. 1966, Jankowska et al. 1967). The appearance of long-latency responses after SCI was proposed to reflect functional changes within the spinal circuitry that enable the emergence of locomotor activity (Viala et al. 1974, Roby-Brami & Bussel 1987, Lavrov et al. 2006). However, others (Dietz et al. 2009) have associated locomotor recovery, assessed during robot-driven air stepping, with the early reflex component. The issue requires further investigation.

Another prominent change in spinal reflex pathways after SCI is a reduction in the efficacy of disynaptic reciprocal inhibition (Dimitrijevic & Nathan 1967, Okuma et al. 2002, Crone et al. 2003, Xia & Rymer 2005). In some patients, disynaptic reciprocal facilitation is evoked, instead of inhibition, which is thought to be mediated by an oligosynaptic excitatory pathway (Okuma et al. 2002, Crone et al. 2003, Xia & Rymer 2005). The rate-dependent depression of the H-reflex is also reduced following incomplete or complete SCI in adult rats (Thompson et al. 1992, Boulenguez et al. 2010) and human patients (Calancie et al. 1993, Schindler-Ivens & Shields 2000). Although downregulation of KCC2 was linked to reduced rate-dependent depression after spinal transection in adult rats (Boulenguez et al. 2010), changes in presynaptic mechanisms have also been suggested (Thompson et al. 1992, Calancie et al. 1993, Schindler-Ivens & Shields 2000). Changes in reflex pathways at rest and during locomotion reflect the new state, or configuration, of the spinal circuitry

Compensation by descending pathways. The compensation by descending pathways may take different forms. Damaged pathways may regenerate while undamaged pathways may sprout or may change the efficacy of their transmission. In doing so, new circuits could result from new anatomical connections (new circuits) or from enhanced connectivity (enhancing existing circuits)

Regeneration and sprouting. After SCI, substantial research efforts have focused on reconnecting supraspinal and spinal levels by promoting the regeneration of damaged pathways or by favoring a takeover of function through collateral sprouting of undamaged pathways. This is a formidable task because powerful inhibitory cues or barriers exist that minimize or limit regeneration. Numerous and active labs have been working on deciphering these molecular inhibitory cues, or on minimizing the glial scar that forms a physical and chemical barrier to regeneration caudal to an SCI, which has been reviewed several times (Bradbury & McMahon 2006, Maier & Schwab 2006, Thuret et al. 2006).

The case has been clearly made (Bradbury & McMahon 2006) that there is a lack of hard evidence that regenerated lesioned axons induce significant functional improvements because of the small number of regenerating axons. This might not be, in itself, a strong argument because function could be sustained by small numbers of spared fibers, provided that the outcome being measured (e.g., hindlimb locomotion) is generated by other mechanisms, such as intrinsic functional changes within the spinal cord. In this case, the remaining descending fibers play more of a triggering role and not a controlling one (**Figure 4*b***). Very limited extension of regenerated fibers past the lesion site may also be a serious argument against a functional role for regenerated lesioned axons. However, it is only a cogent argument if the underlying model requires regenerated descending fibers to access targets located throughout the lumbosacral cord. More rostral segments could be sufficient to organize the rhythmic output of the spinal cord generated by more caudal segments (see section above on propriospinal pathways). As stated, new functional circuits composed of short propriospinal connections are most likely involved in the recovery process.

An extension of collaterals (i.e., sprouting) from damaged and spared systems can make synaptic contacts with new targets, which are themselves connected redundantly to the original target (**Figure 4**). Such new connections could promote or hinder functional recovery. Recent work in primates suggests a key role of corticospinal projections in the recovery of hand function and locomotion in monkeys (Rosenzweig et al. 2010). This work further suggests that the recovery in nonhuman primates may rely more on such regeneration of the corticospinal pathway than in rodents. The formation of new circuits through propriospinal pathways is of great interest (Bareyre et al. 2004, Maier & Schwab 2006). Other descending systems may also serve similar roles (Ballermann & Fouad 2006), and even in the presence of substantive regeneration, the role of these other pathways cannot be minimized.

Are regenerated fibers even necessary for the studied function? In that context, a second lesion is not necessarily convincing because the second lesion may not influence restored function but may affect compensation from spared structures. A critical question is whether regenerated fibers are even functional. Regenerated fibers can conduct impulses because electrical stimulation of the motor cortex after spinal hemisection clearly evokes cord dorsum potentials below the lesion (Bradbury et al. 2002). However, this does not mean that the regenerated fibers can mediate any complex functions, such as interacting with reflexes or the spinal locomotor circuitry in an appropriate manner. After unilateral spinal hemisection, it is tempting to think that contralateral pathways play a major role in functional compensation. However, experiments using staggered spinal hemisections show that the regeneration of long descending pathways is not necessary (Kato 1989). Work by Helgren & Goldberger (1993) in adult cats also suggests that contralateral pathways are not crucial. Experiments on the isolated spinal cord (Kato 1989), consisting of a hemisection at L2 and a longitudinal myelotomy, clearly show that hindlimb locomotion can be generated without crossing fibers, which indicates that locomotor recovery depends more on intrinsic spinal mechanisms and contributions from sensory afferents.

After a unilateral left cervical hemisection at C3/C4 (equivalent to a Brown-Séquard syndrome) in the rat, the forelimb on the right side was predominantly used, which led to an expansion of the left cortical sensorimotor representations of the forelimb, as shown by functional magnetic resonance imaging and voltage-sensitive dyes (Ghosh et al. 2009). Increased collateral sprouting of corticospinal cells from the left cortex that recross at cervical and lumbar levels was observed. Interestingly, the forelimb on the side of the lesion remained impaired, whereas the hindlimbs recovered an almost normal pattern during overground and skilled locomotion (Ghosh et al. 2009). The same work (Ghosh et al. 2009) also showed that after large bilateral T8 lesions that damage both corticospinal tracts, rendering the hindlimbs paretic, sprouting of surviving hindlimb corticospinal tract neurons occurred in the cervical cord, which could account for the greater use of the forelimbs during overground and skilled locomotion. The strategy of transferring body propulsion to the forelimbs was also observed in the cat after large bilateral ventrolateral spinal lesions (Brustein & Rossignol 1998).

New/old circuits. Another important issue stemming from regeneration and sprouting studies is whether new axons represent new circuits or strengthening of existing connections

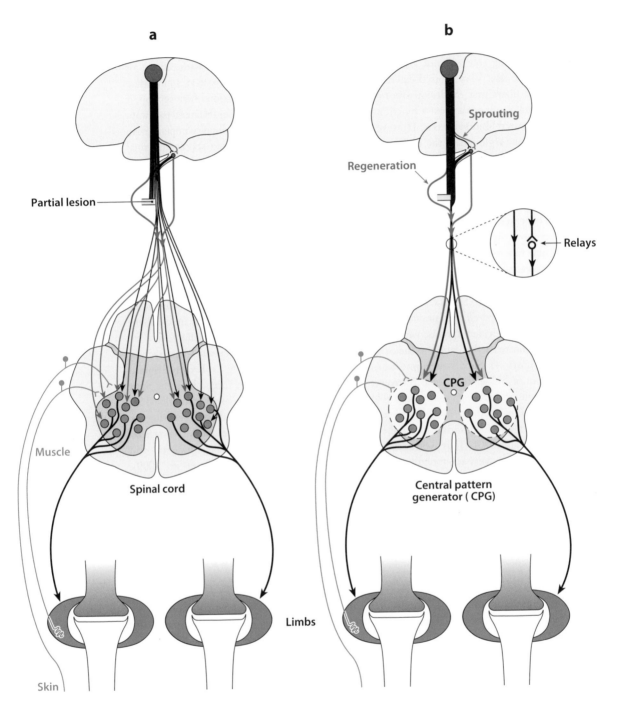

(Bareyre et al. 2004). Serial hemisections and pharmacological stimulation in the in vitro neonatal rat spinal cord (Zaporozhets et al. 2006, Cowley et al. 2008) demonstrated that propriospinal neurons transmit command signals from the brainstem to the spinal cord to initiate rhythmic locomotor-like activity of the hindlimbs. After double staggered spinal hemisections at different levels in the cat (Kato et al. 1985), animals recovered voluntary locomotion, indicating that descending commands reached the lumbar spinal cord through a system of interconnecting propriospinal neurons that formed a link from the brainstem to the lumbar cord. Sherrington & Laslett (1903) described in much detail the organization of these intrinsic spinal cells and showed how they might be responsible for the return of some spinal functions after SCI. An overall scheme of activation of the spinal locomotor CPG by brainstem stimulation was previously proposed (Shik 1983) based on Lloyd's (1941) concept of spinal organization. Similar conclusions were reached in the mouse (Courtine et al. 2008) and rat (Grill et al. 1997, Murray et al. 2010), highlighting the potential role of the propriospinal system after SCI.

There is no doubt that the propriospinal systems (long and short) can reach the CPG, but should we think of the CPG as part of the propriospinal system? The CPG circuits can be viewed as a particular set of interneurons organized to generate rhythmic alternate activity, which can be accessed by various long descending pathways, either directly or indirectly through propriospinal pathways, as well as by sensory afferents. Propriospinal pathways, especially the short intersegmental pathways, probably provide a robust circuit for basic functions such as posture and locomotion, which could bridge supraspinal and intrinsic spinal circuits after SCI. Such an organization explains how lesions of specific pathways or sensory afferents in isolation do not prevent activation of an interconnected neuronal network. What is the number of spinal segmental neurons required to give rise to a given behavior? We showed previously that lesions below L4 prevented hindlimb locomotion in adult cats (Langlet et al. 2005). We also showed that blocking several spinal segments simultaneously above L4 prevented hindlimb locomotion in decerebrate cats, consistent with interference of a multisegmental propriospinal system involved in generating locomotion (Delivet-Mongrain et al. 2008).

Compensation by the Intrinsic Spinal Circuitry

As stated above, the recovery of hindlimb locomotion after a complete SCI in cats, rats, and mice absolutely requires a spinal circuitry capable of generating the basic locomotor pattern independently of descending commands (Grillner 1981; Rossignol 1996, 2006; Rossignol et al. 2006). How important is the spinal CPG for the recovery of hindlimb locomotion after partial spinal lesions, considering that spared descending pathways still access the spinal cord? The essential role of the intrinsic spinal circuitry was clearly demonstrated by a dual spinal lesion paradigm, in which a partial spinal lesion at T10-T11 was followed several

---

Figure 4

Two conceptual models of functional recovery. (a) Descending pathways project to the spinal cord directly or through connections in the brainstem or other descending circuits, including long propriospinal tracts that also reach the spinal cord. In this model, we assume that the spinal circuits are not organized in functional modules. In the model, regenerating fibers or sprouting axons (both in *green*) after spinal cord injury (SCI) will have to reach the spinal cord directly or indirectly to organize the rhythmic spinal locomotor output. New fibers have the complex task of rewiring with spinal components to regain function. (b) Processes of regeneration and sprouting are the same, but the new fibers reach (directly or indirectly through propriospinal pathways) a central pattern generator (CPG) that has been modified to function more autonomously following the partial SCI. The two conceptual models emphasize that recovery of function, such as locomotion, can be simplified if one integrates the CPG circuitry as a major component of the recovery.

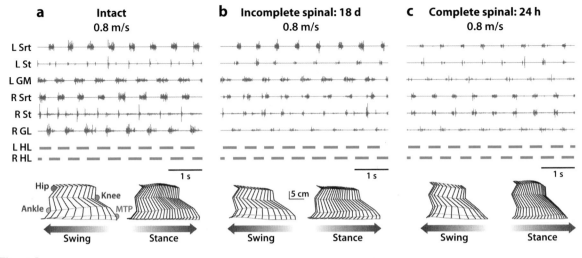

Figure 5

Left hindlimb kinematics and bilateral electromyographic (EMG) activity of a cat walking at 0.8 m s⁻¹ in the (*a*) intact, (*b*) incomplete, and (*c*) complete spinal states. The traces below the EMGs show the associated duty cycles of the stance phases for both hindlimbs (HLs). The stick figures at the bottom illustrate a typical swing and stance for each condition reconstructed from video recordings. Muscles are abbreviated as follows (L, left; R, right): Srt, sartorius (knee extensor, hip flexor); St, semitendinosus (knee flexor/hip extensor); and GL and GM, gastrocnemius lateralis and medialis (ankle extensors and knee flexors). Abbreviation: MTP, metatarsophalangeal. Figure reproduced with permission from Barrière et al. (2008).

weeks later by a spinalization at T13 (i.e., two to three segments below the initial partial lesion and where complete spinal lesions are usually made in the cat). Hindlimb walking was observed within hours following spinalization, demonstrating that important intrinsic changes had already occurred within the locomotor spinal circuitry following the partial spinal lesion (Barrière et al. 2008, 2010). **Figure 5** illustrates sequences of locomotion recorded in the same cat in three states: intact, hemispinal, and 24 h after spinalization. Despite some minor changes in the timing of EMG bursts, the overall pattern in the three conditions was remarkably similar, as would be expected from a pattern generated by an intrinsic spinal network.

Sherrington & Laslett (1903) observed that cats with an initial cervical hemisection (C4 or C6) had brisker reflexes on the side of the lesion following spinalization at T4 or C8 several months later. Basso et al. (1996) showed that rats with an initial spinal contusion had some retention of function after a complete SCI

performed 7 weeks later so that their BBB (Basso et al. 1996) scores tended to be better than those of rats with only a complete SCI. Recent work on the escape swim of a mollusc (*Tritonia diomedea*) showed near immediate changes within the functional connectivity of the swim CPG following a lesion within the intrinsic circuitry that compensated for the loss of long projections and reinstated function in the absence of regeneration (Sakurai & Katz 2009). Intrinsic functional plasticity within spinal circuits after incomplete SCI is a mechanism that has generally been underestimated for motor recovery. Indeed, many studies of partial spinal lesions using various types of regenerative therapies have largely ignored the possibility that a considerable portion of the recovery could be mediated within spinal circuits rather than by a functional takeover by descending pathways. Therefore, it is absolutely crucial to understand what happens to the spinal circuitry after a partial spinal lesion. We propose that the recovery of function by descending or afferent inputs after SCI essentially depends on how the circuitry has adapted to the total or partial absence of descending inputs.

IMPLICATIONS FOR HUMANS WITH SPINAL CORD INJURY

Although most of the initial work on spinal cord lesions aimed at understanding the intrinsic capabilities of the spinal cord and how neurotransmitters could induce different states (Grillner 1973, 1981; Forssberg & Grillner 1973; Forssberg et al. 1980a,b), the principles emerging from these studies were rapidly considered within a clinical context of SCI in humans. For instance, pharmacology was used to improve locomotor capabilities in humans with SCI. Using intrathecal clonidine (Remy-Neris et al. 1999), locomotion improved in some SCI subjects but was adversely affected in others, presumably due to the suppression of spastic-like activity, which might be important to maintain posture during walking. Pharmacotherapy in humans is still hampered by a lack of knowledge regarding changes in receptors within the human spinal cord after SCI and which neurotransmitter systems are more important in initiating and maintaining locomotor activity. In cats, clonidine can trigger locomotion after complete SCI (Forssberg & Grillner 1973; Barbeau et al. 1987; Chau et al. 1998a,b), but the same drug can have different effects after an incomplete SCI (Brustein & Rossignol 1999) or in the intact cat (Giroux et al. 1998). Thus the problem is more complex than a simple species difference and depends on the state of the spinal circuitry after SCI, which involves a myriad of factors. Indeed, the effects of receptor agonists/antagonists after spinal lesions depend on the state of receptor sensitivity, which in turn depends on lesion extent. Recent work by Murray et al. (2010) exemplifies how a better understanding of pharmacological changes after SCI could lead to better therapeutic approaches to treat spasticity and improve locomotion. Targeting intrinsic changes in membrane properties and receptor activities within the spinal circuits responsible for generating locomotion after SCI should be a prime focus for rehabilitation.

Several of the reflex features described earlier have also been incorporated into the understanding of human locomotion and recovery of function after SCI (Yang & Stein 1990, Pang & Yang 2000, Dietz & Harkema 2004, Dobkin et al. 2006, Harkema 2008, Dy et al. 2010). Reflexes are modified after SCI, and these changes could also be targeted as part of combinatorial therapeutic approaches (Thuret et al. 2006). Gossard and colleagues (Côté et al. 2003, Côté & Gossard 2004) showed that treadmill training could normalize load and cutaneous pathways in the cat. This, in itself, could provide easily accessible indices of beneficial consequences of rehabilitation approaches in patients. In human SCI patients, the basis of locomotor training is to provide sensory cues consistent with normal walking (Harkema 2001), which is primarily derived from work on adult cat locomotion. Recent evidence using spinal electromagnetic stimulation suggests that lower thoracic spinal segments (T11-T12) might be important for rhythm generation in humans (Gerasimenko et al. 2010) and could potentially be used as an adjuvant to locomotor training.

Perhaps one of the most important consequences of the experimental work in animals is the notion of a spinal locomotor CPG. The present review highlights the notion that the spinal locomotor CPG is indispensable in understanding locomotor recovery after complete and incomplete SCI. The issue of whether humans possess a spinal locomotor CPG has been debated, and although it can probably never be directly demonstrated, circumstantial evidence for the existence of such circuitry is quite compelling (Lhermitte 1919, Kuhn 1950, Calancie et al. 1994, Bussel et al. 1988, Dimitrijevic et al. 1998, Calancie 2006, Nadeau et al. 2010). Consequently, targeting intrinsic spinal circuits by stimulating or engaging remaining pathways and sensory afferents should be a focus for rehabilitative strategies in humans with SCI (Barbeau & Rossignol 1994, Wirz et al. 2005, Harkema 2008).

DISCLOSURE STATEMENT

The authors are not aware of any affiliations, memberships, funding, or financial holdings that might be perceived as affecting the objectivity of this review.

ACKNOWLEDGMENTS

We acknowledge the support of the Canadian Institutes for Health Research (CIHR) through individual or group grants as well as through the Team Grant to the Multidisciplinary Team on Locomotor Rehabilitation or through a Tier I Chair on the spinal cord. Alain Frigon is funded by a postdoctoral fellowship from the CIHR and by an individual grant from the Wings for Life Foundation.

LITERATURE CITED

Afelt Z. 1974. Functional significance of ventral descending tracts of the spinal cord in the cat. *Acta Neurobiol. Exp.* 34:393–407

Anden NE, Jukes MGM, Lundberg A, Vyklicky L. 1966. The effect of DOPA on the spinal cord. 1. Influence on transmission from primary afferents. *Acta Physiol. Scand.* 67:373–86

Antri M, Mouffle C, Orsal D, Barthe JY. 2003. 5-HT1A receptors are involved in short- and long-term processes responsible for 5-HT-induced locomotor function recovery in chronic spinal rat. *Eur. J. Neurosci.* 18(7):1963–72

Ballermann M, Fouad K. 2006. Spontaneous locomotor recovery in spinal cord injured rats is accompanied by anatomical plasticity of reticulospinal fibers. *Eur. J. Neurosci.* 23(8):1988–96

Barbeau H, Julien C, Rossignol S. 1987. The effects of clonidine and yohimbine on locomotion and cutaneous reflexes in the adult chronic spinal cat. *Brain Res.* 437:83–96

Barbeau H, Rossignol S. 1987. Recovery of locomotion after chronic spinalization in the adult cat. *Brain Res.* 412:84–95

Barbeau H, Rossignol S. 1994. Enhancement of locomotor recovery following spinal cord injury. *Curr. Opin. Neurol.* 7:517–24

Bareyre FM, Kerschensteiner M, Raineteau O, Mettenleiter TC, Weinmann O, Schwab ME. 2004. The injured spinal cord spontaneously forms a new intraspinal circuit in adult rats. *Nat. Neurosci.* 7(3):269–77

Barrière G, Frigon A, Leblond H, Provencher J, Rossignol S. 2010. Dual spinal lesion paradigm in the cat: evolution of the kinematic locomotor pattern. *J. Neurophysiol.* 104:1119–33

Barrière G, Leblond H, Provencher J, Rossignol S. 2008. Prominent role of the spinal central pattern generator in the recovery of locomotion after partial spinal cord injuries. *J. Neurosci.* 28:3976–87

Basso DM, Beattie MS, Bresnahan JC. 1996. Graded histological and locomotor outcomes after spinal cord contusion using the NYU weight-drop device versus transection. *Exp. Neurol.* 139(2):244–56

Bédard P, Barbeau H, Barbeau B, Filion M. 1979. Progressive increase of motor activity induced by 5-HTP in the rat below a complete section of the cord. *Brain Res.* 169:393–97

Belanger M, Drew T, Provencher J, Rossignol S. 1996. A comparison of treadmill locomotion in adult cats before and after spinal transection. *J. Neurophysiol.* 76(1):471–91

Bem T, Gorska T, Majczynski H, Zmyslowski W. 1995. Different patterns of fore-hindlimb coordination during overground locomotion in cats with ventral and lateral spinal lesions. *Exp. Brain Res.* 104:70–80

Bennett DJ, Li Y, Harvey PJ, Gorassini M. 2001. Evidence for plateau potentials in tail motoneurons of awake chronic spinal rats with spasticity. *J. Neurophysiol.* 86(4):1972–82

Bertrand S, Cazalets JR. 2002. The respective contribution of lumbar segments to the generation of locomotion in the isolated spinal cord of newborn rat. *Eur. J. Neurosci.* 16(9):1741–50

Bolton DA, Misiaszek JE. 2009. The contribution of hindpaw cutaneous inputs to the control of lateral stability during walking in the cat. *J Neurophysiol* 102(3):1711–24

Boulenguez P, Liabeuf S, Bos R, Bras H, Jean-Xavier C, et al. 2010. Down-regulation of the potassium-chloride cotransporter KCC2 contributes to spasticity after spinal cord injury. *Nat. Med.* 16(3):302–7

Boulenguez P, Vinay L. 2009. Strategies to restore motor functions after spinal cord injury. *Curr. Opin. Neurobiol.* 19(6):587–600

Bouyer LJG, Rossignol S. 2003a. Contribution of cutaneous inputs from the hindpaw to the control of locomotion: 1. Intact cats. *J. Neurophysiol.* 90:3625–39

Bouyer LJG, Rossignol S. 2003b. Contribution of cutaneous inputs from the hindpaw to the control of locomotion: 2. Spinal cats. *J. Neurophysiol.* 90:3640–53

Bradbury EJ, McMahon SB. 2006. Spinal cord repair strategies: Why do they work? *Nat. Rev. Neurosci.* 7(8):644–53

Bradbury EJ, Moon LD, Popat RJ, King VR, Bennett GS, et al. 2002. Chondroitinase ABC promotes functional recovery after spinal cord injury. *Nature* 416:636–40

Brocard F, Tazerart S, Vinay L. 2010. Do pacemakers drive the central pattern generator for locomotion in mammals? *Neuroscientist* 16(2):139–55

Brownstone RM, Gossard J-P, Hultborn H. 1994. Voltage-dependent excitation of motoneurones from spinal locomotor centres in the cat. *Exp. Brain Res.* 102:34–44

Brustein E, Rossignol S. 1998. Recovery of locomotion after ventral and ventrolateral spinal lesions in the cat. I. Deficits and adaptive mechanisms. *J. Neurophysiol.* 80:1245–67

Brustein E, Rossignol S. 1999. Recovery of locomotion after ventral and ventrolateral spinal lesions in the cat. II. Effects of noradrenergic and serotoninergic drugs. *J. Neurophysiol.* 81:1513–30

Buchanan JT. 2001. Contributions of identifiable neurons and neuron classes to lamprey vertebrate neurobiology. *Prog. Neurobiol.* 63(4):441–66

Bussel BC, Roby-Brami A, Yakovleff A, Bennis N. 1988. Evidences for the presence of a spinal stepping generator in patients with a spinal cord section. In *Posture and Gait: Development, Adaptation and Modulation*, ed. B Amblard, A Berthoz, F Clarac, pp. 273–78. Amsterdam: Elsevier

Cafferty WB, McGee AW, Strittmatter SM. 2008. Axonal growth therapeutics: regeneration or sprouting or plasticity? *Trends Neurosci.* 31(5):215–20

Calancie B. 2006. Spinal myoclonus after spinal cord injury. *J. Spinal Cord. Med.* 29(4):413–24

Calancie B, Broton JG, Klose KJ, Traad M, Difini J, Ayyar DR. 1993. Evidence that alterations in presynaptic inhibition contribute to segmental hypo- and hyperexcitability after spinal cord injury in man. *Electroencephalogr. Clin. Neurophysiol.* 89(3):177–86

Calancie B, Needham-Shropshire B, Jacobs P, Willer K, Zych G, Green BA. 1994. Involuntary stepping after chronic spinal cord injury: evidence for a central rhythm generator for locomotion in man. *Brain* 117:1143–59

Carrier L, Brustein L, Rossignol S. 1997. Locomotion of the hindlimbs after neurectomy of ankle flexors in intact and spinal cats: model for the study of locomotor plasticity. *J. Neurophysiol.* 77:1979–93

Cazalets JR, Borde M, Clarac F. 1995. Localization and organization of the central pattern generator for hindlimb locomotion in newborn rat. *J. Neurosci.* 15:4943–51

Cazalets JR, Borde M, Clarac F. 1996. The synaptic drive from the spinal locomotor network to motoneurons in the newborn rat. *J. Neurosci.* 16(1):298–306

Chau C, Barbeau H, Rossignol S. 1998a. Early locomotor training with clonidine in spinal cats. *J. Neurophysiol.* 79:392–409

Chau C, Barbeau H, Rossignol S. 1998b. Effects of intrathecal α_1- and α_2-noradrenergic agonists and norepinephrine on locomotion in chronic spinal cats. *J. Neurophysiol.* 79:2941–63

Choo AM, Liu J, Lam CK, Dvorak M, Tetzlaff W, Oxland TR. 2007. Contusion, dislocation, and distraction: primary hemorrhage and membrane permeability in distinct mechanisms of spinal cord injury. *J. Neurosurg. Spine* 6(3):255–66

Contamin F. 1983. Sections médullaires incomplètes et locomotion chez le chat. *Bull. Acad. Nat. Med.* 167:727–30

Côté M-P, Gossard J-P. 2004. Step training-dependent plasticity in spinal cutaneous pathways. *J. Neurosci.* 24:11317–27

Côté M-P, Menard A, Gossard J-P. 2003. Spinal cats on the treadmill: changes in load pathways. *J. Neurosci.* 23(7):2789–96

Courtine G, Roy RR, Raven J, Hodgson J, McKay H, et al. 2005. Performance of locomotion and foot grasping following a unilateral thoracic corticospinal tract lesion in monkeys (*Macaca mulatta*). *Brain* 128:2338–58

Courtine G, Song B, Roy RR, Zhong H, Herrmann JE, et al. 2008. Recovery of supraspinal control of stepping via indirect propriospinal relay connections after spinal cord injury. *Nat. Med.* 14:69–74

Cowley KC, Schmidt BJ. 1995. Effects of inhibitory amino acid antagonists on reciprocal inhibitory interactions during rhythmic motor activity in the in vitro neonatal rat spinal cord. *J. Neurophysiol.* 74(3):1109–17

Cowley KC, Zaporozhets E, Schmidt BJ. 2008. Propriospinal neurons are sufficient for bulbospinal transmission of the locomotor command signal in the neonatal rat spinal cord. *J. Physiol.* 586(6):1623–35

Crone C, Johnsen LL, Biering-Sorensen F, Nielsen JB. 2003. Appearance of reciprocal facilitation of ankle extensors from ankle flexors in patients with stroke or spinal cord injury. *Brain* 126(Pt. 2):495–507

Cuellar CA, Tapia JA, Juarez V, Quevedo J, Linares P, et al. 2009. Propagation of sinusoidal electrical waves along the spinal cord during a fictive motor task. *J. Neurosci.* 29(3):798–810

de Leon RD, Hodgson JA, Roy RR, Edgerton VR. 1998a. Full weight-bearing hindlimb standing following stand training in the adult spinal cat. *J. Neurophysiol.* 80(1):83–91

de Leon RD, Hodgson JA, Roy RR, Edgerton VR. 1998b. Locomotor capacity attributable to step training versus spontaneous recovery after spinalization in adult cats. *J. Neurophysiol.* 79(3):1329–40

de Leon RD, Tamaki H, Hodgson JA, Roy RR, Edgerton VR. 1999. Hindlimb locomotor and postural training modulates glycinergic inhibition in the spinal cord of the adult spinal cat. *J. Neurophysiol.* 82(1):359–69

Delivet-Mongrain H, Leblond H, Rossignol S. 2008. Effects of localized intraspinal injections of a noradrenergic blocker on locomotion of high decerebrate cats. *J. Neurophysiol.* 100:907–21

de Seze M, Falgairolle M, Viel S, Assaiante C, Cazalets JR. 2008. Sequential activation of axial muscles during different forms of rhythmic behavior in man. *Exp. Brain Res.* 185(2):237–47

Dietz V, Grillner S, Trepp A, Hubli M, Bolliger M. 2009. Changes in spinal reflex and locomotor activity after a complete spinal cord injury: a common mechanism? *Brain* 132(Pt. 8):2196–205

Dietz V, Harkema SJ. 2004. Locomotor activity in spinal cord-injured persons. *J. Appl. Physiol.* 96(5):1954–60

Dietz V, Michel J. 2009. Human bipeds use quadrupedal coordination during locomotion. *Ann. N. Y. Acad. Sci.* 1164:97–103

Dimitrijevic MR, Gerasimenko Y, Pinter MM. 1998. Evidence for a spinal central pattern generator in humans. *Ann. N. Y. Acad. Sci.* 860:360–76

Dimitrijevic MR, Nathan PW. 1967. Studies of spasticity in man. 2. Analysis of stretch reflexes in spasticity. *Brain* 90(2):333–58

Dobkin B, Apple D, Barbeau H, Basso M, Behrman A, et al. 2006. Weight-supported treadmill versus overground training for walking after acute incomplete SCI. *Neurology* 484–93

Drew T, Prentice S, Schepens B. 2004. Cortical and brainstem control of locomotion. *Prog. Brain Res.* 143:251–61

Duysens J, Pearson KG. 1976. The role of cutaneous afferents from the distal hindlimb in the regulation of the step cycle of thalamic cats. *Exp. Brain Res.* 24:245–55

Dy CJ, Gerasimenko YP, Edgerton VR, Dyhre-Poulsen P, Courtine G, Harkema SJ. 2010. Phase-dependent modulation of percutaneously elicited multisegmental muscle responses after spinal cord injury. *J. Neurophysiol.* 103(5):2808–20

Eidelberg E, Story JL, Walden JG, Meyer BL. 1981a. Anatomical correlates of return of locomotor function after partial spinal cord lesions in cats. *Exp. Brain Res.* 42:81–88

Eidelberg E, Walden JG, Nguyen LH. 1981b. Locomotor control in macaque monkeys. *Brain* 104:647–63

Eidelberg E, Yu J. 1981. Effects of corticospinal lesions upon treadmill locomotion by cats. *Exp. Brain Res.* 43:101–3

Eken T, Hultborn H, Kiehn O. 1989. Possible functions of transmitter-controlled plateau potentials in alpha motoneurones. *Prog. Brain Res.* 80:257–67

Eng JJ, Winter DA, Patla AE. 1994. Strategies for recovery from a trip in early and late swing during human walking. *Exp. Brain Res.* 102:339–49

English AW. 1980. Interlimb coordination during stepping in the cat: effects of dorsal column section. *J. Neurophysiol.* 44:270–79

Forssberg H. 1979. Stumbling corrective reaction: a phase-dependent compensatory reaction during locomotion. *J. Neurophysiol.* 42:936–53

Forssberg H, Grillner S. 1973. The locomotion of the acute spinal cat injected with clonidine i.v. *Brain Res.* 50:184–86

Forssberg H, Grillner S, Halbertsma J. 1980a. The locomotion of the low spinal cat. I. Coordination within a hindlimb. *Acta Physiol. Scand.* 108:269–81

Forssberg H, Grillner S, Halbertsma J, Rossignol S. 1980b. The locomotion of the low spinal cat: II. Interlimb coordination. *Acta Physiol. Scand.* 108:283–95

Forssberg H, Grillner S, Rossignol S. 1975. Phase dependent reflex reversal during walking in chronic spinal cats. *Brain Res.* 85:103–7

Frigon A, Barriere G, Leblond H, Rossignol S. 2009. Asymmetric changes in cutaneous reflexes after a partial spinal lesion and retention following spinalization during locomotion in the cat. *J. Neurophysiol.* 102(5):2667–80

Frigon A, Rossignol S. 2006a. Experiments and models of sensorimotor interactions during locomotion. *Biol. Cybern.* 95:607–27

Frigon A, Rossignol S. 2006b. Functional plasticity following spinal cord lesions. *Prog. Brain Res.* 157(16):231–60

Frigon A, Rossignol S. 2008. Adaptive changes of the locomotor pattern and cutaneous reflexes during locomotion studied in the same cats before and after spinalization. *J. Physiol.* 586:2927–45

Frigon A, Rossignol S. 2009. Partial denervation of ankle extensors prior to spinalization in cats impacts the expression of locomotion and the phasic modulation of reflexes. *Neuroscience* 158(4):1675–90

Frigon A, Sirois J, Gossard JP. 2010. The effects of ankle and hip muscle afferent inputs on rhythm generation during fictive locomotion. *J. Neurophysiol.* 103(3):1591–605

Gerasimenko Y, Gorodnichev R, Machueva E, Pivovarova E, Semyenov D, et al. 2010. Novel and direct access to the human locomotor spinal circuitry. *J. Neurosci.* 30(10):3700–8

Ghosh A, Sydekum E, Haiss F, Peduzzi S, Zorner B, et al. 2009. Functional and anatomical reorganization of the sensory-motor cortex after incomplete spinal cord injury in adult rats. *J. Neurosci.* 29(39):12210–19

Giroux N, Brustein E, Chau C, Barbeau H, Reader TA, Rossignol S. 1998. Differential effects of the noradrenergic agonist clonidine on the locomotion of intact, partially and completely spinalized adult cats. In *Neuronal Mechanisms for Generating Locomotor Activity*, ed. O Kiehn, RM Harris-Warrick, LM Jordan, H Hulborn, N Kudo, pp. 517–20. New York: NY Acad. Sci.

Giroux N, Rossignol S, Reader TA. 1999. Autoradiographic study of α_1-, α_2-noradrenergic and serotonin $_{1A}$ receptors in the spinal cord of normal and chronically transected cats. *J. Comp. Neurol.* 406:402–14

Giuliani CA, Smith JL. 1987. Stepping behaviors in chronic spinal cats with one hindlimb deafferented. *J. Neurosci.* 7(8):2537–46

Goldberger ME. 1977. Locomotor recovery after unilateral hindlimb deafferentation in cats. *Brain Res.* 123:59–74

Gorassini MA, Knash ME, Harvey PJ, Bennett DJ, Yang JF. 2004. Role of motoneurons in the generation of muscle spasms after spinal cord injury. *Brain* 127:2247–58

Gorska T, Bem T, Majczynski H. 1990. Locomotion in cats with ventral spinal lesions: support patterns and duration of support phases during unrestrained walking. *Acta Neurobiol. Exp.* 50:191–200

Gorska T, Bem T, Majczynski H, Zmyslowski W. 1993a. Unrestrained walking in cats with partial spinal lesions. *Brain Res. Bull.* 32:241–49

Gorska T, Chojnicka-Gittins B, Majczynski H, Zmyslowski W. 2007. Overground locomotion after incomplete spinal lesions in the rat: quantitative gait analysis. *J. Neurotrauma* 24(7):1198–218

Gorska T, Majczynski H, Bem T, Zmyslowski W. 1993b. Hindlimb swing, stance and step relationships during unrestrained walking in cats with lateral funicular lesion. *Acta Neurobiol. Exp.* 53:133–42

Gossard J-P, Floeter MK, Degtyarenko AM, Simon ES, Burke RE. 1996. Disynaptic vestibulospinal and reticulospinal excitation in cat lumbosacral motoneurons: modulation during fictive locomotion. *Exp. Brain Res.* 109(2):277–88

Goulding M. 2009. Circuits controlling vertebrate locomotion: moving in a new direction. *Nat. Rev. Neurosci.* 10(7):507–18

Grill R, Murai K, Blesch A, Gage FH, Tuszynski MH. 1997. Cellular delivery of neurotrophin-3 promotes corticospinal axonal growth and partial functional recovery after spinal cord injury. *J. Neurosci.* 17(14):5560–72

Grillner S. 1973. Locomotion in the spinal cat. In *Control of Posture and Locomotion. Advances in Behavioral Biology*, Vol. 7, ed. RB Stein, KG Pearson, RS Smith, JB Redford, pp. 515–35. New York: Plenum

Grillner S. 1981. Control of locomotion in bipeds, tetrapods, and fish. In *Handbook of Physiology. The Nervous System II*, ed. JM Brookhart, VB Mountcastle, pp. 1179–236. Bethesda, MD: Am. Physiol. Soc.

Grillner S. 2003. The motor infrastructure: from ion channels to neuronal networks. *Nat. Rev. Neurosci.* 4:573–86

Grillner S, Georgopoulos AP, Jordan LM. 1997. Selection and initiation of motor behavior. In *Neurons, Networks, and Motor Behavior*, ed. PSG Stein, S Grillner, AI Selverston, DG Stuart, pp. 3–19. Cambridge, MA: MIT Press

Grillner S, Jessell TM. 2009. Measured motion: searching for simplicity in spinal locomotor networks. *Curr. Opin. Neurobiol.* 19(6):572–86

Grillner S, Wallen P. 2002. Cellular bases of a vertebrate locomotor system: steering, intersegmental and segmental co-ordination and sensory control. *Prog. Brain Res.* 40:92–106

Grillner S, Zangger P. 1979. On the central generation of locomotion in the low spinal cat. *Exp. Brain Res.* 34:241–61

Guertin P, Angel MJ, Perreault M-C, McCrea DA. 1995. Ankle extensor group I afferents excite extensors throughout the hindlimb during fictive locomotion in the cat. *J. Physiol.* 487:197–209

Gulino R, Dimartino M, Casabona A, Lombardo SA, Perciavalle V. 2007. Synaptic plasticity modulates the spontaneous recovery of locomotion after spinal cord hemisection. *Neurosci. Res.* 57(1):148–56

Harkema SJ. 2001. Neural plasticity after human spinal cord injury: application of locomotor training to the rehabilitation of walking. *Neuroscientist* 7(5):455–68

Harkema SJ. 2008. Plasticity of interneuronal networks of the functionally isolated human spinal cord. *Brain Res. Rev.* 57(1):255–64

Harvey PJ, Li X, Li Y, Bennett DJ. 2006. 5-HT2 receptor activation facilitates a persistent sodium current and repetitive firing in spinal motoneurons of rats with and without chronic spinal cord injury. *J. Neurophysiol.* 96(3):1158–709

Heckman CJ, Lee RH, Brownstone RM. 2003. Hyperexcitable dendrites in motoneurons and their neuro-modulatory control during motor behavior. *Trends Neurosci.* 26(12):688–95

Helgren ME, Goldberger ME. 1993. The recovery of postural reflexes and locomotion following low thoracic hemisection in adult cats involves compensation by undamaged primary afferent pathways. *Exp. Neurol.* 123:17–34

Hinckley C, Seebach B, Ziskind-Conhaim L. 2005. Distinct roles of glycinergic and GABAergic inhibition in coordinating locomotor-like rhythms in the neonatal mouse spinal cord. *Neuroscience* 131:745–58

Holstege JC, Kuypers HGJM. 1987. Brainstem projections to spinal motoneurons: an update. *Neuroscience* 23:809–21

Hultborn H. 2003. Changes in neuronal properties and spinal reflexes during development of spasticity following spinal cord lesions and stroke: studies in animal models and patients. *J. Rehabil. Med.* 41(Suppl.):46–55

Hultborn H, Brownstone RB, Toth TI, Gossard J-P. 2004. Key mechanisms for setting the input-output gain across the motoneuron pool. *Prog. Brain Res.* 143:77–95

Hultborn H, Conway B, Gossard J-P, Brownstone R, Fedirchuk B, Schomburg ED. 1998. How do we approach the locomotor network in the mammalian spinal cord? *Ann. N. Y. Acad. Sci.* 860:70–82

Jankowska E, Jukes MGM, Lund S, Lundberg A. 1967. The effect of DOPA on the spinal cord. 5. Reciprocal organization of pathways transmitting excitatory action to alpha motoneurones of flexors and extensors. *Acta Physiol. Scand.* 70:369–88

Jiang W, Drew T. 1996. Effects of bilateral lesions of the dorsolateral funiculi and dorsal columns at the level of the low thoracic spinal cord on the control of locomotion in the adult cat: I. Treadmill walking. *J. Neurophysiol.* 76(2):849–66

Jordan LM. 1998. Initiation of locomotion in mammals. *Ann. N. Y. Acad. Sci.* 860:83–93

Juvin L, Simmers J, Morin D. 2005. Propriospinal circuitry underlying interlimb coordination in mammalian quadrupedal locomotion. *J. Neurosci.* 25(25):6025–35

Kanagal SG, Muir GD. 2008. Effects of combined dorsolateral and dorsal funicular lesions on sensorimotor behaviour in rats. *Exp. Neurol.* 214:229–39

Kato M. 1989. Chronically isolated lumbar half spinal cord produced by hemisection and longitudinal myelotomy generates locomotor activities of the ipsilateral hindlimb of the cat. *Neurosci. Lett.* 98:149–53

Kato M, Murakami S, Hirayama H, Hikino K. 1985. Recovery of postural control following chronic bilateral hemisections at different spinal cord levels in adult cats. *Exp. Neurol.* 90:350–64

Kato M, Murakami S, Yasuda K, Hirayama H. 1984. Disruption of fore- and hindlimb coordination during overground locomotion in cats with bilateral serial hemisection of the spinal cord. *Neurosci. Res.* 2:27–47

Kiehn O. 2006. Locomotor circuits in the mammalian spinal cord. *Annu. Rev. Neurosci.* 29:279–306

Kiehn O, Johnson BR, Raastad M. 1996. Plateau properties in mammalian spinal interneurons during transmitter-induced locomotor activity. *Neuroscience* 75:263–73

Kjaerulff O, Kiehn O. 1997. Crossed rhythmic synaptic input to motoneurons during selective activation of the contralateral spinal locomotor network. *J. Neurosci.* 17:9433–47

Kremer E, Lev-Tov A. 1997. Localization of the spinal network associated with generation of hindlimb locomotion in the neonatal rat and organization of its transverse coupling system. *J. Neurophysiol.* 77(3):1155–70

Kuhn RA. 1950. Functional capacity of the isolated human spinal cord. *Brain* 73:1–51

Kuhtz-Buschbeck JP, Boczek-Funcke A, Mautes A, Nacimiento W, Weinhardt C. 1996. Recovery of locomotion after spinal cord hemisection: an X-ray study of the cat hindlimb. *Exp. Neurol.* 137:212–24

Lam T, Wolstenholme C, Van Der LM, Pang MY, Yang JF. 2003. Stumbling corrective responses during treadmill-elicited stepping in human infants. *J. Physiol.* 553(Pt. 1):319–31

Langlet C, Leblond H, Rossignol S. 2005. The mid-lumbar segments are needed for the expression of locomotion in chronic spinal cats. *J. Neurophysiol.* 93:2474–88

Lavrov I, Courtine G, Dy CJ, Van Den Brand R, Fong AJ, et al. 2008. Facilitation of stepping with epidural stimulation in spinal rats: role of sensory input. *J. Neurosci.* 28:7774–80

Lavrov I, Gerasimenko YP, Ichiyama RM, Courtine G, Zhong H, et al. 2006. Plasticity of spinal cord reflexes after a complete transection in adult rats: relationship to stepping ability. *J. Neurophysiol.* 96(4):1699–710

Leblond H, Menard A, Gossard J-P. 2001. Corticospinal control of locomotor pathways generating extensor activities in the cat. *Exp. Brain Res.* 138(2):173–84

Lhermitte J. 1919. *La section totale de la Moelle Dorsale*. Paris: Impr. Vve Tardy-Pigelet Fils

Li X, Murray KC, Harvey PJ, Ballou EW, Bennett DJ. 2007. Serotonin facilitates a persistent calcium current in motoneurons of rats with and without chronic spinal cord injury. *J. Neurophysiol.* 97:1236–46

Li Y, Gorassini MA, Bennett DJ. 2004. Role of persistent sodium and calcium currents in motoneuron firing and spasticity in chronic spinal rats. *J. Neurophysiol.* 91(2):767–83

Li Y, Oskouian RJ, Day Y-J, Kern JA, Linden J. 2006. Optimization of a mouse locomotor rating system to evaluate compression-induced spinal cord injury: correlation of locomotor and morphological injury indices. *J. Neurosurg. Spine* 4:165–73

Liddell EGT, Phillips CG. 1944. Pyramidal section in the cat. *Brain* 67:1–9

Lloyd DPC. 1941. Activity in neurons of the bulbospinal correlation system. *J. Neurophysiol.* 4:115–34

Magnuson DS, Lovett R, Coffee C, Gray R, Han Y, et al. 2005. Functional consequences of lumbar spinal cord contusion injuries in the adult rat. *J. Neurotrauma* 22(5):529–43

Magnuson DS, Trinder TC. 1997. Locomotor rhythm evoked by ventrolateral funiculus stimulation in the neonatal rat spinal cord in vitro. *J. Neurophysiol.* 77(1):200–6

Maier IC, Schwab ME. 2006. Sprouting, regeneration and circuit formation in the injured spinal cord: factors and activity. *Philos. Trans. R. Soc. Lond. B* 361:1611–34

Majczynski H, Maleszak K, Gorska T, Slawinska U. 2007. Comparison of two methods for quantitative assessment of unrestrained locomotion in the rat. *J. Neurosci. Methods* 163:197–207

Marcoux J, Rossignol S. 2000. Initiating or blocking locomotion in spinal cats by applying noradrenergic drugs to restricted lumbar spinal segments. *J. Neurosci.* 20(22):8577–85

McCrea DA, Rybak IA. 2008. Organization of mammalian locomotor rhythm and pattern generation. *Brain Res. Rev.* 57(1):134–46

Metz GA, Curt A, Van de Meent H, Klusman I, Schwab ME, Dietz V. 2000a. Validation of the weight-drop contusion model in rats: a comparative study of human spinal cord injury. *J. Neurotrauma* 17(1):1–17

Metz GA, Merkler D, Dietz V, Schwab ME, Fouad K. 2000b. Efficient testing of motor function in spinal cord injured rats. *Brain Res.* 883(2):165–77

Muir GD, Steeves JD. 1995. Phasic cutaneous input facilitates locomotor recovery after incomplete spinal injury in the chick. *J. Neurophysiol.* 74:358–68

Muir GD, Webb AA, Kanagal S, Taylor L. 2007. Dorsolateral cervical spinal injury differentially affects forelimb and hindlimb action in rats. *Eur. J. Neurosci.* 25(5):1501–10

Murray KC, Nakae A, Stephens MJ, Rank M, D'Amico J, et al. 2010. Recovery of motoneuron and locomotor function after spinal cord injury depends on constitutive activity in 5-HT(2C) receptors. *Nat. Med.* 16(6):694–700

Nadeau S, Jacquemin G, Fournier C, Lamarre Y, Rossignol S. 2010. Spontaneous motor rhythms of the back and legs in a patient with a complete spinal cord transection. *Neurorehabil. Neural Repair* 24(4):377–83

Nathan PW. 1994. Effects on movement of surgical incisions into the human spinal cord. *Brain* 117:337–46

Nielsen JB, Crone C, Hultborn H. 2007. The spinal pathophysiology of spasticity: from a basic science point of view. *Acta Physiol.* 189:171–80

Norreel JC, Pflieger JF, Pearlstein E, Simeoni-Alias J, Clarac F, Vinay L. 2003. Reversible disorganization of the locomotor pattern after neonatal spinal cord transection in the rat. *J. Neurosci.* 23(5):1924–32

Okuma Y, Mizuno Y, Lee RG. 2002. Reciprocal Ia inhibition in patients with asymmetric spinal spasticity. *Clin. Neurophysiol.* 113(2):292–97

Orlovsky GN, Shik ML. 1976. Control of locomotion: a neurophysiological analysis of the cat locomotor system. In *International Review of Physiology: Neurophysiology II*, ed. R Portez, pp. 281–309. Baltimore: Univ. Park Press

Pang MY, Yang JF. 2000. The initiation of the swing phase in human infant stepping: importance of hip position and leg loading. *J. Physiol.* 528(Pt. 2):389–404

Pearson KG. 2008. Role of sensory feedback in the control of stance duration in walking cats. *Brain Res. Rev.* 57(1):222–27

Pearson KG, Misiaszek JE, Fouad K. 1998. Enhancement and resetting of locomotor activity by muscle afferents. *Ann. N. Y. Acad. Sci.* 860:203–15

Pearson KG, Rossignol S. 1991. Fictive motor patterns in chronic spinal cats. *J. Neurophysiol.* 66:1874–87

Perez T, Tapia JA, Mirasso CR, Garcia-Ojalvo J, Quevedo J, et al. 2009. An intersegmental neuronal architecture for spinal wave propagation under deletions. *J. Neurosci.* 29(33):10254–63

Perreault M-C, Rossignol S, Drew T. 1994. Microstimulation of the medullary reticular formation during fictive locomotion. *J. Neurophysiol.* 71:229–45

Petras JM. 1967. Cortical, tectal and segmental fiber connections in the spinal cord of the cat. *Brain Res.* 6:275–324

Philippson M. 1905. L'autonomie et la centralisation dans le système nerveux des animaux. *Trav. Lab. Physiol. Inst. Solvay (Bruxelles)* 7:1–208

Poon PC, Gupta D, Shoichet MS, Tator CH. 2007. Clip compression model is useful for thoracic spinal cord injuries: histologic and functional correlates. *Spine* 32:2853–59

Rank MM, Li X, Bennett DJ, Gorassini MA. 2007. Role of endogenous release of norepinephrine in muscle spasms after chronic spinal cord injury. *J. Neurophysiol.* 97(5):3166–80

Remy-Neris O, Barbeau H, Daniel O, Boiteau F, Bussel B. 1999. Effects of intrathecal clonidine injection on spinal reflexes and human locomotion in incomplete paraplegic subjects. *Exp. Brain Res.* 129(3):433–40

Rho MJ, Lavoie S, Drew T. 1999. Effects of red nucleus microstimulation on the locomotor pattern and timing in the intact cat: a comparison with the motor cortex. *J. Neurophysiol.* 81(5):2297–315

Roberts A, Soffe SR, Wolf ES, Yoshida M, Zhao FY. 1998. Central circuits controlling locomotion in young frog tadpoles. *N. Y. Acad. Sci. Conf.* 860:19–34

Robinson GA, Goldberger ME. 1985. Interfering with inhibition may improve motor function. *Brain Res.* 346(2):400–3

Roby-Brami A, Bussel B. 1987. Long-latency spinal reflex in man after flexor reflex afferent stimulation. *Brain* 110:707–25

Rosenzweig ES, Courtine G, Jindrich DL, Brock JH, Ferguson AR, et al. 2010. Extensive spontaneous plasticity of corticospinal projections after primate spinal cord injury. *Nat. Neurosci.* 13(12):1505–10

Rossignol S. 1996. Neural control of stereotypic limb movements. In *Handbook of Physiology, Section 12. Exercise: Regulation and Integration of Multiple Systems*, ed. LB Rowell, JT Sheperd, pp. 173–216. New York: Oxford Univ. Press

Rossignol S. 2006. Plasticity of connections underlying locomotor recovery after central and or peripheral lesions in the adult mammals. *Philos. Trans. R. Soc. Lond. B* 361:1647–71

Rossignol S, Barriere G, Alluin O, Frigon A. 2009. Re-expression of locomotor function after partial spinal cord injury. *Physiology* 24:127–39

Rossignol S, Bélanger M, Chau C, Giroux N, Brustein E, et al. 2000. The spinal cat. In *Neurobiology of Spinal Cord Injury*, ed. RG Kalb, SM Strittmatter, pp. 57–87. Totowa, NJ: Humana

Rossignol S, Chau C, Brustein E, Bélanger M, Barbeau H, Drew T. 1996. Locomotor capacities after complete and partial lesions of the spinal cord. *Acta Neurobiol. Exp.* 56(1):449–63

Rossignol S, Chau C, Giroux N, Brustein E, Bouyer L, et al. 2002. The cat model of spinal injury. *Prog. Brain Res.* 137:151–68

Rossignol S, Drew T, Brustein E, Jiang W. 1999. Locomotor performance and adaptation after partial or complete spinal cord lesions in the cat. *Prog. Brain Res.* 123(31):349–65

Rossignol S, Dubuc R, Gossard JP. 2006. Dynamic sensorimotor interactions in locomotion. *Physiol Rev.* 86(1):89–154

Rossignol S, Giroux N, Chau C, Marcoux J, Brustein E, Reader TA. 2001. Pharmacological aids to locomotor training after spinal injury in the cat. *J. Physiol.* 533:65–74

Russel DF, Zajac FE. 1979. Effects of stimulating Deiter's nucleus and medial longitudinal fasciculus on the timing of the fictive locomotor rhythm induced in cats by DOPA. *Brain Res.* 177:588–92

Sakurai A, Katz PS. 2009. Functional recovery after lesion of a central pattern generator. *J. Neurosci.* 29(42):13115–25

Saruhashi Y, Young W, Perkins R. 1996. The recovery of 5-HT immunoreactivity in lumbosacral spinal cord and locomotor function after thoracic hemisection. *Exp. Neurol.* 139(2):203–13

Schillings AM, Van Wezel BM, Mulder T, Duysens J. 2000. Muscular responses and movement strategies during stumbling over obstacles. *J. Neurophysiol.* 83(4):2093–102

Schindler-Ivens S, Shields RK. 2000. Low frequency depression of H-reflexes in humans with acute and chronic spinal-cord injury. *Exp. Brain Res.* 133(2):233–41

Schmidt BJ, Jordan LM. 2000. The role of serotonin in reflex modulation and locomotor rhythm production in the mammalian spinal cord. *Brain Res. Bull.* 53(5):689–710

Shefchyk SJ, Jordan LM. 1985. Excitatory and inhibitory post-synaptic potentials in alpha-motoneurons produced during fictive locomotion by stimulation of the mesencephalic locomotor region. *J. Neurophysiol.* 53:1345–55

Sherrington CS, Laslett EE. 1903. Observations on some spinal reflexes and the interconnection of spinal segments. *J. Physiol.* 29:58–96

Shields CB, Zhang YP, Shields LB, Han Y, Burke DA, Mayer NW. 2005. The therapeutic window for spinal cord decompression in a rat spinal cord injury model. *J. Neurosurg. Spine* 3(4):302–7

Shik ML. 1983. Action of the brainstem locomotor region on spinal stepping generators via propriospinal pathways. In *Spinal Cord Reconstruction*, ed. CC Kao, RP Bunge, PJ Reier, pp. 421–34. New York: Raven

Shik ML, Severin FV, Orlovsky GN. 1966. Control of walking and running by means of electrical stimulation of the mid-brain. *Biophysics* 11:756–65

Siegal T. 1995. Spinal cord compression: from laboratory to clinic. *Eur. J. Cancer* 31A(11):1748–53

Smith RR, Shum-Siu A, Baltzley R, Bunger M, Baldini A, et al. 2006. Effects of swimming on functional recovery after incomplete spinal cord injury in rats. *J. Neurotrauma* 23(6):908–19

Suresh BR, Muthusamy R, Namasivayam A. 2000. Behavioural assessment of functional recovery after spinal cord hemisection in the bonnet monkey (*Macaca radiata*). *J. Neurol. Sci.* 178(2):136–52

Tan AM, Petruska JC, Mendell LM, Levine JM. 2007. Sensory afferents regenerated into dorsal columns after spinal cord injury remain in a chronic pathophysiological state. *Exp. Neurol.* 206(2):257–68

Tazerart S, Vinay L, Brocard F. 2008. The persistent sodium current generates pacemaker activities in the central pattern generator for locomotion and regulates the locomotor rhythm. *J. Neurosci.* 28(34):8577–89

Thompson FJ, Reier PJ, Lucas CC, Parmer R. 1992. Altered patterns of reflex excitability subsequent to contusion injury of the rat spinal cord. *J. Neurophysiol.* 68(5):1473–86

Thuret S, Moon LD, Gage FH. 2006. Therapeutic interventions after spinal cord injury. *Nature* 7:628–43

Tillakaratne NJ, de Leon RD, Hoang TX, Roy RR, Edgerton VR, Tobin AJ. 2002. Use-dependent modulation of inhibitory capacity in the feline lumbar spinal cord. *J. Neurosci.* 22(8):3130–43

Tillakaratne NJ, Mouria M, Ziv NB, Roy RR, Edgerton VR, Tobin AJ. 2000. Increased expression of glutamate decarboxylase (GAD_{67}) in feline lumbar spinal cord after complete thoracic spinal cord transection. *J. Neurosci. Res.* 60(2):219–30

Viala D, Valin A, Buser P. 1974. Relationship between the "late reflex discharge" and locomotor movements in acute spinal cats and rabbits treated with DOPA. *Arch. Ital. Biol.* 112:299–306

Viala D, Vidal C. 1978. Evidence for distinct spinal locomotion generators supplying respectively fore- and hindlimbs in the rabbit. *Brain Res.* 155:182–86

Vilensky JA, Moore AM, Eidelberg E, Walden JG. 1992. Recovery of locomotion in monkeys with spinal cord lesions. *J. Motor Behav.* 24:288–96

Wilson P, Kitchener PD. 1996. Plasticity of cutaneous primary afferent projections to the spinal dorsal horn. *Prog. Neurobiol.* 48(2):105–29

Wirz M, Zemon DH, Rupp R, Scheel A, Colombo G, et al. 2005. Effectiveness of automated locomotor training in patients with chronic incomplete spinal cord injury: a multicenter trial. *Arch. Phys Med. Rehabil.* 86(4):672–80

Xia R, Rymer WZ. 2005. Reflex reciprocal facilitation of antagonist muscles in spinal cord injury. *Spinal Cord* 43(1):14–21

Yakovenko S, Drew T. 2009. A motor cortical contribution to the anticipatory postural adjustments that precede reaching in the cat. *J. Neurophysiol.* 102:853–74

Yang JF, Stein RB. 1990. Phase-dependent reflex reversal in human leg muscles during walking. *J. Neurophysiol.* 63:1109–17

Young W. 2002. Spinal cord contusion models. *Prog. Brain Res.* 137:231–55

Zaporozhets E, Cowley KC, Schmidt BJ. 2006. Propriospinal neurons contribute to bulbospinal transmission of the locomotor command signal in the neonatal rat spinal cord. *J. Physiol.* 572(Pt. 2):443–58

Zehr EP, Hundza SR, Vasudevan EV. 2009. The quadrupedal nature of human bipedal locomotion. *Exerc. Sport Sci. Rev.* 37(2):102–8

Zmyslowski W, Gorska T, Majczynski H, Bem T. 1993. Hindlimb muscle activity during unrestrained walking in cats with lesions of the lateral funiculi. *Acta Neurobiol. Exp.* 53:143–53

RELATED RESOURCES

Grillner S, El Manira A, Kiehn O, Rossignol S, Stein PSG, eds. 2008. Special issue: Networks in motion. *Brain Res. Rev.* 57(1):1–269

Grillner S, Stein PSG, Stuart DG, Forssberg H, Herman RM, eds. 1986. *Neurobiology of Vertebrate Locomotion*, Wenner-Gren Int. Symp. Ser. New York: Macmillan. 735 pp.

Herman RM, Grillner S, Stein PSG, Stuart DG, eds. 1976. *Neural Control of Locomotion*, Vol. 18. New York: Plenum. 822 pp.

McKerracher L, Doucet G, Rossignol S, eds. 2002. *Spinal Cord Trauma: Regeneration, Neural Repair and Functional Recovery*. New York: Elsevier

Stein PSG, Grillner S, Selverston AI, Stuart DG, eds. 1997. *Neurons, Networks, and Motor Behavior*. Cambridge, MA: MIT Press

Modulation of Striatal Projection Systems by Dopamine

Charles R. Gerfen[1] and D. James Surmeier[2]

[1]Laboratory of Systems Neuroscience, National Institute of Mental Health, Bethesda, Maryland 20892; email: gerfenc@mail.nih.gov

[2]Department of Physiology, Northwestern University Feinberg School of Medicine, Chicago, Illinois 60611; email: j-surmeier@northwestern.edu

Annu. Rev. Neurosci. 2011. 34:441–66

First published online as a Review in Advance on April 4, 2011

The *Annual Review of Neuroscience* is online at neuro.annualreviews.org

This article's doi:
10.1146/annurev-neuro-061010-113641

0147-006X/11/0721-0441$20.00

Keywords

basal ganglia, synaptic plasticity, movement disorders, Parkinson's disease, motor learning

Abstract

The basal ganglia are a chain of subcortical nuclei that facilitate action selection. Two striatal projection systems—so-called direct and indirect pathways—form the functional backbone of the basal ganglia circuit. Twenty years ago, investigators proposed that the striatum's ability to use dopamine (DA) rise and fall to control action selection was due to the segregation of D_1 and D_2 DA receptors in direct- and indirect-pathway spiny projection neurons. Although this hypothesis sparked a debate, the evidence that has accumulated since then clearly supports this model. Recent advances in the means of marking neural circuits with optical or molecular reporters have revealed a clear-cut dichotomy between these two cell types at the molecular, anatomical, and physiological levels. The contrast provided by these studies has provided new insights into how the striatum responds to fluctuations in DA signaling and how diseases that alter this signaling change striatal function.

Contents

INTRODUCTION

The basal ganglia are part of the forebrain circuitry involved in action selection (Cisek & Kalaska 2010, Redgrave et al. 1999, Wichmann & DeLong 2003). Diseases of the basal ganglia result in a spectrum of movement disorders, ranging from those characterized by hypokinesia, such as Parkinson's disease (PD), to those producing hyperkinesia, such as chorea and dystonia (Mink 2003). A cardinal feature of the striatum—the principal integrator of cortical and thalamic information reaching the basal ganglia—is its dense dopamine (DA) innervation, which arises from the midbrain nigrostriatal pathway (Bolam et al. 2000). This input is critical to normal functioning of the striatum and basal ganglia. One of the major targets of this innervation is the principal neuron of the striatum: GABAergic spiny projection neurons (SPNs). SPNs constitute as much as 90% of the striatal neuron population. SPNs can be divided into two

approximately equally sized populations based on axonal projections. One population—direct pathway SPNs—projects axons to the nuclei at the interface between the basal ganglia and the rest of the brain, whereas the other population—indirect pathway SPNs—projects only indirectly to the interface nuclei.

A second distinguishing feature of direct- and indirect-pathway SPNs is their differential expression of DA receptors. The D_1 DA receptor (Drd1a) is expressed selectively by direct-pathway SPNs, whereas the D_2 (Drd2) is expressed by indirect-pathway SPNs. These two G protein–coupled receptors (GPCRs) have distinctive intracellular signaling cascades and targets, leading to fundamentally different cellular responses to extracellular DA. The dichotomous effect of DA on gene expression in direct- and indirect-pathway SPNs was demonstrated some 20 years ago (Gerfen et al. 1990). However, electrophysiological studies did not yield as clean a picture, leading to a debate about the extent to which these receptors were colocalized in SPNs. The murkiness of the physiological picture has subsequently been shown to be largely a consequence of the complexity of the striatal circuitry, not a consequence of D_1 and D_2 receptor colocalization. The most compelling evidence for this conclusion has come from work with bacterial artificial chromosome (BAC) transgenic mice in which expression of a fluorescent reporter molecule, most commonly enhanced green fluorescent protein (eGFP), is driven by the promoter region of either the D_1 or the D_2 receptor (Gong et al. 2003). In these mice, a precise alignment of D_1 and D_2 receptor expression can be seen with direct and indirect SPN pathways. This review summarizes work in the past few years, primarily with these new mice, that has advanced our understanding of how DA modulates the structure and function of direct- and indirect-pathway SPNs in health and in disease states, such as PD.

ORGANIZATION OF THE BASAL GANGLIA

The largest nucleus of the basal ganglia is the striatum, which comprises the caudate,

PD: Parkinson's disease

DA: dopamine

SPN: spiny projection neuron

BAC: bacterial artificial chromosome

a
Basal ganglia circuits

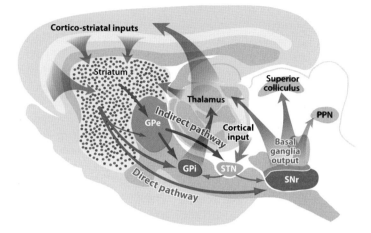

Drd1a BAC-EGFP (direct pathway)

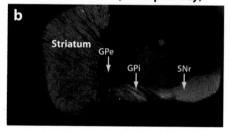

Drd2 BAC-EGFP (indirect pathway)

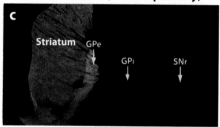

Figure 1

Diagram of select basal ganglia circuits. (*a*) The striatum receives excitatory corticostriatal and thalamic inputs. Outputs of the basal ganglia arise from the internal segment of the globus pallidus (GPi) and substantia nigra pars reticulata (SNr), which are directed to the thalamus, superior colliculus, and pendunculopontine nucleus (PPN). The direct pathway originates from Drd1a-expressing spiny projection neurons (SPNs) that project to the GPi and SNr output nuclei. The indirect pathway originates from Drd2-expressing SPNs that project only to the external segment of the globus pallidus (GPe), which together with the subthalamic nucleus (STN) contain transsynaptic circuits connecting to the basal output nuclei. The direct and indirect pathways provide opponent regulation of the basal ganglia output interface. (*b*) Fluorescent imaging of a brain section from a mouse expressing enhanced green fluorescent protein (eGFP) under regulation of the Drd1a promoter shows Drd1a-expressing SPNs in the striatum that project axons through the GPe, which terminate in the GPi and GPe. (*c*) Fluorescent imaging of a Drd2-eGFP mouse shows that labeled SPNs provide axonal projections that terminate in the GPe but do not extend to the GPi or SNr.

putamen, and ventral striatum, including the nucleus accumbens (**Figure 1**). Essentially, all cortical areas—sensory, motor, and associational—project to the striatum (Bolam et al. 2000, Gerfen et al. 2002). The other major input to the striatum comes from the thalamus, particularly the intralaminar thalamic nuclei (Doig et al. 2010, Smith et al. 2004). Both projections are glutamatergic, forming excitatory synaptic connections with SPNs and four classes of interneuron.

Striatal information flows to the rest of the basal ganglia through GABAergic SPNs of the direct and indirect pathways (Gerfen & Wilson 1996). Direct-pathway SPNs extend axonal projections to the GABAergic output

GPi: globus pallidus
internal segment

GPe: globus pallidus
external segment

nuclei of the basal ganglia, the internal segment of the globus pallidus (GPi), and the substantia nigra pars reticulata (SNr). These nuclei may be considered as one somatotopically organized structure, with the GPi involved in axial and limb movements and the SNr involved in head and eye movements. Direct-pathway neurons also extend an axon collateral to the GABAergic neurons of the external segment of the globus pallidus (GPe). Indirect-pathway SPNs extend axonal projections only to the GPe. GPe neurons project to the glutamatergic neurons of the subthalamic nucleus (STN) and to the output nuclei (GPi/SNr). STN neurons project to the GPe and to both the GPi and the SNr, forming a parallel pathway to the output nuclei. Thus, the indirect pathway forms a multisynaptic circuit between the striatum and the basal ganglia output nuclei. Activity within this indirect-pathway circuit is also modulated by excitatory cortical input to the STN. All the neurons in the GPe, GPi, STN, and SNr are autonomous pacemakers; that is, they generate action potentials on their own, without synaptic input. This mechanism creates a means by which direct- and indirect-pathway GABAergic neurons can bidirectionally modulate the spiking rate of GPi and SNr neurons projecting to the thalamus, superior colliculus, and pedunculopontine nucleus (PPN). Thus, their neuroanatomical connections endow direct- and indirect-pathway SPNs with distinct effects on the activity of the inhibitory output interface of the basal ganglia. The indirect pathway provides excitatory input through disinhibition, whereas the direct pathway provides inhibitory input at this interface. Some GPe neurons also send collaterals to the SNr, which provide an additional source of inhibition to the output nuclei from the indirect-pathway circuit.

STRIATAL DISTRIBUTION OF DOPAMINE RECEPTORS

Five GPCRs mediate DA signaling (D_1–D_5). These receptors are grouped into two classes on the basis of the G-protein to which they couple: D_1 and D_5 (D_1-like) receptors stimulate G_s and G_{olf} proteins, whereas D_2, D_3, and D_4 (D_2-like) receptors stimulate G_o and G_i proteins (Neve et al. 2004). G_s and G_{olf} proteins stimulate adenylyl cyclase, elevating intracellular levels of cyclic adenosine monophosphate (cAMP) and activating protein kinase A (PKA). PKA has a broad array of cellular targets, including transcription factors, voltage-dependent ion channels, and glutamate receptors (Svenningsson et al. 2004). $G_{i/o}$ proteins target voltage-dependent ion channels through a membrane-delimited mechanism, as well as enzymes such as phospholipase C (PLC) isoforms (e.g., Hernandez-Lopez et al. 2000). G_i proteins also inhibit adenylyl cyclase (Stoof & Kebabian 1984). Another important feature of the D_2-like pathway is its potent modulation by RGS (regulators of G-protein signaling) proteins that are robustly expressed in striatal neurons (Geurts et al. 2003).

All five DA receptors are expressed in the striatum, but D_1 and D_2 receptors are by far the most abundant. These two receptors are segregated in direct- and indirect-pathway SPNs: D_1 receptors are expressed by direct-pathway SPNs, whereas D_2 receptors are expressed by indirect-pathway SPNs (Gerfen et al. 1990, Surmeier et al. 1996). This dichotomy was first inferred from the effects of DA depletion on gene expression. In the mid-1980s, investigators discovered that direct-pathway SPNs express high levels of substance P (SP) and dynorphin (DYN), whereas indirect-pathway SPNs express enkephalin (ENK). They subsequently found that after DA-depleting lesions, striatal ENK levels rose and SP levels fell, suggesting that DA was differentially modulating these two populations (Young et al. 1986). D_1 receptor agonists restored striatal SP levels and D_2 receptor agonists restored ENK levels (Gerfen et al. 1990). In situ hybridization studies confirmed that D_1 receptor mRNA and SP colocalized in direct-pathway SPNs, and D_2 receptor mRNA and ENK colocalized in indirect-pathway SPNs. As elegant and simple as this story was, it did not align perfectly with electrophysiological studies that showed that both D_1 and D_2 receptor agonists seemed to have effects on the same striatal neurons (e.g., White & Wang 1986). The simplest explanation for the physiological studies was that D_1 and D_2 receptors were colocalized in SPNs. Although parsimonious, this explanation did not hold up to further experimental study. Single-cell reverse transcription-polymerase chain reaction (RT-PCR) studies found that D_1 and D_2 receptor mRNAs were largely segregated (Surmeier et al. 1996). Because these studies were not quantitative, the modest extent of colocalization found in these studies could be attributable to the detection of low-abundance transcripts that had little or no functional impact. This point has been driven home more recently by the development of BAC transgenics in which eGFP or Cre-recombinase is expressed under control of the D_1 or D_2 promoter (Gong et al. 2003). Examination of these mice has confirmed the segregation of D_1 and D_2 receptor expression in direct- and indirect-pathway SPNs (Gertler et al. 2008, Valjent et al. 2009), confirming the original hypothesis advanced by Gerfen et al. (1990).

What then explains the physiological data? There are two obvious alternatives to colocalization. First, because the pharmacological tools are crude and only allow the D_1- and D_2-like receptors to be discriminated, it could be that SPNs express the less abundant DA receptors (D_5, D_3, D_4). Single-cell RT-PCR profiles of SPNs clearly support this view, showing that direct-pathway SPNs expressed low levels of D_3 receptor mRNA and indirect-pathway SPNs expressed low levels of D_5 mRNA (Surmeier et al. 1996). However, the functional roles of these receptors have been difficult to demonstrate.

Another possibility that has gained more support in recent years is that DA receptor–expressing striatal interneurons play an important role in regulating both direct- and indirect-pathway SPNs, leading to indirect responses to a global DA signal. There are four well-characterized classes of striatal interneuron: cholinergic interneurons, parvalbumin-expressing GABAergic interneurons, calretinin-expressing GABAergic interneurons, and neuropeptide Y (NPY)/nitric oxide–expressing GABAergic interneurons (Tepper et al. 2008). Together, these interneurons constitute ~5%–10% of all striatal neurons. One of them, the cholinergic interneuron, coexpresses D_2 and D_5 receptors (Bergson et al. 1995, Hersch et al. 1995, Yan et al. 1997) and modulates both SPN populations through muscarinic receptors (Bernard et al. 1992, Yan & Surmeier 1996). Two other prominent interneurons, the somatostatin (SOM)/NPY-expressing GABAergic interneuron and the parvalbumin (PV) GABAergic interneuron, express D_5 DA receptors (Centonze et al. 2002, 2003). To complicate matters, the PV interneuron is strongly innervated by globus pallidus neurons that express D_2 receptors (Bevan et al. 1998), creating a microcircuit that is influenced by both D_1- and D_2-class receptors (Wiltschko et al. 2010). As outlined below, there are good reasons to believe that these interneurons play important roles in regulating direct- and indirect-pathway activity, providing a mechanism by which a ligand for a single receptor class can have broad effects.

DOPAMINE MODULATION OF CANONICAL STRIATAL MICROCIRCUITS

The striatal circuitry controlling direct and indirect pathways is complex, and the impact of DA on these circuits is even more so. Nevertheless, in the past decade, considerable progress has been made toward understanding both. To make the review of this work more tractable, we consider five canonical striatal microcircuits (**Figure 2**). These basic circuits are found throughout the striatum.

The Corticostriatal Circuit

The most basic striatal microcircuit is the one formed by glutamatergic cortical pyramidal neurons and SPNs (Bolam et al. 2000). The synapses formed by cortical pyramidal neurons are exclusively on dendritic spines of SPNs. These spines are absent from soma and the most

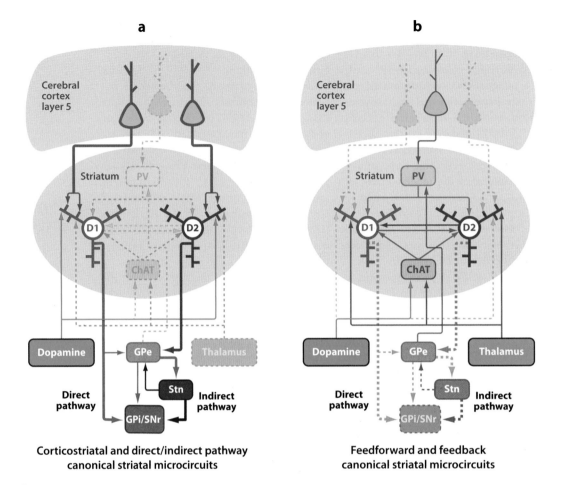

a Corticostriatal and direct/indirect pathway
canonical striatal microcircuits

b Feedforward and feedback
canonical striatal microcircuits

Figure 2

Canonical basal ganglia microcircuits. (*a*) Cortico-striatal and direct/indirect pathway canonical circuits. Layer 5 cortical pyramidal neurons provide excitatory glutamatergic inputs to the spines of Drd1a (D_1) and Drd2 (D_2) spiny projection neurons (SPNs). D_1 SPNs give rise to striatal direct-pathway projections to the output nuclei of the basal ganglia [internal segment of the globus pallidus (GPi)/substantia nigra pars reticulata (SNr)], whereas D_2 SPNs give rise to striatal indirect-pathway projections to basal ganglia output nuclei. Dopamine input through the nigrostriatal pathway is directed to the spine necks of D_1 and D_2 bearing SPNs to modulate corticostriatal inputs. (*b*) Feedforward, feedback, and intrinsic striatal circuits. One feedforward circuit involves fast-spiking, parvalbumin (PV) GABAergic interneurons that provide perisomatic synapses on both D_1 and D_2 SPNs. These PV neurons receive excitatory inputs from layer 5 corticostriatal neurons and are inhibited by the external segment of the globus pallidus (GPe). Intralaminar thalamic nuclei provide inputs to D_1 and D_2 SPNs as well as contribute to a feedforward circuit involving thalamostriatal inputs to cholinergic (ChAT) interneurons that provide input to both D_1 and D_2 SPNs. Cholinergic neuron activity is also affected by dopamine inputs. Feedback striatal microcircuits involve interconnections between local axonal collaterals of D_1 and D_2 SPNs that make synaptic contact with other SPNs.

proximal dendrites, rising to a peak density (1–2 per μm) 50–60 μm from the soma and then falling off very gradually in density to the tips of the sparsely branching dendrites (250–400 μm) (Wilson 1994). Individual cortical axons are sparsely connected to any one SPN, typically making one or two en passant synapses (Parent & Parent 2006). The cortical synapses on the dendritic tree of SPNs have no obvious organization, but this organization may simply be difficult to see because the striatum lacks the lamination characteristic of other regions in

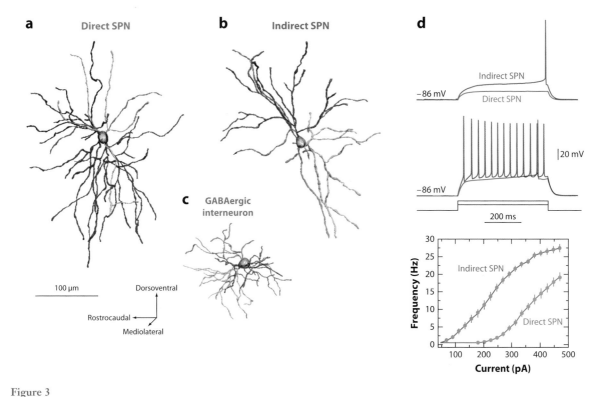

a Direct SPN

b Indirect SPN

c GABAergic interneuron

100 μm

Dorsoventral

Rostrocaudal

Mediolateral

d

Indirect SPN

−86 mV

Direct SPN

20 mV

−86 mV

200 ms

Frequency (Hz)

Indirect SPN

Direct SPN

Current (pA)

Figure 3

Dichotomous anatomy and physiology of direct- and indirect-pathway SPNs. Reconstructions of biocytin-filled direct- (*a*) and indirect-pathway (*b*) SPNs from P35–P45 BAC transgenic mice; a GABAergic interneuron (*c*) is included for comparison. Intrasomatic current injection consistently revealed that indirect-pathway SPNs were more excitable (*d*). Summary of the responses of direct- and indirect-pathway SPNs to a range of intrasomatic current steps; indirect-pathway SPNs were more excitable over a broad range of injected currents. From Gertler et al. (2008).

which this organization is apparent (e.g., cerebral cortex).

New techniques are revealing the extent of differences between direct- and indirect-pathway SPNs (Heiman et al. 2008, Lobo et al. 2006, Meurers et al. 2009). The functional implications of this dichotomy are only beginning to be understood. One dichotomy that is readily interpretable is in the somatodendritic anatomy (Gertler et al. 2008, Fujiyama et al. 2011). In rodents, the total dendritic length of indirect-pathway SPNs is significantly less than that of direct-pathway SPNs (**Figure 3**). This difference is because indirect-pathway SPNs have, on average, one to two fewer primary dendrites than do direct-pathway SPNs. This difference in surface area, in the absence of any obvious difference in spine density, suggests

that direct-pathway SPNs receive more glutamatergic input than do indirect-pathway SPNs. This also makes indirect-pathway SPNs more excitable, as judged by somatic current injection (Kreitzer & Malenka 2007, Gertler et al. 2008, Lobo et al. 2010). Thus, although dendritic organization of SPNs has long been thought to be of a single morphologic type, there are differences between direct- and indirect-pathway SPNs.

The neuromodulatory effects of DA on direct- and indirect-pathway SPNs parallel this dichotomy in intrinsic excitability. Both SPNs have a similar core physiological phenotype. At rest, SPNs are dominated by inwardly rectifying, Kir2 K$^+$ channels that hold the membrane potential near the K$^+$ equilibrium potential ($\sim$−90 mV), far from spike threshold (Shen et al.

2007, Wilson 1993). This is the so-called down state. In response to excitatory glutamatergic synaptic input from the cerebral cortex, SPNs depolarize. If this input lacks spatial or temporal convergence, the constitutively open Kir2 K^+ channels shunt this synaptic current, minimizing the cellular response. However, if this input is highly convergent, the glutamatergic synapses can overwhelm the Kir2 channels and promote their closure (Wilson & Kawaguchi 1996). Closure of dendritic Kir2 K^+ channels and inactivation of neighboring Kv4 (A-type) K^+ channels lead to a dramatic elevation of the input impedance of SPN dendrites and a reduction in their electrotonic length (Day et al. 2008, Wilson 1993). These physiological changes can bring the SPN somatic membrane potential to near spike threshold. This up state can last hundreds of milliseconds. It is during this up state that SPNs spike. These spikes are typically not correlated with the transition to the up state, which suggests that they are driven by an independent synaptic input (Stern et al. 1998).

DA modulates the glutamatergic synapses responsible for the transition to the up state and the ion channels controlling spiking. The qualitative features of the modulation depend on which DA receptor is being stimulated. As originally hypothesized on the basis of changes in gene expression induced by DA depletion (Gerfen et al. 1990), D_2 receptor signaling impedes the up-state transition and diminishes up-state spiking in indirect-pathway SPNs, whereas D_1 receptor signaling does precisely the opposite in direct-pathway SPNs (Surmeier et al. 2007). How this happens is still being unraveled, but some elements of the formula have been defined. In the somatic region where spikes are generated, activation of D_2 receptor signaling reduces inward, depolarizing currents through Cav1 (L-type) Ca^{2+} channels and Nav1 Na^+ channels, while increasing outward, hyperpolarizing K^+ channel currents (Greif et al. 1995, Hernandez-Lopez et al. 2000, Kitai & Surmeier 1993, Olson et al. 2005, Schiffmann et al. 1998, Surmeier et al. 1992). D_2 receptor stimulation also decreases dendritic Ca^{2+} entry through voltage-dependent channels (Day et al. 2008, Higley & Sabatini 2010). Complementing these alterations in the gating of voltage-dependent ion channels is a reduction in glutamatergic signaling. Although it is not clear whether they are located pre- or postsynaptically (Yin & Lovinger 2006), D_2 receptor stimulation diminishes presynaptic release of glutamate (Bamford et al. 2004). Activation of D_2 receptors also decreases α-amino-3-hydroxyl-5-methyl-4-isoxazole-propionate (AMPA) receptor currents of SPNs (Cepeda et al. 1993, Hernandez-Echeagaray et al. 2004). This reduction could be accomplished by D_2 receptor–triggered dephosphorylation of GluR1 subunits, which should promote trafficking of AMPA receptors out of the synaptic membrane (Hakansson et al. 2006). However, this modulation may be specific to a particular type of synapse: Recent work using glutamate uncaging on proximal dendritic spines failed to find any acute effect of D_2 receptor stimulation on AMPA receptor currents (Higley & Sabatini 2010). They did find that D_2 receptor signaling decreases Ca^{2+} entry through N-methyl-D-aspartic acid (NMDA) receptors, keeping with the effect on voltage-dependent Ca^{2+} channels.

D_1 receptor signaling has almost diametrically opposing effects on direct-pathway SPNs. D_1 receptor stimulation and PKA activation increases Cav1 L-type Ca^{2+} channel currents and decreases somatic K^+ currents (Galarraga et al. 1997, Gao et al. 1997, Kitai & Surmeier 1993, Surmeier et al. 1995). In addition, D_1 receptor signaling decreases the opening of Cav2 Ca^{2+} channels that control activation of Ca^{2+}-dependent, small-conductance K^+ (SK) channels that slow repetitive spiking in SPNs (Surmeier et al. 1995, Vilchis et al. 2000). All three of these effects serve to increase spiking of direct-pathway SPNs with somatic depolarization. The only apparently incongruent modulation is the one that was described first. Confirming inferences drawn from earlier work in tissue slices (Calabresi et al. 1987), voltage clamp work has shown that D_1 receptor signaling reduces Na^+ channel availability without altering the voltage dependence of fast activation or inactivation (Surmeier et al. 1992). Subsequent work has shown that PKA phosphorylation of the pore-forming subunit of the Na^+ channel promotes activity-dependent entry into a nonconducting, slow inactivated state that can be reversed only by membrane hyperpolarization (Carr et al. 2003, Chen et al. 2006). The D_1 receptor modulation is likely mediated by phosphorylation of somatic Nav1.1 channels because axon initial segment Nav1.6 channels are not efficiently phosphorylated by PKA (Scheuer & Catterall 2006), and Na^+ channels do not extend any significant distance into the dendrites of SPNs (Day et al. 2008). Given this restricted site of action and its time and voltage dependence, it is reasonable to hypothesize that this effect simply acts as a brake on the spike-promoting effects of D_1 receptors on Ca^{2+} and K^+ channels (as well as on glutamatergic signaling, described below).

A number of studies suggest that D_1 receptor, in contrast to D_2 receptor, signaling has positive effects on AMPA and NMDA receptor function and trafficking. For example, D_1 receptor activation of PKA enhances surface expression of both AMPA and NMDA receptors (Hallett et al. 2006, Snyder et al. 2000, Sun et al. 2005). The precise mechanisms underlying the trafficking are still being pursued, but the tyrosine kinase Fyn and the protein phosphatase STEP (striatal-enriched-phosphatase) appear to be important regulators of glutamate receptor surface expression (Braithwaite et al. 2006). Trafficking and localization may also be affected by a direct interaction between D_1 and NMDA receptors (Dunah et al. 2000, Lee et al. 2002, Scott et al. 2006). Rapid D_1 receptor effects on glutamate receptor gating have been more difficult to see. Although PKA phosphorylation of the NR1 subunit is capable of enhancing NMDA receptor currents (Blank et al. 1997), the presence of this modulation in SPNs is controversial. In neurons where the engagement of dendritic voltage-dependent ion channels has been minimized by dialyzing the cytoplasm with cesium ions, D_1 receptor agonists have little or no discernible effect on AMPA or NMDA receptor–mediated currents (Nicola

& Malenka 1998). However, in SPNs where this has not been done, D_1 receptor stimulation rapidly enhances currents evoked by NMDA receptor stimulation (Cepeda et al. 1993). The difference between these results suggests that the effect of D_1 receptors on NMDA receptor currents is largely indirect and mediated by voltage-dependent dendritic conductances that are negated by blocking K^+ channels and clamping dendritic voltage. Indeed, blocking L-type Ca^{2+} channels, which open in the same voltage range as do NMDA receptors (Mg^{2+} unblock), attenuates the D_1 receptor–mediated enhancement of NMDA receptor currents (Liu et al. 2004).

Taken together, this body of work paints a relatively simple picture of the divergent modulation of direct- and indirect-pathway SPN excitability by DA. However, there are two important caveats to the simplicity of this model. DA receptor–expressing striatal interneurons are important regulators of activity in the striatal circuit. In most of the studies of SPNs, their actions have been blocked to reduce the complexity of the preparation. The other caveat is that virtually nothing is known about the integrative mechanisms of SPN dendrites, where most D_1 and D_2 receptors are located.

Although DA has an important role in modulating the moment-to-moment activity of the corticostriatal network, it also has an important part to play in regulating long-term changes in synaptic strength. Several recent reviews cover this ground (Surmeier et al. 2007, Kreitzer & Malenka 2008, Surmeier et al. 2009, Wickens 2009), so only the major points bearing on the dichotomy between direct and indirect pathways are discussed here. At the outset, we must acknowledge that nearly all of what has been described as corticostriatal synaptic plasticity is truly glutamatergic synaptic plasticity because corticostriatal and thalamostriatal glutamatergic synapses have not been reliably distinguished. With that caveat in mind, we follow the convention and assume that what has been described applies to the corticostriatal synapse formed on a dendritic spine.

The best characterized form of SPN synaptic plasticity is long-term depression (LTD). Unlike the situation at many other synapses, striatal LTD induction requires pairing of postsynaptic depolarization with moderate- to high-frequency afferent stimulation at physiological temperatures (Kreitzer & Malenka 2005, Lovinger et al. 1993). For the induction to be successful, postsynaptic L-type calcium channels and mGluR5 receptors typically need to be coactivated. Both L-type calcium channels and mGluR5 receptors are found near glutamatergic synapses on SPN spines, enabling them to respond to local synaptic events (Carter et al. 2007, Carter & Sabatini 2004, Day et al. 2006, Testa et al. 1994). The induction of LTD requires the postsynaptic generation of endocannabinoids (ECs) (Gerdeman et al. 2002). ECs diffuse retrogradely to activate presynaptic CB1 receptors and decrease glutamate release probability. Ongoing work suggests that both of the abundant striatal ECs, anandamide and 2-arachidonoylglycerol (2-AG), are involved in SPN signaling (Gao et al. 2010, Giuffrida et al. 1999, Lerner et al. 2010, Tanimura et al. 2010). A key question about the induction of striatal LTD is whether activation of D_2 receptors is necessary. Activation of D_2 receptors is a potent stimulus for anandamide production (Giuffrida et al. 1999). However, recent work showing the sufficiency of L-type channel opening in EC-dependent LTD (Adermark & Lovinger 2007) makes it clear that D_2 receptors play a modulatory, not obligatory, role. The real issue is the role of D_2 receptors in LTD induction following synaptic stimulation. Studies have consistently found that in indirect-pathway SPNs, D_2 receptor activation is necessary (Kreitzer & Malenka 2007, Shen et al. 2008, Wang et al. 2006). This could be due to the need to suppress A2a adenosine receptor signaling, which could impede efficient EC synthesis and LTD induction (Fuxe et al. 2007, Shen et al. 2008). Indeed, Lerner et al. (2010) demonstrate quite convincingly that antagonism of A2a receptors promotes EC-dependent LTD induction in indirect-pathway SPNs.

The question, then, is can EC-dependent LTD be induced in direct-pathway SPNs that do not express D_2 receptors? When a minimal local stimulation paradigm is used, LTD does not appear to be induced in these SPNs (Kreitzer & Malenka 2007, Shen et al. 2008). However, using macroelectrode stimulation, EC-dependent LTD is readily inducible in identified direct-pathway SPNs (Wang et al. 2006), consistent with the high probability of SPN induction seen in previous work (Calabresi et al. 2007). The mechanisms controlling LTD induction in the ventral striatum appear to be different (Grueter et al. 2010).

In the dorsal striatum, how could induction of LTD in direct-pathway SPNs be dependent on D_2 receptors? There are a couple of possibilities. One is that D_2 receptor stimulation reduces DA release through a presynaptic mechanism, preferentially reducing stimulation of D_1 receptors that oppose the induction of LTD in direct-pathway SPNs (Shen et al. 2008). The other possibility is that for LTD to be induced in direct-pathway SPNs, acetylcholine release and postsynaptic M1 muscarinic receptor signaling must fall (Calabresi et al. 2007, Wang et al. 2006). This possibility has received additional support recently (Tozzi et al. 2011), emphasizing the importance of cholinergic interneurons in regulation of LTD in both direct- and indirect-pathway SPNs.

Long-term potentiation (LTP) at glutamatergic synapses is less well characterized because it is more difficult to induce in the in vitro preparations typically used to study plasticity. Most of the work describing LTP at glutamatergic synapses has been done with sharp electrodes (either in vivo or in vitro), not with patch clamp electrodes in brain slices, which affords greater experimental control and definition of the cellular and molecular determinants of induction. Previous studies have argued that LTP induced in SPNs by pairing high frequency stimulation (HFS) of glutamatergic inputs and postsynaptic depolarization depends on coactivation of D_1, NMDA, and TrkB receptors (Calabresi et al. 2007, Jia et al.

2010, Kerr & Wickens 2001). The involvement of NMDA receptors in LTP induction is clear. The involvement of TrkB receptors and its ligand, brain-derived neurotrophic factor (BDNF), is less well characterized but plausible given the expression of TrkB receptors in both classes of SPN (Lobo et al. 2010). However, the necessity of D_1 receptors is another matter. If this were the case, there would be no LTP in indirect-pathway SPNs—an unlikely situation. Again, the advent of BAC transgenic mice has provided an invaluable tool for unraveling this puzzle. In direct-pathway SPNs, the induction of LTP at glutamatergic synapses is dependent on D_1 DA receptors (Pawlak & Kerr 2008, Shen et al. 2008). However, this is not the case in indirect-pathway SPNs, where LTP requires activation of A2a receptors (Flajolet et al. 2008, Shen et al. 2008, Schiffmann et al. 2003). These receptors are robustly expressed in the indirect pathway and have an intracellular signaling linkage very similar to that of D_1 receptors; that is, they positively couple to adenylyl cyclase and PKA. Acting through PKA, D_1, and A2a receptor activation leads to the phosphorylation of DARPP-32 and a variety of other signaling molecules, including mitogen-activated protein kinases (MAPKs), linked to synaptic plasticity (Svenningsson et al. 2004, Sweatt 2004).

Although most of the induction protocols for synaptic plasticity that have been used to study striatal plasticity are decidedly unphysiological, involving sustained, strong depolarization and/or high-frequency synaptic stimulation that induces dendritic depolarization, they do make clear the necessity of postsynaptic depolarization. In a physiological setting, which types of depolarization are likely to gate induction? One possibility is that spikes generated in the axon initial segment (AIS) propagate into dendritic regions where synapses are formed. Recent work has shown that spike timing–dependent plasticity (STDP) is present in SPNs (Fino et al. 2005, Pawlak & Kerr 2008, Shen et al. 2008). But there are reasons to believe that this type of plasticity is relevant for only a subset of the synapses formed on SPNs. SPN dendrites are several hundred microns

long, thin, and modestly branched. Their initial 20–30 microns are largely devoid of spines and glutamatergic synapses. Glutamatergic synapse and spine density peak near 50 microns from the soma and then modestly decline with distance (Wilson 2004). Because of their geometry and ion channel expression, AIS-generated spikes rapidly decline in amplitude as they invade SPN dendrites (as judged by their ability to open voltage-dependent calcium channels), producing only a modest depolarization 80–100 microns from the soma (Day et al. 2008). This is less than half the distance to the dendritic tips, arguing that a large portion of the synaptic surface area is not normally accessible to somatic feedback about the outcome of aggregate synaptic activity. High-frequency, repetitive somatic spiking improves dendritic invasion, but distal (>120 microns) synapses remain relatively inaccessible, particularly in direct-pathway SPNs (Day et al. 2008).

In the more distal dendritic regions, what controls plasticity? The situation in SPNs may be very similar to that found in deep-layer pyramidal neurons, in which somatically generated bAPs do not invade the apical dendritic tuft (Golding et al. 2002). In this region, convergent synaptic stimulation is capable of producing a local calcium spike or plateau potential that produces a depolarization strong enough to open NMDA receptors and promote plasticity. Calcium imaging using two photon laser scanning microscopy (2PLSM) has shown that there is robust expression of both low threshold T-(Cav3) and L-type (Cav1) calcium channels—in addition to NMDA receptors—in SPN dendrites (Carter et al. 2007, Carter & Sabatini 2004, Day et al. 2006), a result that has been confirmed using cell-type specific gene profiling (Day et al. 2006). If distal dendrites are capable of regenerative activity, they could also be important to the induction of plasticity in SPNs. SPN dendrites do appear to be active in some circumstances (Carter & Sabatini 2004; Flores-Barrera et al. 2010; Kerr & Plenz 2002, 2004; Vergara et al. 2003), and this activity does not appear to arise from the somatic and proximal dendrites of adult SPNs (Wilson &

Kawaguchi 1996). One possible scenario is that distal SPN dendrites are capable of regenerative activity when NMDA receptors open during synaptic stimulation. If this is the case, spatial convergence of glutamatergic inputs onto a distal dendrite could induce a local plateau potential capable of pulling the rest of the cell into a depolarized state, like an up state. Doing so would collapse the electrotonic structure of SPNs (Wilson 1992), enhancing the impact of excitatory input anywhere on the dendrite. The lack of temporal correlation between up-state transitions and excitatory postsynaptic potential (EPSP)-driven spike generation is consistent with a scenario such as this one (Stern et al. 1998). If this were how SPNs operate, it would significantly increase their pattern-recognition capacity (e.g., Poirazi & Mel 2001) and create a new means of inducing plasticity. Dendritic D_1 and D_2 receptors should play complementary roles in modulating these dendritic events and the induction of synaptic plasticity there.

In vivo studies of striatal synaptic plasticity have provided an important counterpoint to the perspectives based on reduced in vitro preparations. The pioneering work of Charpier and Deniau (Charpier et al. 1999, Charpier & Deniau 1997) demonstrated that with more intact input, LTP was readily inducible in SPNs. More recent studies have shown that the sign of synaptic plasticity in SPNs is influenced by anesthetic and presumably the degree of cortical synchronization in corticostriatal projections (Stoetzner et al. 2010). In particular, in barbiturate-anesthetized rats, 5 Hz stimulation of the motor cortex evokes LTP in the striatum, but in awake animals, the same stimulation induced LTD. A challenge facing the field is how to bridge these observations. Because glutamatergic connections are sparse, it is virtually impossible to stimulate reliably a collection of synapses onto a particular SPN dendrite with an electrode in a brain slice. Optogenetic techniques may provide a feasible alternative strategy. Another strategy would be to employ two-photon laser uncaging (2PLU) of glutamate at visualized synaptic sites (Carter & Sabatini 2004, Higley & Sabatini

2010). These tools are becoming more widely available and should allow the properties of SPN dendrites to be characterized soon.

The Feedforward Corticostriatal Circuit

Fast-spiking (FS), PV GABAergic interneurons receive a prominent glutamatergic input from cortical pyramidal neurons and, in turn, convey this activity through perisomatic synapses to both direct- and indirect-pathway SPNs (Bennett & Bolam 1994, Gittis et al. 2010, Kita 1993, Koos & Tepper 1999, Planert et al. 2010). This feedforward inhibition is thought to contribute to action selection by suppressing SPN activity in circuits associated with unwanted actions (Gage et al. 2010, Kita et al. 1990, Parthasarathy & Graybiel 1997). Although both types of SPNs are targeted in this circuit, paired recordings in BAC mice have found some preferential connectivity of FS interneurons with direct-pathway SPNs (Gittis et al. 2010). More importantly, the dichotomy between direct- and indirect-pathway SPNs contributes to the regulation of this network. One of the major projections to FS interneurons originates from GPe neurons that are preferentially controlled by indirect-pathway SPNs (Bevan et al. 1998, Rajakumar et al. 1994, Spooren et al. 1996). This feedback loop complements the one formed by collateral projections of SPNs. D_2 receptor agonists depress the GABAergic inputs to FS interneurons (Bracci et al. 2002, Centonze et al. 2003, Gage et al. 2010, Sciamanna et al. 2009), which are presumably derived in large measure from GPe neurons [which express D_2 receptors (Hoover & Marshall 2004)]. This dual control of the feedback by D_2 receptors in indirect-pathway and GPe neurons suggests that its function is attenuated by basal DA, but it is capable of rapid facilitation if DA levels fall. This suppression of the GABAergic feedback circuit complements the elevation in FS interneuron excitability mediated by postsynaptic D_5 receptors.

SOM/NPY GABAergic interneurons also form another, less well-studied, part of the feedforward corticostriatal circuit (Tepper et al. 2010). If these interneurons are like the SOM-expressing, Martinotti interneurons of cortex (Wang et al. 2004), their innervation of distal dendrites could make it difficult to judge their importance accurately (Gittis et al. 2010), as with SPN recurrent collaterals. Whether this component of feedforward circuit differentially controls direct- and indirect-pathway SPNs remains to be determined, but their expression of D_5 receptors certainly creates a situation in which D_1 class agonists may influence indirect-pathway SPNs (Centonze et al. 2002).

The Feedback Striatal Circuit

SPNs have a richly branching recurrent axon collateral that arborizes in the neighborhood of its parent cell body (Kawaguchi et al. 1989). This feedback could provide the substrate for lateral inhibition (Groves 1983) and has figured prominently in several models of striatal processing (Beiser et al. 1997). However, the functional significance of this feedback circuit has been controversial, in large measure, because the synapses formed by recurrent collaterals are predominantly onto distal dendrites of other SPNs (Bolam et al. 1983, Wilson & Groves 1980), making their physiological effects difficult to see with a somatic electrode (Jaeger et al. 1994). Using paired patch clamp recordings from neighboring SPNs, investigators have been able to see more reliably the effects of collateral activation (Czubayko & Plenz 2002, Koos et al. 2004, Taverna et al. 2008, Tunstall et al. 2002), but the percentage of synaptically connected neighbors has been small ($\sim$10–15%) in randomly selected SPNs in brain slices. Using D_1 and D_2 BAC transgenic mice to direct sampling, Taverna et al. (2008) found that although indirect-pathway SPNs project to both themselves and direct-pathway SPNs, direct-pathway SPNs connect essentially only with other direct-pathway SPNs. The percentage of SPNs showing demonstrable connectivity doubled when sampling was not random. More recent work using optogenetic approaches to activate SPNs has inferred

an even higher degree of connectivity (Chuhma et al. 2011). Whether these approaches yield a pattern of connectivity consistent with that inferred from paired recordings remains to be determined.

As at their extrastriatal terminals, the release of GABA at recurrent collateral synapses is regulated by DA (Guzman et al. 2003). D_2 receptor stimulation decreases GABA release, whereas D_1 receptor stimulation increases release. Given the differences in the DA affinity in the native membrane (see above), at normal extrastriatal DA levels, the feedback circuit of the indirect pathway should be depressed, disinhibiting dendritic integration. A transient elevation in striatal DA should preferentially activate D_1 receptors, leading to enhanced GABA release at collaterals formed between direct pathway SPNs, enhancing intrapathway lateral inhibition of dendritic integration. Because D_1 receptor stimulation enhances the somatodendritic excitability of direct-pathway SPNs, the collateral modulation should in principle promote the type of population sculpting typically envisioned for lateral inhibition (Beiser et al. 1997). Conversely, a transient drop in striatal DA should enhance GABA release at collaterals formed by the indirect-pathway SPNs. Because this drop elevates the somatodendritic excitability of indirect-pathway SPNs, the collateral modulation should promote a similar type of population sculpting, but one favoring a subset of strongly activated indirect-pathway SPNs. The critical gap in this scenario is the lack of compelling data about how dendritic GABAergic synapses shape synaptic integration in SPNs. Recent work suggests that these processes differ in direct- and indirect-pathway SPNs (Flores-Barrera et al. 2010).

The Feedforward Thalamostriatal Circuit

The other major glutamatergic projection to the striatum originates in the thalamus (Smith et al. 2004). This input targets both direct- and indirect-pathway SPNs (Ding et al. 2008, Doig et al. 2010). The synapses formed by this projection are found on both dendritic shafts and spine heads, in the same regions as those formed by the corticostriatal projection. In contrast to the corticostriatal synapses, those formed by thalamic axons have a high release probability, making them well suited to signaling transient events (Ding et al. 2008). Another major target of this projection is the cholinergic interneuron. Like the corticostriatal feedforward circuit involving FS interneurons, the thalamostriatal projection makes a feedforward connection to SPNs through cholinergic interneurons (Ding et al. 2010) (**Figure 4**). There appear to be two phases to this feedforward system. The first phase is a rapid and transient inhibition of cortically driven activity in SPNs, which is mediated by a presynaptic, M2/M4 receptor-dependent inhibition of glutamate release (Ding et al. 2010) and a postsynaptic, GABAergic inhibition (Witten et al. 2010). Whether this GABAergic inhibition relies on nicotinic receptor activation of PV GABAergic interneurons remains to be determined (Koos & Tepper 2002, Sullivan et al. 2008). The second phase is mediated by postsynaptic M4 and M1 receptors that enhance the somatic excitability of both SPNs (Perez-Rosello et al. 2005, Pisani et al. 2007) but preferentially enhance the dendritic excitability of indirect-pathway SPNs by decreasing the Kir2 K^+ channel opening (Shen et al. 2007). With a burst of thalamic activity like that seen after presentation of a salient stimulus, cholinergic interneurons exhibit a burst-pause pattern of activity that engages both phases of the response, but because the inhibitory effects are fast (milliseconds in duration) and the postsynaptic effects are slow (hundreds of milliseconds), the two modulations do not conflict and lead to a patterned change in SPN activity that could underlie the alerting response.

The role of DA in regulating the thalamostriatal projection to SPNs has not been systematically explored. However, DA potently modulates the activity of cholinergic interneurons through D_2 and D_5 receptors. D_2 receptors suppress the ongoing pacemaking activity of interneurons (Deng et al. 2007, Maurice

et al. 2004) and diminished acetylcholine release (DeBoer et al. 1996, Ding et al. 2006). D_5 receptors complement this modulation by enhancing GABAergic responsiveness (Deng et al. 2007, Yan & Surmeier 1997), possibly to FS interneuron inputs (Sullivan et al. 2008). Thus, by modulating cholinergic interneurons, transient changes in striatal DA release differentially modulate direct- and indirect-pathway SPNs.

DOES THE BAC TRANSGENE ALTER NEURONAL PHENOTYPE?

Concerns have recently been raised about the BAC transgenic mice. Specifically, homozygous D_2-eGFP BAC mice on a Swiss-Webster (SW) background overexpressed D_2 receptors, leading to alterations in physiology and psychomotor stimulant sensitization (Kramer et al. 2011). In contrast, homozygous D_1-eGFP BAC mice appeared to be indistinguishable from wild-type SW mice. These observations certainly warrant additional study, given the widespread use of these mice and other BAC transgenics. However, there are reasons to believe that hemizygous mice on other backgrounds have less significant alterations in gene expression and faithfully mimic wild-type mice. For example, in hemizygous D_2-eGFP BAC mice, D_2 SPNs had intrinsic properties that were indistinguishable from eGFP-negative SPNs in hemizygous D_1-eGFP BAC mice (Gertler et al. 2008). Moreover, psychomotor stimulant sensitization appears to be normal in hemizygous D_2 BAC mice on the C57Bl/6 background (Grueter et al. 2010, Lobo et al. 2010).

NORMAL FUNCTION OF DIRECT AND INDIRECT PATHWAYS

According to the classic model of basal ganglia function, movements occur during pauses in the tonic inhibitory activity of the basal ganglia output interface, generated by activity in the direct pathway. This model had its origins in neurophysiologic studies showing that corticostriatal activation of the direct

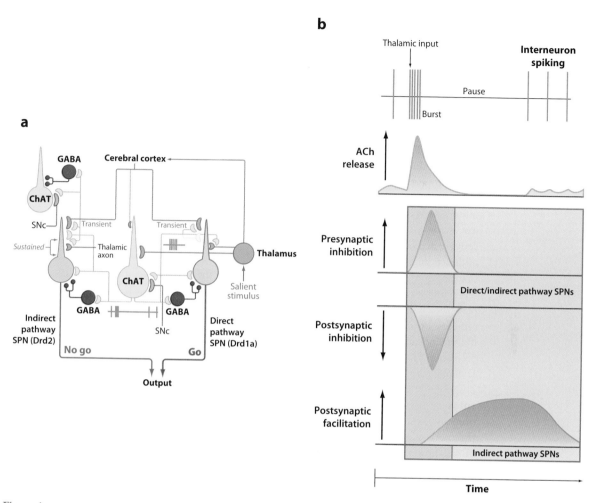

Figure 4

Thalamic gating of glutamatergic signaling in the striatum. (*a*) Schematic illustration of cortico- and thalamostriatal glutamatergic projections. Both direct- and indirect-pathway SPNs receive glutamatergic afferents from the cortex and the thalamus. However, the cholinergic interneuron receives glutamatergic inputs primarily from the thalamus. (*b*) Thalamic inputs efficiently drive cholinergic interneurons and generate a burst-pause firing pattern. By acting at presynaptic M2-class receptors, acetylcholine release transiently suppresses release probability at corticostriatal synapses formed on both direct- and indirect-pathway SPNs. In addition, activation of nicotinic receptors on GABAergic interneurons leads to a transient postsynaptic inhibition of both types of SPN. By acting at postsynaptic M1 receptors, acetylcholine release primarily enhances the responsiveness of D_2 SPNs to corticostriatal input for about a second. The pause in cholinergic interneuron activity ensures that there is not a concomitant presynaptic suppression in this window. Thalamic stimulation should activate neighboring cholinergic interneurons, as well. The pause is generated in part by recurrent collateral or neighboring interneuron activation of nicotinic receptors on dopaminergic terminals. In this way, the burst of thalamic spikes engages cholinergic interneurons to transiently suppress cortical drive of striatal circuits and then create a second long period in which the striatal network is strongly biased toward cortical activation of D_2 SPNs. From Ding et al. (2010).

striatonigral pathway resulted in pauses of the tonic activity of GABAergic neurons in the substantia nigra pars reticulata (Chevalier et al. 1985, Deniau & Chevalier 1985). The demonstration that saccadic eye movements occur during these pauses clearly implicated the direct pathway in movement control (Hikosaka & Wurtz 1983). The opposing role of the indirect pathway in suppressing movements was originally proposed on the basis of studies in animal models of PD (see below). Recent work using optogenetic or genetic approaches supports the general tenets of the model (Hikida et al. 2010, Kravitz et al. 2010). Although this model has proven to be of considerable clinical value, it fails to account for the great diversity and complexity of the decision-making process in action selection (Cisek & Kalaska 2010).

Mink (2003) elaborated the classic model, proposing that the activity of direct and indirect pathways is coordinated to select particular motor programs and to inhibit competing motor programs. This model predicts that during ongoing behavior, there will be increased activity in neuronal ensembles that are part of direct- and indirect-pathway circuits, rather than one or the other. The execution of a movement sequence would then generate a complex pattern of activity in specific neuronal ensembles. Physiologic studies of limb-reaching movements in primates have confirmed this prediction (Turner & Anderson 1997, 2005).

Consistent with such data, current models propose that the striatum performs a computation on sensorimotor, cognitive, and emotional/motivational information provided by the cerebral cortex to facilitate the selection of an appropriate action out of a collection of possibilities (Balleine et al. 2007, Cisek & Kalaska 2010, Nambu 2008). Distinct cortico-basal ganglia loops are thought to perform different aspects of this computation. Distinct cortico-basal ganglia loops may also play different roles in the acquisition and stabilization of context-dependent action selection. For example, work by Costa's group suggests that the dorsomedial striatum-associative cortex loop plays an important role in the early phases of skill acquisition, whereas the dorsolateral striatum–sensorimotor cortex loop is more involved once the skill has been established and the action program becomes more automated and inflexible.

More germane to our topic is the role that DA modulation of direct- and indirect-pathway SPNs may serve in this process. Certainly, DA is thought to play a crucial role in linking the outcome of a particular action choice to the probability that it will be chosen in the future. Given the robust DA innervation of the striatum, it is natural to think that modulation of the striatal circuitry is critical to decision making. How might this happen? As noted above, SNc DA neurons innervating the striatum are autonomous pacemakers, providing a tonic release of DA in the striatum. Rewarding events transiently increase the activity of SNc DA neurons, whereas aversive events transiently decrease it, providing bidirectional signaling to the striatum (Brown et al. 2009; Hikosaka et al. 2008; Schultz 1998, 2007). Although there is debate about precisely what SNc DA neuron activity codes, this is not critical to our discussion. What is important from our standpoint is that the dichotomous modulation of direct and indirect pathways by DA provides a mechanism by which these outcome-driven changes in SNc activity can be translated into changes in corticostriatal networks facilitating action selection. As outlined above, the transient elevation in striatal DA release associated with rewarding events promotes the induction of LTP at corticostriatal synapses formed on direct-pathway SPNs and increases their excitability. In contrast, this same change promotes LTD induction at synapses formed on indirect-pathway SPNs and decreases their excitability. A reasonable conjecture from this data is that reward promotes the ability of a particular cortical ensemble to turn on direct-pathway SPNs linked to action initiation while decreasing the ability of that cortical ensemble to activate indirect-pathway SPNs linked to action suppression (Cohen & Frank 2009). Transient depression of DA release associated with aversive events should have exactly the opposite effect on the corticostriatal circuitry, promoting the ability of a cortical ensemble to turn on indirect-pathway SPNs and reducing its ability to turn on direct-pathway SPNs. Thus, the differential expression of DA receptors by striatal efferent pathways appears to be critical to the striatum's ability to participate in action selection.

Although this model is consistent with most of what we know, we do not really know that much, and a great many questions need to be answered. For example, the role of the striatal microcircuits outlined above is almost completely undefined in this model. Also, it is commonly assumed that direct- and indirect-pathway SPNs receive the same cortical information. This assumption is being debated in the literature (Ballion et al. 2008, Lei et al. 2004, Parthasarathy & Graybiel 1997) but will not be easy to resolve, given the complex architecture of the striatum. If this is not true, it would fundamentally change our ideas about what each pathway was doing to control behavior. It is also far from clear how the timing of activity in the direct and indirect pathways shapes the activity in the circuitry controlling actual movement (Nambu 2008). In diseases of the basal ganglia, such as PD, the timing of activity, rather than the overall level of activity, appears to be the most important determinant of movement choice and initiation (Bevan et al. 2002).

PARKINSON'S DISEASE

The striatum and the basal ganglia have been implicated in a wide variety of psychomotor disorders, ranging from PD to schizophrenia and drug abuse. Altered DA regulation of direct- and indirect-pathway SPNs appears to be a pivotal aspect of most, if not all, of these diseases. The best characterized example of this is in PD, during which DA neurons innervating the basal ganglia degenerate (Hornykiewicz 1966). Studies of PD models provided the first clear indication of the dichotomy in the regulation of the direct and indirect pathways, suggesting that their excitability shifts in opposite directions following the loss of DA, creating an imbalance in the regulation of the motor thalamus favoring suppression of movement (Albin et al. 1989). Specifically, direct-pathway SPNs were posited to spike less in the PD state, whereas indirect-pathway SPNs were thought to spike more. This conjecture was based on changes in gene expression following DA-depleting lesions (Gerfen et al. 1990). The cellular mechanisms underlying the postulated shift in excitability were not known at the time the model was formulated, but they have widely been assumed to reflect changes in intrinsic excitability. In the past two decades, electrophysiological studies have provided strong support for the proposition that following DA depletion, the excitability of direct- and indirect-pathway SPNs moves in opposite directions (Surmeier et al. 2007). Recent work has added a dimension to the classic model, showing that DA depletion alters the induction of long-term plasticity at glutamatergic synapses (Shen et al. 2008). In direct-pathway SPNs, the loss of D_1 receptor signaling biases glutamatergic synapses toward LTD. In contrast, in indirect-pathway SPNs, the loss of D_2 receptor signaling promotes the induction of LTP. Thus, activity-dependent changes in synaptic strength parallel those of intrinsic excitability following DA depletion. Work in vivo examining the responsiveness of antidromically identified SPNs to cortical stimulation following unilateral lesions of the striatal DA innervation is consistent with this broader model (Mallet et al. 2006). Additional support for this model has recently come from an elegant application of optogenetics to selectively activate direct- and indirect-pathway SPNs in mouse models of PD (Kravitz et al. 2010).

What this model does not account for is the propensity of neural circuits to adapt to sustained perturbations in activity, such as those induced by DA depletion. The fact that changes in striatal gene expression take weeks to stabilize after lesioning DA fibers suggests that many adaptations are taking place. Sustained perturbations in synaptic or intrinsic properties that make neurons spike more or less than their set point typically engage homeostatic mechanisms that attempt to bring activity back to the desired level (Marder & Goaillard 2006, Turrigiano 1999). One of the most common mechanisms of homeostatic plasticity is to alter synaptic strength or to scale synapses. In indirect-pathway SPNs, the elevation in activity following DA depletion triggers a dramatic downregulation of glutamatergic synapses formed on spines (Day et al. 2006, Deutch et al. 2007). Like scaling seen in other cell types, the synaptic modification depends on calcium entry through voltage-dependent L-type channels that activates the protein phosphatase calcineurin, which dephosphorylates myocyte enhancer factor 2 (MEF2) (Flavell et al. 2006), increasing its transcriptional activity. In indirect-pathway SPNs, MEF2 upregulation increased the expression of at least two genes linked to synaptic remodeling: Nur77 and Arc (Tian et al. 2010).

Although studies are in their very early stages, DA depletion clearly induces adaptations in the striatal microcircuits shaping direct- and indirect-pathway activity. In the feedforward pathway, the activity of SOM/NOS interneurons is elevated, whereas that of FS interneurons appears unchanged (Dehorter et al. 2009). The strength of the feedback, recurrent collateral system within the striatum appears to be dramatically downregulated (Taverna et al. 2008). These adaptations could be a major factor in the attenuation of differences between direct- and indirect-pathway SPNs to repetitive cortical stimulation in PD models (Flores-Barrera et al. 2010). Complementing these changes, cholinergic signaling is enhanced in the thalamostriatal feedforward circuit following DA depletion (DeBoer et al. 1996, Ding et al. 2006), which could differentially modulate SPNs.

Another example of an asymmetric adaptation in direct- and indirect-pathway SPNs in PD models comes from study of the effects of treatment with DA receptor agonists. In indirect-pathway SPNs, D_2 receptor agonists reverse the effects of DA depletion on gene expression (Gerfen et al. 1990). However, in direct-pathway SPNs, the responses to DA precursors or D_1 receptor agonists fundamentally change, resulting in a loss of both regional and compartmental regulation of gene expression (Gerfen et al. 2002). The signaling changes underlying this shift remain to be fully elucidated, but an important component appears to be the enhanced coupling of D_1 receptors to activation of two MAPKs, extracellular signal–regulated kinases 1/2 and mammalian target of rapamycin (mTOR) (Gerfen et al. 2002; Pavon et al. 2006; Santini et al. 2007, 2009; Westin et al. 2007). This enhancement and the aberrant activation of direct-pathway SPNs is widely thought to be a key step in the induction of dyskinesias following the repeated administration of the DA precursor, L-DOPA.

SUMMARY

Direct and indirect striatal efferent pathways form the anatomical and functional backbone of the basal ganglia. These pathways are formed by two types of SPN: one that projects directly to the basal ganglia interface (the internal segment of the globus pallidus and the substantia nigra pars reticulata) and one that indirectly connects with this interface. These GABAergic SPNs integrate information conveyed by the principal neurons of the cerebral cortex and thalamus to facilitate the selection of appropriate actions. DA released by SNc neurons, whose activity is regulated by the outcome of chosen actions, modulates this process. By virtue of the dichotomous expression of D_1 and D_2 receptors, DA differentially regulates the intrinsic excitability and synaptic connectivity of direct- and indirect-pathway SPNs. Activation of D_1 receptors in direct-pathway SPNs increases their excitability and promotes long-term potentiation of excitatory synapses. In contrast, activation of D_2 receptors in indirect-pathway SPNs decreases their excitability and promotes long-term depression of excitatory synapses. In addition, DA modulates striatal microcircuits formed by either SPN recurrent collaterals or GABAergic interneurons to modulate SPN activity. This dichotomous modulation of SPNs is consistent with their hypothesized roles in action selection. However, many questions remain unanswered. The development of genetic tools that allow the selective expression of transgenes in these two populations of SPNs promises to put many of these questions within our experimental reach in the near future.

DISCLOSURE STATEMENT

The authors are not aware of any affiliations, memberships, funding, or financial holdings that might be perceived as affecting the objectivity of this review.

LITERATURE CITED

Adermark L, Lovinger DM. 2007. Combined activation of L-type Ca^{2+} channels and synaptic transmission is sufficient to induce striatal long-term depression. *J. Neurosci.* 27:6781–87

Albin RL, Young AB, Penney JB. 1989. The functional anatomy of basal ganglia disorders. *Trends Neurosci.* 12:366–75

Balleine BW, Delgado MR, Hikosaka O. 2007. The role of the dorsal striatum in reward and decision-making. *J. Neurosci.* 27:8161–65

Ballion B, Mallet N, Bezard E, Lanciego JL, Gonon F. 2008. Intratelencephalic corticostriatal neurons equally excite striatonigral and striatopallidal neurons and their discharge activity is selectively reduced in experimental parkinsonism. *Eur. J. Neurosci.* 27:2313–21

Bamford NS, Zhang H, Schmitz Y, Wu NP, Cepeda C, et al. 2004. Heterosynaptic dopamine neurotransmission selects sets of corticostriatal terminals. *Neuron* 42:653–63

Beiser DG, Hua SE, Houk JC. 1997. Network models of the basal ganglia. *Curr. Opin. Neurobiol.* 7:185–90

Bennett BD, Bolam JP. 1994. Synaptic input and output of parvalbumin-immunoreactive neurons in the neostriatum of the rat. *Neuroscience* 62:707–19

Bergson C, Mrzljak L, Smiley JF, Pappy M, Levenson R, Goldman-Rakic PS. 1995. Regional, cellular, and subcellular variations in the distribution of D1 and D5 dopamine receptors in primate brain. *J. Neurosci.* 15:7821–36

Bernard V, Normand E, Bloch B. 1992. Phenotypical characterization of the rat striatal neurons expressing muscarinic receptor genes. *J. Neurosci.* 12:3591–600

Bevan MD, Booth PA, Eaton SA, Bolam JP. 1998. Selective innervation of neostriatal interneurons by a subclass of neuron in the globus pallidus of the rat. *J. Neurosci.* 18:9438–52

Bevan MD, Magill PJ, Terman D, Bolam JP, Wilson CJ. 2002. Move to the rhythm: oscillations in the subthalamic nucleus-external globus pallidus network. *Trends Neurosci.* 25:525–31

Blank T, Nijholt I, Teichert U, Kugler H, Behrsing H, et al. 1997. The phosphoprotein DARPP-32 mediates cAMP-dependent potentiation of striatal N-methyl-D-aspartate responses. *Proc. Natl. Acad. Sci. USA* 94:14859–64

Bolam JP, Hanley JJ, Booth PA, Bevan MD. 2000. Synaptic organisation of the basal ganglia. *J. Anat.* 196(Pt. 4):527–42

Bolam JP, Somogyi P, Takagi H, Fodor I, Smith AD. 1983. Localization of substance P-like immunoreactivity in neurons and nerve terminals in the neostriatum of the rat: a correlated light and electron microscopic study. *J. Neurocytol.* 12:325–44

Bracci E, Centonze D, Bernardi G, Calabresi P. 2002. Dopamine excites fast-spiking interneurons in the striatum. *J. Neurophysiol.* 87:2190–94

Braithwaite SP, Paul S, Nairn AC, Lombroso PJ. 2006. Synaptic plasticity: one STEP at a time. *Trends Neurosci.* 29:452–58

Brown MT, Henny P, Bolam JP, Magill PJ. 2009. Activity of neurochemically heterogeneous dopaminergic neurons in the substantia nigra during spontaneous and driven changes in brain state. *J. Neurosci.* 29:2915–25

Calabresi P, Mercuri N, Stanzione P, Stefani A, Bernardi G. 1987. Intracellular studies on the dopamine-induced firing inhibition of neostriatal neurons in vitro: evidence for D1 receptor involvement. *Neuroscience* 20:757–71

Calabresi P, Picconi B, Tozzi A, Di Filippo M. 2007. Dopamine-mediated regulation of corticostriatal synaptic plasticity. *Trends Neurosci.* 30:211–19

Carr DB, Day M, Cantrell AR, Held J, Scheuer T, et al. 2003. Transmitter modulation of slow, activity-dependent alterations in sodium channel availability endows neurons with a novel form of cellular plasticity. *Neuron* 39:793–806

Carter AG, Sabatini BL. 2004. State-dependent calcium signaling in dendritic spines of striatal medium spiny neurons. *Neuron* 44:483–93

Carter AG, Soler-Llavina GJ, Sabatini BL. 2007. Timing and location of synaptic inputs determine modes of subthreshold integration in striatal medium spiny neurons. *J. Neurosci.* 27:8967–77

Centonze D, Bracci E, Pisani A, Gubellini P, Bernardi G, Calabresi P. 2002. Activation of dopamine D1-like receptors excites LTS interneurons of the striatum. *Eur. J. Neurosci.* 15:2049–52

Centonze D, Grande C, Usiello A, Gubellini P, Erbs E, et al. 2003. Receptor subtypes involved in the presynaptic and postsynaptic actions of dopamine on striatal interneurons. *J. Neurosci.* 23:6245–54

Cepeda C, Buchwald NA, Levine MS. 1993. Neuromodulatory actions of dopamine in the neostriatum are dependent upon the excitatory amino acid receptor subtypes activated. *Proc. Natl. Acad. Sci. USA* 90:9576–80

Charpier S, Deniau JM. 1997. In vivo activity-dependent plasticity at cortico-striatal connections: evidence for physiological long-term potentiation. *Proc. Natl. Acad. Sci. USA* 94:7036–40

Charpier S, Mahon S, Deniau JM. 1999. In vivo induction of striatal long-term potentiation by low-frequency stimulation of the cerebral cortex. *Neuroscience* 91:1209–22

Chen Y, Yu FH, Surmeier DJ, Scheuer T, Catterall WA. 2006. Neuromodulation of Na$^+$ channel slow inactivation via cAMP-dependent protein kinase and protein kinase C. *Neuron* 49:409–20

Chevalier G, Vacher S, Deniau JM, Desban M. 1985. Disinhibition as a basic process in the expression of striatal functions. I. The striato-nigral influence on tecto-spinal/tecto-diencephalic neurons. *Brain Res.* 334:215–26

Chuhma N, Tanaka KF, Hen R, Rayport S. 2011. Functional connectome of the striatal medium spiny neuron. *J. Neurosci.* 31:1183–92

Cisek P, Kalaska JF. 2010. Neural mechanisms for interacting with a world full of action choices. *Annu. Rev. Neurosci.* 33:269–98

Cohen MX, Frank MJ. 2009. Neurocomputational models of basal ganglia function in learning, memory and choice. *Behav. Brain Res.* 199:141–56

Czubayko U, Plenz D. 2002. Fast synaptic transmission between striatal spiny projection neurons. *Proc. Natl. Acad. Sci. USA* 99:15764–69

Day M, Wang Z, Ding J, An X, Ingham CA, et al. 2006. Selective elimination of glutamatergic synapses on striatopallidal neurons in Parkinson disease models. *Nat. Neurosci.* 9:251–59

Day M, Wokosin D, Plotkin JL, Tian X, Surmeier DJ. 2008. Differential excitability and modulation of striatal medium spiny neuron dendrites. *J. Neurosci.* 28:11603–14

DeBoer P, Heeringa MJ, Abercrombie ED. 1996. Spontaneous release of acetylcholine in striatum is preferentially regulated by inhibitory dopamine D2 receptors. *Eur. J. Pharmacol.* 317:257–62

Dehorter N, Guigoni C, Lopez C, Hirsch J, Eusebio A, et al. 2009. Dopamine-deprived striatal GABAergic interneurons burst and generate repetitive gigantic IPSCs in medium spiny neurons. *J. Neurosci.* 29:7776–87

Deng P, Zhang Y, Xu ZC. 2007. Involvement of I(h) in dopamine modulation of tonic firing in striatal cholinergic interneurons. *J. Neurosci.* 27:3148–56

Deniau JM, Chevalier G. 1985. Disinhibition as a basic process in the expression of striatal functions. II. The striato-nigral influence on thalamocortical cells of the ventromedial thalamic nucleus. *Brain Res.* 334:227–33

Deutch AY, Colbran RJ, Winder DJ. 2007. Striatal plasticity and medium spiny neuron dendritic remodeling in parkinsonism. *Parkinsonism Relat. Disord.* 13:S251–58

Ding J, Guzman JN, Tkatch T, Chen S, Goldberg JA, et al. 2006. RGS4-dependent attenuation of M4 autoreceptor function in striatal cholinergic interneurons following dopamine depletion. *Nat. Neurosci.* 9:832–42

Ding J, Peterson JD, Surmeier DJ. 2008. Corticostriatal and thalamostriatal synapses have distinctive properties. *J. Neurosci.* 28:6483–92

Ding JB, Guzman JN, Peterson JD, Goldberg JA, Surmeier DJ. 2010. Thalamic gating of corticostriatal signaling by cholinergic interneurons. *Neuron* 67:294–307

Doig NM, Moss J, Bolam JP. 2010. Cortical and thalamic innervation of direct and indirect pathway medium-sized spiny neurons in mouse striatum. *J. Neurosci.* 30:14610–18

Dunah AW, Wang Y, Yasuda RP, Kameyama K, Huganir RL, et al. 2000. Alterations in subunit expression, composition, and phosphorylation of striatal N-methyl-D-aspartate glutamate receptors in a rat 6-hydroxydopamine model of Parkinson's disease. *Mol. Pharmacol.* 57:342–52

Fino E, Glowinski J, Venance L. 2005. Bidirectional activity-dependent plasticity at corticostriatal synapses. *J. Neurosci.* 25:11279–87

Flajolet M, Wang Z, Futter M, Shen W, Nuangchamnong N, et al. 2008. FGF acts as a co-transmitter through adenosine A(2A) receptor to regulate synaptic plasticity. *Nat. Neurosci.* 11:1402–9

Flavell SW, Cowan CW, Kim TK, Greer PL, Lin Y, et al. 2006. Activity-dependent regulation of MEF2 transcription factors suppresses excitatory synapse number. *Science* 311:1008–12

Flores-Barrera E, Vizcarra-Chacon BJ, Tapia D, Bargas J, Galarraga E. 2010. Different corticostriatal integration in spiny projection neurons from direct and indirect pathways. *Front. Syst. Neurosci.* 4:1–15

Fujiyama F, Sohn J, Nakano T, Furuta T, Nakamura KC, et al. 2011. Exclusive and common targets of neostriatofugal projections of rat striosome neurons: a single neuron-tracing study using a viral vector. *Eur. J. Neurosci.* 33:668–77

Fuxe K, Ferre S, Genedani S, Franco R, Agnati LF. 2007. Adenosine receptor-dopamine receptor interactions in the basal ganglia and their relevance for brain function. *Physiol. Behav.* 92:210–17

Gage GJ, Stoetzner CR, Wiltschko AB, Berke JD. 2010. Selective activation of striatal fast-spiking interneurons during choice execution. *Neuron* 67:466–79

Galarraga E, Hernandez-Lopez S, Reyes A, Barral J, Bargas J. 1997. Dopamine facilitates striatal EPSPs through an L-type Ca^{2+} conductance. *Neuroreport* 8:2183–86

Gao T, Yatani A, Dell'Acqua ML, Sako H, Green SA, et al. 1997. cAMP-dependent regulation of cardiac L-type Ca^{2+} channels requires membrane targeting of PKA and phosphorylation of channel subunits. *Neuron* 19:185–96

Gao Y, Vasilyev DV, Goncalves MB, Howell FV, Hobbs C, et al. 2010. Loss of retrograde endocannabinoid signaling and reduced adult neurogenesis in diacylglycerol lipase knock-out mice. *J. Neurosci.* 30:2017–24

Gerdeman GL, Ronesi J, Lovinger DM. 2002. Postsynaptic endocannabinoid release is critical to long-term depression in the striatum. *Nat. Neurosci.* 5:446–51

Gerfen CR, Engber TM, Mahan LC, Susel Z, Chase TN, et al. 1990. D1 and D2 dopamine receptor-regulated gene expression of striatonigral and striatopallidal neurons. *Science* 250:1429–32

Gerfen CR, Miyachi S, Paletzki R, Brown P. 2002. D1 dopamine receptor supersensitivity in the dopamine-depleted striatum results from a switch in the regulation of ERK1/2/MAP kinase. *J. Neurosci.* 22:5042–54

Gerfen CR, Wilson CJ. 1996. The basal ganglia. In *Handbook of Chemical Neuroanatomy*, ed. LW Swanson, A Bjorklund, T Hokfelt, pp. 365–62. Amsterdam: Elsevier

Gertler TS, Chan CS, Surmeier DJ. 2008. Dichotomous anatomical properties of adult striatal medium spiny neurons. *J. Neurosci.* 28:10814–24

Geurts M, Maloteaux JM, Hermans E. 2003. Altered expression of regulators of G-protein signaling (RGS) mRNAs in the striatum of rats undergoing dopamine depletion. *Biochem. Pharmacol.* 66:1163–70

Gittis AH, Nelson AB, Thwin MT, Palop JJ, Kreitzer AC. 2010. Distinct roles of GABAergic interneurons in the regulation of striatal output pathways. *J. Neurosci.* 30:2223–34

Giuffrida A, Parsons LH, Kerr TM, Rodriguez de Fonseca F, Navarro M, Piomelli D. 1999. Dopamine activation of endogenous cannabinoid signaling in dorsal striatum. *Nat. Neurosci.* 2:358–63

Golding NL, Staff NP, Spruston N. 2002. Dendritic spikes as a mechanism for cooperative long-term potentiation. *Nature* 418:326–31

Gong S, Zheng C, Doughty ML, Losos K, Didkovsky N, et al. 2003. A gene expression atlas of the central nervous system based on bacterial artificial chromosomes. *Nature* 425:917–25

Greif GJ, Lin YJ, Liu JC, Freedman JE. 1995. Dopamine-modulated potassium channels on rat striatal neurons: specific activation and cellular expression. *J. Neurosci.* 15:4533–44

Groves PM. 1983. A theory of the functional organization of the neostriatum and the neostriatal control of voluntary movement. *Brain Res.* 286:109–32

Grueter BA, Brasnjo G, Malenka RC. 2010. Postsynaptic TRPV1 triggers cell type-specific long-term depression in the nucleus accumbens. *Nat. Neurosci.* 13:1519–25

Guzman JN, Hernandez A, Galarraga E, Tapia D, Laville A, et al. 2003. Dopaminergic modulation of axon collaterals interconnecting spiny neurons of the rat striatum. *J. Neurosci.* 23:8931–40

Hakansson K, Galdi S, Hendrick J, Snyder G, Greengard P, Fisone G. 2006. Regulation of phosphorylation of the GluR1 AMPA receptor by dopamine D2 receptors. *J. Neurochem.* 96:482–88

Hallett PJ, Spoelgen R, Hyman BT, Standaert DG, Dunah AW. 2006. Dopamine D1 activation potentiates striatal NMDA receptors by tyrosine phosphorylation-dependent subunit trafficking. *J. Neurosci.* 26: 4690–700

Heiman M, Schaefer A, Gong S, Peterson JD, Day M, et al. 2008. A translational profiling approach for the molecular characterization of CNS cell types. *Cell* 135:738–48

Hernandez-Echeagaray E, Starling AJ, Cepeda C, Levine MS. 2004. Modulation of AMPA currents by D2 dopamine receptors in striatal medium-sized spiny neurons: Are dendrites necessary? *Eur. J. Neurosci.* 19:2455–63

Hernandez-Lopez S, Tkatch T, Perez-Garci E, Galarraga E, Bargas J, et al. 2000. D2 dopamine receptors in striatal medium spiny neurons reduce L-type Ca^{2+} currents and excitability via a novel PLC[beta]1-IP3-calcineurin-signaling cascade. *J. Neurosci.* 20:8987–95

Hersch SM, Ciliax BJ, Gutekunst CA, Rees HD, Heilman CJ, et al. 1995. Electron microscopic analysis of D1 and D2 dopamine receptor proteins in the dorsal striatum and their synaptic relationships with motor corticostriatal afferents. *J. Neurosci.* 15:5222–37

Higley MJ, Sabatini BL. 2010. Competitive regulation of synaptic Ca^{2+} influx by D2 dopamine and A2A adenosine receptors. *Nat. Neurosci.* 13:958–66

Hikida T, Kimura K, Wada N, Funabiki K, Nakanishi S. 2010. Distinct roles of synaptic transmission in direct and indirect striatal pathways to reward and aversive behavior. *Neuron* 66:896–907

Hikosaka O, Sesack SR, Lecourtier L, Shepard PD. 2008. Habenula: crossroad between the basal ganglia and the limbic system. *J. Neurosci.* 28:11825–29

Hikosaka O, Wurtz RH. 1983. Visual and oculomotor functions of monkey substantia nigra pars reticulata. III. Memory-contingent visual and saccade responses. *J. Neurophysiol.* 49:1268–84

Hoover BR, Marshall JF. 2004. Molecular, chemical, and anatomical characterization of globus pallidus dopamine D2 receptor mRNA-containing neurons. *Synapse* 52:100–13

Hornykiewicz O. 1966. Dopamine (3-hydroxytyramine) and brain function. *Pharmacol. Rev.* 18:925–64

Jaeger D, Kita H, Wilson CJ. 1994. Surround inhibition among projection neurons is weak or nonexistent in the rat neostriatum. *J. Neurophysiol.* 72:2555–58

Jia Y, Gall CM, Lynch G. 2010. Presynaptic BDNF promotes postsynaptic long-term potentiation in the dorsal striatum. *J. Neurosci.* 30:14440–45

Kawaguchi Y, Wilson CJ, Emson PC. 1989. Intracellular recording of identified neostriatal patch and matrix spiny cells in a slice preparation preserving cortical inputs. *J. Neurophysiol.* 62:1052–68

Kerr JN, Plenz D. 2002. Dendritic calcium encodes striatal neuron output during up-states. *J. Neurosci.* 22:1499–512

Kerr JN, Plenz D. 2004. Action potential timing determines dendritic calcium during striatal up-states. *J. Neurosci.* 24:877–85

Kerr JN, Wickens JR. 2001. Dopamine D-1/D-5 receptor activation is required for long-term potentiation in the rat neostriatum in vitro. *J. Neurophysiol.* 85:117–24

Kita H. 1993. GABAergic circuits of the striatum. *Prog. Brain Res.* 99:51–72

Kita H, Kosaka T, Heizmann CW. 1990. Parvalbumin-immunoreactive neurons in the rat neostriatum: a light and electron microscopic study. *Brain Res.* 536:1–15

Kitai ST, Surmeier DJ. 1993. Cholinergic and dopaminergic modulation of potassium conductances in neostriatal neurons. *Adv. Neurol.* 60:40–52

Koos T, Tepper JM. 1999. Inhibitory control of neostriatal projection neurons by GABAergic interneurons. *Nat. Neurosci.* 2:467–72

Koos T, Tepper JM. 2002. Dual cholinergic control of fast-spiking interneurons in the neostriatum. *J. Neurosci.* 22:529–35

Koos T, Tepper JM, Wilson CJ. 2004. Comparison of IPSCs evoked by spiny and fast-spiking neurons in the neostriatum. *J. Neurosci.* 24:7916–22

Kramer PF, Christensen CH, Hazelwood LA, Dobi A, Bock R, et al. 2011. Dopamine D2 receptor overexpression alters behavior and physiology in Drd2-EGFP mice. *J. Neurosci.* 31:126–32

Kravitz AV, Freeze BS, Parker PR, Kay K, Thwin MT, et al. 2010. Regulation of parkinsonian motor behaviours by optogenetic control of basal ganglia circuitry. *Nature* 466:622–26

Kreitzer AC, Malenka RC. 2005. Dopamine modulation of state-dependent endocannabinoid release and long-term depression in the striatum. *J. Neurosci.* 25:10537–45

Kreitzer AC, Malenka RC. 2007. Endocannabinoid-mediated rescue of striatal LTD and motor deficits in Parkinson's disease models. *Nature* 445:643–47

Kreitzer AC, Malenka RC. 2008. Striatal plasticity and basal ganglia circuit function. *Neuron* 60:543–54

Lee FJ, Xue S, Pei L, Vukusic B, Chery N, et al. 2002. Dual regulation of NMDA receptor functions by direct protein-protein interactions with the dopamine D1 receptor. *Cell* 111:219–30

Lei W, Jiao Y, Del Mar N, Reiner A. 2004. Evidence for differential cortical input to direct pathway versus indirect pathway striatal projection neurons in rats. *J. Neurosci.* 24:8289–99

Lerner TN, Horne EA, Stella N, Kreitzer AC. 2010. Endocannabinoid signaling mediates psychomotor activation by adenosine A2A antagonists. *J. Neurosci.* 30:2160–64

Liu JC, DeFazio RA, Espinosa-Jeffrey A, Cepeda C, de Vellis J, Levine MS. 2004. Calcium modulates dopamine potentiation of N-methyl-D-aspartate responses: electrophysiological and imaging evidence. *J. Neurosci. Res.* 76:315–22

Lobo MK, Covington HE 3rd, Chaudhury D, Friedman AK, Sun H, et al. 2010. Cell type-specific loss of BDNF signaling mimics optogenetic control of cocaine reward. *Science* 330:385–90

Lobo MK, Karsten SL, Gray M, Geschwind DH, Yang XW. 2006. FACS-array profiling of striatal projection neuron subtypes in juvenile and adult mouse brains. *Nat. Neurosci.* 9:443–52

Lovinger DM, Tyler EC, Merritt A. 1993. Short- and long-term synaptic depression in rat neostriatum. *J. Neurophysiol.* 70:1937–49

Mallet N, Ballion B, Le Moine C, Gonon F. 2006. Cortical inputs and GABA interneurons imbalance projection neurons in the striatum of parkinsonian rats. *J. Neurosci.* 26:3875–84

Marder E, Goaillard JM. 2006. Variability, compensation and homeostasis in neuron and network function. *Nat. Rev. Neurosci.* 7:563–74

Maurice N, Mercer J, Chan CS, Hernandez-Lopez S, Held J, et al. 2004. D2 dopamine receptor-mediated modulation of voltage-dependent Na^+ channels reduces autonomous activity in striatal cholinergic interneurons. *J. Neurosci.* 24:10289–301

Meurers BH, Dziewczapolski G, Shi T, Bittner A, Kamme F, Shults CW. 2009. Dopamine depletion induces distinct compensatory gene expression changes in DARPP-32 signal transduction cascades of striatonigral and striatopallidal neurons. *J. Neurosci.* 29:6828–39

Mink JW. 2003. The basal ganglia and involuntary movements: impaired inhibition of competing motor patterns. *Arch. Neurol.* 60:1365–68

Nambu A. 2008. Seven problems on the basal ganglia. *Curr. Opin. Neurobiol.* 18:595–604

Neve KA, Seamans JK, Trantham-Davidson H. 2004. Dopamine receptor signaling. *J. Recept. Signal Transduct. Res.* 24:165–205

Nicola SM, Malenka RC. 1998. Modulation of synaptic transmission by dopamine and norepinephrine in ventral but not dorsal striatum. *J. Neurophysiol.* 79:1768–76

Olson PA, Tkatch T, Hernandez-Lopez S, Ulrich S, Ilijic E, et al. 2005. G-protein-coupled receptor modulation of striatal CaV1.3 L-type Ca^{2+} channels is dependent on a Shank-binding domain. *J. Neurosci.* 25:1050–62

Parent M, Parent A. 2006. Single-axon tracing study of corticostriatal projections arising from primary motor cortex in primates. *J. Comp. Neurol.* 496:202–13

Parthasarathy HB, Graybiel AM. 1997. Cortically driven immediate-early gene expression reflects modular influence of sensorimotor cortex on identified striatal neurons in the squirrel monkey. *J. Neurosci.* 17:2477–91

Pavon N, Martin AB, Mendialdua A, Moratalla R. 2006. ERK phosphorylation and FosB expression are associated with L-DOPA-induced dyskinesia in hemiparkinsonian mice. *Biol. Psychiatry* 59:64–74

Pawlak V, Kerr JN. 2008. Dopamine receptor activation is required for corticostriatal spike-timing-dependent plasticity. *J. Neurosci.* 28:2435–46

Perez-Rosello T, Figueroa A, Salgado H, Vilchis C, Tecuapetla F, et al. 2005. Cholinergic control of firing pattern and neurotransmission in rat neostriatal projection neurons: role of CaV2.1 and CaV2.2 Ca^{2+} channels. *J. Neurophysiol.* 93:2507–19

Pisani A, Bernardi G, Ding J, Surmeier DJ. 2007. Re-emergence of striatal cholinergic interneurons in movement disorders. *Trends Neurosci.* 30:545–53

Planert H, Szydlowski SN, Hjorth JJ, Grillner S, Silberberg G. 2010. Dynamics of synaptic transmission between fast-spiking interneurons and striatal projection neurons of the direct and indirect pathways. *J. Neurosci.* 30:3499–507

Poirazi P, Mel BW. 2001. Impact of active dendrites and structural plasticity on the memory capacity of neural tissue. *Neuron* 29:779–96

Rajakumar N, Elisevich K, Flumerfelt BA. 1994. The pallidostriatal projection in the rat: a recurrent inhibitory loop? *Brain Res.* 651:332–36

Redgrave P, Prescott TJ, Gurney K. 1999. The basal ganglia: a vertebrate solution to the selection problem? *Neuroscience* 89:1009–23

Santini E, Heiman M, Greengard P, Valjent E, Fisone G. 2009. Inhibition of mTOR signaling in Parkinson's disease prevents L-DOPA-induced dyskinesia. *Sci. Signal.* 2:ra36

Santini E, Valjent E, Usiello A, Carta M, Borgkvist A, et al. 2007. Critical involvement of cAMP/DARPP-32 and extracellular signal-regulated protein kinase signaling in L-DOPA-induced dyskinesia. *J. Neurosci.* 27:6995–7005

Scheuer T, Catterall WA. 2006. Control of neuronal excitability by phosphorylation and dephosphorylation of sodium channels. *Biochem. Soc. Transact.* 34:1299–302

Schiffmann SN, Dassesse D, d'Alcantara P, Ledent C, Swillens S, Zoli M. 2003. A2A receptor and striatal cellular functions: regulation of gene expression, currents, and synaptic transmission. *Neurology* 61:S24–29

Schiffmann SN, Desdouits F, Menu R, Greengard P, Vincent JD, et al. 1998. Modulation of the voltage-gated sodium current in rat striatal neurons by DARPP-32, an inhibitor of protein phosphatase. *Eur. J. Neurosci.* 10:1312–20

Schultz W. 1998. Predictive reward signal of DA neurons. *J. Neurophysiol* 80:1–27

Schultz W. 2007. Multiple DA functions at different time courses. *Annu. Rev. Neurosci.* 30:259–88

Sciamanna G, Bonsi P, Tassone A, Cuomo D, Tscherter A, et al. 2009. Impaired striatal D2 receptor function leads to enhanced GABA transmission in a mouse model of DYT1 dystonia. *Neurobiol. Dis.* 34:133–45

Scott L, Zelenin S, Malmersjo S, Kowalewski JM, Markus EZ, et al. 2006. Allosteric changes of the NMDA receptor trap diffusible dopamine 1 receptors in spines. *Proc. Natl. Acad. Sci. USA* 103:762–67

Shen W, Flajolet M, Greengard P, Surmeier DJ. 2008. Dichotomous dopaminergic control of striatal synaptic plasticity. *Science* 321:848–51

Shen W, Tian X, Day M, Ulrich S, Tkatch T, et al. 2007. Cholinergic modulation of Kir2 channels selectively elevates dendritic excitability in striatopallidal neurons. *Nat. Neurosci.* 10:1458–66

Smith Y, Raju DV, Pare JF, Sidibe M. 2004. The thalamostriatal system: a highly specific network of the basal ganglia circuitry. *Trends Neurosci.* 27:520–27

Snyder GL, Allen PB, Fienberg AA, Valle CG, Huganir RL, et al. 2000. Regulation of phosphorylation of the GluR1 AMPA receptor in the neostriatum by dopamine and psychostimulants in vivo. *J. Neurosci.* 20:4480–88

Spooren WP, Lynd-Balta E, Mitchell S, Haber SN. 1996. Ventral pallidostriatal pathway in the monkey: evidence for modulation of basal ganglia circuits. *J. Comp. Neurol.* 370:295–312

Stern EA, Jaeger D, Wilson CJ. 1998. Membrane potential synchrony of simultaneously recorded striatal spiny neurons in vivo. *Nature* 394:475–78

Stoetzner CR, Pettibone JR, Berke JD. 2010. State-dependent plasticity of the corticostriatal pathway. *Neuroscience* 165:1013–18

Stoof JC, Kebabian JW. 1984. Two dopamine receptors: biochemistry, physiology and pharmacology. *Life Sci.* 35:2281–96

Sullivan MA, Chen H, Morikawa H. 2008. Recurrent inhibitory network among striatal cholinergic interneurons. *J. Neurosci.* 28:8682–90

Sun X, Zhao Y, Wolf ME. 2005. Dopamine receptor stimulation modulates AMPA receptor synaptic insertion in prefrontal cortex neurons. *J. Neurosci.* 25:7342–51

Surmeier DJ, Bargas J, Hemmings HC Jr, Nairn AC, Greengard P. 1995. Modulation of calcium currents by a D1 dopaminergic protein kinase/phosphatase cascade in rat neostriatal neurons. *Neuron* 14:385–97

Surmeier DJ, Ding J, Day M, Wang Z, Shen W. 2007. D1 and D2 dopamine-receptor modulation of striatal glutamatergic signaling in striatal medium spiny neurons. *Trends Neurosci.* 30:228–35

Surmeier DJ, Eberwine J, Wilson CJ, Cao Y, Stefani A, Kitai ST. 1992. Dopamine receptor subtypes colocalize in rat striatonigral neurons. *Proc. Natl. Acad. Sci. USA* 89:10178–82

Surmeier DJ, Plotkin J, Shen W. 2009. Dopamine and synaptic plasticity in dorsal striatal circuits controlling action selection. *Curr. Opin. Neurobiol.* 19:621–28

Surmeier DJ, Song WJ, Yan Z. 1996. Coordinated expression of dopamine receptors in neostriatal medium spiny neurons. *J. Neurosci.* 16:6579–91

Svenningsson P, Nishi A, Fisone G, Girault JA, Nairn AC, Greengard P. 2004. DARPP-32: an integrator of neurotransmission. *Annu. Rev. Pharmacol. Toxicol.* 44:269–96

Sweatt JD. 2004. Mitogen-activated protein kinases in synaptic plasticity and memory. *Curr. Opin. Neurobiol.* 14:311–17

Tanimura A, Yamazaki M, Hashimotodani Y, Uchigashima M, Kawata S, et al. 2010. The endocannabinoid 2-arachidonoylglycerol produced by diacylglycerol lipase alpha mediates retrograde suppression of synaptic transmission. *Neuron* 65:320–27

Taverna S, Ilijic E, Surmeier DJ. 2008. Recurrent collateral connections of striatal medium spiny neurons are disrupted in models of Parkinson's disease. *J. Neurosci.* 28:5504–12

Tepper JM, Tecuapetla F, Koos T, Ibanez-Sandoval O. 2010. Heterogeneity and diversity of striatal GABAergic interneurons. *Front. Neuroanat.* 4:1–18

Tepper JM, Wilson CJ, Koos T. 2008. Feedforward and feedback inhibition in neostriatal GABAergic spiny neurons. *Brain Res. Rev.* 58:272–81

Testa CM, Standaert DG, Young AB, Penney JB Jr. 1994. Metabotropic glutamate receptor mRNA expression in the basal ganglia of the rat. *J. Neurosci.* 14:3005–18

Tian X, Kai L, Hockberger PE, Wokosin DL, Surmeier DJ. 2010. MEF-2 regulates activity-dependent spine loss in striatopallidal medium spiny neurons. *Mol. Cell Neurosci.* 44:94–108

Tozzi A, de Iure A, Di Filippo M, Tantucci M, Costa C, et al. 2011. The distinct role of medium spiny neurons and cholinergic interneurons in the D2/A2A receptor interaction in the striatum: implications for Parkinson's disease. *J. Neurosci.* 31:1850–62

Tunstall MJ, Oorschot DE, Kean A, Wickens JR. 2002. Inhibitory interactions between spiny projection neurons in the rat striatum. *J. Neurophysiol.* 88:1263–69

Turner RS, Anderson ME. 1997. Pallidal discharge related to the kinematics of reaching movements in two dimensions. *J. Neurophysiol.* 77:1051–74

Turner RS, Anderson ME. 2005. Context-dependent modulation of movement-related discharge in the primate globus pallidus. *J. Neurosci.* 25:2965–76

Turrigiano GG. 1999. Homeostatic plasticity in neuronal networks: The more things change, the more they stay the same. *Trends Neurosci.* 22:221–27

Valjent E, Bertran-Gonzalez J, Herve D, Fisone G, Girault JA. 2009. Looking BAC at striatal signaling: cell-specific analysis in new transgenic mice. *Trends Neurosci.* 32:538–47

Vergara R, Rick C, Hernandez-Lopez S, Laville JA, Guzman JN, et al. 2003. Spontaneous voltage oscillations in striatal projection neurons in a rat corticostriatal slice. *J. Physiol.* 553:169–82

Vilchis C, Bargas J, Ayala GX, Galvan E, Galarraga E. 2000. Ca^{2+} channels that activate Ca^{2+}-dependent K^+ currents in neostriatal neurons. *Neuroscience* 95:745–52

Wang Y, Toledo-Rodriguez M, Gupta A, Wu C, Silberberg G, et al. 2004. Anatomical, physiological and molecular properties of Martinotti cells in the somatosensory cortex of the juvenile rat. *J. Physiol.* 561:65–90

Wang Z, Kai L, Day M, Ronesi J, Yin HH, et al. 2006. Dopaminergic control of corticostriatal long-term synaptic depression in medium spiny neurons is mediated by cholinergic interneurons. *Neuron* 50:443–52

Westin JE, Vercammen L, Strome EM, Konradi C, Cenci MA. 2007. Spatiotemporal pattern of striatal ERK1/2 phosphorylation in a rat model of L-DOPA-induced dyskinesia and the role of dopamine D1 receptors. *Biol. Psychiatry* 62:800–10

White FJ, Wang RY. 1986. Electrophysiological evidence for the existence of both D-1 and D-2 dopamine receptors in the rat nucleus accumbens. *J. Neurosci.* 6:274–80

Wichmann T, DeLong MR. 2003. Functional neuroanatomy of the basal ganglia in Parkinson's disease. *Adv. Neurol.* 91:9–18

Wickens JR. 2009. Synaptic plasticity in the basal ganglia. *Behav. Brain Res.* 199:119–28

Wilson CJ. 1992. Dendritic morphology, inward rectification and the functional properties of neostriatal neurons. In *Single Neuron Computation*, ed. T McKenna, J Davis, SF Zornetzer, pp. 141–71. San Diego: Academic

Wilson CJ. 1993. The generation of natural firing patterns in neostriatal neurons. *Prog. Brain Res.* 99:277–97

Wilson CJ. 1994. Understanding the neostriatal microcircuitry: high-voltage electron microscopy. *Microsc. Res. Tech.* 29:368–80

Wilson CJ. 2004. Basal ganglia. In *The Synaptic Organization of the Brain*, ed. GM Shepherd, pp. 361–414. Oxford: Oxford Univ. Press

Wilson CJ, Groves PM. 1980. Fine structure and synaptic connections of the common spiny neuron of the rat neostriatum: a study employing intracellular inject of horseradish peroxidase. *J. Comp. Neurol.* 194:599–615

Wilson CJ, Kawaguchi Y. 1996. The origins of two-state spontaneous membrane potential fluctuations of neostriatal spiny neurons. *J. Neurosci.* 16:2397–410

Wiltschko AB, Pettibone JR, Berke JD. 2010. Opposite effects of stimulant and antipsychotic drugs on striatal fast-spiking interneurons. *Neuropsychopharmacology* 35:1261–70

Witten IB, Lin SC, Brodsky M, Prakash R, Diester I, et al. 2010. Cholinergic interneurons control local circuit activity and cocaine conditioning. *Science* 330:1677–81

Yan Z, Song WJ, Surmeier J. 1997. D2 dopamine receptors reduce N-type Ca^{2+} currents in rat neostriatal cholinergic interneurons through a membrane-delimited, protein-kinase-C-insensitive pathway. *J. Neurophysiol.* 77:1003–15

Yan Z, Surmeier DJ. 1996. Muscarinic (m2/m4) receptors reduce N- and P-type Ca^{2+} currents in rat neostriatal cholinergic interneurons through a fast, membrane-delimited, G-protein pathway. *J. Neurosci.* 16:2592–604

Yan Z, Surmeier DJ. 1997. D5 dopamine receptors enhance Zn^{2+}-sensitive GABA(A) currents in striatal cholinergic interneurons through a PKA/PP1 cascade. *Neuron* 19:1115–26

Yin HH, Lovinger DM. 2006. Frequency-specific and D2 receptor-mediated inhibition of glutamate release by retrograde endocannabinoid signaling. *Proc. Natl. Acad. Sci. USA* 103:8251–56

Young WS 3rd, Bonner TI, Brann MR. 1986. Mesencephalic dopamine neurons regulate the expression of neuropeptide mRNAs in the rat forebrain. *Proc. Natl. Acad. Sci. USA* 83:9827–31

How Is the Olfactory Map Formed and Interpreted in the Mammalian Brain?

Kensaku Mori[1] and Hitoshi Sakano[2]

[1] Department of Physiology, Graduate School of Medicine, University of Tokyo, Tokyo 113-0033, Japan; email: moriken@m.u-tokyo.ac.jp

[2] Department of Biophysics and Biochemistry, Graduate School of Science, University of Tokyo, Tokyo 113-0032, Japan; email: sakano@mail.ecc.u-tokyo.ac.jp

Annu. Rev. Neurosci. 2011. 34:467–99

First published online as a Review in Advance on April 4, 2011

The *Annual Review of Neuroscience* is online at neuro.annualreviews.org

This article's doi:
10.1146/annurev-neuro-112210-112917

Keywords

olfactory sensory neurons, glomerular map formation, olfactory bulb, mitral and tufted cells, olfactory cortex

Abstract

Odor signals received by odorant receptors (ORs) expressed by olfactory sensory neurons (OSNs) in the olfactory epithelium (OE) are represented as an odor map in the olfactory bulb (OB). In the mouse, there are ~1,000 different OR species, and each OSN expresses only one functional OR gene in a monoallelic manner. Furthermore, OSN axons expressing the same type of OR converge on a specific target site in the OB, forming a glomerular structure. Because each glomerulus represents a single OR species, and a single odorant can interact with multiple OR species, odor signals received in the OE are converted into a topographic map of multiple glomeruli activated with varying magnitudes. Here we review recent progress in the study of the mammalian olfactory system, focusing on the formation of the olfactory map and the transmission of topographical information in the OB to the olfactory cortex to elicit various behaviors.

Contents

INTRODUCTION

The mammalian olfactory system mediates various responses, including aversive behaviors to spoiled food smells and fear responses to predator odors. Sensory information is spatially encoded in the brain, forming neural maps that are fundamental for higher-order processing of sensory information. Molecular mechanisms of sensory map formation in the visual system in particular have been extensively studied. It is well established that axonal projection of retinal ganglion cells is instructed by multiple pairs of axon-guidance molecules that demonstrate graded expression in the retina and tectum (McLaughlin & O'Leary 2005, Luo & Flanagan 2007). The spatial organization of the projecting neurons in retina is maintained and projected onto the target, preserving the nearest-neighbor relationship. In contrast, in olfactory map formation, projecting axons of spatially dispersed cell bodies with the same neuronal identity converge on one glomerular location in the target olfactory bulb (OB) (Mombaerts et al. 1996) (**Figure 1a**). In the mouse olfactory system, there are ~1,000 functional odorant receptor (OR) genes expressed in the olfactory epithelium (OE) (Buck & Axel 1991). Thus ~1,000 different identities of olfactory sensory neurons (OSNs) determined by expressed OR species are represented in

~1,000 distinct glomeruli in each mirror map on the OB (Mori et al. 2006). Unlike the visual system, much of glomerular map formation appears to occur autonomously by axon-axon interaction, without involving target-derived cues (Sakano 2010). Within the OB, the odor information coded in the olfactory map is processed in the local neuronal circuits and then

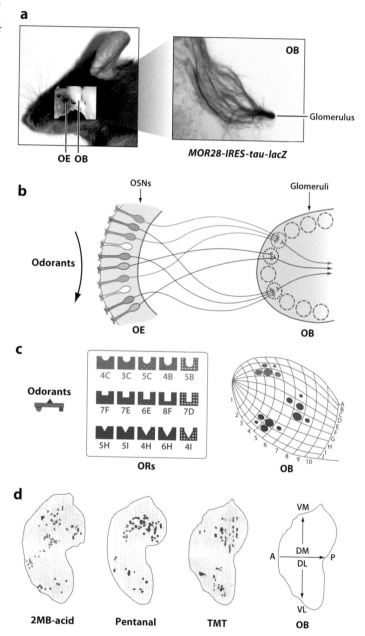

Figure 1

Olfactory map formation in the mouse. (*a*) Odorant receptor (OR)-instructed axonal projection of olfactory sensory neurons (OSNs). MOR28-expressing axons are stained blue with X-gal in the transgenic mouse, MOR28-IRES-tau-lacZ. OSNs expressing the transgene converge their axons to a specific site, forming a glomerulus in the olfactory bulb (OB). (*b*) Axonal segregation and olfactory map formation. OSN axons are guided to approximate destinations in the OB by a combination of dorsal-ventral (D-V) patterning and anterior-posterior (A-P) patterning. D-V projection occurs based on anatomical locations of OSNs in the olfactory epithelium (OE). A-P projection is regulated by OR-derived cAMP signals. The map is further refined in an activity-dependent manner during the early neonatal period. (*c*) Conversion of olfactory signals. In the olfactory epithelium OE, each OSN expresses only one functional OR gene in a monoallelic manner. Furthermore, OSN axons expressing the same OR species converge to a specific glomerulus in the OB. Thus each glomerulus represents one OR species. Because a given odorant activates multiple OR species and a given OR responds to multiple odorants, odorant signals received by OSNs in the OE are converted into a two-dimensional odor map of activated glomeruli with varying magnitudes of activity. (*d*) Unrolled odor maps for aversive odorants. Glomeruli activated by 2-methylbutyric (2MB) acid, pentanal, and trimethyl-thiazoline (TMT) were detected by staining for the immediate-early gene product Zif268. Zif268-positive glomeruli are indicated in the unrolled maps: D_I (*orange*), D_{II} (*blue*), and V (*magenta*). Abbreviations: DL, dorsolateral; DM, dorsomedial; VL, ventrolateral; VM, ventromedial.

conveyed by mitral and tufted cells to various areas of the olfactory cortex (OC) (Mori et al. 1999, Shepherd et al. 2004). Here we review recent progress on the study of OR-instructed axonal projection of OSNs and olfactory map formation in rodents. We discuss how the odor map formed on the OB is interpreted in the brain for various odor responses.

HOW IS NEURONAL IDENTITY ESTABLISHED IN OLFACTORY SENSORY NEURONS?

In the mammalian olfactory system, each OSN chooses for expression only one functional OR gene in a monoallelic manner (Chess et al. 1994, Serizawa et al. 2000, Ishii et al. 2001). Such unique expression forms the genetic basis for the OR-instructed axonal projection of OSNs to the OB. How is the singular OR gene choice regulated, and how is the one neuron–one receptor rule maintained in OSNs? Based on previous studies of other multigene families, three mechanisms have been considered for the choice and activation of OR genes: (*a*) DNA recombination, which brings a promoter and the enhancer region into close proximity; (*b*) gene conversion, which transfers a copy of the gene into the expression cassette; and (*c*) a regulatory DNA region, which interacts with only one promoter site. Irreversible DNA changes, such as recombination and gene conversion, had been attractive explanations for single OR gene expression because of the many parallels between the immune system and the olfactory system. However, these theories were dismissed after two groups independently cloned mice from postmitotic OSN nuclei and determined that the mice showed no irreversible DNA rearrangement in the OR genes (Eggan et al. 2004, Li et al. 2004).

Odorant-Receptor Gene Choice by Locus Control Region

Because the genetic translocation model appeared unlikely, another possibility was explored, namely a locus control region (LCR)

that might regulate the single OR gene choice (**Figure 2**). LCR is defined as a *cis*-acting regulatory region that controls multiple genes clustered at a specific genetic locus. The first example of an LCR was identified in the globin gene locus containing developmentally regulated and related genes (Grosveld et al. 1987). It has been assumed that transcription-activating factors bind to the LCR (holocomplex) that physically interacts with the remote promoter site by looping out the intervening DNA. In the globin gene system, an active chromatin hub structure has been reported, in which the LCR is in close proximity to the gene to be expressed (Carter et al. 2002, Tolhuis et al. 2002). One example of such an LCR in the mouse olfactory system, named *H* for the homology between the mouse and human DNA sequences, was identified upstream of the *MOR28* cluster containing seven murine OR genes (Serizawa et al. 2003). Deletion and mutation analyses of the *H* region further revealed that the 124-bp core-*H* region, which contains two homeodomain sequences and one O/E-like sequence, is sufficient to achieve the enhancer activity (Nishizumi et al. 2007). Both the homeodomain and O/E-like sequences are often found in the OR promoter regions (Vassalli et al. 2002, Hirota & Mombaerts 2004, Rothman et al. 2005, Michaloski et al. 2006). Homeodomain factors, Lhx2 and Emx2, and O/E family proteins are known to bind to their motifs in the OR gene promoter (Wang et al. 2004, Hirota et al. 2007). It is possible that these nuclear factors bind to the *H* region and form a complex that remodels the chromatin structure near the cluster, thereby activating one OR promoter site at a time by physical interaction (Serizawa et al. 2003) (**Figure 2***b*). This model is attractive because it reduces the likelihood of the simultaneous activation of multiple OR genes, from a probability of ~1,000 individual genes to ~50 LCR loci.

In the immune system, it has been reported that the LCR of the *IL-4* gene in naïve T helper cells interacts not only with the adjacent genes on chromosome 11, but also with the alternatively expressed *IFN-γ* gene on chromosome 10

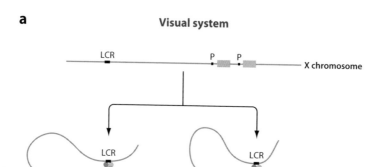

a

Visual system

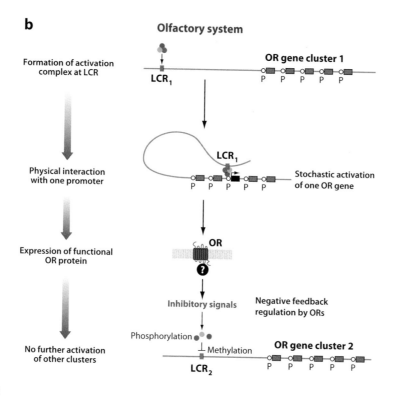

b

Olfactory system

Figure 2

Mutually exclusive expression of related genes. (*a*) In the human visual system, a locus control region (LCR) plays an important role in choosing either the red or green photopigment gene in a mutually exclusive manner in the cone cells of the retina. Stochastic interaction between the LCR and either promoter (P) ensures the mutually exclusive expression of the human photopigment genes encoded on the X chromosome (Smallwood et al. 2002). (*b*) In the odorant receptor (OR) gene system, the LCR-promoter interaction alone would not preclude the activation of a second OR gene located in the other allele or in other OR gene clusters. Once a functional gene is expressed, the OR molecules transmit inhibitory signals to block the further activation of additional OR gene clusters.

(Spilianakis et al. 2005). Such an interchromo-somal association of multiple gene loci may be a common feature of chromosomal organization for the coordinately regulated genes. Does the *H* region act on other OR gene clusters, not only in *cis* but also in *trans*, like the *IL-4* LCR in T cells? Based on fluorescent in situ hybridiza-tion and chromosome conformation capture analyses of OSN nuclei, Lomvardas et al. (2006) reported that the single *trans*-acting enhancer *H* (Serizawa et al. 2003) may allow the stochastic activation of only one OR gene in each OSN. However, recent knockout (KO) studies of *H* contradict the single *trans*-acting LCR model for the OR gene choice (Fuss et al. 2007, Nishizumi et al. 2007). Targeted deletion of *H* abolished the expression of only three proximal OR genes in *cis*, indicating the presence of another LCR in the downstream region to regulate the four distal OR genes in the same *MOR28* cluster. Furthermore, in heterozygous (H^+/H^-) KO mice, the wild-type H^+ allele could not rescue the H^- mutant allele in *trans*, indicating that *H* can act only in *cis* and not in *trans* (Nishizumi et al. 2007).

Negative Feedback Regulation to Maintain the One Neuron–One Receptor Rule

In the human visual system, an LCR plays an important role in choosing either the red or green photopigment gene in a mutually ex-clusive manner in the cone cells of the retina (Wang et al. 1999). It has been shown that stochastic interaction between the LCR and ei-ther of the two promoters ensures the mutually exclusive expression of the human photopig-ment genes that are encoded on the X chro-mosome (Smallwood et al. 2002) (**Figure 2a**). In contrast, in the OR gene system, the LCR-promoter interaction alone would not preclude the activation of a second OR gene located in the other allele or in other OR gene clusters. Therefore, it has been postulated that the func-tional OR proteins have an inhibitory role to prevent further activation of other OR genes (Serizawa et al. 2003, Lewcock & Reed 2004).

Transgenic experiments demonstrated that the mutant OR genes lacking either the entire cod-ing sequence or the start codon can permit a second OR gene to be expressed (Serizawa et al. 2003, Feinstein et al. 2004, Lewcock & Reed 2004, Shykind et al. 2004). Naturally occurring frameshift mutants of OR genes also allow the coexpression of a functional OR gene (Serizawa et al. 2003). It is known that a substantial num-ber of pseudogenes are present in the mam-malian OR gene families (~30% of total OR genes in mice and ~60% in humans). If an ac-tivated LCR has selected a pseudogene and has been trapped by its promoter, other LCRs must undergo a similar process to ensure the activa-tion of an intact OR gene (Roppolo et al. 2007). A pseudogene may help slow the process of OR gene activation and further reduce the likeli-hood of activating two functional OR genes. Thus rate-limited activation of an OR gene by *cis*-acting LCRs and negative feedback regula-tion by the OR gene product together appear to ensure the maintenance of the one neuron–one receptor rule (Serizawa et al. 2003, 2004) (**Figure 2b**). However, the exact nature of the negative feedback signals has not been explored. Targets of the feedback signals are also issues for future studies. Promoters of OR genes, en-hancers of OR gene clusters, and protein factors binding them could be silenced by OR-derived negative feedback signals.

It has been reported that forced expression of an OR using heterologous promoters can suppress the expression of endogenous ORs (Nguyen et al. 2007, Fleischmann et al. 2008). As the forced expression was found to be in-efficient, researchers have proposed that the OR coding sequence has an inhibitory effect on transcriptional regulation. However, it has not been demonstrated that the suppressive effect resulted from the DNA sequence and not the potentially toxic effects of OR protein overex-pression. It is important to identify the essential nucleotide sequences responsible for the sup-pressive effects of functional ORs.

In neonatal mice, a small fraction of OSNs expresses two functional OR species simulta-neously, which are eventually eliminated in an activity-dependent manner (Tian & Ma 2008). This form of negative selection has also been considered to be a fail-safe mechanism to ensure monogenic OR expression (Mombaerts 2006). OR-mediated negative feedback does not require G protein signaling (Imai et al. 2006, Nguyen et al. 2007). The involvement of G proteins in negative feedback regulation was tested using transgenic mice expressing a mutant-type OR, in which the DRY motif, essential for G protein signaling, was changed to RDY. This mutation completely abolished odor-evoked calcium responses in OSNs, indicating the loss of G protein signaling. However, OSNs expressing the mutant OR maintained the one neuron–one receptor rule. Seven-transmembrane receptors often utilize G protein–independent signaling pathways, such as those involving β-arrestin (Shenoy & Lefkowitz 2005). We note that the DRY motif is dispensable for β-arrestin-mediated seven-transmembrane-receptor signaling (Seta et al. 2002). KO studies will clarify whether β-arrestin-mediated signaling is involved in negative feedback regulation.

It has been reported that each OR gene possesses a unique expression area in the OE along the dorsomedial-ventrolateral axis (Ressler et al. 1993, Vassar et al. 1993, Miyamichi et al. 2005). Thus OR gene choice may not be totally stochastic, but rather it may be restricted by the OSN location in the OE. How does this positional information within the OE regulate OR gene choice? It is possible that this regulation is determined by cell lineage, resulting in the use of zone-specific transcription factors, e.g., Msx1 and Foxg1 (Norlin et al. 2001, Duggan et al. 2008).

HOW IS ODORANT-RECEPTOR-INSTRUCTED AXONAL PROJECTION REGULATED?

The olfactory map in the OB comprises discrete glomeruli, each representing a single OR species (Ressler et al. 1994, Vassar et al. 1994, Mombaerts et al. 1996). This is another rule for olfactory map formation, which is referred

to as the one glomerulus–one receptor rule. Coding-swap experiments of OR genes demonstrated the instructive role of the OR protein in OSN projection (Mombaerts et al. 1996, Wang et al. 1998, Feinstein & Mombaerts 2004). Because OSNs expressing the same OR are scattered in the OE for anterior-posterior projection, topographic organization must occur during the process of axonal projection of OSNs. Unlike axonal projection in other sensory systems in which relative positional information is preserved between the periphery and the brain (Lemke & Reber 2005, McLaughlin & O'Leary 2005, Petersen 2007), there is no such correlation for the projection along the anterior-posterior axis in the mouse olfactory system. Although OR molecules have been known to play an instructive role in projecting OSN axons to form the glomerular map, it has remained entirely unclear how this occurs at the molecular level. Intriguingly, OR molecules are detected in axon termini by tagging with green fluorescent protein (Feinstein & Mombaerts 2004) or by immunostaining with anti-OR antibodies (Barnea et al. 2004, Strotmann et al. 2004). On the basis of these observations, it was suggested that the OR protein itself may recognize positional cues in the OB and also may mediate homophilic interaction of similar axons (Mombaerts 2006). Although these models were attractive, recent studies argue against them. Instead of directly acting as guidance receptors or adhesion molecules, ORs regulate transcription levels of axon-guidance and axon-sorting molecules by OR-derived cAMP signals with levels uniquely determined by the OR species (**Figure 3a**) (Imai et al. 2006, Serizawa et al. 2006).

Anterior-Posterior Positioning of Glomeruli is Determined by Odorant-Receptor-Derived cAMP Signals

In odor detection, binding signals of odorants to ORs are converted into neuronal activities via cAMP. The olfactory-specific G protein G_{olf} activates adenylyl cyclase type III (ACIII), generating cAMP, which opens cyclic nucleotide gated (CNG) channels (Wong et al. 2000). The CNG channel, together with the chloride channels (Stephan et al. 2009), depolarizes the plasma membrane to generate the action potential. Targeted KOs of G_{olf} and CNGA2 cause severe anosmia (Brunet et al. 1996). Despite their essential role in odor signal transduction, KOs of the genes did not demonstrate major defects in glomerular map formation (Belluscio et al. 1998, Lin et al.

2000, Zheng et al. 2000). It was therefore assumed that OR-derived cAMP signals are not required for OSN projection.

Despite these observations, it was possible that an alternate G protein mediates OR-instructed OSN projection. To examine this possibility, Imai et al. (2006) generated a mutant OR whose G protein coupling motif, DRY, was mutated to RDY. The authors found that axons expressing the mutant OR remained in the anterior region of the OB and failed to

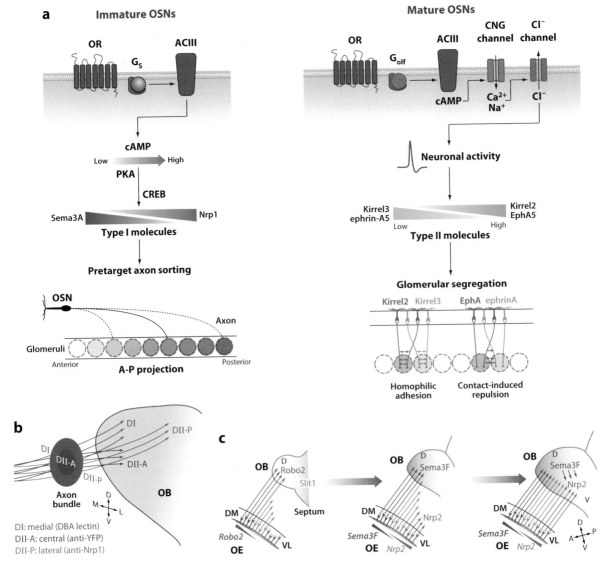

converge to a specific glomerulus in the OB. Interestingly, coexpression of the constitutively active G_s mutant restored axonal convergence and glomerular formation. Partial rescue was also observed with the constitutively active mutants of protein kinase PKA and transcription factor CREB. Thus PKA-mediated transcriptional regulation appears to be involved in OSN projection. Furthermore, constitutively active G_s results in a posterior shift of glomeruli when expressed with the wild-type OR, whereas dominant-negative PKA results in an anterior shift of glomeruli (Imai et al. 2006). These findings suggest that it is the OR-derived cAMP signals, rather than the direct action of OR molecules themselves, that determine the target destination of OSNs along the anterior-posterior axis in the OB (**Figure 3a**).

The above results provide insight into some puzzling observations about OSN targeting that cannot be explained by a model that assumes a direct role of OR molecules in axonal projection and fasciculation. It has been reported that the β_2-adrenergic receptor (β_2AR), but not the vomeronasal receptor Vlrb2, can substitute for an OR in OR-instructed axonal outgrowth and glomerular formation (Feinstein et al. 2004). The explanation for this observation may be that β_2AR can couple to G_s, but Vlrb2 cannot. This explanation is consistent with the idea that G_s-mediated cAMP levels set by the receptors determine the target sites of OSN axons. Another puzzling observation is that alterations in OR expression levels can affect OSN projection and cause glomerular segregation (Serizawa et al. 2000, Feinstein et al. 2004). Both the OR identity and the amount of OR protein within the cell may affect cAMP signaling levels, which in turn affect transcription or translation parameters.

Pretarget Axon Sorting Establishes the Anterior-Posterior Topography

We assume that each OR species generates a unique level of cAMP signals that regulates the transcription of axon-guidance molecules, e.g., the Nrp1 receptor in a positive manner and Sema3A (the repulsive ligand of Nrp1) in a negative manner (Imai et al. 2006). As mentioned above, OSNs producing high levels of cAMP signaling project their axons to the posterior OB, whereas those producing low levels target the anterior OB. When protein levels of

---

Figure 3

Axon sorting of olfactory sensory neurons (OSNs) in olfactory map formation. (*a*) Differential utilization of cAMP signals for global versus local axon sorting. In the mouse olfactory system, axon-sorting molecules can be categorized into two different types: type I (*left*) and type II (*right*). Type I molecules are expressed at axon termini of immature OSNs in a graded manner along the anterior-posterior (A-P) axis in the olfactory bulb (OB) and regulate A-P positioning of glomeruli. In contrast, type II molecules are expressed at axon termini of mature OSNs, showing a mosaic pattern in the glomerular map, and are involved in glomerular segregation. Expression of both type I and type II genes is regulated by odorant receptor (OR)-derived cAMP signals. Each OR generates a unique level of cAMP with the aid of G protein and ACIII. The level of cAMP signals in immature OSNs is converted into a relative expression level of type I molecules, e.g., Nrp1 and Sema3A, via cAMP-dependent PKA and CREB (Imai et al. 2006). In mature OSNs, different ORs generate different neuronal activities that determine the expression levels of axon-sorting molecules necessary for glomerular segregation, e.g., Kirrel2/3 and EphA/ephrin-A (Serizawa et al. 2006). (*b*) Pretarget axon sorting of OSNs (Imai et al. 2009). The axons of dorsal-zone OSNs course through the bundle on the dorsal roof of the olfactory epithelium before projecting to the D domain of the OB. The D domain comprises D_I and D_{II} subdomains in the OB. D_I is located in the most dorsal part of the OB; D_{II} is just ventral to D_I and is further divided into D_{II-P} (Nrp1-high, posterior) and D_{II-A} (Nrp1-low, anterior). Three types of axons projecting to the D_I, D_{II-A}, and D_{II-P} regions in the OB are intermingled in the bundle near the olfactory epithelium (OE) but segregate to form a tripartite organization as they approach the OB. D_I, D_{II-A}, and D_{II-P} axons are sorted to the medial, central, and lateral areas, respectively. (*c*) A model for the dorsal-ventral (D-V) projection of OSN axons (Takeuchi et al. 2010). Axonal extension of OSNs occurs sequentially along the dorsomedial-ventrolateral (DM-VL) axis of the OE as the OB grows ventrally during development. (*Left panel*) Dorsal-zone OSNs express Robo2 and project their axons to the prospective dorsal domain of the embryonic OB. Repulsive interactions between Robo2 and Slit1 are needed to restrict early OSN projection to the embryonic OB. (*Middle panel*) In the OE, *Nrp2* and *Sema3F* genes are expressed in a complementary manner. (*Right panel*) Sema3F is deposited at the anterodorsal region of the OB by early arriving dorsal-zone axons and prevents the late-arriving Nrp2$^+$ axons from invading the dorsal region of the OB.

Nrp1 were measured in axon termini of OSNs, Nrp1 was found in an anterior-low/posterior-high gradient in the OB. Increases or decreases of Nrp1 expression in OSNs caused posterior or anterior glomerular shifts, respectively (Imai et al. 2009). Furthermore, the anterior-posterior topography of the glomerular map was perturbed in mice deficient for Nrp1 or Sema3A. How then do guidance molecules regulate olfactory map formation?

Map order emerges in axon bundles, well before they reach the target (Satoda et al. 1995). It has been reported that pretarget axon sorting plays an important role in the organization of the olfactory map (Imai et al. 2009). Nrp1 and Sema3A are both expressed in OSNs, but in a complementary manner. Within the axon bundles, Nrp1^{low}/Sema3A^{high} axons are sorted into the central compartment of the bundle, whereas Nrp1^{high}/Sema3A^{low} axons are confined to the outer-lateral compartment. OSN-specific KOs of Nrp1 and Sema3A not only perturbed axon sorting within the bundle, but also caused an anterior shift of glomeruli in the OB (Imai et al. 2009). These results indicate that pretarget axon sorting within the bundle contributes to olfactory map formation along the anterior-posterior axis (**Figure 3b**). Although OSN axons may be sorted autonomously within the bundle, they still require an extrinsic cue for orientation along the correct axis before projecting onto the OB. Sema3A, which is also detected in ensheathing glial cells along the medial side of the axon bundles, may act as the extrinsic cue that orients the axon bundle, pushing Nrp1$^+$ axons toward the lateral side of the bundle (Imai et al. 2009).

HOW IS DORSAL-VENTRAL POSITIONING REGULATED?

In contrast to anterior-posterior projection, the OB projection sites of OSNs along the dorsal-ventral axis are correlated with the anatomical locations of the OSNs in the OE. On the basis of the expression patterns of zone-specific markers, e.g., O-MACS (Oka et al. 2003), NQO1 (Gussing & Bohm 2004), and OCAM (Yoshihara et al. 1997), the OE can be divided into two nonoverlapping zones: dorsal and ventral (Kobayakawa et al. 2007, Bozza et al. 2009). Two sets of repulsive ligands/receptors, Slits/Robo2 and Sema3F/Nrp2, have been shown to participate in OSN projection along the dorsal-ventral axis (Norlin et al. 2001, Cho et al. 2007, Nguyen-Ba-Charvet et al. 2008, Takeuchi et al. 2010).

Positional Information of Olfactory Sensory Neurons in the Olfactory Epithelium Regulates Dorsal-Ventral Projection

In the OE, OR genes expressed by OSNs that project to the dorsal domain of the OB are distributed throughout the dorsal zone (Tsuboi et al. 2006). However, ventral-zone-specific OR genes exhibit spatially limited expression along the dorsomedial-ventrolateral axis of the OE (Miyamichi et al. 2005). The relationship between the dorsal-ventral positioning of glomeruli and the locations of OSNs in the OE has been demonstrated by retrograde DiI staining of OSN axons (Astic et al. 1987, Miyamichi et al. 2005). These observations suggest that the anatomical locations of OSNs in the OE contribute to the dorsal-ventral positioning of glomeruli in the OB. How is this positional information of OSNs in the OE translated to their target site during olfactory map formation? Nrp2 is expressed on OSN axons in such a way that it forms a gradient in the OB along the dorsal-ventral axis. Loss-of-function and gain-of-function experiments demonstrated that Nrp2 indeed regulates the axonal projection of OSNs along this axis (Takeuchi et al. 2010). Based on the visual system, the repulsive ligand Sema3F was expected to show a gradient in the target organ, the OB. Curiously, however, the *Sema3F* transcript was not detected in the OB, and animals in which Sema3F was knocked out specifically in OSNs showed mistargeting of Nrp2$^+$ axons along the dorsal-ventral axis. In the olfactory system, an axon-guidance receptor, Nrp2, and its repulsive ligand, Sema3F, are both expressed by OSN axons in a complementary manner to regulate dorsal-ventral projection (Takeuchi et al. 2010).

Expression levels of dorsal-ventral guidance molecules, such as Nrp2 and Sema3F, are closely correlated with the expressed OR species. However, unlike Nrp1 and Sema3A, which are involved in anterior-posterior positioning, the transcription of *Nrp2* and *Sema3F* is not downstream of OR signaling. This notion was supported by the analyses of some transgenic mice in which the expression areas of particular ORs were genetically altered. When the expression areas of ORs were shifted, the projection sites changed accordingly along the dorsal-ventral axis in the OB (Nakatani et al. 2003, Miyamichi et al. 2005).

It turns out that OR gene choice is not purely stochastic and is affected by location in the OE. This idea was demonstrated using transgenic mice in which the coding sequence of the transgenic OR gene was deleted and replaced with *EGFP*. In these mice, the choice of the secondary OR gene in *EGFP*-positive OSNs was not random and was primarily limited to a group of OR genes whose expression areas and transcription levels of *Nrp2* were comparable with those of the coding-deleted OR gene (Serizawa et al. 2003). If dorsal-ventral guidance molecules are not regulated by OR-derived signals, how are their expression levels determined and correlated with the expressed OR species? It appears that both OR gene choice and *Nrp2* expression levels are commonly regulated by positional information within the OE. This regulation is likely determined by cell lineage, resulting in the use of specific sets of transcription factors, which can explain the anatomical correlation along the dorsomedial-ventrolateral axis of the OE.

Complementary expression of Nrp2 and Sema3F in OSN axons suggested that repulsive interactions to establish map order occurred among OSN axons before they reach the target, as described above for anterior-posterior positioning. This, however, could not be true because dorsal-zone axons and ventral-zone axons are segregated in separate bundles. Not surprisingly, axonal segregation in the bundles was not noticeably affected in the OSN-specific Sema3F KO (Takeuchi et al. 2010). Furthermore, using molecular markers for mature OSNs, Sullivan et al. (1995) showed that OSNs in the dorsal zone mature earlier than those in the ventral zone during embryonic development. Glomerular structures first emerge in the anterodorsal domain of the OB (Bailey et al. 1999). The olfactory map appears to expand ventrally as the OB grows during development. Where and how, then, does Sema3F interact with Nrp2$^+$ OSN axons? By using the inducible Tet-ON system with Doxycycline to enhance *Sema3F* expression, Takeuchi et al. (2010) successfully detected secreted Sema3F protein in the outer olfactory nerve layer of the anterodorsal OB. The olfactory nerve layer has been reported to serve as an axon-sorting area in which OSN axons defasciculate before converging at their final destination (Royal & Key 1999, Treloar et al. 2002). These observations point toward an intriguing possibility that a repulsive ligand, Sema3F, is secreted by early arriving dorsal-zone axons and is deposited in the anterodorsal OB to serve as a guidance cue to repel late-arriving ventral-zone axons that express the Nrp2 receptor.

In *Drosophila*, a similar situation has been reported for axon targeting of olfactory receptor neurons: Early arriving antennal axons expressing a repulsive ligand, Sema1A, are required for the targeting fidelity of late-arriving maxillary pulp axons that express Plxn-A, a receptor for Sema1A (Sweeney et al. 2007). Thus the same family of axon-guidance molecules plays a crucial role in neural map formation in both the mouse and fly olfactory systems (Miyamichi & Luo 2009). Unlike the mouse Sema3F, the *Drosophila* Sema1A is a transmembrane molecule that is not secreted; therefore, it mediates direct interactions among axons.

What guides pioneer OSN axons to the anterodorsal area? It has been reported that Robo2$^+$ dorsal-zone axons project to the dorsal region of the OB by repulsive interactions with secreted ligands (Cho et al. 2007, Nguyen-Ba-Charvet et al. 2008). One of the Robo2 ligands, Slit1, is detected in the septum and ventral OB during early developmental stages. These observations suggest that Robo2/Slit1 signaling also plays an important role in dorsal-ventral projection of OSNs. In the total KO for the Robo/Slit system, OSN axons can mistarget to surrounding non-OB tissues (Nguyen-Ba-Charvet et al. 2008). Repulsive interactions between Robo2 and Slit1 are probably needed to restrict the first wave of OSN projection to the anterodorsal OB. It has been reported that the embryonic OB represents the prospective dorsal OB because mitral cells at E14 are all positive for OCAM (a dorsal-region mitral cell marker) (Takeuchi et al. 2010). Over the course of embryonic development, the OB map appears to expand ventrally. Axonal projection of OSNs also occurs sequentially from the dorsomedial to the ventrolateral area, which may help to establish the map order in the OB along the dorsal-ventral axis. Spatiotemporal regulation of axonal projection of OSNs aided by Robo2 and Slit1, and the graded and complementary expression of Nrp2 and Sema3F, contributes to olfactory map formation along the dorsal-ventral axis (**Figure 3c**).

HOW DOES THE OLFACTORY MAP CONVERT FROM CONTINUOUS TO DISCRETE?

As mentioned above, a coarse olfactory map is generated by a combination of dorsal-ventral patterning, based on anatomical locations of OSNs, and anterior-posterior patterning, based on OR-derived cAMP signals. Developmental studies have shown that neighboring glomerular structures are intermingled before birth, and the discrete glomeruli emerge during the neonatal period (Conzelman et al. 2001, Potter et al. 2001, Sengoku et al. 2001). After OSN axons reach their approximate destinations in the OB, further refinement of the glomerular map needs to occur through fasciculation and segregation of axon termini in an activity-dependent manner.

Activity-Dependent Glomerular Segregation

To study how OR-specific glomerular segregation is controlled, investigators searched to find a group of genes with expression profiles that correlate with the expressed OR species. Using a transgenic mouse in which the majority of OSNs express a particular OR, they were indeed able to identify such genes (Serizawa et al. 2006): These genes include ones that code for homophilic adhesive molecules, e.g., Kirrel2 and Kirrel3. Mosaic gain of function and loss of function of these genes generate duplicated glomeruli even though the expressed OR species are identical (Serizawa et al. 2006; H. Takeuchi and H. Sakano, unpublished results), suggesting that they play a role in the attraction of similar axons. Yoshihara's group (Kaneko-Goto et al. 2008) recently reported another adhesive molecule, BIG2, which is expressed in axon termini of OSNs in an OR-specific manner and facilitates local sorting with unknown heterophilic binding partners. Repulsive molecules, such as ephrin-As and EphAs, are also expressed in a complementary manner in each subset of OSNs (Cutforth et al. 2003, Serizawa et al. 2006). Therefore, interactions between two subsets of axons, one that is ephrin-A^{high}/EphA^{low} and another that is ephrin-A^{low}/EphA^{high}, may be important for the segregation of dissimilar OSN axons. Although OR molecules at axon termini have been thought to mediate homophilic interactions of similar axons (Feinstein & Mombaerts 2004), ORs may not directly act as adhesion molecules. We assume that a specific set of adhesive and repulsive molecules, whose expression levels are determined by OR molecules, regulates the axonal fasciculation of OSNs (**Figure 3a**) (Serizawa et al. 2006). It is not clear at this point how many sets of sorting molecules are involved in glomerular segregation. However, several sets of adhesion/repulsion molecules should be enough to segregate neighboring glomerular structures.

Unlike the global targeting of OSN axons in embryos, local sorting appears to occur in an activity-dependent manner in neonatal animals.

Blocking neuronal activity by the expression of an inward rectifying potassium channel, Kir2.1, severely affects axonal convergence (Yu et al. 2004). Mice that are mosaic KOs for CNGA2 reveal the segregation of CNGA2-positive and -negative glomeruli for the same OR (Zheng et al. 2000, Serizawa et al. 2006). The expression levels of OR-correlated axon-sorting molecules are affected by the *CNGA2* mutation in OSNs. In the *CNGA2* KO, *Kirrel2* was downregulated, whereas *Kirrel3* was upregulated, indicating that these genes are transcribed in an activity-dependent manner (Serizawa et al. 2006).

The elimination of ectopic glomeruli is an important process for olfactory map refinement. Minor satellite glomeruli found in young animals disappear with age in an activity-dependent manner. In mice deficient for CNGA2 or whose naris is surgically occluded, ectopic glomeruli persist (Nakatani et al. 2003, Zou et al. 2004). In mosaic female mice (*CNGA2^{+/−}$*), CNGA2-negative OSN axons are eliminated in a competitive manner but survive in a noncompetitive manner provided by naris occlusion (Zhao & Reed 2001). It seems that both the elimination and segregation of glomerular structures occur in an activity-dependent manner and contribute to the refinement of the olfactory map.

The olfactory map is established by a combination of genetically determined targeting and activity-dependent refinement processes (Chen & Flanagan 2006). Activity-dependent refinement, which follows initial targeting, plays an important role in many other sensory systems during development (Goodman & Shatz 1993). In the visual system, initial mapping by graded guidance molecules is followed by Hebbian activity-dependent refinement. In the olfactory system, after OSN axons reach the OB, further refinement occurs through fasciculation and segregation of axon termini in an activity-dependent manner. How is it then that map formation is regulated in a stepwise fashion?

In the mouse olfactory system, axon-sorting molecules whose expression levels are regulated by OR-derived cAMP signals can be categorized into two types (**Figure 3a**). Type I molecules, including Nrp1 and Plexin-A1, are expressed at axon termini in a graded manner along the anterior-posterior axis in the OB. Type II molecules, including Kirrel2 and Kirrel3, are expressed at axon termini showing a mosaic pattern in the OB. In the ACIII-deficient mouse, Nrp1 and Kirrel2 are downregulated, whereas Plexin-A1 and Kirrel3 are upregulated, suggesting that both type I and type II genes are under the control of cAMP signals (Col et al. 2007; H. Takeuchi and H. Sakano, unpublished results). How is the same second messenger, cAMP, capable of regulating two types of genes that are seemingly independent of one another?

We provide the following explanation. cAMP signals for type I and type II genes appear to be derived from distinct sources. One demonstration of this idea is naris occlusion, which affects the expression of type II, but not of type I, genes. It is generally thought that naris occlusion reduces the inhalation of odorants and, consequently, the signal intensity elicited by them. Type I genes are likely to be driven by intrinsic signals such as ligand-independent basal signaling by ORs (Imai & Sakano 2008). Differential regulation of type I and type II molecules may also result from the different subcellular localization of ORs, namely, cilia (for type II) versus axons (for type I). How are the cAMP signals independently processed within the OSNs for type I and type II genes? Both types are regulated by cAMP signals at distinct stages of OSN maturation. Both G_s and G_{olf} activate ACIII; however, the expression profiles of these G proteins are quite different: G_s is predominantly expressed in immature OSNs, whereas G_{olf} is expressed in mature OSNs (Menco et al. 1994, Chesler et al. 2007). As G_s and G_{olf} are biochemically quite similar, the distinction may derive from different subcellular distributions of the stimulatory subunits. Furthermore, downstream signaling components also differ for the expression of type I and type II genes (Imai et al. 2006, Serizawa et al. 2006). Type I genes are affected by dominant-negative PKA. In contrast, CNGA2 deficiency affects type II genes. Therefore, type I axon-guidance events, which form a course map, are dependent on PKA signaling, whereas type II events, which refine the final map, are dependent on changes in membrane potential or ion flux. It is therefore possible that OR-mediated cAMP signals may influence early guidance decisions via PKA, whereas OR-mediated activity may influence later events via calcium signaling (**Figure 3a**).

Because type I and type II genes are differentially expressed in immature and mature OSNs, respectively, olfactory map formation might occur in a stepwise fashion during development. A coarse continuous map is formed first, and axons are later sorted locally to make the map discrete. For anterior-posterior projection, pre-target sorting in axon bundles contributes to glomerular map formation (Imai et al. 2009). For dorsal-ventral projection, sequential arrival and repulsive interactions of OSN axons play important roles in the formation of the map order in the OB (Takeuchi et al. 2010). After the arrival of OSN axons to the target OB, the local sorting occurs in an activity-dependent manner (Serizawa et al. 2006).

Neuronal Activity that Regulates Glomerular Segregation

The stepwise regulation of axon guidance appears to be a general feature of neural map formation for both continuous and discrete maps. During olfactory development, the glomerular map is refined after axons are guided to their approximate destination. Similarly, the topographic map in the visual system is established by a combination of genetically determined targeting and activity-dependent refinement processes (Goodman & Shatz 1993, McLaughlin & O'Leary 2005, Huberman et al. 2008).

What is the nature of neuronal activity in neural map formation? In the visual system, spontaneous activity is needed for the refinement of the retinotopic map (McLaughlin et al. 2003, McLaughlin & O'Leary 2005). Prior to the onset of visual experience, the vertebrate retina generates spontaneous waves of action potentials, which spread across the retina, correlating activities among the neighboring reti-

nal ganglion cells. When the retinal waves are blocked in the early neonatal animal, retinal ganglion cell axons cannot form the dense termination zone (McLaughlin et al. 2003). In the mouse olfactory system, expression of type I molecules for anterior-posterior projection seems to be regulated by intrinsic signals or the basal activity of OR molecules (Imai & Sakano 2008). It has been reported that CNGA2 KO affects the expression of type II, but not type I, molecules (Serizawa et al. 2006), suggesting that glomerular segregation is regulated by neuronal activity driven by CNG channels. Curiously, glomerular structures still form for some ORs even in CNGA2 KOs (Lin et al. 2000). This may suggest that additional neuronal activities are involved in the glomerular formation.

It has been reported that spontaneous firing is recorded in CNGA2 KO mice (Brunet et al. 1996). When the inward rectifying potassium channel Kir2.1 is overexpressed in OSNs to prevent spontaneous activity, axonal convergence is severely affected (Yu et al. 2004). Interestingly, axonal convergence is not perturbed in the mutant mouse in which neurotransmitter release has been blocked with tetanus-toxin light chain (Yu et al. 2004). These observations suggest that activity-dependent axonal fasciculation of OSNs does not require the postsynaptic activity of mitral and tufted cells in the OB. How are the spontaneous activities of OSNs correlated with expressed OR species? In the visual system, spontaneous retinal waves propagate across spatially so that neighboring cells can correlate firing. However, OSNs expressing the same OR species are randomly distributed across the OE. It is unlikely that similar spontaneous waves are utilized in the olfactory system. In *Drosophila*, expressed ORs regulate the spontaneous firing rate (Hallem et al. 2004). It is possible that mouse OSNs exhibit variable rates of spontaneous activities, depending on the expressed OR species. Because activity and calcium influx regulate gene expression in other systems (Itoh et al. 1995, Dolmetsch et al. 1997, Chang & Berg 2001, Borodinsky et al. 2004, Hanson & Landmesser 2004), mouse

OSNs might also convert the firing rate of spontaneous activities to expression profiles of axon-guidance/sorting molecules.

HOW IS THE OLFACTORY MAP ORGANIZED FOR VARIOUS ODORANTS?

As mentioned above, an individual glomerulus in the OB represents a single OR species, as OSN axons expressing a given OR converge to a fixed projection site in each olfactory map (Mombaerts 2006, Imai & Sakano 2007). Within glomeruli, OSN axons generate excitatory synaptic connections on primary dendrites of mitral and tufted cells (**Figure 4a**) (Shepherd et al. 2004). Because these second-order neurons project a single primary dendrite to a neighboring glomerular structure, each glomerulus and its associated mitral and tufted cells form a single OR channel or OR module in the OB. In this section, we discuss the functional significance of the compartmental organization of OR channels as it relates to the processing of odor information in the OB and OC.

Each Glomerular Map is Subdivided into Three Distinct Domains: D_I, D_{II}, and V

The right and left OBs each have two mirror-symmetric maps: a lateral map and a medial map (**Figure 4c**). The lateral map is located at the rostro-dorso-lateral half of the OB and receives olfactory axon inputs from OSNs in the rostro-dorso-lateral part of the OE. The medial map is situated at the caudo-ventro-medial half of the OB and receives olfactory axon inputs from the caudo-ventro-medial part of the OE (Astic & Saucier 1986, Saucier & Astic 1986, Schoenfeld et al. 1994, Nagao et al. 2000). The lateral and medial maps are precisely and topographically linked via the axon collaterals of tufted cells (Schoenfeld et al. 1985, Liu & Shipley 1994, Belluscio et al. 2002, Lodovichi et al. 2003, Zhou & Belluscio 2008). Therefore, intimate functional interactions have been suggested between the pair of glomerular channels representing the same OR species.

Why does each OB possess duplicated maps? One plausible explanation is that the lateral and medial maps process different aspects of OR signals. For example, the two maps may be differentiated to separately handle rapidly adapting and slowly adapting signals, similar to the parallel pathways in the somatic sensory system (Kaas et al. 1981). Alternatively, it is possible that the two maps are specialized to handle odor information at different intervals of the respiration cycle: One map processes odor signals primarily from orthonasal inputs during inhalation, and the other processes signals from retronasal inputs during exhalation. A third possibility is that a simultaneous activation of the pair of glomeruli representing the same OR is necessary for OR-specific behavioral responses. It is also possible that the two redundant maps are needed in case one does not function due to the transient local abnormality of the corresponding OE. Functionally, a single map alone may perform odor-information processing perfectly well.

Grouping of glomeruli into specific compartments, such as domains at stereotypical positions in the OB, has been reported based on the OR species (Zhang & Firestein 2002). **Figure 4b** shows a schematic diagram of the domain arrangement in the lateral view of the OB: Glomeruli are grouped into two domains, dorsal and ventral, arranged in parallel with the anterior-posterior axis of the OB. Dorsal-domain glomeruli receive signals from dorsal-zone OSNs in the OE and represent dorsal-zone ORs, whereas ventral-domain glomeruli receive inputs from ventral-zone OSNs and represent ventral-zone ORs (Yoshihara et al. 1997). Over the course of embryonic development, the dorsal domain is formed first in the OB, and the glomerular map expands ventrally (Takeuchi et al. 2010).

The dorsal domain in the OB is further divided into two domains, D_I and D_{II}, according to the expressed OR classes (**Figure 4b**) (Tsuboi et al. 2006, Kobayakawa et al. 2007, Bozza et al. 2009). The D_I and D_{II} domains are also arranged in parallel with the anterior-posterior axis of the OB. D_I glomeruli represent

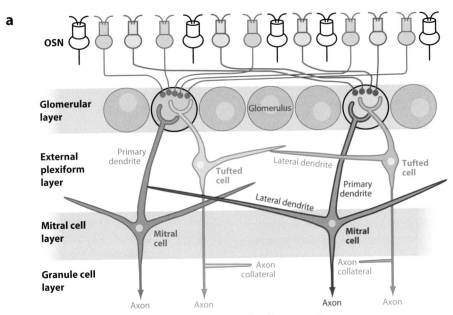

a

OSN

Glomerular layer

Glomerulus

External plexiform layer

Primary dendrite

Tufted cell

Lateral dendrite

Lateral dendrite

Primary dendrite

Tufted cell

Mitral cell layer

Mitral cell

Mitral cell

Granule cell layer

Axon collateral

Axon collateral

Axon

Axon

Axon

Axon

Axonal projection to the olfactory cortex

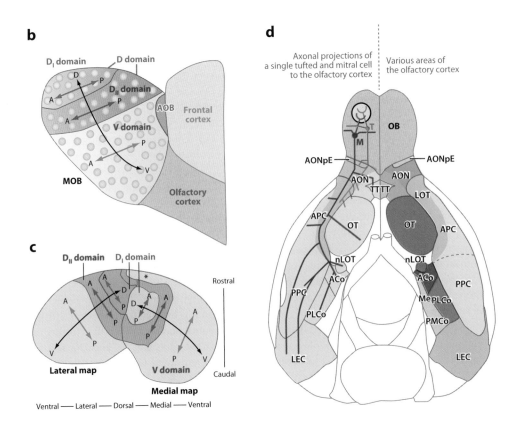

b

D_I domain

D domain

D_{II} domain

AOB

Frontal cortex

MOB

V domain

Olfactory cortex

c

D_{II} domain

D_I domain

Rostral

Lateral map

V domain

Medial map

Caudal

Ventral — Lateral — Dorsal — Medial — Ventral

d

Axonal projections of a single tufted and mitral cell to the olfactory cortex

Various areas of the olfactory cortex

OB

AONpE

AONpE

AON

AON

LOT

APC

OT

OT

APC

nLOT

nLOT

ACo

ACo

PPC

PPC

MePLCo

PLCo

PMCo

LEC

LEC

class I (fish-type) ORs, whereas D_{II} glomeruli represent class II (terrestrial-type) ORs. It is interesting that OSNs expressing class I ORs and those expressing class II ORs are intermingled in the dorsal zone of the OE. However, the two subsets of OSNs project their axons to separate dorsal domains. In the ventral zone of the OE, most OSNs express class II ORs. Thus glomeruli in the ventral domain mostly represent class II ORs, with the exception of two class I ORs (Kobayakawa et al. 2007). It has been proposed that the ventral domain consists of at least three subdomains (Mori et al. 1999, Takahashi et al. 2010) that receive inputs from previously reported zones 2–4 of the OE (Ressler et al. 1993, Vassar et al. 1993). However, owing to the lack of appropriate molecular markers, it is not clear how the subdomains are spatially arranged in the ventral domain of the glomerular map. It has been reported that the sequential arrival of OSN axons from the dorsal to the ventral zone in the OE plays a key role in arranging glomeruli along the dorsal-ventral axis in the

OB (Takeuchi et al. 2010). Like the lateral map, the medial map is also divided into three domains (D_I, D_{II}, and V) arranged in parallel to the anterior-posterior axis. A comparison between the lateral and medial maps using the unrolled maps (**Figure 4c**) indicates that each map has its own anterior-posterior/dorsal-ventral axis coordinates, although the two maps are roughly arranged in a mirror-symmetric manner.

Molecular Feature Clusters of Glomeruli in the Olfactory Bulb

Another way of grouping glomeruli in the olfactory map is by the odorant selectivity, or molecular receptive range (MRR), of individual glomeruli. Mapping the spatial representation of odorants and the MRR properties of glomeruli demonstrated that individual glomeruli respond to a range of odorants sharing common molecular features. Furthermore, glomeruli with similar MRR properties gather to form a molecular feature cluster

---

Figure 4

Structure and organization of the olfactory bulb (OB) and the olfactory cortex (OC). (*a*) Glomerular modules in the mammalian OB. Each glomerulus receives converging axonal inputs from olfactory sensory neurons (OSNs) expressing the same species of odorant receptor (OR). Within each glomerulus, olfactory axons make excitatory synaptic terminals on the primary dendrites of mitral cells and tufted cells, the two types of projection neurons in the OB. Each glomerulus together with OSN axon inputs and associated mitral and tufted cells form an OR channel. Tufted cells project lateral dendrites in the superficial half of the external plexiform layer, whereas mitral cells extend long lateral dendrites to the deeper half of the external plexiform layer. Mitral cells and tufted cells project axons to the OC. (*b*) A lateral view of the domain organization of glomeruli in the lateral map of the rodent main olfactory bulb (MOB). Double-headed colored arrows indicate the anterior-posterior (A-P) axis of each domain. Black arrows indicate the dorsal-ventral (D-V) axis of the lateral map. Abbreviations: AOB, accessory olfactory bulb; D_I, class I part of the D domain; D_{II}, class II part of the D domain. (*c*) A dorsal centered view of an unrolled flattened map of glomerular layer of the OB. Double-headed colored arrows indicate the A-P axis of each domain of the lateral and medial maps. Black arrows indicate the D-V axis of the lateral and medial maps. In mice, an individual OR is typically represented by a pair of glomeruli: one in the lateral map and the other in the medial map. However, for a small subset of ORs, each OR is represented only by a single glomerulus. Some of these glomeruli are located in the tongue-shaped region indicated by an asterisk. (*d*) Schematic diagrams illustrating the axonal projection of mitral and tufted cells (*left*) and various areas of the OC (*right*). The left diagram shows the axonal projection pattern of a single mitral cell (M) and a single tufted cell (T) to the OC. The OC is shown by the light-blue shaded area. A single mitral cell projects axons to nearly all areas of the OC, whereas a single tufted cell projects axons only to restricted parts of the anterior areas of the OC. The right diagram indicates various regions of the OC that receive axonal inputs from the MOB and AOB. Abbreviations: ACo, anterior cortical amygdaloid nucleus; AON, anterior olfactory nucleus pars principalis; AONpE, anterior olfactory nucleus pars externa; APC, anterior piriform cortex; LEC, lateral entorhinal cortex; lot, lateral olfactory tract; Me, medial amygdaloid nucleus; nLOT, nucleus of lateral olfactory tract; OT, olfactory tubercle; PLCo, postero-lateral cortical amygdaloid nucleus; PMCo, postero-medial cortical amygdaloid nucleus; PPC, posterior piriform cortex; TT, tenia tecta. Figure (*d*) adapted from Luskin & Price (1983).

at a stereotyped position in the glomerular sheet (Takahashi et al. 2004a, Mori et al. 2006, Johnson & Leon 2007, Johnson et al. 2009). Molecular feature clusters of glomeruli have been studied in mice by combining optical imaging of intrinsic signals and OCAM immunohistochemistry to determine the domain organization of the OB (Matsumoto et al. 2010).

As shown in **Figure 6a**, cluster A, which responds to fatty acids and alkylamines, is located in the lateral part of the D_I domain. Cluster B, which responds to aliphatic alcohols, aliphatic ketones, and phenyl ethers, is located in the anterior part of the D_{II} domain. Glomeruli in cluster C, which respond to phenols and phenyl ethers, are located just posterior to cluster B in the D_{II} domain. Cluster D glomeruli are activated by aliphatic and aromatic ketones and are located posterior to cluster C. In addition, cluster J glomeruli for thiazoles and thiazolines are located in the most caudal part of the D_{II} domain. Thus clusters B, C, D, and J are invariably arranged from the anterior to the posterior region in the D_{II} domain. Molecular feature clusters of glomeruli are also found in the V domain of the OB (Igarashi & Mori 2005). The stereotyped arrangement of the molecular feature clusters suggests that they represent specific subsets of ORs that have similar MRR properties (Matsumoto et al. 2010). Functional molecular feature clusters appear to be organized in a genetically determined domain arrangement in the OB. The clustering of glomeruli with similar MRR properties implies that glomeruli in the same cluster represent ORs that have evolved from a common ancestor. Duplication and diversification of an ancestral OR gene may have created new glomeruli in a nearby region of the OB, which correspond to distinct but related odor ligands. We assume that newly evolved glomeruli such as the one responsible for fear responses to predators underwent positive selection during evolution to ensure an animal's survival.

The arrangement of molecular feature clusters (B, C, D, and J in the D_{II} domain) may be regulated by glomerular positioning in the anterior-posterior axis using Neuropilin-1 (Nrp1): D_{II} glomeruli in the posterior area receive Nrp1[high] OSN axons, whereas those in the anterior area receive Nrp1[low] axons (Taniguchi et al. 2003, Imai et al. 2009). Glomerular positioning may also regulate their more precise spatial arrangement within the molecular feature cluster. For example, in cluster A, glomeruli for fatty acids and aliphatic aldehydes with different carbon chain lengths are arranged with a gradual shift along the anterior-posterior axis within the D_I domain (Meister & Bonhoeffer 2001, Takahashi et al. 2004a, Uchida et al. 2000). Nrp2-regulated dorsal-ventral positioning of glomeruli (Takeuchi et al. 2010) may also be responsible for cluster arrangement.

Interactions Among Odorant-Receptor Channels via Local Neuronal Circuits in the Olfactory Bulb

Mitral and tufted cells associated with a particular OR channel interact with those associated with other OR channels via interneurons such as periglomerular cells, granule cells, and short axon cells (Ramón y Cajal 1955, Shepherd et al. 2004, Willhite et al. 2006). Thus the spatial arrangement of glomeruli within the molecular feature clusters and domains must have implications for interactions among OR channels. Mitral cells have long lateral dendrites in the deeper half of the external plexiform layer (**Figure 4a**) and have dendrodendritic reciprocal synaptic connections with a subset of granule cells. In contrast, tufted cells project their lateral dendrites to the superficial half of the external plexiform layer and have dendrodendritic reciprocal connections with a different subset of granule cells that project apical dendrites to the superficial half of the layer. The laminar segregation of mitral and tufted cell dendrites suggests that mitral cells preferentially interact with other mitral cells via dendrodendritic synapses with granule cells, whereas tufted cells mainly interact with other tufted cells via dendrodendritic synapses (Mori 1987).

Mitral cells associated with glomeruli for trimethyl-thiazoline (TMT), located in cluster

J of the D_{II} domain, project their lateral dendrites not only to neighboring cluster D, but also to other clusters within the D_{II} domain (H. Matsumoto, K. Igarashi, K. Mori, unpublished results). Some of the lateral dendrites project to the D_I domain and even to the ventral domain. These observations suggest that mitral cells associated with other OR channels within reach can modify mitral cell activity via dendrodendritic synapses with granule cells. Dendrodendritic synaptic interactions between mitral cell's lateral dendrites and granule cell's dendrites enhance the odorant selectivity of the output neurons by lateral inhibition (Yokoi et al. 1995, Tan et al. 2010). In addition, the dendrodendritic synaptic interaction causes synchronized discharges in the mitral and tufted cells associated with different glomeruli (Kashiwadani et al. 1999, Schoppa 2006). This facilitates synaptic activation of OC pyramidal cells that receive axonal input from different OR channels. Studies on the interactions among mitral and tufted cells associated with different OR channels will help us understand the neuronal logics in the OC, which integrate odor signals derived from different ORs.

HOW IS THE OLFACTORY MAP ORGANIZED FOR VARIOUS FUNCTIONS AND BEHAVIORS?

In mammals, neural maps in the sensory cortices encode an object's essential information to elicit the proper behavioral responses (Kaas et al. 1981). In primate visual cortices, the dorsal parietal pathway handles the motion of objects, whereas the ventral inferior-temporal pathway processes the form of objects (Merigan & Maunsell 1993). A major challenge of OB mapping is to understand the functional significance of the spatial organization of molecular feature clusters in the OR map with regard to the translation of odor signals into behavioral responses by the central olfactory system (Doty 1986). Olfactory inputs to a given set of OR channels in the OB appear to drive specific behavioral responses (Mainen 2007, Mori et al. 2009, Sakano 2010). In this section we

discuss how the olfactory map is organized for different functions and behaviors.

Innate Versus Learned Odor Processing in the Olfactory Bulb

The mammalian olfactory system mediates various innate responses, including aversive behaviors to spoiled foods and fear responses to predator odors. Because a particular odorant interacts with many different OR species, multiple sets of glomeruli are activated in both the dorsal and ventral domains of the OB (Xu et al. 2003, Johnson & Leon 2007) (**Figure 1d**). However, little is known about how topographic information in the OB is transmitted to and interpreted in the brain to elicit various behaviors. To address these questions, Kobayakawa et al. (2007) generated mutant mice with OSNs in a specific area of the OE ablated with diphtheria toxin (**Figure 5a**). They demonstrated that, in dorsal-zone-depleted mice (ΔD), the dorsal domain of the OB was devoid of glomerular structures. The mutant mice failed to show innate responses to aversive odorants, such as 2-methylbutyric acid from spoiled foods and TMT from the fox anal gland (**Figure 5b**). Interestingly, the ΔD mutant mice were capable of detecting these odorants and could be conditioned for aversion using the remaining glomeruli. It was once thought that the dorsal- and ventral-domain glomeruli contribute equally to odor-information processing in the glomerular map. However, the main olfactory system in the mouse seems to comprise at least two functional modalities: one for innate odor responses with hard-wired circuits and another for learned responses based on memory (**Figure 5c**) (Kobayakawa et al. 2007, Mainen 2007).

Functional compartmentalization of the glomerular map is also seen in fish OB, suggesting that it is a general principle conserved among different vertebrate species. In fish, food odors and social odors (bile salts) (Li et al. 2002) are represented in different domains of the OB (Døving et al. 1980, Hara & Zhang 1998, Nikonov & Caprio 2001, Nikonov et al.

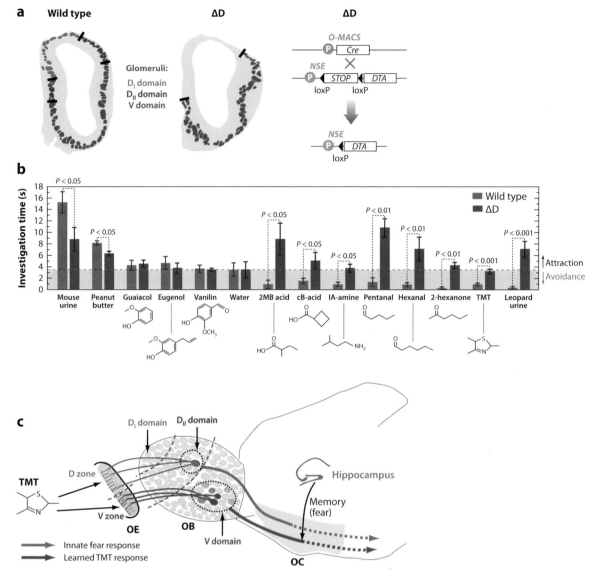

Figure 5

Two distinct neural pathways in the mouse olfactory system (Kobayakawa et al. 2007). (*a*) Coronal sections of the olfactory bulbs (OBs) isolated from the wild-type mouse (*left*) and ΔD mutant (*middle*). Distributions of glomeruli are schematically shown. In the ΔD mutant, olfactory sensory neurons in the dorsal (D) zone are ablated by targeted expression of the diphtheria toxin gene (*right*). In the ΔD mutant, the dorsal domain of the OB was devoid of glomerular structures. (*b*) Innate olfactory preference tests. The duration for which the mouse investigated the scented filter paper was measured for various odorants. Mean investigation times (± s.e.m.) are shown for each odorant during the 3-min test period with the wild-type (*brown*) and ΔD mutant (*magenta*) mice. The investigation time for water was used as a criterion for attraction versus avoidance responses (*dashed red line*). Investigation times less than this criterion (avoidance responses) are marked by the gray shaded area. (*c*) Schematic diagram of the neural pathways in the mouse. A predator's odorant, trimethyl-thiazoline (TMT), is detected by two separate sets of ORs: One is expressed in the D zone, and the other is in the ventral (V) zone of the olfactory epithelium (OE). TMT activates two different neuronal pathways: one for the innate fear response (*red*) and the other for the learned fear response based on memory (*blue*). Abbreviation: OC, olfactory cortex.

2005, Sato et al. 2005, Koide et al. 2009) and are processed in distinct regions of the forebrain (Nikonov et al. 2005).

Despite the complete absence of glomeruli, the dorsal domain of the ΔD mutant OB showed otherwise normal cytoarchitecture with distinct layers (Kobayakawa et al. 2007). Individual mitral cells in the dorsal domain of the ΔD mutant emitted several dendrites into the external plexiform layer but did not form dendritic terminal tufts, suggesting that these mitral cells lacked synaptic inputs from OSN axons. Ventral-zone axons of OSNs did not form ectopic contacts with the dorsal-domain mitral cells in the ΔD mutant. No glomeruli were found in the dorsal domain of the ΔD mutant during the course of embryonic development. These results indicate that mitral cells are specified as dorsal-domain and ventral-domain subsets before the axonal projection of OSNs occurs. The dorsal-domain mitral cells seem to be committed to receiving olfactory inputs exclusively from dorsal-zone OSNs. These findings are unexpected because in other sensory systems, such as retino-tectal projection (Horder 1971) and barrel formation (Van der Loos & Woolsey 1973), competing axons eventually occupy vacant projection sites. In the olfactory system, the OB may not simply be a projection screen to form a glomerular map, but may instead have region-specific functions that are genetically predetermined.

Functional Modules in the Olfactory Map

As mentioned above, individual odorants activate specific sets of glomeruli distributed in various areas of the OB (**Figure 1d**) (Xu et al. 2003, Mori et al. 2006, Johnson & Leon 2007, Johnson et al. 2010). Furthermore, each object emits scores of different odorants with a unique combination and different concentrations. It has been difficult to understand the functional significance of compartmental organization of the glomerular map with respect to the coding of an object's odor information. However, for some innate responses, such as to predators'

odors and spoiled food smells, only one or a few core odorants are sufficient to induce specific behavioral responses (Novotny 2003). Functional compartmentalization of the glomerular map suggests that the spatial organization of the OB has evolved to adapt the map to an animal's behavioral responses to the core odorants.

Predators of rodents emit specific chemical classes of odorants. These predator odors signify danger and induce fear responses, such as freezing (Dielenberg & McGregor 2001; Hebb et al. 2002, 2004). Although a fox emits numerous odorants, TMT alone is capable of inducing fear responses in mice (**Figure 6b**). TMT activates glomeruli in cluster J of the D_{II} domain and glomeruli in the ventral domain. Ablation of glomeruli in the D_{II} domain abolished TMT-induced fear responses in mice (Kobayakawa et al. 2007). We therefore assume that a specific set of ORs has been selected to detect the fear quality of TMT during evolution and that these glomerular locations confined to cluster J contain a part, or all, of the functional domain for innate fear responses to TMT. Because the ΔD mutant mice also fail to demonstrate fear responses to the cat odor, another core odorant that induces fear responses to the cat may have the corresponding glomeruli in the dorsal domain.

Many odorants in mouse urine are known to activate glomeruli in clusters J and D, and these clusters appear to contain glomeruli responsive for innate fear and social behaviors. In addition to responding to TMT, glomeruli in cluster J respond to the odorant 2-sec-butyl-dihydrothiazole (SBT) found in mouse urine. SBT has a molecular structure similar to TMT and is known to induce male-male aggressive behaviors (Novotny et al. 1985). Another set of cluster J glomeruli, responsive to dehydro-exo-brevicomine that has a molecular feature distinct from TMT and SBT, also induces intermale aggressive behaviors (Novotny et al. 1985). These observations suggest that cluster J contains glomeruli responsible not only for predator odor–induced fear responses, but also for urine odor–induced aggressive responses (**Figure 6b**).

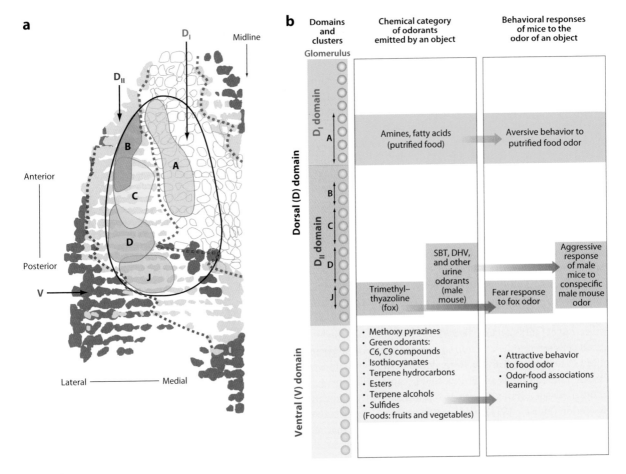

Figure 6

Molecular feature clusters of glomeruli. (*a*) The spatial arrangement of molecular feature clusters (A–D and J) in the D_I and D_{II} domains of the mouse olfactory bulb (OB). The schematic presents a dorsal view of the left OB: glomeruli in the D_I domain (*open circles*), in the D_{II} domain (*light gray circles*), and in the V domain (*dark gray circles*). Molecular feature cluster A is located in the lateral region of the D_I domain. Clusters B, C, D, and J are located in order from the anterior to posterior regions of the D_{II} domain. Red broken lines indicate the boundary between the D_I and D_{II} domains. Figure adapted from Matsumoto et al. (2010). (*b*) Domains and clusters of the odorant receptor (OR) map of the mouse OB. (*Left column*) Domain and cluster arrangement of OR-representing glomeruli (*gradient spheres*). Molecular feature clusters of glomeruli in the dorsal domain are represented by arrows. (*Middle column*) The chemical category of odorants that activate the domain and cluster. The object that emits these odorants is shown in parentheses. (*Right column*) The behavioral responses of mice to the odor of an object. Each compartment of the OR map appears to play a key role in relating specific molecular features of odorants to behavioral responses of the mouse to the odors of the object. The relationships are shown by thick arrows with compartment-specific colors. Because the compartmental organization within the ventral domain is not well understood, the ventral domain is shown in a relatively small scale. Green odorants include *cis*-3-hexenol and *trans*-2-hexenal. Abbreviations: DHB, dehydro-exo-brevicomine; SBT, 2-sec-butyl-dihydrothiazole.

Glomeruli that Elicit Aversive Responses to Spoiled Food Smells

Putrid odors of spoiled foods and putrefied objects serve as important warning signals. A variety of amines are produced in spoiled foods by decarboxylation of free amino acids with bacterial enzymes. Spoiled foods also produce fatty acids and aliphatic aldehydes by lipid oxidation. Amines and fatty acids induce innate aversive responses in rats and mice (Dielenberg & McGregor 2001) and activate multiple glomeruli in various areas of the OB. Interestingly, only cluster A glomeruli in the D_I domain are responsible for detecting the innate aversion-inducing quality of amines and fatty acids (Kobayakawa et al. 2007). Although amines and fatty acids possess different molecular features, many amine-responsive glomeruli are also activated by fatty acids (Takahashi et al. 2004b), suggesting that cluster A glomeruli can be activated either by a carboxyl group or by an amine group. Because the glomeruli in the D_I domain represent fish-type class I ORs, ancestral class I ORs used to detect water-soluble amino acids may have evolved in mammals to detect amines and carboxylic acids in the air. At least for the dorsal domain in the OB, molecular feature clusters appear to have evolved not only by the clustering glomeruli that represent ORs with similar MRR properties, but also by adopting those glomeruli that represent functionally related ORs. Functional compartmentalization of the OB implies that distinct domains in the OB give rise to distinct flows of olfactory information in the central olfactory system, each mediating a specific behavioral response. However, whether similar compartmentalization is also found in the ventral domain in the glomerular map has not been studied.

Projection of Mitral and Tufted Cells to the Olfactory Cortex

How does the OC interpret odor maps in the OB and translate odor information into behavioral responses? Two types of projection neurons within each glomerulus, the mitral and tufted cells, receive synaptic inputs from OSNs (**Figure 4a** and **d**). Because each mitral/tufted cell sends a single primary dendrite to a specific glomerulus, each cell appears to be specialized to handle olfactory inputs from a single OR species. Mitral cells have somata in the mitral cell layer, whereas somata of tufted cells are scattered in the external plexiform layer (**Figure 4a**). Tufted cells are further classified into three subtypes: internal tufted cells (which have somata in the deeper one-third of the external plexiform layer), middle tufted cells

(with somata in the superficial two-thirds of the external plexiform layer), and external tufted cells (with somata at the border between the external plexiform layer and the glomerular layer) (Ramón y Cajal 1955, Macrides & Schneider 1982, Hayar et al. 2004).

Because mitral and tufted cells project their axons to the OC, odor information is conveyed from the OB to the OC via two parallel projection neuron pathways (Haberley & Price 1977). By labeling axons of individual mitral and tufted cells with known functions, Igarashi et al. (K. Igarashi, N. Ieki, M. An, Y. Yamaguchi, S. Nagayama, et al., unpublished results) demonstrated that the parallel pathways of mitral and tufted cells transmit distinct aspects of odor information to different targets in the OC (Nagayama et al. 2004, Griff et al. 2008). Compared with mitral cells, tufted cells demonstrate odor responses with a lower concentration threshold, wider dynamic range with regard to odor concentrations, higher firing frequency, and higher temporal resolution (K. Igarashi, N. Ieki, M. An, Y. Yamaguchi, S. Nagayama, et al., unpublished results). Thus, for low-concentration odorants, only the tufted cell pathway conveys sensory information to the OC. The parallel pathways to the OC suggest that they give rise to distinct streams of informational processing in the higher cortical areas and subcortical regions beyond the OC.

It is interesting that projection neuron axons demonstrate pretarget axon sorting. Within the bundles of the lateral olfactory tract, axons of mitral cells are confined to the middle compartment, whereas those of immature mitral cells are sorted into the superficial compartment (Inaki et al. 2004). Axons of middle tufted cells are found in the deep-medial compartment (Matsutani et al. 1989). As is the case for OSN projection (Imai et al. 2009), pretarget axon sorting appears to play an important role in the projection of second-order neurons to different targets in the OC. Various areas of the rat OC are schematically shown in **Figure 4d** (Heimer 1968, Haberly & Price 1977, Neville & Haberly 2004). Mitral/tufted cells in the accessory OB and projection neurons in the main

OB demonstrate segregated axonal projection to the OC. OC areas receiving axons from the accessory OB include the medial amygdaloid nucleus (Me) and posteromedial cortical amygdaloid nucleus (PMCo). OC areas receiving axons from the main OB include anterior olfactory nucleus pars externa (AONpE), anterior olfactory nucleus pars principalis (AON), tenia tecta (TT), anterior piriform cortex (APC), olfactory tubercle (OT), posterior piriform cortex (PPC), nucleus of lateral olfactory tract (nLOT), anterior cortical amygdaloid nucleus (ACo), posterolateral cortical amygdaloid nucleus (PLCo), and lateral entorhinal cortex (LEC) (Neville & Haberly 2004).

Both mitral and tufted cells project their axons to the anterior areas of the OC, including the anterior olfactory nucleus pars externa (AONpE), anterior olfactory nucleus pars principalis (AON), anterior piriform cortex, and olfactory tubercle (OT). In each anterior area, mitral and tufted cells differentially project their axons to separate parts (K. Igarashi, N. Ieki, M. An, Y. Yamaguchi, S. Nagayama, et al., unpublished results). In contrast, posterior areas of the OC receive inputs only from mitral cells. Single-cell labeling demonstrated that a TMT-responsive mitral cell projected its axon to virtually all areas of the OC. Unlike mitral cells, a single TMT-responsive tufted cell projected its axon to focal targets only in the AONpE, AON, anterior piriform cortex, and OT. These observations suggest that TMT-derived information carried by tufted cell axons is transmitted to specific target neurons in the anterior areas of the OC.

Differential Processing of Odor Information in Distinct Areas of the Olfactory Cortex

Each area of the OC is thought to be involved in different attributes of informational processing (Neville & Haberly 2004). Namely, different areas of the OC may interpret the odor map of the OB in distinct manners. Recent development of trans-synaptic tracing has begun to reveal the manner of axonal projection of OB neurons to the OC (Miyamichi et al. 2011). However, the connectivity pattern of mitral and tufted cells to each area of the OC remains largely unknown in different domains and clusters of the glomerular map. To elucidate the axonal connectivity of OB neurons to numerous areas in the OC, it is necessary to selectively label the mitral and tufted cell axons in a specific compartment of the OB, or only those neurons that project to a specific target in the OC (Miyamichi et al. 2011). Here we discuss the differential functions of three areas with respect to odor-information processing: the AONpE, piriform cortex (PC), and OT.

The anterior olfactory nucleus is the most rostral area in the OC. Anatomical studies have revealed highly topographic axonal projection of mitral and tufted cells to the rostral part (pars externa) of the AON (Schoenfeld & Macrides 1984, Yan et al. 2008) in terms of the dorsal-ventral axis of the glomerular map. A recent study demonstrated that neurons in the AONpE detect the localization of odor sources by comparing ipsi-nostril and contra-nostril inputs of the same odorant category (Kikuta et al. 2010). This observation suggests that topographic axonal projection may serve as the basis for the binostril comparison of the same odorant category. TMT-responsive tufted cells were found to project their axons to a specific part of AONpE. This may indicate that the directional localization of the fox odor is detected by the neuronal circuit in that part of the AONpE.

In contrast to the highly topographic projection described above, axonal projection of mitral cells to the PC appears to be sparsely distributed and highly overlapping (Illig & Haberly 2003, Neville & Haberly 2004, Stettler & Axel 2009). Convergence of signals from various OR channels onto individual cortical neurons may be important in processing odor information within the PC (Haberly 2001). Based on these observations, we speculate the following: (a) For the detailed olfactory discrimination of individual objects, integration of information is necessary from a specific combination of odorant categories (Yoshida & Mori 2007). (b) Connectivity patterns of mitral cell axons to the PC may reveal a key logic for the integration of odor signals from many different OR channels, although the detailed connectivity patterns of individual mitral cells to the PC are not well understood. Individual pyramidal cells in the PC give rise to massive axon collaterals (association fibers) that form synapses on other pyramidal cells, not only in the vicinity of the parent neuron, but also across wide areas of the PC (Neville & Haberly 2004). To elucidate odor signal processing in the PC, one should study the logics for the axonal connections of mitral cells, as well as those for the association fibers. Understanding the coordination of mitral afferent inputs and associational fiber inputs onto individual pyramidal cells may be critical in the functional analysis of the PC.

The OT, together with the accumbens referred to as the ventral striatum, is assumed to be involved in the induction of appetitive and fearful motivational behaviors (Ikemoto 2007). Output neurons of the OT project their axons to the ventral pallidum, whose neurons in turn send axons to dopaminergic neurons in the ventral tegmental area. Therefore, food odor information processed in the OT appears to influence the dopaminergic circuits for reward expectation. The OT comprises two cytoarchitectually distinct subregions: the cortical region and the cap region. Layer II of the cortical region is packed with medium-sized spiny neurons, whereas that of the cap region contains dwarf cells and small pyramidal cells. TMT-responsive tufted cells send their axons preferentially to the cap region, whereas TMT-responsive mitral cells project to the cortical region (K. Igarashi, N. Ieki, M. An, Y. Yamaguchi, S. Nagayama, et al., unpublished results). These results suggest that the cortical region and the cap region of the OT are functionally distinct, processing different aspects of odor information related to motivational behaviors.

CONCLUSION AND FUTURE DIRECTIONS

In this review, we discuss how the glomerular map is formed and how the odor map is interpreted in the brain to elicit various behaviors. In the mouse olfactory system, much of dorsal-ventral and anterior-posterior positioning, as well as glomerular segregation, appears to occur autonomously, without the involvement of target-derived cues (Sakano 2010). These autonomous mapping strategies may be quite advantageous for sensory development in embryos, and may help propel evolution, as it would not require the codevelopment or coevolution of the target organ. During evolution, the OR gene family expanded rapidly in amphibians to respond to a large variety of volatile ligands (Niimura & Nei 2005, Nei et al. 2008). In contrast, a significant number of OR genes became pseudogenes when the animals became more reliant on the visual system or went back to the water. Target-independent axonal projection would allow the system to adapt easily to sudden changes in the OR gene repertoire and their ligand qualities.

Although axon-derived guidance molecules alone could organize a topographic map, the map would still require proper connections to second-order neurons for the appropriate conversion of sensory stimuli to functional and behavioral responses mediated by higher cortical neurons. At least for the fox odor, TMT appears to be detected by two separate sets of ORs, one expressed in the dorsal zone and the other in the ventral zone of the OE, and processed by dorsal-domain glomeruli and ventral-domain glomeruli in the OB for innate and learned circuits, respectively (Kobayakawa et al. 2007). As for the dorsal domain of the OB, there seem to be multiple functional modalities for distinct innate responses, such as fearful, aversive, and attractive behaviors mediated by genetically programmed, hard-wired circuits (Kobayakawa et al. 2007). Also several subsets of OSNs have been reported in the ventral zone of the OE, which respond to specific ligands such as CO_2 and induce specific innate behaviors (Lin et al. 2005, Liberles & Buck 2006, Hu et al. 2007, Cockerham et al. 2009).

Are mitral and tufted cells naïve with respect to projections that drive innate behaviors? At least for the hard-wired circuits, proper connections are required between specialized OSNs and specific projection neurons. Are mitral and tufted cells instructed by OSNs? How is the axonal projection of these cells regulated in the OC? Answering these questions seems to be essential to further explore the neural circuits in the central olfactory system. As for the learned processing of olfactory information, how is odor quality judged based on memories? How are the neuronal circuits reorganized during the postlearning sleep period? For flavor perception of foods (Shepherd 2006), how is olfactory information integrated with gustatory and somatic sensory information? These interesting questions are expected to be answered in the future. The olfactory system will continue to serve as an excellent model system for the study of axon wiring and neural circuit formation in the mammalian brain.

DISCLOSURE STATEMENT

The authors are not aware of any affiliations, memberships, funding, or financial holdings that might be perceived as affecting the objectivity of this review.

ACKNOWLEDGMENTS

This work was supported by the CREST Program of the Japan Science and Technology Agency (K.M.) and a Specially Promoted Research Grant from the Japanese Government (H.S.).

LITERATURE CITED

Astic L, Saucier D. 1986. Anatomical mapping of the neuroepithelial projection to the olfactory bulb in the rat. *Brain Res. Bull.* 16:445–54

Astic L, Saucier D, Holley A. 1987. Topographical relationships between olfactory receptor cells and glomerular foci in the rat olfactory bulb. *Brain Res.* 424:144–52

Bailey MS, Puche AC, Shipley MT. 1999. Development of the olfactory bulb: evidence for glia-neuron interactions in glomerular formation. *J. Comp. Neurol.* 415:423–48

Barnea G, O'Donnell S, Mancia F, Sun X, Nemes A, et al. 2004. Odorant receptors on axon termini in the brain. *Science* 304:1468

Belluscio L, Gold GH, Nemes A, Axel R. 1998. Mice deficient in G(olf) are anosmic. *Neuron* 20:69–81

Belluscio L, Lodovichi C, Feinstein P, Mombaerts P, Katz LC. 2002. Odorant receptors instruct functional circuitry in the mouse olfactory bulb. *Nature* 419:296–300

Borodinsky LN, Root CM, Cronin JA, Sann SB, Gu X, Spitzer NC. 2004. Activity-dependent homeostatic specification of transmitter expression in embryonic neurons. *Nature* 429:523–30

Bozza T, McGann J, Mombaerts P, Wachowiak M. 2004. In vivo imaging of neuronal activity by targeted expression of a genetically encoded probe in the mouse. *Neuron* 42:9–21

Brunet LJ, Gold GH, Ngai J. 1996. General anosmia caused by a targeted disruption of the mouse olfactory cyclic nucleotide-gated cation channel. *Neuron* 17:681–93

Buck L, Axel R. 1991. A novel multigene family may encode odorant receptors: a molecular basis for odor recognition. *Cell* 65:175–87

Carter D, Chakalova L, Osborne CS, Dai YF, Fraser P. 2002. Long-range chromatin regulatory interaction in vivo. *Nat. Genet.* 32:623–26

Chang KT, Berg DK. 2001. Voltage-gated channels block nicotinic regulation of CREB phosphorylation and gene expression in neurons. *Neuron* 32:855–65

Chen Y, Flanagan JG. 2006. Follow your nose: axon pathfinding in olfactory map formation. *Cell* 127:881–84

Chesler AT, Zou DJ, Le Pichon CE, Peterlin ZA, Matthews GA, et al. 2007. A G protein/cAMP signal cascade is required for axonal convergence into olfactory glomeruli. *Proc. Natl. Acad. Sci. USA* 104:1039–44

Chess A, Simon I, Cedar H, Axel R. 1994. Allelic inactivation regulates olfactory receptor gene expression. *Cell* 78:823–34

Cho JH, Lepine M, Andrews W, Parnavelas J, Cloutier JF. 2007. Requirement for Slit-1 and Robo-2 in zonal segregation of olfactory sensory neuron axons in the main olfactory bulb. *J. Neurosci.* 27:9094–104

Cockerham RE, Leinders-Zufall T, Munger SD, Zufall F. 2009. Functional analysis of the guanylyl cyclase type D signaling system in the olfactory epithelium. *Ann. N. Y. Acad. Sci.* 1170:173–76

Col JA, Matsuo T, Storm DR, Rodriguez I. 2007. Adenylyl cyclase-dependent axonal targeting in the olfactory system. *Development* 134:2481–89

Conzelman S, Malun D, Strotmann J. 2001. Brain targeting and glomerulus formation of two olfactory neuron populations expressing related receptor types. *Eur. J. Neurosci.* 14:1623–32

Cutforth T, Moring L, Mendelsohn M, Nemes A, Shah NM, et al. 2003. Axonal ephrin-As and odorant receptors: coordinate determination of the olfactory sensory map. *Cell* 114:311–22

Dielenberg R, McGregor I. 2001. Defensive behavior in rats towards predatory odors: a review. *Neurosci. Biobehav. Rev.* 25:597–609

Dolmetsch RE, Lewis RS, Goodnow CC, Healy JI. 1997. Differential activation of transcription factors induced by Ca^{2+} response amplitude and duration. *Nature* 386:855–58

Doty RL. 1986. Odor-guided behavior in mammals. *Experientia* 42:257–71

Døving K, Selset R, Thommesen G. 1980. Olfactory sensitivity to bile acids in salmonid fishes. *Acta Physiol. Scand.* 108:123–31

Duggan CD, Demaria S, Baudhuin A, Stafford D, Ngai J. 2008. Foxg1 is required for development of the vertebrate olfactory system. *J. Neurosci.* 28:5229–39

Eggan K, Baldwin K, Tackett M, Osborne J, Gogos J, et al. 2004. Mice cloned from olfactory sensory neurons. *Nature* 428:44–49

Feinstein P, Bozza T, Rodriguez I, Vassalli A, Mombaerts P. 2004. Axon guidance of mouse olfactory sensory neurons by odorant receptors and the β2 adrenergic receptor. *Cell* 117:833–46

Feinstein P, Mombaerts P. 2004. A contextual model for axonal sorting into glomeruli in the mouse olfactory system. *Cell* 117:817–31

Fleischmann A, Shykind BM, Sosulski DL, Franks KM, Glinka ME, et al. 2008. Mice with a "monoclonal nose": perturbations in an olfactory map impair odor discrimination. *Neuron* 60:1068–81

Fuss SH, Omura M, Mombaerts P. 2007. Local and *cis* effects of the H element on expression of odorant receptor genes in mouse. *Cell* 130:373–84

Goodman CS, Shatz CJ. 1993. Developmental mechanisms that generate precise patterns of neuronal connectivity. *Cell* 72(Suppl.):77–98

Griff ER, Mafhouz M, Chaput MA. 2008. Comparison of identified mitral and tufted cells in freely breathing rats: II. Odor-evoked responses. *Chem. Senses* 33:793–802

Grosveld F, van Assendelft DB, Greaves DR, Kollias G. 1987. Position-independent, high-level expression of the human β-globin gene in transgenic mice. *Cell* 51:975–85

Gussing F, Bohm S. 2004. NQO1 activity in the main and the accessory olfactory systems correlates with the zonal topography of projection maps. *Eur. J. Neurosci.* 19:2511–18

Haberly LB. 2001. Parallel-distributed processing in olfactory cortex: new insights from morphological and physiological analysis of neuronal circuitry. *Chem. Senses* 26:551–76

Haberly LB, Price JL. 1977. The axonal projection patterns of the mitral and tufted cells of the olfactory bulb in the rat. *Brain Res.* 129:152–57

Hallem EA, Ho MG, Carlson JR. 2004. The molecular basis of odor coding in the *Drosophila* antenna. *Cell* 117:965–79

Hanson MG, Landmesser LT. 2004. Normal patterns of spontaneous activity are required for correct motor axon guidance and the expression of specific guidance molecules. *Neuron* 43:687–701

Hara T, Zhang C. 1998. Topographic bulbar projections and dual neural pathways of the primary olfactory neurons in salmonid fishes. *Neurosc.* 82:301–13

Hayar A, Karnup S, Shipley MT, Ennis M. 2004. Olfactory bulb glomeruli: External tufted cells intrinsically burst at theta frequency and are entrained by patterned olfactory input. *J. Neurosci.* 24:1190–99

Hebb A, Zacharko R, Dominguez H, Trudel F, Laforest S, Drolet G. 2002. Odor-induced variation in anxiety-like behavior in mice is associated with discrete and differential effects on mesocorticolimbic cholecystokinin mRNA expression. *Neuropsychopharmacology* 27:744–55

Hebb A, Zacharko R, Gauthier M, Trudel F, Laforest S, Drolet G. 2004. Brief exposure to predator odor and resultant anxiety enhances mesocorticolimbic activity and enkephalin expression in CD-1 mice. *Eur. J. Neurosci.* 20:2415–29

Heimer L. 1968. Synaptic distribution of centripetal and centrifugal nerve fibres in the olfactory system of the rat: an experimental anatomical study. *J. Anat.* 103:413–32

Hirota J, Mombaerts P. 2004. The LIM-homeodomain protein Lhx2 is required for complete development of mouse olfactory sensory neurons. *Proc. Natl. Acad. Sci. USA* 101:8751–55

Hirota J, Omura M, Mombaerts P. 2007. Differential impact of Lhx2 deficiency on expression of class I and class II odorant receptor genes in mouse. *Mol. Cell Neurosci.* 34:679–88

Horder TJ. 1971. The course of recovery of the retinotectal projection during regeneration of the fish optic nerve. *J. Physiol.* 217:P53–54

Hu J, Zhong C, Ding C, Chi Q, Walz A, et al. 2007. Detection of near-atmospheric concentrations of CO_2 by an olfactory subsystem in the mouse. *Science* 317:953–57

Huberman AD, Feller MB, Chapman B. 2008. Mechanisms underlying development of visual maps and receptive fields. *Annu. Rev. Neurosci.* 31:479–509

Igarashi K, Mori K. 2005. Spatial representation of hydrocarbon odorants in the ventrolateral zones of the rat olfactory bulb. *J. Neurophysiol.* 93:1007–19

Ikemoto S. 2007. Dopamine reward circuitry: two projection systems from the ventral midbrain to the nucleus accumbens-olfactory tubercle complex. *Brain Res. Rev.* 56:27–78

Illig KR, Haberly LB. 2003. Odor-evoked activity is spatially distributed in piriform cortex. *J. Comp. Neurol.* 457:361–73

Imai T, Sakano H. 2007. Roles of odorant receptors in projecting axons in the mouse olfactory system. *Curr. Opin. Neurobiol.* 17:507–15

Imai T, Sakano H. 2008. Odorant receptor-mediated signaling in the mouse. *Curr. Opin. Neurobiol.* 18:251–60

Imai T, Suzuki M, Sakano H. 2006. Odorant receptor-derived cAMP signals direct axonal targeting. *Science* 314:657–61

Imai T, Yamazaki T, Kobayakawa R, Kobayakawa K, Abe T, et al. 2009. Pre-target axon sorting establishes the neural map topography. *Science* 325:585–90

Inaki K, Nishimura S, Nakashiba T, Itohara S, Yoshihara Y. 2004. Laminar organization of the developing lateral olfactory tract revealed by differential expression of cell recognition molecules. *J. Comp. Neurol.* 479:243–56

Ishii T, Serizawa S, Kohda A, Nakatani H, Shiroishi T, et al. 2001. Monoallelic expression of the odourant receptor gene and axonal projection of olfactory sensory neurons. *Genes Cells* 6:71–78

Itoh K, Stevens B, Schachner M, Fields RD. 1995. Regulated expression of the neural cell adhesion molecule L1 by specific patterns of neural impulses. *Science* 270:1369–72

Johnson BA, Leon M. 2007. Chemotopic odorant coding in a mammalian olfactory system. *J. Comp. Neurol.* 503:1–34

Johnson BA, Ong J, Leon M. 2010. Glomerular activity patterns evoked by natural odor objects in the rat olfactory bulb are related to patterns evoked by major odorant components. *J. Comp. Neurol.* 518:1542–55

Johnson BA, Xu Z, Ali SS, Leon M. 2009. Spatial representations of odorants in olfactory bulbs of rats and mice: similarities and differences in chemotopic organization. *J. Comp. Neurol.* 514:658–73

Kaas J, Nelson R, Sur M, Merzenich M. 1981. Organization of somatosensory cortex in primates. In *The Organization of the Cerebral Cortex*, ed. F Schmitt, F Worden, G Adelman, S Dennis, pp. 237–61. Cambridge, MA: MIT Press

Kaneko-Goto T, Yoshihara S, Miyazaki H, Yoshihara Y. 2008. BIG-2 mediates olfactory axon convergence to target glomeruli. *Neuron* 57:834–46

Kashiwadani H, Sasaki Y, Uchida N, Mori K. 1999. Synchronized oscillatory discharges of mitral/tufted cells with different molecular receptive ranges in the rabbit olfactory bulb. *J. Neurophysiol.* 82:1786–92

Kikuta S, Sato K, Kashiwadani H, Tsunoda K, Yamasoba T, Mori K. 2010. Neurons in the anterior olfactory nucleus pars externa detect right or left localization of odor sources. *Proc. Natl. Acad. Sci. USA* 107:12363–68

Kobayakawa K, Kobayakawa R, Matsumoto H, Oka Y, Imai T, et al. 2007. Innate versus learned odour processing in the mouse olfactory bulb. *Nature* 450:503–8

Koide T, Miyasaka N, Morimoto K, Asakawa K, Urasaki A, et al. 2009. Olfactory neural circuitry for attraction to amino acids revealed by transposon-mediated gene trap approach in zebrafish. *Proc. Natl. Acad. Sci. USA* 106:9884–89

Lemke G, Reber M. 2005. Retinotectal mapping: new insights from molecular genetics. *Annu. Rev. Cell Dev. Biol.* 21:551–80

Lewcock JW, Reed RR. 2004. A feedback mechanism regulates monoallelic odorant receptor expression. *Proc. Natl. Acad. Sci. USA* 101:1069–74

Li J, Ishii T, Feinstein P, Mombaerts P. 2004. Odorant receptor gene choice is reset by nuclear transfer from mouse olfactory sensory neurons. *Nature* 428:393–99

Li W, Scott A, Siefkes M, Yan H, Liu Q, et al. 2002. Bile acid secreted by male sea lamprey that acts as a sex pheromone. *Science* 296:138–41

Liberles S, Buck LB. 2006. A second class of chemosensory receptors in the olfactory epithelium. *Nature* 442:645–50

Lin DM, Wang F, Lowe G, Gold GH, Axel R, et al. 2000. Formation of precise connections in the olfactory bulb occurs in the absence of odorant-evoked neuronal activity. *Neuron* 26:69–80

Lin DY, Zhang SZ, Block E, Katz LC. 2005. Encoding social signals in the mouse main olfactory bulb. *Nature* 434:470–77

Liu W, Shipley M. 1994. Intrabulbar associational system in the rat olfactory bulb comprises cholecystokinin-containing tufted cells that synapse onto the dendrites of GABAergic granule cells. *J. Comp. Neurol.* 346:541–58

Lodovichi C, Belluscio L, Katz L. 2003. Functional topography of connections linking mirror-symmetric maps in the mouse olfactory bulb. *Neuron* 38:265–76

Lomvardas S, Barnea G, Pisapia DJ, Mendelsohn M, Kirkland J, Axel R. 2006. Interchromosomal interactions and olfactory receptor choice. *Cell* 126:403–13

Luo L, Flanagan JG. 2007. Development of continuous and discrete neural maps. *Neuron* 56:284300

Luskin MB, Price JL. 1983. The topographic organization of association fibers of the olfactory system in the rat, including centrifugal fibers to the olfactory bulb. *J. Comp. Neurol.* 216:264–91

Macrides F, Schneider S. 1982. Laminar organization of mitral and tufted cells in the main olfactory bulb of the adult hamster. *J. Comp. Neurol.* 208:419–30

Mainen ZF. 2007. The main olfactory bulb and innate behavior: different perspectives on an olfactory scene. *Nat. Neurosci.* 10:1511–12

Matsumoto H, Kobayakawa K, Kobayakawa R, Tashiro T, Mori K, Sakano H. 2010. Spatial arrangement of glomerular molecular-feature clusters in the odorant-receptor class domains of the mouse olfactory bulb. *J. Neurophysiol.* 103:3490–500

Matsutani S, Senba E, Tohyama M. 1989. Terminal field of cholecystokinin-8-like immunoreactive projection neurons of the rat main olfactory bulb. *J. Comp. Neurol.* 285:73–82

McLaughlin T, O'Leary DD. 2005. Molecular gradients and development of retinotopic maps. *Annu. Rev. Neurosci.* 28:327–55

McLaughlin T, Torborg CL, Feller MB, O'Leary DD. 2003. Retinotopic map refinement requires spontaneous retinal waves during a brief critical period of development. *Neuron* 40:1147–60

Meister M, Bonhoeffer T. 2001. Tuning and topography in an odor map on the rat olfactory bulb. *J. Neurosci.* 21:1351–60

Menco BP, Tekula FD, Farbman AI, Danho W. 1994. Developmental expression of G-proteins and adenylyl cyclase in peripheral olfactory systems: light microscopic and freeze-substitution electron microscopic immunocytochemistry. *J. Neurocytol.* 23:708–27

Merigan W, Maunsell J. 1993. How parallel are the primate visual pathways? *Annu. Rev. Neurosci.* 16:369–402

Michaloski JS, Galante PA, Malnic B. 2006. Identification of potential regulatory motifs in odorant receptor genes by analysis of promoter sequences. *Genome Res.* 16:1091–98

Miyamichi K, Amat F, Moussavi F, Wang C, Wickersham I, et al. 2011. Cortical representations of olfactory input by trans-synaptic tracing. *Nature*. In press

Miyamichi K, Luo L. 2009. Brain wiring by presorting axons. *Science* 325:544–45

Miyamichi K, Serizawa S, Kimura HM, Sakano H. 2005. Continuous and overlapping expression domains of odorant receptor genes in the olfactory epithelium determine the dorsal/ventral positioning of glomeruli in the olfactory bulb. *J. Neurosci.* 25:3586–92

Mombaerts P. 2006. Axonal wiring in the mouse olfactory system. *Annu. Rev. Cell Dev. Biol.* 22:713–37

Mombaerts P, Wang F, Dulac C, Chao SK, Nemes A, et al. 1996. Visualizing an olfactory sensory map. *Cell* 87:675–86

Mori K. 1987. Membrane and synaptic properties of identified neurons in the olfactory bulb. *Prog. Neurobiol.* 29:275–320

Mori K, Matsumoto H, Tsuno Y, Igarashi KM. 2009. Dendrodendritic synapses and functional compartmentalization in the olfactory bulb. *Ann. N. Y. Acad. Sci.* 1170:255–58

Mori K, Nagao H, Yoshihara Y. 1999. The olfactory bulb: coding and processing of odor molecule information. *Science* 286:711–15

Mori K, Takahashi YK, Igarashi KM, Yamaguchi M. 2006. Maps of odorant molecular features in the mammalian olfactory bulb. *Physiol. Rev.* 86:409–33

Nagao H, Yoshihara Y, Mitsui S, Fujisawa H, Mori K. 2000. Two mirror-image sensory maps with domain organization in the mouse main olfactory bulb. *NeuroReport* 11:3023–27

Nagayama S, Takahashi Y, Yoshihara Y, Mori K. 2004. Mitral and tufted cells differ in the decoding manner of odor maps in the rat olfactory bulb. *J. Neurophysiol.* 91:2532–40

Nakatani H, Serizawa S, Nakajima M, Imai T, Sakano H. 2003. Developmental elimination of ectopic projection sites for the transgenic OR gene that has lost zone specificity in the olfactory epithelium. *Eur. J. Neurosci.* 18:2425–32

Nei M, Niimura Y, Nozawa M. 2008. The evolution of animal chemosensory receptor gene repertoires: roles of chance and necessity. *Nat. Rev. Genet.* 9:951–63

Neville K, Haberly L. 2004. Olfactory cortex. In *The Synaptic Organization of the Brain*, ed. G Shepherd, pp. 415–54. New York: Oxford Univ. Press

Nguyen-Ba-Charvet KT, Di Meglio T, Fouquet C, Chedotal A. 2008. Robos and slits control the pathfinding and targeting of mouse olfactory sensory axons. *J. Neurosci.* 28:4244–49

Nguyen MQ, Zhou Z, Marks CA, Ryba NJ, Belluscio L. 2007. Prominent roles for odorant receptor coding sequences in allelic exclusion. *Cell* 131:1009–17

Niimura Y, Nei M. 2005. Evolutionary dynamics of olfactory receptor genes in fishes and tetrapods. *Proc. Natl. Acad. Sci. USA* 102:6039–44

Nikonov A, Caprio J. 2001. Electrophysiological evidence for a chemotopy of biologically relevant odors in the olfactory bulb of the channel catfish. *J. Neurophysiol.* 86:1869–76

Nikonov A, Finger T, Caprio J. 2005. Beyond the olfactory bulb: an odotopic map in the forebrain. *Proc. Natl. Acad. Sci. USA* 102:18688–93

Nishizumi H, Kumasaka K, Inoue N, Nakashima A, Sakano H. 2007. Deletion of the core-H region in mice abolishes the expression of three proximal odorant receptor genes in *cis*. *Proc. Natl. Acad. Sci. USA* 104:20067–72

Norlin EM, Alenius M, Gussing F, Hagglund M, Vedin V, Bohm S. 2001. Evidence for gradients of gene expression correlating with zonal topography of the olfactory sensory map. *Mol. Cell Neurosci.* 18:283–95

Novotny M. 2003. Pheromones, binding proteins and receptor responses in rodents. *Biochem. Soc. Trans.* 31:117–22

Novotny M, Harvey S, Jemiolo B, Alberts J. 1985. Synthetic pheromones that promote inter-male aggression in mice. *Proc. Natl. Acad. Sci. USA* 82:2059–61

Oka Y, Kobayakawa K, Nishizumi H, Miyamichi K, Hirose S, et al. 2003. O-MACS, a novel member of the medium-chain acyl-CoA synthetase family, specifically expressed in the olfactory epithelium in a zone-specific manner. *Eur. J. Biochem.* 270:1995–2004

Petersen CC. 2007. The functional organization of the barrel cortex. *Neuron* 56:339–55

Potter SM, Zheng C, Koos DS, Feinstein P, Fraser SE, Mombaerts P. 2001. Structure and emergence of specific olfactory glomeruli in the mouse. *J. Neurosci.* 21:9713–23

Ramón y Cajal S. 1955. *Studies on the Cerebral Cortex*. Transl. LM Kraft. Chicago: Year Book

Ressler KJ, Sullivan SL, Buck LB. 1993. A zonal organization of odorant receptor gene expression in the olfactory epithelium. *Cell* 73:597–609

Ressler KJ, Sullivan SL, Buck LB. 1994. Information coding in the olfactory system: evidence for a stereotyped and highly organized epitope map in the olfactory bulb. *Cell* 79:1245–55

Roppolo D, Vollery S, Kan CD, Luscher C, Broillet MC, Rodriguez I. 2007. Gene cluster lock after pheromone receptor gene choice. *EMBO J.* 26:3423–30

Rothman A, Feinstein P, Hirota J, Mombaerts P. 2005. The promoter of the mouse odorant receptor gene M71. *Mol. Cell Neurosci.* 28:535–46

Royal SJ, Key B. 1999. Development of P2 olfactory glomeruli in P2-internal ribosome entry site-tau-LacZ transgenic mice. *J. Neurosci.* 19:9856–64

Sakano H. 2010. Neural map formation in the mouse olfactory system. *Neuron* 67:530–42

Sato Y, Miyasaka N, Yoshihara Y. 2005. Mutually exclusive glomerular innervation by two distinct types of olfactory sensory neurons revealed in transgenic zebrafish. *J. Neurosci.* 25:4889–97

Satoda M, Takagi S, Ohta K, Hirata T, Fujisawa H. 1995. Differential expression of two cell surface proteins, neuropilin and plexin, in *Xenopus* olfactory axon subclasses. *J. Neurosci.* 15:942–55

Saucier D, Astic L. 1986. Analysis of the topographical organization of olfactory epithelium projections in the rat. *Brain Res. Bull.* 16:455–62

Schoenfeld T, Clancy A, Forbes W, Macrides F. 1994. The spatial organization of the peripheral olfactory system of the hamster. Part I: receptor neuron projections to the main olfactory bulb. *Brain Res. Bull.* 34:183–210

Schoenfeld T, Macrides F. 1984. Topographic organization of connections between the main olfactory bulb and pars externa of the anterior olfactory nucleus in the hamster. *J. Comp. Neurol.* 227:121–35

Schoenfeld T, Marchand J, Macrides F. 1985. Topographic organization of tufted cell axonal projections in the hamster main olfactory bulb: an intrabulbar associational system. *J. Comp. Neurol.* 235:503–18

Schoppa N. 2006. Synchronization of olfactory bulb mitral cells by precisely timed inhibitory inputs. *Neuron* 49:271–83

Sengoku S, Ishii T, Serizawa S, Nakatani H, Nagawa F, et al. 2001. Axonal projection of olfactory sensory neurons during the developmental and regeneration processes. *NeuroReport* 12:1061–66

Serizawa S, Ishii T, Nakatani H, Tsuboi A, Nagawa F, et al. 2000. Mutually exclusive expression of odorant receptor transgenes. *Nat. Neurosci.* 3:687–93

Serizawa S, Miyamichi K, Nakatani H, Suzuki M, Saito M, et al. 2003. Negative feedback regulation ensures the one receptor–one olfactory neuron rule in mouse. *Science* 302:2088–94

Serizawa S, Miyamichi K, Sakano H. 2004. One neuron–one receptor rule in the mouse olfactory system. *Trends Genet.* 20:648–53

Serizawa S, Miyamichi K, Takeuchi H, Yamagishi Y, Suzuki M, Sakano H. 2006. A neuronal identity code for the odorant receptor-specific and activity-dependent axon sorting. *Cell* 127:1057–69

Seta K, Nanamori M, Modrall JG, Neubig RR, Sadoshima J. 2002. AT1 receptor mutant lacking heterotrimeric G protein coupling activates the Src-Ras-ERK pathway without nuclear translocation of ERKs. *J. Biol. Chem.* 277:9268–77

Shenoy SK, Lefkowitz RJ. 2005. Seven-transmembrane receptor signaling through β-arrestin. *Sci. STKE* 2005:cm10

Shepherd G. 2006. Smell images and the flavour system in the human brain. *Nature* 444:316–21

Shepherd G, Chen WR, Greer CA. 2004. Olfactory bulb. In *The Synaptic Organization of the Brain*, ed. GM Shepherd, pp. 165–216. New York: Oxford Univ. Press

Shykind BM, Rohani SC, O'Donnell S, Nemes A, Mendelsohn M, et al. 2004. Gene switching and the stability of odorant receptor gene choice. *Cell* 117:801–15

Smallwood PM, Wang Y, Nathans J. 2002. Role of a locus control region in the mutually exclusive expression of human red and green cone pigment genes. *Proc. Natl. Acad. Sci. USA* 99:1008–11

Spilianakis CG, Lalioti MD, Town T, Lee GR, Flavell RA. 2005. Interchromosomal associations between alternatively expressed loci. *Nature* 435:637–45

Stephan AB, Shum EY, Hirsh S, Cygnar KD, Reisert J, Zhao H. 2009. ANO2 is the cilial calcium-activated chloride channel that may mediate olfactory amplification. *Proc. Natl. Acad. Sci. USA* 106:11776–81

Stettler D, Axel R. 2009. Representations of odor in the piriform cortex. *Neuron* 63:854–64

Strotmann J, Levai O, Fleischer J, Schwarzenbacher K, Breer H. 2004. Olfactory receptor proteins in axonal processes of chemosensory neurons. *J. Neurosci.* 24:7754–61

Sullivan SL, Bohm S, Ressler KJ, Horowitz LF, Buck LB. 1995. Target-independent pattern specification in the olfactory epithelium. *Neuron* 15:779–89

Sweeney LB, Couto A, Chou YH, Berdnik D, Dickson BJ, et al. 2007. Temporal target restriction of olfactory receptor neurons by Semaphorin-1a/PlexinA-mediated axon-axon interactions. *Neuron* 53:185–200

Takahashi H, Yoshihara S, Nishizumi H, Tsuboi A. 2010. Neuropilin-2 is required for the proper targeting of ventral glomeruli in the mouse olfactory bulb. *Mol. Cell Neurosci.* 44:233–45

Takahashi Y, Kurosaki M, Hirono S, Mori K. 2004a. Topographic representation of odorant molecular features in the rat olfactory bulb. *J. Neurophysiol.* 92:2413–27

Takahashi Y, Nagayama S, Mori K. 2004b. Detection and masking of spoiled food smells by odor maps in the olfactory bulb. *J. Neurosci.* 24:8690–94

Takeuchi H, Inokuchi K, Aoki M, Suto F, Tsuboi A, et al. 2010. Sequential arrival and graded secretion of Sema3F by olfactory neuron axons specify map topography at the bulb. *Cell* 141:1056–67

Tan J, Savigner A, Ma M, Luo M. 2010. Odor information processing by the olfactory bulb analyzed in gene-targeted mice. *Neuron* 65:912–26

Taniguchi M, Nagao H, Takahashi Y, Yamaguchi M, Mitsui S, et al. 2003. Distorted odor maps in the olfactory bulb of semaphorin 3A-deficient mice. *J. Neurosci.* 23:1390–97

Tian H, Ma M. 2008. Activity plays a role in eliminating olfactory sensory neurons expressing multiple odorant receptors in the mouse septal organ. *Mol. Cell Neurosci.* 38:484–88

Tolhuis B, Palstra RJ, Splinter E, Grosveld F, de Laat W. 2002. Looping and interaction between hypersensitive sites in the active β-globin locus. *Mol. Cell* 10:1453–65

Treloar HB, Feinstein P, Mombaerts P, Greer CA. 2002. Specificity of glomerular targeting by olfactory sensory axons. *J. Neurosci.* 22:2469–77

Tsuboi A, Miyazaki T, Imai T, Sakano H. 2006. Olfactory sensory neurons expressing class I odorant receptors converge their axons on an antero-dorsal domain of the olfactory bulb in the mouse. *Eur. J. Neurosci.* 23:1436–44

Uchida N, Takahashi YK, Tanifuji M, Mori K. 2000. Odor maps in the mammalian olfactory bulb: domain organization and odorant structural features. *Nat. Neurosci.* 3:1035–43

Van der Loos H, Woolsey TA. 1973. Somatosensory cortex: structural alterations following early injury to sense organs. *Science* 179:395–98

Vassalli A, Rothman A, Feinstein P, Zapotocky M, Mombaerts P. 2002. Minigenes impart odorant receptor-specific axon guidance in the olfactory bulb. *Neuron* 35:681–96

Vassar R, Chao SK, Sitcheran R, Nunez JM, Vosshall LB, Axel R. 1994. Topographic organization of sensory projections to the olfactory bulb. *Cell* 79:981–91

Vassar R, Ngai J, Axel R. 1993. Spatial segregation of odorant receptor expression in the mammalian olfactory epithelium. *Cell* 74:309–18

Wang F, Nemes A, Mendelsohn M, Axel R. 1998. Odorant receptors govern the formation of a precise topographic map. *Cell* 93:47–60

Wang SS, Lewcock JW, Feinstein P, Mombaerts P, Reed RR. 2004. Genetic disruptions of O/E2 and O/E3 genes reveal involvement in olfactory receptor neuron projection. *Development* 131:1377–88

Wang Y, Smallwood PM, Cowan M, Blesh D, Lawler A, Nathans J. 1999. Mutually exclusive expression of human red and green visual pigment-reporter transgenes occurs at high frequency in murine cone photoreceptors. *Proc. Natl. Acad. Sci. USA* 96:5251–56

Willhite D, Nguyen KT, Masurkar AV, Greer CA, Shepherd GM, Chen WR. 2006. Viral tracing identifies distributed columnar organization in the olfactory bulb. *Proc. Natl. Acad. Sci. USA* 103:12592–97

Wong ST, Trinh K, Hacker B, Chan GC, Lowe G, et al. 2000. Disruption of the type III adenylyl cyclase gene leads to peripheral and behavioral anosmia in transgenic mice. *Neuron* 27:487–97

Xu F, Liu N, Kida I, Rothman D, Hyder F, Shepherd G. 2003. Odor maps of aldehydes and esters revealed by functional MRI in the glomerular layer of the mouse olfactory bulb. *Proc. Natl. Acad. Sci. USA* 100:11029–34

Yan Z, Tan J, Qin C, Lu Y, Ding C, Luo M. 2008. Precise circuitry links bilaterally symmetric olfactory maps. *Neuron* 58:613–24

Yokoi M, Mori K, Nakanishi S. 1995. Refinement of odor molecule tuning by dendrodendritic synaptic inhibition in the olfactory bulb. *Proc. Natl. Acad. Sci. USA* 92:3371–75

Yoshida I, Mori K. 2007. Odorant category profile selectivity of olfactory cortex neurons. *J. Neurosci.* 27:9105–14

Yoshihara Y, Kawasaki M, Tamada A, Fujita H, Hayashi H, et al. 1997. OCAM: a new member of the neural cell adhesion molecule family related to zone-to-zone projection of olfactory and vomeronasal axons. *J. Neurosci.* 17:5830–42

Yu CR, Power J, Barnea G, O'Donnell S, Brown HE, et al. 2004. Spontaneous neural activity is required for the establishment and maintenance of the olfactory sensory map. *Neuron* 42:553–66

Zhang X, Firestein S. 2002. The olfactory receptor gene superfamily of the mouse. *Nat. Neurosci.* 5:124–33

Zhao H, Reed RR. 2001. X inactivation of the OCNC1 channel gene reveals a role for activity-dependent competition in the olfactory system. *Cell* 104:651–60

Zheng C, Feinstein P, Bozza T, Rodriguez I, Mombaerts P. 2000. Peripheral olfactory projections are differentially affected in mice deficient in a cyclic nucleotide-gated channel subunit. *Neuron* 26:81–91

Zhou Z, Belluscio L. 2008. Intrabulbar projecting external tufted cells mediate a timing-based mechanism that dynamically gates olfactory bulb output. *J. Neurosci.* 28:9920–28

Zou DJ, Feinstein P, Rivers AL, Mathews GA, Kim A, et al. 2004. Postnatal refinement of peripheral olfactory projections. *Science* 304:1976–79

Vestibular Hair Cells and Afferents: Two Channels for Head Motion Signals

Ruth Anne Eatock and Jocelyn E. Songer

Department of Otology and Laryngology, Department of Neurobiology, Harvard Medical School, Eaton-Peabody Laboratories, Massachusetts Eye and Ear Infirmary, Boston, Massachusetts 02114; email: eatock@meei.harvard.edu, jocelyn_songer@meei.harvard.edu

Annu. Rev. Neurosci. 2011. 34:501–34

First published online as a Review in Advance on April 4, 2011

The *Annual Review of Neuroscience* is online at neuro.annualreviews.org

This article's doi: 10.1146/annurev-neuro-061010-113710

Keywords

mechanoelectrical transduction, adaptation, low-voltage-activated K channels, type I hair cells, calyx terminals, spike regularity

Abstract

Vestibular epithelia of the inner ear detect head motions over a wide range of amplitudes and frequencies. In mammals, afferent nerve fibers from central and peripheral zones of vestibular epithelia form distinct populations with different response dynamics and spike timing. Central-zone afferents are large, fast conduits for phasic signals encoded in irregular spike trains. The finer afferents from peripheral zones conduct more slowly and encode more tonic, linear signals in highly regular spike trains. The hair cells are also of two types, I and II, but the two types do not correspond directly to the two afferent populations. Zonal differences in afferent response dynamics may arise at multiple stages, including mechanoelectrical transduction, voltage-gated channels in hair cells and afferents, afferent transmission at calyceal and bouton synapses, and spike generation in regular and irregular afferents. In contrast, zonal differences in spike timing may depend more simply on the selective expression of low-voltage-activated ion channels by irregular afferents.

Contents

INTRODUCTION

The vestibular sense allows animals to maintain orientation and stabilize visual input in the face of environmental and self motions and is therefore critical to all other interactions with the world. Because we rarely attend to this necessary aspect of our lives, however, vestibular sensation has attracted less public and scientific notice than its sister sense, hearing. Vestibular hair cells work in the background to transmit information about head tilt and head motion to reflex pathways that control gaze, head position, and body posture and to cortical centers that sense heading and orientation. Slow head motions may also engage visual and somatosensory signals that contribute to compensatory responses, but vestibular hair cells provide the only signals fast enough to drive compensation for rapid head motions. In vivo animal studies show that head motion signals emerge from each organ in the vestibular labyrinth in two channels with different response dynamics and spike timing. In this review, we focus on properties of hair cells and afferent terminals that may differentiate the afferent-fiber channels, particularly in mammals.

Evolutionary and Functional Context

Multicelled organisms share the key ability to detect which way is up: their orientation relative to the gravity vector. Gravity sensing arises when weighty structures (starch grains in plants, calcareous crystals in vertebrates) in fluid-filled sacs fall onto mechanosensitive cells. Because gravity is physically indistinguishable from linear acceleration, gravity sensors in animals also detect translational motions. The vestibular labyrinth of the vertebrate inner ear (**Figure 1a**) has between one and three linear accelerometers; each of these otolith ("earstone") organs includes an innervated sensory epithelium overlaid with otoconia (calcareous crystals). Vertebrates also have two or three semicircular canal organs for detecting head rotations. The primitive agnathic fishes, hagfish and lampreys, have one otolithic organ and two vertical canals. Advanced vertebrates added one or more otolith organs plus a horizontal canal, all by expressing a single additional transcription factor (Morsli et al. 1999, Fritzsch et al. 2001, Hammond & Whitfield 2006). Placental mammals have three canals and the otolithic utricle and saccule; monotremes, birds, and reptiles also have an otolithic lagena.

Evolution from stem tetrapods to amniotes (reptiles, birds, and mammals) included big changes in the inner ear as animals adapted to life fully out of the water. From its saccular end, the inner ear sprouted a specialized hearing organ sensitive to airborne sounds. At the same time, a novel combination of sensory receptor cell (type I hair cell) and afferent terminal (calyceal ending) arose in the vestibular labyrinth. Although the value of the hearing

---

Afferent: (adjective) flow of signals from sensory receptors toward the brain; (noun) primary neuron that carries afferent signals

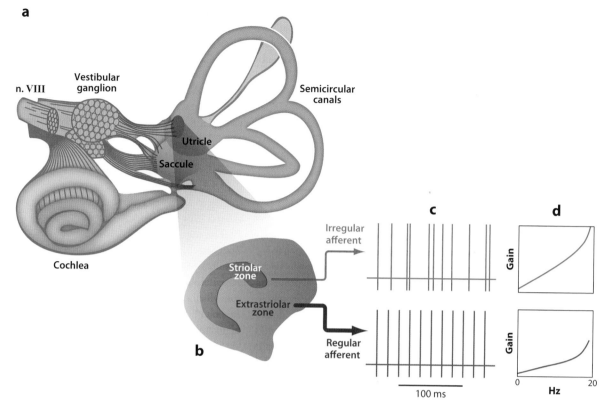

a

n. VIII

Vestibular
ganglion

Semicircular
canals

Utricle

Saccule

Cochlea

b

Striolar
zone

Extrastriolar
zone

Irregular
afferent

Regular
afferent

c

d

Gain

Gain

100 ms

0   20
Hz

Figure 1

Two afferent channels emerge from different zones of vestibular sensory epithelia. (*a*) Schematic of mammalian inner ear. (*b*) Schematic of the utricular sensory epithelium, showing the striolar zone (*red*) surrounded by the blue extrastriolar zone. Arrows represent striolar and extrastriolar afferent channels, which have different spike regularities (*c*) and response dynamics (*d*). The extrastriolar arrow is thicker because more afferents are extrastriolar. (*c*) Schematized background firing for irregular otolith afferent (top) and a regular otolith afferent (bottom). (*d*) Response gain (average spike rate per stimulus unit; linear axes) as a function of stimulus frequency, based on data for semicircular canal afferents from Sadeghi et al. (2007a). Otolith afferent data show similar trends over the same frequency range (K.E. Cullen, personal communication). Gain increases more with frequency for irregular afferents than for regular afferents. Irregular afferent activity also peaks earlier in each stimulus cycle (has a greater phase lead re: head motion) than does regular afferent activity (not shown).

organ is obvious, we do not yet understand the driving forces behind the evolution of the type I–calyx complex. Certainly, the move from water to land brought new challenges for postural and gaze control as well as new sensory opportunities. For example, the elongation of the neck in land vertebrates allows larger independent head motions. Involuntary head motions can be very rapid: Although the dominant frequency of head motion for a running or walking man is just several Hz, significant higher harmonics are present (Grossman et al. 1988), and

when the head stops moving or changes direction, frequencies as high as 80 Hz can be measured (Armand & Minor 2001). Some species evolved high-acuity retinal centers (foveas) and three-dimensional depth perception (stereopsis), which would intensify the need for rapid and accurate eye motions to compensate for head motions. The need is most acute between ~1 Hz (Paige 1983) and the frequency above which flickering visual images fuse (~50 Hz in humans). Below 1 Hz, visual feedback is effective at preventing retinal slip of images; above

VOR: vestibulo-ocular reflex

Crista: sensory epithelium of a semicircular canal organ

Macula: sensory epithelium of an otolith organ (utricle, saccule, or lagena)

the flicker-fusion frequency, the slowness of visual transduction filters out retinal motion. Visual field stability in this important intermediate range of head motion frequencies is largely provided by vestibulo-ocular reflexes (VORs): Signals from vestibular hair cells drive eye muscles to move the eye oppositely to the head motion, cancelling the effect of head motion on the direction of gaze. VORs are likely the fastest reflexes in the body and are highly accurate: In primates, the latency is as low as 5–6 ms and eye motion is nearly perfectly compensatory for head motion frequencies at least up to 25 Hz (Huterer & Cullen 2002).

We are usually unaware of this remarkable compensatory system until it malfunctions. First experiences of dizziness and vertigo may arise from alcohol consumption, which alters the relative densities of accessory structures that stimulate vestibular hair cells. In middle age, some people experience attacks of vertigo following the dislocation of otoconia from the utricle into the posterior canal (benign paroxysmal positional vertigo). Bilateral loss of vestibular inner-ear function prevents gaze stabilization during rapid head motions, producing a bouncing visual field (oscillopsia), which renders ordinary tasks such as walking and driving difficult. The most common vestibular pathology may be a gradual loss of function as sensory cells and other elements of the inner ear age. In humans, the incidence of balance dysfunction of apparently vestibular origin increases steeply after 40 years, exceeding 80% of people older than 80 years (Agrawal et al. 2009). Bilateral damage can also arise acutely from pharmaceuticals such as anticancer drugs (cisplatin) or powerful antibiotics (gentamicin[1]). Ménière's disease or vestibular neuritis can produce errant afferent signals and cause disorientation so severe that patients are treated by ablating vestibular hair cells (and therefore the aberrant signals) with gentamicin. Patients accommodate to unilateral loss of vestibular afferent signals in a gradual process referred to as "vestibular compensation" (Dutia 2010). The compensatory changes, which occur in the brain over several weeks, may reflect the unmasking of proprioceptive signals (Sadeghi et al. 2010) and visual signals that accompany low-frequency head motions. Compensation is incomplete (Halmagyi et al. 2010), however, particularly for impulsive head motions with high-frequency content for which other senses are too slow to provide useful feedback.

The significance of vestibular function in daily activities, widespread incidence of balance dysfunction arising in the inner ear, and lack of selective inner-ear therapies together provide strong clinical motivation for studying the vestibular inner ear. For neurobiologists, the vestibular inner ear offers an unusual hair cell, the type I cell, and a unique calyceal synapse, which may yield insights into high-volume excitatory transmission at large synaptic contacts. Vestibular epithelia are also attractive models in which to investigate developmental mechanisms and sensory processing because they are both relatively simple and highly ordered. In the next section, we lay out the relationships among epithelial zones, afferent classes, and hair cell types in mammalian vestibular epithelia in preparation for considering specific mechanisms for zonal differences in afferent physiology.

Mapping of Afferent Physiological Properties on Vestibular Epithelia

In each vestibular epithelium, thousands of hair cells and supporting cells are arrayed in roughly ovoid sheets. The maculae (epithelia of the otolith organs) are relatively flat; the cristae (epithelia of canals) are each draped over a saddle-shaped ridge. There are ~10,000 vestibular hair cells in each mouse labyrinth (extrapolated from Desai et al. 2005a,b; Li et al. 2008), distributed nearly equally between otolith and canal functions. A mouse utricle has about as many hair

---

[1] For the impact of loss of vestibular hair cells from aminoglycoside overdose on two individuals, see "In the Balance," broadcasted by *Dateline NBC*, on July 31, 2006 (available at http://www.bing.com/videos/watch/video/in-the-balance/6hv0i0a) and the autobiographical case study by J.C. ("Living Without a Balancing Mechanism," *New England Journal of Medicine*, March 20, 1952, p. 458).

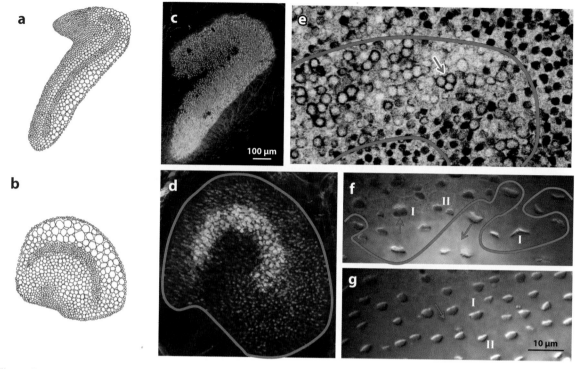

Figure 2

Zones of otolithic epithelia. (*a,b*) The otoconial masses (calcium carbonate crystals) of the saccule (*a*) and utricle (*b*) have marked zonal differences in otoconial size (guinea pig; reproduced in modified form from Lindeman 1973, with permission from the publisher, S. Karger, AG Basel). (*c*) The rat saccular macula, viewed with confocal microscopy after removal of the otoconial layer. Red stripe, approximate location of striola. Postnatal day 2. (*d*) Calcium binding protein expression reveals zones in the adult mouse utricle. Calretinin (*red*) is expressed by calyx-only afferent terminals in the striola and type II hair cells everywhere. Calbindin-D28k (*green*) is expressed by striolar afferents. Yellow indicates coexpression of calretinin and calbindin-D28k by many calyx-only afferent endings. From E.H. Peterson, unpublished results. (*e*) The striola (~50–100-μm wide, within the *red lines*) in a postnatal day 8 saccular macula at the approximate location of the box in panel *c*. Calretinin antibody labels type II hair cells (*filled black circles*) and calyx-only afferent endings (*black rings*), many of which form complex calyces. Red arrow: a triple calyx (*black*), around three type I hair cells (*light circles*). (*f,g*) Hair bundles differ between the striola (*f*) and extrastriola (*g*); rat saccular macula, postnatal day 8, viewed from above with Nomarski optics. Note the denser spacing, smaller type I bundles, and uniform bundle orientations of the extrastriola. I, II: hair cell types. Red line: line of polarity reversal of bundle orientations. Arrows: orientations of select bundles, pointing from the short edge of the bundle to the kinocilium at the tall edge. (*g*) I, II: the same bundles are labeled in **Figure 4a–b**.

cells (~3,000) as a mouse cochlea (extrapolated from Cheatham et al. 2009).

Mammalian cristae have central and peripheral zones. The maculae have corresponding zones known as the striola, a central stripe that follows the curve of the macular outline, and the surrounding extrastriola (**Figure 1b**, **Figure 2**). In the maculae, hair bundles reverse orientation at a line of polarity reversal (LPR) running through or next to the striola (**Figure 2f**). The striola has sometimes been equated with the LPR. More commonly and in this review, the striola is a swath of epithelium, 10%–20% of the macula in rodents (Desai et al. 2005b) (**Figure 2a–d**). In many ways, the striola and extrastriola resemble the central and peripheral zones, respectively, of cristae. The genetic mechanisms of zonal differentiation are not understood (Fekete & Wu 2002), but a role is likely for differential expression of neurotrophins (Sugawara et al. 2007, Gómez-Casati et al. 2010), which are required

LPR: line of polarity reversal of hair bundle orientation in the maculae

for development and stabilization of synapses between hair cells and afferents.

Afferent physiology varies with zone. Primary afferent populations of the vestibular nerves of diverse species are distinctive for the range of spike timing, from highly irregular to highly regular, which varies with epithelial zone (reviewed for several species in Lysakowski & Goldberg 2004). In mammals, afferents from peripheral zones of cristae and extrastriolar zones of maculae have strikingly regular spike timing, in contrast to the highly irregular timing of afferents from the smaller central and striolar zones (**Figure 1**) (Baird et al. 1988, Goldberg et al. 1990b). In the nerve to each organ, the coefficient of variation (CV) of interspike interval has a bimodal distribution (when normalized by rate; Goldberg et al. 1984), with modal values that differ by an order of magnitude (Baird et al. 1988, Goldberg et al. 1990a). Regularly firing afferents are the most numerous, in keeping with the large territories of the peripheral crista and extrastriolar macula.

The difference in spike regularity is so striking and easily measured that vestibular primary afferents are universally classified by their spike regularity. Spike timing may reflect encoding strategy (as discussed below) and also covaries with other properties that arise at various stages and are clearly important for signaling head motions (reviewed in Goldberg 1991, 2000). Compared with regular afferents, irregular afferents have

1. Larger diameters and higher conduction velocities (Goldberg & Fernández 1977);
2. Greater directional asymmetry favoring excitatory head motions over inhibitory head motions (Fernández & Goldberg 1976a);
3. Larger, faster responses to the efferent feedback mediated by neural projections from the brain stem to the inner ear (Goldberg & Fernández 1980, Sadeghi et al. 2009); and
4. More adapting (phasic, high-pass filtering) response dynamics (**Figure 1d**), making them more sensitive to motion

transients (reviewed in Goldberg 2000; see also Hullar & Minor 1999, Hullar et al. 2005, Sadeghi et al. 2007a, Yang & Hullar 2007, Lasker et al. 2008).

Anatomical and molecular markers of zones. As the two kinds of firing pattern and response dynamics became apparent, it was appealing to hypothesize that irregular and regular afferents were separately supplied by type I and type II hair cells (Goldberg & Fernández 1977). This idea still has some traction, despite conclusive evidence that both hair cell types supply both irregular and regular afferents (Baird et al. 1988, Goldberg et al. 1990b) at least in mammals, where type I and type II cells are both found in both zones. In the vestibular epithelia of some reptiles and birds, in contrast, type I cells are confined to the striola (Si et al. 2003, Xue & Peterson 2006).

Even in mammals, however, hair cells and synapses of a given type differ markedly between zones, and such differences may contribute to the specialization of afferent signals. Our classification schemes based on the form of the afferent ending may be overly simple. A striolar type I hair cell looks very different from an extrastriolar type I hair cell and likewise for type II cells and dimorphic afferents. Type I hair cells differ between zones in hair-bundle structure (see Hair Bundles, below) and synaptic specializations (**Figure 3**). Calyces differ between zones: The calyces of peripheral dimorphic afferents tend to be simple (around single hair cells), whereas ~50% of central calyces are complex, enveloping 2–4 hair cells (Desai et al. 2005a,b). Terminal arbors are more compact in central and striolar zones than in peripheral and extrastriolar zones (Fernández et al. 1988, 1990). Bouton-only afferents innervate just the periphery and extrastriola; like thin afferents in other systems, they express the intermediate filament, peripherin (Leonard & Kevetter 2002). Calyx-only afferents innervate just the central and striolar zones.

These anatomical features provide simple zonal markers, which can facilitate comparisons across species, organs, and studies. In the

Nomarski optics of electrophysiological experiments (**Figures 2f,g** and **7c,d**), the rodent striola can be identified by the wider spacing and larger size of hair bundles and especially by the prominence of complex calyces (Li et al. 2008). With fixed tissue, immunocytochemistry for Ca^{2+}-binding proteins defines the zones. Afferent endings and hair cells in striolar and central zones express high levels of fast Ca^{2+}-binding proteins. Calbindin antibodies stain striolar calyx-only and striolar dimorphic afferents (Hume et al. 2007) (**Figure 2d**), calretinin antibodies selectively stain striolar calyx-only terminals (Leonard & Kevetter 2002; Desai et al. 2005a,b) (**Figure 2e**), and antibody to oncomodulin stains striolar type I hair cells (Simmons et al. 2010). The functional significance of the striolar-zone Ca^{2+}-binding proteins is not established. As described below, transduction and synaptic currents may be especially large in this zone and the accompanying Ca^{2+} influx may require robust, fast Ca^{2+} buffering.

Functional significance of zones/afferent channels. Ideas and experiments addressing the functional significance of the two afferent streams were thoroughly reviewed by Goldberg (2000). The question is still not resolved. We briefly note two recent investigations that have yielded relevant results by examining vestibular activity up to higher stimulus frequencies.

In one approach, Cullen and colleagues considered information transfer at a single-afferent level over the frequency range from 0 to 20 Hz (Sadeghi et al. 2007a). For stimulus frequencies <15 Hz, regular afferent responses have greater mutual information density (bits/spike/Hz) than do irregular afferents, providing more sensitive detection thresholds and more detailed information about stimulus waveform. When average rate is the measured quantity, as noted earlier, irregular afferents have the advantage in gain (spikes/s/stimulus input) for mid-high frequencies (>2 Hz,[2] see

Figure 1d) and greater phase leads, which should reduce the response time of reflexes that they drive. These results suggested that irregular and regular firing patterns are suited to spike rate and spike timing codes, respectively.

In a quite different approach, Straka and colleagues have explored the electrical responses of single central vestibular neurons to frequency sweeps of injected current over a wide range of frequencies (Pfanzelt et al. 2008). They found two categories of central neurons that are reminiscent of the regular (tonic) and irregular (phasic) classes of primary afferents and propose that regular and irregular afferent streams persist through central vestibulo-motor transformations (Straka et al. 2009).

The functional significance of the two afferent populations depends on how their signals are differentially distributed to central targets. On this critical issue we know surprisingly little for several reasons. First, regularity as a signature feature is not preserved in central neurons, where highly regular firing is not reported (Goldberg 2000). This does not, by itself, imply mingling of the afferent streams; even if all inputs to a secondary neuron are regular, they would have to be correlated in time to evoke regular firing. Second, the Ca^{2+}-binding proteins that distinguish the afferents in the periphery are too widely distributed in neurons to be useful in brain tissue. Third, the central territories of primary vestibular afferents can be very large and overlapping. Nevertheless, evidence indicates that irregular afferents target different parts of the secondary neurons (Sato & Sasaki 1993). Irregular afferents tend to form synapses on large cell bodies, consistent with secure, rapid transmission. Regular afferents are more likely to synapse on the dendrites of smaller neurons and are therefore more likely to be integrated with other inputs. Thus, some data on central projections are consistent with the specialization of irregular, central-zone, afferents for speed and reliability of transmission.

---

[2]On the basis of the responses of mammalian canal afferents at 2 Hz (Baird et al. 1988), calyx-only, irregular, afferents have been called "low gain." This terminology can be misleading, however, because at higher frequencies these afferents have the highest gain.

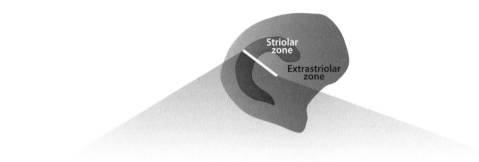

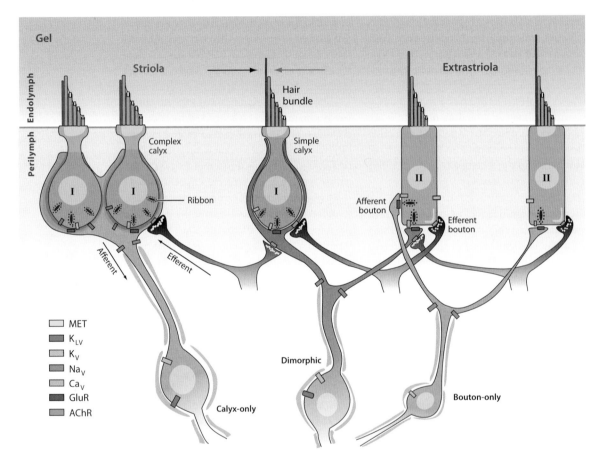

Figure 3

A slice through a rodent utricular macula is illustrated in cross-section (schematic modified from Eatock et al. 2008), showing part of the striola and part of the extrastriola. Calyceal and bouton terminals of primary afferents define two types of hair cell (I and II) and three types of afferent (calyx-only in the striola, bouton-only in the extrastriola, and dimorphic in both zones). Head motion or tilt moves the gel in contact with the hair cells' apical hair bundles. Deflection of the bundle toward its tall edge (+ deflections, *gray arrow*) opens MET channels at the stereociliary tips, depolarizing the cell and modulating current through voltage-gated ion channels (K_V, K_{LV}, Ca_V), which feeds back on the receptor potential. Synaptic vesicles arrayed around synaptic ribbons fuse with the hair-cell membrane and release glutamate onto GluR receptor channels in the postsynaptic afferent membranes, triggering spikes in the afferent nerve fibers, which are wrapped in myelin. Negative deflections (*black arrow*) close MET channels and so reduce afferent activity. Efferent nerve fibers modulate afferent signals by releasing acetylcholine (ACh) onto receptor (AChR) channels in the hair-cell and afferent-terminal membranes.

To summarize, discrete zones in mammalian vestibular epithelia give rise to two afferent streams with different spontaneous and evoked spiking properties. Because the strikingly different type I and II hair cells contribute to both afferent streams, we must look for more subtle factors in the physiological differences between regular and irregular afferents. Progress will be faster if data are specified in terms of zone; to that end, we have described simple ways to recognize zones in either unfixed or fixed tissue. The remainder of this review focuses on recent analyses of afferent stages—those controlling information flow from the hair bundle to the brain—with the potential to set differences in spike regularity and response dynamics (see sidebar, Transduction and Transmission in a Mammalian Otolith Organ, and **Figure 3** for an overview of afferent stages).

MECHANISMS UNDERLYING DIFFERENCES AMONG AFFERENT NERVE POPULATIONS

Mechanical Input to Transduction Channels Varies with Zone

The rise time of afferent signals is determined in part by the mechanical properties of structures that communicate head motions to the mechanosensitive channels: the accessory structures (otolithic membranes in otolith organs and cupulae in semicircular canals) and the hair bundles themselves. The otolithic membrane and hair bundles are strongly differentiated among zones of the maculae, suggesting that differences in mechanics contribute to differences in afferent response dynamics. Zonal differentiation of the corresponding structures in cristae is less well studied. The cupula may be more homogeneous than the otolithic membrane, but some zonal differences in crista hair bundles resemble those described for maculae (Peterson et al. 1996). Here we focus on the better-studied maculae.

TRANSDUCTION AND TRANSMISSION IN A MAMMALIAN OTOLITH ORGAN

Head motion or tilt moves extracellular matrices [gel layer (**Figure 3**) and overlying otoconia] in contact with the hair cells' apical hair bundles: arrays of linked microvilli (stereocilia) with a single true cilium (kinocilium) centered on the tall edge of the bundle. Deflection of the bundle toward its tall edge (+ deflections; gray arrow, **Figure 3**) opens mechano-electrical transduction (MET) channels at the stereociliary tips, allowing positive ions (transduction current) to enter stereocilia and depolarize the hair cell membrane (receptor potential).

The hair bundle is bathed in endolymph, a special medium with a high K^+ content (reviewed in Lang et al. 2007). Most of the transduction current is K^+, but Ca^{2+} entry is also significant. The basolateral membrane has many voltage-gated, K^+-selective (K_V) channels and is bathed in perilymph, a more standard extracellular medium with low K^+ content. Some channels (low-voltage-activated, K_{LV}) are open at resting membrane potential, and others open as the transduction current depolarizes the cell membrane, allowing K^+ to leave the hair cell. In this way, K^+ moves into and out of the hair cell along its concentration gradient.

A depolarizing receptor potential opens voltage-dependent calcium (Ca_V) channels, allowing Ca^{2+} influx that stimulates synaptic vesicles to fuse with the cell membrane, each releasing a stereotyped bolus or quantum of excitatory transmitter (glutamate) into the synaptic cleft. The synaptic vesicles are organized around electron-dense structures, synaptic ribbons, adjacent to the hair cell membrane and opposite afferent terminals. Glutamate released into the cleft opens glutamate-receptor channels (GluR) in the afferent terminal. The transient influx of positive ions [excitatory postsynaptic current (EPSC)] depolarizes the afferent membrane [excitatory postsynaptic potential (EPSP)]. EPSPs excite voltage-gated Na (Na_V) channels in the nerve fiber membrane, triggering action potentials (spikes). Spikes propagate down the myelinated afferent process, through the bipolar cell body located in the ganglion, and along the central afferent process toward secondary neurons in the brain stem and cerebellum. Calyces may also support nonquantal modes of transmission.

Some central neurons extend axons out to the vestibular epithelia, where they form efferent synaptic terminals on hair cells and afferent terminals. When stimulated by arriving spikes, efferent terminals release acetylcholine (ACh) and possibly other transmitters, modulating receptor potentials and/or afferent spike activity.

(*Continued*)

Vestibular hair cells can report motions in two directions because some MET, K_{LV}, and Ca_V channels are open at the resting bundle position. Negative deflections (black arrow, **Figure 3**) close MET channels, reducing the influx of cations, hyperpolarizing the hair cell, closing K_{LV} and Ca_V channels, and reducing transmitter release and afferent spike activity.

Accessory structures. Under linear acceleration force, the dense otoconial mass lags the underlying epithelium, deflecting the hair bundles with a predicted rise time of <50 μs in humans. For a comprehensive review of how vestibular accessory structures respond to head motions, see Rabbitt et al. (2004). Natural stimuli include head tilt, which changes the otolith angle relative to gravity, translation of the head, and substrate vibrations. Irregular, striolar afferents are also sensitive to high-frequency stimuli delivered as bone vibrations, which excite the utricle (Curthoys 2010), or loud airborne sounds, which excite the saccule.[3] In the cat, irregular saccular afferents are tuned to ~800 Hz with spike rate thresholds of ~90 dB SPL (decibels of sound pressure level referenced to 20 microPascals)—well above conversational sound level but not so loud as to be painful or damaging (McCue & Guinan 1994). Factors in the sound sensitivity of the saccule may include its proximity to the cochlea, mechanical resonance of the otolith (De Vries 1950), and micromechanical and electrical tuning at the level of striolar hair cells (below).

Although often modeled as a simple uniform mass, the otolithic accessory structure has multiple layers (Lins et al. 2000) that differ sharply between striolar and extrastriolar zones. At the top is a mass of crystalline otoconia, with diameters of 100 nm or less in rat, organic cores, and calcium carbonate shells (reviewed in Lundberg et al. 2006). Striolar otoconia are particularly small (**Figure 2a,b**) (Lindeman 1973), possibly because macular growth displaces them from extrastriolar sites of production (Kim et al. 2010). Supporting the otoconia is a gel layer, sometimes perforated with holes (Lim 1984, Jacobs & Hudspeth 1990) and connected in different ways to hair bundles. Below the gel layer, a gauzy column filament layer surrounds hair bundles and extends to the apical surface of the epithelium. Because the gel layers are secreted by supporting cells around the hair cells, extra-large perforations of the striolar gel (Lim 1984) may, in part, reflect the large apical surfaces of striolar hair cells. Zonal differences in gel thickness may influence the deflection of hair bundles that extend into the gel layer (Davis et al. 2007).

Hair bundles. Orientation of the hair bundle determines response directionality: A hair cell detects only that stimulus component parallel to a line bisecting the bundle from its short to its tall edge. In canal cristae, all hair bundles are aligned with the direction of fluid motion in the canal. Because translational head motions and gravity move the otolithic mass in multiple directions, macular hair bundles have diverse orientations, with an abrupt reversal at the LPR (**Figure 2f**).

Until recently the LPR was thought to bisect the striola. This may be approximately true for the saccular macula (J.E. Songer, unpublished observations in the rat; **Figure 2f**), but the utricular striola in both turtles and rodents lies almost entirely medial to the LPR (Xue & Peterson 2006, Li et al. 2008, Schweizer et al. 2009). This observation has functional and technical implications. Functionally, it implies that utricular striolar afferents favor a subset of motions: They are excited by head translations toward the opposite side of the head and tilt toward the same side (Li et al. 2008). Conversely, other head motions are represented by just one afferent stream: the regular, extrastriolar afferents. From a technical standpoint,

---

[3]This sensitivity has given rise to clinical tests of vestibular and brain stem function, which measure VEMPs (vestibular evoked myogenic potentials) either from neck muscle (cervical VEMP) or eye muscles (ocular VEMP). In humans, the utricle is thought to drive ocular VEMPS in response to bone-conducted vibration of the mastoid at 500 Hz, and the saccule is thought to drive cervical VEMPs in response to loud sounds of similar frequency. Controversy and agreement in this developing methodology are reviewed in Curthoys (2010) and Welgampola & Carey (2010).

experimentalists should note that in mouse and rat utricles, used increasingly for biophysical and genetic experiments, the striola does not straddle the LPR. For example, a comparison of transduction properties between mouse utricular zones (in Vollrath & Eatock 2003) probably mixed extrastriolar and striolar bundles because the utricular striola at that time was thought to straddle the LPR.

Maps of individual bundle orientations in both utricular and saccular maculae reveal tortuous LPRs and a local disorder in bundle orientations (**Figure 2f**) (Rowe & Peterson 2006), which may contribute to a comparable disorder in stimulus direction preferences of otolith afferents (Goldberg et al. 1990b, figure 4). Genetic control of hair bundle orientation includes a planar cell polarity gene complex, which may set up an underlying pattern without specifying reversal at the LPR (Deans et al. 2007). Control of bundle reversal may involve Emx3, a homeodomain protein (Holley et al. 2010). GATA3, a transcription factor with multiple actions in inner ear development, colocalizes with the LPR in the avian utricle and lagena, but not in the saccule (Warchol & Speck 2007).

The structure of hair bundles differs between zones in ways that are likely to influence afferent sensitivity and dynamics, as carefully analyzed by Peterson and colleagues in the turtle utricular macula. Here type I cells occupy a narrow belt within the striola, and type II cells differ strongly between striolar and extrastriolar zones. The numerous stereocilia of type I bundles (~100 versus 50–60 in type II bundles; Moravec & Peterson 2004) allow larger maximal transduction currents (Rennie et al. 2004). Large currents should speed up quantal transmission by depolarizing the membrane faster and may also enhance nonquantal transmission (below) by increasing current flow and K^+ flux into the calyceal synaptic cleft. Striolar bundles of both hair cell types have kinocilia close to the height of the tallest stereocilia (Peterson et al. 1996, Moravec & Peterson 2004, Rowe & Peterson 2006, Xue & Peterson 2006), which may make the bundles stiffer (Spoon et al. 2005), and wide stereociliary arrays (as viewed from above) which may increase the fluid-coupled stimulus component. Extrastriolar bundles have long kinocilia, which can extend deeply into the overlying layers, and very narrow arrays of short, fine stereocilia. In such bundles, stimulation may focus on the upper part of the kinocilium. Simulations by Nam et al. (2005) illustrate possible interactions between stimulation modes and bundle shapes.

Hair bundles in rodent otolith organs share some of the zonal differences of the turtle utricular macula—e.g., striolar kinocilia are similar in height to the tallest stereocilia (Li et al. 2008). Stereocilia tend to be thicker in type I bundles than in type II bundles, with the greatest difference in the striola (Rüsch et al. 1998). Viewed from above (**Figure 2f,g**), the broad, often wing-shaped outlines of striolar type I bundles contrast with the compact round or wedge-shaped outlines of type II bundles. A cytoskeletal structure in the apical part of the hair cell, the striated organelle, is particularly prominent in type I cells (Vranceanu & Lysakowski 2009). Although its function is not yet known, it is well positioned to interact with the cytoskeleton of the hair bundle.

In summary, striolar type I bundles have relatively numerous and thick stereocilia, may couple differently to gel layers, and may present large surface areas to endolymph motion parallel to the epithelial surface. Such differences may affect stimulation mode, bundle stiffness, and transduction current amplitude in ways that contribute to the more phasic response dynamics of striolar afferents.

Transducer Adaptation Affects Frequency Filtering and Operating Range

According to the gating-spring model of transduction in hair bundles, force is applied to MET channels by "gating springs," elastic elements that are stretched by positive bundle deflections (Corey & Hudspeth 1983). Possible gating springs include elastic parts of the MET channel (Howard & Bechstedt 2004) or even the stereociliary membrane (Kung 2005). Once

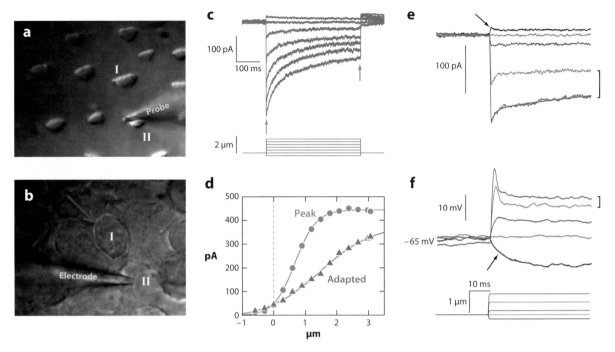

Figure 4

Transduction and adaptation in a hair cell of the rat saccular macula. (*a*) A probe deflected the hair bundle of an extrastriolar type II hair cell (bundle II in **Figure 2g**). (*b*) An electrode on the hair cell body recorded transduction current or receptor potential. Hair cells are identified as type I by the calyces that surround them (*arrow in panel* b). (*c,e*) Transduction currents evoked by families of bundle displacements (*shown below*) adapt markedly but not fully, leaving substantial steady-state responses. (*d*) Current-displacement [*I(X)*] relations measured at two time points (*see arrows in panel* c). X, the probe displacement viewed from above. Adapted I(X) relations are broader than peak I(X) relations. (*e*) More transduction currents from the same cell, expanded to show the early time course of step responses. A double-exponential fit (*red curve*) to the response to a +1-μm step has time constants of 2 ms and 70 ms. (*f*) Receptor potentials recorded from the same cell to the same bundle deflections as in panel e. Comparing (*e*) with (*f*) reveals the effects of membrane capacitance and voltage-gated conductances on the sensory signal. Membrane charging time slows the receptor potential, as shown by comparing the rise times of current (*e*) and voltage (*f*) responses to similar negative bundle deflections (*black traces, black arrows*). Red curve in panel f: monoexponential fit yielding τ_m of 7 ms. K_V channel activation compresses the voltage response to positive (depolarizing) displacements, as shown by (1) the smaller voltage response to a positive step (*maroon trace*) compared to the comparable negative step (*black trace*) and (2) the decrease in voltage per unit displacement (*square bracket*) for stimuli that evoke nearly linear increases in current (*square bracket, panel* e).

force is applied, activation of MET channels takes tens of microseconds (Corey & Hudspeth 1979), a negligible delay on the time scale of vestibular signals. As soon as MET channels start to open, however, they begin to close (adapt). This transducer adaptation proceeds by at least two mechanisms that are relevant to the time course, sensitivity, and operating range of vestibular signals.

To illustrate transducer adaptation, we show transduction currents and receptor potentials evoked by sustained (step) deflections of the hair

Adaptation: the decline with time in a sensory response to a steady stimulus

bundle of a rat saccular hair cell (**Figure 4**). The bundle is moved by a rigid probe (**Figure 4a**), while a glass electrode sealed on the basolateral membrane (**Figure 4b**) records whole-cell currents (**Figure 4c,e**) or voltages (**Figure 4f**). Positive steps evoke inward currents that rise rapidly and then decay (adapt) strongly but not completely, leaving substantial steady responses (**Figures 4c,e**). Negative steps close MET channels that are open at rest, reducing the small resting inward current (black arrow, **Figure 4e**).

Adaptation broadens the bundle's operating range and decreases sensitivity (Vollrath & Eatock 2003), as shown by comparing how peak and steady-state currents (I, in pA) vary with bundle displacement (X, in μm) as viewed from above (**Figure 4d**). These effects likely contribute to the low gains and broad operating ranges of vestibular afferents at low stimulus frequencies (Fernández & Goldberg 1976a, Baird et al. 1988), where transduction currents are partly or fully adapted. Adaptation also shifts the instantaneous operating range in the direction of the bundle displacement (Vollrath & Eatock 2003), allowing sensitivity to transient disturbances in the face of steady otolith displacements imposed by gravity (Eatock et al. 1987). At the same time, the incompleteness of adaptation provides for sustained signaling of head tilt.

In adaptation, Ca^{2+} entering the stereocilium through open MET channels reduces the channels' open probability. Although the Ca^{2+} level of vestibular endolymph is low (~100 μM), Ca^{2+} influx is still significant because MET channels have a high Ca^{2+} permeability (Jørgensen & Kroese 1995), and low intracellular free Ca^{2+} creates a strong driving force into the cell. The two time scales of adaptation (reviewed in Eatock 2000, Holt & Corey 2000) may reflect intracellular sites of Ca^{2+} action at different distances from the MET channels. In mouse utricular hair cells (Vollrath & Eatock 2003) and rat saccular hair cells (Songer & Eatock 2011), fast and slow components of adaptation have time constants (τ_A) on the order of 1–10 ms and 10–100 ms, respectively. Fast adaptation dominates responses to small stimuli, slow adaptation dominates responses to large stimuli, and both are evident at intermediate stimulus levels (**Figure 4e**) (Vollrath & Eatock 2003, Stauffer et al. 2005). Slow adaptation requires myosin 1c, a motor protein that may connect the gating springs to the actin core of stereocilia (Howard & Hudspeth 1987, Holt et al. 2002). Fast adaptation may result from Ca^{2+} binding to the MET channel protein (Howard & Hudspeth 1987, Fettiplace & Ricci 2003, Cheung & Corey 2006) or to

calmodulin associated with myosin 1c (Bozovic & Hudspeth 2003, Stauffer et al. 2005).

Figure 5 illustrates how adaptation operates in the frequency domain for better comparison with in vivo responses of vestibular afferents to sinusoidal head motions. By selectively decreasing responses at low stimulus frequencies, adaptation acts as a high-pass frequency filter. The corner frequencies for the fast and slow adaptation filters [$f_A = (2\pi\tau_A)^{-1}$; f_A in Hz, τ_A in seconds] are about an order of magnitude apart: tens of Hz and several Hz or less in both mouse utricular hair cells (Vollrath & Eatock 2003) and rat saccular hair cells (Songer & Eatock 2011). For natural stimuli, which are less rigid than the stimulus probe of **Figures 4** and **5**, adaptation's effects on MET channels or myosin 1c may amplify stereociliary motion over a narrow frequency range (Howard & Hudspeth 1987, Le Goff et al. 2005). Thus, hair cell adaptation processes can both reduce sensitivity at low frequencies (high-pass filter) and enhance sensitivity at specific frequencies (tune).

How transducer adaptation manifests in afferent responses depends on where stimulus frequencies lie relative to f_A values. In early studies that focused on low stimulus frequencies or slow stimulus trapezoids, regular afferents showed little evidence of adaptation. By extending the stimulus frequency range to higher frequencies, however, recent studies have revealed high-pass filtering even in regular afferents (Hullar et al. 2005).

Transducer adaptation as it occurs in vivo may differ quantitatively from its properties as measured in vitro. Adaptation data have been collected at room temperature and in ~1–5 mM external Ca^{2+}; bringing temperature and Ca^{2+} to physiological values may have opposing effects. The use of a rigid stimulus probe with a fast rise time (**Figure 4**) maximizes transduction-current decay and therefore the apparent adaptation. Using a fluid jet slows the decay as predicted by simple mechanical modeling of adaptation (Vollrath & Eatock 2003). Otolithic gel layers may deliver a stimulus that is intermediate between the displacement step of a rigid probe and the force step of a fluid jet.

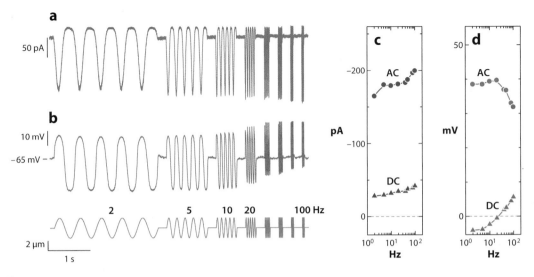

Figure 5

Effects of transducer adaptation and K_V channels on frequency dependency of the hair-cell response. (*a*) Transduction currents and (*b*) receptor potentials evoked by sinusoidal bursts of hair-bundle deflections (gray, *bottom trace*) at frequencies between 2 and 100 Hz. (*c,d*) Sinusoidal (AC, *circles*) and average (DC, *triangles*) components of the transduction current (*c*; note the inverted y-axis for better comparison with voltage in panel *d*) and receptor potential (*d*). AC current (*c*) increased with frequency with two exponential components corresponding to fast and slow adaptation processes. AC receptor potential (*d*) decreased above 20 Hz, as a consequence of membrane charging time. DC transduction current (*c*) was attenuated at low frequencies by transducer adaptation. DC receptor potential (*d*) was positive above 20 Hz, as expected from the large inward DC current (*a,c*), but reversed sign at frequencies below 20 Hz, where depolarization was slow enough to activate K_V current. Same hair cell as in **Figure 4**.

Despite these experimental differences, the reported time course of adaptation of mammalian vestibular hair cells is appropriate to contribute to the response dynamics of vestibular afferents (Songer & Eatock 2011). In addition, adaptation processes are considered essential to hair-cell mechanosensitivity (Hudspeth et al. 2000, Le Goff et al. 2005). For these reasons, we hypothesize that differences in transducer adaptation contribute to different response dynamics of striolar and extrastriolar afferents.

Basolateral K_V Channels Shape the Receptor Potential

The transduction current initiates the receptor potential, recruiting multiple voltage-dependent ion channels in the basolateral hair cell membrane to shape the time course, gain, and linearity of the afferent signal. The translation from current to voltage differs between hair cells largely through systematic variation in the expression of voltage-dependent ion channels. The most striking difference is between the K_V channels of type I and type II hair cells. For type II cells, K_V channel properties vary with zone. Whether type I cell properties also vary with zone has not been established.

The number of open ion channels (membrane conductance) determines the gain of the receptor potential ($\Delta V / \Delta I$) and its time course, quantified by the membrane time constant, τ_m. The more open channels there are, the smaller and faster the voltage response is. In **Figure 4f**, for example, the asymmetry of the size of the voltage response to negative and positive bundle deflections (black and maroon traces) arises from a difference in the number of voltage-gated ion channels opened by the two stimuli. The negative step (black traces) elicited a small fast transduction current (black arrow, **Figure 4e**), which in turn evoked a relatively large, slow hyperpolarizing receptor potential

Membrane time constant: The product of the input resistance (inversely proportional to membrane conductance) and membrane capacitance (proportional to membrane surface area)

(black arrow, **Figure 4*f***). The positive step (maroon traces) evoked a larger inward transduction current but a smaller, faster depolarizing receptor potential because it activated more voltage-gated ion channels (see below). In the frequency domain, membrane charging time introduces a low-pass filter, attenuating responses above the corner frequency f_m $(2\pi\tau_m^{-1}$, f_m in Hz, τ_m in seconds) by tenfold per tenfold increase in frequency. The input conductances of rodent vestibular hair cells vary markedly, from 1 nS to 50 nS, corresponding to τ_m values from 10 ms to 200 μs and f_m values from 15 Hz to 750 Hz.

The wide range of reported input conductances reflects systematic variation between hair cell types and zones in the expression of ion channels that are gated directly or indirectly by voltage: inward rectifying and outward rectifying K channels, HCN (hyperpolarization-activated cyclic-nucleotide modulated) channels, Ca_V channels, and, in young animals, Na_V channels. Some of the variation is developmental. Over the first postnatal month, rat and mouse hair cells tend to add K_V channels and lose Na_V and Ca_V channels, with the overall effect of making them less excitable [reviewed in Eatock & Hurley (2003) and Goodyear et al. (2006); see Géléoc et al. (2004), Hurley et al. (2006), Wooltorton et al. (2007), Schweizer et al. (2009), and Li et al. (2010) for recent observations on developmental changes in K_V, Na_V, and K(Ca) channel expression specific to vestibular hair cells]. Changes may continue past the first month; the electrophysiological properties of mouse primary canal afferents continue to differentiate past seven weeks (Lasker et al. 2008). Nevertheless, the main reason for the unusually wide range of input conductances is the differential expression of outward rectifying K_V channels by type I and type II hair cells, a mature property but one that can be detected in mouse hair cells even before birth (Géléoc et al. 2004).

K_V channels in type II cells.

In type II cells, outward-rectifying K_V channels have a low open probability at resting potential and are activated by depolarizing receptor potentials. The channel types include rapidly inactivating (A-type) channels and slowly inactivating (delayed rectifier) channels in different proportions, depending on hair cell type and zone. A current family evoked by a typical voltage protocol is illustrated for a type II cell in **Figure 6*a***. These data are used to generate activation or $G(V)$ curves (**Figure 6*c***), showing how membrane conductance (G) varies with voltage.

Following a depolarizing current step or bundle deflection, K^+ currents exit the hair cell through activated K_V channels, repolarizing the membrane (*red arrow*, **Figure 6*d***, inset) and contributing to adaptation of the receptor potential (**Figure 4*f***). The voltage response typically decays with at most a slight rebound or weak oscillation (e.g., Angelaki & Correia 1991, Rennie et al. 1996, Rüsch & Eatock 1996), indicating that the hair cell membrane has broad electrical tuning, consistent with the broad tuning of mammalian (and avian; Dickman & Correia 1989) afferent spike rates to head-motion stimuli. These data contrast with the sharp electrical tuning of frog saccular hair cells (Hudspeth & Lewis 1988, Rutherford & Roberts 2009), which is likely to enhance afferent sensitivity to substrate vibrations and sounds. Neonatal rat saccular hair cells, however, may have sharper electrical resonances than do more mature cells (J.E. Songer, unpublished results), which could benefit development of the immature vestibular system by boosting transmitter release and afferent spike activity.

K_V channel activation selectively compresses the steady-state response to depolarizing currents, as shown by steady-state voltage changes evoked by current steps (bracket, **Figure 6*d***) or displacement steps (compare **Figures 4*e,f***). Steady-state $I(X)$ and $V(I)$ relations have complementary asymmetries: Positive bundle displacements evoke larger depolarizing currents (**Figure 4*d***), but depolarizing currents evoke smaller voltages (**Figures 4*f*** and **6*f***), together producing a more linear $V(X)$ relation (Holt et al. 1999). This effect has a frequency dependency

Low-pass filter: filters out frequencies above a corner frequency

Outward rectifying K channels: voltage-gated potassium channels that tend to be open at voltages for which K^+ current flows out of the cell

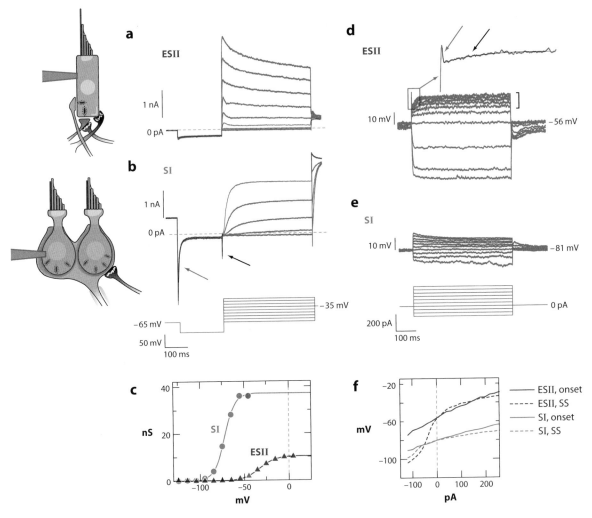

Figure 6

The K_V channels of type I and type II hair cells shape distinctive voltage responses to current. (*a,b*) A standard voltage-clamp protocol (below *b*) evokes different whole-cell currents in an extrastriolar type II cell (ESII) and a striolar type I cell (SI). The cells were clamped to −65 mV, then stepped to −125 mV to deactivate low-voltage activated currents, and then stepped positively to show how currents activate with increasing depolarization. (*a*) K_V current in the ESII cell activates positive to resting potential (−65 mV), in this case the same as the holding potential. (*b*) K_{LV} currents in the SI cell are activated at the holding potential, producing a positive current. They are deactivated by stepping to −125 mV (*red arrow*) and then reactivated by stepping positive to −90 mV. Fast inward current (*black arrow*) is Na_V current. The red traces in panels *a* and *b* highlight responses to the same −45 mV step. (*c*) Conductance-voltage, $G(V)$, relations for the data in panels *a* and *b*. Boltzmann (sigmoidal) fits have activation midpoints of −72 mV (type I) and −32 mV (type II). Red symbols show values at −45 mV, from the red traces in panels *a* and *b*. (*d,e*) Voltage responses to injected current steps (shown below panel *e*) differ between hair cell types because they have different complements of K_V channels. For the type II cell (*d*), strong depolarization rapidly activates K_V channels, causing a fast voltage decay (inset, $\tau \sim 1$ ms, *red arrow*) and compressing the gain ($\Delta V/\Delta I$) (*bracket*). Inactivation of K_V channels then produces a gradual depolarization (inset, $\tau \sim 20$ ms, *black arrow*). For the type I cell (*e*), voltage responses are small because of the large K_{LV} conductance at resting potential. (*f*) Voltage-current, $V(I)$, relations for onset and steady-state responses from the data in panels *d* and *e*. Onset gain (slope) is smaller in the type I cell than the type II cell because the type I cell has a high resting conductance. The steady-state $V(I)$ relation of the type II cell changes slope (rectifies) around the 0-current (resting potential) value because positive currents activate the large K_V conductance. In contrast, the steady-state $V(I)$ relation of the type I cell is linear around the 0-current value because K_{LV} channels are already activated at resting potential. Rat saccular macula, postnatal days 7-8.

determined by the kinetics of the K_V channels. For the type II cell of **Figures 4** and **5**, $I(X)$ and $V(I)$ asymmetries cancelled each other at 20 Hz (**Figure 5b**), producing a symmetric receptor potential (zero "DC" component at 20 Hz; **Figure 5d**). This mechanism for compensating the asymmetry of transduction could account for the symmetric linearity of regular afferents (Hullar & Minor 1999).

K_V currents of type II cells differ with zones (Masetto & Correia 1997, Holt et al. 1999, Weng & Correia 1999, Brichta et al. 2002) in ways that affect gain, frequency filtering, and response symmetry (linearity). Because peripheral/extrastriolar K_V channels are faster to activate, they will influence the receptor potential up to higher stimulus frequencies. They also show a greater tendency to inactivate, which gradually relieves their negative feedback on depolarizing responses (**Figure 6d**, black arrow)—an effect that could oppose the slow transducer adaptation and help make regular afferent responses more tonic (less phasic) than the responses of irregular afferents.

K_V channels in type I cells. The outward rectifying K^+ currents of type I hair cells (**Figure 6b**) differ markedly from those of type II cells, as originally described in birds (Correia & Lang 1990) and confirmed in mammals and reptiles (reviewed in Eatock & Lysakowski 2006). Type I K_V channels include one or more channel types that activate at least 30 mV more negatively than do K_V channels of type II cells (**Figure 6c**). The conductances have been called g_{KI}, for type I–specific conductance, or $g_{K,L}$, for low-voltage-activated conductance. Here we refer to them as K_{LV} channels to emphasize their similarity to low-voltage-activated channels in primary afferent neurons.

The K_{LV} channels of type I cells activate more slowly than K_V channels of type II cells, inactivate little, and can be much more numerous. As a consequence of their large resting K^+ conductance, type I hair cells have relatively negative resting potentials, high input conductances, and short τ_m values. These properties yield small, fast voltage responses to

input currents (**Figure 6e,f**) and high corner frequencies for low-pass filtering (Rennie et al. 1996, Holt et al. 1998). The more negative resting potentials may contribute to the low background firing rates of some irregular afferents by reducing transmitter release at resting potential. Because they reduce membrane charging time, K_{LV} channels are thought to enhance temporal fidelity, for example, in neurons in auditory timing pathways (Rothman & Manis 2003). K_{LV} channels also make cell membranes more linear by reducing voltage responses (compare the slopes of the onset $V(I)$ relations for the type I and type II cells in **Figure 6f**), thereby reducing the activation of voltage-sensitive (nonlinear) conductances.

The K_{LV} conductance of type I hair cells may comprise more than one channel type, with subunits from the KCNQ ($K_V7.x$) and erg ($K_V11.x$) families (Kharkovets et al. 2000, Rennie et al. 2001, Hurley et al. 2006, Holt et al. 2007b, Rocha-Sanchez et al. 2007), the same K_V channel families that form neuronal M-currents (Selyanko et al. 2002). Like neuronal M-currents, K_{LV} currents of type I cells can be modulated by second messengers (Behrend et al. 1997, Chen & Eatock 2000, Rennie 2002, Hurley et al. 2006), which may be triggered by efferent or retrograde signals from the calyx. For example, efferents or calyces might, in response to their own activation, release nitric oxide (Lysakowski & Singer 2000), which could close K_{LV} channels (Chen & Eatock 2000) and make the receptor potential larger and slower, more like that of a type II hair cell. This kind of action could contribute to the large excitatory effect of efferents on irregular afferents (Sadeghi et al. 2009).

K_{LV} channels may also be important for the flow of K^+ through the hair cell during transduction. Cochlear outer hair cells also have a substantial K_{LV} conductance, which appears to be an important part of the K^+ circulation pathway (Lang et al. 2007). Similarly, K_{LV} conductances may help support relatively large standing currents through the type I hair cells. Such currents may increase K^+ concentration in the thin synaptic cleft surrounding most of the

Low-voltage-activated (LV): ion channels that are open at or near resting membrane potential

EPSC: excitatory
postsynaptic current

EPSP: excitatory
postsynaptic potential

basolateral hair cell membrane (Lim et al. 2011), dynamically altering the K^+ reversal potential and the current through K_{LV} channels.

In summary, the size, voltage dependency, and time course of K_V conductances differ among hair cell types and zones in ways that are relevant to vestibular afferent physiology. In particular, the different voltage dependencies of K_V channels in type I and type II cells make type I receptor potentials faster and smaller than type II voltage responses. More modest zonal variations in type II K_V channels could contribute to differences in response dynamics and linearity between regular afferents, which contact peripheral type II cells, and irregular dimorphic afferents, which contact central type II cells.

Filtering by the Synapse

The receptor potential drives chemical transmission from ribbon synapses in the hair cells onto afferent endings. **Figure 7** shows examples of excitatory postsynaptic currents (EPSCs) and excitatory postsynaptic potentials (EPSPs) recorded from a rat saccular complex calyx while one of the input type I hair bundles was deflected by a rigid probe driven at 20 Hz. Each cycle of bundle motion triggered a response at a particular phase in the cycle; such phase-locking illustrates the temporal precision of stimulus processing summed over all processing stages preceding the EPSC. Hair cells are evidently very good at minimizing this processing time: In the cochlea, the delay between basilar membrane motion and afferent response is $\sim$1 ms (Temchin et al. 2005). The brief latency of the monkey VOR (Huterer & Cullen 2002) and the ability of cat saccular afferents to phase-lock to 800-Hz tones (McCue & Guinan 1994) indicate that vestibular synapses are also specialized for speed and reliability of transfer. Adaptation of vesicular release (Furukawa & Matsuura 1978, Avissar et al. 2007) causes EPSC and EPSP size to decrease with successive cycles during a sinusoidal burst (**Figure 7**).

Here we consider emerging information about processes that may influence delay,

precision, reliability, and adaptation of transmission at vestibular hair cell synapses. Processes that slow transmission act as low-pass filters, whereas synaptic adaptation processes act as high-pass filters.

Calcium channels and calcium sensing. Depolarization opens Ca_V channels in hair cells, and the ensuing Ca^{2+} influx triggers transmitter release (exocytosis). The exocytosis evoked by voltage steps is faster in type I cells than in type II cells, and the greater speed depends in part on a Ca^{2+}-binding protein, otoferlin (Dulon et al. 2009)[4], which is also critical for afferent transmission in the cochlea. Otoferlin may be the hair cell's Ca^{2+}-sensing intermediary between Ca^{2+} influx and exocytosis, assuming the role played by synaptotagmin in other synapses (Roux et al. 2006).

Exocytosis in cochlear hair cells is exclusively mediated by $Ca_V 1.x$ (L-type) channels (Spassova et al. 2001). Experiments on null mutant mice show that $Ca_V 1.3$ channels are a smaller percentage of all Ca_V channels in utricular hair cells ($\sim$50%; Dou et al. 2004) than in cochlear inner hair cells ($\sim$90%; Platzer et al. 2000), mostly because inner hair cells have a very large number of $Ca_V 1.3$ channels. All Ca_V channels in rodent vestibular hair cells and other hair cells share biophysical properties that are critical to function at these highly active and specialized synapses. (*a*) Channel activation and deactivation are rapid, with time constants of $\sim$1 ms at room temperature (Bao et al. 2003), ensuring that the Ca_V channel response time does not slow down the transmitted signal. (*b*) Ca_V channels are located immediately below the ribbons (Roberts et al. 1990, Frank et al. 2009), allowing for rapid, large jumps in Ca^{2+} next to docked vesicles, reducing the

---

[4]Dulon et al. (2009) found that knocking out otoferlin affected gross vestibular potentials evoked by linear acceleration transients (VsEPs, Jones et al. 2002). Previously no vestibular phenotype was detected in otoferlin-null mice, perhaps because testing was of motor activities that compensate with input from other senses. This result suggests that we should use VsEP testing to reevaluate vestibular function in other deafness mutants.

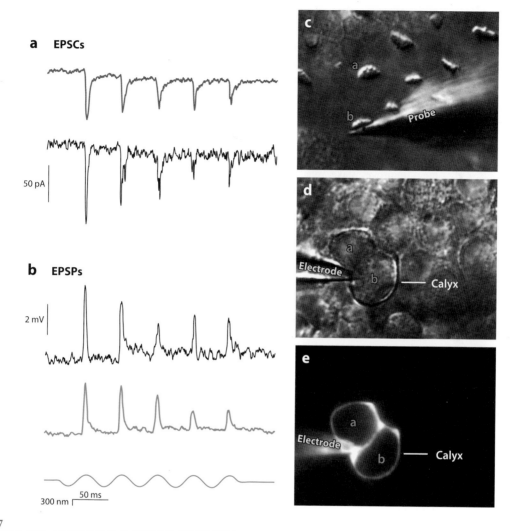

Figure 7

Transmission from a type I hair cell onto a complex calyx shows phase-locking and synaptic adaptation. (*a*) Excitatory postsynaptic currents (EPSCs) and (*b*) excitatory postsynaptic potentials (EPSPs) were driven by 20-Hz bundle deflections with a rigid probe (*c*). Individual sweeps (*black traces*) and averages of several sweeps (*colored traces*) are shown. The diminution of responses following the first stimulus cycle arises from synaptic adaptation because it is not seen in the receptor potential (*not shown*). This complex calyx, visible in panel *d*, surrounded two hair cells (cells a and b; bundles shown in panel *c*) and was filled with fluorescent dye from the pipette (*e*). Rat saccular striola, postnatal day 4.

delay between depolarization and transmitter release. (*c*) The channels have a relatively negative voltage range of activation and low tendency to inactivate (Bao et al. 2003, Dou et al. 2004), creating enough Ca^{2+} influx at resting potential to drive high background firing rates: 40–60 spikes/s in mouse canal afferents (Yang & Hullar 2007) and 90 spikes/s in primate canal afferents (Goldberg & Fernández 1971). Background spiking is essential to the bidirectionality and linearity of vestibular afferents. Some irregular (central-zone) afferents do have a low background rate, which can be silenced by a small negative stimulus—a phenomenon called "inhibitory cut-off" in the vestibular afferent literature (Hullar et al. 2005).

Synaptic ribbons. The presynaptic ribbon is likely to play a critical role in holding "readily releasable" synaptic vesicles at high concentration near the synaptic active zone (Wittig & Parsons 2008). In cochlear hair cells, ribbons promote short-latency, reliable transmission, possibly by bringing synaptic vesicles and Ca_V channels into close proximity. Experiments on mutant animals in which ribbons are not anchored in place show that normal ribbon placement improves spike reliability and reduces response latency. It is not critical, however, for temporal precision as measured by phase-locking (Buran et al. 2010). In normal cochlear hair cells, many vesicles (up to 25, on average 7) are released simultaneously in a process called multivesicular release, producing large, fast EPSCs that rapidly drive the afferent terminal to spike threshold (Grant et al. 2010). Emerging results from vestibular afferent terminals (Rennie & Streeter 2006, Dulon et al. 2009) suggest a narrower distribution of EPSC sizes, more like that of amphibian papillar hair cells (Li et al. 2009), for which the usual number of simultaneously released vesicles is 2–3 and the maximum number is 5–6. As in the cochlea, glutamate released from both type I and type II vestibular hair cells activates fast AMPA (α-amino-3-hydroxy-5-methyl-4-isoxazolepropionic acid) receptors (Bonsaquet et al. 2006, Holt et al. 2006b, Holt et al. 2007a, Dulon et al. 2009).

The number and size of synaptic ribbons may also influence the availability of readily releasable vesicles and therefore the size and speed of the onset afferent response and the kinetics of synaptic adaptation. Ribbon numbers and size vary with zone in the mature chinchilla crista (Lysakowski & Goldberg 1997). Type I cells tend to have small, round ribbons, whereas type II cells have ribbon-shaped and barrel-shaped structures that are larger, especially in central zones. The effect of ribbon size is not understood; in mammalian cochlear afferents, ribbons facing low-spontaneous-rate afferents are, counterintuitively, larger and more complex than those facing high-spontaneous-rate fibers (Merchan-Perez &

Liberman 1996); perhaps small ribbons keep all associated vesicles closer to vesicle docking sites at the plasma membrane. In the cochlea, each afferent receives input from ~1 ribbon. In the chinchilla crista, in contrast, each central-zone calyx ending receives inputs from ~20 ribbons in each type I cell it contacts (Lysakowski & Goldberg 1997). Because head motions coactivate nearby hair cells of similar orientation, a complex calyx with three hair cells is barraged with glutamate from ~60 closely spaced ribbons, which should produce large, fast EPSCs. Each regular afferent also sums input from tens of ribbons (see tables in Fernández et al. 1988, 1990), but the greater distances between ribbons, together with the finer caliber of the fibers, may spread out EPSP arrival times at the spike initiation zone.

Synaptic adaptation. Synaptic adaptation of auditory hair cells can be fit with multiple exponential decays, modeled as depletion of successive pools of synaptic vesicles at ever-greater radii from the active zone (Schnee et al. 2005). We know less about such adaptation mechanisms in vestibular afferents, both because it is hard to deliver sufficiently fast head motions and because of a long-standing emphasis on frequency domain analysis, which quantifies the steady-state (fully adapted) responses to sinusoidal stimuli, ignoring onset effects.

Preliminary comparisons of hair cell receptor potentials and EPSCs in an in vitro preparation of the rat saccule suggest that synaptic adaptation, by high-pass filtering of the receptor potential, sharpens tuning in the vestibular frequency range (Songer & Eatock 2011).

Differential expression of synaptic transmitters and receptors may help diversify afferent response dynamics, as suggested by Highstein et al. (2005) based on data from the toadfish crista, where certain hair cells are highly immunoreactive for both GABA and glutamate (Holstein et al. 2004a). A slow adaptation component of the afferent response to canal indentation is mediated by GABA-B receptors, possibly activated by hair cell release of GABA (Holstein et al. 2004b).

Nonquantal transmission. In addition to quantal transmission, Type I–calyceal transmission may include an unconventional, nonvesicular mode. In a semi-intact preparation of the chick crista (Yamashita & Ohmori 1990), mechanical stimulation evoked EPSPs from bouton afferents, but the voltage responses of calyx afferents looked like low-pass-filtered versions of the hair cell response. There are no gap junctions at type I–calyx synapses, however, to mediate conventional electrical transmission (Gulley & Bagger-Sjöbäck 1979, Ginzberg 1984, Yamashita & Ohmori 1991). Intracellular recordings from turtle crista afferents can be dissected into quantal and nonquantal components; the nonquantal components are largest in afferents that contact type I hair cells (Holt et al. 2007a). Candidate mechanisms include K^+ accumulation in the narrow synaptic cleft (Goldberg 1996), which would depolarize membranes by shifting the K^+ equilibrium potential, and ephaptic transmission, which occurs when currents flowing out of one cell create extracellular voltages that polarize a neighboring cell. Such transmission is proposed for the triad synapse in the retina (Fahrenfort et al. 2009). Like the triad synapse, the calyx has a thin, extensive synaptic cleft that might support an appreciable voltage.

Nonquantal transmission may be two-way. The inner calyx membrane, facing the cleft, is studded with KCNQ channels at a high density (Kharkovets et al. 2000, Hurley et al. 2006, Sousa et al. 2009), especially in the central and striolar zones (Lysakowski & Price 2003). Thus, spikes in the calyx will lead to strong K^+ efflux into the cleft from the postsynaptic side and could also change cleft voltage. Nonquantal components may be largest in central-zone complex calyces, where transduction and K_V currents of hair cells are large, calyceal KCNQ channels are dense, and contributions from several hair cells and postsynaptic surfaces sum together. They also have many more "invaginations," where the calyx pushes slightly into the hair cell and the cleft narrows (Lysakowski & Goldberg 1997). These structures could, like punctae in the calyx of Held (Sätzler et al. 2002),

fasten pre- and postsynaptic surfaces together (Goldberg 1996), and their greater number in the striola may be an adaptation to the larger currents that flow through striolar synapses.

Nonquantal transmission may boost signaling speed and improve reflex responses to fast head motions because it avoids the delays of vesicle fusion, transmitter diffusion across the cleft, and binding of the transmitter to postsynaptic receptors. In this way, it could serve a similar function to electrical transmission components of other fast pathways, such as the Mauthner escape reflex of fish (Sillar 2009). In excised preparations of mature mouse cristae (Lee et al. 2005) and turtle cristae (Bonsaquet et al. 2006), however, various methods of blocking quantal transmission eliminated all afferent spiking, suggesting that spiking requires vesicular release and that nonquantal transmission plays at most a modulatory role.

In summary, numerous pre- and postsynaptic specializations vary with epithelial zone and so may play a role in zonal differences in afferent firing patterns and response dynamics. In general, central/striolar hair cells tend to have more of everything: more ribbons, more invaginations, more K_V channels. On the postsynaptic side, central/striolar-zone calyces receive inputs from larger numbers of ribbons and express more K_{LV} channels (Lysakowski & Price 2003) and Na_V channels (Wooltorton et al. 2007) than do peripheral/extrastriolar calyces. The intense transmitter barrage and ion channel expression of central-zone calyces should reduce their membrane resistance and therefore the rise time of EPSPs. It remains to be determined whether unconventional modes of transmission, involving K^+ accumulation or ephaptic voltage changes in the calyceal cleft, significantly affect signaling at calyx synapses and help differentiate transmission onto irregular (central) and regular (peripheral) afferents.

Filtering at the Spike Generator Stage

EPSPs of sufficient size and rate of voltage change trigger spiking in the afferent terminal, between the synaptic zone and the first wrapping of myelin (internode). Aspects of spike generation that may affect spike latencies include electrotonic distance from the active zone to the spike initiation zone, charging of the afferent's membrane capacitance, spike threshold mechanisms, and spiking at preferred intervals. Each of these attributes differs between regular and irregular afferents and, therefore, has the potential to contribute to their different response dynamics. The striking difference in Ca^{2+}-binding protein expression by regular and irregular afferents (**Figure 2d**) may also affect afferent activity. All afferents express high levels of parvalbumin, but only irregular afferents also express high levels of calretinin and/or calbindin D-28. These proteins could act to mop up Ca^{2+} signals rapidly, either as protection from excitotoxicity or as required for good temporal resolution (Edmonds et al. 2000).

To address the contribution of the spike-generating stage to afferent response dynamics, Goldberg et al. (1982) drove firing in chinchilla vestibular afferents with sinusoidal currents (<4 Hz) supplied by an extracellular electrode. The stimulated spike rates were insensitive to the frequency of the injected currents, indicating that spike accommodation is not likely to contribute to high-pass filtering of irregular afferents. Spike-generating mechanisms may affect the timing and tuning of vestibular afferents in other ways, however. Here we raise several possibilities: spike thresholding, electrotonic differences, and electrical resonances or preferred intervals.

Spike generation can sharpen tuning by simply not transmitting subthreshold signals, such that the frequency range over which spikes are triggered is narrower than the frequency range that evokes EPSPs. This effect has been seen in preliminary results from immature afferents (Songer & Eatock 2011). Because spike thresholds determine which signals pass the spike generator filter, threshold differences arising from different ion channel complements (Kalluri et al. 2010) may affect spike rate tuning. In particular, expression of K_{LV} channels by irregular afferents may

increase spike thresholds and cause them to reject more EPSP input. This effect could play a role in the greater high-pass filtering of irregular afferents relative to regular afferents.

The different morphologies of irregular and regular afferent terminals likely affect the delay and rise time of generator potentials at the spike initiation zone. Central/striolar afferents sum inputs from closely spaced ribbons at a nearby spike initiation zone. Peripheral/extrastriolar afferents sum inputs from a great many ribbons distributed over many hair cells covering larger epithelial territories. The greater average distance from active zones to the common spike-initiating zone—presumably adjacent to the first internode—may delay spike initiation but encourage summation and integration, thereby enhancing sensitivity.

The preference for specific interspike intervals of regular afferents may specifically enhance their sensitivity at the corresponding head motion frequencies. In chinchilla and mouse, preferred intervals range from 10 to 30 ms (Baird et al. 1988, Goldberg et al. 1990a, Lasker et al. 2008), yielding preferred frequencies (preferred interval^{-1}) of 30 to 100 Hz. Young mouse ganglion cells respond to sinusoidal current injections with broad tuning centered on 5–10 Hz at room temperature (Risner & Holt 2006). At body temperature, the preferred frequencies would likely be 15–30 Hz, similar to mean background frequencies recorded from immature rat canal afferents in vivo (Curthoys 1979). Regular spiking in frog saccular afferents is driven by high-quality electrical resonances in the hair cells (Rutherford & Roberts 2009) at frequencies that match best frequencies for mechanical stimuli. In mammals, however, mechanisms intrinsic to the vestibular afferents are important in setting regularity.

Mechanisms of Spike Regularity

In vivo experiments in which spiking was driven by extracellular current injections suggested that intrinsic electrical properties of afferents were responsible for differences in spike regularity (Goldberg et al. 1984, Smith & Goldberg 1986). With the advent of patch clamp methods, we can identify which ion channels regulate the excitability of vestibular afferents. Initial characterizations have been of afferent cell bodies isolated from the vestibular ganglion (reviewed in Eatock et al. 2008), a more accessible preparation for patch clamping than the distal terminals where spike patterns originate. On the basis of several studies (Limón et al. 2005, Risner & Holt 2006, Iwasaki et al. 2008, Kalluri et al. 2010), the cell bodies fall into two groups. Transient neurons respond to positive current steps with single onset spikes and have relatively large somata, negative resting potentials, high current thresholds for spiking, a broad frequency range (as determined from sinusoidal current injections), and small spike afterhyperpolarizations (AHPs). In many respects, this group fits the expected profile for irregular afferents. Sustained neurons generate many spikes in response to current steps and have other properties expected of regular afferents, including smaller cell bodies and prominent AHPs.

Shorn of their synaptic inputs, isolated vestibular cell bodies do not fire spontaneously. To obtain measures of regularity that could be compared with in vivo data, Kalluri et al. (2010) drove spiking with synthetic EPSCs delivered in a fixed series with pseudorandom timing. As predicted, these stimuli evoke irregular firing from transient neurons and regular firing from sustained neurons (**Figure 8a**). Both spike irregularity and the transient firing pattern depend on the expression of K_{LV} channels, including members of the K_V1 (dendrotoxin-sensitive) and K_V7 (M-like KCNQ) families (Iwasaki et al. 2008, Kalluri et al. 2010, Pérez et al. 2010). Blocking both channel types converts current-step responses from the transient pattern to the sustained pattern and converts spike timing of pseudo-EPSC-driven responses from irregular to regular (**Figure 8b**) (Kalluri et al. 2010). By reducing the membrane time constant, K_{LV} channels may also help

differentiate the time course of EPSPs, which are brief and temporally isolated in irregular afferents and more rounded and summating in regular afferents (Schessel et al. 1991, Kalluri et al. 2010).

Thus, the pattern of K_{LV} channel expression can account for differences in excitability and spike timing of ganglion cell bodies. Whether this difference accounts for spike timing at the afferents' spike initiation zones in the vestibular epithelia remains to be determined. Immunocytochemistry shows greater expression of $K_V7.4$ subunits in central/striolar (irregular) calyces than in peripheral/extrastriolar calyces (Hurley et al. 2006), consistent with an important role for K_{LV} channels. But differences in other channels such as Na_V and HCN channels may also affect input resistance and excitability at afferent terminals.

Note that physiological and immunocytochemical results place M-like K_{LV} channels on both sides of the calyx, in the type I membrane and the inner face of the calyx. Both membranes are therefore low-resistance pathways for K^+ flow and may be jointly modulated by cleft substances, such as K^+ or glutamate, or by nitric oxide released from afferent or efferent terminals (Chen & Eatock 2000, Lysakowski & Singer 2000).

Efferent Feedback from the Brain to Vestibular Hair Cells and Afferents

A modest number of efferent neurons (~300) on each side of the brain stem send axonal processes to both vestibular labyrinths to terminate on type II hair cells and on calyceal endings and other terminals (**Figure 3**) (see Sadeghi et al. 2009 for connectivity). The terminals release acetylcholine (ACh) and possibly other transmitters and modulators such as GABA, CGRP (calcitonin gene-related peptide), and nitric oxide (reviewed in Eatock & Lysakowski 2006).

Cullen and Minor and colleagues have recently refuted two attractive hypotheses for the function of efferent feedback to the vestibular inner ear. In alert primates, efferents to the

ACh: acetylcholine

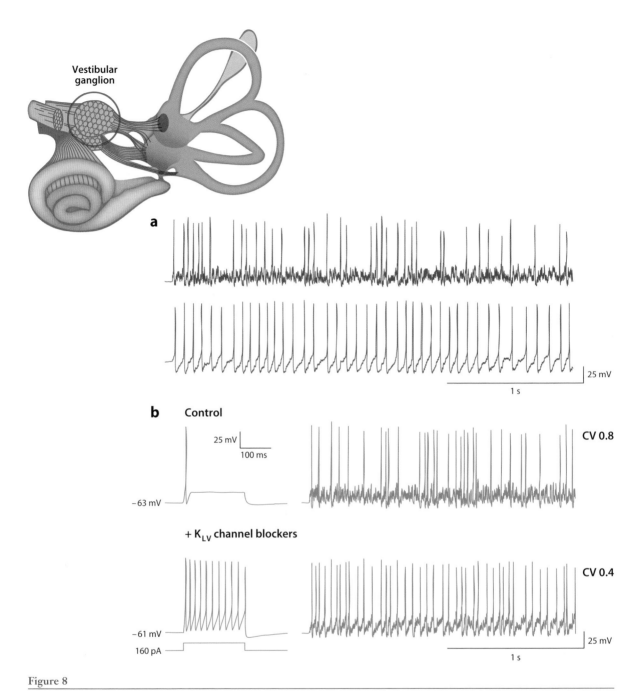

Figure 8

Spike regularity in vestibular afferent neurons is affected by neuronal expression of K_{LV} channels (modified from Kalluri et al. 2010). (*Inset*) Vestibular afferent cell bodies were isolated from the rodent vestibular ganglion then studied with the patch clamp method in current clamp mode. (*a*) Firing driven by the same temporal pattern of pseudo-EPSCs was irregular in some isolated cell bodies (*top*) and regular in others (*bottom*). (*b*) K_{LV} channels control the firing pattern. Control: Neurons that respond with a single onset spike to a current step (*left*) generate irregular firing in response to pseudo-EPSC trains (*right*). +K_{LV} channel blockers: Blocking two classes of K_{LV} channels converted the step response to sustained (*left*) and made firing twice as regular (*right*), as measured by coefficient of variation (CV). Effects were robust across different mean firing rates.

inner ear do not reduce the vestibular afferents' gain during active head movements (Cullen & Minor 2002, Jamali et al. 2009), although an "efference copy" mechanism of this sort is in place at the secondary neuron level in the brain stem (Cullen et al. 2001). Nor do efferents drive compensation for unilateral loss of vestibular afferent activity (Sadeghi et al. 2007b). Results of Sadeghi et al. (2009) suggest that vestibular efferents participate in a fast, positive feedback loop: High activity in vestibular afferents stimulates efferents to further stimulate the afferents, with the greatest effects on central, irregular afferents.

Work from the turtle crista (Holt et al. 2006a) suggests a mechanism for the excitation. Efferent terminals release ACh onto calyceal terminals, opening nicotinic receptor-channels (AChRs), depolarizing the terminals and increasing afferent spike rates. Other possible mechanisms for excitation include second-messenger-mediated inhibition of K_{LV} channels in the calyx endings. $K_V7.x$ channels in vestibular ganglion somata can be modulated by ACh via muscarinic receptors (Pérez et al. 2010); the greater $K_V7.4$ immuno-labeling in central/striolar calyces could play a role in the greater efferent actions on central, irregular afferents. If efferents do suppress $K_V7.x$ channel activity, the results of Kalluri et al. (2010) suggest that efferent-mediated excitation of irregular afferents could be accompanied by increased regularity of spike timing.

CONCLUDING REMARKS

Afferent properties vary with location in vestibular epithelia. In mammals, central (striolar) and peripheral (extrastriolar) zones have sharp morphological and molecular boundaries. In central/striolar zones, the large sizes of hair bundles, afferent terminals, and axons are compelling indicators that irregular afferents are specialized for rapid signaling. The more distributed and finer arbors of

peripheral/extrastriolar afferents may allow for more spatial and temporal integration, improving the signal-to-noise ratio at the cost of speed.

We have reviewed how such biophysical properties as transducer adaptation, expression of low-voltage-activated K channels, synaptic-cleft resistance, and spike thresholds could contribute to diversity of afferent response dynamics. With a few exceptions, we lack systematic comparisons of these properties across zones, whether because hair cells or neurons were dissociated, zonal boundaries were unclear, or investigators were more focused on type I versus type II comparisons. Given a renewed attention to intact epithelial preparations and new clarity on zonal boundaries, we anticipate obtaining the appropriate comparisons.

The functional significance of the zonal difference in spike regularity could also relate to differences in how stimulus frequency is mapped on the two zones. In their information-theoretic approach, Sadeghi et al. (2007a) found that mutual information density of regular afferents greatly exceeds that of irregular afferents for frequencies below 15 Hz. In contrast, the average rates of irregular afferents have greater gains and phase leads at mid to high frequencies. Thus, the new emphasis on testing neuronal and behavioral responses at stimulus frequencies above several Hz may better discriminate the roles of central/striolar and peripheral/extrastriolar afferents. Expanding the upper end of the tested frequency range also focuses on head motion frequencies for which vestibular signals are more robust and timely than visual and somatosensory signals. Below 2 Hz, both mice and primates—our most popular behavioral models—may manage peripheral vestibular loss well because of functional redundancy with the slower visual and somatosensory systems. By challenging the nervous system with high-frequency head motions, we can better appreciate how the vestibular inner ear allows us to navigate a rapidly moving sensory environment.

SUMMARY POINTS

1. Mammalian vestibular epithelia are organized in zones that give rise to different streams of afferent nerve fibers. Central hair cells of canal organs and striolar hair cells of otolith organs transmit signals to large afferents with high conduction speeds and phasic (adapting) response dynamics. Peripheral and extrastriolar afferents have complementary properties. Neither the function of these different streams nor the mechanisms giving rise to their differences is fully understood.

2. The zones are crisply delineated by various anatomical and molecular markers, which can be seen at multiple levels, from the accessory structures to the primary afferent terminals.

3. Although there are two hair cell types (I and II), two types of afferent terminal (calyx and bouton), and two types of afferent nerve fiber (irregular and regular), there is no simple correspondence between the classifications.

4. In the past decade, the known frequency range of vestibular function in mammals has increased at its upper end to 30 Hz or more, as measured by head motions, afferent sensitivities, and reflexes. The fast central/striolar afferents appear best suited for signaling at high frequencies and for driving high-speed reflexes.

5. Mechanoelectrical transduction in hair cells adapts via two Ca^{2+}-dependent mechanisms that appear to be intrinsic to hair cell transduction and critical to its high sensitivity. The time course and other properties of these mechanisms influence the frequency dependency, sensitivity, and operating range of vestibular afferent signals.

6. Basolateral K_V conductances differ between hair cell types and zones in ways that may contribute to the dynamics, sensitivity, linearity, and operating range of afferent signals. Type I hair cells have large numbers of low-voltage-activated (K_{LV}) channels that are activated at resting potential and do not inactivate. Type II hair cells have K_V conductances that activate positive to resting potential and differ with zone, being larger and less inactivating in central and striolar zones.

7. Type I hair cells are enveloped by large calyceal afferent terminals, and type II hair cells are contacted by smaller bouton endings. At both types of synapse, glutamate is released from synaptic vesicles organized around presynaptic ribbons. Significant nonquantal transmission of mechanically evoked signals has been recorded at calyceal synapses, which have an extensive, thin synaptic cleft that may accumulate K^+ or sustain extracellular voltage. Specializations in central-zone calyces may enhance the size and speed of both quantal and nonquantal transmission.

8. Afferent spike timing differs by zone, being highly irregular in central/striolar afferents and highly regular in peripheral/extrastriolar afferents; this difference may reflect different stimulus-encoding strategies. Analysis of ion channels expressed by vestibular afferent cell bodies suggests that expression of K_{LV} channels confers irregularity. Such channels are present at afferent terminals, where spikes are initiated, and could be a target for efferent transmitters to modulate spike rate or timing.

FUTURE ISSUES

1. What are the downstream targets of the two channels of vestibular afferent signals?

2. How does transmission work at type I–calyx synapses? What is the functional significance of synaptic differences between zones?

3. How do hair bundles and accessory structures work together, and how do zonal differences in their morphology influence the afferent signal?

4. How are hair cell types, afferent types, and zones set up during development?

5. To what extent does loss of vestibular function (e.g., with age) originate in deterioration of hair cells, synapses, and neurons, and are such changes reparable?

DISCLOSURE STATEMENT

R.A.E. is a member of the scientific advisory board to Sensorion Pharmaceuticals, a company dedicated to generating therapies for vestibular disorders.

ACKNOWLEDGMENTS

The authors' research is supported by NIDCD R01 DC02290 (R.A.E. and J.E.S.) and NSBRI through NASA NCC 9-58 (J.E.S.). We thank Radha Kalluri, Xiao-Ping Liu, Gang Li, Michaela Meyer, Will McLean, and Jingbing Xue for inspiring discussions and Ellengene Peterson for sharing unpublished material.

LITERATURE CITED

Agrawal Y, Carey JP, Della Santina CC, Schubert MC, Minor LB. 2009. Disorders of balance and vestibular function in US adults: data from the National Health and Nutrition Examination Survey, 2001–2004. *Arch. Intern. Med.* 169(10):938–44

Angelaki DE, Correia MJ. 1991. Models of membrane resonance in pigeon semicircular canal type II hair cells. *Biol. Cybern.* 65:1–10

Armand M, Minor LB. 2001. Relationship between time- and frequency-domain analyses of angular head movements in the squirrel monkey. *J. Comput. Neurosci.* 11(3):217–39

Avissar M, Furman AC, Saunders JC, Parsons TD. 2007. Adaptation reduces spike-count reliability, but not spike-timing precision, of auditory nerve responses. *J. Neurosci.* 27(24):6461–72

Baird RA, Desmadryl G, Fernández C, Goldberg JM. 1988. The vestibular nerve of the chinchilla. II. Relation between afferent response properties and peripheral innervation patterns in the semicircular canals. *J. Neurophysiol.* 60:182–203

Bao H, Wong WH, Goldberg JM, Eatock RA. 2003. Voltage-gated calcium channel currents in type I and type II hair cells isolated from the rat crista. *J. Neurophysiol.* 90(1):155–64

Behrend O, Schwark C, Kunihiro T, Strupp M. 1997. Cyclic GMP inhibits and shifts the activation curve of the delayed-rectifier (I_{K1}) of type I mammalian vestibular hair cells. *NeuroReport* 8:2687–90

Bonsaquet J, Brugeaud A, Compan V, Desmadryl G, Chabbert C. 2006. AMPA type glutamate receptor mediates neurotransmission at turtle vestibular calyx synapse. *J. Physiol.* 576:63–71

Bozovic D, Hudspeth AJ. 2003. Hair-bundle movements elicited by transepithelial electrical stimulation of hair cells in the sacculus of the bullfrog. *Proc. Natl. Acad. Sci. USA* 100:958–63

Brichta AM, Aubert A, Eatock RA, Goldberg JM. 2002. Regional analysis of whole cell currents from hair cells of the turtle posterior crista. *J. Neurophysiol.* 88:3259–78

Buran BN, Strenzke N, Neef A, Gundelfinger ED, Moser T, Liberman MC. 2010. Onset coding is degraded in auditory nerve fibers from mutant mice lacking synaptic ribbons. *J. Neurosci.* 30(22):7587–97

Cheatham MA, Low-Zeddies S, Naik K, Edge R, Zheng J, et al. 2009. A chimera analysis of prestin knock-out mice. *J. Neurosci.* 29:12000–8

Chen JWY, Eatock RA. 2000. Major potassium conductance in type I hair cells from rat semicircular canals: characterization and modulation by nitric oxide. *J. Neurophysiol.* 84:139–51

Cheung EL, Corey DP. 2006. Ca^{2+} changes the force sensitivity of the hair-cell transduction channel. *Biophys. J.* 90:124–39

Corey DP, Hudspeth AJ. 1979. Response latency of vertebrate hair cells. *Biophys. J.* 26:499–506

Corey DP, Hudspeth AJ. 1983. Kinetics of the receptor current in bullfrog saccular hair cells. *J. Neurosci.* 3:962–76

Correia MJ, Lang DG. 1990. An electrophysiological comparison of solitary type I and type II vestibular hair cells. *Neurosci. Lett.* 116:106–11

Cullen KE, Minor LB. 2002. Semicircular canal afferents similarly encode active and passive head-on-body rotations: implications for the role of vestibular efference. *J. Neurosci.* 22(11):RC226

Cullen KE, Roy JE, Sylvestre PA. 2001. Signal processing by vestibular nuclei neurons is dependent on the current behavioral goal. *Ann. NY Acad. Sci.* 942:345–63

Curthoys IS. 1979. The development of function of horizontal semicircular canal primary neurons in the rat. *Brain Res.* 167(1):41–52

Curthoys IS. 2010. A critical review of the neurophysiological evidence underlying clinical vestibular testing using sound, vibration and galvanic stimuli. *Clin. Neurophysiol.* 121(2):132–44

Davis JL, Xue J, Peterson EH, Grant JW. 2007. Layer thickness and curvature effects on otoconial membrane deformation in the utricle of the red-ear slider turtle: static and modal analysis. *J. Vestib. Res.* 17(4):145–62

Deans MR, Antic D, Suyama K, Scott MP, Axelrod JD, Goodrich LV. 2007. Asymmetric distribution of Prickle-like 2 reveals an early underlying polarization of vestibular sensory epithelia in the inner ear. *J. Neurosci.* 27:3139–47

Desai SS, Ali H, Lysakowski A. 2005a. Comparative morphology of rodent vestibular periphery. II. Cristae ampullares. *J. Neurophysiol.* 93(1):267–80

Desai SS, Zeh C, Lysakowski A. 2005b. Comparative morphology of rodent vestibular periphery. I. Saccular and utricular maculae. *J. Neurophysiol.* 93(1):251–66

De Vries HL. 1950. Mechanics of the labyrinth organs. *Acta Otolaryngol.* 38:262–73

Dickman JD, Correia MJ. 1989. Responses of pigeon horizontal semicircular canal afferent fibers. II. High-frequency mechanical stimulation. *J. Neurophysiol.* 62:1102–343

Dou H, Vazquez AE, Namkung Y, Chu H, Cardell EL, et al. 2004. Null mutation of alpha1D Ca^{2+} channel gene results in deafness but no vestibular defect in mice. *J. Assoc. Res. Otolaryngol.* 5(2):215–26

Dulon D, Safieddine S, Jones SM, Petit C. 2009. Otoferlin is critical for a highly sensitive and linear calcium-dependent exocytosis at vestibular hair cell ribbon synapses. *J. Neurosci.* 29(34):10474–87

Dutia MB. 2010. Mechanisms of vestibular compensation: recent advances. *Curr. Opin. Otolaryngol. Head Neck Surg.* 18(5):420–24

Eatock RA. 2000. Adaptation in hair cells. *Annu. Rev. Neurosci.* 23:285–314

Eatock RA, Corey DP, Hudspeth AJ. 1987. Adaptation of mechanoelectrical transduction in hair cells of the bullfrog's sacculus. *J. Neurosci.* 7:2821–36

Eatock RA, Fay RR, Popper AN. 2006. *Vertebrate Hair Cells.* New York: Springer

Eatock RA, Hurley KM. 2003. Functional development of hair cells. In *Development of the Auditory and Vestibular Systems 3: Molecular Development of the Inner Ear*, ed. R Romand, I Varela-Nieto, pp. 389–448. San Diego, CA: Academic

Eatock RA, Lysakowski A. 2006. Mammalian vestibular hair cells. See Eatock et al. 2006, pp. 348–442

Eatock RA, Xue J, Kalluri R. 2008. Ion channels in mammalian vestibular afferents may set regularity of firing. *J. Exp. Biol.* 211(Pt. 11):1764–74

Edmonds B, Reyes R, Schwaller B, Roberts WM. 2000. Calretinin modifies presynaptic calcium signaling in frog saccular hair cells. *Nat. Neurosci.* 3(8):786–90

Fahrenfort I, Steijaert M, Sjoersma T, Vickers E, Ripps H, et al. 2009. Hemichannel-mediated and pH-based feedback from horizontal cells to cones in the vertebrate retina. *PLoS ONE* 4(6):e6090

Fekete DM, Wu DK. 2002. Revisiting cell fate specification in the inner ear. *Curr. Opin. Neurobiol.* 12(1):35–42

Fernández C, Baird RA, Goldberg JM. 1988. The vestibular nerve of the chinchilla. I. Peripheral innervation patterns in the horizontal and superior semicircular canals. *J. Neurophysiol.* 60(1):167–81

Fernández C, Goldberg JM. 1976a. Physiology of peripheral neurons innervating otolith organs of the squirrel monkey. II. Directional selectivity and force-response relations. *J. Neurophysiol.* 39:985–95

Fernández C, Goldberg JM. 1976b. Physiology of peripheral neurons innervating otolith organs of the squirrel monkey. III. Response dynamics. *J. Neurophysiol.* 39:996–1008

Fernández C, Goldberg JM, Baird RA. 1990. The vestibular nerve of the chinchilla. III. Peripheral innervation patterns in the utricular macula. *J. Neurophysiol.* 63(4):767–80

Fettiplace R, Ricci AJ. 2003. Adaptation in auditory hair cells. *Curr. Opin. Neurobiol.* 13(4):446–51

Frank T, Khimich D, Neef A, Moser T. 2009. Mechanisms contributing to synaptic Ca^{2+} signals and their heterogeneity in hair cells. *Proc. Natl. Acad. Sci. USA* 106(11):4483–88

Fritzsch B, Signore M, Simeone A. 2001. Otx1 null mutant mice show partial segregation of sensory epithelia comparable to lamprey ears. *Dev. Genes Evol.* 211(8–9):388–96

Furukawa T, Matsuura S. 1978. Adaptive rundown of excitatory post-synaptic potentials at synapses between hair cells and eighth nerve fibres in the goldfish. *J. Physiol.* 276:193–209

Géléoc GS, Risner JR, Holt JR. 2004. Developmental acquisition of voltage-dependent conductances and sensory signaling in hair cells of the embryonic mouse inner ear. *J. Neurosci.* 24:11148–59

Ginzberg RD. 1984. Freeze-fracture morphology of the vestibular hair cell-primary afferent synapse in the chick. *J. Neurocytol.* 13:393–405

Goldberg JM. 1991. The vestibular end organs: morphological and physiological diversity of afferents. *Curr. Opin. Neurobiol.* 1:229–35

Goldberg JM. 1996. Theoretical analysis of intercellular communication between the vestibular type I hair cell and its calyx ending. *J. Neurophysiol.* 76:1942–57

Goldberg JM. 2000. Afferent diversity and the organization of central vestibular pathways. *Exp. Brain Res.* 130(3):277–97

Goldberg JM, Desmadryl G, Baird RA, Fernández C. 1990a. The vestibular nerve of the chinchilla. IV. Discharge properties of utricular afferents. *J. Neurophysiol.* 63:781–90

Goldberg JM, Desmadryl G, Baird RA, Fernández C. 1990b. The vestibular nerve of the chinchilla. V. Relation between afferent discharge properties and peripheral innervation patterns in the utricular macula. *J. Neurophysiol.* 63:791–804

Goldberg JM, Fernández C. 1971. Physiology of peripheral neurons innervating semicircular canals of the squirrel monkey. I. Resting discharge and response to constant angular accelerations. *J. Neurophysiol.* 34:635–60

Goldberg JM, Fernández C. 1977. Conduction times and background discharge of vestibular afferents. *Brain Res.* 122:545–50

Goldberg JM, Fernández C. 1980. Efferent vestibular system in the squirrel monkey: anatomical location and influence on afferent activity. *J. Neurophysiol.* 43(4):986–1025

Goldberg JM, Fernández C, Smith CE. 1982. Responses of vestibular-nerve afferents in the squirrel monkey to externally applied galvanic currents. *Brain Res.* 252(1):156–60

Goldberg JM, Smith CE, Fernández C. 1984. Relation between discharge regularity and responses to externally applied galvanic currents in vestibular nerve afferents of the squirrel monkey. *J. Neurophysiol.* 51:1236–56

Gómez-Casati ME, Murtie JC, Rio C, Stankovic K, Liberman MC, Corfas G. 2010. Nonneuronal cells regulate synapse formation in the vestibular sensory epithelium via erbB-dependent BDNF expression. *Proc. Natl. Acad. Sci. USA* 107:17005–10

Goodyear RJ, Kros CJ, Richardson GP. 2006. The development of hair cells in the inner ear. See Eatock et al. 2006, pp. 20–94

Grant L, Yi E, Glowatzki E. 2010. Two modes of release shape the postsynaptic response at the inner hair cell ribbon synapse. *J. Neurosci.* 30(12):4210–20

Grossman GE, Leigh RJ, Abel LA, Lanska DJ, Thurston SE. 1988. Frequency and velocity of rotational head perturbations during locomotion. *Exp. Brain Res.* 70:470–76

Gulley RL, Bagger-Sjöbäck D. 1979. Freeze-fracture studies on the synapse between the type I hair cell and the calyceal terminal in the guinea-pig vestibular system. *J. Neurocytol.* 8:591–603

Halmagyi GM, Weber KP, Curthoys IS. 2010. Vestibular function after acute vestibular neuritis. *Restor. Neurol. Neurosci.* 28(1):37–46

Hammond KL, Whitfield TT. 2006. The developing lamprey ear closely resembles the zebrafish otic vesicle: otx1 expression can account for all major patterning differences. *Development* 133(7):1347–57

Highstein SM, Popper AN, Fay RR, eds. 2004. *Anatomy and Physiology of the Central and Peripheral Vestibular System*. New York: Springer-Verlag

Highstein SM, Rabbitt RD, Holstein GR, Boyle RD. 2005. Determinants of spatial and temporal coding by semicircular canal afferents. *J. Neurophysiol.* 93(5):2359–70

Holley M, Rhodes C, Kneebone A, Herde MK, Fleming M, Steel KP. 2010. Emx2 and early hair cell development in the mouse inner ear. *Dev. Biol.* 340(2):547–56

Holstein GR, Martinelli GP, Henderson SC, Friedrich VL Jr, Rabbitt RD, Highstein SM. 2004a. Gamma-aminobutyric acid is present in a spatially discrete subpopulation of hair cells in the crista ampullaris of the toadfish Opsanus tau. *J. Comp. Neurol.* 471(1):1–10

Holstein GR, Rabbitt RD, Martinelli GP, Friedrich VL Jr, Boyle RD, Highstein SM. 2004b. Convergence of excitatory and inhibitory hair cell transmitters shapes vestibular afferent responses. *Proc. Natl. Acad. Sci. USA* 101(44):15766–71

Holt JC, Chatlani S, Lysakowski A, Goldberg JM. 2007a. Quantal and nonquantal transmission in calyx-bearing fibers of the turtle posterior crista. *J. Neurophysiol.* 98(3):1083–101

Holt JC, Lysakowski A, Goldberg JM. 2006a. Mechanisms of efferent-mediated responses in the turtle posterior crista. *J. Neurosci.* 26(51):13180–93

Holt JC, Xue JT, Brichta AM, Goldberg JM. 2006b. Transmission between type II hair cells and bouton afferents in the turtle posterior crista. *J. Neurophysiol.* 95(1):428–52

Holt JR, Corey DP. 2000. Two mechanisms for transducer adaptation in vertebrate hair cells. *Proc. Natl. Acad. Sci. USA* 97(22):11730–35

Holt JR, Gillespie SK, Provance DW, Shah K, Shokat KM, et al. 2002. A chemical-genetic strategy implicates myosin-1c in adaptation by hair cells. *Cell* 108(3):371–81

Holt JR, Rüsch A, Vollrath MA, Eatock RA. 1998. The frequency dependence of receptor potentials in hair cells of the mouse utricle. *Prim. Sens. Neuron* 2:233–41

Holt JR, Stauffer EA, Abraham D, Géléoc GS. 2007b. Dominant-negative inhibition of M-like potassium conductances in hair cells of the mouse inner ear. *J. Neurosci.* 27(33):8940–51

Holt JR, Vollrath MA, Eatock RA. 1999. Stimulus processing by type II hair cells in the mouse utricle. *Ann. NY Acad. Sci.* 871:15–26

Howard J, Bechstedt S. 2004. Hypothesis: a helix of ankyrin repeats of the NOMPC-TRP ion channel is the gating spring of mechanoreceptors. *Curr. Biol.* 14(6):R224–26

Howard J, Hudspeth AJ. 1987. Mechanical relaxation of the hair bundle mediates adaptation in mechanoelectrical transduction by the bullfrog's saccular hair cell. *Proc. Natl. Acad. Sci. USA* 84:3064–68

Hudspeth AJ, Choe Y, Mehta AD, Martin P. 2000. Putting ion channels to work: mechanoelectrical transduction, adaptation, and amplification by hair cells. *Proc. Natl. Acad. Sci. USA* 97(22):11765–72

Hudspeth AJ, Lewis RS. 1988. A model for electrical resonance and frequency tuning in saccular hair cells of the bull-frog, *Rana catesbeiana*. *J. Physiol.* 400:275–97

Hullar TE, Della Santina CC, Hirvonen T, Lasker DM, Carey JP, Minor LB. 2005. Responses of irregularly discharging chinchilla semicircular canal vestibular-nerve afferents during high-frequency head rotations. *J. Neurophysiol.* 93(5):2777–86

Hullar TE, Minor LB. 1999. High-frequency dynamics of regularly discharging canal afferents provide a linear signal for angular vestibuloocular reflexes. *J. Neurophysiol.* 82(4):2000–5

Hume CR, Bratt DL, Oesterle EC. 2007. Expression of LHX3 and SOX2 during mouse inner ear development. *Gene Expr. Patterns* 7(7):798–807

Hurley KM, Gaboyard S, Zhong M, Price SD, Wooltorton JR, et al. 2006. M-like K^+ currents in type I hair cells and calyx afferent endings of the developing rat utricle. *J. Neurosci.* 26(40):10253–69

Huterer M, Cullen KE. 2002. Vestibuloocular reflex dynamics during high-frequency and high-acceleration rotations of the head on body in rhesus monkey. *J. Neurophysiol.* 88(1):13–28

Iwasaki S, Chihara Y, Komuta Y, Ito K, Sahara Y. 2008. Low-voltage-activated potassium channels underlie the regulation of intrinsic firing properties of rat vestibular ganglion cells. *J. Neurophysiol.* 100(4):2192–204

Jacobs RA, Hudspeth AJ. 1990. Ultrastructural correlates of mechanoelectrical transduction in hair cells of the bullfrog's internal ear. *Cold Spring Harb. Symp. Quant. Biol.* 55:547–61

Jamali M, Sadeghi SG, Cullen KE. 2009. Response of vestibular nerve afferents innervating utricle and saccule during passive and active translations. *J. Neurophysiol.* 101(1):141–49

Jones SM, Subramanian G, Avniel W, Guo Y, Burkard RF, Jones TA. 2002. Stimulus and recording variables and their effects on mammalian vestibular evoked potentials. *J. Neurosci. Methods* 118(1):23–31

Jørgensen F, Kroese AB. 1995. Ca selectivity of the transduction channels in the hair cells of the frog sacculus. *Acta Physiol. Scand.* 155(4):363–76

Kalluri R, Xue J, Eatock RA. 2010. Ion channels set spike timing regularity of mammalian vestibular afferent neurons. *J. Neurophysiol.* 104:2034–51

Kharkovets T, Hardelin J-P, Safieddine S, Schweizer M, El-Amraoui A, et al. 2000. KCNQ4, a K^+ channel mutated in a form of dominant deafness, is expressed in the inner ear and the central auditory pathway. *Proc. Natl. Acad. Sci. USA* 97:4333–38

Kim E, Hyrc KL, Speck J, Lundberg YW, Salles FT, et al. 2010. Regulation of cellular calcium in vestibular supporting cells by Otopetrin 1. *J. Neurophysiol.* 104:3439–50

Kung C. 2005. A possible unifying principle for mechanosensation. *Nature* 436:647–54

Lang F, Vallon V, Knipper M, Wangemann P. 2007. Functional significance of channels and transporters expressed in the inner ear and kidney. *Am. J. Physiol. Cell Physiol.* 293(4):C1187–208

Lasker DM, Han GC, Park HJ, Minor LB. 2008. Rotational responses of vestibular-nerve afferents innervating the semicircular canals in the C57BL/6 mouse. *J. Assoc. Res. Otolaryngol.* 9(3):334–48

Le Goff L, Bozovic D, Hudspeth AJ. 2005. Adaptive shift in the domain of negative stiffness during spontaneous oscillation by hair bundles from the internal ear. *Proc. Natl. Acad. Sci. USA* 102(47):16996–7001

Lee HY, Camp AJ, Callister RJ, Brichta AM. 2005. Vestibular primary afferent activity in an in vitro preparation of the mouse inner ear. *J. Neurosci. Methods* 145:73–87

Leonard RB, Kevetter GA. 2002. Molecular probes of the vestibular nerve. I. Peripheral termination patterns of calretinin, calbindin and peripherin containing fibers. *Brain Res.* 928(1–2):8–17

Li A, Xue J, Peterson EH. 2008. Architecture of the mouse utricle: macular organization and hair bundle heights. *J. Neurophysiol.* 99(2):718–33

Li GL, Keen E, Andor-Ardo D, Hudspeth AJ, von Gersdorff H. 2009. The unitary event underlying multiquantal EPSCs at a hair cell's ribbon synapse. *J. Neurosci.* 29(23):7558–68

Li GQ, Meredith FL, Rennie KJ. 2010. Development of K^+ and Na^+ conductances in rodent postnatal semicircular canal type I hair cells. *Am. J. Physiol. Regul. Integr. Comp. Physiol.* 298(2):R351–58

Lim DJ. 1984. Otoconia in health and disease. A review. *Ann. Otol. Rhinol. Laryngol. Suppl.* 112:17–24

Lim R, Kindig AE, Donne SW, Callister RJ, Brichta AM. 2011. Potassium accumulation between type I hair cells and calyx terminals in mouse crista. *Exp. Brain Res.* 210:607–21

Limón A, Pérez C, Vega R, Soto E. 2005. Ca^{2+}-activated K^+-current density is correlated with soma size in rat vestibular-afferent neurons in culture. *J. Neurophysiol.* 94(6):3751–61

Lindeman HH. 1973. Anatomy of the otolith organs. *Adv. Otorhinolaryngol.* 20:405–33

Lins U, Farina M, Kurc M, Riordan G, Thalmann R, et al. 2000. The otoconia of the guinea pig utricle: internal structure, surface exposure, and interactions with the filament matrix. *J. Struct. Biol.* 131(1):67–78

Lundberg YW, Zhao X, Yamoah EN. 2006. Assembly of the otoconia complex to the macular sensory epithelium of the vestibule. *Brain Res.* 1091(1):47–57

Lysakowski A, Goldberg JM. 1997. Regional variations in the cellular and synaptic architecture of the chinchilla cristae. *J. Comp. Neurol.* 389:419–43

Lysakowski A, Goldberg JM. 2004. Morphophysiology of the vestibular sensory periphery. See Highstein et al. 2004, pp. 57–152

Lysakowski A, Price SD. 2003. Potassium channel localization in sensory epithelia of the rat inner ear. *Abstr. Assoc. Res. Otolaryngol.* 26:1534

Lysakowski A, Singer M. 2000. Nitric oxide synthase localized in a subpopulation of vestibular efferents with NADPH diaphorase histochemistry and nitric oxide synthase immunohistochemistry. *J. Comp. Neurol.* 427:508–21

Masetto S, Correia MJ. 1997. Ionic currents in regenerating avian vestibular hair cells. *Int. J. Dev. Neurosci.* 15:387–99

McCue MP, Guinan JJ Jr. 1994. Acoustically responsive fibers in the vestibular nerve of the cat. *J. Neurosci.* 14(10):6058–70

Merchan-Perez A, Liberman MC. 1996. Ultrastructural differences among afferent synapses on cochlear hair cells: correlations with spontaneous discharge rate. *J. Comp. Neurol.* 371(2):208–21

Moravec WJ, Peterson EH. 2004. Differences between stereocilia numbers on type I and type II vestibular hair cells. *J. Neurophysiol.* 92(5):3153–60

Morsli H, Tuorto F, Choo D, Postiglione MP, Simeone A, Wu DK. 1999. Otx1 and Otx2 activities are required for the normal development of the mouse inner ear. *Development* 126(11):2335–43

Nam JH, Cotton JR, Grant JW. 2005. Effect of fluid forcing on vestibular hair bundles. *J. Vestib. Res.* 15(5–6):263–78

Paige GD. 1983. Vestibuloocular reflex and its interactions with visual following mechanisms in the squirrel monkey. I. Response characteristics in normal animals. *J. Neurophysiol.* 49(1):134–51

Pérez C, Vega R, Soto E. 2010. Phospholipase C-mediated inhibition of the M-potassium current by muscarinic-receptor activation in the vestibular primary-afferent neurons of the rat. *Neurosci. Lett.* 468(3):238–42

Peterson EH, Cotton JR, Grant JW. 1996. Structural variation in ciliary bundles of the posterior semicircular canal: quantitative anatomy and computational analysis. *Ann. NY Acad. Sci.* 781:85–102

Pfanzelt S, Rossert C, Rohregger M, Glasauer S, Moore LE, Straka H. 2008. Differential dynamic processing of afferent signals in frog tonic and phasic second-order vestibular neurons. *J. Neurosci.* 28(41):10349–62

Platzer J, Engel J, Schrott-Fischer A, Stephan K, Bova S, et al. 2000. Congenital deafness and sinoatrial node dysfunction in mice lacking class D L-type Ca^{2+} channels. *Cell* 102(1):89–97

Rabbitt RD, Damiano ER, Grant JW. 2004. Biomechanics of the semicircular canals and otolith organs. See Highstein et al. 2004, pp. 153–201

Rennie KJ. 2002. Modulation of the resting potassium current in type I vestibular hair cells by cGMP. In *Hair Cell Micromechanics and Otoacoustic Emissions*, ed. CI Berlin, LJ Hood, AJ Ricci, pp. 79–89. San Diego, CA: Singular

Rennie KJ, Manning KC, Ricci AJ. 2004. Mechano-electrical transduction in the turtle utricle. *Biomed. Sci. Instrum.* 40:441–46

Rennie KJ, Ricci AJ, Correia MJ. 1996. Electrical filtering in gerbil isolated type I semicircular canal hair cells. *J. Neurophysiol.* 75:2117–23

Rennie KJ, Streeter MA. 2006. Voltage-dependent currents in isolated vestibular afferent calyx terminals. *J. Neurophysiol.* 95(1):26–32

Rennie KJ, Weng T, Correia MJ. 2001. Effects of KCNQ channel blockers on K(+) currents in vestibular hair cells. *Am. J. Physiol. Cell Physiol.* 280(3):C473–80

Risner JR, Holt JR. 2006. Heterogeneous potassium conductances contribute to the diverse firing properties of postnatal mouse vestibular ganglion neurons. *J. Neurophysiol.* 96(5):2364–76

Roberts WM, Jacobs RA, Hudspeth AJ. 1990. Colocalization of ion channels involved in frequency selectivity and synaptic transmission at presynaptic active zones of hair cells. *J. Neurosci.* 10:3664–84

Rocha-Sanchez SM, Morris KA, Kachar B, Nichols D, Fritzsch B, Beisel KW. 2007. Developmental expression of Kcnq4 in vestibular neurons and neurosensory epithelia. *Brain Res.* 1139:117–25

Rothman JS, Manis PB. 2003. The roles potassium currents play in regulating the electrical activity of ventral cochlear nucleus neurons. *J. Neurophysiol.* 89(6):3097–113

Roux I, Safieddine S, Nouvian R, Grati M, Simmler MC, Bahloul A, et al. 2006. Otoferlin, defective in a human deafness form, is essential for exocytosis at the auditory ribbon synapse. *Cell* 127:277–89

Rowe MH, Peterson EH. 2006. Autocorrelation analysis of hair bundle structure in the utricle. *J. Neurophysiol.* 96(5):2653–69

Rüsch A, Eatock RA. 1996. Voltage responses of mouse utricular hair cells to injected currents. *Ann. NY Acad. Sci.* 781:71–84

Rüsch A, Lysakowski A, Eatock RA. 1998. Postnatal development of type I and type II hair cells in the mouse utricle: acquisition of voltage-gated conductances and differentiated morphology. *J. Neurosci.* 18(18):7487–501

Rutherford MA, Roberts WM. 2009. Spikes and membrane potential oscillations in hair cells generate periodic afferent activity in the frog sacculus. *J. Neurosci.* 29(32):10025–37

Sadeghi SG, Chacron MJ, Taylor MC, Cullen KE. 2007a. Neural variability, detection thresholds, and information transmission in the vestibular system. *J. Neurosci.* 27(4):771–81

Sadeghi SG, Goldberg JM, Minor LB, Cullen KE. 2009. Efferent-mediated responses in vestibular nerve afferents of the alert macaque. *J. Neurophysiol.* 101(2):988–1001

Sadeghi SG, Minor LB, Cullen KE. 2007b. Response of vestibular-nerve afferents to active and passive rotations under normal conditions and after unilateral labyrinthectomy. *J. Neurophysiol.* 97(2):1503–14

Sadeghi SG, Minor LB, Cullen KE. 2010. Neural correlates of motor learning in the vestibulo-ocular reflex: dynamic regulation of multimodal integration in the macaque vestibular system. *J. Neurosci.* 30(30):10158–68

Sato F, Sasaki H. 1993. Morphological correlations between spontaneously discharging primary vestibular afferents and vestibular nucleus neurons in the cat. *J. Comp. Neurol.* 333(4):554–66

Sätzler K, Söhl LF, Bollmann JH, Borst JG, Frotscher M, et al. 2002. Three-dimensional reconstruction of a calyx of Held and its postsynaptic principal neuron in the medial nucleus of the trapezoid body. *J. Neurosci.* 22(24):10567–79

Schessel DA, Ginzberg R, Highstein SM. 1991. Morphophysiology of synaptic transmission between type I hair cells and vestibular primary afferents. An intracellular study employing horseradish peroxidase in the lizard, *Calotes versicolor. Brain Res.* 544:1–16

Schnee ME, Lawton DM, Furness DN, Benke TA, Ricci AJ. 2005. Auditory hair cell-afferent fiber synapses are specialized to operate at their best frequencies. *Neuron* 47(2):243–54

Schweizer FE, Savin D, Luu C, Sultemeier DR, Hoffman LF. 2009. Distribution of high-conductance calcium-activated potassium channels in rat vestibular epithelia. *J. Comp. Neurol.* 517(2):134–45

Selyanko AA, Delmas P, Hadley JK, Tatulian L, Wood IC, et al. 2002. Dominant-negative subunits reveal potassium channel families that contribute to M-like potassium currents. *J. Neurosci.* 22(5):RC212

Si X, Zakir MM, Dickman JD. 2003. Afferent innervation of the utricular macula in pigeons. *J. Neurophysiol.* 89(3):1660–77

Sillar KT. 2009. Mauthner cells. *Curr. Biol.* 19(9):R353–55

Simmons DD, Tong B, Schrader AD, Hornak AJ. 2010. Oncomodulin identifies different hair cell types in the mammalian inner ear. *J. Comp. Neurol.* 518(18):3785–802

Smith CE, Goldberg JM. 1986. A stochastic afterhyperpolarization model of repetitive activity in vestibular afferents. *Biol. Cybern.* 54:41–51

Songer JE, Eatock RA. 2011. Transduction, tuning and spike timing in the rat saccular macula. *Assoc. Res. Otolaryngol. Abstr.* 34:344

Sousa AD, Andrade LR, Salles FT, Pillai AM, Buttermore ED, et al. 2009. The septate junction protein caspr is required for structural support and retention of KCNQ4 at calyceal synapses of vestibular hair cells. *J. Neurosci.* 29(10):3103–8

Spassova M, Eisen MD, Saunders JC, Parsons TD. 2001. Chick cochlear hair cell exocytosis mediated by dihydropyridine-sensitive calcium channels. *J. Physiol.* 535(Pt. 3):689–96

Spoon C, Peterson E, Grant JW. 2005. Bundle mechanics depends on bundle structure in turtle utricle. *Assoc. Res. Otolaryngol. Abstr.* 28:300

Stauffer EA, Scarborough JD, Hirono M, Miller ED, Shah K, et al. 2005. Fast adaptation in vestibular hair cells requires myosin-1c activity. *Neuron* 47(4):541–53

Straka H, Lambert FM, Pfanzelt S, Beraneck M. 2009. Vestibulo-ocular signal transformation in frequency-tuned channels. *Ann. NY Acad. Sci.* 1164:37–44

Sugawara M, Murtie JC, Stankovic KM, Liberman MC, Corfas G. 2007. Dynamic patterns of neurotrophin 3 expression in the postnatal mouse inner ear. *J. Comp. Neurol.* 501(1):30–37

Temchin AN, Recio-Spinoso A, van Dijk P, Ruggero MA. 2005. Wiener kernels of chinchilla auditory-nerve fibers: verification using responses to tones, clicks, and noise and comparison with basilar-membrane vibrations. *J. Neurophysiol.* 93(6):3635–48

Vollrath MA, Eatock RA. 2003. Time course and extent of mechanotransducer adaptation in mouse utricular hair cells: comparison with frog saccular hair cells. *J. Neurophysiol.* 90(4):2676–89

Vranceanu F, Lysakowski A. 2009. A re-examination of the striated organelle in vestibular endorgans. *Abstr. Assoc. Res. Otolaryngol.* 32:274

Warchol ME, Speck JD. 2007. Expression of GATA3 and tenascin in the avian vestibular maculae: normative patterns and changes during sensory regeneration. *J. Comp. Neurol.* 500(4):646–57

Welgampola MS, Carey JP. 2010. Waiting for the evidence: VEMP testing and the ability to differentiate utricular versus saccular function. *Otolaryngol. Head Neck Surg.* 143(2):281–83

Weng T, Correia MJ. 1999. Regional distribution of ionic currents and membrane voltage responses of type II hair cells in the vestibular neuroepithelium. *J. Neurophysiol.* 82(5):2451–61

Wittig JH Jr, Parsons TD. 2008. Synaptic ribbon enables temporal precision of hair cell afferent synapse by increasing the number of readily releasable vesicles: a modeling study. *J. Neurophysiol.* 100(4):1724–39

Wooltorton JR, Gaboyard S, Hurley KM, Price SD, Garcia JL, et al. 2007. Developmental changes in two voltage-dependent sodium currents in utricular hair cells. *J. Neurophysiol.* 97(2):1684–704

Xue J, Peterson EH. 2006. Hair bundle heights in the utricle: differences between macular locations and hair cell types. *J. Neurophysiol.* 95(1):171–86

Yamashita M, Ohmori H. 1990. Synaptic responses to mechanical stimulation in calyceal and bouton type vestibular afferents studied in an isolated preparation of semicircular canal ampullae of chicken. *Exp. Brain Res.* 80:475–88

Yamashita M, Ohmori H. 1991. Synaptic bodies and vesicles in the calix type synapse of chicken semicircular canal ampullae. *Neurosci. Lett.* 129:43–46

Yang A, Hullar TE. 2007. The relationship of semicircular canal size to vestibular-nerve afferent sensitivity in mammals. *J. Neurophysiol.* 98:3197–205

RELATED RESOURCES

Angelaki DE, Cullen KE. 2008. Vestibular system: the many facets of a multimodal sense. *Annu. Rev. Neurosci.* 31:125–50

Dateline NBC. 2006. In the Balance. *Dateline NBC*, July 31. **http://www.bing.com/videos/watch/video/in-the-balance/6hv0i0a**

J.C. 1952. Living without a balancing mechanism. *New Engl. J. Med.*, March 20, p. 458

Mechanisms of Inhibition within the Telencephalon: "Where the Wild Things Are"

Gord Fishell[1,2] and Bernardo Rudy[1,3]

[1]Smilow Neuroscience Program; Departments of [2]Cell Biology, Neural Science, [3]Physiology and Neuroscience, and Biochemistry; Smilow Research Center, New York University School of Medicine, New York, New York 10016; email: fishell@saturn.med.nyu.edu, rudyb01@med.nyu.edu

Annu. Rev. Neurosci. 2011. 34:535–67

First published online as a Review in Advance on April 4, 2011

The *Annual Review of Neuroscience* is online at neuro.annualreviews.org

This article's doi: 10.1146/annurev-neuro-061010-113717

Keywords

interneurons, GABA, cortex, subpallial, ganglionic eminences

Abstract

In this review, we first provide a historical perspective of inhibitory signaling from the discovery of inhibition through to our present understanding of the diversity and mechanisms by which GABAergic interneuron populations function in different parts of the telencephalon. This is followed by a summary of the mechanisms of inhibition in the CNS. With this as a starting point, we provide an overview describing the variations in the subtypes and origins of inhibitory interneurons within the pallial and subpallial divisions of the telencephalon, with a focus on the hippocampus, somatosensory, paleo/piriform cortex, striatum, and various amygdala nuclei. Strikingly, we observe that marked variations exist in the origin and numerical balance between GABAergic interneurons and the principal cell populations in distinct regions of the telencephalon. Finally we speculate regarding the attractiveness and challenges of establishing a unifying nomenclature to describe inhibitory neuron diversity throughout the telencephalon.

Contents

INTRODUCTION

Excitation and inhibition are the two fundamental modes of signaling within the CNS. Although several neurotransmitters exist within the nervous system, glutamate and γ-aminobutyric acid (GABA), two neurotransmitters with surprisingly similar structures, mediate the majority of excitatory and inhibitory signals within much of the CNS including the telencephalon. With regards to inhibition, the discovery of GABA in 1950 (Roberts & Frankel 1950, Udenfriend 1950, Williams 1950) determined the identity of the primary inhibitory transmitter in the CNS (**Figure 1**). Another decade would pass, however, before work by Kravitz, Kuffler, and colleagues definitively showed that it fulfilled the criteria to act as an endogenous neurotransmitter (Kravitz et al. 1963a,b). These experiments immediately suggested that the high levels of GABA in the mammalian brain were not as then thought merely metabolites (Roberts 1963) but indicative of neuronal populations that mediate inhibition. Work over the past three decades has borne out this hypothesis and demonstrated

that GABAergic neurons mediate most of the inhibition throughout the CNS.

With the notable exception of stellate neurons in neocortex, most excitatory cells within the telencephalon are projection neurons. In contrast, inhibitory signaling is mediated by both projection and interneuron populations, distinguished primarily by whether their axons project locally or outside the area where the cell body is located. Although such a distinction may on occasion seem semantic, there is little doubt that interneurons or "local projection neurons" subserve a unique function in gating signaling within the nervous system. Although the functional properties of a diverse set of "short-axon" cells in the cortex were unknown until recently, their identification was already noted by Ramón y Cajal more than 100 years ago (Ramón y Cajal 1891). Subsequent to the realization that these were GABAergic neurons and hence inhibitory (Dreifuss et al. 1969; Fonnum & Storm-Mathisen 1969; Hendry et al. 1984, 1986; Storm-Mathisen & Fonnum 1971), the recognition that the considerable heterogeneity in this population had functional significance took a further two decades (reviewed in Ascoli et al. 2008, Whittington & Traub 2003). During the mid-1980s a number of groups first demonstrated that, in addition to morphological diversity, interneurons throughout the telencephalon had both immunochemical (reviewed in Jones 1986) and electrophysiological (reviewed in Buzsaki 1984, Freund & Buzsaki 1996) diversity that likely reflected the diverse roles they played in both cortical and subcortical networks. In this review we examine these populations in different regions of the telencephalon from a historical and functional perspective. Our aim in doing so is not foremost to determine a unified classification for these cells, but rather to provide a context for how this diversity impacts signaling in various regions throughout the telencephalon. Although this survey of interneuron subtypes does not resolve the daunting problem of how best to compare different interneuron classes in distinct brain areas, it does serve to illustrate the remarkable computational

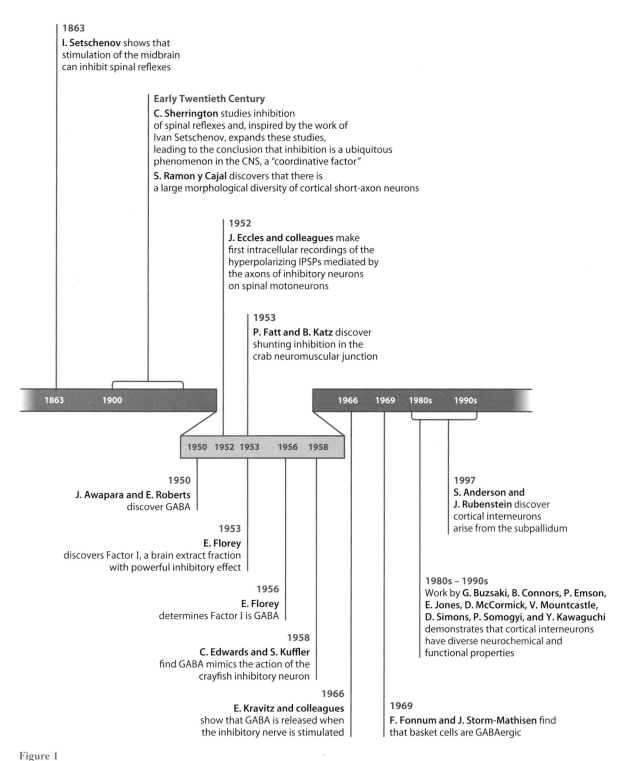

Figure 1

Key events in the discovery of inhibition.

power this population provides to neural circuitry.

DISCOVERY OF INHIBITION

Universally, nervous systems in higher organisms include two types of neurons: excitatory neurons that ensure the transmission of signals through various stages of processing and inhibitory neurons that control this transfer of information. These two, however, are intrinsically linked, as the nervous system depends on both: one cannot function without the other. Inhibition was discovered early in the investigations of the nervous system (**Figure 1**), when studies in both invertebrate and vertebrate preparations showed that stimulation of certain neurons (or fibers from those neurons) produced inhibition of stimulatory responses. Inhibition had a prominent place in the research of Charles Sherrington, a pioneer of neurophysiology. In fact, when Sherrington received the Nobel Prize with Edgar Douglas Adrian in 1932 for their "discoveries regarding the functions of neurons," inhibition was a major topic of his lecture (Sherrington 1932). Sherrington spent a great deal of effort studying "reciprocal innervation," the observation that stimuli evoking contraction in a given muscle will provoke inhibition of the antagonist muscle, leading to the view that "inhibitory" afferents conduct impulses that "at certain central loci cause, directly or indirectly, inhibition." Inhibition also became important in his studies of brain control of spinal reflexes, having stated in his Nobel lecture that "[f]urther study of central nervous action, however, finds central inhibition too extensive and ubiquitous to make it likely that it is confined solely to the taxis of antagonistic muscles." Sherrington cited as inspiration earlier studies including the 1863 work of Ivan Setschenov, perhaps the first to show that brain activity is linked to electric currents, demonstrating that stimulation of the midbrain could inhibit spinal reflexes. Sherrington concluded his Nobel lecture with the following:

> The role of inhibition in the working of the central nervous system has proved to be more and more extensive and more and more

fundamental as experiment has advanced in examining it. Reflex inhibition can no longer be regarded merely as a factor specially developed for dealing with the antagonism of opponent muscles acting at various hinge-joints. Its role as a coordinative factor comprises that, and goes beyond that. In the working of the central nervous machinery inhibition seems as ubiquitous and as frequent as is excitation itself. The whole quantitative grading of the operations of the spinal cord and brain appears to rest upon mutual interaction between the two central processes "excitation" and "inhibition," the one no less important than the other. (Sherrington 1932)

It was not until much later, however, that the action of an inhibitory neuron was studied directly, when Eccles and his colleagues using intracellular electrodes recorded the effect of the axon of an inhibitory neuron on spinal motoneurons (Brock et al. 1952). The inseparability of excitation and inhibition is reflected in the fact that inhibitory neurons are found throughout the brain. Functional units of the mammalian CNS generally consist of both excitatory and inhibitory neurons. Indeed, in areas in which this is not the case, such as the striatum or central or medial amygdala, the inhibitory cells are the ones that can exist in the absence of excitatory neurons rather than vice versa. This no doubt reflects the fundamental instability of solely excitatory networks and serves to emphasize the ubiquitous need for inhibition in the CNS. It would be impossible to cover in one review all the inhibitory neurons of the mammalian nervous system. So, following a discussion of aspects of inhibition that apply to all CNS areas, we focus in more detail on the inhibitory neurons of the mammalian telencephalon.

INHIBITORY NEUROTRANSMITTERS

Two known neurotransmitters mediate ionotropic inhibitory responses: GABA and glycine. GABA is an inhibitory neurotransmitter in invertebrates and vertebrates and is the major inhibitory neurotransmitter in the brain (Jentsch et al. 2002). Glycine is the major inhibitory neurotransmitter in the spinal cord. Strychnine-sensitive glycine receptors (GlyRs) have also been long recognized as important mediators of synaptic inhibition in the brainstem and medulla (Betz & Laube 2006, Betz 1991). However, recent evidence suggests that glycine also serves as an inhibitory neurotransmitter in other CNS areas including in the forebrain (Betz & Laube 2006, Hernandes & Troncone 2009, Trombley & Shepherd 1994). High densities of GlyRs have been found in the striatum and the hippocampus (Zarbin et al. 1981, Araki et al. 1988, Friauf et al. 1997). Moreover, as receptors to both neurotransmitters mediate their influence predominantly by altering chloride conductances, discussing them in conjunction is appropriate.

MECHANISMS OF INHIBITION

The activation of ionotropic GABA (GABA$_A$) and glycine receptors produces inhibition by at least two major mechanisms: hyperpolarizing and shunting. The existence of these two forms of inhibition was revealed by the very earliest investigations of synaptic inhibition. Hyperpolarizing inhibition was discovered by Eccles and colleagues (Brock et al. 1952a,b) in cat spinal motoneurons. The discovery of the inhibitory postsynaptic potential (IPSP) was the finding that finally led Eccles to abandon his electrical hypothesis for synaptic transmission. Inhibition in these cells is blocked by strychnine and simulated by the application of glycine, suggesting it is mediated by glycine receptors (Young & MacDonald 1983).

Hyperpolarizing inhibition occurs when the resting membrane potential (E_m) is more positive than the equilibrium potential for Cl$^-$ (E_{Cl}). In mature neurons, [Cl$^-$] is actively maintained at values lower than the electrochemical equilibrium for Cl$^-$ by specific transporters, resulting in an inward Cl$^-$ gradient (see Control of Intracellular Cl$^-$ and the Functions of Inhibitory Neurotransmitters, sidebar above). GABA$_A$ and glycine ionotropic receptors are

CONTROL OF INTRACELLULAR Cl$^-$ AND THE FUNCTIONS OF INHIBITORY NEUROTRANSMITTERS

In adult CNS neurons, the ionotropic actions of GABA and glycine are generally inhibitory. However, it is believed they both have excitatory actions during embryonic development and early postnatal ages. This is due to elevated intracellular concentrations of Cl$^-$ and hence a positive chloride equilibrium potential of the postsynaptic cell in young neurons, resulting in chloride efflux upon receptor activation and, hence, membrane depolarization (Ben-Ari 2002, Plotkin et al. 1997, Reichling et al. 1994). These excitatory actions of GABA and glycine are believed to be important for proliferation, migration, synaptogenesis, neuronal differentiation, and neuronal network stability (Kirsch & Betz 1998, Wester et al. 2008). The shift from excitatory to inhibitory at later stages is due to a shift in the chloride equilibrium potential to more negative values as a result of the expression of the K$^+$/Cl$^-$ cotransporter KCC2 (**Figure 2**). This transporter produces active chloride extrusion, thus reducing intracellular Cl$^-$ concentrations (Alvarez-Leefmans 1990, Alvarez-Leefmans & Delpire 2009). As a result of this shift in the Cl$^-$ equilibrium potential, receptor activation produces hyperpolarization. Na$^+$K$^+$/Cl$^-$ (NKCCs) cotransporters dominate in immature neurons of cortex and hippocampus, whereas KCC2 expression is induced only after birth (compare **Figure 2**, left and right).

Cation-chloride cotransporters (CCCs) constitute a family of transporters that include Na-Cl (NCCs), Na/K-Cl (NKCCs), and K$^+$-Cl$^-$ (KCCs) cotransporters, which differ in tissue and cellular distribution and have a variety of physiological roles (reviewed in Delpire & Mount 2002; Payne et al. 2003; Alvarez-Leefmans 1990, Alvarez-Leefmans & Delpire 2009). Under physiological conditions, NCCs and NKCCs provide the main route for Cl$^-$ uptake, whereas KCCs are responsible for Cl$^-$ extrusion. NKCC and KCC are of particular interest because they are critically involved in Cl$^-$ homeostasis in the brain.

ligand-gated anionic channels. The opening of the channels following association with GABA or glycine leads in mature neurons to chloride influx that hyperpolarizes the membrane. This hyperpolarization increases the difference between the membrane potential and spike threshold, thereby decreasing the effectiveness of excitatory inputs, i.e., inhibiting neuronal excitability.

However, in many instances in which the resting membrane potential is close to the chloride equilibrium potential, the activation of GABA$_A$ and glycine receptors can still produce inhibition through a mechanism known as "shunting inhibition." In these cases, the increase in anionic conductance does not produce significant changes in membrane potential. However, the increase in membrane conductance decreases the membrane resistance, i.e., the synaptic conductance "shunts" or short-circuits the membrane thus reducing the impact of the currents generated by concurrent excitatory inputs resulting in excitatory postsynaptic potentials of smaller amplitude. Excitatory inputs are less effective because the anionic conductance transiently clamps the membrane potential to the E$_{IPSP}$, resembling the effect of inward-rectifying K$^+$ channels that clamp the membrane potential to E$_K$. Shunting inhibition occurs when the membrane potential (V$_m$) is close to the chloride equilibrium potential (E$_{Cl}$); that is, V$_m \sim$ E$_{Cl} =$ E$_{IPSP}$.

If the chloride equilibrium potential is more positive than the resting potential, the opening of the chloride channels produces depolarization. However, as long as the depolarization is well below spike threshold, the shunting effect dominates the effect on excitability, resulting in inhibition of excitability (Alger & Nicoll 1979, Andersen et al. 1980, Gulledge & Stuart 2003). However, early in development, in certain pathological situations and apparently also in certain subcellular compartments such as the axon initial segment, the chloride gradient is such that the depolarization is large and dominates over the shunting effect. Thus, GABA becomes excitatory (Ben-Ari 2002) (see Control of Intracellular Cl$^-$ and the Functions of Inhibitory Neurotransmitters, sidebar above, and Chandelier Cells May Be Excitatory, sidebar below) (see also **Figure 2**).

Fatt & Katz (1953) discovered shunting inhibition in the crab neuromuscular junction. They concluded that "the main effect of inhibitory impulses is to attenuate the 'end-plate potentials,' i.e., to diminish the local depolarization produced by [excitatory] motor impulses." Shunting inhibition is believed to be a frequent inhibitory mechanism in central neurons. Recently, Mann & Paulsen (2007) and Jonas and colleagues (Vida et al. 2006) discussed the importance of this form of GABAergic inhibition in the control of spike timing of excitatory neurons and the generation of network oscillations.

Ionotropic anionic receptors may also produce inhibition by a third, and much less studied, mechanism of "depolarizing inhibition," which has been suggested to operate in GABA-mediated presynaptic inhibition. Some neurons receive axo-axonic GABAergic inputs and contain GABA$_A$ receptors (sometimes in addition to GABA$_B$ receptors) in their terminals, where the intracellular Cl$^-$ concentration is higher than E$_{Cl}$ (Howard et al. 2005). The activation of the GABA$_A$ receptors depolarizes the terminal and yet reduces neurotransmitter release. It has been suggested that the depolarization reduces transmitter release by blocking the invasion of action

CHANDELIER CELLS MAY BE EXCITATORY

It was traditionally believed that Chandelier cells might exert a particularly powerful inhibitory control of spike generation by virtue of the location of their synapses to the axon initial segment (AIS), the site of spike generation. A recent report questioned this classic conceptualization of chandelier cells as inhibitory, showing that chandelier cell activity can actually drive spikes in target pyramidal cells in layer 2/3 of the human and rodent neocortex (Szabadics et al. 2006). The basis of this phenomenon was suggested to be due to a relative absence of the chloride transporter KCC2 at the AIS of pyramidal neurons. As a result, chloride concentration is elevated in this subcellular compartment, and the reversal potential of the GABA-mediated postsynaptic potential produced by chandelier cell activity becomes depolarizing and its action excitatory (see also Control of Intracellular Cl$^-$ and the Functions of Inhibitory Neurotransmitters, sidebar above, as well as **Figure 2**). Additional data indicate that a high internal chloride concentration at the AIS is produced by chloride influx by the chloride transporter NKCC1 (Khirug et al. 2008). However, this view of chandelier cells remains controversial (Glickfeld et al. 2009).

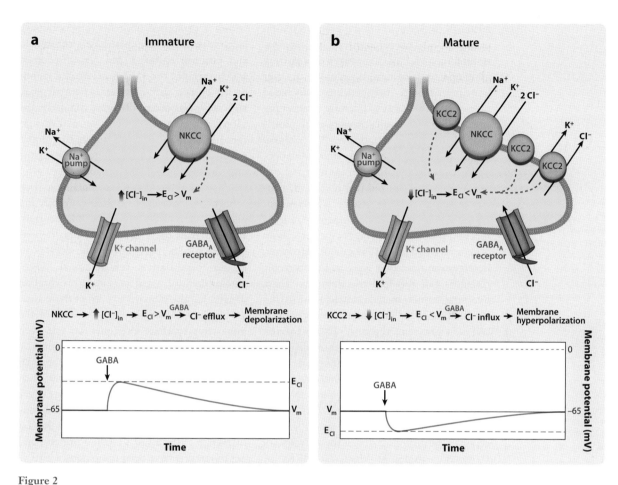

Figure 2

(*a*) Immature neuron expressing the NKCC (sodium-potassium chloride cotransporter) transporter. NKCC uses the inward Na^+ gradient maintained by the Na^+ pump to transport Cl^- into the cell. This increases the intracellular Cl^- concentration, resulting in a chloride equilibrium potential (E_{Cl}) that is more positive than the membrane potential (V_m). GABA-mediated activation of $GABA_A$ receptors at V_m produces Cl^- efflux, resulting in membrane depolarization. (*b*) Mature neuron coexpressing NKCC (sodium-potassium chloride) and KCC2 (sodium-potassium) cotransporters. KCC2 dominates over NKCC. KCC2 extrudes Cl^-, lowering the intracellular Cl^- concentration, resulting in a chloride equilibrium potential (E_{Cl}) that is more negative than the membrane potential (V_m). GABA-mediated activation of $GABA_A$ receptors at V_m produces Cl^- influx, resulting in membrane hyperpolarization.

potentials into the terminals by a mixture of Na^+-channel inactivation and membrane shunting (reviewed in Alvarez-Leefmans & Delpire 2009). This mechanism may thus be responsible for the GABA-mediated presynaptic inhibition of primary afferent terminals in the spinal cord. This mechanism may also explain GABA-mediated inhibition of release from secretory nerve terminals of the posterior pituitary. In this case, researchers were able to record directly from the terminal and show that

GABA depolarizes the terminal via $GABA_A$ receptors and that AP invasion is blocked, thus inhibiting neurosecretion (Zhang & Jackson 1993). Zhang & Jackson (1993) suggested that the depolarization of the terminal produces inactivation of Na^+ channels, thus blocking action potential invasion. However, in the case of the spinal cord, although there is evidence that GABA depolarizes primary sensory neurons, there is no evidence, to our knowledge, that this blocks spike invasion of the terminals, and even less that the depolarization produces Na$^+$-channel inactivation. Presynaptic inhibition via GABA-mediated depolarizing inhibition has also been observed in invertebrate preparations. In a modeling study based on studies in sensory neurons innervating spider VS-3 slit sensilla, French et al. (2006) showed that either shunting or Na$^+$ channel inactivation are sufficient to produce inhibition.

When speaking of inhibitory neurons, one usually thinks of neurons that release the neurotransmitter GABA or glycine. These neurotransmitters indeed activate ionotropic receptors that produce the changes described above leading to inhibition of excitability. However, neurotransmitters that activate metabotropic receptors can also lead to inhibition. These neurotransmitters include acetylcholine (ACh), serotonin (5HT), dopamine, neuropeptides, and even glutamate, the most prevalent excitatory neurotransmitter in the mammalian central nervous system (Lee & Sherman 2009). GABA can also activate metabotropic receptors (known as GABA$_B$ receptors). Not all the actions of metabotropic receptors are inhibitory, but one widespread example of such action in the nervous system is the activation of G-protein-activated K$^+$ channels (GIRKs) by metabotropic receptors that activate the appropriate G proteins (Padgett & Slesinger 2010). The resulting increase in K$^+$ conductance produces membrane hyperpolarization resulting in inhibition that can be as potent and effective as that mediated by ionotropic GABA and glycine receptors and often more long lasting. GABA action via GABA$_B$ receptors is widespread in the CNS and in some cases may be the main effect of the neurotransmitter.

INTERNEURON CLASSES WITHIN DIFFERENT REGIONS OF THE TELENCEPHALON

The telencephalon can be broadly divided into pallial and subpallial structures (Rubenstein et al. 1998). The pallium can be divided into regions based on their presumed phylogenic origin. According to this nomenclature, the cortex is comprised of archi- (hippocampus), paleo- (periamygdala, perirhinal, entorhinal, and piriform cortices), and neo- (isocortex, which includes the various areal subdivisions of the cortex, such as visual, motor, frontal, and auditory) cortical divisions. Similarly, the subpallium or pallidum can be divided into the neostriatum (caudate-putamen), globus pallidus, and the amygdala nuclei, to name only the most prominent divisions.

Each of these structures contains a wide cohort of specialized inhibitory projecting neurons and local inhibitory interneurons. If one defines these populations on the basis of their specific connectivity, the differences in the organization and neuronal classes within each of these regions by definition dictate that the subtypes in each of these areas are unique. Indeed, in the CA1 area of the hippocampus where this question has been most extensively addressed, the inhibitory neuronal population consists of more than 20 distinct classes (Somogyi & Klausberger 2005, Klausberger & Somogyi 2008, Klausberger 2009). Alternatively, if one uses a broad constellation of characteristics such as molecular markers (of which Ca^{2+}-binding proteins and neuropeptides have been prominently used as a means to distinguish interneuron subtypes), morphology, and firing pattern, it is possible to compare interneurons across different telencephalic regions (cf. Benes & Berretta 2001, Tepper & Bolam 2004, Wonders & Anderson 2006, Batista-Brito & Fishell 2009). Indeed, if one excludes the special features of interneuron populations residing in each of these structures, one discovers that there are common subtypes that share a number of salient features, most prominently their site of origin within the brain (summarized in **Figure 3** and **Table 1**).

In addition to their site of origin, molecular marker expression, firing pattern, and morphology (Ascoli et al. 2008), the classical means of discerning whether interneurons are distinct include (*a*) whether they innervate excitatory or inhibitory neurons or both (Freund & Gulyas 1997), (*b*) the cellular subdomain they target for innervation (somatic, proximal or distal dendrites, or axonal) (Mittmann et al. 2004), and (*c*) whether their connectivity is confined within or between particular structures or nuclei (Klausberger & Somogyi 2008). Unfortunately for the vast majority of telencephalic interneurons subtypes detailed information on their connectivity and their synaptic ultrastructure is incomplete (Freund & Gulyas 1997, Meglas et al. 2001, Karube et al. 2004, Kubota et al. 2009).

Another criterion that has been employed for the classification of interneuron populations in the hippocampus is in vivo temporal dynamics. In this region (Cobb et al. 1995, Klausberger et al. 2004, Somogyi & Klausberger 2005, Kwag & Paulsen 2009, Ellender & Paulsen 2010), specific interneuron subtypes fire at particular phases of specific network oscillations. These observations provided evidence that hippocampal functional networks are not hardwired, but instead are recruited dynamically by these oscillations. If so, rather than being an innate property of given interneuron populations, the circuitry in which they are embedded may dictate the context-specific role of interneuron subtypes. Alternatively, one may speculate that specific interneuron populations have evolved to influence particular aspects of the temporal dynamics of brain rhythms regardless of context. Hence, it will be of great interest to determine if particular interneuron subtypes predict their contribution to temporal dynamics in different structures. However, this can be addressed only after in vivo recordings of interneurons in relationship to the rhythms they participate in are completed.

The task of whether interneurons can be meaningfully classified across structures is presently hampered by a lack of data (see **Table 2**). Even with the criteria suggested above, the extent to which different telencephalic regions have been examined is very uneven, varying from the hippocampus (most particularly the CA1 region), which has been studied extensively, to the medial amygdala, which remains much less well characterized. To focus on the region most familiar to us, we use the somatosensory (S1) cortex (also known as barrel cortex in rodents), an area that shares features with all the telencephalic regions examined, as the standard with which to compare each of these structures. Nonetheless, this is done acknowledging at the outset that if one were to consider distinct subclasses of interneurons based on their afferent/efferent connections, then even the most archetypal interneuron subtype, the fast-spiking (FS) basket cells, would be considered distinct cell types in different cortical layers (Tremblay et al. 2010).

Somatosensory Cortex

Although some evidence exists that interneuron diversity may vary according to areal division (see Piriform Cortex section below), the repertoire within the barrel cortex is likely fairly representative of those found throughout the neocortex, albeit perhaps with slightly different ratios among subtypes. Indeed, such variances in subtype numbers have been reported in analyses comparing interneuron diversity in somatosensory (Kubota et al. 1993; Cauli et al. 1997; Gupta et al. 2000; Kawaguchi 2001; Kawaguchi & Kondo 2002; Butt et al. 2005; Miyoshi et al. 2007, 2010; Xu & Callaway 2009; Batista-Brito & Fishell 2009; Xu et al. 2010), visual (Gonchar & Burkhalter 1997, Gonchar et al. 2007, Xu et al. 2010) and prefrontal/frontal cortex (Lund & Lewis 1993, Kawaguchi & Kubota 1997, Kawaguchi & Kondo 2002, Lewis et al. 2002, Zaitsev et al. 2009). Depending on the criteria used, anywhere up to ~20 different interneuron subtypes have been described in the barrel cortex. Nonetheless, the most common subtypes can be broadly classified into six distinct categories: parvalbumin (PV)+ (*a*) chandelier or (*b*) basket cells, (*c*) somatostatin (SST)+ Martinotti cells, (*d*) vasoactive intestinal protein (VIP)+ interneurons (bipolar, bitufted and multipolar cells), (*e*) Reelin+ (nonsomatostatin) interneurons, and (*f*) other poorly defined CGE-derived interneurons that do not express VIP or reelin (reviewed in Batista-Brito & Fishell 2009, Lee et al. 2010, Rudy et al. 2010). The most stereotypical features of each of

Table 1 Diversity of neocortical interneurons[a,b]

	Origin	Molecular marker	Functional properties	Laminar distribution	Morphology	Diversity
FS basket cells	MGE	PV	FS firing pattern[c]. Low-input resistance and fast-membrane time constant. Mediate fast, powerful, and precise IPSPs	Layers II-VI; highest proportion in layer IV	Mostly multipolar, occasionally bitufted dendrites; dense local axon often extending to nearby columns and layers, targeting the perisomatic domain (forming "basket" terminals) of principal cells and interneurons, including other FS cells	1. Variations in expression of channels and receptors, e.g., Erg1 K$^+$ channels in cingulate CX and hippocampus 2. Layer-specific differences in threshold firing: delayed or onset spike and in degree of adaptation during large depolarizations 3. Variations in axonal transcolumnar and translaminar extent and somatodendritic size 4. Facilitating inputs from corticothalamic axons, other excitatory inputs are depressing 5. Fire early or late during UP states
Chandelier cells (axo-axonic cells)	MGE	PV	Firing pattern resembles FS basket cells, with higher input resistance and slower membrane time constant[d]	Layers II-VI. In rodents observed most often in layers II/III	Multipolar or bitufted dendrites. Preterminal axon branches form short vertical rows of boutons resembling candlesticks making synapses on the axonal initial segment of pyramidal cells	

Cell type	Origin	Markers	Physiology	Layers	Morphology	Notes
Martinotti cells	MGE	SST	Often called LTS cells (low-threshold spiking). Some fire two or more spikes on slow-depolarizing humps from hyperpolarized potentials, while others have an adapting regular-spiking firing pattern. Often show rebound spike(s) on repolarization. Strongly facilitating excitatory inputs. Strong excitation by muscarinic agonists	Layers II–VI	Multipolar, bitufted, or bipolar dendrites; axon arborizes locally and then ascends to layer 1, where it usually produces a dense axonal arborization. Target distal and tuft dendrites	1. Calretinin positive and negative 2. Variations in expression of other markers: reelin, NPY? 3. Variations in location and extent of distal axonal tuft
X-94-like SST neurons	MGE	SST	Lower input resistance and spikes of shorter duration than Martinotti cells (MCs), approaching those of FS cells. Often have a stuttering firing pattern during intermediate current injections. Capable of firing at higher frequencies than MCs but in contrast to FS neurons exhibit spike-frequency adaptation. Strongly facilitating excitatory inputs like MCs	Layers IV and Vb	Multipolar dendrites. Axon ramifies extensively in layer IV	
Neurogliaform cells	CGE	5HT3aR, reelin	Late-spiking firing pattern: delayed firing preceded by a slow depolarizing ramp at threshold and low-current injections. Regular adapting firing during large-current injections. Mediate combined slow $GABA_A$ and $GABA_B$ synaptic responses. Target dendritic spines but also produce nonsynaptic "volume" GABA release	Layers I–VI	Multipolar, short highly branched dendritic and axonal arbors near the cell body	
LS2	CGE	5HT3aR, reelin	Late-spiking less robust than neurogliaform cells, no discernable delay to first spike upon firing more than one spike		Multipolar. Wider and less branched dendritic and axonal arborization than neurogliaform cells	
CCK-expressing interneurons	CGE (at least for VIP+)	CCK. Two populations VIP+ and VIP-. CB1 receptors[e]	BSNP or RSNP	Layers I–VI; predominantly in upper layers	At least some of these populations are basket cells. Large CCK cells are VIP-; small CCK cells are VIP+	At least two populations VIP+ and VIP-

(Continued)

Table 1 (*Continued*)

	Origin	Molecular marker	Functional properties	Laminar distribution	Morphology	Diversity
Bitufted irregular spiking	CGE	5HT3aR, VIP, CR +	Irregular spiking at low current injections and adapting regular spiking with amplitude accommodation during larger depolarizations. High membrane resistance.	Layers I-VI; mainly layers II/III	Bipolar/bitufted dendrites, descending interlaminar axon. May preferentially target other INs	
Bipolar fast adapting	CGE	5HT3aR, VIP, CR negative	Fast-adapting cells are unable to maintain continuous firing during 500 ms suprathreshold depolarizations and show a large sAHP following the depolarizing step. High input resistance like other VIP cells	Mainly layers II/III	Bipolar and sometimes tripolar or multipolar dendrites	
Bursting bipolar cells (bNA2)	CGE	5HT3aR, VIP	High input resistance like other VIP cells	Mainly layers II/III	Bipolar dendritic morphology	
Other VIP interneurons	CGE	VIP, 5HT3aR		Mainly layers II/III	Arcade cells: multipolar or bitufted dendrites; axonal arcades, with vertical arborizations and descending collaterals. Small-basket cells	
Other CGE-derived INs lacking VIP or reelin	CGE	5HT3aR				
IS multipolar	PoA	5HT3aR			Multipolar	

[a]Abbreviations: BSNP, burst spiking nonpyramidal cell; CCK, cholecystokinin; CGE, caudal ganglionic eminence; CR, calretinin; FS, fast-spiking; IN, interneuron; IS, irregular spiking; MGE, medial ganglionic eminence; PoA, preoptic area; RSNP, regular spiking nonpyramidal cell; sAHP, slow after hyperpolarization; SST, somatostatin; VIP, vasoactive intestinal peptide.

[b]Based mainly on data in somatosensory cortex.

[c]Brief spikes, large fast AHP, high-frequency repetitive firing with little adaptation. At threshold, these cells fire abrupt trains of action potentials after a delay or after an initial spike(s) followed by a pause.

[d]Woodruff et al. 2009 (but see Xu & Callaway 2009).

[e]It is controversial whether both populations or only the VIP- express CB1 receptors.

Table 2 Comparison of GABAergic interneuron populations across telencephalic structures[a]

Neocortex	Hippocampus	Paleocortex/Piriform Cortex	Striatum	Basolateral amygdala
FS basket cells	Basket PV cells	FS multipolar cells	FS cells	FS cells with diverse firing patterns resembling the diversity observed in the neocortex
Chandelier cells (axo-axonic cells).	Chandelier or axo-axonic cells	Chandelier cells		Chandelier cells
Martinotti cells	OLM cells	Regular-spiking multipolar cells (somatostatin-expressing neurons)	LTS or pLTS SST+ interneurons	SST+ BLA interneurons
X-94-like SST neurons				
Neurogliaform cells (CGE derived)	Neurogliaform and Ivy cells. Most express NOS and are MGE derived	Neurogliaform cells		Neurogliaform cells
CCK+ interneurons	CCK basket cells (VIP+ and VIP-/VGLUT3+)			
LS2		Horizontal cells that resemble the interneurons described in somatosensory cortex as LS2		
Bitufted irregular spiking	CR+ VIP+ interneuron targeting INs			CGE-derived VIP neurons are present in BLA but have not been characterized
Bipolar fast adapting		Bitufted cell. VIP+ bitufted neurons with a fast-adapting firing pattern		
Bursting bipolar cells (bNA2)				
Projecting GABAergic neurons	Projecting GABAergic neurons: back projection cells; hippocampus to septum projecting cells; double projection; oriens-retrohippocampal projection		Medium spiny neurons	
	Bistratified cells			
	Other CCK INs: Schaffer collateral associated cells, /LM-PP-associated cells, and the LM-R-PP-associated cells			
	Trilaminar cells			
	CR (VIP-) IN targeting cells			
	Large calbindin			

[a]Abbreviations: BLA, basolateral amygdala; CCK, cholecystokinin; CGE, caudal ganglionic eminence; CR, calretinin; FS, fast-spiking; IN, interneuron; LM, lacunosum-moleculare; LS, late-spiking; LTS, low-threshold spiking; MGE, medial ganglionic eminence; NOS, nitric oxide synthase; OLM, oriens-lacunosum molecular; pLTS, persistent low-threshold spiking; PP, perforant pathway; PV, parvalbumin; R-PP, radiatum/perforant pathway; SST, somatostatin; VGLUT3, vesicular glutamate transporter; VIP, vasoactive intestinal peptide.

these classes as they are manifested within the somatosensory cortex is summarized in **Table 1**.

In total, the two most numerous populations of interneurons are the PV+ FS cells and the Martinotti cells, both of which originate embryonically within the medial ganglionic eminence (MGE) (Butt et al. 2005; Fishell 2007; Xu et al. 2004, 2008). Comprising ∼40% of all GABAergic interneurons, FS cells are the most prominent subtype across all cortical layers. These cells are characterized by their narrow spike width (∼0.3 ms), large and fast afterhyperpolarization, high maximal firing rate (>150 spikes/sec), and minimal firing frequency adaption during sustained depolarization (Connors & Gutnick 1990, McCormick et al. 1985) (see **Table 1** for a summary). They can be divided into two broad classes, basket cells and chandelier cells, both of which express the Ca^{2+}-binding protein PV (Celio 1986; reviewed in Kisvarday et al. 1990). Basket cells target the perisomatic domain of both excitatory and inhibitory neurons and are thought to be the strongest source of inhibition. On the other hand, chandelier cells target the axon initial segment (hence, they are often referred to as axoaxonic cells), have been observed to target only pyramidal neurons, and have recently been suggested to elicit depolarizing rather than hyperpolarizing postsynaptic responses (see Chandelier Cells May Be Excitatory, sidebar above). Despite these important differences, without visualizing their morphology the two cell types are difficult to distinguish because their firing properties are similar, although not identical (Woodruff et al. 2009; however, see also Xu & Callaway 2009).

FS basket cells are by far the most common of the interneuron subtypes and populate all layers except layer 1 (L1). Despite their relative abundance, in superficial cortical layers where they are still numerous, they are nevertheless collectively outnumbered by cells derived from the caudal ganglionic eminence (CGE) (Lee et al. 2010). Moreover, despite their common classification as a single interneuron subtype, their connectivity varies in accordance with

their laminar position (Dantzker & Callaway 2000, Xu & Callaway 2009, Tremblay et al. 2010). In L5/6 this population innervates the pyramidal neurons that provide all cortical output (Kawaguchi & Kubota 1993). In L4 this population receives thalamic input, is thought to mediate thalamocortical (TC) feed-forward inhibition, and connects to the stellate cell population (Gibson et al. 1999, 2004; Cruikshank et al. 2007). In more superficial layers, basket cells provide inhibition to both L5 projecting pyramidal cells as well as commissural and associative projecting pyramidal neurons (Dantzker & Callaway 2000). They also appear to possess a delay to firing at threshold, although the physiological significance of this property remains poorly understood. FS basket cells also innervate other interneurons including other FS cells (Gibson et al. 1999, Galarreta & Hestrin 2002).

FS basket cells also vary in size; their expression of channels, receptors, and other molecules; and the dynamics of their excitatory inputs (usually depressing, except for FS cell populations in L6). Furthermore, the details of their firing patterns are variable (Chow et al. 1999; Saganich et al. 1999, 2001; Beierlein & Connors 2002; West et al. 2006; Ascoli et al. 2008) (see **Table 1**).

The second most prevalent inhibitory cell type within the barrel cortex can be loosely classified as Martinotti cells on the basis of their ascending axon and their expression of the neuropeptide SST (Wang et al. 2004). Although it seems inevitable that all six groups described here will be further subdivided, perhaps nowhere is there a need for more reliable classifiers than with regards to this subtype (cf. Kawaguchi & Kubota 1997, Ma et al. 2006, Xu et al. 2006, McGarry et al. 2010). Classically the term Martinotti cell was used for neurons with somas located in L5/6 that send axons to L1. However, if we use the broadly defined criteria described here, the cell bodies of this population, although biased toward deeper layers, are present in all cortical layers except L1. This population is also characterized by their targeting of distal dendrites and by the fact that excitatory synapses onto these cells are strongly facilitating, a property that substantially influences their function (Oviedo & Reyes 2002, Silberberg & Markram 2007, Kapfer et al. 2007, Fanselow et al. 2008, Hull et al. 2009). Martinotti cells either have a bursting firing pattern (responding to stimulation from slightly hyperpolarized potentials with a burst of 1–3 spikes) or are characterized as adapting regular-spiking nonpyramidal cells (Kawaguchi & Kubota 1997, Gibson et al. 1999, Beierlein et al. 2003, Xu et al. 2006, Miyoshi et al. 2007).

Recently, Ma et al. (2006) characterized a population of SST+ interneurons in a mouse line called X94 that differed from classical Martinotti cells in a number of properties. These cells were located in L4 and L5b and had axonal projections that profusely innervated L4 instead of targeting L1. The cells also differed from Martinotti cells in a number of electrophysiological properties. X94 cells had a much lower input resistance, approaching that of FS cells. They had spikes of shorter duration and often produced a stuttering firing pattern. They fired at higher frequencies than did Martinotti cells, but in contrast to FS neurons, they exhibited spike frequency adaptation. However, as in the case of Martinotti cells, these SST-expressing cells received strongly facilitating excitatory synapses.

In addition, variability in SST neurons in the neocortex has been observed in their intrinsic firing properties, their expression of molecular markers, and their connectivity (Butt et al. 2005, Xu et al. 2006, Gonchar et al. 2007, Miyoshi et al. 2007, McGarry et al. 2010). For example, approximately one-third of SST interneurons in the mouse frontal, somatosensory (S1), and visual cortex (V1) contain calretinin (Xu et al. 2006). Although both SST/calretinin (CR)+ cells and SST/CR-cells exhibit similar Martinotti cell anatomical features and have similar adapting spike-firing patterns, they differ in the horizontal extension of their dendritic fields and in their number of primary processes (Xu et al. 2006). In addition, SST/CR- cells have narrower spikes with faster after-hyperpolarizing potentials. Moreover, in L2/3, where SST/CR+ cells are concentrated,

the two subtypes of SST interneurons have different connectivity (Xu & Callaway 2009). Whereas SOM+/CR– Martinotti cells receive strong excitatory input from both L2/3 and L4, SOM+/CR+ Martinotti cells receive excitatory input mainly from L2/3. Further arguing for a true delineation between SOM+/CR– and SOM+/CR+ Martinotti cells is evidence suggesting that they are derived from different areas during development, with the SOM+/CR+ population mainly arising from the dorsal Nkx6-2-positive region of the MGE, while the SOM+/CR– population is derived preferentially from the more ventrally positioned portion of the MGE (Fogarty et al. 2007, Sousa et al. 2009, Xu et al. 2008). It seems likely that some of the diversity within the SST population is functionally significant.

Most of the remainder of the interneuron populations within the somatosensory cortex are derived from the caudal ganglionic eminence (CGE) and are concentrated in the more superficial layers (L1–3) where they collectively represent the largest interneuron population (60% of L1–3 and approximately 30% of all interneurons) (Miyoshi et al. 2010, Lee et al. 2010). Between six and nine electrophysiological and morphological subtypes of CGE-derived interneurons have been described (Lee et al. 2010, Miyoshi et al. 2010). They have a similar expression of the ionotropic 5-HT receptor 5HT3a, which is expressed in the cortex exclusively in these interneurons. Also similar is their fast responsiveness to serotonin and ACh (Lee et al. 2010, Vucurovic et al. 2010). Analysis of these neurons in a mouse line expressing enhanced green fluorescent protein in 5HT3aR+ cells has demonstrated that PV-, SST-, and 5HT3aR+ interneurons account for nearly 100% of all glutamate amino-decarboylase-67-expressing (GAD67, an enzyme essential for the synthesis of GABA) neurons in S1 cortex (Lee et al. 2010).

CGE-derived interneurons include all the interneurons that express the neuropeptide VIP, which account for approximately 40% of the CGE-derived population. Another ~40–50% of the CGE-derived interneurons express reelin, which is also expressed in a fraction of SST+ interneurons. Both of these subpopulations are heterogeneous. Each includes two major subtypes, each representing approximately ~5% of the entire cortical interneuron population. These include two subtypes of reelin+ late-spiking cells, the CR+ VIP+ bipolar/bitufted neurons and the CR- VIP+ neurons. The two forms of late-spiking cells, one of which is the neurogliaform cells, can be distinguished by electrophysiology and morphology. Recent work has suggested that neurogliaform cells mediate tonic inhibition (see Tonic Inhibition, sidebar below) via nonsynaptic release of GABA (referred to as volume transmission) (Olah et al. 2009). Most of the CR/VIP+ bitufted cells have been described as irregular spiking, while most the VIP CR- cells are bitufted and multipolar neurons with a strongly adapting firing pattern, and have been termed "fast-adapting" (fAD) cells. An as yet undetermined subpopulation of the VIP+ cells apparently preferentially targets other interneurons (Staiger et al. 2004, David et al. 2007, Caputi et al. 2009).

Finally, a number of reports have suggested that a population of cortico-cortical projection GABAergic neurons may exist, particularly during development. Although these populations are not well described, these findings do suggest that populations akin to those seen in the hippocampus may also exist (Jinno et al. 2007, Jinno 2009, Higo et al. 2009).

TONIC INHIBITION

Recently, evidence of a new form of GABA$_A$ receptor-mediated inhibition has been reported and termed "tonic" inhibition to distinguished it from the classical synaptic GABA$_A$ receptor-mediated "phasic" (transient) inhibition (Otis et al. 1991, Salin & Prince 1996). This tonic inhibition is mediated by extrasynaptic or perisynaptic GABA$_A$ receptors that are activated by bulk ("volume") release of GABA to produce a long lasting or "persistent" contribution to the resting membrane potential and membrane leak (Soltesz et al. 1990, Brickley et al. 1996, Mody 2001, Nusser & Mody 2001, Stell & Mody 2002; reviewed in Semyanov et al. 2004, Farrant & Nusser 2005, Glykys & Mody 2007, Belelli et al. 2009). GABA$_A$ receptor-mediated tonic inhibition can be revealed by the depolarization of the membrane and the increase in membrane resistance following application of GABA$_A$ receptor inhibitors.

The GABA mediating this signaling was thought to exclusively originate through spillover from the synaptic cleft; however, Tamas and colleagues (Oláh et al. 2009) recently suggested that, in the cortex, neurogliaform cells may make a particularly strong contribution to tonic inhibition by producing "volume" transmission of GABA. Originally studied in cerebellar granule cells and the hippocampus, where it has important roles in modulating neuronal and network excitability (Mitchell & Silver 2003, Rothman et al. 2009), tonic inhibition has now been observed in many other structures including the neocortex, thalamus, hypothalamus, striatum, and spinal cord (Belelli et al. 2009) and is considered to provide a significant contribution to several physiological and pathological processes.

Following synaptic GABA release, the concentration of GABA in the extracellular fluid is much lower than that reached in the synaptic cleft. Extrasynaptic GABA$_A$ receptors are able to respond to low concentrations of GABA because the subunit composition of these channels bestow on them a higher affinity for GABA. Extrasynaptic receptors also tend to desensitize less than synaptic receptors. As a result of these properties, activation of these channels tends to produce more persistent actions than are produced by synaptic receptors (Saxena & Macdonald 1994, 1996; Haas & Macdonald 1999; Bianchi & Macdonald 2002; Brown et al. 2002).

Paleocortex (Piriform Cortex)

Analysis of the GABAergic interneurons of the primary olfactory or piriform cortex, a region of paleocortex, although less studied than the neocortex, reveals similar interneuron subtypes to those found in somatosensory cortex. The piriform cortex is a three-layered paleocortex with a simpler anatomy than that of the neocortex. In recent studies, Suzuki & Bekkers (2007, 2010a,b) classified the interneurons in the anterior piriform cortex into five classes based on electrophysiological properties, morphology, and expression of molecular markers. These included some of the most common CGE-derived neocortical interneurons including neurogliaform cells and VIP+ bitufted neurons with a fast-adapting firing pattern (Lee et al. 2010, Miyoshi et al. 2010).

They also observed FS basket cells and a group of "regular-spiking" multipolar somatostatin-expressing neurons resembling in morphology and electrophysiological properties neocortical Martinotti cells. Finding homologies among these five classes of interneurons in the aPC and interneuron subtypes in neocortex is not difficult. Nonetheless, some intriguing differences support the idea that interneuron classes can have specializations that allow them to adapt to the requirements of specific microcircuits.

Neurogliaform cells account for ~34% of the interneuron population in L1 of the piriform cortex and possess the delayed firing (late-spiking) characteristic of these neurons in neocortex and hippocampus. The short-term dynamics of excitatory inputs onto these cells appears to be variable. According to Suzuki & Bekkers (2010b), synapses from lateral olfactory tract (LOT) afferents onto layer Ia neurogliaform cells showed strong paired-pulse facilitation, whereas L1b, L2, and L3 intracortical inputs onto neurogliaform cells in those layers showed little or no facilitation. In the hippocampus, excitatory inputs onto neurogliaform cells are typically depressing (Price et al. 2005).

Excitatory inputs onto FS cells in the aPC also showed layer-specific short-term dynamics. Layer-specific dynamics of excitatory inputs is also observed on neocortical FS cells. Excitatory synapses on FS cells are usually depressing; however, facilitating excitatory synapses have been observed on layer 6 FS cells (Beierlein & Connors 2002, West et al. 2006). Synaptic dynamics depends on presynaptic and postsynaptic mechanisms (Reyes et al. 1998), and several still to be investigated mechanisms are likely responsible for the variability of this property among interneurons of the same subtype in different locations.

In the hippocampus, striatum and neocortex somatic targeting FS basket cells mediate powerful feed-forward inhibition of principal cells (Pouille & Scanziani 2001, 2004; Gabernet et al. 2005; Cruikshank et al. 2007). From this, one might feel justified in believing that FS interneurons are truly an interneuron subtype with a dedicated function to which they are recruited whenever the need for fast feed-forward inhibition arises. However, examination of FS cells within the piriform cortex challenges this notion. A recent study of the olfactory cortex illustrates that the function of a specific interneuron subtype is not dependent solely on its intrinsic properties. Instead, it is also a function of the circuit in which it is incorporated.

In the neocortex, TC axons synapse on both FS cells and excitatory principal neurons in thalamo-recipient layers (mainly layer 4). Activation of both excitatory and inhibitory cells establishes a simple disynaptic circuit that provides powerful, local feed-forward inhibition. The latency between the onset of the TC excitation of excitatory cells and the onset of the disynaptic feed-forward inhibition from perisomatic-targeting FS cells results in a brief time window during which the excitatory neurons can integrate TC inputs. This window is critical for the processing of sensory information in the neocortex (Miller et al. 2001, Swadlow 2002, Gabernet et al. 2005, Wilent & Contreras 2005, Cruikshank et al. 2007). A similar circuit exists in the CA1 area of the hippocampus, where basket cells mediate feed-forward inhibition of CA1 pyramidal cells. As in the neocortex, the duration of the time window within which Schaffer collateral excitatory inputs on CA1 pyramidal cells can summate to reach spike threshold is determined by the basket cell-mediated feed-forward inhibition (Pouille & Scanziani 2001). FS cells mediate powerful feed-forward inhibition in other structures, such as L2/3 of somatosensory cortex and the neostriatum (Helmstaedter et al. 2008, Tepper & Bolam 2004). In contrast, in the piriform cortex, the input layer is L1a, which receives the LOT afferents containing the axons of projecting mitral and tufted cells of the olfactory bulb. L1a lacks FS cells. Instead, LOT axons synapse on L1a interneurons that target the apical dendrites of pyramidal cells with cell bodies in deeper layers (Stokes & Isaacson 2010). LOT axons also target the dendrites of the excitatory cells. As a result, similar to what is observed in the neocortex and hippocampus, bursts of mitral and tufted cell activity mediate a short-latency feed-forward disynaptic inhibition of the pyramidal cells, except that this inhibition occurs on the distal apical dendrite, instead of the soma, of the pyramidal cells and is mediated by non-FS layer 1a INs.

On the other hand, FS cells in the piriform cortex are localized mainly in L3 where they are recruited by recurrent excitation from the pyramidal cells. As in other structures, the FS cells in the piriform cortex target the somatic domain of the pyramidal cells and their inputs are depressing. However, the excitatory inputs on the pyramidal cells are facilitating. Hence, during bursts of inputs to the piriform cortex, perisomatic-targeting FS cells are recruited late and provide "feedback" somatic inhibition of the pyramidal cells. Thus, in the olfactory cortex, inhibition shifts from the dendrite to the soma, the opposite of what is observed in the neocortex and hippocampus.

Hippocampus

Interneurons have been studied in all regions of the hippocampal formation (Ceranik et al. 1997, McBain & Fisahn 2001, Lawrence & McBain 2003, Hefft & Jonas 2005, Bartos et al. 2007, Fuentealba et al. 2010, Tricoire et al. 2010). As a result of the CA1 region's relatively simple cellular architecture, limited afferent connectivity, and extensive examination by numerous groups, the connectivity of interneuron subtypes within this region has been characterized most extensively. Somogyi and colleagues have made a special effort to characterize interneuron diversity in the CA1 subfield and have defined at last count 21 distinct classes of GABAergic neurons in this area (Somogyi & Klausberger 2005, Klausberger & Somogyi 2008). However, the reliance on connectivity to define these populations makes extrapolation of the cell types found in this region difficult to compare with those found elsewhere. Even though certain subtypes may be specific to the hippocampus, or even more specifically to particular layers of this structure,

many of the interneuron subtypes in CA1 seem to share both functional and lineal relationships (Tricoire et al. 2010) to those found in other parts of the telencephalon, most notably the neocortex (see **Table 2** for comparisons).

The easiest populations to compare with those found in the cortex are the PV+ basket and chandelier (axo-axonic) cells. The location of their output synapses, perisomatic and the axon initial segment, respectively, intrinsic biophysical properties, properties of input and output synapses, the expression of the Ca^{2+}-binding protein PV, as well as the fact that they are all MGE-derived suggest that these interneuron subtypes are closely homologous to those found in the cortex.

Despite their expression of PV, oriens-lacunosum moleculare cells (known as OLM cells) also appear to be related to populations seen in the cortex. Their biophysical properties, long axon targeting distal apical dendrites, and marker expression (somatostatin) suggest they are homologous to the neocortical Martinotti cells. Additional populations that have close homologs within the hippocampus and the cortex are CR- and/or VIP+ interneuron populations that preferentially innervate other inhibitory interneuronal populations (Staiger et al. 2004, David et al. 2007, Caputi et al. 2009).

Even though neurogliaform cells would also appear to exist in both the hippocampus and cortex, the relationship between those within the cortex and hippocampus seems more complicated. Neurogliaform cells in the two structures are very similar in morphology and they share a number of unique functional properties (Fuentealba et al. 2008, 2010; Tricoire et al. 2010). Furthermore, there is a neurogliaform population in the hippocampus that like those in neocortex arises from the CGE. However, the majority of neurogliaform cells in the hippocampus express nitric-oxidase synthetase- (nNOS-) and are MGE-derived (Tricoire et al. 2010). Moreover, another MGE-derived (Tricoire et al. 2010) nNOS+ hippocampal population of interneurons termed Ivy cells has been identified. Although their somas are positioned in the pyramidal rather than the lacunosum-moleculare layer, the Ivy cells strongly resemble the nNOS+ neurogliaform cells (Fuentealba et al. 2008, Szabadics & Soltesz 2009). Interestingly, the nNOS$^+$ Ivy interneuron population appears to be numerically the largest interneuron population within the hippocampus. Whether there are any MGE-derived cortical homologs of the nNOS+ neurogliaform and Ivy interneurons within the neocortex remains to be determined.

Another population of interneurons that exists in both the neocortex and the hippocampus expresses the neuropeptide cholecystokinin (CCK) (Nunzi et al. 1985). In the CA1 area, at least five classes of CCK interneurons have been identified (Klausberger et al. 2004), including two types of CCK+ basket cells (with and without VIP) (Somogyi et al. 2004), the Schaffer collateral-associated cells, lacunosum-moleculare/perforant pathway (LM-PP)-associated cells (Vida et al. 1998), and the lacunosum-moleculare/radiatum/perforant pathway (LM-R-PP)-associated cells (Cossart et al. 1998, Vida et al. 1998). At least two types of CCK interneurons have been described in the neocortex, those expressing VIP and those that do not (Wang et al. 2002, Gonchar et al. 2007, Xu et al. 2010). The latter type closely resembles the CCK basket cells in the hippocampus on the basis of both morphology and multiple functional properties including modulation by cannabinoid (CB1) receptors and many other neuromodulators (Freund & Katona 2007). However, much less is known about their connectivity. Moreover, immunostaining within the cortex has shown a sparse CCK+ population, raising the question of whether CCK basket cells are much more abundant in hippocampus. Indeed, CCK mRNA expression was identified in a recent microarray analysis of developing neocortical interneurons (Batista-Brito et al. 2008). While the origin of the CCK+ cortical interneuron population is at present unknown, the coexpression of 5HT3aR in a substantial population of CCK+ interneurons in the hippocampus (Somogyi et al. 2004) and neocortex (Lee et al. 2010, Vucurovic et al. 2010) argues that

this population may be CGE derived. This suggestion fits well with the fact that there is as yet no identified marker available for a portion of the CGE-derived cortical interneurons.

The hippocampus, like the neocortex, also contains projection GABAergic neurons. In CA1, projection GABAergic neurons include subiculum-projecting trilaminar cells; back-projecting CA1 cells that provide widespread innervation to CA1, CA3, and the dentate hilus (Sik et al. 1994, 1995); and the hippocampal-septal cells (Freund & Buzsaki 1996).

Among the populations of hippocampal interneurons for which there does not appear to be a parallel population in the neocortex, the cells called bistratified stand out. These interneurons first described in the CA1 area by Somogyi and colleagues (Buhl et al. 1994, Pawelzik et al. 1999, Maccaferri et al. 2000) have axons targeting the dendrites of CA1 pyramidal cells in stratum oriens and stratum radiatum (hence the term bistratified). They are coaligned with the Schaffer collateral input onto those dendrites and receive both Schaffer collateral input and recurrent input from CA1 pyramidal cells. They are PV+ and have a FS firing pattern (Pawelzik et al. 2002, Fujiwara-Tsukamoto et al. 2010). Bistratified cells make synapses with small dendritic shafts of the pyramidal cells, and they fire at the trough of the theta oscillation, similar to oriens-lacunosum molecular cells. However, in contrast to oriens-lacunosum molecular cells, which are silenced during ripple episodes, bistratified cells fired strongly during these events (Klausberger et al. 2004).

Striatal Interneurons

The basal ganglia is comprised of several brain structures, most notably the striatum and the globus pallidus. In both these structures, the only projecting neuronal population, the medium spiny neurons (MSNs) of the striatum, and the principal projection cells of the globus pallidus are inhibitory GABAergic neurons (Difiglia et al. 1982, Wilson 2007). Of the various basal ganglia, only the interneurons within the striatum have been extensively studied (Kawaguchi 1997, Tepper & Bolam 2004). The striatum contains several subtypes of interneurons, including GABAergic and cholinergic neurons (the sole excitatory striatal population). Given the diversity of interneurons that have been described in the neocortex and hippocampus, it is surprising how few subtypes have been observed to date within the basal ganglia and even there the focus of investigators has been almost entirely limited to the striatum. Over 15 years ago, Kawaguchi (1993) described three major interneuron subtypes within the striatum and these collectively are thought to comprise less than 3% of all the neurons within this structure (Tepper et al. 2004, Tepper & Bolam 2004) (see **Figure 3** for a comparison between structures). These

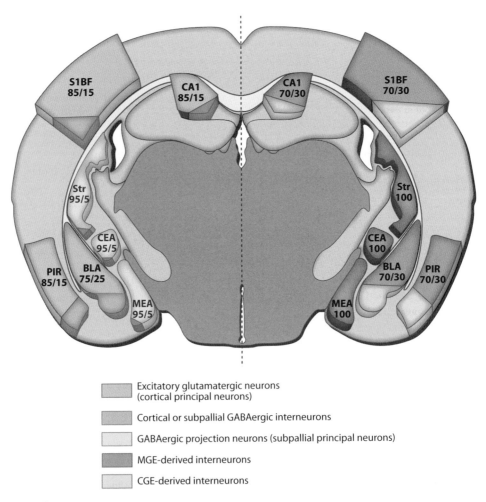

Figure 3

Cross section of an adult brain: (*left*) the relative proportion of excitatory (*green*) or inhibitory (*light red*) principal projection neurons compared with the percentage of inhibitory interneurons in each of the illustrated structures; (*right*) the relative proportion of inhibitory interneurons within these same structures that are thought to be derived from the medial ganglionic eminence (MGE) (*blue*) or the caudal ganglionic eminence (CGE) (*yellow*). Abbreviations: CA1, CA1 region of the hippocampus; S1BF, primary somatosensory cortex/barrel field; PIR, piriform cortex; Str, striatum; BLA, basal-lateral amygdala; CEA, central amygdala nuclei; MEA, medial amygdala nuclei.

are the large aspiny cholinergic interneurons, so called tonically active neurons (TAN), two types of GABAergic interneurons, PV+ FS basket cells and low-threshold spiking (LTS or pLTS, persistent low-threshold spiking owing to their long calcium-mediated plateau) SST+ interneurons. This later population is heterogeneous and includes NPY+ and NOS+ subpopulations. Beyond this, the only additional population that has been reported is a small population of CR+ neurons that are provisionally thought to possess LTS-like physiological features, suggesting that, although they are SST-negative, they may be related to the SST+ populations. Finally, a few references, most notably from Ramón y Cajal (1911), have suggested that a small population of neurogliaform cells reside within the striatum (Fox & Rafols 1971, DiFiglia et al. 1976, Chang et al. 1982).

The interneuron subtypes within the striatum are quite similar to, albeit apparently less diverse than, subtypes within the neocortex and hippocampus. However, the principal neuronal population in the striatum is GABAergic (the medium spiny neurons), and as such, in contrast to interneuron populations in the pallium, striatal interneurons (with the exception of those targeting the ACh interneuron population) target inhibitory GABAergic neurons. Interestingly, the interneuron marker expression within the striatum is consistent with most, if not all, interneurons within this structure being MGE-derived (Marin et al. 2000). If the striatum does not receive CGE-derived interneuron populations, this could explain the apparently lower level of diversity of interneurons within this structure.

In terms of the connectivity and function of the interneurons within the striatum, whereas the cholinergic populations are thought to get input primarily from the thalamus, the FS and LTS populations are thought to receive afferents mainly from the neocortex (although to some extent all these cell types receive afferent inputs from both structures). Functionally, these latter two interneuron subtypes form synapses on MSNs, thus providing feed-forward inhibition (Vuillet et al.

1989, Sidibe & Smith 1999, Gertler et al. 2008). In addition to their extremely small numbers, the FS interneurons give sparse input to the MSNs, as compared with the extensive excitatory input provided to MSNs by afferent pyramidal neurons (Koos & Tepper 1999, Kreitzer 2009, Gittis et al. 2010). However, like in the neocortex, these populations possess FS properties, PV expression, and high release probability. They also produce strongly temporally coordinated inhibition. Indeed, although a typical MSN receives on the order of ten times more excitatory cortical inputs (100–125 compared with 15–25 synapses from FS cells), the FS population is thought to be extremely effective at reducing MSN excitability (Koos et al. 2004, reviewed in Kreitzer 2009). By contrast, the LTS population, which also provides weak input to MSN neurons, likely mediates much of its impact on striatal function through neuromodulation mediated through release of NPY or NOS (Chesselet & Graybiel 1986, Kubota et al. 1993). Indeed, a primary function of LTS cells may be related to the control of blood flow (Aoki & Pickel 1990).

Neuromodulation by dopamine and ACh plays an extremely important role in regulating the function of the striatum, as evidenced by Parkinson's and Huntington's diseases where imbalances in neuromodulation are strongly associated with the dysfunction seen in these syndromes (Shen et al. 2008, reviewed in Pisani et al. 2001). Indeed, studies examining striatal function belie the simplistic notion that specific neuron's contribution to signaling can be reduced to providing net excitation or inhibition. Hence, the ability of striatal interneurons to regulate complex interactions between all cell types in this structure including other interneurons, combined with their secretion of neuromodulators suggest that they are central to the control and timing of striatal activity.

The Amygdala Nuclei

In many regards, the amygdala nuclei represent a hybrid of cortical and subcortical structures (reviewed in Sah et al. 2003). Whereas lateral amygdala nuclei such as the lateral and basal nuclei (collectively termed BLA) have a cortical-like architecture, the medial nuclei such as the central nucleus (which can be subdivided into the lateral and medial divisions) and the medial amygdala (MeA) all more closely resemble the GABAergic basal ganglia. Linking these portions of the amygdala are the intercalated cell masses, which largely consist of GABAergic populations that appear more specialized and have recently been suggested to have arisen from a unique origin near the pallial-subpallial boundary (Stenman et al. 2003, Waclaw et al. 2010; J.G. Corbin, personal communication). Functionally, the amygdala can be divided into the more laterally positioned nuclei including the BLA and central nucleus, which are more involved in fear circuits, and the MeA, which processes innate behaviors such as aggression and mating and is closely linked to the hypothalamus (MacDonald 1984, reviewed in Ehrlich et al. 2009). Nonetheless, all structures within the amygdala are characterized by the presence of at least some inhibitory interneuron populations, which resemble those found in other portions of the telencephalon but may differ in their electrophysiological characteristics, connectivity, and site of origin.

Our understanding of the contribution of interneurons within each of these set of nuclei is varied; the BLA is the most studied and MeA the least. Within the BLA, immunological and electrophysiological analyses indicate that the diversity of interneurons closely resemble that seen in the neocortex or hippocampus with PV-, SST-, VIP-, CCK-, CR-, and NOS-expressing populations all having been reported (Mascagni & McDonald 2003). As in the cortex, interneurons make up a percentage of the total neuronal population of the BLA (only approximately 25%) (McDonald & Augustine 1993) (see **Figure 3** for comparisons between structures). Similarly, as in the cortex, approximately 50% of all interneurons within these regions appear to be PV positive. This population is primarily comprised of FS basket cells (although chandelier neurons have also been reported in this region) (McDonald

1982); however, a variety of physiological firing patterns and some variation in morphology has been used to suggest that some heterogeneity exists within this population (Woodruff & Sah 2007a,b). This heterogeneity is similar to that observed for FS cells in the neocortex (Ascoli et al. 2008, Tremblay et al. 2010). In addition, the function of this population has been speculated to involve both feed-forward and feedback regulation of signaling (Woodruff et al. 2006). Nonetheless, many of these variations have also been observed within the neocortex and hippocampus; hence, these cells may appropriately be classified as belonging to the FS basket cell subtype. A further similarity with the cortex is that interneurons within the BLA (and to a lesser extent the MeA) are under the strong control of a variety of neuromodulators. The PV and VIP populations receive pronounced serotonin modulation, and at least the latter of these also possesses CB1-receptors as well as significant cannabinoid modulation (Morales et al. 2004). The similarities of the interneurons within the BLA and cortex in part reflect their common origin. As in the cortex, both our work and that of others have implicated the MGE as the source of both PV and SST+ BLA interneurons (Nery et al. 2002, Xu et al. 2008, Hirata et al. 2009). In addition, consistent with the presence of 5HT3aR+ interneurons in this region (Mascagni & McDonald 2007, Muller et al. 2007), recent unpublished genetic fate-mapping efforts (G. Miyoshi and J. Corbin, personal communication) have reported that CGE-derived interneurons populate the BLA (as well as the CeA).

Central and Medial Amygdala Nuclei

As noted above, the central nuclei of the amygdala is striatal-like in organization and provides the major output for fear (although unlike the majority of cells in the medial amygdala, these cells are glutamatergic). Despite being overwhelmingly comprised of GABAergic neurons, it does not appear to contain a major interneuron population. Like the striatum, the central and medial amygdala nuclei possess only sparse PV, VIP, and CCK populations and a somewhat larger somatostatin-expressing cohort particularly in the medial amygdala (McDonald 1989, Pare & Smith 1993, McDonald et al. 1995, Kemppainen & Pitkänen 2000, Równiak et al. 2008). The intrinsic physiological properties and morphologies of cells within these subdivisions have yet to be thoroughly analyzed. However, consistent with the MGE origin of populations within this region (Xu et al. 2008), both FS and LTS burst spiking (with hyperpolarizing rebound burst properties) have been reported. These observations coupled with the general structure of these nuclei indicate that the central and medial amygdala will ultimately prove reasonably similar in its interneuron subtypes to those seen in the striatum (**Figure 3**).

Least well studied are the interneuron populations associated with innate behaviors and located within the medial divisions of the amygdala. Recent fate-mapping experiments indicate that this region receives cells that developmentally express *Shh* or related genes, including *Gli1 and Nkx2.1* (Carney et al. 2010). In addition, it has been reported that this region possesses large numbers of cells that express the 5HT3a receptor (Mascagni & McDonald 2007). Although the interneuron subtypes and their origins have yet to be sorted out thoroughly, this structure appears to receive its interneuron repertoire from the same set of eminences that supply the rest of the telencephalon (**Figure 3**).

PERSPECTIVE

This survey of the cellular substrates and mechanisms mediating inhibition within the telencephalon, although far from complete, provides a sense of the breadth of cell types and mechanisms that mediate inhibition. It also highlights commonalities in inhibitory function within this brain region. Although a growing cohort of scientists have called for a standardization of nomenclature for interneuron subtypes (Ascoli et al. 2008), the challenges to this effort remain daunting. Despite the undeniable appeal of providing a common vernacular for specific cell types, the variance in anatomy both across and within structures dictates that such identifiers are uncomfortably restricting. An alternative approach is to be less precise but capture some of the salient features of particular subsets. For instance, referring to FS, perisomatic-targeting basket cells as a common category based on their shared site of origin within the MGE, coupled with their anatomical and physiological similarities, is tacitly accepted. This is true despite the fact that depending on structure they primarily target either excitatory (cortex, hippocampus, BLA) or inhibitory projection neurons (striatum, CeA, MeA) or receive depressing (most neocortical layers and hippocampus) versus facilitating (layer 6) excitatory synapses as a result of their afferent connectivity. On balance, we would argue that using an interneuron developmental origin and molecular signature provides a guide to the identification of commonalities that provides a compelling argument for acceptance of certain classes as orthologs, if not homologs. There is no doubt that we are at the cusp of a deluge of new findings that will greatly enhance our understanding of the connectome and transcriptome of all classes of interneurons. The foremost goal in this effort is to organize the wealth of information addressing the similarities and differences in inhibitory neural signaling across the telencephalon, and there seems little doubt that the points of view of both the "lumpers" (those individuals who focus on the commonalities between different interneuron subtypes) and "splitters" (those who focus on the differences) will aid in this effort. In the words of Maurice Sendak, "Let the wild rumpus begin."

DISCLOSURE STATEMENT

The authors are not aware of any affiliations, memberships, funding or financial holding that might be perceived as affecting the objectivity of this review.

ACKNOWLEDGMENTS

We are grateful to Michael Long, Jens Hjerling-Leffler, Theofanis Karayannis, and Soohyun Lee for suggestions and comments on the manuscript and Drs. David Anderson and Joshua Corbin for help with understanding the amygdala. Both Drs. Fishell and Rudy are generously supported by grant funding from the NIH. Dr. Fishell is specifically supported by the National Institute of Health grants (5RO1MH068469-08 and 2R01MH071679-09), National Institute of Neurological Disorders and Stroke grant (5R01NS039007-1), Simons Foundation and New York Stem Cell Science State grant (NGSG-130), and Dr. Rudy by National Institute of Neurological Disorders and Stroke grants (5R01NS045217-08, 2R01NS030989-18, and 3R01NS045217-06A1S1) and the Alzheimer's Association.

LITERATURE CITED

Alger BE, Nicoll RA. 1979. GABA-mediated biphasic inhibitory responses in hippocampus. *Nature* 281:315–17

Alvarez-Leefmans FJ. 1990. Intracellular Cl⁻ regulation and synaptic inhibition in vertebrate and invertebrate neurons. In *Chloride Channels and Carriers in Nerve, Muscle and Glial Cells*, ed. FJ Alvarez-Leefmans, JM Russell. pp. 109–58. New York: Plenum

Alvarez-Leefmans FJ, Delpire E. 2009. Thermodynamics and kinetics of chloride transport in neurons: an outline. In *Physiology and Pathology of Chloride Transporters and Channels in the Nervous System*, ed. FJ Alvarez-Leefmans, E Delpire, pp. 81–108. New York: Academic

Andersen P, Dingledine R, Gjerstad L, Langmoen IA, Laursen AM. 1980. Two different responses of hippocampal pyramidal cells to application of gamma-amino butyric acid. *J. Physiol.* 305:279–96

Aoki C, Pickel VM. 1990. Neuropeptide Y in cortex and striatum. Ultrastructural distribution and coexistence with classical neurotransmitters and neuropeptides. *Ann. NY Acad. Sci.* 611:186–205

Araki T, Yamano M, Murakami T, Wanaka A, Betz H, Tohyama M. 1988. Localization of glycine receptors in the rat central nervous system: an immunocytochemical analysis using monoclonal antibody. *Neuroscience* 25:613–24

Ascoli GA, Alonso-Nanclares L, Anderson SA, Barrionuevo G, Benavides-Piccione R, et al. 2008. Petilla terminology: nomenclature of features of GABAergic interneurons of the cerebral cortex. *Nat. Rev. Neurosci.* 9:557–68

Bartos M, Vida I, Jonas P. 2007. Synaptic mechanisms of synchronized gamma oscillations in inhibitory interneuron networks. *Nat. Rev. Neurosci.* 8:45–56

Batista-Brito R, Fishell G. 2009. The developmental integration of cortical interneurons into a functional network. *Curr. Top. Dev. Biol.* 87:81–118

Batista-Brito R, Machold R, Klein C, Fishell G. 2008. Gene expression in cortical interneuron precursors is prescient of their mature function. *Cereb. Cortex* 18:2306–17

Beierlein M, Connors BW. 2002. Short-term dynamics of thalamocortical and intracortical synapses onto layer 6 neurons in neocortex. *J. Neurophysiol.* 88:1924–32

Beierlein M, Gibson JR, Connors BW. 2003. Two dynamically distinct inhibitory networks in layer 4 of the neocortex. *J. Neurophysiol.* 90:2987–3000

Belelli D, Harrison NL, Maguire J, Macdonald RL, Walker MC, Cope DW. 2009. Extrasynaptic GABAA receptors: form, pharmacology, and function. *J. Neurosci.* 29:12757–63

Ben-Ari Y. 2002. Excitatory actions of GABA during development: the nature of the nurture. *Nat. Rev. Neurosci.* 3:728–39

Benes FM, Berretta S. 2001. GABAergic interneurons: implications for understanding schizophrenia and bipolar disorder. *Neuropsychopharmacology* 25:1–27

Betz H. 1991. Glycine receptors: heterogeneous and widespread in the mammalian brain. *Trends Neurosci.* 14:458–61

Betz H, Laube B. 2006. Glycine receptors: recent insights into their structural organization and functional diversity. *J. Neurochem.* 97:1600–10

Bianchi MT, Macdonald RL. 2002. Slow phases of GABAA receptor desensitization: structural determinants and possible relevance for synaptic function. *J. Physiol.* 544:3–18

Brickley SG, Cull-Candy SG, Farrant M. 1996. Development of a tonic form of synaptic inhibition in rat cerebellar granule cells resulting from persistent activation of GABAA receptors. *J. Physiol.* 497:753–59

Brock LG, Coombs JS, Eccles JC. 1952a. The nature of the monosynaptic excitatory and inhibitory processes in the spinal cord. *Proc. R. Soc. Lond. B Biol. Sci.* 140:170–76

Brock LG, Coombs JS, Eccles JC. 1952b. The recording of potentials from motoneurones with an intracellular electrode. *J. Physiol.* 117:431–60

Brown N, Kerby J, Bonnert TP, Whiting PJ, Wafford KA. 2002. Pharmacological characterization of a novel cell line expressing human α4β3δ GABAA receptors. *Br. J. Pharmacol.* 136:965–74

Buhl EH, Halasy K, Somogyi P. 1994. Diverse sources of hippocampal unitary inhibitory postsynaptic potentials and the number of synaptic release sites. *Nature* 368:823–28

Butt SJ, Fuccillo M, Nery S, Noctor S, Kriegstein A, Corbin JG, Fishell G. 2005. The temporal and spatial origins of cortical interneurons predict their physiological subtype. *Neuron* 48:591–604

Buzsáki G. 1984. Feed-forward inhibition in the hippocampal formation. *Prog. Neurobiol.* 22:131–53

Caputi A, Rozov A, Blatow M, Monyer H. 2009. Two calretinin-positive GABAergic cell types in layer 2/3 of the mouse neocortex provide different forms of inhibition. *Cereb. Cortex* 19:1345–59

Carney RS, Mangin JM, Hayes L, Mansfield K, Sousa VH, et al. 2010. Sonic hedgehog expressing and responding cells generate neuronal diversity in the medial amygdala. *Neural Dev.* 5:14

Cauli B, Audinat E, Lambolez B, Angulo MC, Ropert N, et al. 1997. Molecular and physiological diversity of cortical nonpyramidal cells. *J. Neurosci.* 17:3894–906

Celio MR. 1986. Parvalbumin in most gamma-aminobutyric acid-containing neurons of the rat cerebral cortex. *Science* 231:995–97

Ceranik K, Bender R, Geiger JR, Monyer H, Jonas P, et al. 1997. A novel type of GABAergic interneuron connecting the input and the output regions of the hippocampus. *J. Neurosci.* 17:5380–94

Chang HT, Wilson CJ, Kitai ST. 1982. A Golgi study of rat neostriatal neurons: light microscopic analysis. *J. Comput. Neurol.* 208:107–26

Chow A, Erisir A, Farb C, Nadal MS, Ozaita A, et al. 1999. K+ channel expression distinguishes subpopulations of parvalbumin- and somatostatin-containing neocortical interneurons. *J. Neurosci.* 19:9332–45

Chesselet MF, Graybiel AM. 1986. Striatal neurons expressing somatostatin-like immunoreactivity: evidence for a peptidergic interneuronal system in the cat. *Neuroscience* 17:547–71

Cobb SR, Buhl EH, Halasy K, Paulsen O, Somogyi P. 1995. Synchronization of neuronal activity in hippocampus by individual GABAergic interneurons. *Nature* 378:75–78

Connors BW, Gutnick MJ. 1990. Intrinsic firing patterns of diverse neocortical neurons. *Trends Neurosci.* 13:99–104

Cossart R, Esclapez M, Hirsch JC, Bernard C, Ben-Ari Y. 1998. GluR5 kainate receptor activation in interneurons increases tonic inhibition of pyramidal cells. *Nat. Neurosci.* 1:470–78

Cruikshank SJ, Lewis TJ, Connors BW. 2007. Synaptic basis for intense thalamocortical activation of feedforward inhibitory cells in neocortex. *Nat. Neurosci.* 10:462–68

Dantzker JL, Callaway EM. 2000. Laminar sources of synaptic input to cortical inhibitory interneurons and pyramidal neurons. *Nat. Neurosci.* 3:701–7

Dávid C, Schleicher A, Zuschratter W, Staiger JF. 2007. The innervation of parvalbumin-containing interneurons by VIP-immunopositive interneurons in the primary somatosensory cortex of the adult rat. *Eur. J. Neurosci.* 25:2329–40

Delpire E, Mount DB. 2002. Human and murine phenotypes associated with defects in cation-chloride cotransport. *Annu. Rev. Physiol.* 64:803–43

DiFiglia M, Pasik P, Pasik T. 1976. A Golgi study of neuronal types in the neostriatum of monkeys. *Brain Res.* 114:245–56

DiFiglia M, Pasik P, Pasik T. 1982. A Golgi and ultrastructural study of the monkey globus pallidus. *J. Comp. Neurol.* 212:53–75

Dreifuss JJ, Kelly JS, Krnjević K. 1969. Cortical inhibition and gamma-aminobutyric acid. *Exp. Brain Res.* 9:137–54

Ellender TJ, Paulsen O. 2010. The many tunes of perisomatic targeting interneurons in the hippocampal network. *Front. Cell Neurosci.* 4:ii, 26

Ehrlich I, Humeau Y, Grenier F, Ciocchi S, Herry C, Lüthi A. 2009. Amygdala inhibitory circuits and the control of fear memory. *Neuron* 62:757–71

Fanselow EE, Richardson KA, Connors BW. 2008. Selective, state-dependent activation of somatostatin-expressing inhibitory interneurons in mouse neocortex. *J. Neurophysiol.* 100:2640–52

Farrant M, Nusser Z. 2005. Variations on an inhibitory theme: phasic and tonic activation of GABA(A) receptors. *Nat. Rev. Neurosci.* 6:215–29

Fatt P, Katz B. 1953. The effect of inhibitory nerve impulses on a crustacean muscle fibre. *J. Physiol.* 121:374–89

Fishell G. 2007. Perspectives on the developmental origins of cortical interneuron diversity. *Novartis Found. Symp.* 288:21–35; discuss. 35–44, 96–98

Fogarty M, Grist M, Gelman D, Marín O, Pachnis V, Kessaris N. 2007. Spatial genetic patterning of the embryonic neuroepithelium generates GABAergic interneuron diversity in the adult cortex. *J. Neurosci.* 27:10935–46

Fonnum F, Storm-Mathisen J. 1969. GABA synthesis in rat hippocampus correlated to the distribution of inhibitory neurons. *Acta Physiol. Scand.* 76:35A–36A

Fox CA, Rafols JA. 1971. Observations on the oligodendroglia in the primate striatum. Are they Ramón y Cajal's "dwarf" or "neurogliaform" neurons? *J. Hirnforsch.* 13:331–40

French AS, Panek I, Torkkeli PH. 2006. Shunting vs inactivation: simulation of GABAergic inhibition in spider mechanoreceptors suggests that either is sufficient. *Neurosci. Res.* 55:189–96

Freund TF, Buzsáki G. 1996. Interneurons of the hippocampus. *Hippocampus* 6:347–470

Freund TF, Gulyás AI. 1997. Inhibitory control of GABAergic interneurons in the hippocampus. *Can. J. Physiol. Pharmacol.* 75:479–87

Freund TF, Katona I. 2007. Perisomatic inhibition. *Neuron* 56:33–42

Friauf E, Hammerschmidt B, Kirsch J. 1997. Development of adult-type inhibitory glycine receptors in the central auditory system of rats. *J. Comp. Neurol.* 385:117–34

Fuentealba P, Begum R, Capogna M, Jinno S, Márton LF, et al. 2008. Ivy cells: a population of nitric-oxide-producing, slow-spiking GABAergic neurons and their involvement in hippocampal network activity. *Neuron* 57:917–29

Fuentealba P, Klausberger T, Karayannis T, Suen WY, Huck J, et al. 2010. Expression of COUP-TFII nuclear receptor in restricted GABAergic neuronal populations in the adult rat hippocampus. *J. Neurosci.* 30:1595–609

Fujiwara-Tsukamoto Y, Isomura Y, Imanishi M, Ninomiya T, Tsukada M, et al. 2010. Prototypic seizure activity driven by mature hippocampal fast-spiking interneurons. *J. Neurosci.* 30:13679–89

Gabernet L, Jadhav SP, Feldman DE, Carandini M, Scanziani M. 2005. Somatosensory integration controlled by dynamic thalamocortical feed-forward inhibition. *Neuron* 48:315–27

Galarreta M, Hestrin S. 2002. Electrical and chemical synapses among parvalbumin fast-spiking GABAergic interneurons in adult mouse neocortex. *Proc. Natl. Acad. Sci. USA* 99:12438–43

Gertler TS, Chan CS, Surmeier DJ. 2008. Dichotomous anatomical properties of adult striatal medium spiny neurons. *J. Neurosci.* 28:10814–24

Gibson JR, Beierlein M, Connors BW. 1999. Two networks of electrically coupled inhibitory neurons in neocortex. *Nature* 402:75–79

Gibson JR, Beierlein M, Connors BW. 2004. Functional properties of electrical synapses between inhibitory interneurons of neocortical layer 4. *J. Neurophysiol.* 93:467–80

Gittis AH, Nelson AB, Thwin MT, Palop JJ, Kreitzer AC. 2010. Distinct roles of GABAergic interneurons in the regulation of striatal output pathways. *J. Neurosci.* 30:2223–34

Glickfeld LL, Roberts JD, Somogyi P, Scanziani M. 2009. Interneurons hyperpolarize pyramidal cells along their entire somatodendritic axis. *Nat. Neurosci.* 12:21–23

Glykys J, Mody I. 2007. Activation of GABAA receptors: views from outside the synaptic cleft. *Neuron* 56:763–70

Gonchar Y, Burkhalter A. 1997. Three distinct families of GABAergic neurons in rat visual cortex. *Cereb. Cortex* 7:347–58

Gonchar Y, Wang Q, Burkhalter A. 2007. Multiple distinct subtypes of GABAergic neurons in mouse visual cortex identified by triple immunostaining. *Front. Neuroanat.* 1:3

Gulledge AT, Stuart GJ. 2003. Excitatory actions of GABA in the cortex. *Neuron* 37(2):299–309

Gupta A, Wang Y, Markram H. 2000. Organizing principles for a diversity of GABAergic interneurons and synapses in the neocortex. *Science* 287:273–78

Haas KF, Macdonald RL. 1999. GABAA receptor subunit gamma2 and delta subtypes confer unique kinetic properties on recombinant GABAA receptor currents in mouse fibroblasts. *J. Physiol.* 514:27–45

Hefft S, Jonas P. 2005. Asynchronous GABA release generates long-lasting inhibition at a hippocampal interneuron-principal neuron synapse. *Nat. Neurosci.* 8:1319–28

Helmstaedter M, Staiger JF, Sakmann B, Feldmeyer D. 2008. Efficient recruitment of layer 2/3 interneurons by layer 4 input in single columns of rat somatosensory cortex. *J. Neurosci.* 28:8273–84

Hendry SH, Jones EG, DeFelipe J, Schmechel D, Brandon C, Emson PC. 1984. Neuropeptide-containing neurons of the cerebral cortex are also GABAergic. *Proc. Natl. Acad. Sci. USA* 81:6526–30

Hendry SH, Jones EG, DeFelipe J, Schmechel D, Brandon C, Emson PC. 1986. Neuropeptide-containing neurons of the cerebral cortex are also GABAergic. *Science* 231:995–97

Hernandes MS, Troncone LR. 2009. Glycine as a neurotransmitter in the forebrain: a short review. *J. Neural. Transm.* 116:1551–60

Higo S, Akashi K, Sakimura K, Tamamaki N. 2009. Subtypes of GABAergic neurons project axons in the neocortex. *Front Neuroanat.* 3:25

Hirata T, Li P, Lanuza GM, Cocas LA, Huntsman MM, Corbin JG. 2009. Identification of distinct telencephalic progenitor pools for neuronal diversity in the amygdala. *Nat. Neurosci.* 12:141–49

Howard A, Tamas G, Soltesz I. 2005. Lighting the chandelier: new vistas for axo-axonic cells. *Trends Neurosci.* 28:310–16

Hull C, Adesnik H, Scanziani M. 2009. Neocortical disynaptic inhibition requires somatodendritic integration in interneurons. *J. Neurosci.* 29:8991–95

Jentsch TJ, Stein V, Weinreich F, Zdebik AA. 2002. Molecular structure and physiological function of chloride channels. *Physiol. Rev.* 82:503–68

Jinno S, Klausberger T, Marton LF, Dalezios Y, Roberts JD, et al. 2007. Neuronal diversity in GABAergic long-range projections from the hippocampus. *J. Neurosci.* 27:8790–804

Jinno S. 2009. Structural organization of long-range GABAergic projection system of the hippocampus. *Front. Neuroanat.* 3:13

Jones EG. 1986. Neurotransmitters in the cerebral cortex. *J. Neurosurg.* 65:135–53

Kapfer C, Glickfeld LL, Atallah BV, Scanziani M. 2007. Supralinear increase of recurrent inhibition during sparse activity in the somatosensory cortex. *Nat. Neurosci.* 10:743–53

Karube F, Kubota Y, Kawaguchi Y. 2004. Axon branching and synaptic bouton phenotypes in GABAergic nonpyramidal cell subtypes. *J. Neurosci.* 24:2853–65

Kawaguchi Y, Kubota Y. 1993. Correlation of physiological subgroupings of nonpyramidal cells with parvalbumin- and calbindinD28k-immunoreactive neurons in layer V of rat frontal cortex. *J. Neurophysiol.* 70:387–96

Kawaguchi Y, Kubota Y. 1997. GABAergic cell subtypes and their synaptic connections in rat frontal cortex. *Cereb. Cortex* 7:476–86

Kawaguchi Y. 1993. Physiological, morphological, and histochemical characterization of three classes of interneurons in rat neostriatum. *J. Neurosci.* 13:4908–23

Kawaguchi Y. 1997. Neostriatal cell subtypes and their functional roles. *Neurosci. Res.* 27:1–8

Kawaguchi Y. 2001. Distinct firing patterns of neuronal subtypes in cortical synchronized activities. *J. Neurosci.* 21:7261–72

Kawaguchi Y, Kondo S. 2002. Parvalbumin, somatostatin and cholecystokinin as chemical markers for specific GABAergic interneuron types in the rat frontal cortex. *J. Neurocytol.* 31:277–87

Kemppainen S, Pitkänen A. 2000. Distribution of parvalbumin, calretinin, and calbindin-D(28k) immunoreactivity in the rat amygdaloid complex and colocalization with gamma-aminobutyric acid. *J. Comput. Neurol.* 426:441–67

Khirug S, Yamada J, Afzalov R, Voipio J, Khiroug L, Kaila K. 2008. GABAergic depolarization of the axon initial segment in cortical principal neurons is caused by the Na-K-2Cl cotransporter NKCC1. *J. Neurosci.* 28:4635–39

Kirsch J, Betz H. 1998. Glycine-receptor activation is required for receptor clustering in spinal neurons. *Nature* 392:717–20

Kisvárday ZF, Gulyas A, Beroukas D, North JB, Chubb IW, Somogyi P. 1990. Synapses, axonal and dendritic patterns of GABA-immunoreactive neurons in human cerebral cortex. *Brain* 113:793–812

Klausberger T. 2009. GABAergic interneurons targeting dendrites of pyramidal cells in the CA1 area of the hippocampus. *Eur. J. Neurosci.* 30:947–57

Klausberger T, Márton LF, Baude A, Roberts JD, Magill PJ, Somogyi P. 2004. Spike timing of dendrite-targeting bistratified cells during hippocampal network oscillations in vivo. *Nat. Neurosci.* 7:41–47. Erratum. 2006. *Nat. Neurosci.* 9:979

Klausberger T, Somogyi P. 2008. Neuronal diversity and temporal dynamics: the unity of hippocampal circuit operations. *Science* 321:53–57

Koós T, Tepper JM. 1999. Inhibitory control of neostriatal projection neurons by GABAergic interneurons. *Nat. Neurosci.* 2:467–72

Koos T, Tepper JM, Wilson CJ. 2004. Comparison of IPSCs evoked by spiny and fast-spiking neurons in the neostriatum. *J. Neurosci.* 24:7916–22

Kravitz EA, Kuffler SW, Potter DD. 1963a. Gamma-aminobutric acid and other blocking compounds in crustacea. III. Their relative concentrations in separated motor and inhibitory axons. *J. Neurophysiol.* 26:739–51

Kravitz EA, Kuffler SW, Potter DD, Vangelder NM. 1963b. Gamma-aminobutyric acid and other blocking compounds in crustacea. II. Periperal nervous system. *J. Neurophysiol.* 26:729–38

Kreitzer AC. 2009. Physiology and pharmacology of striatal neurons. *Annu. Rev. Neurosci.* 32:127–47

Kubota Y, Hatada SN, Kawaguchi Y. 2009. Important factors for the three-dimensional reconstruction of neuronal structures from serial ultrathin sections. *Front. Neural Circuits* 3:4

Kubota Y, Mikawa S, Kawaguchi Y. 1993. Neostriatal GABAergic interneurones contain NOS, calretinin or parvalbumin. *Neuroreport* 5:205–8

Kwag J, Paulsen O. 2009. The timing of external input controls the sign of plasticity at local synapses. *Nat. Neurosci.* 12:1219–21

Lawrence JJ, McBain CJ. 2003. Interneuron diversity series: containing the detonation—feedforward inhibition in the CA3 hippocampus. *Trends Neurosci.* 26:631–40

Lee CC, Sherman SM. 2009. Glutamatergic inhibition in sensory neocortex. *Cereb. Cortex* 19:2281–89

Lee S, Hjerling-Leffler J, Zagha E, Fishell G, Rudy B. 2010. The largest group of superficial neocortical GABAergic interneurons expresses ionotropic serotonin receptors. *J. Neurosci.* 30:16796–808

Lewis DA, Melchitzky DS, Burgos GG. 2002. Specificity in the functional architecture of primate prefrontal cortex. *J. Neurocytol.* 31:265–76

Lund JS, Lewis DA. 1993. Local circuit neurons of developing and mature macaque prefrontal cortex: Golgi and immunocytochemical characteristics. *J. Comput. Neurol.* 328:282–312

Ma Y, Hu H, Berrebi AS, Mathers PH, Agmon A. 2006. Distinct subtypes of somatostatin-containing neocortical interneurons revealed in transgenic mice. *J. Neurosci.* 26:5069–82

Maccaferri G, Roberts JD, Szucs P, Cottingham CA, Somogyi P. 2000. Cell surface domain specific postsynaptic currents evoked by identified GABAergic neurones in rat hippocampus in vitro. *J. Physiol.* 524(Pt. 1):91–116

Mann EO, Paulsen O. 2007. Role of GABAergic inhibition in hippocampal network oscillations. *Trends Neurosci.* 30:343–49

Marin O, Anderson SA, Rubenstein JL. 2000. Origin and molecular specification of striatal interneurons. *J. Neurosci.* 20(16):6063–76

Mascagni F, McDonald AJ. 2003. Immunohistochemical characterization of cholecystokinin containing neurons in the rat basolateral amygdala. *Brain Res.* 976:171–84

Mascagni F, McDonald AJ. 2007. A novel subpopulation of 5-HT type 3A receptor subunit immunoreactive interneurons in the rat basolateral amygdala. *Neuroscience* 144:1015–24

McBain CJ, Fisahn A. 2001. Interneurons unbound. *Nat. Rev. Neurosci.* 2:11–23

McCormick DA, Connors BW, Lighthall JW, Prince DA. 1985. Comparative electrophysiology of pyramidal and sparsely spiny stellate neurons of the neocortex. *J. Neurophysiol.* 54:782–806

McDonald AJ. 1982. Neurons of the lateral and basolateral amygdaloid nuclei: a Golgi study in the rat. *J. Comput. Neurol.* 212:293–312

McDonald AJ. 1984. Neuronal organization of the lateral and basolateral amygdaloid nuclei in the rat. *J. Comput. Neurol.* 222:589–606

McDonald AJ. 1989. Coexistence of somatostatin with neuropeptide Y, but not with cholecystokinin or vasoactive intestinal peptide, in neurons of the rat amygdala. *Brain Res.* 500:37–45

McDonald AJ, Augustine JR. 1993. Localization of GABA-like immunoreactivity in the monkey amygdala. *Neuroscience* 52:281–94

McDonald AJ, Mascagni F, Augustine JR. 1995. Neuropeptide Y and somatostatin-like immunoreactivity in neurons of the monkey amygdala. *Neuroscience* 66:959–82

McGarry LM, Packer AM, Fino E, Nikolenko V, Sippy T, Yuste R. 2010. Quantitative classification of somatostatin-positive neocortical interneurons identifies three interneuron subtypes. *Front. Neural Circuits* 14(4):12

Megías M, Emri Z, Freund TF, Gulyás AI. 2001. Total number and distribution of inhibitory and excitatory synapses on hippocampal CA1 pyramidal cells. *Neuroscience* 102:5275–40

Mihály A. 1982. Histochemistry and functional significance of putative neocortical transmitter substances: a review. *Z. Mikrosk. Anat. Forsch.* 96:916–36

Miller KD, Pinto DJ, Simons DJ. 2001. Processing in layer 4 of the neocortical circuit: new insights from visual and somatosensory cortex. *Curr. Opin. Neurobiol.* 11:488–97

Mitchell SJ, Silver RA. 2003. Shunting inhibition modulates neuronal gain during synaptic excitation. *Neuron* 38:433–45

Mittmann W, Chadderton P, Häusser M. 2004. Neuronal microcircuits: frequency-dependent flow of inhibition. *Curr. Biol.* 14:R837–39

Miyoshi G, Butt SJ, Takebayashi H, Fishell G. 2007. Physiologically distinct temporal cohorts of cortical interneurons arise from telencephalic Olig2-expressing precursors. *J. Neurosci.* 27:7786–98

Miyoshi G, Hjerling-Leffler J, Karayannis T, Sousa VH, Butt SJ, et al. 2010. Genetic fate mapping reveals that the caudal ganglionic eminence produces a large and diverse population of superficial cortical interneurons. *J. Neurosci.* 30:1582–94

Mody I. 2001. Distinguishing between GABA(A) receptors responsible for tonic and phasic conductances. *Neurochem. Res.* 26:907–13

Morales M, Wang SD, Diaz-Ruiz O, Jho DH. 2004. Cannabinoid CB1 receptor and serotonin 3 receptor subunit A (5-HT3A) are co-expressed in GABA neurons in the rat telencephalon. *J. Comput. Neurol.* 468:205–16

Muller JF, Mascagni F, McDonald AJ. 2007. Serotonin-immunoreactive axon terminals innervate pyramidal cells and interneurons in the rat basolateral amygdala. *J. Comput. Neurol.* 505:314–35

Nery S, Fishell G, Corbin JG. 2002. The caudal ganglionic eminence is a source of distinct cortical and subcortical cell populations. *Nat. Neurosci.* 5:1279–87

Nicoll RA, Alger BE. 1979. Presynaptic inhibition: transmitter and ionic mechanisms. *Int. Rev. Neurobiol.* 21:217–58

Nunzi MG, Gorio A, Milan F, Freund TF, Somogyi P, Smith AD. 1985. Cholecystokinin-immunoreactive cells form symmetrical synaptic contacts with pyramidal and nonpyramidal neurons in the hippocampus. *J. Comput. Neurol.* 237:485–505

Nusser Z, Mody I. 2002. Selective modulation of tonic and phasic inhibitions in dentate gyrus granule cells. *J. Neurophysiol.* 87:2624–28

Oláh S, Füle M, Komlósi G, Varga C, Báldi R, Barzó P, Tamás G. 2009. Regulation of cortical microcircuits by unitary GABA-mediated volume transmission. *Nature* 461:1278–81

Otis TS, Staley KJ, Mody I. 1991. Perpetual inhibitory activity in mammalian brain slices generated by spontaneous GABA release. *Brain Res.* 545:142–50

Oviedo H, Reyes AD. 2002. Boosting of neuronal firing evoked with asynchronous and synchronous inputs to the dendrite. *Nat. Neurosci.* 5:261–66

Padgett CL, Slesinger PA. 2010. GABAB receptor coupling to G-proteins and ion channels. *Adv. Pharmacol.* 58:123–47

Paré D, Smith Y. 1993. Distribution of GABA immunoreactivity in the amygdaloid complex of the cat. *Neuroscience* 57:1061–76

Pawelzik H, Bannister AP, Deuchars J, Ilia M, Thomson AM. 1999. Modulation of bistratified cell IPSPs and basket cell IPSPs by pentobarbitone sodium, diazepam and Zn2+: dual recordings in slices of adult rat hippocampus. *Eur. J. Neurosci.* 11:3552–64

Pawelzik H, Hughes DI, Thomson AM. 2002. Physiological and morphological diversity of immunocyto-chemically defined parvalbumin- and cholecystokinin-positive interneurones in CA1 of the adult rat hippocampus. *J. Comput. Neurol.* 443:346–67

Payne JA, Rivera C, Voipio J, Kaila K. 2003. Cation-chloride co-transporters in neuronal communication, development and trauma. *Trends Neurosci.* 26:199–206

Pisani A, Bonsi P, Picconi B, Tolu M, Giacomini P, Scarnati E. 2001. Role of tonically-active neurons in the control of striatal function: cellular mechanisms and behavioral correlates. *Prog. Neuropsychopharmacol. Biol. Psychiatry* 25:211–30

Plotkin MD, Snyder EY, Hebert SC, Delpire E. 1997. Expression of the Na-K-2Cl cotransporter is devel-opmentally regulated in postnatal rat brains: a possible mechanism underlying GABA's excitatory role in immature brain. *J. Neurobiol.* 33:781–95

Pouille F, Scanziani M. 2001. Enforcement of temporal fidelity in pyramidal cells by somatic feed-forward inhibition. *Science* 293:1159–63

Pouille F, Scanziani M. 2004. Routing of spike series by dynamic circuits in the hippocampus. *Nature* 429:717–23

Price CJ, Cauli B, Kovacs ER, Kulik A, Lambolez B, et al. 2005. Neurogliaform neurons form a novel inhibitory network in the hippocampal CA1 area. *J. Neurosci.* 25:6775–86

Ramón y Cajal S. 1891. Sur la structure de l'écorce cérébrale de quelques mammifères. *Cellule* 7:123–76

Ramón y Cajal S. 1911. *Histologie du Système Nerveux de l'Homme et des Vertébrés*, transl. L Azoulay, Vol. 2, pp. 504–18. Paris: Maloine

Reichling DB, Kyrozis A, Wang J, MacDermott AB. 1994. Mechanisms of GABA and glycine depolarization-induced calcium transients in rat dorsal horn neurons. *J. Physiol.* 476:411–21

Reyes A, Lujan R, Rozov A, Burnashev N, Somogyi P, Sakmann B. 1998. Target-cell-specific facilitation and depression in neocortical circuits. *Nat. Neurosci.* 1:279–85

Roberts E. 1963. Gamma-aminobutyric acid (gamma-ABA)-vitamin B6 relationships in the brain. *Am. J. Clin. Nutr.* 12:291–307

Roberts E, Frankel S. 1950. γ-Aminobutyric acid in brain: its formation from glutamic acid. *J. Biol. Chem.* 187:55–63

Rothman JS, Cathala L, Steuber V, Silver RA. 2009. Synaptic depression enables neuronal gain control. *Nature* 457:1015–18

Równiak M, Robak A, Bogus-Nowakowska K, Kolenkiewicz M, Bossowska A, et al. 2008. Somatostatin-like immunoreactivity in the amygdala of the pig. *Folia Histochem. Cytobiol.* 46:229–38

Rubenstein JL, Shimamura K, Martinez S, Puelles L. 1998. Regionalization of the prosencephalic neural plate. *Annu. Rev. Neurosci.* 21:445–77

Rudy B, Fishell G, Lee S, Hjerling-Leffler J. 2010. Three groups of interneurons account for nearly 100% of neocortical GABAergic neurons. *J. Dev. Neurobiol.* 71:45–61

Saganich MJ, Vega-Saenz de Miera E, Nadal MS, Baker H, Coetzee WA, Rudy B. 1999. Cloning of compo-nents of a novel subthreshold-activating K+ channel with a unique pattern of expression in the cerebral cortex. *J. Neurosci.* 19:10789–802

Saganich MJ, Machado E, Rudy B. 2001. Differential expression of genes encoding subthreshold-operating voltage-gated K+ channels in brain. *J. Neurosci.* 21:4609–24

Sah P, Faber ES, Lopez De Armentia M, Power J. 2003. The amygdaloid complex: anatomy and physiology. *Physiol. Rev.* 83:803–34

Salin PA, Prince DA. 1996. Spontaneous GABAA receptor-mediated inhibitory currents in adult rat so-matosensory cortex. *J. Neurophysiol.* 75:1573–88

Saxena NC, Macdonald RL. 1994. Assembly of GABAA receptor subunits: role of the delta subunit. *J. Neurosci.* 14:7077–86

Saxena NC, Macdonald RL. 1996. Properties of putative cerebellar gamma-aminobutyric acid A receptor isoforms. *Mol. Pharmacol.* 49:567–79

Semyanov A, Walker MC, Kullmann DM, Silver RA. 2004. Tonically active GABA A receptors: modulating gain and maintaining the tone. *Trends Neurosci.* 27:262–69

Shen W, Flajolet M, Greengard P, Surmeier DJ. 2008. Dichotomous dopaminergic control of striatal synaptic plasticity. *Science* 321:848–51

Sherrington C. 1932. *Inhibition as a coordinative factor.* http://nobelprize.org/nobel_prizes/medicine/laureates/ 1932/sherrington-lecture.html

Sidibé M, Smith Y. 1999. Thalamic inputs to striatal interneurons in monkeys: synaptic organization and co-localization of calcium binding proteins. *Neuroscience* 89:1189–208

Sik A, Ylinen A, Penttonen M, Buzsáki G. 1994. Inhibitory CA1-CA3-hilar region feedback in the hippocampus. *Science* 265:1722–24

Sik A, Penttonen M, Ylinen A, Buzsáki G. 1995. Hippocampal CA1 interneurons: an in vivo intracellular labeling study. *J. Neurosci.* 15:6651–65

Silberberg G, Markram H. 2007. Disynaptic inhibition between neocortical pyramidal cells mediated by Martinotti cells. *Neuron* 53:735–46

Soltesz I, Roberts JD, Takagi H, Richards JG, Mohler H, Somogyi P. 1990. Synaptic and nonsynaptic localization of benzodiazepine/GABAA receptor/cl- channel complex using monoclonal antibodies in the dorsal lateral geniculate nucleus of the cat. *Eur. J. Neurosci.* 2:414–29

Somogyi J, Baude A, Omori Y, Shimizu H, El Mestikawy S, et al. 2004. GABAergic basket cells expressing cholecystokinin contain vesicular glutamate transporter type 3 (VGLUT3) in their synaptic terminals in hippocampus and isocortex of the rat. *Eur. J. Neurosci.* 19:552–69

Somogyi P, Klausberger T. 2005. Defined types of cortical interneurone structure space and spike timing in the hippocampus. *J. Physiol.* 562:9–26

Sousa VH, Miyoshi G, Hjerling-Leffler J, Karayannis T, Fishell G. 2009. Characterization of Nkx6-2-derived neocortical interneuron lineages. *Cereb. Cortex* 19(Suppl. 1):1–10

Staiger JF, Masanneck C, Schleicher A, Zuschratter W. 2004. Calbindin-containing interneurons are a target for VIP-immunoreactive synapses in rat primary somatosensory cortex. *J. Comput. Neurol.* 468:179–89

Stell BM, Mody I. 2002. Receptors with different affinities mediate phasic and tonic GABA(A) conductances in hippocampal neurons. *J. Neurosci.* 22:RC223

Stenman J, Yu RT, Evans RM, Campbell K. 2003. Tlx and Pax6 co-operate genetically to establish the pallio-subpallial boundary in the embryonic mouse telencephalon. *Development* 130:1113–22

Stokes CC, Isaacson JS. 2010. From dendrite to soma: dynamic routing of inhibition by complementary interneuron microcircuits in olfactory cortex. *Neuron* 67:452–65

Storm-Mathisen J, Fonnum F. 1971. Quantitative histochemistry of glutamate decarboxylase in the rat hippocampal region. *J. Neurochem.* 18:1105–11

Suzuki N, Bekkers JM. 2007. Inhibitory interneurons in the piriform cortex. *Clin. Exp. Pharmacol. Physiol.* 34:1064–69

Suzuki N, Bekkers JM. 2010a. Inhibitory neurons in the anterior piriform cortex of the mouse: classification using molecular markers. *J. Comput. Neurol.* 518:1670–87

Suzuki N, Bekkers JM. 2010b. Distinctive classes of GABAergic interneurons provide layer-specific phasic inhibition in the anterior piriform cortex. *Cereb. Cortex* 20:2971–84

Swadlow HA. 2002. Thalamocortical control of feed-forward inhibition in awake somatosensory 'barrel' cortex. *Philos. Trans. R. Soc. Lond. B* 357:1717–27

Szabadics J, Soltesz I. 2009. Functional specificity of mossy fiber innervation of GABAergic cells in the hippocampus. *J. Neurosci.* 29:4239–51

Szabadics J, Varga C, Molnár G, Oláh S, Barzó P, Tamás G. 2006. Excitatory effect of GABAergic axo-axonic cells in cortical microcircuits. *Science* 311:233–35

Tepper JM, Koós T, Wilson CJ. 2004. GABAergic microcircuits in the neostriatum. *Trends Neurosci.* 27:662–69

Tepper JM, Bolam JP. 2004. Functional diversity and specificity of neostriatal interneurons. *Curr. Opin. Neurobiol.* 14:685–92

Tremblay R, Clark BD, Rudy B. 2010. Layer-specific organization within the fast-spiking interneuron population of mouse barrel cortex. *Soc. Neurosci.* Abstr.

Tricoire L, Pelkey KA, Daw MI, Sousa VH, Miyoshi G, et al. 2010. Common origins of hippocampal ivy and nitric oxide synthase expressing neurogliaform cells. *J. Neurosci.* 30:2165–176

Trombley PQ, Shepherd GM. 1994. Glycine exerts potent inhibitory actions on mammalian olfactory bulb neurons. *J. Neurophysiol.* 71:761–67

Udenfriend S. 1950. Indentification of γ-aminobutyric acid in brain by the isotope derivative method. *J. Biol. Chem.* 187:65–69

Vida I, Halasy K, Szinyei C, Somogyi P, Buhl EH. 1998. Unitary IPSPs evoked by interneurons at the stratum radiatum-stratum lacunosum-moleculare border in the CA1 area of the rat hippocampus in vitro. *J. Physiol.* 506(Pt. 3):755–73

Vida I, Bartos M, Jonas P. 2006. Shunting inhibition improves robustness of gamma oscillations in hippocampal interneuron networks by homogenizing firing rates. *Neuron* 49:107–17

Vucurovic K, Gallopin T, Ferezou I, Rancillac A, Chameau P, et al. 2010. Serotonin 3A receptor subtype as an early and protracted marker of cortical interneuron subpopulations. *Cereb. Cortex* 20:2333–47

Vuillet J, Kerkerian L, Salin P, Nieoullon A. 1989. Ultrastructural features of NPY-containing neurons in the rat striatum. *Brain Res.* 477:241–51

Wang Y, Gupta A, Toledo-Rodriguez M, Wu CZ, Markram H. 2002. Anatomical, physiological, molecular and circuit properties of nest basket cells in the developing somatosensory cortex. *Cereb. Cortex* 12:395–410

Wang Y, Toledo-Rodriguez M, Gupta A, Wu C, Silberberg G, Luo J, Markram H. 2004. Anatomical, physiological and molecular properties of Martinotti cells in the somatosensory cortex of the juvenile rat. *J. Physiol.* 561:65–90

Wester MR, Teasley DC, Byers SL, Saha MS. 2008. Expression patterns of glycine transporters (xGlyT1, xGlyT2, and xVIAAT) in *Xenopus laevis* during early development. *Gene Expr. Patterns* 8:261–70

West DC, Mercer A, Kirchhecker S, Morris OT, Thomson AM. 2006. Layer 6 cortico-thalamic pyramidal cells preferentially innervate interneurons and generate facilitating EPSPs. *Cereb. Cortex* 16:200–11

Whittington MA, Traub RD. 2003. Interneuron diversity series: inhibitory interneurons and network oscillations in vitro. *Trends Neurosci.* 26:676–82

Wilent WB, Contreras D. 2005. Dynamics of excitation and inhibition underlying stimulus selectivity in rat somatosensory cortex. *Nat. Neurosci.* 8:1364–70

Williams JN Jr. 1950. The effects of folic acid and aminopterin upon enzyme systems in vitro. *J. Biol. Chem.* 187:47–54

Wilson CJ. 2007. GABAergic inhibition in the neostriatum. *Prog. Brain Res.* 160:91–110

Woodruff AR, Sah P. 2007a. Networks of parvalbumin-positive interneurons in the basolateral amygdala. *J. Neurosci.* 27:553–63

Woodruff AR, Sah P. 2007b. Inhibition and synchronization of basal amygdala principal neuron spiking by parvalbumin-positive interneurons. *J. Neurophysiol.* 98:2956–61

Woodruff AR, Monyer H, Sah P. 2006. GABAergic excitation in the basolateral amygdala. *J. Neurosci.* 26:11881–87

Woodruff A, Xu Q, Anderson SA, Yuste R. 2009. Depolarizing effect of neocortical chandelier neurons. *Front. Neural. Circuits* 3:15

Wonders CP, Anderson SA. 2006. The origin and specification of cortical interneurons. *Nat. Rev. Neurosci.* 7:687–96

Xu Q, Cobos I, De La Cruz E, Rubenstein JL, Anderson SA. 2004. Origins of cortical interneuron subtypes. *J. Neurosci.* 2426:2612–22

Xu Q, Guo L, Moore H, Waclaw RR, Campbell K, Anderson SA. 2010. Sonic hedgehog signaling confers ventral telencephalic progenitors with distinct cortical interneuron fates. *Neuron* 65:328–40

Xu Q, Tam M, Anderson SA. 2008. Fate mapping Nkx2.1-lineage cells in the mouse telencephalon. *J. Comput. Neurol.* 506:16–29

Xu X, Callaway EM. 2009. Laminar specificity of functional input to distinct types of inhibitory cortical neurons. *J. Neurosci.* 29:70–85

Xu X, Roby KD, Callaway EM. 2006. Mouse cortical inhibitory neuron type that coexpresses somatostatin and calretinin. *J. Comput. Neurol.* 499:144–60

Xu X, Roby KD, Callaway EM. 2010. Immunochemical characterization of inhibitory mouse cortical neurons: three chemically distinct classes of inhibitory cells. *J. Comput. Neurol.* 518:389–404

Waclaw RR, Ehrman LA, Pierani A, Campbell K. 2010. Developmental origin of the neuronal subtypes that comprise the amygdalar fear circuit in the mouse. *J. Neurosci.* 30:6944–53

Young AB, MacDonald RL. 1983. Glycine as a spinal cord neurotransmitter. In *Handbook of the Spinal Cord: Pharmacology*, Vol. I, ed. RA Davidoff, pp. l–44. New York: Marcel Dekker

Zaitsev AV, Povysheva NV, Gonzalez-Burgos G, Rotaru D, Fish KN, et al. 2009. Interneuron diversity in layers 2–3 of monkey prefrontal cortex. *Cereb. Cortex* 19:1597–615

Zarbin MA, Wamsley JK, Kuhar MJ. 1981. Glycine receptor: light microscopic autoradiographic localization with [3H]strychnine. *J. Neurosci.* 1:532–47

Zhang SJ, Jackson MB. 1993. GABA-activated chloride channels in secretory nerve endings. *Science* 259:531–34

Spatial Neglect and Attention Networks

Maurizio Corbetta[1,2,3] and Gordon L. Shulman[1]

Departments of [1] Neurology, [2] Radiology, and [3] Anatomy and Neurobiology, Washington University School of Medicine, St. Louis, Missouri 63110; email: mau@npg.wustl.edu, gordon@npg.wustl.edu

Annu. Rev. Neurosci. 2011. 34:569–99

The *Annual Review of Neuroscience* is online at neuro.annualreviews.org

This article's doi:
10.1146/annurev-neuro-061010-113731

Copyright © 2011 by Annual Reviews.
All rights reserved

0147-006X/11/0721-0569$20.00

Keywords

hemispheric dominance, stroke, neuroimaging, functional connectivity, arousal, white matter

Abstract

Unilateral spatial neglect is a common neurological syndrome following predominantly right hemisphere injuries and is characterized by both spatial and non-spatial deficits. Core spatial deficits involve mechanisms for saliency coding, spatial attention, and short-term memory and occur in conjunction with nonspatial deficits that involve reorienting, target detection, and arousal/vigilance. We argue that neglect is better explained by the dysfunction of distributed cortical networks for the control of attention than by structural damage of specific brain regions. Ventral lesions in right parietal, temporal, and frontal cortex that cause neglect directly impair nonspatial functions partly mediated by a ventral frontoparietal attention network. Structural damage in ventral cortex also induces physiological abnormalities of task-evoked activity and functional connectivity in a dorsal frontoparietal network that controls spatial attention. The anatomy and right hemisphere dominance of neglect follow from the anatomy and laterality of the ventral regions that interact with the dorsal attention network.

Contents

INTRODUCTION

L.J. is a 58-year-old schoolteacher, who suddenly developed malaise and confusion while at school. He was brought to the hospital for evaluation. On first appearance, his affect was flat and his vigilance reduced; his responses were grammatically accurate but delayed, and when asked "what was the matter" he replied he was not sure why he had been brought in. On exam, his visual fields were normal but his gaze tended to deviate to the right spontaneously when looking ahead. When presented with two objects, one in each visual field, he always looked first to the right object and denied the presence of the object on the left. However, when asked, he was able to move his eyes to the left and reported seeing the object. Overall, he was very slow in reporting stimuli even in his right visual field. When searching blindfolded for objects scattered on a table with either the left or right hand, he explored mainly the right side of the table, and his search progressed to the center only when objects on the right were removed. Touches on the right hand were easily reported and localized, whereas touches to the left hand were inconsistently detected and reported as belonging to the examiner. Motor function was normal in terms of strength, coordination, and dexterity on both sides, but L.J. was reluctant to use his left hand unless asked to do so. Head MRI showed a diffusion restriction in the right inferior and middle frontal gyrus and anterior insula, consistent with an acute ischemic stroke.

This clinical history illustrates some key features of unilateral spatial neglect: a reduction of arousal and speed of processing; an inability to attend to and report stimuli on the side opposite the lesion (contralesional) despite apparently normal visual perception; a spatial bias for directing actions toward the hemi-space or hemi-body on the same side as the lesion (ipsilesional); and several disorders of awareness, including a degree of obliviousness toward being ill and confabulation about body ownership.

Spatial neglect is caused by lesions, typically strokes, in a number of different cortical and subcortical areas. Although acutely both left and right hemisphere lesions can cause neglect, only right hemisphere lesions cause severe and persistent deficits (Stone et al. 1993), which is the primary basis for the widely held view that the right hemisphere is dominant for attention.

Spatial neglect is unique among the behavioral disorders resulting from focal lesions because its severity can be modulated by behavioral interventions over very short timescales (e.g., seconds). Deficits in attending to and reporting objects in contralesional space can be lessened by (a) encouraging a patient to attend to the previously ignored stimuli using verbal cues (Riddoch & Humphreys 1983); (b) presenting salient sensory stimuli, such as noises (Robertson et al. 1998); (c) asking the patient to perform hand movements controlled by the injured hemisphere (Robertson & North 1992); or (d) training patients to increase their alertness (Robertson et al. 1995). These observations suggest that the neural mechanisms underlying the spatial deficit can be dynamically modulated by signals from other parts of the brain reflecting endogenous or exogenous attention, movement, and arousal. Moreover, in a matter of days or weeks, most patients with spatial neglect recover from the more obvious spatial impairments, which continue to negatively influence their ability to return to a productive life (Denes et al. 1982, Paolucci et al. 2001).

Spatial neglect has attracted tremendous interest as a model for understanding the neurological basis of awareness, cerebral lateralization, spatial cognition, and recovery of function. Yet its neural bases remain poorly understood. Here, we provide a selective review of the vast literature on neglect and present a framework for understanding the disorder.

We propose that neglect is mediated by the abnormal interaction between brain networks that control attention to the environment in the healthy brain (see Heilman et al. 1985 and Mesulam 1999 for other network formulations). Although many authors have emphasized dissociations and subtypes of spatial neglect, we argue for a core set of spatial and nonspatial deficits that match the physiological properties of these networks. The core spatial deficit, a bias in spatial attention and salience mapped in an egocentric coordinate frame, is caused by the dysfunction of a dorsal frontoparietal network that controls attention and eye movements and represents stimulus saliency. Core nonspatial deficits of arousal, reorienting, and detection reflect structural damage to more ventral regions that partly overlap with a right hemisphere dominant ventral frontoparietal network recruited during reorienting and detection of novel behaviorally relevant events.

Next, we emphasize that the ventral lesions that result in neglect alter the physiology of structurally undamaged dorsal frontoparietal regions, consistent with the fact that dorsal and ventral attention regions interact in the healthy brain. Physiological dysfunction in dorsal frontoparietal regions is empirically observed not only during task performance but also at rest, and it correlates with the severity of the egocentric spatial bias. Moreover, this dysfunction decreases the top-down modulation of visual cortex, reducing its responsiveness, which can also contribute to neglect. The highly dynamic and plastic nature of the spatial deficits in neglect strongly argues that they are mediated by parts of the brain that still function, even if abnormally. Measurements of the physiology of brain regions, not just of structural damage, are essential for understanding neglect (Deuel & Collins 1983), and we emphasize neuroimaging methods that provide a window on physiological function.

Finally, perhaps the least understood clinical feature of spatial neglect is its right hemisphere lateralization. Brain imaging studies have shown that dorsal frontoparietal regions controlling spatial attention and eye movements are largely symmetrically organized, with each hemisphere predominantly representing the contralateral side of space. In contrast, ventral regions that underlie the core nonspatial deficits observed in neglect patients are strongly

fMRI: functional magnetic resonance imaging

right hemisphere dominant. We argue that lateralization of these latter functions, and their interaction with dorsal regions, rather than asymmetries of spatial attention per se, primarily accounts for the hemispheric asymmetry of neglect.

THE CORE SPATIAL DEFICIT: EGOCENTRIC BIAS

A large body of neuropsychological research has tried to characterize the nature of spatial deficits (i.e., involve predominantly one side of space) in neglect. Factor analytic studies have consistently isolated at least one factor associated with impairments in attending/searching/responding to targets in contralateral space, but have yielded inconsistent conclusions with regard to other factors (Azouvi et al. 2002, Halligan et al. 1989, Kinsella et al. 1993, Verdon et al. 2010). Other studies have classified patients based on their performance on different behavioral tasks thought to isolate different subtypes, such as perceptual/attention versus motor/intention deficits (Coslett et al. 1990), but the consistency (Hamilton et al. 2008) and relevance to recovery of these distinctions have been disappointing (Farne et al. 2004, Rengachary et al. 2011). Most patients suffer from perceptual/attention deficits that recover, albeit incompletely, in parallel with recovery from spatial neglect (Rengachary et al. 2011). We argue that at its core, spatial neglect represents a deficit of spatial attention and stimulus saliency that is mapped in an egocentric reference frame.

Gradients of Spatial Attention and Stimulus Saliency

Virtually all patients with spatial neglect manifest a lateralized bias in visual information processing that is evident both clinically and experimentally as a gradient across space (Behrmann et al. 1997, Pouget & Driver 2000). Sensitivity and responsiveness to behaviorally relevant stimuli improve as one moves from contralesional to ipsilesional locations along a continuous spatial gradient. This bias does not reflect abnormal early visual mechanisms, as indicated by normal measures of contrast sensitivity (Spinelli et al. 1990) and image segmentation based on low-level features in the neglected visual field (Driver & Mattingley 1998) and responses in occipital cortex based on visually evoked potentials (Di Russo et al. 2008, Watson et al. 1977) and functional magnetic resonance imaging (fMRI)(Rees et al. 2000).

In contrast, the saliency of objects in the neglected visual field is impaired. Saliency refers to the sensory distinctiveness and behavioral relevance of an object relative to other objects. In a recent study, the saliency of objects in the ipsilesional or contralesional visual field of neglect subjects was measured by indexing the patients' tendency to look at distinctive but task-irrelevant stimuli or at task-relevant stimuli (Bays et al. 2010). Both kinds of stimuli produced an increased probability of eye movements in their direction along a similar spatial gradient from contralesional to ipsilesional locations, suggesting that exogenous (automatic) and goal-driven components of spatial attention were equally affected (**Figure 1***a*). The abnormally high salience of ipsilesional stimuli may prevent them from being filtered when they are task-irrelevant (Bays et al. 2010, Shomstein et al. 2010, Snow & Mattingley 2006) or lead to repeated refixations during search tasks (Husain et al. 2001).

Importantly, a spatially lateralized bias is observed even in the absence of a stimulus. When neglect subjects searched for a nonexistent object in complete darkness, search patterns as measured by eye or head position were strongly biased toward the ipsilesional field (Hornak 1992). Moreover, gaze deviations are observed tonically at rest, similar to those observed during task performance (**Figure 1***b*) (Fruhmann Berger et al. 2008).

Spatial biases in the dark during target search could reflect a reduced salience of contralesional spatial locations during task performance, but the biases observed at rest also suggest an indwelling imbalance in the mechanisms controlling gaze. These motor biases are likely

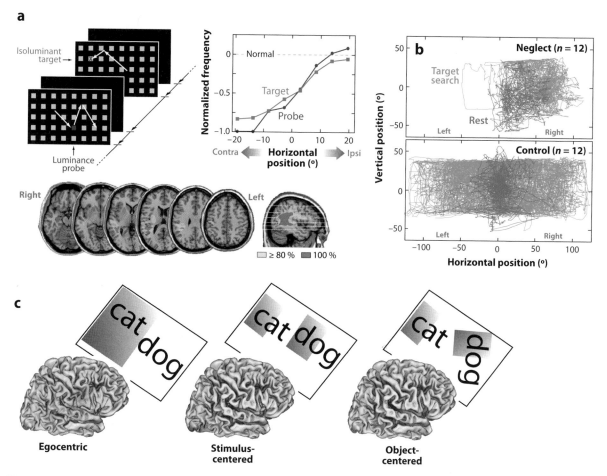

Figure 1

Behavioral and lesion analyses of the egocentric spatial bias in neglect patients. (*a*) Effects of stimulus salience on eye movements of neglect patients. Patients were instructed to saccade to a target letter in an array. A sequence of arrays was presented, some containing a target, some containing a distinctive probe stimulus of a higher luminance or different orientation than the distracter squares, and some containing both a target and a probe. The right graph shows that targets (*red line*) and irrelevant but distinctive luminance probes (*blue line*) produced more saccades in the ipsilesional than contralesional visual field with a similar linear gradient along the horizontal position. This indicates a similar ipsilesional bias for automatic and goal-directed orienting (Bays et al. 2010). Lesions are localized in right temporoparietal junction. (*b*) Neglect patients show an ipsilesional gaze bias while searching for a target letter in a letter array (*blue traces*) and at rest (*green traces*). Non-neglect patients (*bottom*) showed no bias (Fruhmann Berger et al. 2008). (*c*) Lesions classically associated with neglect involve an ipsilesional bias within an egocentric reference frame, but not biases observed within stimulus-centered or object-centered frames. Each image shows the voxelwise lesion distribution associated with deficits within a particular reference frame (Medina et al. 2008). Panel *a* is adapted from figures 1, 2d, and 4b of Bays et al. 2010, with permission from the Society for Neuroscience. Panel *b* is adapted from figure 1 of Fruhmann Berger et al. 2008, with permission from the American Psychological Association. Lesion data in panel *c* are courtesy of Drs. Medina, Pavlak, and Hillis, Johns Hopkins and University of Pennsylvania, from Medina et al. 2008.

associated with attentional biases because of the functional relationship between the corresponding neural systems (Corbetta et al. 1998, Rizzolatti et al. 1987).

An Egocentric Frame of Reference

Spatial deficits can be separated based on the reference frame in which stimuli are coded. Neglect is often egocentric (viewer-centered), with left and right hemi-spaces based on the observer's midline (**Figure 1c**). Coding of the spatial midline can be further fractionated based on eye, head, or body position. Although these factors have been shown to modulate the severity of neglect in individual cases, no consistent dissociation has emerged between eye-, head-, or body-centered neglect (Behrmann & Geng 2002). These null results have traditionally been explained by the large volume and heterogeneity of lesions in humans. Computational studies and physiological analyses in monkeys, however, indicate an additional factor, namely that activity in many areas reflects combinations of different egocentric reference frames (Chang & Snyder 2010, Pouget & Sejnowski 2001).

Neglect can also be allocentric, where the midline is defined from the central axis of a stimulus, irrespective of its position in the environment (stimulus-centered, **Figure 1c**) or of both its position and orientation (object-centered, **Figure 1c**). The great majority of patients with spatial neglect, however, suffer from egocentric deficits. Marsh & Hillis studied 100 consecutive cases of right hemisphere acute stroke, and found that while 17% and 34% showed visual and tactile egocentric neglect, respectively, the corresponding figures for allocentric neglect were only 4% and 2% (Marsh & Hillis 2008). Allocentric and egocentric neglect rarely co-occur clinically and are dissociated anatomically (Medina et al. 2008, Verdon et al. 2010). While egocentric neglect is associated with regions classically damaged in spatial neglect, i.e., inferior parietal lobule (IPL), superior temporal gyrus (STG), and inferior frontal gyrus (IFG), stimulus- and object-centered neglect is associated with damage of inferior temporal regions (**Figure 1c**).

Therefore, egocentric spatial deficits in neglect correspond much more closely to the clinical syndrome both behaviorally and anatomically than do allocentric deficits, and represent core features of the syndrome.

Does Spatial Neglect Involve Only Attention/Salience?

An important question is whether an attention/saliency account of the spatial component of neglect leaves out other impaired perceptual-cognitive functions, particularly visuospatial short-term memory (VSTM) and spatial cognition. An influential account proposes that neglect is a deficit in forming, storing, or manipulating the left side of mental images or information in VSTM, termed representational neglect (i.e., representation within VSTM) (Bisiach & Luzzatti 1978, Della Sala et al. 2004). Most or all of the empirical findings that support an attentional/saliency interpretation of the spatial deficit in neglect patients can also be explained within a representational framework. This duality reflects the fact that mechanisms for spatial attention, VSTM, and imagery are closely related, as shown by psychological and physiological studies (Awh & Jonides 2001, Kosslyn et al. 2001), and entry into VSTM is often considered a normal consequence of conscious perception. Not surprisingly, representational neglect is nearly always observed in association with perceptual neglect (Bartolomeo et al. 1994). The few cases in which a dissociation has been reported (e.g., Guariglia et al. 1993, Ortigue et al. 2001) used pencil-and-paper tasks with lower sensitivity than computerized tasks (Rengachary et al. 2009) and did not control for differences in eye movements (Fruhmann Berger et al. 2008, Hornak 1992) and cognitive load or time-on-task, which affect the direction of spatial attention (Dodds et al. 2008, Rizzolatti et al. 1987). Regardless, both the low frequency of dissociations between perceptual and representational neglect and the likelihood that each is mediated by overlapping neural systems suggest that this distinction, while very important theoretically, may not be critical for first-order identification of the psychological and neural mechanisms that are damaged in the large majority of patients with spatial neglect. Accordingly, in this review, we do not distinguish representational and spatial attention formulations of the core egocentric deficit underlying spatial neglect.

Some neglect patients exhibit deficits in VSTM that may not show a contralesional-to-ipsilesional gradient and could be separate from a contralesional attention/salience/VSTM deficit. VSTM deficits are observed for stimuli presented along a central, vertical axis (Malhotra et al. 2005), while a more specialized deficit has also been reported in trans-saccadic spatial memory (Mannan et al. 2005), possibly reflecting saccadic remapping mechanisms (Vuilleumier et al. 2007). When coupled with a bias to attend to ipsilesional locations, full-field VSTM and trans-saccadic deficits can exacerbate contralesional neglect by leading to multiple refixations of already-searched, ipsilesional objects (Husain et al. 2001, Mannan et al. 2005). However, a recent study reported that both VSTM deficits and visual search performance were worse in the contralesional visual field. This result is consistent with the functional similarity and neural overlap of mechanisms for attention/perception and working memory (Kristjansson & Vuilleumier 2010).

In addition to spatial attention, spatial cognition is involved in many tasks used to assess spatial neglect. Line bisection, for example, requires fine judgments of spatial extent or position. Patients with neglect typically show rightward biases in line bisection (Bisiach et al. 1983) or underestimate the size of objects placed on the left side of space (Milner et al. 1998). These perceptual deficits are sometimes thought to reflect a distortion of the horizontal dimension of space (Bisiach et al. 2002). However, extensive investigations of eye movements during line bisection clearly indicate that neglect patients fail to explore the left side of lines, most often judging as midpoint the leftmost position that was fixated (Ishiai et al. 2006). Moreover, their subjective midline judgment is strongly biased by the position of the rightward line endpoint (McIntosh et al. 2005). These findings suggest a more parsimonious explanation of line bisection errors based on the relative saliency of the right versus left side of lines (**Figure 1a**) and tonic oculomotor biases (**Figure 1b**). Another aspect of spatial cognition, processing of global stimulus structure, is also impaired in at least some patients (Delis et al. 1986). A resulting local bias could increase the tendency to search near the current focus of attention, exacerbating a bias to attend to ipsilesional locations (Robertson & Rafal 2000).

Overall, the egocentric spatial deficit in neglect reflects impairments in a set of related mechanisms for spatial attention, salience, and VSTM, and perhaps spatial cognition. A recent study has shown that this deficit, when assessed using even a very simple task, is highly associated with clinical judgments of neglect. In a paradigm involving simple reaction time to a cued target, performance differences for left- and right-field targets discriminated acute and chronic neglect patients from healthy controls with better accuracy than a variety of standard neuropsychological tests of neglect (Rengachary et al. 2009). Notably, this task did not involve spatial imagery, spatial cognition (line bisection, clock-drawing, copying tasks), or shape identification in a cluttered field (cancellation tasks), and involved minimal VSTM demands.

Summary

Spatial neglect is characterized by a spatial gradient of impaired attention/saliency/representation within an egocentric reference frame. The saliency deficit reflects both task and sensory factors and is linked with indwelling motor imbalances that produce resting ipsilesional deviations in eye, head, and body movements. Abnormal interhemispheric interactions likely play a role in producing the spatial gradient. The gradient fluctuates depending on arousal and task instructions, suggesting that the underlying neural mechanisms are modulated by signals from other parts of the brain and are dysfunctional rather than obliterated by structural damage.

ANATOMY AND PHYSIOLOGY OF THE EGOCENTRIC SPATIAL BIAS

We next discuss anatomical studies of neglect. The most commonly damaged brain regions do not contain physiological signals that can mediate the egocentric spatial bias. In contrast, physiological studies of spatial attention in healthy adults have highlighted the importance of a dorsal frontoparietal attention network that typically is not structurally damaged in neglect patients. The mismatch between anatomical lesions and physiology may be explained by recent physiological studies of the dorsal network in neglect patients.

Structural Damage Does Not Explain the Egocentric Spatial Bias

Spatial neglect was first associated with damage to parietal cortex (Critchley 1953), especially IPL (Mort et al. 2003, Vallar & Perani 1987). However, subsequent studies have also emphasized STG (Karnath et al. 2001) and IFG (Husain & Kennard 1996), and convincing evidence exists that damage to other regions, including anterior insula and middle frontal gyrus (MFG), sometimes produces neglect. Interestingly, the distribution of cortical damage is sim-ilar irrespective of the behavioral criteria (e.g., neglect severity, clinical diagnosis, comparison of neglect versus no-neglect individuals) used to group the lesions (**Figure 2a**). Importantly, neglect patients, especially in severe cases, have white matter fiber damage, which can disconnect frontal, temporal, and parietal cortex (Bartolomeo et al. 2007, Gaffan & Hornak 1997, He et al. 2007, Thiebaut de Schotten et al. 2005). White matter damage most commonly involves a dorsal region lateral to the ventricle where arcuate and superior longitudinal fasciculi (II and III) run parallel in an anterior-to-posterior direction (**Figure 2c**) (Doricchi & Tomaiuolo 2003). Finally, neglect also can be caused by damage to subcortical nuclei (pulvinar, caudate, putamen) (Karnath et al. 2002a, Vallar & Perani 1987) that cause cortical hypoactivation of regions important for the genesis of neglect (Karnath et al. 2005, Perani et al. 1987).

While the detailed profile of behavioral deficits undoubtedly depends on the site of the lesion (Medina et al. 2008, Verdon et al. 2010), and some lesion sites are more likely than others to produce neglect, the striking fact remains that neglect of the left field can be caused by many different right hemisphere lesions.

Figure 2

Relationship between anatomical distribution of lesions associated with neglect, attentional networks, and damage to fiber tracts. (a) Anatomical regions associated with neglect, as shown by lesion-symptom mapping (*left panel*) (Karnath et al. 2011), overlap of lesions in patients diagnosed with neglect (*middle panel*) (A.R. Carter, G.L. Shulman, and M. Corbetta, unpublished data), and comparisons of groups of patients with severe neglect versus no neglect (*right panel*) (Verdon et al. 2010). The three cortical distributions are quite similar and emphasize ventral regions in IPL (angular and supramarginal gyri), TPJ, STG, and VFC. (b) The dorsal (*left panel*) and ventral (*right panel*) attention networks as determined by resting state functional connectivity in 25 healthy controls. The yellow-orange voxels indicate regions with strong and significant positive temporal correlation of spontaneous activity at rest. The blue voxels in the left panel indicate regions negatively correlated with the dorsal network, corresponding to the default network. The anatomical distributions shown in **Figure 2a** match the distribution of the ventral but not dorsal attention networks. Seed regions for the connectivity analysis were determined by meta-analyses of task activation paradigms (shown in **Figures 3a** and **6a**). Dorsal seed regions were determined from the activations evoked by a central cue to shift attention. Ventral seed regions were determined from comparisons of activations to targets presented at unattended and attended locations. (c) Slice representations from the anatomical distributions shown in the left (Karnath et al. 2011) and middle panels (A.R. Carter, G.L. Shulman, and M. Corbetta, unpublished data) of **Figure 2a** indicate that anatomical damage includes both gray and white matter. White matter tracts corresponding to the arcuate fasciculus (AF) and superior longitudinal fasciculus (SLF) II and III, as determined by diffusion tensor imaging (DTI) in 30 healthy controls, are shown in the right panel (M. Glasser and K. Patel, unpublished data). The regions of maximal damage in neglect patients, shown in the left and middle panels, are outlined in green and blue. Lesion damage overlaps most strongly with the AF, but also with SLF II and III. These tracts connect regions within and across networks. Panels *a* and *c*, left panels, are adapted from figure 2 of Karnath et al. 2011, with permission from Oxford University Press. Panel *a*, right panel, is adapted from figure 2 of Verdon et al. 2010, with permission from Oxford University Press.

Attempts to identify a critical region based on structural damage alone inevitably must explain away a large number of reported lesions that produce neglect and yet do not involve the supposedly critical region.

Moreover, regions that are commonly damaged in spatial neglect do not contain the physiological signals expected based on deficits in spatial attention, eye movements, and coding of salience. These regions are not usually recruited in neuroimaging experiments that isolate those processes, and none have yet been shown in humans to contain maps of space. Right ventral frontal cortex (VFC), for example, has been associated with target detection (Stevens et al. 2005), task control and error detection (Dosenbach et al. 2006), and response inhibition (Aron et al. 2004).

To resolve this paradox, we propose that the heterogeneity of lesions in spatial neglect patients masks a greater uniformity at the level of physiology, with common physiological abnormalities in remote neural systems specialized for spatial processing (Corbetta et al. 2005, He et al. 2007). Next, we review evidence in healthy subjects for a dorsal frontoparietal attention network that houses the physiological signals impaired in spatial neglect.

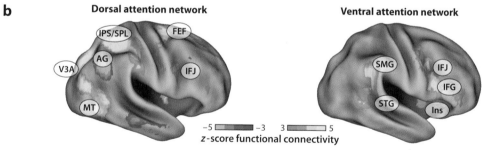

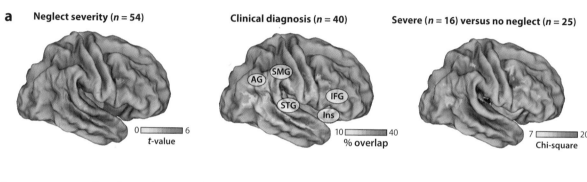

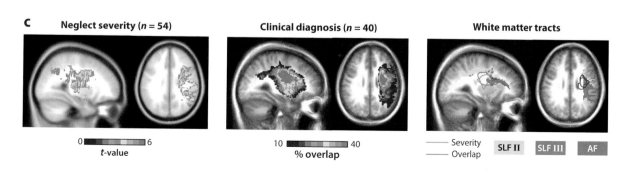

A Dorsal Frontoparietal Network for Spatial Attention, Stimulus Salience, and Eye Movements

SPL: superior parietal lobule

FEF: frontal eye field

IPS: intraparietal sulcus

Regions in dorsal frontal and parietal cortex, including bilateral medial intraparietal sulcus (mIPS), SPL, precuneus, supplementary eye field (SEF), and frontal eye field (FEF), respond to symbolic cues to shift attention voluntarily to a location (e.g., Corbetta et al. 2000, Hopfinger et al. 2000, Kastner et al. 1999) (**Figure 3a**). In some studies, additional responses are reported in lateral prefrontal cortex, including the inferior frontal sulcus/junction (IFS/IFJ) (Sylvester et al. 2007) and MFG (Hopfinger et al. 2000). The standard regions are also recruited when attention is shifted to salient objects based on task relevance and sensory distinctiveness (Shulman et al. 2009) (**Figure 3b**), consistent with the idea that these regions are involved in coding the saliency of objects under both goal-driven and stimulus-driven conditions. Dorsal frontoparietal regions (mIPS, precuneus/SPL,

FEF, SEF, DLPFC) are also recruited during visually and memory-guided saccades, with almost complete overlap of attention- and eye movement–related activations (Corbetta et al. 1998), and in some regions sensory signals are remapped during eye movements (Merriam et al. 2003). Both body-centered and stimulus-centered coding has been reported in IPS/SPL (Galati et al. 2010).

We previously proposed that these frontoparietal regions constituted a dorsal cortical network for the control of spatial and featural attention and stimulus-response mapping (Corbetta & Shulman 2002). Subsequent work demonstrated that at rest, many of these regions show highly correlated activity (**Figure 2b**), consistent with the notion that they represent a separate functional-anatomical network analogous to sensory and motor systems (Fox et al. 2006, He et al. 2007). Importantly, these dorsal frontoparietal regions are not generally damaged in neglect patients,

---

Figure 3

Physiology of spatial attention in healthy adults. (*a*) Dorsal frontoparietal regions are activated following a central cue to shift attention. The statistical map shows the z-map from a meta-analysis of four experiments ($n = 58$) in which BOLD activity was measured following a central cue to shift attention to a peripheral location (Astafiev et al. 2003, Corbetta et al. 2000, Kincade et al. 2005). The time course of the response to the cue shows bilateral activity from right (R) IPS with a contralateral preponderance indicating spatial selectivity. (*b*) Occipital and dorsal frontoparietal regions show spatially selective attentional modulations following a stimulus-driven shift of attention [from a meta-analysis of two experiments ($n = 47$) (Shulman et al. 2009; A. Tosoni, G.L. Shulman, D.L.W. Pope, M.P. McAvoy, and M. Corbetta, unpublished data)]. Subjects were cued to attend left or right to detect targets in a rapid-serial-visual-presentation (RSVP) stream presented among distracter streams. The z-map indicates voxels showing contralateral activity > ipsilateral BOLD activity following a shift of attention to the peripheral cue (*red square*). Note the strong spatially selective response in right IPS and visual cortex for shifting and attending to contralateral rather than ipsilateral stimulus streams. Also, maps for purely goal-driven (Panel *a*) versus stimulus-driven (Panel *b*) shifts of attention are very consistent with a role of these regions in coding stimulus saliency under both conditions. (*c*) Contralateral topographic maps in dorsal parietal cortex. The left image shows five contralateral polar angle maps along R IPS. The right image shows the activations in these maps during a VSTM task in which subjects remembered the orientation and location of target lines presented among distracters. The bottom graph shows a comparison of the magnitude of contralateral and ipsilateral activations in left and right IPS maps as a function of VSTM load. Although left and right IPS contains contralateral polarangle maps (left IPS not shown), right IPS was equally activated by VSTM load in the contralateral and ipsilateral hemifields and left IPS was only modulated by load in the contralateral hemifield (Sheremata et al. 2010). This pattern of activity matches that postulated by the standard model for neglect (Mesulam 1981). (*d*) Interhemispheric coding of spatial attention. BOLD activity was measured following an auditory cue to attend to a peripheral location. The top right graph shows the magnitude of activity in left (L) and R FEF on a trial-to-trial basis following leftward (*blue dots*) and rightward (*red dots*) cues. Activity in L and R FEF is highly correlated across trials, but a contralateral signal is superimposed on the positively correlated noise (i.e., *blue dots plot above red dots*). This correlated noise is partly explained by the presence of strong correlations at rest (*bottom graph*) between homologous regions (e.g., left-right FEF) or parts of maps (e.g., left-right fovea in V1). The locus of attention is only weakly predicted (AUC value = ∼.60, chance = 0.50) by reading out activity only from the portion of the map in visual cortex or area (e.g., FEF) coding for the attended location. The prediction increases significantly (AUC = ∼.80) when subtracting activity from the attended-minus-unattended homologous portion of map or area in the two hemispheres (Sylvester et al. 2007). **Figure 3c** is adapted from figures 3a, 3b, and 5b of Sheremata et al. 2010, with permission from the Society for Neuroscience.

as shown in the comparison of **Figure 2a** and **2b**. Recent physiological studies of these dorsal frontoparietal regions, discussed next, have suggested two mechanisms for coding the locus of behaviorally relevant stimuli that may provide insights into the pathogenesis of spatial neglect.

Topographic maps of contralateral space. Computational theories show that the co-occurrence of topographic maps, eye/attention, and saliency signals is useful for stimulus selection (Koch & Ullman 1985). Correspondingly, many dorsal frontoparietal regions modulated

by spatial attention, eye movements, and salience also contain polar angle maps of the contralateral hemifield (**Figure 3c**, left image) (Hagler & Sereno 2006, Sereno et al. 2001, Swisher et al. 2007): medial IPS, precuneus, medial parieto-occipital cortex, SPL, FEF, and IFS/IFJ. Polar angle maps of the ipsilateral hemifield, however, have not been reported, indicating no evidence for separate topographic representations of both visual fields within a hemisphere, a longstanding explanation for the right hemisphere dominance of neglect (see below).

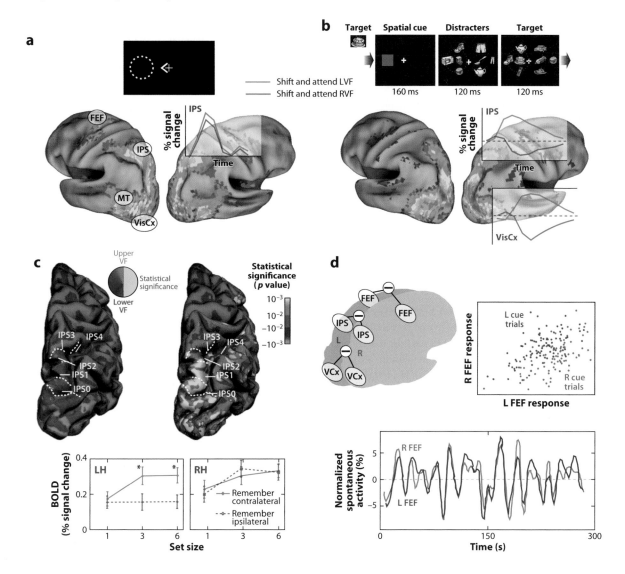

Conversely, topographic maps have not been reported in ventral regions typically damaged in neglect patients, consistent with the lack of evidence for their involvement in spatial attention. However, this null result should be treated cautiously. Maps in these regions may only be imaged with appropriate tasks or may be masked by a larger-scale organization involving other variables, such as eye position (Siegel et al. 2003). Moreover, the absence of a topographic map does not imply null spatial coding, as in rat hippocampus and entorhinal cortex (O'Keefe 2006). Spatial coding in ventral regions may occur at a scale and organization that are amenable to multivariate analyses of activation patterns, but not to standard retinotopic mapping procedures.

Interhemispheric control of the locus of attention. Interhemispheric competition may play a key role in the efficient control of spatial attention by the dorsal network (Kinsbourne 1987). Sylvester and colleagues (Sylvester et al. 2007) found that attention-related blood oxygenation level–dependent (BOLD) activity in dorsal frontoparietal and occipital regions contralateral to an attended location was only modestly predictive of whether a left- or right-field location had been cued on that trial. However, predictability was greatly increased by subtracting the activity of homologous left and right hemisphere regions, largely because activations in the two hemispheres showed strong positive trial-to-trial correlations (**Figure 3d**, right graph). Differencing the signals between the two hemispheres eliminated this common noise (**Figure 3d**, left). High interhemispheric correlation of activity also occurs at rest (**Figure 3d**, bottom), indicating tonic interhemispheric interactions.

Computational studies show that when signals in two brain regions are negatively correlated, as is the case with spatially selective signals in the two hemispheres, positive correlations in the noise increase the amount of information that is encoded by the corresponding neurons (Averbeck et al. 2006). The locus of attention may be efficiently coded by the two hemispheres through a difference signal implemented by interactions via either the corpus callosum (Innocenti 2009) or subcortical routes. Enhanced prediction of the locus of attention based on an interhemispheric difference signal also has been observed with electroencephalography (Thut et al. 2006) and single-unit recordings in monkey posterior parietal cortex (Bisley & Goldberg 2003).

A prediction of this model is that abnormalities in the computation of the locus of attention should correlate with abnormal interhemispheric interactions or response imbalances between the left and right hemisphere dorsal attention network. The next section discusses the physiological activity of these regions in neglect patients.

Physiological Correlates of the Egocentric Spatial Bias in Neglect

Patients with spatial neglect following lesions to right ventral cortex and underlying white matter have shown two types of physiological abnormalities in the structurally intact, dorsal attention network. First, at three weeks poststroke, a widespread cortical hypoactivation during a spatial attention task was observed in both right and left hemispheres (**Figure 4a**), a finding consistent with other studies in the literature (Pizzamiglio 1998, Thimm et al. 2008). The cortical hypoactivation was associated with a large interhemispheric imbalance of activity in dorsal parietal cortex (**Figure 4b**). The interhemispheric imbalance normalized at the chronic stage (9 months poststroke) in parallel with an overall improvement of cortical activity and spatial neglect (**Figure 4a,b**) (Corbetta et al. 2005). Interestingly, activity in ipsilesional occipital visual cortex was also imbalanced, showing reductions in magnitude and spatial selectivity (Corbetta et al. 2005) that were particularly marked under high attentional loads (Vuilleumier et al. 2008). These impairments in sensory-evoked activity, possibly reflecting abnormal top-down control from

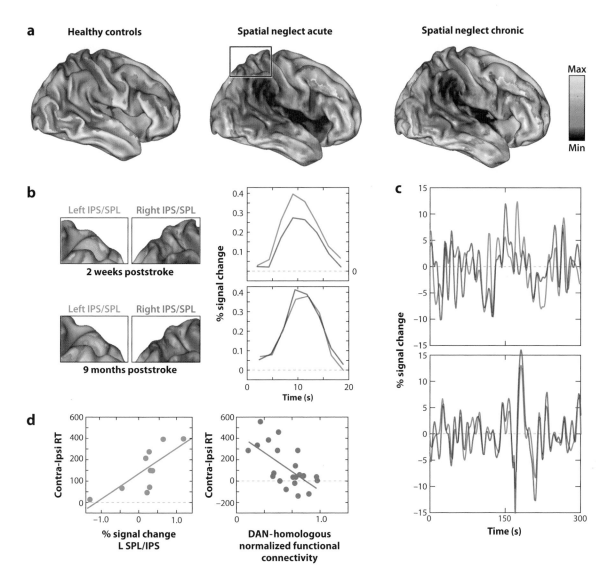

Figure 4

Physiology of egocentric spatial bias in neglect patients. (*a*) Statistical map of BOLD activations during a Posner spatial attention task (same as in **Figure 3a**), in which subjects are cued to a peripheral location and detect a subsequent target. Right hemisphere acute neglect patients show hypoactivation of both hemispheres (right>left) that partly recovers at the chronic stage. The dark shading in the anatomical image indicates the distribution of structural damage (Corbetta et al. 2005). (*b*) As a result of right hemisphere hypoactivation, acute patients show a large imbalance of BOLD activity in IPS/SPL, with relatively greater left than right hemisphere activity, even though activity in both hemispheres is lower than normal. This imbalance normalizes at the chronic stage. The left columns show statistical maps of activity in parietal cortex, the right column shows the averaged time course of activity time-locked to the presentation of the cue over a trial of the Posner task in left (*blue line*) and right (*red line*) parietal cortex (Corbetta et al. 2005). (*c*) Acute neglect patients (*top graph*) show low correlations in BOLD spontaneous activity between homologous regions of left (*blue lines*) and right (*red lines*) parietal cortex, but the correlation recovers at the chronic stage (*bottom graph*) (He et al. 2007). (*d*) Abnormal physiological signals in the dorsal attention network of acute neglect patients are functionally significant. (*Left graph*) Left parietal activity was stronger in subjects with more severe neglect, as indexed by longer response times (RTs) to contralesional versus ipsilesional visual targets. (*Right graph*) Reduced interhemispheric correlation within frontoparietal regions of the dorsal attention network correlates with the severity of neglect of the left visual field, as indexed by longer RTs to contralesional versus ipsilesional visual targets (Carter et al. 2010).

Functional
connectivity: the
correlation over time
of the spontaneous
activity between
different brain regions

frontoparietal cortex (Bressler et al. 2008), may further lessen the saliency of contralesional stimuli.

Second, neglect patients have shown anomalies in the pattern of spontaneous activity fluctuations within the dorsal attention network (**Figure 4c**). Coherence between left and right parietal regions was disrupted three weeks post-stroke and improved over time in parallel with the improvement of neglect (**Figure 4c**) (He et al. 2007).

Dorsal frontoparietal abnormalities in both task-evoked activity and resting coherence are functionally significant. Left parietal activity is stronger in subjects with more severe neglect, as indexed by the difference in response times to contralesional versus ipsilesional visual targets (**Figure 4d**) (Corbetta et al. 2005). Stronger left than right hemisphere activation of parietal and occipital regions may reflect a biased representation of stimulus salience and the locus of spatial attention (Bisley & Goldberg 2003, Sylvester et al. 2007). This interpretation is consistent with trans-cranial magnetic stimulation studies, in which inactivation of left posterior parietal cortex reduced left-field neglect (Brighina et al. 2003, Koch et al. 2008). At rest, significant correlations are found throughout the dorsal attention network between reductions in interhemispheric coherence and the magnitude of the spatial bias (**Figure 4d**) (Carter et al. 2010, He et al. 2007). The anomalies in spontaneous activity may underlie the lateral rotation of the eyes, head, and body in neglect patients at rest (Fruhmann Berger et al. 2008), which likely reflect biased coding of the locus of attention and salience, and contribute to the observed abnormalities in task-evoked responses.

Therefore, in structurally intact regions of the dorsal attention network, ventral lesions induce changes in BOLD resting functional connectivity and task-evoked activity that reflect abnormal interhemispheric interactions and response balances (Kinsbourne 1987), which plausibly explain the egocentric spatial bias in neglect. Recent studies of resting blood perfusion, which assess physiological function,

have associated spatial deficits in neglect patients with hypoperfusion of right IPL (Hillis et al. 2005), STG (Zopf et al. 2009), and IFG/anterior insula (Medina et al. 2008), but not of dorsal frontoparietal regions. However, these studies did not assess the functional relationship or coherence between regions, corresponding to the resting BOLD measure of the dorsal network that shows abnormalities, and did not measure task-evoked signal changes, corresponding to the other BOLD measure of dorsal network function that shows abnormalities. Moreover, the task-evoked changes measured in BOLD studies are quite small, well under 1% (Corbetta et al. 2005), and may not be detected with less-sensitive perfusion methods.

Summary

The extant literature strongly supports the presence of physiological signals (spatial attention, eye movement, saliency, maps of contralateral space, interhemispheric interactions, egocentric frame of reference) in the dorsal frontoparietal network that are likely highly relevant to the pathogenesis of spatial neglect. Recent neuroimaging results potentially resolve the paradox that ventral regions traditionally associated with neglect do not contain physiological signals that could account for an egocentric spatial bias, whereas dorsal frontoparietal regions that contain these signals are not typically damaged in strokes that cause neglect. However, only a few BOLD imaging studies have been conducted, involving relatively small patient samples. Further studies will be necessary to determine if similar changes in dorsal network physiology are observed across the different lesions that cause spatial neglect.

The above neuroimaging results, moreover, do not explain why lesions that cause neglect occur ventrally and in the right hemisphere. After more than 50 years of research, the right hemisphere dominance of spatial neglect remains the most puzzling aspect of this syndrome. These issues are considered next.

RIGHT HEMISPHERE DOMINANCE OF SPATIAL NEGLECT

Right Hemisphere Lateralization of Spatial Deficits

The right hemisphere dominance of neglect might reflect a corresponding asymmetry for spatial attention, the primary mechanism underlying the egocentric spatial deficit, and for related functions of VSTM and spatial cognition. Perhaps the most widely accepted, standard theory of neglect postulates that the right hemisphere controls shifts of attention to both sides of space. while the left hemisphere only controls attention to the right side (Mesulam 1981). Damage to the right hemisphere impairs attention to the left hemifield, whereas damage to the left hemisphere can be compensated. A second, opponent-process theory proposes that each hemisphere promotes orienting in a contralateral direction, but the strength of this bias is stronger in the left than right hemisphere (Kinsbourne 1987). Left hemisphere lesions cause only mild right spatial neglect because the unopposed orienting bias generated by the right hemisphere is relatively weak.

Empirical support for each theory from neuroimaging studies is surprisingly modest. Mapping studies have reported contralateral retinotopic maps in both hemispheres but no evidence of ipsilateral maps in the right hemisphere. Most neuroimaging studies of cueing paradigms have reported largely bilateral dorsal frontoparietal activations for directing spatial attention to locations in either visual field (e.g., Hopfinger et al. 2000, Kastner et al. 1999, Shulman et al. 2010, Sylvester et al. 2007). Several studies, however, have reported larger activations for contralateral than ipsilateral stimuli (contralateral bias) in some left than right dorsal regions (e.g., Szczepanski et al. 2010, Vandenberghe et al. 2005). These asymmetries in contralateral bias are consistent with either the standard or opponent process theory, depending on whether the bias in the right hemisphere was completely absent

(standard theory) or was simply present to a lesser degree (opponent process). Given the large number of discrepant findings, however, it will be important to identify the factors controlling when the specific hemispheric asymmetry postulated by either theory is observed.

A recent study suggested that one factor might involve high loading of VSTM (Sheremata et al. 2010). Under conditions that involved high VSTM load and spatial filtering of distracters, left and right hemisphere activations in regions of IPS, which contained strictly contralateral polar angle maps, nevertheless showed the postulated visual field profile (**Figure 3c**). Despite this intriguing result, however, the presence of the standard visual field organization under high VSTM load does not satisfactorily explain the laterality of neglect. Contralesional neglect is correlated with conditions that do not involve high VSTM loads, such as simple detection of a single visual target (Rengachary et al. 2009). Similarly, gaze biases that correlate with contralesional neglect are present even at rest.

Finally, the laterality of contralesional neglect could partly reflect a similar laterality for aspects of spatial cognition. Variants of line bisection tasks qualitatively produce right-dominant activity in IPS/SPL after they are referenced to control tasks that largely subtract out activations from shifting and maintaining attention (Fink et al. 2001). Similarly, Sack and colleagues observed right hemisphere dominance during variants of a clock task, in which subjects judge the angle formed by hour and minute hands (Sack 2009). To explain the laterality of contralesional neglect, however, mechanisms of spatial cognition must also show the visual field organization postulated by the standard attention theory, for which there is only limited evidence (Kukolja et al. 2006).

Most importantly, right lateralization of spatial attention, VSTM, and spatial cognition has primarily been observed in dorsal parietal regions, leaving unanswered the critical question of why ventral (e.g., IFG or IPL) rather than dorsal (e.g., IPS/SPL) right hemisphere lesions cause neglect.

Right Hemisphere Lateralization and Anatomy of Core Nonspatial Deficits

TPJ: temporoparietal junction

An alternative explanation that addresses this question, and is fully compatible with either the presence or absence of hemispheric asymmetries in dorsal visuospatial mechanisms, links the laterality of neglect to several nonspatial behavioral deficits that are commonly observed in neglect patients (Heilman et al. 1987, Husain & Rorden 2003, Robertson 2001). Nonspatial refers to deficits that are nominally present across the visual field (e.g., the putative full-field VSTM and global processing deficits discussed above). In practice, a nonspatial deficit is rarely if ever shown to be of equal severity in the two hemifields, and measurements across the visual field are often not made, mainly because of the difficulty of separating nonspatial deficits from the egocentric spatial bias (but see (Duncan et al. 1999). More commonly, conclusions are based on finding a deficit relative to control subjects for stimuli located centrally or in the ipsilesional field, although this procedure does not control for spatial gradients that extend across the visual field (e.g., **Figure 1a**).

Below, we briefly review three core nonspatial deficits consistently observed in neglect patients. These deficits are of particular interest because, unlike the egocentric spatial mechanisms discussed above, their physiology maps closely onto the anatomy of neglect, with clear involvement of ventral right hemisphere regions. We discuss evidence for the interaction of these ventral regions, damaged in neglect, with the dorsal attention network, a critical link for understanding the pathophysiology and right hemisphere dominance of the neglect syndrome.

Reorienting of attention. Michael Posner and colleagues reported that neglect patients are impaired in reorienting to unexpected events (Posner et al. 1984). Patients showed especially large deficits in detecting contralesional targets when they were expecting an ipsilesional target, suggesting a deficit in disengaging attention from the ipsilesional field.

A comparison of temporoparietal junction (TPJ)/STG versus SPL patients originally localized disengagement/reorienting deficits to right TPJ/STG (Friedrich et al. 1998). A recent study confirmed that a reorienting deficit was stronger for contralesional than ipsilesional targets following TPJ lesions, but the magnitude of the ipsilesional deficit was substantially increased following VFC lesions (Rengachary et al. 2011) (**Figure 5a**), indicating a bilateral deficit in reorienting.

Detection of behaviorally relevant stimuli. Right hemisphere and neglect patients show deficits in target detection in even the simplest paradigms. Simple auditory reaction time (RT), for example, is much slower following right than left hemisphere damage (Howes & Boller 1975). Related differences have been reported for auditory stimuli presented at ipsilesional locations in right hemisphere patients with neglect versus those without neglect (Samuelsson et al. 1998), indicating that the RT slowing may reflect damage to right hemisphere brain regions specifically associated with neglect (**Figure 5b**). Impairments are sometimes reported even for accurate detection of suprathreshold stimuli presented centrally (Malhotra et al. 2009) or ipsilesionally [although only acutely (Rengachary et al. 2011)], indicating that RT slowing likely does not reflect motor difficulties with the ipsilesional or good hand. RT slowing could reflect deficits in arousal and processing capacity (Duncan et al. 1999, Husain et al. 1997), although the latter effects occur following both left and right hemisphere damage (Peers et al. 2005).

Arousal and vigilance. Reduced arousal and vigilance is an important component of the neglect syndrome following right hemisphere injury. Clinically, patients with neglect and right hemisphere injuries suffer from lower arousal than patients with similar lesions in the left hemisphere. Arousal refers to the combination of autonomic, electrophysiological, and behavioral activity that is associated with an alert

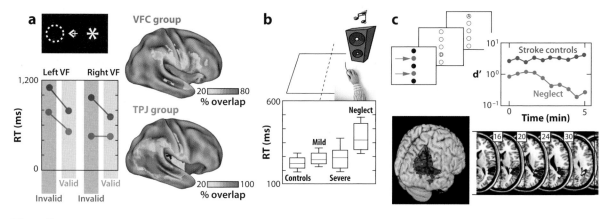

Figure 5

Behavioral analyses of nonspatial deficits in neglect patients. (*a*) Reorienting deficits in neglect patients with VFC and TPJ lesions (Rengachary et al. 2011). Patients detected a visual target (*asterisk*) that occurred in a validly cued location (*dotted circle*) or at an invalidly cued location (shown in the figure). Both TPJ and VFC patients showed large contralesional deficits in reorienting, as indexed by longer response times (RTs) to unattended (invalid) than attended (valid) targets. The VFC group additionally showed reorienting deficits in the ipsilesional field and larger overall detection deficits. Similar results were observed for accuracy (not shown), but the TPJ group showed evidence of a small reorienting deficit in the ipsilesional field. (*b*) Detection deficits in neglect patients. Neglect patients show abnormally slow simple RTs to an ipsilesional auditory stimulus (Samuelsson et al. 1998). Controls were healthy age- and gender-matched subjects. The mild and severe groups consisted of non-neglect patients with minor and major right hemisphere strokes. (*c*) Arousal deficits in neglect patients. Parietal neglect patients show a vigilance decrement (*red curve*) in a task that involved detection of letter targets in two locations (*arrows*) within a central column. No deficit is observed in right hemisphere stroke controls without neglect (*blue curve*). The anatomical images show the association of damaged voxels in right TPJ with the vigilance decrement, with darker areas indicating a weaker association (Malhotra et al. 2009). Panel *b* is adapted from Samuelsson et al. 1998, with permission from Taylor & Francis. Panel *c* is adapted from figures 5a, 6c, and 7 of Malhotra et al. 2009, with permission from Oxford University Press.

state, whereas vigilance refers to the ability to sustain this state over time.

Kenneth Heilman and colleagues have argued that neglect patients have decreased arousal due to hypoactivation of the right hemisphere (**Figure 4a**) (Corbetta et al. 2005, Heilman et al. 1987). For instance, patients with right as opposed to left hemisphere damage do not show the typical slowing of heart rate following a cue that signals a subsequent target (Yokoyama et al. 1987) and show reduced galvanic skin responses to electrical stimulation (Heilman et al. 1978). Lesions studies have associated right frontal damage with decreased arousal (Wilkins et al. 1987) and a decrement over time in sustaining attention and detecting targets (vigilance decrement) (Rueckert & Grafman 1996).

Importantly, there is evidence for interaction between arousal/vigilance and spatial deficits. A nonspeeded, auditory counting test

of arousal discriminated neglect from nonneglect patients in a right hemisphere group with heterogeneous lesions, indicating a strong linkage between arousal and spatial deficits in neglect (Robertson et al. 1997). A recent study reported a specific association between damage to right TPJ cortex and vigilance decrements, but only when attention had to be sustained to a spatial location, not for targets presented at random locations (**Figure 5c**) (Malhotra et al. 2009). The interaction between mechanisms underlying nonspatial and spatial deficits is likely critical for the pathogenesis and right hemisphere lateralization of spatial neglect (see below).

Right Hemisphere Lateralization and Physiology of Core Nonspatial Deficits

The above results indicate that neglect patients show nonspatial deficits in reorienting, target detection, and arousal that are likely right hemisphere dominant. We next show that, correspondingly, physiological signals mediating these functions in the healthy brain are right lateralized and occur in ventral regions typically damaged in neglect (IPL, STG, and IFG). Interestingly, the lateralization of these processes is supported by similar findings in other species (see Right Hemisphere Dominance in Vertebrates sidebar).

RIGHT HEMISPHERE DOMINANCE IN VERTEBRATES

While right lateralization in humans is often considered a byproduct of left hemisphere dominance for language, comparative studies suggest the basic specialization of each hemisphere may have been present early in vertebrate evolution. MacNeilage and colleagues (2009), summarizing this work, write "... the right hemisphere, the primary seat of emotional arousal, was at first specialized for detecting and responding to unexpected stimuli in the environment."

The latter sentence describes the human ventral attention network (Corbetta et al. 2008, Corbetta & Shulman 2002). When visual input is confined to the left eye/right hemisphere of chicks, their behavior is greatly affected by salient or novel stimuli (Rogers & Anson 1979). Feeding behavior, for example, is disrupted by the presence of brightly colored pebbles scattered among grains (Rogers et al. 2007). Similarly, during feeding, a simulated hawk is detected faster when it appears in the left than right monocular field (Rogers 2000).

Behavioral asymmetries in chicks arise partly from asymmetric light exposure prior to hatching. Following exposure, connections from the ipsilateral visual field innervate more the right than left hyperstriatum (Rogers & Sink 1988), reminiscent of the standard neglect theory. This physiology, however, varies widely across avian species. In pigeons, the asymmetry is reversed and is mediated by tectofugal rather than thalamofugal pathways (Valencia-Alfonso et al. 2009).

In mammals, right hemisphere dominance for several nonspatial functions may partly reflect asymmetric brainstem projections. The locus coeruleus/noradrenergic system in rats shows an asymmetric organization (Robinson 1985) that Posner & Petersen (1990) have linked to right lateralization of arousal. Recent evidence also points toward locus coeruleus/noradrenergic involvement in reorienting and target detection, two other lateralized, nonspatial functions associated with neglect (Aston-Jones & Cohen 2005, Bouret & Sara 2005, Corbetta et al. 2008).

The similar right hemisphere specializations in animals and humans may reflect convergent evolution rather than homology, but raise the possibility that the lateralization underlying human neglect is of longstanding origin.

Reorienting of attention. Neuroimaging studies of healthy adults have shown that reorienting to stimuli in either visual field that are presented outside the focus of attention (stimulus-driven reorienting) recruits a right lateralized ventral attention network in TPJ (including separate foci in SMG and STG) (Shulman et al. 2010) and VFC (Insula, IFG, MFG), in conjunction with the dorsal network (**Figure 6a**) (Corbetta & Shulman 2002). While TPJ is uniformly activated by stimulus-driven reorienting, VFC is mainly activated when reorienting is unexpected and requires cognitive control (Shulman et al. 2009) or is coupled to a response. Because one or both conditions usually apply in real-world situations, the two regions are typically coactivated and a similar network is observed in the resting state (**Figure 2b**) (Fox et al. 2006, He et al. 2007).

Importantly, the cortical anatomy of the ventral attention network includes the primary regions damaged in neglect (**Figure 2b, Figure 6a**) and matches the localization of the reorienting/disengagement deficit (**Figure 5a**), indicating a clear convergence between studies in neglect patients and healthy adults (Fox et al. 2006, Friedrich et al. 1998, He et al. 2007, Rengachary et al. 2011, Shomstein et al. 2010, Shulman et al. 2010).

Detection of behaviorally relevant and novel stimuli. The ventral (and dorsal) attention network is activated by detection of behaviorally relevant stimuli that are unattended or unexpected with respect to a wide range of attributes, not just their location, as demonstrated by oddball tasks in which subjects report infrequent targets that differ in some feature (e.g., color, tone frequency) from a standard stimulus (Corbetta & Shulman 2002). Oddball detection, however, activates not only the ventral attention network but also additional regions

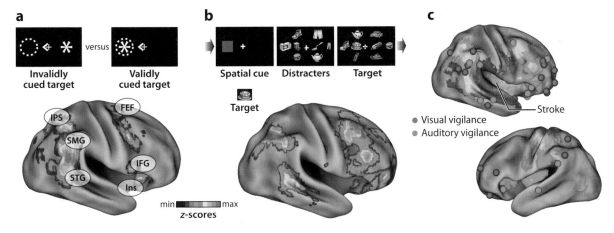

Figure 6

Physiology of nonspatial attention in healthy adults. (*a*) Physiology of reorienting attention in healthy adults. BOLD activity was compared for visual targets that occurred in invalidly cued (*left panel*) versus validly cued (*right panel*) locations. The statistical map shows the z-map from a meta-analysis (*n* = 58) of four experiments (Astafiev et al. 2003, Corbetta et al. 2000, Kincade et al. 2005). Strong activations are observed in right TPJ, extending from posterior temporal cortex (STG) to ventral IPL (SMG), and in right IFG/insula. These ventral regions are coactivated with dorsal attention regions in IPS and FEF. (*b*) Physiology of detection in healthy adults. A spatial cue directed subjects to attend to a rapid-serial-visual-presentation (RSVP) stream in the left or right visual fields that might contain a target. The z-map shows all voxels with greater right than left hemisphere BOLD activity to detected targets. The map includes ventral voxels that showed significant activations from reorienting in panel *a*, but also additional regions in prefrontal cortex. The dorsal attention network, however, does not show an asymmetry. (*c*) Physiology of arousal in healthy adults. A meta-analysis of foci from five experiments reporting activations that index arousal/vigilance, visual (*red spheres*) or auditory (*green spheres*) (Coull et al. 1998, Foucher et al. 2004, Paus et al. 1997, Sturm et al. 1999, Sturm et al. 2004). The right hemisphere foci are superimposed on the average of z-maps for reorienting and detection-related hemispheric asymmetries from panels *a* and *b*. The dark shading shows the distribution of neglect lesions from a recent neglect study (Rengachary et al. 2009). Many more foci are observed in the right than left hemisphere. Foci are clustered around the TPJ and insula/frontal operculum regions that are activated by reorienting and target detection, but many prefrontal foci are anterior to both distributions. Few foci occur in dorsal frontoparietal regions, indicating that arousal-related activations overlap more the ventral than dorsal attention network.

in frontal, parietal, and temporal cortex that show right hemisphere dominance in direct interhemispheric voxelwise comparisons (Stevens et al. 2005). Similar, although not identical, right hemisphere regions are activated by novel stimuli that are task irrelevant (Stevens et al. 2005).

Right hemisphere dominance during target detection is observed in regions that are frequently associated with neglect (IPL, STG, IFG) and for visual targets in both left and right hemifield (Shulman et al. 2010), consistent with the nonspatial deficit in neglect (**Figure 6***b*). Unlike the simple detection tasks studied in neglect patients, however, the above paradigms presented both targets and nontargets, and the frequency of targets was relatively low.

Arousal and vigilance. Neuroimaging studies of arousal and vigilance using simpler auditory and visual detection paradigms, more similar to those adopted in neglect subjects, have qualitatively reported right hemisphere dominance (direct interhemispheric comparisons were not conducted), usually in lateral prefrontal, insula/frontal operculum, and TPJ regions (Coull et al. 1998; Foucher et al. 2004; Pardo et al. 1991; Paus et al. 1997; Sturm et al. 1999, 2004). **Figure 6***c* shows that arousal-related activations are recorded more frequently in ventral cortex of the right than left hemisphere, and are not frequently reported in dorsal frontoparietal cortex, indicating a much stronger overlap with ventral than dorsal attention networks. Importantly, arousal-related

Ventral attention network: regions centered around TPJ and ventral frontal cortex, involved in reorienting to salient stimuli outside the focus of attention

activations in the TPJ and insula/frontal operculum overlap with regions that are damaged in spatial neglect, and recruited during reorienting and target detection, although prefrontal activations are localized more anteriorly.

Summary

The widespread view that attention in healthy adults is right hemisphere dominant is more consistent with the evidence for right lateralization in ventral frontoparietal cortex of reorienting, detection, and arousal than the more modest evidence for right lateralization in dorsal frontoparietal cortex of the mechanisms controlling spatial attention. The main conclusion is that the right hemisphere ventral regions (IPL, STG, IFG/insula) associated with neglect underlie the above nonspatial functions impaired in patients.

VENTRAL AND DORSAL ATTENTION NETWORKS AND SPATIAL NEGLECT

What remains to be considered is how damage to these ventral regions produces the documented abnormalities in the dorsal attention network that likely underlie the egocentric spatial bias. We next consider behavioral evidence for interactions between nonspatial and spatial functions that map to ventral and dorsal regions, and physiological studies that have attempted to identify the specific pathways that might mediate these interactions. Finally, we present a novel physiological model of spatial neglect based on the structural-physiological interaction of ventral and dorsal attention networks.

Interactions Between Ventral and Dorsal Mechanisms

Recent behavioral evidence in healthy adults indicates that arousal, a right lateralized, nonspatial function impaired in neglect patients, interacts with spatial attention, the primary spatial function impaired in neglect patients. Healthy adults show a slight tendency to attend to the left side of an object (Nicholls et al. 1999), but this bias is reduced or shifted to the right under conditions of low arousal (Bellgrove et al. 2004, Manly et al. 2005, Matthias et al. 2009). The specific form of the interaction between arousal and spatial attention predicts that increases in arousal should bias attention to the left visual field, ameliorating left-field neglect.

Robertson and colleagues observed this result in two important studies. Increases in either phasic (Robertson et al. 1998) or sustained arousal (Robertson et al. 1995) decreased neglect of the contralesional field, consistent with a direct effect of the activation of ventral right hemisphere mechanisms on the dorsal attention network. Biases in spatial attention have also been observed immediately following target detection, another right lateralized, nonspatial function damaged in neglect patients (Perez et al. 2009). Perez and colleagues suggest that conditions that reduce processing capacity, such as the attentional blink and low arousal, bias spatial attention to the right.

There is currently only limited evidence of the anatomy and physiology underlying the interactions between nonspatial and spatial mechanisms observed in behavioral studies of healthy adults, and the effects of ventral lesions on dorsal physiology observed in neglect patients. Both may involve similar pathways. Studies in healthy adults have suggested possible linkages between ventral and dorsal regions that run through lateral frontal cortex. A region near IFJ, for example, that shows resting-state connectivity with both dorsal and ventral networks (**Figure 2b**) (He et al. 2007) and shares task-evoked properties with each network (Asplund et al. 2010) may act as a pivot point. Ventral frontal lesions that include IFJ produce greater deficits in spatial attention than temporoparietal lesions (**Figure 5a**) (Rengachary et al. 2011).

Putative ventral-dorsal linkages have also been evaluated by measuring the relationship between ventral to pivot-region functional connectivity and connectivity within the dorsal network. Impaired functional connectivity between STG/TPJ and MFG, for example, is correlated with impaired interhemispheric connectivity between left and right posterior IPS/SPL, which in turn relates to the magnitude of the spatial deficit (He et al. 2007). The same study also assessed structural damage to a white matter tract that putatively connected ventral/pivot regions to the dorsal network, and observed the resulting effects on dorsal network connectivity and spatial behavioral biases. Neglect patients who suffered damage to the superior longitudinal fasciculus showed reduced interhemispheric functional connectivity in posterior parietal cortex and more severe spatial neglect. This may explain why patients with more severe neglect have more involvement of the white matter tracts connecting parietal, temporal, and frontal regions (Gaffan & Hornak 1997, Karnath et al. 2011, Verdon et al. 2010) (**Figure 2c**). Consistent with this latter result, stimulating a white matter tract that connects right frontal and parietal cortex produced rightward deviations in bisection performance (Thiebaut de Schotten et al. 2005). Finally, the neural mechanisms behind ventral-dorsal interactions are unknown but may depend on synchronization of neural activity in ventral and dorsal regions that is time-locked to behaviorally relevant events (Daitch et al. 2010).

These results are provisional and exploratory but suggest that in neglect patients, both the severity of spatial biases and the interhemispheric functional connectivity of dorsal regions are related to the functional connectivity or integrity of ventral/pivot regions and to the integrity of white matter tracts that likely connect ventral and dorsal regions.

A Physiological Framework for Understanding Spatial Neglect

Our review suggests the following account of spatial neglect, as observed in L.J.'s case report. L.J.'s reduced vigilance and slowness in responding to targets, even in his right visual field, reflected impairments in arousal, reorienting, and the detection of novel and behaviorally important stimuli. These nonspatial processes are directly damaged by stroke and other focal brain injuries in neglect patients (in L.J. a ventral frontal stroke involving the anterior insula), and correspondingly, all involve in healthy adults right hemisphere ventral regions that are commonly associated with neglect, including superior temporal cortex, TPJ, IPL, and VFC/insula. These nonspatial mechanisms directly interact with spatial attention mechanisms, providing a link between damage to ventral regions and the abnormal physiology of dorsal regions. Damage to right hemisphere ventral regions, which impairs arousal, reorienting, and detection, hypoactivates the right hemisphere, reducing interactions between the ventral and dorsal attention networks and between regions of the ipsilesional (right) dorsal network. The result is unbalanced interhemispheric physiological activity in the dorsal network, both at rest and during a task, in a direction that favors the left hemisphere. As the locus of attention is coded by mechanisms that take into account activity from both sides of the brain, this imbalance drives spatial attention and eye movements to the right visual field (**Figure 7a** and **7b**). This spatial bias explains L.J.'s tendency to look at and explore first the right visual field, and his inability to detect stimuli in the left visual field.

The right hemisphere dominance of neglect follows from the specific biases produced by right lateralized nonspatial mechanisms on the direction of spatial attention, reflecting the interhemispheric balance of activity in the dorsal network. Right hemisphere lesions may also impair mechanisms that do not directly affect the interhemispheric balance of activity within the dorsal attention network, but increase the behavioral effects produced by the extant stroke-induced imbalance. Full-field deficits in VSTM, trans-saccadic memory, and global perception may act in this fashion (Husain et al. 2001, Robertson & Rafal 2000). Finally, while physiological evidence that the right but not left hemisphere directs spatial attention to both visual fields has not been reported in the majority of studies, the predicted

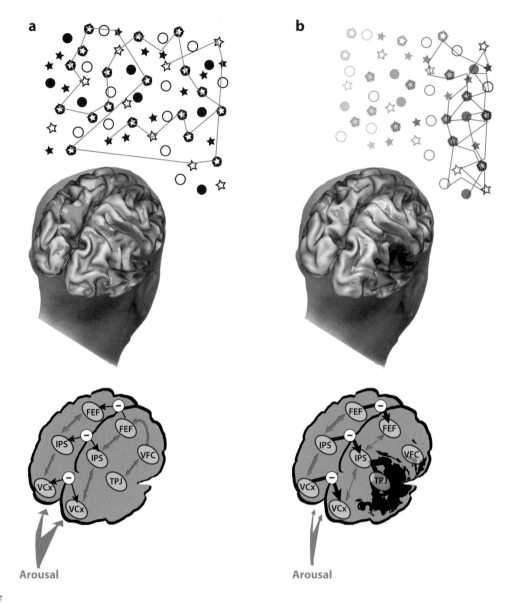

Figure 7

Pathophysiology of spatial neglect. (*a*) In the healthy brain, activity during visual search is symmetric, and interhemispheric interactions between left and right dorsal attention and visual occipital areas are balanced. Each side of the dorsal attention network directs shifts of attention and eye movements contralaterally, and the locus of spatial attention is coded by a differencing mechanism that takes into account activity from both hemispheres, as described in **Figure 3*d*** and by Sylvester et al. (2007). Balanced interhemispheric activity results in a normal eye movement search pattern, shifts of attention, and coding of stimulus saliency. The ventral network is lateralized to the right hemisphere due to a slight asymmetric (right > left) arousal input from the brainstem locus coeruleus/norepinephrine (LC/NE) system, and interacts with the dorsal network (right > left). Accordingly, decreases in arousal shifts spatial attention rightward because of greater left than right activity in the dorsal attention network, while under normal conditions spatial attention shows a slight leftward bias due to slightly greater right than left dorsal activity. (*b*) In a patient with a ventral stroke, direct damage of ventral regions causes a reduction of arousal, target detection, and reorienting that leads to a bilateral visual field impairment. Abnormal ventral-to-dorsal interactions cause an interhemispheric imbalance in the dorsal attention network and visual cortex, leading to tonic and task-dependent rightward spatial biases in attention, eye movements, and stimulus salience.

responses are sometimes observed (Sheremata et al. 2010), suggesting that hemispheric asymmetries in dorsal attentional mechanisms may contribute to the laterality of neglect even if they do not account for its ventral anatomy.

An important future goal is to identify the basis of right hemisphere lateralization of nonspatial mechanisms underlying reorienting, detection, and arousal. Several authors (Corbetta et al. 2008, Posner & Petersen 1990) have argued that this lateralization reflects asymmetries in cortical modulation from the locus coeruleus/norepinephrine (LC/NE) system (Robinson 1985). Lesions to right ventral cortex may damage mechanisms normally receiving strong LC/NE inputs, such as IPL (Morrison & Foote 1986), and cause widespread cortical hypoactivation.

Problems and Omissions

While we propose that the dorsal frontoparietal network underlies control of overt and covert spatial orienting, unilateral lesions of IPS and SPL in humans are traditionally associated with optic ataxia, i.e., difficulties in pointing rather than a general egocentric spatial bias. However, carefully placed lesions in monkey dorsal areas (e.g., Lateral Intraparietal area) cause contralesional deficits in visual search and memory-guided saccades (Wardak et al. 2004). Similarly, recent lesion studies in humans involving careful psychophysical testing found that lesions centered in SPL or in the white matter connecting IPS/SPL to FEF (superior longitudinal fasciculus) cause deficits in goal-driven shifts of attention (Shomstein et al. 2010) and abnormal capture by irrelevant distracters (Ptak & Schnider 2010, Shomstein et al. 2010), consistent with the proposed role of the dorsal attention network in directing spatial attention and coding of stimulus saliency.

The more important point, however, is that damage to dorsal regions alone is insufficient to produce the full neglect syndrome. We argue that the full syndrome depends on a combination of bilateral hypoactivation but greater on the right, associated with damage to mechanisms for reorienting, detection, and arousal in ventral frontoparietal regions and a resulting imbalance in the dorsal network, generating an egocentric spatial bias. The latter emphasis on the dysfunction of dorsal frontoparietal cortex is necessarily provisional because it partly reflects the lack of evidence that in the healthy human brain ventral regions control spatial attention and eye movements or contain topographic maps.

An important topic that we have omitted owing to space limitation concerns the role of subcortical nuclei in spatial neglect. Similarly, the current review does not consider how visuospatial deficits interact with body representations, which may underlie some of the striking symptoms shown by neglect patients, such as their denial of symptoms and confabulations about body ownership (e.g., L.J.).

SUMMARY POINTS

1. The primary spatial impairment in neglect patients is a failure to attend to the contralesional side of space within a reference frame centered on the observer.

2. This spatial deficit is observed both at rest and during task performance.

3. The spatial deficit reflects tonic and task-evoked interhemispheric imbalances of activity within the dorsal attention network.

4. The dorsal attention network may be physiologically impaired across the wide variety of right hemisphere ventral frontal and temporoparietal lesions that can produce neglect.

5. The right hemisphere dominance of neglect primarily reflects the laterality of nonspatial mechanisms for reorienting, detection, and arousal in right ventral frontoparietal cortex, rather than the laterality of mechanisms for spatial attention within dorsal frontoparietal cortex.

6. The activation of nonspatial mechanisms directly biases spatial attention, corresponding to the interaction of ventral and dorsal frontoparietal regions.

7. Damage to right ventral frontoparietal cortex in neglect patients impairs nonspatial functions, hypoactivates the right hemisphere, and unbalances the activity of the dorsal attention network.

8. Ventral-dorsal interactions link the ventral lesions that cause neglect to the egocentric spatial bias that is the hallmark of the neglect syndrome.

FUTURE ISSUES

1. What are the anatomy and physiology of the interaction between dorsal and ventral networks? Are there critical frontal regions and fiber tracts that link the two networks? How is activity in the two networks synchronized?

2. Is the right hemisphere dominance of mechanisms for arousal, reorienting, and detection related to asymmetries in the locus coeruleus/noradrenaline system?

3. Are interhemispheric imbalances in the attention-related activity of dorsal frontoparietal regions present across the wide variety of right hemisphere lesions that can cause neglect?

4. Do the ventral frontoparietal regions associated with neglect contain spatial maps that are involved in controlling attention, either independently or in conjunction with the dorsal network?

5. Under what conditions does the dorsal attention network consistently show hemispheric asymmetries in visual field organization?

6. What nonspatial manipulations bias spatial attention to the right hemifield in healthy adults? Are saliency and eye movement deficits related to arousal deficits?

DISCLOSURE STATEMENT

The authors are not aware of any affiliations, memberships, funding, or financial holdings that might be perceived as affecting the objectivity of this review.

ACKNOWLEDGMENTS

We acknowledge the help of many colleagues who generously made their data available for this review: Drs. Bays and Driver, University College London, for **Figure 1a**; Drs. Karnath and Fruhmann-Berger, University of Tubingen, for **Figure 1b**; Drs. Medina, Pavlak, and Hillis, Johns Hopkins and University of Pennsylvania, for **Figure 1c**; Drs. Karnath, University of Tubingen, Verdon and Vuilleumier, University of Geneva, for **Figure 2a**; Dr. Karnath, University of Tubingen, and M. Glasser and K. Patel, Washington University in St. Louis, for **Figure 2c**; Drs. Sheremata and Somers, Boston University, for **Figure 3c**; Dr. Samuelsson, University of

Goteborg, for **Figure 5b**; Drs. Malhotra and Husain, University College London, for **Figure 5c**. Drs. Jon Driver, David Somers, and Patrick Vuilleumier, as well as Dr. Michael Posner of University of Oregon, and Dr. John Duncan of the MRC, Cambridge, also made useful suggestions concerning the manuscript. Finally, we thank Dr. Serguei Astafiev for help in preparing the figures.

This work was supported by the National Institute of Child Health and Human Development (NICHD) RO1 HD061117-05A2 and the National Institute of Mental Health (NIMH) RO1 MH 71920-06.

LITERATURE CITED

Aron AR, Robbins TW, Poldrack RA. 2004. Inhibition and the right inferior frontal cortex. *Trends Cogn. Sci.* 8:170–77

Asplund CL, Todd JJ, Snyder AP, Marois R. 2010. A central role for the lateral prefrontal cortex in goal-directed and stimulus-driven attention. *Nat. Neurosci.* 13:507–12

Astafiev SV, Shulman GL, Stanley CM, Snyder AZ, Van Essen DC, Corbetta M. 2003. Functional organization of human intraparietal and frontal cortex for attending, looking, and pointing. *J. Neurosci.* 23:4689–99

Aston-Jones G, Cohen JD. 2005. An integrative theory of locus coeruleus-norepinephrine function: adaptive gain and optimal performance. *Annu. Rev. Neurosci.* 28:403–50

Averbeck BB, Latham PE, Pouget A. 2006. Neural correlations, population coding and computation. *Nat. Rev. Neurosci.* 7:358–66

Awh E, Jonides J. 2001. Overlapping mechanisms of attention and spatial working memory. *Trends Cogn. Sci.* 5:119–26

Azouvi P, Samuel C, Louis-Dreyfus A, Bernati T, Bartolomeo P, et al. 2002. Sensitivity of clinical and behavioural tests of spatial neglect after right hemisphere stroke. *J. Neurol. Neurosurg. Psychiatry* 73:160–66

Bartolomeo P, D'Erme P, Gainotti G. 1994. The relationship between visuospatial and representational neglect. *Neurology* 44:1710–14

Bartolomeo P, Thiebaut de Schotten M, Doricchi F. 2007. Left unilateral neglect as a disconnection syndrome. *Cereb. Cortex* 17:2479–90

Bays PM, Singh-Curry V, Gorgoraptis N, Driver J, Husain M. 2010. Integration of goal- and stimulus-related visual signals revealed by damage to human parietal cortex. *J. Neurosci.* 30:5968–78

Behrmann M, Geng JJ. 2002. What is 'left' when all is said and done? Spatial coding and hemispatial coding. See Karnath et al. 2002b, pp. 85–100

Behrmann M, Watt S, Black S, Barton J. 1997. Impaired visual search in patients with unilateral neglect: an oculographic analysis. *Neuropsychologia* 35:1445–58

Bellgrove MA, Dockree PM, Aimola L, Robertson IH. 2004. Attenuation of spatial attentional asymmetries with poor sustained attention. *NeuroReport* 15:1065–69

Bisiach E, Bulgarelli C, Sterzi R, Vallar G. 1983. Line bisection and cognitive plasticity of unilateral neglect of space. *Brain Cogn.* 2:32–38

Bisiach E, Luzzatti C. 1978. Unilateral neglect of representational space. *Cortex* 14:129–33

Bisiach E, Neppi-Modona M, Ricci R. 2002. Space anisometry in unilateral neglect. See Karnath et al. 2002b, pp. 145–52

Bisley JW, Goldberg ME. 2003. Neuronal activity in the lateral intraparietal area and spatial attention. *Science* 299:81–86

Bouret S, Sara SJ. 2005. Network reset: a simplified overarching theory of locus coeruleus noradrenaline function. *Trends Neurosci.* 28:574–82

Bressler SL, Tang W, Sylvester CM, Shulman GL, Corbetta M. 2008. Top-down control of human visual cortex by frontal and parietal cortex in anticipatory visual spatial attention. *J. Neurosci.* 28:10056–61

Brighina F, Bisiach E, Oliveri M, Piazza A, La Bua V, et al. 2003. 1 Hz repetitive transcranial magnetic stimulation of the unaffected hemisphere ameliorates contralesional visuospatial neglect in humans. *Neurosci. Lett.* 336:131–33

Carter AR, Astafiev SV, Lang CE, Connor LT, Rengachary J, et al. 2010. Resting interhemispheric functional magnetic resonance imaging connectivity predicts performance after stroke. *Ann. Neurol.* 67:365–75

Chang SW, Snyder LH. 2010. Idiosyncratic and systematic aspects of spatial representations in the macaque parietal cortex. *Proc. Natl. Acad. Sci. USA* 107:7951–56

Corbetta M, Akbudak E, Conturo TE, Snyder AZ, Ollinger JM, et al. 1998. A common network of functional areas for attention and eye movements. *Neuron* 21:761–73

Corbetta M, Kincade JM, Ollinger JM, McAvoy MP, Shulman GL. 2000. Voluntary orienting is dissociated from target detection in human posterior parietal cortex. *Nat. Neurosci.* 3:292–97

Corbetta M, Kincade MJ, Lewis C, Snyder AZ, Sapir A. 2005. Neural basis and recovery of spatial attention deficits in spatial neglect. *Nat. Neurosci.* 8:1603–10

Corbetta M, Patel G, Shulman GL. 2008. The reorienting system of the human brain: from environment to theory of mind. *Neuron* 58:306–24

Corbetta M, Shulman GL. 2002. Control of goal-directed and stimulus-driven attention in the brain. *Nat. Rev. Neurosci.* 3:201–15

Coslett HB, Bowers D, Fitzpatrick E, Haws B, Heilman KM. 1990. Directional hypokinesia and hemispatial inattention in neglect. *Brain* 113:475–86

Coull JT, Frackowiak RS, Frith CD. 1998. Monitoring for target objects: activation of right frontal and parietal cortices with increasing time on task. *Neuropsychologia* 36:1325–34

Critchley M. 1953. *The Parietal Lobes*. London: Edward Arnold

Daitch AL, Snyder AZ, Astafiev SV, Bundy D, Freudenburg Z, et al. 2010. Temporal dynamics of stimulus-driven attention shifts as studied through the combined use of ECoG and fMRI. *Soc. Neurosci. Meet. Plann.* No. 372.19

Delis DC, Robertson LC, Efron R. 1986. Hemispheric specialization of memory for visual hierarchical stimuli. *Neuropsychologia* 24:205–14

Della Sala S, Logie RH, Beschin N, Denis M. 2004. Preserved visuo-spatial transformations in representational neglect. *Neuropsychologia* 42:1358–64

Denes G, Semenza C, Stoppa E, Lis A. 1982. Unilateral spatial neglect and recovery from hemiplegia: a follow-up study. *Brain* 105:543–52

Deuel RM, Collins RC. 1983. Recovery from unilateral neglect. *Exp. Neurol.* 81:733–48

Di Russo F, Aprile T, Spitoni G, Spinelli D. 2008. Impaired visual processing of contralesional stimuli in neglect patients: a visual-evoked potential study. *Brain* 131:842–54

Dodds CM, van Belle J, Peers PV, Dove A, Cusack R, et al. 2008. The effects of time-on-task and concurrent cognitive load on normal visuospatial bias. *Neuropsychology* 22:545–52

Doricchi F, Tomaiuolo F. 2003. The anatomy of neglect without hemianopia: a key role for parietal-frontal disconnection? *NeuroReport* 14:2239–43

Dosenbach NU, Visscher KM, Palmer ED, Miezin FM, Wenger KK, et al. 2006. A core system for the implementation of task sets. *Neuron* 50:799–812

Driver J, Mattingley JB. 1998. Parietal neglect and visual awareness. *Nat. Neurosci.* 1:17–22

Duncan J, Bundesen C, Olson A, Humphreys G, Chavda S, Shibuya H. 1999. Systematic analysis of deficits in visual attention. *J. Exp. Psychol. Gen.* 128:450–78

Farne A, Buxbaum LJ, Ferraro M, Frassinetti F, Whyte J, et al. 2004. Patterns of spontaneous recovery of neglect and associated disorders in acute right brain-damaged patients. *J. Neurol. Neurosurg. Psychiatry* 75:1401–10

Fink GR, Marshall JC, Weiss PH, Zilles K. 2001. The neural basis of vertical and horizontal line bisection judgments: an fMRI study of normal volunteers. *NeuroImage* 14:S59–67

Foucher JR, Otzenberger H, Gounot D. 2004. Where arousal meets attention: a simultaneous fMRI and EEG recording study. *NeuroImage* 22:688–97

Fox MD, Corbetta M, Snyder AZ, Vincent JL, Raichle ME. 2006. Spontaneous neuronal activity distinguishes human dorsal and ventral attention systems. *Proc. Natl. Acad. Sci. USA* 103:10046–51

Friedrich FJ, Egly R, Rafal RD, Beck D. 1998. Spatial attention deficits in humans: a comparison of superior parietal and temporal-parietal junction lesions. *Neuropsychology* 12:193–207

Fruhmann Berger M, Johannsen L, Karnath HO. 2008. Time course of eye and head deviation in spatial neglect. *Neuropsychology* 22:697–702

Gaffan D, Hornak J. 1997. Visual neglect in the monkey. Representation and disconnection. *Brain* 120(Pt. 9):1647–57

Galati G, Pelle G, Berthoz A, Committeri G. 2010. Multiple reference frames used by the human brain for spatial perception and memory. *Exp. Brain Res.* 206:109–20

Guariglia C, Padovani A, Pantano P, Pizzamiglio L. 1993. Unilateral neglect restricted to visual imagery. *Nature* 364:235–37

Hagler DJ Jr, Sereno MI. 2006. Spatial maps in frontal and prefrontal cortex. *NeuroImage* 29:567–77

Halligan PW, Marshall JC, Wade DT. 1989. Visuospatial neglect: underlying factors and test sensitivity. *Lancet* 2:908–11

Hamilton RH, Coslett HB, Buxbaum LJ, Whyte J, Ferraro MK. 2008. Inconsistency of performance on neglect subtype tests following acute right hemisphere stroke. *J. Int. Neuropsychol. Soc.* 14:23–32

He BJ, Snyder AZ, Vincent JL, Epstein A, Shulman GL, Corbetta M. 2007. Breakdown of functional connectivity in frontoparietal networks underlies behavioral deficits in spatial neglect. *Neuron* 53:905–18

Heilman KM, Schwartz HD, Watson RT. 1978. Hypoarousal in patients with the neglect syndrome and emotional indefference. *Neurology* 28:229–32

Heilman KM, Watson RT, Valenstein E. 1985. Neglect and related disorders. In *Clinical Neuropsychology*, ed. KM Heilman, E Valenstein, pp. 243–93. New York: Oxford

Heilman KM, Watson RT, Valenstein E, Goldberg ME. 1987. Attention: behaviour and neural mechanisms. In *The Handbook of Physiology. Section 1: The Nervous System. Volume V. Higher Functions of the Brain Part 2*, ed. F Plum, VB Mountcastle, ST Geiger, pp. 461–81. Bethesda, MD: Am. Physiol. Soc.

Hillis AE, Newhart M, Heidler J, Barker PB, Herskovits EH, Degaonkar M. 2005. Anatomy of spatial attention: insights from perfusion imaging and hemispatial neglect in acute stroke. *J. Neurosci.* 25:3161–67

Hopfinger JB, Buonocore MH, Mangun GR. 2000. The neural mechanisms of top-down attentional control. *Nat. Neurosci.* 3:284–91

Hornak J. 1992. Ocular exploration in the dark by patients with visual neglect. *Neuropsychologia* 30:547–52

Howes D, Boller F. 1975. Simple reaction time: evidence for focal impairment from lesions of the right hemisphere. *Brain* 98:317–32

Husain M, Kennard C. 1996. Visual neglect associated with frontal lobe infarction. *J. Neurol.* 243:652–57

Husain M, Mannan S, Hodgson T, Wojciulik E, Driver J, Kennard C. 2001. Impaired spatial working memory across saccades contributes to abnormal search in parietal neglect. *Brain* 124:941–52

Husain M, Rorden C. 2003. Non-spatially lateralized mechanisms in hemispatial neglect. *Nat. Rev. Neurosci.* 4:26–36

Husain M, Shapiro K, Martin J, Kennard C. 1997. Abnormal temporal dynamics of visual attention in spatial neglect patients. *Nature* 385:154–56

Innocenti GM. 2009. Dynamic interactions between the cerebral hemispheres. *Exp. Brain Res.* 192:417–23

Ishiai S, Koyama Y, Seki K, Hayashi K, Izumi Y. 2006. Approaches to subjective midpoint of horizontal lines in unilateral spatial neglect. *Cortex* 42:685–91

Karnath HO, Ferber S, Himmelbach M. 2001. Spatial awareness is a function of the temporal not the posterior parietal lobe. *Nature* 411:950–53

Karnath HO, Himmelbach M, Rorden C. 2002a. The subcortical anatomy of human spatial neglect: putamen, caudate nucleus and pulvinar. *Brain* 125(Pt. 2):350–60

Karnath HO, Milner D, Vallar G, eds. 2002b. *The Cognitive and Neural Bases of Spatial Neglect*. New York: Oxford Univ. Press

Karnath HO, Zopf R, Johannsen L, Fruhmann Berger M, Nèagele T, Klose U. 2005. Normalized perfusion MRI to identify common areas of dysfunction: patients with basal ganglia neglect. *Brain* 128(Pt. 10):2462–69

Karnath OH, Rennig J, Johannsen L, Rorden C. 2011. The anatomy underlying acute versus chronic spatial neglect. *Brain* 134(Pt. 3):903–12

Kastner S, Pinsk MA, De Weerd P, Desimone R, Ungerleider LG. 1999. Increased activity in human visual cortex during directed attention in the absence of visual stimulation. *Neuron* 22:751–61

Kincade JM, Abrams RA, Astafiev SV, Shulman GL, Corbetta M. 2005. An event-related functional magnetic resonance imaging study of voluntary and stimulus-driven orienting of attention. *J. Neurosci.* 25(18):4593–604

Kinsbourne M. 1987. Mechanisms of unilateral neglect. In *Neurophysiological and Neuropsychological Aspects of Spatial Neglect*, ed. M Jeannerod, pp. 69–86. Amsterdam: Elsevier Sci.

Kinsella G, Olver J, Ng K, Packer S, Stark R. 1993. Analysis of the syndrome of unilateral neglect. *Cortex* 29:135–40

Koch C, Ullman S. 1985. Shifts in visual attention: towards the underlying circuitry. *Hum. Neurobiol.* 4:219–27

Koch G, Oliveri M, Cheeran B, Ruge D, Lo Gerfo E, et al. 2008. Hyperexcitability of parietal-motor functional connections in the intact left-hemisphere of patients with neglect. *Brain* 131:3147–55

Kosslyn SM, Ganis G, Thompson WL. 2001. Neural foundations of imagery. *Nat. Rev. Neurosci.* 2:635–42

Kristjansson A, Vuilleumier P. 2010. Disruption of spatial memory in visual search in the left visual field in patients with hemispatial neglect. *Vision Res.* 50:1426–35

Kukolja J, Marshall JC, Fink GR. 2006. Neural mechanisms underlying spatial judgements on seen and imagined visual stimuli in the left and right hemifields in men. *Neuropsychologia* 44:2846–60

MacNeilage PF, Rogers LJ, Vallortigara G. 2009. Origins of the left & right brain. *Sci. Am.* 301:60–67

Malhotra P, Coulthard EJ, Husain M. 2009. Role of right posterior parietal cortex in maintaining attention to spatial locations over time. *Brain* 132:645–60

Malhotra P, Jager HR, Parton A, Greenwood R, Playford ED, et al. 2005. Spatial working memory capacity in unilateral neglect. *Brain* 128:424–35

Manly T, Dobler VB, Dodds CM, George MA. 2005. Rightward shift in spatial awareness with declining alertness. *Neuropsychologia* 43:1721–28

Mannan SK, Mort DJ, Hodgson TL, Driver J, Kennard C, Husain M. 2005. Revisiting previously searched locations in visual neglect: role of right parietal and frontal lesions in misjudging old locations as new. *J. Cogn. Neurosci.* 17:340–54

Marsh EB, Hillis AE. 2008. Dissociation between egocentric and allocentric visuospatial and tactile neglect in acute stroke. *Cortex* 44:1215–20

Matthias E, Bublak P, Costa A, Muller HJ, Schneider WX, Finke K. 2009. Attentional and sensory effects of lowered levels of intrinsic alertness. *Neuropsychologia* 47:3255–64

McIntosh RD, Schindler I, Birchall D, Milner AD. 2005. Weights and measures: a new look at bisection behaviour in neglect. *Brain Res. Cogn. Brain Res.* 25:833–50

Medina J, Kannan V, Pawlak MA, Kleinman JT, Newhart M, et al. 2008. Neural substrates of visuospatial processing in distinct reference frames: evidence from unilateral spatial neglect. *J. Cogn. Neurosci.* 21:2073–84

Merriam EP, Genovese CR, Colby CL. 2003. Spatial updating in human parietal cortex. *Neuron* 39:361–73

Mesulam MM. 1981. A cortical network for directed attention and unilateral neglect. *Ann. Neurol.* 10:309–25

Mesulam MM. 1999. Spatial attention and neglect: parietal, frontal and cingulate contributions to the mental representation and attentional targeting of salient extrapersonal events. *Philos. Trans. R. Soc. Lond. B Biol. Sci.* 354:1325–46

Milner AD, Harvey M, Pritchard CL. 1998. Visual size processing in spatial neglect. *Exp. Brain. Res.* 123:192–200

Morrison JH, Foote SL. 1986. Noradrenergic and serotoninergic innervation of cortical, thalamic and tectal visual structures in Old and New World monkeys. *J. Comp. Neurol.* 243:117–28

Mort DJ, Malhotra P, Mannan SK, Rorden C, Pambakian A, et al. 2003. The anatomy of visual neglect. *Brain* 126:1986–97

Nicholls ME, Bradshaw JL, Mattingley JB. 1999. Free-viewing perceptual asymmetries for the judgement of brightness, numerosity and size. *Neuropsychologia* 37:307–14

O'Keefe J. 2006. Hippocampal neurophysiology in the behaving animal. In *The Hippocampus Book*, ed. P Andersen, R Morris, D Amaral, T Bliss, J O'Keefe, pp. 475–548. New York: Oxford Univ. Press

Ortigue S, Viaud-Delmon I, Annoni JM, Landis T, Michel C, et al. 2001. Pure representational neglect after right thalamic lesion. *Ann. Neurol.* 50:401–4

Paolucci S, Antonucci G, Grasso MG, Pizzamiglio L. 2001. The role of unilateral spatial neglect in rehabilitation of right brain-damaged ischemic stroke patients: a matched comparison. *Arch. Phys. Med. Rehabil.* 82:743–49

Pardo JV, Fox PT, Raichle ME. 1991. Localization of a human system for sustained attention by positron emission tomography. *Nature* 349:61–64

Paus T, Zatorre RJ, Hofle N, Zografos C, Gotman J, et al. 1997. Time-related changes in neural systems underlying attention and arousal during the performance of an auditory vigilance task. *J. Cogn. Neurosci.* 9(3):392–408

Peers PV, Ludwig CJ, Rorden C, Cusack R, Bonfiglioli C, et al. 2005. Attentional functions of parietal and frontal cortex. *Cereb. Cortex* 15(10):1469–84

Perani D, Vallar G, Cappa S, Messa C, Fazio F. 1987. Aphasia and neglect after subcortical stroke: a clinical/cerebral perfusion correlation study). *Brain* 110:1211–29

Perez A, Peers PV, Valdes-Sosa M, Galan L, Garcia L, Martinez-Montes E. 2009. Hemispheric modulations of alpha-band power reflect the rightward shift in attention induced by enhanced attentional load. *Neuropsychologia* 47:41–49

Pizzamiglio L. 1998. Recovery of neglect after right hemispheric damage: $H_2^{15}O$ positron emission tomographic activation study. *Arch. Neurol.* 55:561–68

Posner MI, Petersen SE. 1990. The attention system of the human brain. *Annu. Rev. Neurosci.* 13:25–42

Posner MI, Walker JA, Friedrich FJ, Rafal RD. 1984. Effects of parietal injury on covert orienting of attention. *J. Neurosci.* 4:1863–74

Pouget A, Driver J. 2000. Relating unilateral neglect to the neural coding of space. *Curr. Opin. Neurobiol.* 10:242–49

Pouget A, Sejnowski TJ. 2001. Simulating a lesion in a basis function model of spatial representations: comparison with hemineglect. *Psychol. Rev.* 108:653–73

Ptak R, Schnider A. 2010. The dorsal attention network mediates orienting toward behaviorally relevant stimuli in spatial neglect. *J. Neurosci.* 30:12557–65

Rees G, Wojciulik E, Clarke K, Husain M, Frith C, Driver J. 2000. Unconscious activation of visual cortex in the damaged right hemisphere of a parietal patient with extinction. *Brain* 123(Pt. 8):1624–33

Rengachary J, d'Avossa G, Sapir A, Shulman GL, Corbetta M. 2009. Is the Posner reaction time test more accurate than clinical tests in detecting left neglect in acute and chronic stroke? *Arch. Phys. Med. Rehabil.* 90:2081–88

Rengachary J, He BJ, Shulman GL, Corbetta M. 2011. A behavioral analysis of spatial neglect and its recovery after stroke. *Front. Hum. Neurosci.* 5:29. doi: 10.3389/fnhum.2011.00029. In press

Riddoch MJ, Humphreys GW. 1983. The effect of cueing on unilateral neglect. *Neuropsychologia* 21:589–99

Rizzolatti G, Riggio L, Dascola I, Umiltá C. 1987. Reorienting attention across the horizontal and vertical meridians: evidence in favor of a premotor theory of attention. *Neuropsychologia* 25:31–40

Robertson I, North N. 1992. Spatio-motor cueing in unilateral left neglect: the role of hemispace, hand and motor activation. *Neuropsychologia* 30:553–63

Robertson I, Tegner R, Tham K, Lo A, Nimmo-Smith I. 1995. Sustained attention training for unilateral neglect: theoretical and rehabilitation implications. *J. Clin. Exp. Neuropsychol.* 17:416–30

Robertson IH. 2001. Do we need the "lateral" in unilateral neglect? Spatially nonselective attention deficits in unilateral neglect and their implications for rehabilitation. *Neuroimage* 14:S85–90

Robertson IH, Manly T, Beschin N, Daini R, Haeske-Dewick H, et al. 1997. Auditory sustained attention is a marker of unilateral spatial neglect. *Neuropsychologia* 35:1527–32

Robertson IH, Mattingley JB, Rorden C, Driver J. 1998. Phasic alerting of neglect patients overcomes their spatial deficit in visual awareness. *Nature* 395:169–72

Robertson LC, Rafal RD. 2000. Disorders of visual attention. In *The New Cognitive Neurosciences*, ed. MS Gazzaniga, pp. 633–50. Cambridge: MIT Press

Robinson RG. 1985. Lateralized behavior and neurochemical consequences of unilateral brain injury in rats. In *Cerebral Lateralization in Nonhuman Species*, ed. SD Glick, pp. 135–56. New York: Academic

Rogers LJ. 2000. Evolution of hemispheric specialization: advantages and disadvantages. *Brain Lang.* 73:236–53

Rogers LJ, Andrew RJ, Johnston AN. 2007. Light experience and the development of behavioural lateralization in chicks. III. Learning to distinguish pebbles from grains. *Behav. Brain Res.* 177:61–69

Rogers LJ, Anson JM. 1979. Lateralisation of function in the chicken fore-brain. *Pharmacol. Biochem. Behav.* 10:679–86

Rogers LJ, Sink HS. 1988. Transient asymmetry in the projections of the rostral thalamus to the visual hyperstriatum of the chicken, and reversal of its direction by light exposure. *Exp. Brain Res.* 70:378–84

Rueckert L, Grafman J. 1996. Sustained attention deficits in patients with right frontal lesions. *Neuropsychologia* 34:953–63

Sack AT. 2009. Parietal cortex and spatial cognition. *Behav. Brain Res.* 202:153–61

Samuelsson H, Hjelmquist EK, Jensen C, Ekholm S, Blomstrand C. 1998. Nonlateralized attentional deficits: an important component behind persisting visuospatial neglect? *J. Clin. Exp. Neuropsychol.* 20:73–88

Sereno MI, Pitzalis S, Martinez A. 2001. Mapping of contralateral space in retinotopic coordinates by a parietal cortical area in humans. *Science* 294:1350–54

Sheremata SL, Bettencourt KC, Somers DC. 2010. Hemispheric asymmetry in visuotopic posterior parietal cortex emerges with visual short-term memory load. *J. Neurosci.* 30:12581–88

Shomstein S, Lee J, Behrmann M. 2010. Top-down and bottom-up attentional guidance: investigating the role of the dorsal and ventral parietal cortices. *Exp. Brain Res.* 206(2):197–208

Shulman GL, Astafiev SV, Franke D, Pope DL, Snyder AZ, et al. 2009. Interaction of stimulus-driven reorienting and expectation in ventral and dorsal frontoparietal and basal ganglia-cortical networks. *J. Neurosci.* 29:4392–407

Shulman GL, Pope DL, Astafiev SV, McAvoy MP, Snyder AZ, Corbetta M. 2010. Right hemisphere dominance during spatial selective attention and target detection occurs outside the dorsal frontoparietal network. *J. Neurosci.* 30:3640–51

Siegel RM, Raffi M, Phinney RE, Turner JA, Jando G. 2003. Functional architecture of eye position gain fields in visual association cortex of behaving monkey. *J. Neurophysiol.* 90:1279–94

Snow JC, Mattingley JB. 2006. Goal-driven selective attention in patients with right hemisphere lesions: how intact is the ipsilesional field? *Brain* 129:168–81

Spinelli D, Guariglia C, Massironi M, Pizzamiglio L, Zoccolotti P. 1990. Contrast sensitivity and low spatial frequency discrimination in hemi-neglect patients. *Neuropsychologia* 28:727–32

Stevens MC, Calhoun VD, Kiehl KA. 2005. Hemispheric differences in hemodynamics elicited by auditory oddball stimuli. *Neuroimage* 26:782–92

Stone SP, Halligan PW, Greenwood RJ. 1993. The incidence of neglect phenomena and related disorders in patients with an acute right or left hemisphere stroke. *Age Ageing* 22:46–52

Sturm W, de Simone A, Krause BJ, Specht K, Hesselmann V, et al. 1999. Functional anatomy of intrinsic alertness: evidence for a fronto-parietal-thalamic-brainstem network in the right hemisphere. *Neuropsychologia* 37:797–805

Sturm W, Longoni F, Fimm B, Dietrich T, Weis S, et al. 2004. Network for auditory intrinsic alertness: a PET study. *Neuropsychologia* 42:563–68

Swisher JD, Halko MA, Merabet LB, McMains SA, Somers DC. 2007. Visual topography of human intraparietal sulcus. *J. Neurosci.* 27:5326–37

Sylvester CM, Shulman GL, Jack AI, Corbetta M. 2007. Asymmetry of anticipatory activity in visual cortex predicts the locus of attention and perception. *J. Neurosci.* 27(52):14424–33

Szczepanski SM, Konen CS, Kastner S. 2010. Mechanisms of spatial attention control in frontal and parietal cortex. *J. Neurosci.* 30:148–60

Thiebaut de Schotten M, Urbanski M, Duffau H, Volle E, Levy R, et al. 2005. Direct evidence for a parietal-frontal pathway subserving spatial awareness in humans. *Science* 309:2226–28

Thimm M, Fink GR, Sturm W. 2008. Neural correlates of recovery from acute hemispatial neglect. *Restor. Neurol. Neurosci.* 26:481–92

Thut G, Nietzel A, Brandt SA, Pascual-Leone A. 2006. Alpha-band electroencephalographic activity over occipital cortex indexes visuospatial attention bias and predicts visual target detection. *J. Neurosci.* 26:9494–502

Valencia-Alfonso CE, Verhaal J, Gunturkun O. 2009. Ascending and descending mechanisms of visual lateralization in pigeons. *Philos. Trans. R. Soc. Lond. B Biol. Sci.* 364:955–63

Vallar G, Perani D. 1987. The anatomy of spatial neglect in humans. In *Neurophysiological and Neuropsychological Aspects of Spatial Neglect*, ed. M Jeannerod, pp. 235–58. North-Holland: Elsevier Sci.

Vandenberghe R, Geeraerts S, Molenberghs P, Lafosse C, Vandenbulcke M, et al. 2005. Attentional responses to unattended stimuli in human parietal cortex. *Brain* 128:2843–57

Verdon V, Schwartz S, Lovblad KO, Hauert CA, Vuilleumier P. 2010. Neuroanatomy of hemispatial neglect and its functional components: a study using voxel-based lesion-symptom mapping. *Brain* 133:880–94

Vuilleumier P, Schwartz S, Verdon V, Maravita A, Hutton C, et al. 2008. Abnormal attentional modulation of retinotopic cortex in parietal patients with spatial neglect. *Curr. Biol.* 18:1525–29

Vuilleumier P, Sergent C, Schwartz S, Valenza N, Girardi M, et al. 2007. Impaired perceptual memory of locations across gaze-shifts in patients with unilateral spatial neglect. *J. Cogn. Neurosci.* 19:1388–406

Wardak C, Olivier E, Duhamel JR. 2004. A deficit in covert attention after parietal cortex inactivation in the monkey. *Neuron* 42:501–8

Watson RT, Miller BD, Heilman KM. 1977. Evoked potential in neglect. *Arch. Neurol.* 34:224–27

Wilkins AJ, Shallice T, McCarthy R. 1987. Frontal lesions and sustained attention. *Neuropsychologia* 25:359–65

Yokoyama K, Jennings R, Ackles P, Hood P, Boller F. 1987. Lack of heart rate changes during an attention-demanding task after right hemisphere lesions. *Neurology* 37:624–30

Zopf R, Berger MF, Klose U, Karnath HO. 2009. Perfusion imaging of the right perisylvian neural network in acute spatial neglect. *Front. Hum. Neurosci.* 3:15

General Anesthesia and Altered States of Arousal: A Systems Neuroscience Analysis

Emery N. Brown,[1,2,3] Patrick L. Purdon,[1,2] and Christa J. Van Dort[1,2]

[1] Department of Anesthesia, Critical Care and Pain Medicine, Massachusetts General Hospital, Harvard Medical School, Boston, Massachusetts 02114; email: enb@neurostat.mit.edu, patrickp@nmr.mgh.harvard.edu, vandortc@mit.edu

[2] Department of Brain and Cognitive Sciences, [3] Harvard-MIT Division of Health Sciences and Technology, Massachusetts Institute of Technology, Cambridge, Massachusetts 02139

Annu. Rev. Neurosci. 2011. 34:601–28

The *Annual Review of Neuroscience* is online at neuro.annualreviews.org

This article's doi:
10.1146/annurev-neuro-060909-153200

Keywords

dexmedetomidine, droperidol, ketamine, opioids, propofol

Abstract

Placing a patient in a state of general anesthesia is crucial for safely and humanely performing most surgical and many nonsurgical procedures. How anesthetic drugs create the state of general anesthesia is considered a major mystery of modern medicine. Unconsciousness, induced by altered arousal and/or cognition, is perhaps the most fascinating behavioral state of general anesthesia. We perform a systems neuroscience analysis of the altered arousal states induced by five classes of intravenous anesthetics by relating their behavioral and physiological features to the molecular targets and neural circuits at which these drugs are purported to act. The altered states of arousal are sedation-unconsciousness, sedation-analgesia, dissociative anesthesia, pharmacologic non-REM sleep, and neuroleptic anesthesia. Each altered arousal state results from the anesthetic drugs acting at multiple targets in the central nervous system. Our analysis shows that general anesthesia is less mysterious than currently believed.

Contents

INTRODUCTION

The practice of administering general anesthesia, which began nearly 165 years ago, revolutionized medicine by transforming surgery from trauma and butchery into a humane therapy (Bigelow 1846, Long 1849, Fenster 2001, Boland 2009). In the United States, 21 million patients receive general anesthesia each year for surgical procedures alone (Jt. Comm. 2004). General anesthesia is also essential for conducting many nonsurgical diagnostic and therapeutic procedures. Significant improvements have been made in anesthetic drugs (Schuttler & Schwilden 2008), monitoring standards (Eichhorn et al. 1986), monitoring devices (Ali et al. 1970, 1971; Tremper & Barker 1989; Kearse et al. 1994), and delivery systems (Wuesten 2001, Sloan 2002). Nevertheless, debate continues in anesthesiology over how best to define general anesthesia (Eger 1993, Kissin 1993). How anesthetic drugs create the state of general anesthesia is considered one of the biggest mysteries of modern medicine (Kennedy & Norman 2005).

Studies of the mechanisms of general anesthesia focus on characterizing the actions of anesthetic drugs at molecular targets in the central nervous system (Hemmings et al. 2005, Grasshoff et al. 2006, Franks 2008). This work has been crucial for identifying molecular and pharmacological principles that underlie anesthetic drugs and for establishing the existence of multiple mechanisms of anesthetic action. Studies of the pharmacokinetic and pharmacodynamic properties of anesthetics provide the main guidelines for drug dosing (Shafer et al. 2009). Neurophysiological (Angel 1991, 1993; Gredell et al. 2004; Velly et al. 2007; Bieda et al. 2009; Breshears et al. 2010) and functional imaging studies of general anesthesia (Alkire et al. 1995, Veselis et al. 1997, Bonhomme et al. 2001, Purdon et al. 2009, Mhuircheartaigh et al. 2010) are helping to identify the neural circuits involved in creating the states of general anesthesia. Furthermore, the behavioral and physiological changes observed when anesthetics are administered in clinical practice provide important insights into their mechanisms. Deciphering the mechanisms of general anesthesia requires a systems neuroscience analysis—that is, relating the actions of the drugs at specific molecular targets in specific neural circuits to the behavioral and physiological states that comprise general anesthesia (Brown 2007).

To develop such an analysis we define general anesthesia as a drug-induced, reversible condition composed of the behavioral states of unconsciousness, amnesia, analgesia, and immobility along with physiological stability (Brown et al. 2010). Of the behavioral states, unconsciousness is perhaps the most fascinating. Patients often admit that the unknown of being unconscious is their biggest fear about undergoing general anesthesia. Unconsciousness can be achieved by altering a patient's level of arousal and/or cognition. Anesthesiologists use different drugs to alter arousal in order to create the state of general anesthesia. We review the altered arousal states of five classes of intravenous anesthetic drugs: gamma-amino butyric acid type A (GABA$_A$) receptor agonists, opioid receptor agonists, N-methyl D-aspartate receptor (NMDA) antagonists, alpha 2 receptor agonists, and dopamine receptor antagonists. We demonstrate that each altered arousal state can be characterized in terms of its specific behavioral and physiological features, as well as the molecular targets and neural circuits in which the drugs are purported to act. For each drug, we summarize these features in a table at the end of each section.

SEDATION AND UNCONSCIOUSNESS: GABA$_A$ RECEPTOR AGONISTS

Clinical Features of Altered Arousal of GABA$_A$ Agonists

The anesthetic drugs that act as agonists at GABA$_A$ receptors are used primarily to induce sedation or unconsciousness. These drugs, which include propofol, sodium thiopental, methoxital, and etomidate, belong to different drug classes. Propofol is 2,6 di-isopropyl-phenol (James & Glen 1980), sodium thiopental and methohexital are barbiturates (Russo & Bressolle 1998), and etomidate is a carboxylated imidazole derivative (Vanlersberghe & Camu 2008). All act at GABA$_A$ receptors to enhance inhibition. Their effects depend on how much and how rapidly they are administered.

Administration of small doses of a GABA$_A$ hypnotic induces sedation. If the dose is increased slowly, the patient can enter a state of paradoxical excitation or disinhibition defined by euphoria or dysphoria, incoherent speech, purposeless or defensive movements, and increased electroencephalogram (EEG) oscillations in the beta range (13–25 Hz).

The excitatory state is termed paradoxical because a drug given to induce sedation or unconsciousness results in excitation (Brown et al. 2010). Paradoxical excitation frequently occurs when GABA$_A$ agonists are administered as sedatives for colonoscopies, dental extractions, or radiological procedures (Fulton & Mullen 2000, Tung et al. 2001). Methohexital is distinct from other hypnotics because small doses of it can induce seizures (Rockoff & Goudsouzian 1981). Therefore, methohexital is used as an induction agent during electroconvulsive therapy because the therapeutic benefit of this procedure is related to the quality of the seizure activity it induces (Avramov et al. 1995). Low-dose methohexital is also used during cortical mapping studies to help identify seizure foci (Hufnagel et al. 1992).

If the hypnotic is administered as a bolus dose over 5 to 10 seconds, as is typical for induction of general anesthesia, then within 20 to 30 seconds, the patient becomes unresponsive to verbal commands, regular breathing transitions to apnea, and ventilation by bag and mask must be started to support breathing (Brown et al. 2010). Loss of consciousness can be easily tracked during drug administration on induction by having the patient execute a cognitive task such as counting backward or performing smooth eye pursuit of the anesthesiologist's finger. Patients counting backward rarely go beyond 10 digits. With loss of consciousness, the smooth pursuit excursions decrease then stop; blinking increases; nystagmus may appear; and the eyelash, corneal and oculocephalic reflexes are lost. The pupillary response to light remains intact (Brown et al. 2010). With propofol, a concomitant loss of skeletal muscle tone occurs (Brown et al. 2010), whereas with etomidate, the loss of muscle tone may be preceded by myoclonus (Van Keulen & Burton 2003). The EEG typically shows alpha and delta oscillations or burst suppression. The vasodilatory and myocardial depressant effects of the hypnotic drugs generally cause a decrease in blood pressure that leads to a baroreceptor reflex-mediated increase in heart rate. An opioid is frequently administered prior to or during

Sedation: a pharmacologically induced state of decreased movement, anxiolysis, decreased arousal with slow and/or incoherent responses to verbal commands

Hypnotic: any anesthetic agent used to induce unconsciousness in a patient

Electroencephalogram (EEG) oscillations: EEG oscillatory patterns are characterized in terms of 5 different frequency bands: delta (0–4 Hz), theta (4–8 Hz), alpha (8–12 Hz), beta (13–25 Hz), gamma (40–80 Hz)

Induction: initiation of general anesthesia achieved usually by administering a bolus dose of a hypnotic to render a patient unconscious

Smooth pursuit: the ability of the eyes to track closely a moving object with an uninterrupted change in gaze

Burst suppression: an EEG oscillation pattern composed of high-frequency activity interspersed with periods of isoelectricity (flatlines)

Myoclonus: brief involuntary twitching in a muscle or group of muscles

induction to help mitigate the rise in heart rate, and vasopressors may be administered to maintain blood pressure. After the anesthesiologist administers a muscle relaxant, the patient is intubated, and maintenance of general anesthesia is achieved by a combination of inhalational agents, opioids, hypnotic agents, muscle relaxants, sedatives, and cardiovascular medications with ventilation and thermoregulatory support.

Mechanisms of GABA$_A$-Mediated Altered Arousal

GABA$_A$ receptors are widely distributed throughout the brain (Bowery et al. 1987). Molecular studies have identified GABA$_A$ receptors as the targets of anesthetic drugs (Hemmings et al. 2005, Grasshoff et al. 2006). Binding of a GABA$_A$ hypnotic to the GABA receptors helps maintain postsynaptic chloride channels in the open position, thereby enhancing the inward chloride current, which hyperpolarizes the pyramidal neurons (Bai et al. 1999). For example, a pyramidal neuron in the cortex receives excitatory inputs from each of the major cholinergic, monoaminergic, and orexinergic arousal pathways as well as inhibitory inputs from local inhibitory interneurons (Figure 1a). During wakefulness, there is a balance between the pyramidal neuron's excitatory and inhibitory inputs. Because small numbers of inhibitory interneurons control large numbers of excitatory pyramidal neurons, hypnotic-induced enhancement of GABA$_A$ inhibition can efficiently inactivate large brain regions (Figure 1b) (Markram et al. 2004). Several neurophysiological and imaging studies

support the idea that cortical sites are targets of GABAergic hypnotics (Angel 1991, Alkire et al. 1995, Velly et al. 2007, Purdon et al. 2009, Breshears et al. 2010, Ferrarelli et al. 2010).

When administered as a bolus for induction of general anesthesia, the hypnotic rapidly reaches the GABAergic neurons in the respiratory centers in the pons and medulla (Feldman et al. 2003) and the arousal centers in the pons, midbrain, hypothalamus and basal forebrain (Saper et al. 2005) (Figure 2a). The clinical signs are consistent with inhibitory actions in the brainstem. Loss of the oculocephalic and corneal reflexes is a nonspecific indicator of impaired brain-stem function due to the actions of the hypnotic agent on the nuclei that control eye movements in the midbrain and pons (Posner et al. 2007). Basic science studies have shown that unconsciousness can result from direct injection of a barbiturate into the mesopontine tegmental area (Devor & Zalkind 2001). Brain injury studies show that brainstem coma typically involves bilateral lesions in either the midbrain or pontine tegmentum (Parvizi & Damasio 2003). The preoptic area (POA) of the hypothalamus provides GABAergic inhibition to the principal arousal centers in the hypothalamus, midbrain, and pons (Saper et al. 2005) (Figure 1a). These GABAergic synapses onto pyramidal cells in the arousal centers are likely sites of action of the GABAergic hypnotics (Figure 1b). Inhibition of these ascending arousal centers decreases their input to the cortex and hence decreases cortical activation. The actions of the hypnotic on GABA$_A$ interneurons in the respiratory control network in the ventral

Figure 1

GABAergic signaling during the awake state (a) and during propofol administration (b). (a) A cortical interneuron during wakefulness mediating control of pyramidal neurons and being modulated by ascending arousal centers. Also shown are inhibitory projections from the POA to the arousal-promoting nuclei. (b) Propofol enhances GABAergic transmission in the cortex and at the inhibitory projections from the POA to the arousal centers. Abbreviations: 5HT, serotonin; ACh, acetylcholine; DA, dopamine; DR, dorsal raphe; GABA, gamma aminobutyric acid; Gal, galanin; His, histamine; LC, locus coeruleus; LDT, laterodorsal tegmental area; LH, lateral hypothalamus; NE, norepinephrine; POA, preoptic area; PPT, pedunculopontine tegmental area; TMN, tuberomamillary nucleus; vPAG, ventral periaquaductal gray.

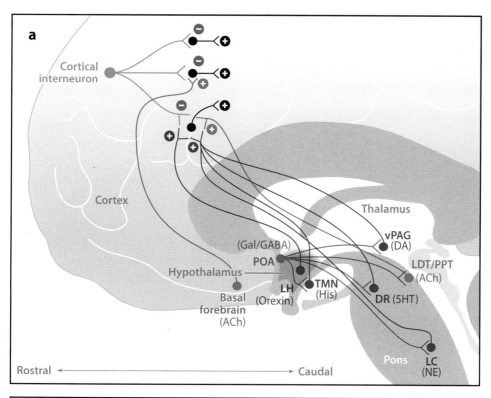

a

Cortical interneuron

Cortex

Thalamus

Hypothalamus

(Gal/GABA)
POA

Basal forebrain (ACh)

LH (Orexin)

TMN (His)

vPAG (DA)

LDT/PPT (ACh)

DR (5HT)

Pons

LC (NE)

Rostral ← → Caudal

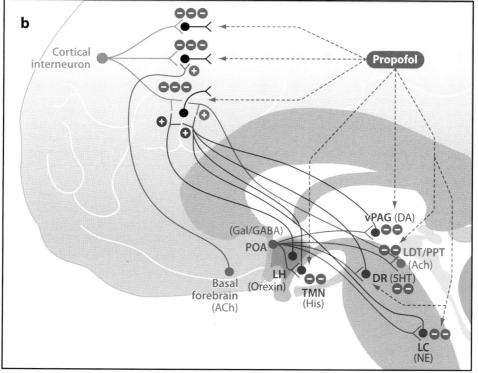

b

Cortical interneuron

Propofol

(Gal/GABA)
POA

Basal forebrain (ACh)

LH (Orexin)

TMN (His)

vPAG (DA)

LDT/PPT (Ach)

DR (5HT)

LC (NE)

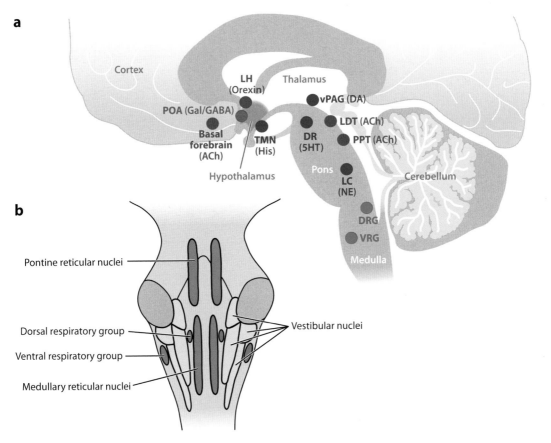

Figure 2

Sites of major nuclei that regulate arousal and respiration. (*a*) A sagittal diagram of a human brain with cholinergic nuclei (*green*), monoaminergic nuclei (*dark blue*), GABAergic and galanergic nuclei (*red*), orexin nuclei (*pink*), and the respiratory nuclei (*light blue*). (*b*) Sites of major respiratory and motor relay nuclei in the pons and medulla. Abbreviations: 5HT, serotonin; ACh, acetylcholine; DA, dopamine; DRG, dorsal respiratory group; GABA, gamma aminobutyric acid; Gal, galanin; His, histamine; LC, locus coeruleus; LDT, laterodorsal tegmental area; LH, lateral hypothalamus; NE, norepinephrine; POA, preoptic area; PPT, pedunculopontine tegmental area; TMN, tuberomamillary nucleus; VAG, ventral periaquaductal gray; VRG, ventral respiratory group.

Interscalene block: regional anesthesia for the brachial plexus achieved by injecting a local anesthetic through a needle inserted between the anterior and middle interscalene muscles at C6

medulla are most likely responsible for the apnea (Feldman et al. 2003).

The atonia observed following bolus administration of propofol may be attributed to its actions on GABAergic circuits in the spinal cord (Kungys et al. 2009) and in the pontine and medullary reticular nuclei that control the antigravity muscles (Brown et al. 2010) (**Figure 2b**). This latter hypothesis is consistent with reports of rapid atonia following inadvertent subarachnoid space and basilar artery injection of local anesthetics during interscalene blocks (Durrani & Winnie 1991, Dutton et al. 1994), direct injection of a barbiturate into the mesopontine tegmental area (Devor & Zalkind 2001), and atonia in patients suffering from pontine strokes and locked-in syndromes (Posner et al. 2007).

The output pathways from the thalamus to the cortex are regulated by major GABAergic inhibitory inputs from the thalamic reticular nucleus (Jones 2002). Most certainly, GABAergic hypnotics contribute to sedation

Table 1 Summary of the behaviors, physiological responses, and neural circuits for the actions of GABA$_A$ agonists[a]

Drugs	Behaviors and physiological responses	Possible neural circuit mechanism	Receptors
Propofol, Etomidate, Thiopental, Methohexital	Unconsciousness and sedation	Potentiation of GABAergic interneurons in the cortex, the thalamic reticular nucleus, and the arsousal centers in the midbrain and pons	GABA$_A$
	Seizures	Potentiation of GABAergic interneurons locally in the cortex	
	Apnea	GABAergic inhibition of brain stem respiratory control areas (ventral medulla)	
	Atonia	Inhibition of the pontine and medullary reticular nuclei that control antigravity muscles and inhibition of spinal motor neurons	
	Myoclonus	Possible effects on primary motor pathways	
	Loss of corneal and oculocephalic reflexes	GABAergic inhibition in the brain stem of the oculomotor, abducens and trochlear nuclei, the trigeminal nuclei, and the motor nuclei of the seventh cranial nerve	

[a]Abbreviations: GABA$_A$, gamma aminobutyric acid type A.

and unconsciousness by enhancing inhibitory activity at these thalamic reticular synapses.

Table 1 summarizes the behavioral and physiological responses of the GABA agonists along with possible neural circuit mechanisms for these responses.

ANALGESIA AND SEDATION: OPIOID RECEPTOR AGONISTS

Opioids can be divided into three categories: natural (morphine, codeine, papaverine), synthetic (methadone, meperidine, fentanyl, alfentanil, sufentanil, remifentanil), and semisynthetic (hydromorphone) (Fukuda 2010). Because the primary clinical feature of opioids is analgesia, i.e., a relief of pain, these drugs are widely used to treat postoperative pain and to provide analgesia and to serve as an adjunct to maintain unconsciousness during general anesthesia (Fukuda 2010). Opioids are frequently combined with a benzodiazepine to provide analgesia and sedation for nonsurgical procedures (Jamieson 1999, Dionne et al. 2001). An awake patient receiving an opioid to treat pain reports analgesia and sedation (Shapiro et al. 1995, Bowdle 1998). Opioids have a number of side effects, including respiratory depression, bradycardia, miosis, constipation, nausea,

vomiting, euphoria, and dysphoria (Bowdle 1998).

The principal molecular targets of opioids are the μ, κ, and δ opioid receptors. These three types of G protein–coupled receptors are expressed in many brain regions, including the periaqueductal gray (PAG), the rostral ventral medulla (RVM), the basal ganglia (Burn et al. 1995), the amygdala, and the spinal cord (Stein 1995, Dowlatshahi & Yaksh 1997). Activation of the opioid receptor leads to hyperpolarization of the nerve cell membrane by inhibiting adenyl cyclase, decreasing conductance of voltage-gated calcium channels, and opening inward-rectifying potassium channels that allow potassium efflux (Fukuda 2010).

A nociceptive stimulus, such as a surgical incision, activates the free nerve endings of C-fibers and/or A-delta fibers, which make excitatory synapses onto projection neurons in the dorsal horn of the spinal cord (**Figure 3a**). The axons of the projection neurons cross the midline of the spinal cord and ascend in the anterolateral fasiculus to synapse in the RVM, the PAG, the thalamus, the amygdala, and the primary and secondary somatosensory cortices (Millan 2002). These are the principal components of the ascending nociceptive pathway. Nociceptive stimulation

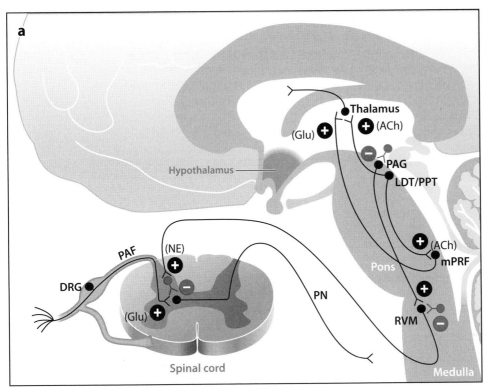

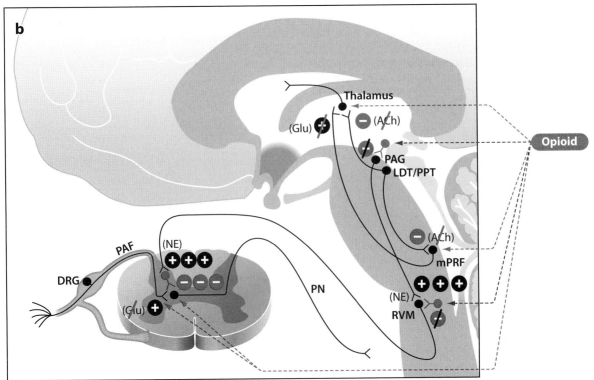

of the PAG and the RVM initiates descending pathways that modulate the nociceptive signaling through a combination of descending inhibition and descending facilitation (Millan 2002). These are the major components of the descending nociceptive pathways. These sites are also targets of the endogenous opioids, endorphins, and enkephalins (Millan 2002).

Because inhibition of nociceptive processing leads to decreased arousal, the several mechanisms through which the opioids achieve their analgesic and antinociceptive effects are the principal mechanisms through which these drugs alter arousal. Opioids binding to opioid receptors in the PAG and the RVM leads to activation of the descending nociceptive pathways, which produces an overall inhibitory effect on nociceptive transmission (Millan 2002) (**Figure 3b**). One mechanism for activating descending inhibition is through opioid receptor mediated hyperpolarization of GABAergic inhibitory interneurons in the PAG (Heinricher & Morgan 1999). These inhibitory neurons synapse onto excitatory neurons that project to the RVM. Excitatory stimulation of the RVM by the PAG leads to increased firing of RVM off-cells and decreased firing of RVM on-cells (Heinricher & Morgan 1999). This pattern of activation in the RVM is associated with descending inhibition of nociception at the level of peripheral afferent neurons and of the projection neurons in the spinal cord.

Opioids applied directly to the spinal cord produce analgesia by at least two mechanisms (**Figure 3b**). Binding of opioids to presynaptic opioid receptors on peripheral afferent neurons inhibits adenylate cyclase and suppresses voltage-dependent calcium channels and subsequent release of glutamate and neuropeptides such as substance P and calcitonin (Meunier et al. 1995). Postsynaptic activation of the opioid receptors on the projection neurons in the dorsal horn initiates an increase in potassium conductance, leading to hyperpolarization of the nerve cell membrane and, hence, to decreased firing. Opioids may also produce analgesia by binding to opioid receptors on primary sensory neurons (Stein 1995).

Opioids also alter arousal through their anticholinergic effects. For example, during normal waking, the cholinergic projections from the lateral dorsal tegmental nucleus (LDT) and the peduculopontine tegmental nucleus (PPT) activate the medial pontine reticular formation (mPRF) and thalamus (Lydic & Baghdoyan 2005) (**Figure 3a**). The mPRF in turn provides excitatory glutamatergic inputs to the thalamus, which sends excitatory inputs to the cortex. The basal forebrain provides another important source of excitatory cholinergic inputs to the cortex (McCarley 2007). Fentanyl decreases arousal by decreasing acetylcholine in the mPRF, whereas morphine decreases arousal by inhibiting the neurons in the LDT, the mPRF, and the basal forebrain (Mortazavi et al. 1999, Lydic & Baghdoyan 2005) (**Figure 3b**).

Actions of opioids in the cortex and the limbic system also lead to altered arousal. Functional imaging studies of pain processing in humans have shown that low-dose morphine induces euphoria associated with positive blood oxygen level dependent (BOLD) signal changes in the nucleus accumbens along with a pattern of negative BOLD signal changes in the cortex similar to that seen with GABA$_A$ agonists (Becerra et al. 2006).

Figure 3

Sites of opioid receptor effects during the awake state (*a*) and during opioid administration (*b*). (*a*) Normal cholinergic signaling during wakefulness and normal nociceptive signaling from the spinal cord to brain stem. (*b*) Illustrates opioid-induced decrease in arousal through a decrease in cholinergic signaling and the mechanisms of opioid-induced analgesia. Abbreviations: ACh, acetylcholine; DRG, dorsal root ganglia; Glu, glutamate; LDT, laterodorsal tegmental area; mPRF, medial pontine reticular formation; NE, norepinephrine; PAF, peripheral afferent fiber; PAG, periaquaductal gray; PN, projection neuron; PPT, pedunculopontine tegmental area; RVM, rostal ventral medulla.

The altered arousal state induced by opioids does not reliably produce complete unconsciousness even when high doses are administered (Bailey et al. 1985). This is evidenced by the high incidence of awareness during cardiac surgery for which high-dose opioids had been the standard anesthetic regimen for many years because of their combined effects of analgesia, decreased arousal, and cardiovascular stability (Ranta et al. 1996, Silbert & Myles 2009).

Respiratory depression, the most clinically significant side effect of opioids, is mediated through opioid binding to μ_2 receptors in the medulla (Weil et al. 1975). Reversal of respiratory depression by using an opioid antagonist also reverses analgesia as well because both work through the same mechanism (Greer & Ren 2009). Ampakines, AMPA receptor antagonists, have been shown in rats to counter fentanyl-induced respiratory depression without significantly altering analgesia and sedation (Greer & Ren 2009, Ren et al. 2009). These drugs may offer a way of maintaining adequate respiration while providing analgesia with opioids. Opioids acting at opioid receptors in the gut help explain the decreased motility and constipation seen with opioid use (Stewart et al. 1978, Greenwood-Van Meerveld et al. 2004). Nausea and emesis are due in part to binding of these drugs to opioid receptors in the chemotactic trigger zone (CMTZ) in the area postrema (Takahashi et al. 2007). Other side effects of opioids are likely to be mediated cholinergically. Bradycardia is likely due to GABA$_A$-mediated disinhibition, leading to activation of cholinergic projections from the nucleus ambiguus to the sino-atrial node of the heart (Griffoen et al. 2004). Miosis may result from the direct or indirect anticholinergic effects of the opioids on the Edinger-Westfall nuclei in the midbrain.

Table 2 summarizes the behavioral and physiological responses of the opioid agonists along with possible neural circuit mechanisms for these responses.

ANALGESIA, HALLUCINATIONS, AND DISSOCIATIVE ANESTHESIA: NMDA RECEPTOR ANTAGONISTS

Ketamine, an arylcyclohexylamine, is a congener of phencyclidines that anesthesiologists use as an analgesic and a hypnotic (Bergman 1999). It is also used as an adjunct to help maintain general anesthesia and as an alternative to opioids for the management of perioperative and neuropathic pain.

When a small dose of ketamine is administered, immediate, intense analgesia is the most apparent clinical feature (Kohrs & Durieux 1998). For this reason, low-dose ketamine is used commonly in anesthesiology to treat labor pain (Mercier & Benhamou 1998), to position a patient with lower extremity fracture for placement of a regional block, and to perform dressing changes in burn patients (Kohrs & Durieux 1998). The patient enters a dissociative state in which he/she perceives nociceptive stimuli but not as pain (Garfield et al. 1972, Kohrs & Durieux 1998, Bergman 1999). Auditory and visual hallucinations that resemble those seen in schizophrenia are common with ketamine (Seamans 2008). This is why ketamine is used as a pharmacological model of schizophrenia in animals (Olney et al. 1999, Kehrer et al. 2008) and why it is a widely used drug of abuse (Morgan et al. 2009). Respiratory function is generally preserved at sedative doses of ketamine. In contrast to the GABA$_A$ hypnotics, under ketamine, the EEG (Hayashi et al. 2007, Tsuda et al. 2007) is active, and both cerebral metabolic rates and blood flow increase (Cavazzuti et al. 1987, Strebel et al. 1995, Vollenweider et al. 1997) rather than decrease. Other key physiological effects of ketamine are horizontal nystagmus, pupillary dilation, salivation, lacrimation, tachycardia, and bronchodilation (Kohrs & Durieux 1998, Reves et al. 2009).

Ketamine's principal molecular target is the NMDA receptor, a major postsynaptic, ionotropic receptor for the excitatory

Table 2 Summary of the behaviors, physiological responses, and neural circuits for the actions of opioid agonists[a]

Drugs	Behavioral and physiological responses	Possible neural circuit mechanism	Receptors
Morphine, Fentanyl, Hydromorphone, Remifentanil	Analgesia, Antinociception	Activation of descending inhibitory pathways through disinhibition of GABAergic interneurons in the periacequeductal gray and rostal ventral medulla Inhibition of ascending pain pathways in the spinal cord pre- and postsynaptically in peripheral neurons and ascending projection neurons, respectively	Opioid
	Respiratory depression	μ_2-opioid mediated inhibition of brain stem respiratory control areas (ventral medulla)	
	Nausea and emesis	δ-opioid receptor mediated activity in chemotactic trigger zone in area postrema	
	Sedation	Reduced arousal due to antinociception	
	Bradycardia	Activation of projections from nucleus ambiguous to sino-atrial node in heart	
	Sedation	Anticholinergic activity in lateral dorsal tegmental nucleus, pedunculopontine tegmental nucleus, median pontine reticular formation, and basal forebrain	ACh
	Miosis	Anticholinergic activity possibly in Edinger-Westfall nuclei	
	Insomnia	Reduction in brain levels of adenosine, required to initiate sleep	Adenosine
	Catalepsy	Catalepsy induced by dopaminergic antagonist activity (see Dopamine Antagonist section below)	Dopamine

[a]Abbreviations: ACh, acetylcholine; GABA$_A$, gamma aminobutyric acid type A.

neurotransmitter glutamate (Sinner & Graf 2008). These pharmacologically defined receptors have an obligatory NR1 subunit and a modulatory NR2 subunit. Channel opening requires that an agonist (glutamate or NMDA) bind to the NR2 subunit while the coagonist glycine binds to the NR1 subunit (Purves et al. 2008). Experimental evidence suggests that ketamine blocks NMDA receptors by uncompetitive binding at a location other than the glutamate or glycine sites (Kim et al. 2002).

An understanding of how ketamine alters arousal is provided by strong experimental evidence that shows that ketamine binds preferentially to NMDA receptors on GABAergic inhibitory interneurons (Olney et al. 1999, Seamans 2008) (**Figure 4**). By selectively down-regulating GABAergic inhibition, ketamine disinhibits pyramidal neurons, leading to an altered arousal state consisting of abberant excitatory activity in multiple brain areas. Hallucinations and the dissociative state most likely result because multiple brain areas such as the cortex, the hippocampus and the limbic system are allowed to communicate through abberant activity that lacks normal spatial and temporal coordination. Benzodiazepines, GABA$_A$ agonists, are commonly administered with ketamine to help mitigate the hallucinations by possibly helping to restore some of the inhibitory activity in the affected brain regions. The increased pyramidal activity helps explain the increased EEG activity, the cerebral metabolic rate, and the cerebral blood flow seen under ketamine.

The analgesic effects of ketamine are due in part to its dissociative effects and to its blockade of NMDA receptors in the spinal cord, most notably on peripheral afferent neurons in the nociceptive pathways (**Figure 4b**) (Oye et al. 1992, Sinner & Graf 2008). Ketamine may also act at opioid receptors (Finck & Ngai 1982). Because certain types of neuropathic pain may be mediated in part by hyperactivity in NMDA circuits (Hocking & Cousins 2003), ketamine is widely used as a treatment for this disorder.

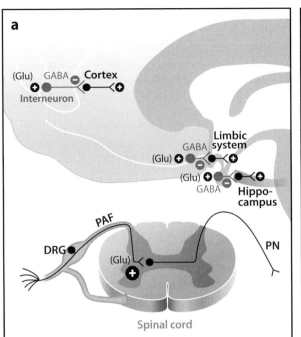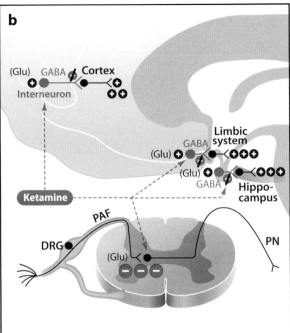

Figure 4

NMDA receptor signaling during the awake state (*a*) and during ketamine administration (*b*). (*a*) Normal glutamatergic regulation of GABAergic interneuron signaling during wakefulness and normal pain pathway signaling. (*b*) Ketamine blockage of GABAergic interneuron inhibition and ketamine-induced mechanism of analgesia. Abbreviations: DRG, dorsal root ganglia; GABA, gamma aminobutyric acid; Glu, glutamate; PAF, peripheral afferent fiber; PN, projection neuron.

Chronic regional pain syndrome (CRPS) is a type of neuropathic pain that may develop following trauma to an extremity that results in nerve damage (Kiefer et al. 2008). Although the initial injury may not be severe, with time, the pain associated with the injury becomes more intense and extends beyond the area that was originally affected. Sufferers from severe CRPS complain of constant burning sensations across large parts of their body.

An experimental therapy, the ketamine coma, is being used to treat severe CRPS (Kiefer et al. 2008, Becerra et al. 2009). The ketamine coma, which is not currently approved in the United States, is conducted by inducing and maintaining general anesthesia with a continuous intravenous infusion of ketamine, intubation, and mechanical ventilation for five to seven days. Following cessation of ketamine, some patients report dramatic reductions in

pain symptoms that can last from a few days to several months. Although the mechanism of the effect remains unclear, it is possible that the weeklong central nervous system blockade of the NMDA receptors helps break the wind-up and central sensitization cascade believed to underlie neuropathic pain syndromes (Costigan et al. 2009). Patients are unresponsive during the ketamine coma. However, unlike the coma period resulting from a brain injury, or more standard general anesthesia, about which patients typically have no recall, patients emerging from ketamine coma recall vivid and often disturbing hallucinations (Kiefer et al. 2008).

Recent reports that low-dose ketamine can provide immediate symptom relief in patients suffering from chronic bipolar disorders and chronic depression have stimulated significant interest in another way this drug alters arousal (Zarate et al. 2006). In contrast to standard

Table 3 Summary of the behaviors, physiological responses, and neural circuits for the actions of ketamine (NMDA receptor antagonist)[a]

Drugs	Behavioral and physiological responses	Possible neural circuit mechanisms	Receptors
Ketamine	Analgesia	Blockade of NMDA-mediated nociceptive stimuli in the dorsal horn of the spinal cord	NMDA
	Dissociative state hallucinations	Preferential binding to NMDA receptors on GABAergic inhibitory interneurons in cerebral cortex, hippocampus, and limbic structures, resulting in disinhibition and aberrant excitatory activity in these areas. Hallucinations are treated with benzodiazepines, which is consistent with a GABA-mediated mechanism	
	Antidepressant effect	Increased neurogenesis and synaptogenesis Change in the NMDA to AMPA receptor activity ratio NMDA-mediated cortical activation by inhibition of GABAergic inhibitory interneurons (chemical ECT induced by disinhibition)	
	Lacrimation and salivation	Increased parasympathetic stimulation of inferior and superior salivatory nuclei due to NMDA-mediated inhibition of GABAergic interneurons (disinhibition)	
	Pupillary dilation, tachycardia, bronchodilation	Increased sympathetic output from nucleus tractus solitarius due to NMDA-mediated inhibition of GABAergic interneurons (disinhibition)	
	Nystagmus	Increased activity in cortical areas generating saccades (disinhibition via NMDA inhibition of GABA interneurons) combined with decreased activity in cerebellum and brain stem reticular formation and medial vestibular nuclei (direct inhibition of NMDA neurons)	

[a]Abbreviations: AMPA, 2-amino-3-(5-methyl-3-oxo-1,2-oxazol-4-yl)propanoic acid; ECT, electroconvulsive therapy; GABA$_A$, gamma aminobutyric acid type A; NMDA: N-methyl D-aspartate.

therapies for these disorders, which may require weeks to months to produce an effect, ketamine's effects can begin within 40 minutes and last for up to 7 days. This effect may be due to a change in the AMPA to NMDA receptor activity ratio (Maeng et al. 2008), to ketamine-induced synaptogenesis (Li et al. 2010), or to the possibility that the excitatory state induced by ketamine, through its actions on inhibitory interneurons, may be providing chemically what electroconvulsive therapy (ECT) provides electrically.

Ketamine-induced nystagmus appears to be the consequence of the drug's differential effects in the frontal cortex, the cerebellum, and the vestibular nuclei (Porro et al. 1999). Lacrimation and salivation seen with ketamine possibly reflect increased parasympathetically mediated activity in the inferior and superior salivatory nuclei (Purves et al. 2008),

whereas pupillary dilation, tachycardia, and bronchodilation most likely reflect increased sympathetic output from the nucleus tractus solitarius (Ogawa et al. 1993, Freeman 2008). These symptoms, we speculate, may be due to NMDA-mediated disinhibition of GABAergic interneurons in these areas that leads to coactivation of the parasympathetic and sympathetic systems.

Table 3 summarizes the behavioral and physiological responses of the ketamine along with possible neural circuit mechanisms for these responses.

PHARMACOLOGICAL NON-REM SLEEP: ALPHA ADRENERGIC RECEPTOR AGONISTS

Dexmedetomidine and clonidine belong to the imidazole subclass of α_2 adrenergic receptor agonists (Coursin & Maccioli 2001). Dexmedetomidine is a sedative whose indicated uses are for short-term sedation (<24 h) for mechanically ventilated patients in the intensive care unit (Gerlach & Dasta 2007) and for sedation of nonintubated patients for surgical and medical procedures (Hospira 2009). It is now being used off-label as an adjunct for general anesthesia (Carollo et al. 2008, Chrysostomou & Schmitt 2008). Sedation with dexmedetomidine is noticeably different from sedation with $GABA_A$ agonists because patients are readily arousable and have little to no respiratory depression (Hsu et al. 2004).

Basic science studies have demonstrated that a primary target at which dexmedetomidine acts to alter arousal is the α_2 receptors on neurons emanating from the locus coeruleus (LC) (Correa-Sales et al. 1992, Chiu et al. 1995, Mizobe et al. 1996). Binding of dexmedetomidine to this G protein–coupled receptor initiates inwardly rectifying potassium currents that allow potassium efflux and inhibition of voltage-sensitive calcium channels. The net hyperpolarization of the LC neurons leads to a decrease in norepinephrine release from these neurons (Jorm & Stamford 1993, Nacif-Coelho et al. 1994, Nelson et al. 2003).

The behavioral effects of dexmedetomidine are consistent with its proposed mechanisms of action in the LC. Basic science studies suggest that, during the wake state, the LC provides norepinephrine-mediated inhibition to the POA in the hypothalamus (**Figure 5a**) (Osaka & Matsumura 1994, Saper et al. 2005, Lu et al. 2006). The LC also provides key adrenergically mediated excitatory inputs to the basal forebrain, intralaminar nucleus of the thalamus, and the cortex (España & Berridge 2006). POA neurons releasing GABA and the inhibitory neurotransmitter galanin inhibit the ascending arousal centers in the midbrain, upper pons, and hypothalamus (Sherin et al. 1998, Saper et al. 2005). Thus inhibition of POA neurons promotes wakefulness. At the onset of sleep, the LC is inhibited, and its noradrenergically mediated inhibition of the POA ceases. As a consequence, the POA becomes active and its GABAergic and galaninergic neurons inhibit the ascending arousal centers to provide a possible mechanism for initiating non-REM sleep.

It is plausible that decreased norepinephrine release from the LC induced by dexmedetomidine disinhibits the POA (Nelson et al. 2002, 2003). Therefore, the disinhibited POA inhibits the ascending arousal pathways, and sedation ensues (**Figure 5b**). Blocking norepinephrine release from the LC would also contribute to sedation due to decreased excitatory inputs to the basal forebrain, to the intralaminar nucleus of the thalamus, and to the cortex (España & Berridge 2006). A clinical study has demonstrated a close resemblance between EEG spindles observed in dexmedetomidine-induced sedation and those observed during stage 2 non-REM sleep (Huuponen et al. 2008). Subjects were easily arousable from dexmedetomidine-induced sedation and from non-REM sleep. These basic science and clinical studies help us understand why the altered arousal state of dexmedetomidine-induced sedation closely resembles non-REM sleep.

Clonidine, injected intrathecally, is commonly used as an adjunct in postoperative pain

---

Figure 5

Alpha adrenergic signaling during the awake state (*a*) and during dexmedetomidine or clonidine administration (*b*). (*a*) Normal adrenergic signaling from the LC during wakefulness and normal pain signaling during wakefulness from the spinal cord to the brain stem. (*b*) Dexmedetomidine-induced loss of consciousness through NE mediated inhibition of the POA and decreased noradrenergic signaling in the thalamus and cortex. Clonidine-induced analgesia through enhanced inhibitory activity in the descending pain pathway. Abbreviations: 5HT, serotonin; ACh, acetylcholine; DA, dopamine; DRG, dorsal root ganglia; $GABA_A$, gamma aminobutyric acid receptor subtype A; Gal, galanin; His, histamine; ILN, intralaminar nucleus of the thalamus; LC, locus coeruleus; LDT, laterodorsal tegmental area; NE, norepinephrine; PAF, peripheral afferent fiber; PN, projection neuron; POA, preoptic area; PPT, pedunculopontine tegmental area; RVM, rostral ventral medulla; TMN, tuberomamillary nucleus; vPAG, ventral periaquaductal gray.

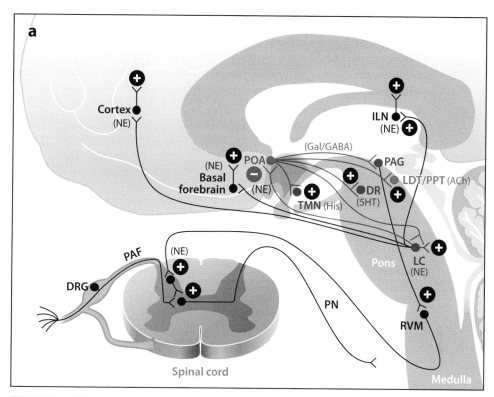

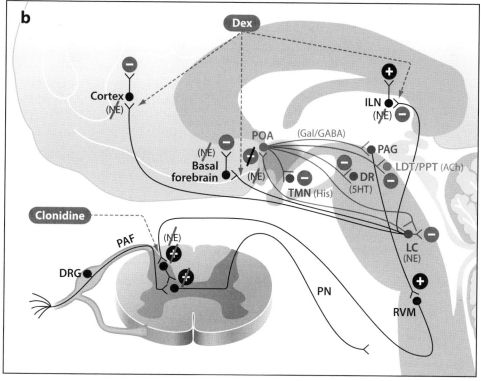

Table 4 Summary of the behaviors, physiological responses, and neural circuits for the actions of alpha receptor agonists[a]

Drugs	Behavioral and physiological responses	Possible neural circuit mechanisms	Receptors
Dexmedetomidine	Sedation mimicking non-REM sleep	Hyperpolarization of LC reducing inhibitory output to POA, allowing POA to inhibit multiple ascending arousal systems (PPT/LDT, vPAG, TMN, Raphe, LC) by GABA and galanin projections. Reduction of NE-mediated activation of the cortex, the ILN of the thalamus, and the BF	α_2 adrenergic
Clonidine	Analgesia	Inhibition of descending pain facilitation pathway achieved by blocking presynaptic NE release in excitatory interneuron in the spinal cord	

[a]Abbreviations: BF, basal forebrain; GABA$_A$, gamma aminobutyric acid type A; ILN, intralaminar nucleus; LC, locus coeruleus; LDT, laterodorsal tegmental area; NE, norepinephrine; PPT, pedunculopontine tegmental area; TMN, tuberomamillary nucleus; POA preoptic area; vPAG, ventral periaquaductal gray.

therapy (Andrieu et al. 2009). Clonidine acts at α_2 receptors in the descending nociceptive pathways in the spinal cord to enhance antinociception (**Figure 5b**). The α_2 receptor agonist xylazine is frequently administered with ketamine to provide general anesthesia for animal surgeries (Rodrigues et al. 2006).

Table 4 summarizes the behavioral and physiological responses of the α_2 agonists along with possible neural circuit mechanisms for these responses.

NEUROLEPTIC ANESTHESIA: DOPAMINE RECEPTOR ANTAGONISTS

The butyrophenones, haloperidol and droperidol, are dopamine antagonists that have been used in anesthesiology (Hardman 2001) as sedatives, anesthetic adjuncts, and antiemetics. The altered arousal state induced by these drugs is insufficient for their use as sole anesthetic agents. Butyrophenones have been used in combination with opioids—haloperidol with phenoperidine and droperidol with fentanyl—to create a state of neuroleptic anesthesia characterized by unresponsiveness with eyes open, analgesia, and decreased mobility (catalepsy) with maintenance of ventilation and cardiovascular stability (Corssen et al. 1964). Neuroleptic anesthesia is no longer used because not infre-

quently patients complained of feeling locked in: being aware of what transpired during the surgery with substantial pinned-up emotion that they could not express (Klausen et al. 1983, Linnemann et al. 1993, Klafta et al. 1995).

The butyrophenones, used almost exclusively now as antiemetics (Apfel et al. 2009), can cause Parkisonian symptoms (bradykinesia, tremor, muscle rigidity, blunted affect) at both low (Rivera et al. 1975, Melnick 1988) and high (Lieberman et al. 2008) doses. In the United States, droperidol carries a controversial black label warning from the US FDA because of rare reports of cardiac dysrhythmias following its use in high doses (FDA 2001, Dershwitz 2003). Use of droperidol at any dose requires perioperative electrocardiogram monitoring.

Basic science studies have shown that the butyrophenones are dopamine receptor antagonists. The five types of dopamine receptors are all 7-transmembrane, G protein–coupled receptors that are divided into two classes: D_1 and D_2 (Girault & Greengard 2004). The D_1 receptors mediate excitation by activating G-proteins that stimulate cyclic AMP (cAMP) synthesis. In contrast, the D_2 receptors mediate inhibition by activation of G-proteins that inhibit cAMP synthesis, suppress Ca^{2+} currents, and activate receptor-operated potassium currents.

The brain has three principal dopaminergic pathways (**Figure 6a**). The nigrostrial pathway,

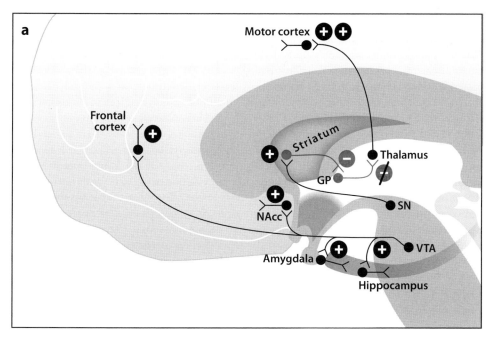

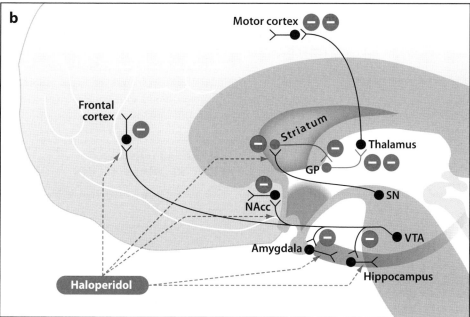

Figure 6

Dopamine signaling during the awake state (*a*) and during haloperidol administration (*b*). (*a*) Normal dopamine signaling during wakefulness. (*b*) Haloperidol effects on dopamine signaling. Abbreviations: GP, globus pallidus; NAcc, nucleus accumbens; SN, substantia nigra; VTA, ventral tegmental area.

which projects from the pars compacta of the substantia nigra to the striatum, is a component of the basal ganglia that is crucial for movement control (Graybiel 1991, Graybiel et al. 1994). Patients suffering from Parkinson's disease have slow movements, a resting tremor, rigidity in all extremities, and minimal facial expressions due to a lack of dopamine production in the substantia nigra (Obeso et al. 2008). The mesolimbic pathway that projects from the ventral tegmentum to the nucleus accumbens, amygdala, and hippocampus plays a key role in processing reward, motivation, emotion, and reinforcement. The mesocortical pathway, which projects from the ventral tegmentum to the frontal cortex, complements the function of the mesolimbic pathway and aids in cognition (Obeso et al. 2008).

The cataleptic state induced by the butyrophenones can be attributed to their binding to D_1 and D_2 receptors in the striatum, leading to a net increase in inhibitory output from the globus pallidus to the thalamus and, as a consequence, to reduced excitation of the motor cortex (Lieberman et al. 2008) (**Figure 6b**). Although, the mechanism may not be as simple as stated, these results are consistent with the direct- and indirect-pathway models of Parkinson's disease. The muted affect and emotion observed in patients receiving a butyrophenone may be due to loss of facial muscle control induced by the drugs' dopamine antagonistic actions in the nigrostriatal pathways (Tickle-Degnen & Lyons 2004, Lieberman et al. 2008). Sedation most likely results from the action of the drugs in the mesolimbic and mesocortical pathways (**Figure 6b**). The antiemetic effects of the butyrophenones are most likely multifactorial: blocking dopamine receptors in the CMTZ of the area postrema (Apfel 2009), blocking histaminergic inputs to the nucleus of the tractus solitarius (NTS), blocking serotonergic inputs to the NTS and CMTZ, and blocking cholinergic effects in the dorsal motor nucleus of the vagus (Peroutka & Synder 1980, 1982; Scuderi 2003).

Although not broadly appreciated in anesthesiology, the opioids are perhaps the most widely used dopaminergic antagonist in clinical practice. Basic science studies have shown that opioid binding to μ and κ opioid receptors in the striatum and substantia nigra inhibit dopamine release (Havemann et al. 1982, Burn et al. 1995). This decrease in striatal dopamine levels can contribute to a state of decreased mobility similar to that seen in Parkinson's disease. This observation offers insight into why awareness has been more likely under high-dose fentanyl anesthesia. Fentanyl binds to opioid receptors in the rostral ventral medulla to provide both analgesia (Yaksh 1997) and activation of parasympathetic outflow (Griffioen et al. 2004). In high dose, its antidopaminergic effects are more likely to be present also. This combination of analgesia, catalepsy, parasympathetic activation (sympathetic quiescence), and muted stress response is likely to be part of the drug-induced locked-in state about which patients who received Innovar complained (Klafta et al. 1995). Hence, if high-dose fentanyl is used with few or no additional anesthetic agents that have cortical effects, a patient can be comfortable, immobile, show little to no stress response, yet remain aware.

Table 5 summarizes the behavioral and physiological responses of the dopaminergic antagonist along with possible neural circuit mechanisms for these responses.

IMPLICATIONS AND FUTURE DIRECTIONS

Five altered states of arousal induced by intravenous anesthetic drugs can be understood by analyzing the behavioral and physiological effects of the drugs in relation to the molecular targets in specific neural circuits at which they are believed to act. In each case, we can suggest how the altered arousal state is created by the drug's actions at multiple sites in the central nervous system. Hence, these states are neither mysterious nor indecipherable. This systems neuroscience paradigm suggests a framework for improving research, practice, and education in anesthesiology and for relating general anesthesia to other altered states of arousal.

Table 5 Summary of the behaviors, physiological responses, neural circuits, and receptors for the actions of dopamine antagonists

Drugs	Behavioral and physiological responses	Possible neural circuit mechanisms	Receptors
Haloperidol, Droperidol	Antinausea, antiemesis	Blockade of D_2 receptors in chemotactic trigger zone in area postrema	Dopamine
	Catalepsy, tremor, rigidity	Blockade of D_1 and D_2 receptors in striatum leading to a net increase in inhibition from globus pallidus to thalamus, leading to reduced activity in motor cortex	
	Loss of affect and emotional expression	Loss of facial muscle control due to antidopaminergic activity in nigrostriatal pathway and decreased affect from actions in mesolimbic and mesocortical pathways	
	Sedation	Combined inhibition of arousal, mesolimbic and mesocortical dopaminergic pathways	
	Antinausea, antiemesis	Antihistaminergic activity in nucleus tractus solitarius	Histamine
	Antinausea, antiemesis	Antiserotinergic activity in nucleus tractus solitarius and chemotactic trigger zone	Serotonin

Anesthesiology: Research, Practice, and Education

Further clinical and basic systems neuroscience studies of general anesthesia's altered arousal states that use the latest functional neuroimaging, neurophysiological, behavioral and molecular techniques will have important implications for anesthesiology and for neuroscience. A clearer understanding of the molecular targets and the neural circuits is needed to design novel site-specific anesthetic drugs that produce desired behavioral and physiological changes in a controlled manner while obviating side effects.

For example, postoperative nausea and vomiting remain two of the most vexing side effects following general anesthesia. Postoperative cognitive dysfunction is now recognized as a major concern following surgery with general anesthesia for at least 30% of patients (Monk et al. 2008). For the elderly, the incidence can be higher than 40% (Monk et al. 2008, Price et al. 2008). Although the mechanisms are multifactorial and some are not related to the anesthetic drugs, evidence is accruing to suggest how anesthetics contribute to postoperative cognitive dysfunction.

Opioids are strongly associated with postoperative delirium, in part because of their anticholinergic effects (Marcantonio et al. 1994), which may be worst in the elderly (Hshieh et al. 2008, Campbell et al. 2009). Similarly, prolonged exposure of intensive care unit patients to benzodiazepines is associated with delirium, prolonged intensive care unit and hospital stays, and increased mortality (Pandharipande et al. 2007). These findings suggest that prolonged exposure to the nonphysiological inhibitory state created by $GABA_A$ agonists in the central nervous system may be deleterious. In contrast, the recently demonstrated benefits of dexmedetomidine as an intensive care unit sedative compared with lorazepam can be attributed, in part, to the fact that the sedative actions of the former are more site specific (Pandharipande et al. 2007).

How certain anesthetics may contribute to the high prevalence of postoperative delirium in children (Vlajkovic & Sindjelic 2007) and the potentially adverse effect of repeated anesthetic exposures in children on central nervous system development are becoming increasingly important concerns of the United States Food and Drug Administration and the International Anesthesia Research Society (Int. Anesth. Res. Soc. 2010). Ketamine and other NMDA antagonists contribute to apoptosis in the newborn brains of animals, leading to recent recommendations to reconsider the use of these drugs in neonates (Anand 2007, Mellon et al. 2007). Similarly, benzodiazepines are frequently given as sedatives to children who require multiple reconstructive surgeries following burn injuries (Stoddard et al. 2002), yet the same drugs are used in experimental models of nervous system development to probe the role of GABA in excitatory-inhibitory balance (Hensch 2004).

Successful use of systems neuroscience concepts in the daily practice of anesthesiology will require more in-depth training of anesthesiologists in neuroanatomy, neurophysiology, and neuropharmacology to bridge the current gap between clinical management of general anesthesia and the understanding of the neurophysiological and molecular mechanisms that underlie those management decisions. This improved understanding will allow more astute interpretations of findings from clinical examinations of patients under general anesthesia, which when coupled with results from ongoing systems neuroscience studies, should lead to improved, neurophysiologically based approaches to producing general anesthesia and to monitoring the states of the brain under general anesthesia. In this way, anesthesiologists can avoid even rare though potentially traumatic events, such as awareness under general anesthesia (Avidan et al. 2008, Errando et al. 2008).

General Anesthesia and Other Altered Arousal States

We began our systems neuroscience analysis by defining general anesthesia as changes in behavioral states with maintenance of homeostasis because there is not a universally accepted definition of general anesthesia. This is due in part to the difficulty anesthesiologists have of being able to state accurately their well-formed empirical understanding of this condition. For example, confusion arises because anesthesiologists use the term sleep as a nonthreatening description of general anesthesia when speaking with patients. A level of general anesthesia appropriate for surgery is not sleep but rather a

coma (Brown et al. 2010). However, like sleep, general anesthesia is reversible and can allow dreaming (Leslie & Skrzypek 2007, Errando et al. 2008, Samuelsson et al. 2008). The concept of coma is less comforting and harder for patients to understand than the notion of sleep. Despite several descriptions of the differences between sleep and general anesthesia (Lydic & Baghdoyan 2005, Brown et al. 2010), uses of the term sleep to describe the altered arousal states of general anesthesia appear in anesthesiology (Reves et al. 2009), medical (Gawande et al. 2008), and legal (Dershwitz & Henthorn 2008) articles.

Using the systems neuroscience framework to define altered arousal states in terms of behavioral and physiological responses, molecular targets and neural circuits can facilitate cross-disciplinary communication and research on how altered arousal states induced by anesthetic agents relate to the fundamental questions of consciousness (Crick & Koch 2003, Mashour 2006) and how they compare with other altered arousal states such as sleep (Lydic & Baghdoyan 2005, McCarley 2007, Alkire et al. 2008, Franks 2008), sleep aided by pharmacologic agents (NIH Consens. Dev. Conf. 1984), the stages of coma (Giacino et al. 2002), schizophrenia (Olney et al. 1999), seizures (Blumenfeld & Taylor 2003), locked-in states (Posner et al. 2007), drug-induced high or paradoxical excitation (Brown et al. 2010), meditation (Lazar et al. 2005), hypnosis (Vanhaudenhuyse et al. 2009), hibernation (Revel et al. 2007), and suspended animation (Blackstone et al. 2005). Many of these altered arousal states have analogs in pharmacologic states induced by anesthetic drugs.

SUMMARY

General anesthesia is a nonphysiological, drug-controlled condition created so that surgical and medical therapies can be provided safely and humanely. The fundamental question for anesthesiology research is how can this state be created by making physiologically sound, reversible manipulations of the neural circuits in the central nervous system? The systems neuroscience paradigm we used to analyze general anesthesia–induced states of altered arousal offers an integrated framework for formulating and answering this question.

DISCLOSURE STATEMENT

The authors are not aware of any affiliations, memberships, funding, or financial holdings that might be perceived as affecting the objectivity of this review.

ACKNOWLEDGMENTS

Research was supported by the Massachusetts General Hospital Department of Anesthesia, Critical Care and Pain Medicine and by the National Institutes of Health (NIH) Director's Pioneer Award (DP1OD003646 to E.N.B), the NIH New Innovator Award (DP2OD006454 to P.L.P.), the NIH K25 (NS057580 to P.L.P.), and the NIH-sponsored Training Program in Sleep, Circadian and Respiratory Neurobiology (HL07901 to C.J.V.). **Figures 1b, 3b, 4b,** and **5b** were redrawn from Brown et al. (2010). We thank Kirsten Fraser and Helena Yardley for research assistance in preparing the manuscript.

LITERATURE CITED

Ali HH, Utting JE, Gray TC. 1970. Stimulus frequency in the detection of neuromuscular block in humans. *Br. J. Anaesth.* 42(11):967–78

Ali HH, Utting JE, Gray TC. 1971. Quantitative assessment of residual antidepolarizing block. II. *Br. J. Anaesth.* 43(5):478–85

Alkire MT, Haier RJ, Barker SJ, Shah NK, Wu JC, Kao YJ. 1995. Cerebral metabolism during propofol anesthesia in humans studied with positron emission tomography. *Anesthesiology* 82(2):393–403; discussion 27A

Alkire MT, Hudetz AG, Tononi G. 2008. Consciousness and anesthesia. *Science* 322(5903):876–80

Anand KJ. 2007. Anesthetic neurotoxicity in newborns: Should we change clinical practice? *Anesthesiology* 107(1):2–4

Andrieu G, Roth B, Ousmane L, Castaner M, Petillot P, et al. 2009. The efficacy of intrathecal morphine with or without clonidine for postoperative analgesia after radical prostatectomy. *Anesth. Analg.* 108(6):1954–57

Angel A. 1991. The G. L. Brown lecture. Adventures in anaesthesia. *Exp. Physiol.* 76(1):1–38

Angel A. 1993. Central neuronal pathways and the process of anaesthesia. *Br. J. Anaesth.* 71(1):148–63

Apfel CC. 2010. Post-operative nausea and vomiting. See Miller et al. 2010, pp. 2729–56

Apfel CC, Cakmakkaya OS, Frings G, Kranke P, Malhotra A, et al. 2009. Droperidol has comparable clinical efficacy against both nausea and vomiting. *Br. J. Anaesth.* 103(3):359–63

Avidan MS, Zhang L, Burnside BA, Finkel KJ, Searleman AC, et al. 2008. Anesthesia awareness and the bispectral index. *N. Engl. J. Med.* 358(11):1097–108

Avramov MN, Husain MM, White PF. 1995. The comparative effects of methohexital, propofol, and etomidate for electroconvulsive therapy. *Anesth. Analg.* 81(3):596–602

Bai D, Pennefather PS, MacDonald JF, Orser BA. 1999. The general anesthetic propofol slows deactivation and desensitization of GABA(A) receptors. *J. Neurosci.* 19(24):10635–46

Bailey PL, Wilbrink J, Zwanikken P, Pace NL, Stanley TH. 1985. Anesthetic induction with fentanyl. *Anesth. Analg.* 64(1):48–53

Becerra L, Harter K, Gonzalez RG, Borsook D. 2006. Functional magnetic resonance imaging measures of the effects of morphine on central nervous system circuitry in opioid-naive healthy volunteers. *Anesth. Analg.* 103(1):208–16

Becerra L, Schwartzman RJ, Kiefer RT, Rohr P, Moulton EA, et al. 2009. CNS measures of pain responses pre- and post-anesthetic ketamine in a patient with complex regional pain syndrome. *Pain Med.* doi: 10.1111/j.1526-4637.2009.00559.x

Benthuysen JL, Smith NT, Sanford TJ, Head N, Dec-Silver H. 1986. Physiology of alfentanil-induced rigidity. *Anesthesiology* 64(4):440–46

Bergman SA. 1999. Ketamine: review of its pharmacology and its use in pediatric anesthesia. *Anesth. Prog.* 46(1):10–20

Bieda MC, Su H, Maciver MB. 2009. Anesthetics discriminate between tonic and phasic gamma-aminobutyric acid receptors on hippocampal CA1 neurons. *Anesth. Analg.* 108(2):484–90

Bigelow H. 1846. Insensibility during surgical operations produced by inhalation. *Boston Med. Surg. J.* XXXV(16):309–17

Blackstone E, Morrison M, Roth MB. 2005. H2S induces a suspended animation-like state in mice. *Science* 308:518

Blumenfeld H, Taylor J. 2003. Why do seizures cause loss of consciousness? *Neuroscientist* 9(5):301–10

Boland F. 2009. *The First Anesthetic: The Story of Crawford Long.* Athens: Univ. Ga. Press

Bonhomme V, Fiset P, Meuret P, Backman S, Plourde G, et al. 2001. Propofol anesthesia and cerebral blood flow changes elicited by vibrotactile stimulation: a positron emission tomography study. *J. Neurophysiol.* 85(3):1299–308

Bowdle TA. 1998. Adverse effects of opioid agonists and agonist-antagonists in anaesthesia. *Drug Saf.* 19(3):173–89

Bowery NG, Hudson AL, Price GW. 1987. GABAA and GABAB receptor site distribution in the rat central nervous system. *Neuroscience* 20(2):365–83

Breshears JD, Roland JL, Sharma M, Gaona CM, Freudenburg ZV, et al. 2010. Stable and dynamic cortical electrophysiology of induction and emergence with propofol anesthesia. *Proc. Natl. Acad. Sci. USA* 107:21170–75

Brown EN. 2007. *A systems neuroscience approach for the study of general anesthesia.* NIH Director's Pioneer Award Symp., Bethesda, MD

Brown EN, Lydic R, Schiff ND. 2010. General anesthesia, sleep and coma. *N. Engl. J. Med.* 363(27):2638–50

Burn DJ, Rinne JO, Quinn NP, Lees AJ, Marsden CD, Brooks DJ. 1995. Striatal opioid receptor binding in Parkinson's disease, striatonigral degeneration and Steele-Richardson-Olszewski syndrome, a [11C]diprenorphine PET study. *Brain* 118(Pt. 4):951–58

Campbell N, Boustani M, Limbil T, Ott C, Fox C, et al. 2009. The cognitive impact of anticholinergics: a clinical review. *Clin. Interv. Aging* 4(1):225–33

Carollo DS, Nossaman BD, Ramadhyani U. 2008. Dexmedetomidine: a review of clinical applications. *Curr. Opin. Anaesthesiol.* 21(4):457–61

Cavazzuti M, Porro CA, Biral GP, Benassi C, Barbieri GC. 1987. Ketamine effects on local cerebral blood flow and metabolism in the rat. *J. Cereb. Blood Flow Metab.* 7(6):806–11

Chiu TH, Chen MJ, Yang YR, Yang JJ, Tang FI. 1995. Action of dexmedetomidine on rat locus coeruleus neurones: intracellular recording in vitro. *Eur. J. Pharmacol.* 285(3):261–68

Chou TC, Bjorkum AA, Gaus SE, Lu J, Scammell TE, Saper CB. 2002. Afferents to the ventrolateral preoptic nucleus. *J. Neurosci.* 22(3):977–90

Chrysostomou C, Schmitt CG. 2008. Dexmedetomidine: sedation, analgesia and beyond. *Expert Opin. Drug Metab. Toxicol.* 4(5):619–27

Correa-Sales C, Rabin BC, Rabin BC, Maze M. 1992. A hypnotic response to dexmedetomidine, an alpha 2 agonist, is mediated in the locus coeruleus in rats. *Anesthesiology* 76(6):948–52

Corssen G, Domino EF, Sweet RB. 1964. Neuroleptanalgesia and anesthesia. *Anesth. Analg.* 43:748–63

Costigan M, Scholz J, Woolf CJ. 2009. Neuropathic pain: a maladaptive response of the nervous system to damage. *Annu. Rev. Neurosci.* 32:1–32

Coursin DB, Maccioli GA. 2001. Dexmedetomidine. *Curr. Opin. Crit. Care* 7(4):221–26

Crick F, Koch C. 2003. A framework for consciousness. *Nat. Neurosci.* 6(2):119–26

Dershwitz M. 2003. There should be a threshold dose for the FDA black-box warning on droperidol. *Anesth. Analg.* 97(5):1542–43; author reply p. 1543

Dershwitz M, Henthorn TK. 2008. The pharmacokinetics and pharmacodynamics of thiopental as used in lethal injection. *Fordham. Urban Law J.* 35:931–56

Devor M, Zalkind V. 2001. Reversible analgesia, atonia, and loss of consciousness on bilateral intracerebral microinjection of pentobarbital. *Pain* 94(1):101–12

Dionne RA, Yagiela JA, Moore PA, Gonty A, Zuniga J, Beirne OR. 2001. Comparing efficacy and safety of four intravenous sedation regimens in dental outpatients. *J. Am. Dent. Assoc.* 132(6):740–51

Dowlatshahi P, Yaksh TL. 1997. Differential effects of two intraventricularly injected alpha 2 agonists, ST-91 and dexmedetomidine, on electroencephalogram, feeding, and electromyogram. *Anesth. Analg.* 84(1):133–38

Durrani Z, Winnie AP. 1991. Brainstem toxicity with reversible locked-in syndrome after intrascalene brachial plexus block. *Anesth. Analg.* 72(2):249–52

Dutton RP, Eckhardt WF 3rd, Sunder N. 1994. Total spinal anesthesia after interscalene blockade of the brachial plexus. *Anesthesiology* 80(4):939–41

Eger EI. 1993. What is general anesthetic action? *Anesth. Analg.* 77(2):408–9

Eichhorn JH, Cooper JB, Cullen DJ, Maier WR, Philip JH, Seeman RG. 1986. Standards for patient monitoring during anesthesia at Harvard Medical School. *JAMA* 256(8):1017–20

Errando CL, Sigl JC, Robles M, Calabuig E, Garcia J, et al. 2008. Awareness with recall during general anaesthesia: a prospective observational evaluation of 4001 patients. *Br. J. Anaesth.* 101(2):178–85

España RA, Berridge CW. 2006. Organization of noradrenergic efferents to arousal-related basal forebrain structures. *J. Comp. Neurol.* 496(5):668–83

FDA. 2001. *Inapsine (droperidol) Dear Healthcare Professional Letter Dec 2001.* **http://www.fda.gov/Safety/MedWatch/SafetyInformation/SafetyAlertsforHumanMedicalProducts/ucm173778.htm**

Feldman JL, Mitchell GS, Nattie EE. 2003. Breathing: rhythmicity, plasticity, chemosensitivity. *Annu. Rev. Neurosci.* 26:239–66

Fenster J. 2001. *Ether Day: The Strange Tale of America's Greatest Medical Discovery and the Haunted Men Who Made It.* New York: Harper Collins

Ferrarelli F, Massimini M, Sarasso S, Casali A, Riedner BA, et al. 2010. Breakdown in cortical effective connectivity during midazolam-induced loss of consciousness. *Proc. Natl. Acad. Sci. USA* 107(6):2681–86

Finck AD, Ngai SH. 1982. Opiate receptor mediation of ketamine analgesia. *Anesthesiology* 56(4):291–97

Franks NP. 2008. General anaesthesia: from molecular targets to neuronal pathways of sleep and arousal. *Nat. Rev. Neurosci.* 9(5):370–86

Freeman R. 2008. Clinical practice. Neurogenic orthostatic hypotension. *N. Engl. J. Med.* 358(6):615–24

Fukuda K. 2010. Opioids. See Miller et al. 2010, pp. 769–824

Fulton SA, Mullen KD. 2000. Completion of upper endoscopic procedures despite paradoxical reaction to midazolam: a role for flumazenil? *Am. J. Gastroenterol.* 95(3):809–11

Garfield JM, Garfield FB, Stone JG, Hopkins D, Johns LA. 1972. A comparison of psychologic responses to ketamine and thiopental–nitrous oxide–halothane anesthesia. *Anesthesiology* 36(4):329–38

Gawande A, Denno DW, Truog RD, Waisel DM. 2008. Physicians and execution-highlights from a dicussion of lethal injection. *N. Engl. J. Med.* 358(5):448–51

Gerlach AT, Dasta JF. 2007. Dexmedetomidine: an updated review. *Ann. Pharmacother.* 41(2):245–52

Giacino JT, Ashwal S, Childs N, Cranford R, Jennett B, et al. 2002. The minimally conscious state: definition and diagnostic criteria. *Neurology* 58(3):349–53

Girault JA, Greengard P. 2004. The neurobiology of dopamine signaling. *Arch. Neurol.* 61(5):641–44

Glass PS, Shafer SL, Reves JG. 2010. Intravenous drug delivery systems. See Miller et al. 2010, pp. 825–58

Grasshoff C, Drexler B, Rudolph U, Antkowiak B. 2006. Anaesthetic drugs: linking molecular actions to clinical effects. *Curr. Pharm. Des.* 12(28):3665–79

Graybiel AM. 1991. Basal ganglia—input, neural activity, and relation to the cortex. *Curr. Opin. Neurobiol.* 1(4):644–51

Graybiel AM, Aosaki T, Flaherty AW, Kimura M. 1994. The basal ganglia and adaptive motor control. *Science* 265(5180):1826–31

Gredell JA, Turnquist PA, Maciver MB, Pearce RA. 2004. Determination of diffusion and partition coefficients of propofol in rat brain tissue: implications for studies of drug action in vitro. *Br. J. Anaesth.* 93(6):810–17

Greenwood-Van Meerveld B, Gardner CJ, Little PJ, Hicks GA, Dehaven-Hudkins DL. 2004. Preclinical studies of opioids and opioid antagonists on gastrointestinal function. *Neurogastroenterol. Motil.* 16(Suppl. 2):46–53

Greer JJ, Ren J. 2009. Ampakine therapy to counter fentanyl-induced respiratory depression. *Respir. Physiol. Neurobiol.* 168(1–2):153–57

Griffioen KJ, Venkatesan P, Huang ZG, Wang X, Bouairi E, et al. 2004. Fentanyl inhibits GABAergic neurotransmission to cardiac vagal neurons in the nucleus ambiguus. *Brain Res.* 1007:109–15

Hardman JG, Limbird L E, Gilman AG, eds. 2001. *Goodman & Gilman's: The Pharmacological Basis of Therapeutics*. New York: McGraw Hill. 13th ed.

Havemann U, Turski L, Kuschinsky K. 1982. Role of opioid receptors in the substantia nigra in morphine-induced muscular rigidity. *Life Sci.* 31(20–21):2319–22

Hayashi K, Tsuda N, Sawa T, Hagihira S. 2007. Ketamine increases the frequency of electroencephalographic bicoherence peak on the alpha spindle area induced with propofol. *Br. J. Anaesth.* 99(3):389–95

Heinricher MM, Morgan MM. 1999. Supraspinal mechanisms of opioid analgesia. In *Opioids in Pain Control: Basic and Clinical Aspects*, ed. C Stein, pp. 46–69. Cambridge, UK: Cambridge Univ. Press

Hemmings HC Jr, Akabas MH, Goldstein PA, Trudell JR, Orser BA, Harrison NL. 2005. Emerging molecular mechanisms of general anesthetic action. *Trends Pharmacol. Sci.* 26(10):503–10

Hensch TK. 2004. Critical period regulation. *Annu. Rev. Neurosci.* 27:549–79

Hocking G, Cousins MJ. 2003. Ketamine in chronic pain management: an evidence-based review. *Anesth. Analg.* 97(6):1730–39

Hospira. 2009. *Precedex dosing guidelines.* **http://www.precedex.com/wp-content/uploads/2010/02/Dosing_Guide.pdf**

Hshieh TT, Fong TG, Marcantonio ER, Inouye SK. 2008. Cholinergic deficiency hypothesis in delirium: a synthesis of current evidence. *J. Gerontol. A Biol. Sci. Med. Sci.* 63(7):764–72

Hsu YW, Cortinez LI, Robertson KM, Keifer JC, Sum-Ping ST, et al. 2004. Dexmedetomidine pharmacodynamics: part I: crossover comparison of the respiratory effects of dexmedetomidine and remifentanil in healthy volunteers. *Anesthesiology* 101(5):1066–76

Hufnagel A, Burr W, Elger CE, Nadstawek J, Hefner G. 1992. Localization of the epileptic focus during methohexital-induced anesthesia. *Epilepsia* 33(2):271–84

Huupponen E, Maksimow A, Lapinlampi P, Sarkela M, Saastamoinen A, et al. 2008. Electroencephalogram spindle activity during dexmedetomidine sedation and physiological sleep. *Acta Anaesthesiol. Scand.* 52(2):289–94

Int. Anesth. Res. Soc. (IARS). 2010. *SmartTots.* from **http://www.iars.org/smarttots/default.asp**

James R, Glen JB. 1980. Synthesis, biological evaluation, and preliminary structure-activity considerations of a series of alkylphenols as intravenous anesthetic agents. *J. Med. Chem.* 23:1350–57

Jamieson J. 1999. Anesthesia and sedation in the endoscopy suite? (influences and options). *Curr. Opin. Anaesthesiol.* 12(4):417–23

Jones EG. 2002. Thalamic circuitry and thalamocortical synchrony. *Philos. Trans. R. Soc. Lond. B Biol. Sci.* 357(1428):1659–73

Jorm CM, Stamford JA. 1993. Actions of the hypnotic anaesthetic, dexmedetomidine, on noradrenaline release and cell firing in rat locus coeruleus slices. *Br. J. Anaesth.* 71(3):447–49

Jt. Comm. 2004. *Sentinel event alert: preventing, and managing the impact of anesthesia awareness.* **http://www.jointcommission.org/sentinel_event_alert_issue_32_preventing_and_managing_the_impact_of_anesthesia_awareness/**

Kearse LA Jr, Manberg P, Chamoun N, deBros F, Zaslavsky A. 1994. Bispectral analysis of the electroencephalogram correlates with patient movement to skin incision during propofol/nitrous oxide anesthesia. *Anesthesiology* 81(6):1365–70

Kehrer C, Maziashvili N, Dugladze T, Gloveli T. 2008. Altered excitatory-inhibitory balance in the NMDA-hypofunction model of schizophrenia. *Front Mol. Neurosci.* 1:6, doi: 10.3389/neuro.02.006.2008

Kennedy D, Norman C. 2005. What don't we know? *Science* 309(5731):75–102

Kiefer RT, Rohr P, Ploppa A, Dieterich HJ, Grothusen J, et al. 2008. Efficacy of ketamine in anesthetic dosage for the treatment of refractory complex regional pain syndrome: an open-label phase II study. *Pain Med.* 9(8):1173–201

Kim A, Kerchner G, Choi D. 2002. Blocking excitotoxicity. In *CNS Neuroprotection*, ed. FW Marcoux, DW Choi, pp. 3–36. New York: Springer

Kissin I. 1993. General anesthetic action: an obsolete notion? *Anesth. Analg.* 76(2):215–18

Klafta JM, Zacny JP, Young CJ. 1995. Neurological and psychiatric adverse effects of anaesthetics: epidemiology and treatment. *Drug Saf.* 13(5):281–95

Klausen NO, Wiberg-Jorgensen F, Jorgensen B. 1983. Psychomimetic reactions after low-dose ketamine infusion. Comparison with neuroleptanaesthesia. *Br. J. Anaesth.* 55(4):297–301

Kohrs R, Durieux ME. 1998. Ketamine: teaching an old drug new tricks. *Anesth. Analg.* 87(5):1186–93

Kungys G, Kim J, Jinks SL, Atherley RJ, Antognini JF. 2009. Propofol produces immobility via action in the ventral horn of the spinal cord by a GABAergic mechanism. *Anesth. Analg.* 108(5):1531–37

Lazar SW, Kerr CE, Wasserman RH, Gray JR, Greve DN, et al. 2005. Meditation experience is associated with increased cortical thickness. *Neuroreport* 16(17):1893–97

Leslie K, Skrzypek H. 2007. Dreaming during anaesthesia in adult patients. *Best Pract. Res. Clin. Anaesthesiol.* 21(3):403–14

Li N, Lee B, Liu RJ, Banasr M, Dwyer JM, et al. 2010. mTOR-dependent synapse formation underlies the rapid antidepressant effects of NMDA antagonists. *Science* 329(5994):959–64

Lieberman JA, Bymaster FP, Meltzer HY, Deutch AY, Duncan GE, et al. 2008. Antipsychotic drugs: comparison in animal models of efficacy, neurotransmitter regulation, and neuroprotection. *Pharmacol. Rev.* 60(3):358–403

Linnemann MU, Guldager H, Nielsen J, Ibsen M, Hansen RW. 1993. Psychomimetic reactions after neurolept and propofol anaesthesia. *Acta Anaesthesiol. Scand.* 37(1):29–32

Long C. 1849. An account of the first use of sulphur ether by inhalation as an anaesthetic in surgical operations. *South Med. Surg. J.* 5:705–13

Lu J, Sherman D, Devor M, Saper CB. 2006. A putative flip-flop switch for control of REM sleep. *Nature* 441(7093):589–94

Lydic R, Baghdoyan HA. 2005. Sleep, anesthesiology, and the neurobiology of arousal state control. *Anesthesiology* 103(6):1268–95

Maeng S, Zarate CA Jr, Du J, Schloesser RJ, McCammon J, et al. 2008. Cellular mechanisms underlying the antidepressant effects of ketamine: role of alpha-amino-3-hydroxy-5-methylisoxazole-4-propionic acid receptors. *Biol. Psychiatry* 63(4):349–52

Marcantonio ER, Juarez G, Goldman L, Mangione CM, Ludwig LE, et al. 1994. The relationship of postoperative delirium with psychoactive medications. *JAMA* 272(19):1518–22

Markram H, Toledo-Rodriguez M, Wang Y, Gupta A, Silberberg G, Wu C. 2004. Interneurons of the neocortical inhibitory system. *Nat. Rev. Neurosci.* 5(10):793–807

Mashour GA. 2006. Integrating the science of consciousness and anesthesia. *Anesth. Analg.* 103(4):975–82

McCarley RW. 2007. Neurobiology of REM and NREM sleep. *Sleep Med.* 8(4):302–30

Mellon RD, Simone AF, Rappaport BA. 2007. Use of anesthetic agents in neonates and young children. *Anesth. Analg.* 104(3):509–20

Melnick BM. 1988. Extrapyramidal reactions to low-dose droperidol. *Anesthesiology* 69(3):424–26

Mercier FJ, Benhamou D. 1998. Promising non-narcotic analgesic techniques for labour. *Baillieres Clin. Obstet. Gynaecol.* 12(3):397–407

Meunier JC, Mollereau C, Toll L, Suaudeau C, Moisand C, et al. 1995. Isolation and structure of the endogenous agonist of opioid receptor-like ORL1 receptor. *Nature* 3776549:532–35

Mhuircheartaigh RN, Rosenorn-Lanng D, Wise R, Jbabdi S, Rogers R, Tracey I. 2010. Cortical and subcortical connectivity changes during decreasing levels of consciousness in humans: a functional magnetic resonance imaging study using propofol. *J. Neurosci.* 30:9095–102

Millan MJ. 2002. Descending control of pain. *Prog. Neurobiol.* 66(6):355–474

Miller RD, Eriksson LI, Fleischer LA, Wiener-Kronish JP, Young WL. 2010. *Miller's Anesthesia*. Philadelphia: Churchhill Livingstone

Mizobe T, Maghsoudi K, Sitwala K, Tianzhi G, Ou J, Maze M. 1996. Antisense technology reveals the alpha2A adrenoceptor to be the subtype mediating the hypnotic response to the highly selective agonist, dexmedetomidine, in the locus coeruleus of the rat. *J. Clin. Invest.* 98(5):1076–80

Monk TG, Weldon BC, Garvan CW, Dede DE, van der Aa MT, et al. 2008. Predictors of cognitive dysfunction after major noncardiac surgery. *Anesthesiology* 108(1):18–30

Morgan CJ, Muetzelfeldt L, Curran HV. 2009. Ketamine use, cognition and psychological wellbeing: a comparison of frequent, infrequent and ex-users with polydrug and non-using controls. *Addiction* 104(1):77–87

Mortazavi S, Thompson J, Baghdoyan HA, Lydic R. 1999. Fentanyl and morphine, but not remifentanil, inhibit acetylcholine release in pontine regions modulating arousal. *Anesthesiology* 90(4):1070–77

Nacif-Coelho C, Correa-Sales C, Chang LL, Maze M. 1994. Perturbation of ion channel conductance alters the hypnotic response to the alpha 2-adrenergic agonist dexmedetomidine in the locus coeruleus of the rat. *Anesthesiology* 81(6):1527–34

Nelson LE, Guo TZ, Lu J, Saper CB, Franks NP, Maze M. 2002. The sedative component of anesthesia is mediated by GABA(A) receptors in an endogenous sleep pathway. *Nat. Neurosci.* 5(10):979–84

Nelson LE, Lu J, Guo T, Saper CB, Franks NP, Maze M. 2003. The alpha2-adrenoceptor agonist dexmedetomidine converges on an endogenous sleep-promoting pathway to exert its sedative effects. *Anesthesiology* 98(2):428–36

NIH Consens. Dev. Conf. 1984. Drugs and insomnia. The use of medications to promote sleep. *JAMA* 251(18):2410–14

Obeso JA, Marin C, Rodriguez-Oroz C, Blesa J, Benitez-Temino B, et al. 2008. The basal ganglia in Parkinson's disease: current concepts and unexplained observations. *Ann. Neurol.* 64(Suppl. 2):S30–46

Ogawa A, Uemura M, Kataoka Y, Ol K, Inokuchi T. 1993. Effects of ketamine on cardiovascular responses mediated by N-methyl-D-aspartate receptor in the rat nucleus tractus solitarius. *Anesthesiology* 78(1):163–67

Olney JW, Newcomer JW, Farber NB. 1999. NMDA receptor hypofunction model of schizophrenia. *J. Psychiatr. Res.* 33(6):523–33

Osaka T, Matsumura H. 1994. Noradrenergic inputs to sleep-related neurons in the preoptic area from the locus coeruleus and the ventrolateral medulla in the rat. *Neurosci. Res.* 19(1):39–50

Oye I, Paulsen O, Maurset A. 1992. Effects of ketamine on sensory perception: evidence for a role of N-methyl-D-aspartate receptors. *J. Pharmacol. Exp. Ther.* 260(3):1209–13

Pandharipande PP, Pun BT, Herr DL, Maze M, Girard TD, et al. 2007. Effect of sedation with dexmedetomidine versus lorazepam on acute brain dysfunction in mechanically ventilated patients: the MENDS randomized controlled trial. *JAMA* 298(22):2644–53

Parvizi J, Damasio AR. 2003. Neuroanatomical correlates of brainstem coma. *Brain* 126(Pt. 7):1524–36

Peroutka SJ, Synder SH. 1980. Relationship of neuroleptic drug effects at brain dopamine, serotonin, alpha-adrenergic, and histamine receptors to clinical potency. *Am. J. Psychiatry* 137(12):1518–22

Peroutka SJ, Snyder SH. 1982. Antiemetics: neurotransmitter receptor binding predicts therapeutic actions. *Lancet* 1(8273):658–59

Porro CA, Biral GP, Benassi C, Cavazzuti M, Baraldi P, et al. 1999. Neural circuits underlying ketamine-induced oculomotor behavior in the rat: 2-deoxyglucose studies. *Exp. Brain Res.* 124(1):8–16

Posner J, Saper C, Schiff ND, Plum F. 2007. *Diagnosis of Stupor and Coma.* Oxford, UK: Oxford Univ. Press

Price CC, Garvan CW, Monk TG. 2008. Type and severity of cognitive decline in older adults after noncardiac surgery. *Anesthesiology* 108(1):8–17

Purdon PL, Pierce ET, Bonmassar G, Walsh J, Harrell PG, et al. 2009. Simultaneous electroencephalography and functional magnetic resonance imaging of general anesthesia. *Ann. N. Y. Acad. Sci.* 1157:61–70

Purves D, Augustine GJ, Hall WC, Lamantia A, McNamara JO, White LE. 2008. *Neuroscience.* Sunderland, MA: Sinauer

Ranta S, Jussila J, Hynynen M. 1996. Recall of awareness during cardiac anaesthesia: influence of feedback information to the anaesthesiologist. *Acta Anaesthesiol. Scand.* 40(5):554–60

Ren J, Ding X, Funk GD, Greer JJ. 2009. Ampakine CX717 protects against fentanyl-induced respiratory depression and lethal apnea in rats. *Anesthesiology* 110(6):1364–70

Revel FG, Herwig A, Garidou ML, Dardente H, Menet JS, et al. 2007. The circadian clock stops ticking during deep hibernation in the European hamster. *Proc. Natl. Acad. Sci. USA* 104(34):13816–20

Reves JG, Glass PS, Lubarsky DA, McEvoy MD, Martinez-Ruiz R. 2010. Intravenous anesthetics. See Miller et al. 2010, pp. 719–68

Rivera VM, Keichian AH, Oliver RE. 1975. Persistent Parkinsonism following neuroleptanalgesia. *Anesthesiology* 42(5):635–37

Rockoff MA, Goudsouzian NG. 1981. Seizures induced by methohexital. *Anesthesiology* 54(4):333–35

Rodrigues SF, de Oliveira MA, Martins JO, Sannomiya P, de Cassia Tostes R, et al. 2006. Differential effects of chloral hydrate- and ketamine/xylazine-induced anesthesia by the s.c. route. *Life Sci.* 79(17):1630–37

Russo H, Bressolle F. 1998. Pharmacodynamics and pharmacokinetics of thiopental. *Clin. Pharmacokinet.* 35(2):95–134

Samuelsson P, Brudin L, Sandin RH. 2008. Intraoperative dreams reported after general anaesthesia are not early interpretations of delayed awareness. *Acta Anaesthesiol. Scand.* 52(6):805–9

Saper CB, Scammell TE, Lu J. 2005. Hypothalamic regulation of sleep and circadian rhythms. *Nature* 437(7063):1257–63

Schuttler J, Schwilden H. 2008. *Modern Anesthetics: Handbook of Experimental Pharmacology*. Berlin: Springer

Scuderi PE. 2003. Pharmacology of antiemetics. *Int. Anesthesiol. Clin.* 41(4):41–66

Seamans J. 2008. Losing inhibition with ketamine. *Nat. Chem. Biol.* 4(2):91–93

Shafer S, Flood P, Schwinn D. 2010. Basic principles of pharmacology. See Miller et al. 2010, pp. 479–514

Shapiro BA, Warren J, Egol AB, Greenbaum DM, Jacobi J, et al. 1995. Practice parameters for intravenous analgesia and sedation for adult patients in the intensive care unit: an executive summary. Society of Critical Care Medicine. *Crit. Care Med.* 23(9):1596–600

Sherin JE, Elmquist JK, Torrealba F, Saper CB. 1998. Innervation of histaminergic tuberomammillary neurons by GABAergic and galaninergic neurons in the ventrolateral preoptic nucleus of the rat. *J. Neurosci.* 18(12):4705–21

Silbert BS, Myles PS. 2009. Is fast-track cardiac anesthesia now the global standard of care? *Anesth. Analg.* 108(3):689–91

Sinner B, Graf BM. 2008. Ketamine. *Handb. Exp. Pharmacol.* 182:313–33

Sloan I. 2002. The new anesthetic machines. *Clin. J. Anaesth.* 47:201–4

Stein C. 1995. The control of pain in peripheral tissue by opioids. *N. Engl. J. Med.* 332(25):1685–90

Stewart JJ, Weisbrodt NW, Burks TF. 1978. Central and peripheral actions of morphine on intestinal transit. *J. Pharmacol. Exp. Ther.* 205(3):547–55

Stoddard FJ, Sheridan RL, Saxe GN, King BS, King BH, et al. 2002. Treatment of pain in acutely burned children. *J. Burn Care Rehabil.* 23(2):135–56

Strebel S, Kaufmann M, Maitre L, Schaefer HG. 1995. Effects of ketamine on cerebral blood flow velocity in humans. Influence of pretreatment with midazolam or esmolol. *Anaesthesia* 50(3):223–28

Takahashi T, Tsuchida D, Pappas TN. 2007. Central effects of morphine on GI motility in conscious dogs. *Brain Res.* 1166:29–34

Tickle-Degnen L, Lyons KD. 2004. Practitioners' impressions of patients with Parkinson's disease: the social ecology of the expressive mask. *Soc. Sci. Med.* 58(3):603–14

Tremper KK, Barker SJ. 1989. Pulse oximetry. *Anesthesiology* 70(1):98–108

Tsuda N, Hayashi K, Hagihira S, Sawa T. 2007. Ketamine, an NMDA-antagonist, increases the oscillatory frequencies of alpha-peaks on the electroencephalographic power spectrum. *Acta Anaesthesiol. Scand.* 51(4):472–81

Tung A, Tadimeti L, Caruana-Montaldo B, Atkins PM, Mion LC, et al. 2001. The relationship of sedation to deliberate self-extubation. *J. Clin. Anesth.* 13(1):24–29

Vanhaudenhuyse A, Boly M, Laureys S, Faymonville M. 2009. Neurophysiological correlates of hypnotic analgesia. *Contemp. Hypn.* 26(1):15–23

Van Keulen SG, Burton JH. 2003. Myoclonus associated with etomidate for ED procedural sedation and analgesia. *Am. J. Emerg. Med.* 21(7):556–58

Vanlersberghe C, Camu F. 2008. Etomidate and other non-barbiturates. *Handb. Exp. Pharmacol.* 182:267–82

Velly LJ, Rey MF, Bruder NJ, Gouvitsos FA, Witjas T, et al. 2007. Differential dynamic of action on cortical and subcortical structures of anesthetic agents during induction of anesthesia. *Anesthesiology* 107(2):202–12

Veselis RA, Reinsel RA, Beattie BJ, Mawlawi OR, Feshchenko VA, et al. 1997. Midazolam changes cerebral blood flow in discrete brain regions: an H2(15)O positron emission tomography study. *Anesthesiology* 87(5):1106–17

Vlajkovic GP, Sindjelic RP. 2007. Emergence delirium in children: many questions, few answers. *Anesth. Analg.* 104(1):84–91

Vollenweider FX, Leenders KL, Oye I, Hell D, Angst J. 1997. Differential psychopathology and patterns of cerebral glucose utilisation produced by (S)- and (R)-ketamine in healthy volunteers using positron emission tomography (PET). *Eur. Neuropsychopharmacol.* 7(1):25–38

Weil JV, McCullough RE, Kline JS, Sodal IE. 1975. Diminished ventilatory response to hypoxia and hypercapnia after morphine in normal man. *N. Engl. J. Med.* 292(21):1103–6

Wuesten R, Van Aken H, Glass PS, Buerkle H. 2001. Assessment of depth of anesthesia and postoperative respiratory recovery after remifentanil- versus alfentanil-based total intravenous anesthesia in patients undergoing ear-nose-throat surgery. *Anesthesiology* 94(2):211–17

Yaksh TL. 1997. Pharmacology and mechanisms of opioid analgesic activity. *Acta Anaesthesiol. Scand.* 41(1 Pt. 2):94–111

Zarate CA Jr, Singh JB, Carlson PJ, Brutsche NE, Ameli R, et al. 2006. A randomized trial of an N-methyl-D-aspartate antagonist in treatment-resistant major depression. *Arch. Gen. Psychiatry* 63(8):856–64

Cumulative Indexes

Contributing Authors, Volumes 25–34

Chapter Titles, Volumes 25–34